BLACKJACK ATTACK

BLACKJACK ATTACK

Playing the Pros' Way

Don Schlesinger

RGE Publishing, Ltd.
Las Vegas, Nevada

RGE Publishing, Ltd.
Las Vegas, NV 89123
(702) 798-7743
(702) 798-8743 fax
www.rge21.com
books@rge21.com

First edition 1997. Third edition, revised 2005.
Printed in the United States of America
20 19 18 17 16 15 14 13 3 4 5 6

ISBN 0-910575-20-7
Library of Congress Control Number 2004195404

Publisher
Viktor Nacht

Production Editor
Bethany Paige

Cover Design
David Adler, Canyon Creative

Cover Photo Credits
Caesars Palace, exterior view, c. 1960s — Las Vegas News Bureau Collection; The George shot, part of Operation Greenhouse, was fired May 8, 1951 at Enewetak (Marshall Islands) — U.S. Department of Energy Collection.

 Portions of this book have previously appeared in different form in *Blackjack Forum* magazine, 1984–2000.

This book is printed on acid-free paper.

This book is dedicated to all of my family members, past and present, whose love and support have encouraged me to pursue my studies of the game of casino 21. Their patience and understanding, throughout my career of playing, teaching, and writing about blackjack, have certainly contributed, in large measure, to any success I may have achieved in these fields.

The book is also dedicated to the memory of five dear blackjack-playing friends: Kenny Feldman, Lester White, Joel Waller, Paul Keen, and Chris Cummings, whom I miss very much.

Table of Contents

List of Tables

Foreword

Don Schlesinger is a teacher of blackjack. Probably because of his background as a teacher of mathematics and foreign language, he has a knack for explaining difficult concepts in a manner that can be understood by most people.

Schlesinger not only knows blackjack well enough to write about it, but he also has a long track record of big casino wins at the game. Though he is a busy person and has a family, he manages several trips a year to Las Vegas and other casino destinations. He generally takes his family on those trips, but he also hits the blackjack tables hard, making big bets for long hours.

For years Schlesinger has been the angel sitting on the shoulders of blackjack experts. He has given generously of his time and expertise, reading each new book carefully and sending lengthy critiques to authors and publishers. All of my own books have benefited considerably from Schlesinger's talents.

Schlesinger also is a frequent contributor of original material to blackjack periodicals, primarily Arnold Snyder's *Blackjack Forum*. Some of his articles, such as the "Illustrious 18" and "Floating Advantage," have become legendary.

On the Internet, Schlesinger has a reputation as one of the most helpful of blackjack experts, patiently answering questions about arcane aspects of the game.

For years, I have been telling Schlesinger that I would love to have a book that included all of his published and privately disseminated blackjack material. With this book, I am getting my wish.

Stanford Wong
March 1997

Introduction

I have more than a hundred books on casino blackjack in my library. Of these, a relative handful stand out as "classics." These are the books written by the original thinkers in this field, those writers whose works broke new ground and contributed something unique and valuable to our knowledge of the game.

Blackjack Attack: Playing the Pros' Way, by Don Schlesinger, is one of those few really important books on this subject that serious players will come to revere as an indispensable classic. I have no doubt in my mind that this will come to pass, as Don's writings on this subject are already revered by those who have been following his writing career in the pages of *Blackjack Forum*.

Blackjack Attack is the "answer book," the book that every professional player will pack in his suitcase, the book that will solve problems, and settle arguments. For the truly dedicated player, hitherto unfamiliar with Don Schlesinger's work, this book will be a light in the darkness, the fundamental factbook that puts a dollars-and-cents value on the fine points of play.

Don started writing this book in 1984. He didn't know he was starting to write a book at the time, nor did I. He was simply answering players' questions. There was no outline for a book-to-be, no synopsis of what was to come, no formula; this book was never planned.

Blackjack players had questions, lots of questions, complicated questions that were not being answered in the myriad books on card counting. So, for a period of some 13 years, Don Schlesinger answered those questions, one at a time, thoughtfully, diligently, and comprehensively. His audience was small but grateful, consisting of the true aficionados of the game, those who subscribed to *Blackjack Forum* magazine.

As the popularity of casino gambling spread across the country and throughout the world, more and more books on casino blackjack were published. But still, these new books — many of them fine and reputable card-counting guides — continued to ignore the fine points of play. Some years back, I noticed that these new players were asking many of the same questions that Don had already answered. I no longer had to forward these questions to Don for his analyses; I simply referred these players to whichever back issues of *Blackjack Forum* contained Don's pertinent answers.

As Don has contributed columns, letters, and articles to more than 50 different issues of *Blackjack Forum,* this has not always been quite as simple a process as it sounds. I don't really want to think about how many (hundreds of?) hours, in the past ten years, I've spent leafing through back issues of *Blackjack Forum,* just so I could refer a reader to the precise issue where Don had already answered his question. Suffice it to say that the publication of this book makes my life easier. Now, when a player asks me one of those complicated questions that Don has already answered in his trademark elegant and comprehensive style, I can simply refer the player to this book, knowing that should this player get bitten really seriously by the blackjack bug, then this same book will also likely answer the next 16 complicated questions that pop into his head.

Don Schlesinger is a book author by default. He spent 13 years answering players' questions, writing by "popular request," never envisioning a book, yet, here it is, a book that is truly one of the most important texts on the game in print! I am humbly honored to be the publisher of *[the first two editions of]* this work. The amount of information contained within these pages is frankly astonishing. I can see now why all the other blackjack experts left this stuff out of their books. Aside from the fact that most had not figured out about the game much of what Don had, the fact is there wouldn't have been room for this material in any "standard" text on how to count cards.

A word of caution to the reader: This is not a book that will teach you how to play blackjack, or how to count cards. This is the book you should read *after* you learn how to do those things. Almost any dedicated player, possessing average memory and intelligence, can learn how to count cards, but very few ultimately become successful card counters. Of those who make it, you can bet a good number of them have been reading Don Schlesinger's articles for the past 13 years. If you haven't been, you've got some catching up to do!

Arnold Snyder
March 1997

Publisher's Introduction

On behalf of RGE, I am very proud to introduce the third edition of *Blackjack Attack: Playing the Pros' Way — The Ultimate Weapon*. More than just the third edition of a venerable blackjack reference, *The Ultimate Weapon* tips the balance of power back into the hands of the player by bridging the snapper gap, TARGETing misinformation, camouflaging your strategic stockpile, and, if all else fails, supplying the numbers required to initiate assured positive expectation.

And while this newest and perhaps final edition is arguably the most extensive and comprehensive research reference work on the subject of blackjack ever published, what is truly "ultimate" about this work is how it stands as the most tangible legacy of Don Schlesinger's utter selflessness toward the blackjack community.

First, the lineage of this book begins in the auspicious pages of Arnold Snyder's *Blackjack Forum* where, for almost twenty years, Don answered reader-submitted questions through diligent research and collaboration, ultimately building those findings into a de facto manuscript that became *Blackjack Attack*. Given the diminutive scale of *Blackjack Forum*'s budget and audience, it's apparent that it was Don's love of the game and his innate drive to share knowledge that first created this book.

Next, the universe gave us Don, blackjack, AND the Internet, and we still don't know what hit us! Since only moments after the birth of the new information age, Don has been there almost daily, replying to post-after-post-after-post of questions, comments, and "new" discoveries, on virtually every site, from players of every skill level. His posts alone number in the thousands, and one can only imagine how many hours of exhaustive research and authoring, all uncompensated, have been put into those words of wisdom.

To his friends and even casual associates, Don is no less generous. On a personal level, it was Don who helped me through one of the most difficult periods of my life, and he did so with immense care, dominated by reason and fairness. Meanwhile, over the years, Don has fastidiously proofread and critiqued books and products for numerous authors and developers — well-known and not-so-well-known — with generosity and the tireless pursuit of truth and accuracy as his only motivations.

Arnold Snyder wrote in his introduction to the first edition of this book that "Don Schlesinger is a book author by default." More so, Don is a teacher by providence, and a prodigious one at that: wise, sometimes gruff, certainly pedantic, but always in search of new knowledge and the opportunity to share it. With all this said, it would hardly be a surprise to know that Don was in fact a New York City schoolteacher!

There is no one around today who has given as much to the blackjack community as Don has, while continuing to commit as much time, care, and passion as his body and mind will allow, to the game we all love.

It is, therefore, an honor to introduce and publish this book for my friend, Don Schlesinger, who is not only a master of the game, but also a master of life.

Viktor Nacht
Las Vegas, March 2004

Acknowledgments

This list of acknowledgments is quite long, simply because, over the course of these nearly thirty years, so many people have been instrumental in contributing to my knowledge of blackjack. And so, I would like to thank:

The one and only **Stanford Wong.** He, perhaps more than any other individual, has furthered our knowledge of the game of blackjack and its subtleties. I'm proud to call him my friend. His help, throughout the years, has been invaluable to me. In particular, *Professional Blackjack* and the *Blackjack Count Analyzer* will forever stand, in my opinion, as two of the most important resource tools to have ever been created for our study of the game. I thank him, as well, for his Foreword to this book.

Arnold Snyder, who gave me a "forum" in which to publish my blackjack articles, and who finally convinced me, after many years of trying, to write this book. Arnold, perhaps more than for any other individual, this one's for you!

Peter Griffin, our foremost theoretician and expert on the mathematics of the game. On more than one occasion, when I was "stuck" in my pursuit of an answer to a thorny problem, Peter came to my rescue. I will always treasure our pages of written correspondence and our phone calls, during which I never failed to learn from the "master." Since the publication of the first edition of this book, the blackjack community mourned the passing of this legendary inspiration to us all. Peter's self-deprecating wit and graciousness served as a model of humility and reminded us that, often, the bigger they are, the nicer they are. We shall miss him very much.

Seven "pioneers" of blackjack: **Edward O. Thorp, Julian Braun, Lawrence Revere, Allan Wilson, John Gwynn, Lance Humble,** and **Ken Uston.** It is safe to say that I would have never become involved in the game of blackjack were it not for these gentlemen's monumental contributions to our knowledge. In particular, it was from Revere's classic, *Playing Blackjack as a Business,* that I patterned my approach to the game.

Three fellow blackjack authors and experts, notably **George C.,** whose friendship and help over the years have been of paramount importance to me. George's contributions to the game have broken ground in many areas, and his insights and innovations have opened my eyes to several new and original approaches to playing blackjack. **Bryce Carlson,** one of the game's brightest authorities, whose outstanding work, *Blackjack for Blood,* and whose software, the *Omega II Blackjack Casino,* have proven to be valuable reference resources; and **Michael Dalton,** for his quarterly, *Blackjack Review,* and his monumental tome, *Blackjack: A Professional Reference,* which is truly "The Encyclopedia of Casino Twenty-One."

Three computer programmers, for their unparalleled contributions to the world of blackjack simulation: the indefatigable **John Imming,** whose *Universal Blackjack Engine* was the first of the truly great multi-purpose pieces of simulation software. John's creation has unlocked the doors of blackjack research to the point where virtually all questions can now be answered, thanks to his incredible *UBE;* **Norm Wattenberger,** whose equally brilliant *Casino Vérité* has set the standard for the ultimate in blackjack game simulation and practice software. How I wish *CV* had been around when I was practicing and learning how to count cards! And last, but certainly not least, **Karel Janecek,** a relative newcomer on the blackjack scene. Karel's *Statistical Blackjack Analyzer* was used to produce the original simulations for Chapters 9 and 10, and it is safe to say that *SBA* has fast become one of the most respected and reliable simulators in the industry today.

Howard Schwartz, Edna Luckman, and **Peter Ruchman,** of the fabulous Gambler's Book Club, in Las Vegas. No trip to Vegas would ever be complete without a stop in their unique store. The vast majority of the books in my blackjack library were purchased at the GBC, and their annual catalog is a gold mine of casino gaming literature and software. Since the publication of the second edition of this book, Edna Luckman passed away. I always considered her the "grande dame" of Las Vegas, and I shall miss her welcoming smile and hospitality whenever I visit the GBC in the future.

Chris Cummings and **Bill Margrabe,** two former colleagues, and each a brilliant mathematician. When the sledding got a bit rough, and I needed some help developing a formula or mathematical concept, they were there to furnish the crucial pieces of the puzzle.

My many newfound friends on the Internet, whose contributions to the various newsgroups and whose correspondence, through e-mail, have been most enlightening. In particular, I am grateful for the wisdom of **Ralph Stricker,** a gentleman, a world-class player, and truly our "elder statesman" of the game today; **Steve Jacobs,** whom I have anointed the "Godfather" of "rec.gambling.blackjack," and whose advice I have sought on many occasions; **Abdul Jalib,** one of r.g.bj's most brilliant contributors, whose insights are always accurate and illuminating; and, especially, the tireless

John Auston, whose simulations for the original Chapter 10 may have set an all-time record for number of hands played. *[Note: A new record was established when generating the charts for this third edition. — D.S.]* I am also thankful for the guidance furnished by **Kim Lee, Peter Carr, Richard Reid, "MathProf,"** and **Bob Fisher** in preparing new material for the revised versions and in indicating a few errors in the first edition. Finally, a special debt of gratitude to **"Cacarulo,"** for his energetic work in providing the magnificent materials for the new Appendixes of this third edition, and to **"Zenfighter,"** whose complete revision and expansion of Peter Griffin's effects of removal tables now constitutes the new gold standard for research in this area.

All my wonderful friends and blackjack-playing buddies, too numerous to mention. A few, of course, deserve special recognition: **Kenny Feldman,** a great friend, a great player, and my partner in our blackjack-teaching days. Kenny's advice has proven invaluable over our thirty-five years of friendship, and his companionship, on many a blackjack trip, has enhanced my enjoyment of the game; **Les Appel,** another great friend and player who also taught with Kenny and me. His easygoing approach to the game showed me how to relax and not sweat the short-term swings.

My former teammates, many of whom have remained wonderful friends, over the years. For obvious reasons, they shall remain nameless, but you know who you are. We had some great times, guys; thanks for the memories!

My new editors for this third edition, the indefatigable **Viktor Nacht** and **Bettie Paige,** of RGE. Their dedication to every last detail, from preparation of the manuscript, to formatting, layout, cover design, and proofreading, was the epitome of professionalism. If this is the best *BJA* of all, then surely the quality is due in large part to their tireless efforts.

Last, but certainly not least, the seven proofreaders of my original manuscript, who turned the tables on me and provided dozens of insightful comments, corrections, and suggestions. This is certainly a better book because of the remarkable attention to detail of: **John Auston, Stanley Dinnelaw, Ken Feldman, James Grosjean, Olaf Vancura, Ron Wieck,** and **Stanford Wong.** (Special thanks to Ken and Ron, for reading all three editions!) A final proofreader, **John Wargat,** who joined our crew after the publication of the hardcover version. His painstaking enumerations of typos that had eluded us all, and his talent for pointing out several stylistic inconsistencies have made this softcover edition of *BJA3* the most accurate one in print. I thank John for his truly amazing effort.

Preface

First Edition

This is, most definitely, *not* your average blackjack book! Now, that is not to imply, in some haughty manner, that *Blackjack Attack: Playing the Pros' Way* is superior to most of the books that have been written on casino 21; it's simply that this book is substantially *different* from all the others.

Permit me to explain. It might be a good idea to begin by telling you what *Blackjack Attack* is not. It is not a "how to learn to count cards" book (although readers will learn much about card counting). Nor is it a work in which you will find standard basic strategy charts *[Note: This is no longer the case, as these have been added to the third edition. — D.S.],* the rules of the game, or a glossary of its terms — all staple features of the typically thorough blackjack book. Instead, this book assumes a more advanced reader — one who is already familiar with the above concepts and seeks a higher level of sophistication. Indeed, were it not for the publisher's desire for a "snappy," attention-grabbing title, I might have called this book, *Blackjack: The Finer Points.* For, it is my hope that even the most seasoned veteran will find at least some new and fertile ground, while the less experienced player is apt to find much food for thought in *Blackjack Attack.*

I have often been asked, "Where did it all begin?" My involvement with the game of blackjack goes back some 35 years. An avid student of probability theory and, especially, its applications to the field of casino gaming and sports wagering, I read, starting in my early teens, all I could get my hands on concerning gambling. In 1962, when Professor Thorp published his seminal work, *Beat the Dealer,* my fascination with the game of blackjack and, at the time, simply its theoretical aspects, was kindled.

It wasn't until a full 13 years later that I began to actually *play* blackjack, rather than just read about it. And, as they say, the rest is, more or less, history! I read everything I could on the subject and, at the time, the big names, in addition to Thorp, were Lawrence Revere, Stanford Wong, Julian Braun, and, a little later, Peter Griffin, Ken Uston, and Arnold Snyder. Mentors all, they helped me to form a rock-solid foundation for my ensuing blackjack knowledge.

I began to correspond with many of the aforementioned "biggies," and soon gained a reputation, through the long letters I would write to authors, as "the conscience of the blackjack-writing community" (as one expert put it). And, although *playing* blackjack was (and remains) certainly one of my great passions, it seemed that writing about it and, eventually, teaching others to play, was equally fulfilling. Many of my early observations on the game were published in Wong's *Blackjack Newsletter,* but it wasn't until 1984 that my first feature article appeared in Snyder's *Blackjack Forum* quarterly review.

Fortunately, that initial piece was well received by Arnold's readers, and I was encouraged to continue writing. And write I did! The work you are about to read is, in part, a compilation of 13 years of articles published in *Blackjack Forum,* as well as much brand-new material that has never before appeared in print. In particular, recent improvements in our ability to simulate, not only accurately, but also at incredible speeds, have permitted the kind of hitherto impossible, exhaustive study that forms the content of our Chapter 10, The "World's Greatest Blackjack Simulation." It is, in some respects, a mini-book all its own.

I am indebted to all those I recognize in the "Acknowledgments," for, clearly, this book would have never been brought to fruition without their input and guidance. Above all, I would like to thank Stanford Wong and Arnold Snyder who, for many years, urged me to put all of my articles in book form, and especially Snyder, who agreed to publish the finished product.

One brief technical note: For the sake of clarity, within any given chapter, original references to other *articles* from *Blackjack Forum* have been changed to the appropriate *chapter* in the present book. It was felt that this notation would greatly facilitate cross-referencing for the reader.

Finally, a request. Whereas every conceivable effort has been made to assure the accuracy of both the numerical calculations and the text of this book, errors in a work of this size and complexity are, nonetheless, inevitable. Should you find any, I would be most grateful if you would convey those findings to the publisher, or to me directly. *[I may be reached at Viktor Nacht's* ***AdvantagePlayer.com*** *Web site (see, especially, "Don's Domain"), or on that site's free "Parker Pages," where, routinely, I "hang out" as one of the resident "gurus." — D.S.]*

Thank you, in advance, for your continued support and for the interest you have shown in my work over the past years.

Don Schlesinger
New York, March 1997

Second Edition

Since I wrote the above lines, over three years ago, much has happened in the world of casino gaming, in general, and the dissemination of blackjack information, in particular. Among the many developments, as mentioned, I now have my own Web site, "Don's Domain," on the pages of *AdvantagePlayer.com.* So, it is all the easier for you to get in touch with me should you have a question requiring an immediate answer. Of course, with the rapid expansion of the Web have come further projects and innovations in the fields of blackjack research and software creation. There seems to be no end to the creativity and resourcefulness of today's students of the game.

This revised edition of *Blackjack Attack* is my attempt to keep the material in the book fresh, current, and as accurate as possible. To this extent, here are some of the improvements and new features you will find in the second edition: a very thorough revision of the entire text, with special care taken to correct any typos, errors, or omissions from the first edition; inclusion, in the highly regarded Chapter 10, of the entire set of 8-deck charts, to complete the "World's Greatest Blackjack Simulation"; addition of a new Chapter 11 *[now Chapter 9 — D.S.]* — the SCORE article — which first appeared on Arnold Snyder's Web site in March 1999; a brand-new Chapter 12 *[now Chapter 13 — D.S.],* reflecting the latest cutting-edge research in the areas of optimal shoe-departure points and risk-averse strategies; chapter titles added to the tops of all odd-numbered pages, for easier referencing; a revised bibliography and an expanded, more complete, index; finally, a completely new cover, of the highest quality, sporting an entirely new look and design!

I hope you will enjoy this second edition of *Blackjack Attack: Playing the Pros' Way.*

Don Schlesinger
New York, June 2000

Third Edition — Hardcover

There is much to be excited about in this third, "Ultimate Weapon," edition of *Blackjack Attack.* As the subtitle suggests, I have tried to cram in the pages of this one volume an incredible amount of highly sophisticated blackjack information. To that end, I have eagerly sought the advice and collaboration of some of the keenest minds in the area of blackjack research, and together, we have broken new ground in several ways.

Here, then, are the major additions and improvements to be found in this third edition: First and foremost is the complete revision of the Chapter 10 charts, by Norm Wattenberger, to include optimal betting ramps, unit sizes, SCOREs, and much, much more; a light piece, "More on SCORE," appended to the SCORE chapter, as we seek ways to distinguish between the "original" concept of SCORE and the "generic" meaning the term has taken on; the complete version of the Optimal Departure study, which began in the second edition of *BJA,* but was completed in a subsequent issue of *Blackjack Forum;* four brand-new risk of ruin equations, encompassing goal-reaching and time-constraint concepts, and an enlightening testing of the formulas' accuracy, through the computer simulations of Norm Wattenberger; an entirely new Chapter 12, as a mythical blackjack team takes a "Random Walk Down the Strip," utilizing John Auston's genial software application, *BJRM 2002;* a three-part Appendix, courtesy of the amazing Cacarulo, with more than 100 pages (!) of precise expectation tables, basic strategy charts (the most accurate in print), and effects of rules variations on basic strategy expectations; finally, a new cover design, an enlargement of the book's format, and, for collectors, an elegant hardcover edition, which will be followed by a softcover version.

We learn to "never say never," but, with this "Ultimate Weapon," I believe I have furnished virtually all of the tools that the traditional card counter must have at his disposal to levy a full-blown attack on the casinos. I'm sure you'll let me know if I've left anything out, but, for now, this third edition of *Blackjack Attack* contains just about everything I know about the game of blackjack. I suggest, as one reviewer observed (speaking of the second edition), that you "don't leave home without it." Enjoy!

Don Schlesinger
New York, February 2004

Third Edition — Softcover

It's been eight months since the publication of the hardcover edition of *BJA3,* and, frankly, I couldn't be more pleased with the excitement that the book has generated in the blackjack community. I wrote, in my Preface to that third edition, that it "contains just about everything I know about the game of blackjack," and intimated that there might not be any further revisions of the book. But, there was a softcover version of *BJA3* in the offing, and I am pleased to announce that it contains a major addition of historic proportions: a brand-new Appendix D, featuring completely revised and expanded tables of effects of removal, for playing and betting, which improve on the work of the legendary Peter Griffin. We have Zenfighter (more about him in the appendix) and Cacarulo to thank for this monumental study, which I trust will be of immense value to blackjack players and researchers everywhere.

The book you now hold in your hands is the most voluminous reference work on casino 21 in existence, comprising 576 total pages and over 400 tables and charts. Needless to say, its creation involved a prodigious amount of work, not only for the author and his collaborators, but also for the publishers and proofreaders, who, once again, have made a Herculean effort to get everything "just right." It would be hard to describe the level of detail into which we all entered as we fine-tuned the text of the hardcover *BJA3,* searching for the inevitable gremlins that are typos, omissions, and stylistic inconsistencies. Doubtless, we haven't found them all, but it is unquestionably true that the present incarnation of *BJA3* is the most accurate and most painstakingly proofread version of them all, and I am extremely proud to sign my name to what is perhaps the final edition of *Blackjack Attack: Playing the Pros' Way.* I hope that you will enjoy it.

Don Schlesinger
New York, January 2005

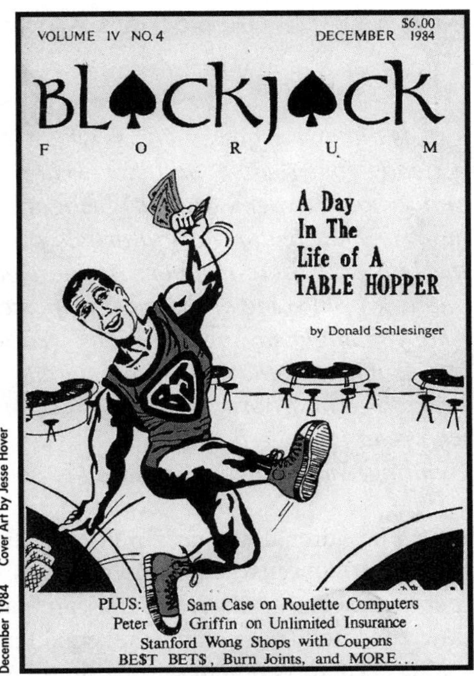

December 1984 Cover Art by Jesse Hover

Chapter 1

Back-Counting the Shoe Game

"Enjoyed Don Schlesinger's fine article."
— *Ken Uston, legendary player, and famed author of many blackjack books, including* ***Million Dollar Blackjack***

"I can provide an assurance that Don Schlesinger really is as good, as successful and as disciplined as he claims."
— *Stanford Wong, noted blackjack author of several landmark works, including* ***Professional Blackjack***

"Schlesinger's article is the best that has ever appeared re: 'Casino Comportment.'"
— *Peter Griffin, author of* ***The Theory of Blackjack***

As the saying goes, "This is where it all began." I had been, for several years prior, a steady contributor and letter-writer to both Stanford Wong's blackjack newsletters and Arnold Snyder's quarterly magazine, ***Blackjack Forum.*** *I proposed to Snyder a more formal, feature-length article on back-counting, and it appeared as "A Day in the Life of a Table-Hopper," in the December 1984 issue. The response (see above quotes), from some of the most respected authorities in the field, was quite flattering. And, you might say, my blackjack-writing career was officially "launched."*

As might be the case with a parent, who always holds a special place in his or her heart for a first-born child, the following article will always be special for me. Many who read it wrote back to say that it was the best depiction of "Wonging," or back-counting, that had ever appeared in print. I'll let you be the judge. Enjoy!

The adrenaline starts pumping the night before the trip down. The dedicated perfectionist leaves no chance for sloppy or inaccurate play. I can recite 165 index numbers in my sleep and can count down a 52-card deck starting from a face-down position (no scanning several clumps of cards at once) in under 14 seconds. No matter. There is practice to be done. Hands are dealt at lightning speed. Cards are flipped over. Indices are recited. This is a *discipline* thing. You either do it right or you don't do it at all. At least that is my approach to the game. The practice completed, I get a good night's sleep. It's going to be a *very* long day.

I have eschewed the junket approach for my entire nine-year playing career. I am very much a loner by nature and I have an infinite capacity for playing the game. I don't care a lot about the freebies if accepting them cramps my style. I like getting the money *and* doing it on my terms. I have never uttered my real name in a casino. I have never established credit anywhere. Central Credit Agency doesn't know I'm alive. I use different names in A.C. from those in Las Vegas and, for various reasons, have established more than one pseudonym, according to the casino in question. Thus, Caesars Palace can't cross-reference a name with Caesars A.C.; ditto for the two Trops, the two Sandses, etc. It pays to be careful.

I have tried to impart to all of my students that the cardinal rule in this game is not to win as much money as possible. Rather, it is to win as much as you can, consistent with being welcome back in the casino the next time. There is a very big difference. If improper money management is the greatest destroyer of potentially successful card counters, then certainly greed and impatience are close runners-up for that top honor. If you can't learn to win (and *lose,* for that matter) with both style and grace, then a) you probably won't last in this game, and b) you will eat yourself up inside while trying.

And so, I drive down the Garden State Parkway. Two and three-quarter hours later, I'm in my first casino, Caesars Boardwalk Regency. It is 12 noon. The battle begins!

The bus would have been cheaper and much more relaxing. So why drive? Because the bus tells you when you must go home. I like to decide for myself. Not that I'll refuse to quit if I'm losing. You lose too often to refuse to stop playing. It's simply that I like to be in control of as much of my own destiny as possible. Thus, the car.

The Back-Counting Approach

Readers of *Blackjack Forum* have questioned the practicality and feasibility of the back-counting approach. No one promised you a rose garden! Sure, I'd rather be in Las Vegas playing the Riviera's two-deck surrender game dealt to the 75% level. Or at Caesars Palace or the Trop with its 85% 4-deck game. But I live in New York and a lot of my play is going to be in A.C. whether I like it or not. The game *can* be beaten, but probably *not* the way you are playing it. Read on and follow me from casino to casino. Warning: If you don't have a good pair of thick rubber-soled shoes, forget it. I'm going to leave you in the dust! Second warning: If you've come down to have fun playing lots of blackjack, stay home and play with your family. You've come to the wrong place. I come to win money. I use blackjack as the vehicle to achieve that goal. We might play 15 minutes out of the hour; we might play less. This doesn't appeal to you? Then you're a loser already, and they haven't even dealt the first card.

The better your eyesight, the easier it is to back-count. Of course, we're already assuming you can count seven hands and the dealer's upcard in two or three seconds. If your concept of back-counting is literally standing two inches behind a player's back and riveting your eyes on each card as it falls, then you've got this thing all wrong. Look, there's a dealer shuffling at the corner table, the one across the aisle from the craps table. I position myself in between the two. I'll be looking at the craps action almost as frequently as at the blackjack table. And I'll be no closer to one than to the other. I'm looking for true counts of +1 or higher to enter the game. Zero is still a minimum bet. Why do I want to play when *they* have the edge? And zeros have a way (almost 50% of the time) of turning into negative counts. Is this any way to start a game? Not for *me,* it isn't!

I get the true of +1, but there are five players at the table. I'll be the sixth hand. Would you play? I pass. And here's why. A true count of +1 with one deck dealt out in an 8-deck game (we'll assume Hi-Lo even though I personally play Revere) is a running count of +7. On the next deal, if I play, approximately 19 cards will be used. Can 19 cards produce a running count of –7 or lower? Of course they can. It happens all the time. So where does that leave me now? I've played one hand, the count is negative, and I have a choice: a) leave the table (and look like a horse's ass!), or b) keep playing into the negative shoe. You don't like either choice? Well, neither do I.

I have a motto: "If it's good, it'll keep." I want to enter the game where I'm reasonably assured of a little action before conditions deteriorate. I might miss a few advantageous hands, but remember, we don't want to win every dollar possible; we want to win what we can while looking *normal* doing it. You *have* to believe this. And what is even harder is that you have to dedicate yourself to playing in a manner that reflects this philosophy.

And so I wait. Two people leave and the count gets better. I'm in! Now, a word about my style of betting. It is Kelly Criterion with several modifications. You won't like most of the constraints, because you want to win *all* of the money. Remember — *I don't.* I want to win and have them as happy with me as I am with them. Are you getting tired of my hammering home this point? Well then, stop being greedy and try the "right" way!

Getting back to the bet scheme: one unit from +1 to +2; two units from +2 to +3; four units from +3 to +4; six units from +4 to +5; and two hands of six units from +5 to anything higher. Why two hands instead of the one-hand, eight-unit bet at +5? Because a) eight units piled up start to look a bit too conspicuous, and b) two hands of six units (thus 12 units) increases my spread and thus my hourly expectation. Yes, it also increases my standard deviation. But it doesn't change the probability of *losing.* It just makes the expected win greater with a commensurate increase in the magnitude of the "swings." Such is life. I'm capitalized properly. And you had better be, too. At the A.C. game, you *work* for your money — they don't give it away!

I lose the first five bets. A lovely greeting! In the process, the true count shoots up to +2. I guess you'd double up, eh? Well, I bet the same one unit. In nine years of play, I can count the number of times I have increased my bet after a loss on the fingers of one hand and still have several fingers left over. I told you there would be constraints you wouldn't like. My rationale: Winners celebrate by parlaying bets. It is the logical and acceptable thing to do. *Card counters* jump bets regardless of the outcome of the previous hand. They make the *mathematically* correct play. If you want to play the single-deck drunken slob routine where the erratic betting scheme bewilders personnel, then go to Reno. That bit doesn't get it in A.C. First of all, the cocktail service is so pathetic, you couldn't get drunk if you wanted to; second, that approach is completely unsuitable for a table-hopping, back-counting style. This is a science, not a freak show.

Hello, I Must Be Going!

I look at my watch constantly. I want everyone to think I'm on the verge of leaving at any moment. In fact, I am; but if they think it's because I'm late for an appointment or because (later in the day) the bus is leaving, my departure from the table is expected and appears more natural. A little common sense goes a long way. I have no hard-and-fast rule for how long to play in one casino. But I am sure of one thing: Most amateur card counters — win or lose — overstay their welcome. If I win a lot — say 30 units or so — I'll be out the door. I consider it poor taste to shove it down the casinos' throats. Remember — OK, OK, you *do* remember the "welcome" bit! I won't mention it again.

And so, on the win side, I let amount rather than time dictate my departure. However, I do have an hour or so limit. Even if the win is meager, I don't show my face for too long. And, on the loss side, it is naïve to think that just because you've been losing, you can play forever. I've been formally barred from one casino in my life — Bally's Park Place — back when they had the right to do so. And do you know what? I was losing 25 units at the time. If someone is skillful enough to determine that you are a card counter, do you think it matters to him whether you are winning or losing? If you think it does, you are quite simply wrong.

Round one goes to Caesars. They beat me rather convincingly. No time to feel sorry for myself. Time is money. Get to the next casino. Trump Plaza is enormous. Dealers are inexperienced. Cut card position varies greatly. They don't know where the hell to put it! They'll learn, but while they're learning, I'll exploit the deep cuts. Every little edge helps. I go down the tubes again. They want to offer me the casino — meals, show, everything. "If there's anything we can do for you, Mr. S. (no, not Schlesinger!), just let us know." I thank them graciously, decline, and leave. I don't want to stay another hour, win back all of my loss and ruin their happiness. I'll get the money back, but it will be at another casino. At least, that's my plan.

You're no more than a 55% favorite to beat any one casino during any one playing session. If you make it a crusade to stay until you beat them all (you can't, no matter how much you want to, anyway), you're making a bad mistake. I have won 62% of the total sessions I have ever played in my life. This is an empirical fact. (I guess I've been a little lucky!) So what am I supposed to do, cry when a casino beats me? Technical ability comes through dedication and practice. But most of all, this game takes an incredible amount of *heart.* It takes an iron will and a fierce determination to succeed. It takes physical stamina, nerves of steel, and an inordinate amount of discipline and self-control. Without all of the latter, the former (technical skill) is meaningless.

Long Memories ...

It's 2:30, I'm a 28-unit loser, and a bit hungry. I grab a quick bite on the boardwalk (you have to understand my aversion to lengthy, drawn-out meals) and decide to honor Playboy (now Atlantis) with my presence. I like the third-floor, posh *salon privé.* Players bet fortunes up there. Nothing I put on the table can upset them. The tuxedoed pit bosses are accustomed to huge action. I find a good count and, as I move in, a young man practically knocks me over getting to the table. He hasn't even played a hand yet, but I already *know* he's a counter. Suggestion: As you spot a good situation, *walk* to the table. Don't you think it's a bit gauche to *charge?!*

I shouldn't have played, because two counters at the same table is deadly. You start orchestrating your bets in unison as the count rises and, to a skilled eye, it looks ridiculous. But the running count goes to +20 (Hi-Lo), and I'd like to be a part of it. I play. And win. But the guy next to me makes an ass of himself. He also happens to win a fortune, but as you know, I'm not impressed by that, because he can never play again at the Atlantis … and I can.

What was his crime? The count escalated so fast, he went from two hands of $200 to two hands of $600 with nothing in between. Result: blackjacks on both hands ($1,800) and several more winning plays before this great shoe ends. Also, *three* pit bosses, two calls to the "eye in the sky," several huddles in the pit, and numerous glares. In short, I hope the guy is satisfied with his score, because that's the last money he wins on day shift at Atlantis for a long, long time. Maybe forever. These bosses have *long* memories.

When the MGM Grand in Vegas changed from four to five badly cut decks a couple of years ago, I stopped playing there. Then, I was told (alas, erroneously) that the cut card had gotten better. After a more than two-year hiatus, I ventured back and played a couple of shoes even though the cut was mediocre. Enter a pit boss: "Oh, hi, Mr. S. Good to see you again. It's been quite some time." The problem was, I had no relationship with this guy. I knew him well by sight and am sure that somewhere along the line he had asked for my name, but I was really surprised. Moral: They don't forget for a long time!

And so, you must parlay your bets when you win. You win, the count goes up, you let the winnings (or a portion) ride. Eventually, you win again at a higher count and you get more money on the table. Yes, mathematically, another constraint. But it's a necessary one. People naturally parlay when they win. I simply consider it very risky to raise a bet after a loss or jump a bet (more than a parlay) no matter what the count is. Remember, survival is the name of the game.

The kid leaves the table and cashes out. Of course, I stay. There is no way in the world I'm going to leave a table at the end of the shoe with him. I mean, you didn't need me writing this article to teach you that, did you? I shouldn't have been at the table in the first place, but you just know I have to stay for a while now.

I pray for another high count, but the shoe is uneventful. The French have a proverb: *Les jours se suivent mais ne se ressemblent pas* — "The days follow one another but don't resemble one another." Substitute the word "shoes" for "days" and you've got the picture. If you think there's a pattern or an exploitable rhythm to this game, if you think there are "biases" or "dumping tables" or *predictable* hot and cold dealers, you'd better save your blackjack playing for Disney World, for as sure as a twenty beats a twelve, you're playing in Fantasyland. TARGET players — it's not too late to play this game properly — while you still have a bankroll. But I digress.

A Narrow Escape

It's 3:45, I'm still losing, but I've narrowed the gap. On to the Tropicana, where I dodge a very big bullet. Come along with me. The Trop is the best technical game in town, but that doesn't make it the best place to win money. After all, if they make it very difficult to play, then what good are the 76 well-cut 6-deck games? For a while I had a hard time playing there. I've never claimed that with a good act it is *impossible* to be detected in a casino. And there isn't a pro in the world who, sometime in his career, hasn't been spotted somewhere by somebody. After all, if I worked for a casino, do you think that there's a counter anywhere whom I couldn't spot in five minutes flat?

Well, then, it's conceivable that if a casino wants to go to the trouble, it can hire the proper personnel to spot me. And that's exactly what the Trop has done. I think they have more counter-catchers than the rest of the city combined. On this day, however, something unexplainable happens. They walk right by me. They let me play. Can it be that I've stayed away long enough (only six months or so) for them to forget? I can't believe that. Does it have anything to do with my being 35 units down? No, as you will learn shortly.

The 35-unit loss exceeds the 30-unit stop loss I use as a guide. In my system, one "session bankroll" equals 30 units and ten such "session bankrolls" (12 to 15 would be even better) constitute the total bankroll. So why have I permitted myself to lose more? Because I reached the limit in the middle of a very high-count shoe, and there were more hands to be dealt. You simply don't walk away from such a situation no matter how badly you're losing.

The "streak" system players will tell you you're throwing good money after bad and that there's no sense being stubborn and getting clobbered even further by finishing an obviously cold, "dealer-biased" shoe. The streak players are full of it!! The count is high and so you keep on playing. Period. If you don't agree with this, then stop reading, close the issue, and write to Arnold for a refund. He can't help you win and neither can I. You don't *want* to win. God bless you and I wish you luck. You'll need plenty of it, for surely that's the *only* way you'll ever win.

And so I play on and finish the shoe. I lose a little more. I did the right thing. In blackjack, you are right when you play correctly and wrong when you don't. Winning and losing have absolutely nothing to do with it. I change tables. After all, if you get your brains beaten in, you have a right to move on, no? Of course, you realize this is what I assume *they'll* be thinking. It's my excuse to leave a table where I no longer have an edge. Losing lets you get away with a few things in a casino. Walking around is one of them. "Let me see if I can find a table where the dealer pays the player once in a while" will do!

I get the dream-come-true situation — the ultimate in a shoe game. We start out as five players. The count skyrockets. And do you know what? Two people get up and leave! Usually, it's the other way around. What's more, the two other remaining players are bigger bettors than I. No matter what I put out, the pit will be more concerned with their play than with mine. It's helpful not to be the "big shot" at the table. Deal the cards, it's get-even time! I win back the 35 units and 18 more. That's right. I run this one shoe for 53 units. Forgive me, Pit Boss in the sky, for I have sinned. I have already told you that I don't approve of winning 53 units at one time. But a) I made sure the whole world knew that I was "almost even" after the bundle I had dropped at the other table, and b) what's a fella to do, quit in the middle of a shoe? I couldn't help myself.

I play a few camouflage hands off the top of the next shoe and make sure I lose the last one. The throwaway line goes something like this: "I've worked too hard to get even. I don't want to give it all back." I color up and leave. I know I promised, but — it's not the 18-unit win (after coming back from the dead!) that makes me happy. You guessed it: I'm *welcome* at the Trop again! Nothing else matters — certainly not the money.

It's 6:30. My feet hurt, my legs hurt, and worst of all, my eyes burn. I loathe smoke. I don't permit it in my house or where I work. But once inside a casino, I am helpless against it. Call it an occupational hazard. God, I hate it so. Well, I'm winning a little now. No big deal, but it's good to be in the black for the first time all day. Unfortunately, it's not going to last. Next stop, Golden Nugget.

The Nugget has won more money the past two months (May and June) than any other casino in A.C. So what, you say? Things like that can actually have an effect on your play, and I'll explain why. They're in a good mood there these days. They're loose and win-happy. The place is crawling with high rollers, and the casino is winning tons of money. That's a good atmosphere for playing. Also, despite the 8-deck game, the cut is excellent — average about 1 1/2 decks. There's money to be made here — unfortunately, not by me on this night!

I walk for 45 minutes and never play a hand. Are *you* capable of doing that? You have to be. Remember, you're in the casino to win money, not to play for the sake of playing. I don't sit down, because I can't find the right conditions. And believe me, it's not for lack of trying. Put a pedometer on me and I bet I've racked up a mile in the Nugget alone! Here's the frustrating part of the A.C. game. You finally find a good shoe, the count is super, and you lose anyway. I make a little comeback, but the net result is that I'm once again losing for the trip.

It's 8 p.m., I left home at 9 a.m., I've decided to drive back the same night (another three hours on the road), and so far the whole trip is for naught. When this happens, many players have a hard time justifying their actions to their families and to themselves. So maybe they press a little. Maybe they increase their stakes, or play negative shoes, just to have a *chance* at winning. You have to watch out for this. You're in this for the *long run.* Day trips (even weekends) are artificial divisions of time that have no real meaning in what is just an ongoing and continuous process. If you are destined to win 15 units in ten hours of play (about the average for the A.C. back-counting approach at 6- and 8-deck games combined), what does it matter if, in two five-hour days, you lose 15 then win 30; lose 5, win 20; win 20, lose 5; win 10, win 5; or any other combination? You have to think this way, or the game will drive you crazy. I'll now describe another way that blackjack will test your mettle.

The Hand

I decide to give it one more shot. Night shift begins at 8 p.m. on Fridays and Saturdays, so it's OK to return to Trump Plaza, as the personnel have all changed since the afternoon. Well, the personnel may have changed, but not the outcome. I'm winning just enough to be even when THE HAND arrives.

Now, before I set it up for you, let's review a few mathematical facts. The true count equals or exceeds +5 about 1.64% of the time in the 8-deck game. For the Revere or Halves counts, the frequencies are slightly higher and thus, correspondingly, so are the hourly win rates. You average around 25–27 hands *played* per hour (based on being able to see and count about 100 rounds per hour). If you put in six hours per day (150 hands played), you will be placing a top bet of two hands of six units each an average of only nine to ten times. And since it is at these counts that the largest contribution to your win is accomplished, they become very important. Win your fair share of them and you'll probably be a winner for the day. Lose them, and it's tough. And when the two hands turn into three, or even four, that can be the whole ball of wax for the day. Now, let's get back to the game.

The count is astronomical. I work up to the max bet. Dealer shows a 6. I make 20 on the first hand and the second hand is a pair of 3s. Already, the count has gotten even higher. I split the 3s and get a 6 on the first. Where are the big cards? The double down produces 19. I turn the other 3 into 18. There are 24 units on the table, and I've got 18, 19 (doubled), and 20. I teach all of my students the number one tenet of the game: *Never* celebrate early! You know you've won a hand when the dealer pays you — not a second before. The count is so high I can't believe it. Normally, a dealer's 6 breaks about 43% of the time. With this count, I'm sure it's closer to 50%. She flips a 3 in the hole. Although the entire process happens in a flash, I nonetheless have time to think: "I push the 19, lose the 18, win the 20. No catastrophe." Yet. The next card is a deuce. My heart sinks. *Still* no big cards. The rest is history. You're not really interested in *which* 10 it was, are you? I lose the 24 units I should have won. To me, this is a 48-unit swing. The dealer, a new young girl, actually apologizes to me: "I'm awfully sorry, sir." I try to console her and make a joke at the same time: "It's not your fault. On the other hand, it sure as hell isn't *my* fault, either!"

You have to play this game like a machine. What would a computer do now? It would play the next hand — after all, the shoe isn't over. If you can't do this, if you're devastated by the sad occurrence, you're not cut out for this yet. If you *do* play on but lose your concentration and keep returning in your mind to the "tragedy," you're not cut out for this. Blackjack will test your soul, your character, and the very fiber of your being. You have to sit there and take it. Otherwise, you're going to be playing this game on the funny farm! There will be better times. I win some back, but the final result of the trip (–32 units) has been sealed by the one hand.

Don't shed any tears for me. I've got a little lead on them! I'm in the car by 10 and home at 1 a.m. It's been a 16-hour day. Some restful way to spend a Saturday! When I write my first book, the title won't be *Blackjack for Fun.* And although it could very well be *Blackjack for Profit,* Arnold has already beaten me to that one!

But it's late now. I've got to get a good night's sleep. I'm going to A.C. tomorrow. There's just no way I'd rather spend a Saturday!

[Editor's note: As Don said, don't shed any tears for him. We understand he got the 32 units back ... and a few more!]

From "The Gospel According to Don," *Blackjack Forum*, June 1990:

Q. *After rereading your table-hopping article, it is clear to me that back-counting alone is a tedious process that takes a great deal of patience and discipline. It has also occurred to me that if I were to enlist the aid of one or more confederates who would act as "spotters" (back-counting different areas of the casino, but never playing), I could certainly enhance my hourly win rate. My question involves the calculation of the increase in profits these spotters would produce. I know, for example, that there would be some "overlapping," but I'm not sure how to do the math. Also, can you suggest an equitable manner for compensating the spotters for their time? I would appreciate any help you can offer.*

A. Although the concept of using spotters to increase back-counting hourly win rates is not new, I doubt that the kind of analysis I'm about to present, in response to our reader's question, has ever been published. I hope it will be useful to players who are contemplating using this style of play.

First, let's define the nature of the activity. Obviously, a spotter must walk in a separate region of the casino from where the primary player finds himself. What good are two people if they're back-counting the same tables? The problem of "overlapping" occurs when the player is already involved in a good shoe and, simultaneously, the non-playing spotter finds a second opportunity. It is possible that this second positive shoe will still be playable after the first opportunity is exhausted, but until that happens, there is a temporary period of time during which the spotter's efforts cannot be exploited. Simply put, the player can't be at two tables at the same time.

Fortunately, with only one spotter, this overlapping does not occur too frequently, and the math involved in calculating the effect is relatively easy. Let's assume, for the sake of simplicity, that a back-counter sees 100 hands per hour, of which he actually plays 25. In practice, these are, in fact, very realistic numbers. Thus, the player plays 1/4 of the hands seen, and the spotter finds another 1/4. Just adding the two, 1/4 + 1/4 = 1/2, or 50 hands, produces the wrong answer, since it does not account for the "interference" described above. Here's how we alter the incorrect answer. Multiply the probabilities that both will find a table simultaneously and then subtract from the 50 the number of hands this probability implies. We get 1/4 x 1/4 = 1/16. Rounded to the nearest whole number, 1/16 of 100 hands is 6. 50 – 6 = 44. The correct number of hands played per hour is 44.

Adding a second spotter can complicate the math, but I'm going to show you a shortcut that often simplifies probability calculations. Before we consider the two-spotter problem, let's go back to the original example. You'll like this approach. Each participant does *not* play 3/4 of the hands. Since, in these instances, probabilities are multiplicative, together, the two do not play 3/4 x 3/4 = 9/16 of the hands. Therefore, they *do* play 1 – 9/16 = 7/16 of the hands. (In probability theory, the totality of all the outcomes is expressed as 100% or, in fractional form, 1.) Now, 7/16 of 100 is approximately 44, and although you may not think this method is much of a shortcut in calculating the first answer, it becomes a very valuable technique when multiple spotters are involved.

Let's add a second spotter. The three now don't play 3/4 x 3/4 x 3/4 = 27/64 of the time. So they do play 1 – 27/64 = 37/64 of the hands or, roughly, 58 betting opportunities. See how simple that was?! While we're having fun, let's examine one more situation, this time with three spotters. $(3/4)^4$ = 81/256. So, 1 – 81/256 = 175/256, or about 68 hands per hour.

We are now in a position to analyze the percent increase in profits that can accrue to the back-counting player who uses spotters. With one spotter, 44 – 25, or 19 extra hands are played. 19/25 means a 76% increase in revenues. Two spotters yield 58 – 25 = 33 extra hands, thus a 33/25 = 132% increase. Finally, three spotters add 68 – 25 = 43 more hands, and so a 43/25 = 172% increase.

Now let's carry this one step further in an attempt to answer your second question. How should the spotters be compensated? Clearly, in my opinion, the actual results of play should have nothing to do with it. Rather, the theoretical "value" of the spotters should be calculated. How many extra dollars, on average, will their presence produce? Next, how should this surplus be divided? I suggest an equal split. After all, the spotters (who, presumably, don't have the bankroll to play themselves) need the player's money. In return, the player needs the spotters to enhance his revenue. I can foresee an objection. Suppose one player is a low-stakes bettor while another is a very high roller. A spotter who hooks up with the former will be paid much less for his efforts than if he were to team up with our well-heeled friend. Yet, in each situation, the spotter's efforts are identical.

Permit me to digress. If you ask a waitress to bring you a hamburger and she does, you tip her 15% of the bill. If the burger costs $3, she gets 45 cents. If, instead, you ask her for filet mignon and her trip to the kitchen (same as for the burger) produces a $30 piece of meat on your plate, you tip her $4.50. Now, I've never found this to be a very rational process, but that's the way it goes. So it really isn't unreasonable that the spotter who hooks up with the "filet mignon" will be paid more for his services than the one who works for "hamburgers."

Now, how will the original player make out if he shares equally with his spotter(s) the extra theoretical revenue that is produced? Well, with one spotter, the extra 76% is split 50-50, so 38% goes to the player. The extra 132% from two spotters is divided three ways, so the player gets 44% more. With three spotters, something interesting occurs. There is a 172% increase in profits, but a four-way split yields only an additional 43% for the player. Obviously, "diminishing returns" have set in, and it does not pay to add the third spotter under this arrangement. Now, I'll explain why, for practical reasons, I don't think a second spotter is worth the trouble either.

Casinos are, more often than not, noisy and crowded (particularly, those in which you are most likely to back-count). An attempt, on the part of the spotter, to get the player's attention by any kind of audio signal will undoubtedly fall on deaf ears. So, a visual call-in must be used. Suppose the player is busy looking down at a prospective table? Suppose he does manage to see the spotter immediately, but can't navigate his way through the casino until one or two hands have been dealt? The bottom line is that, in reality, hands will be missed. If the idea for two spotters is to add 33 extra hands, it is not at all unreasonable to estimate that five of these hands will go unplayed. But this reduces the extra edge to 28/25 = 112%. Split three ways, it becomes 37.3%, or less than the two-way 38% split. And even if only four hands are missed (and I'm certain they would be), 29/25 = 116% and, consequently, an extra 38.7% for the player. Surely, compared to the one spotter 38%, it isn't worth the extra effort. Conclusion: If you intend to share additional revenues equally, play with exactly *one* spotter to maximize your back-counting profits.

Of course, my profit-sharing suggestion is not the only conceivable method for compensating spotters. Indeed, I know of a team in operation now that pays spotters a fixed, hourly wage. Let's say, in our above situation, spotters earned $25 per hour, no matter how many were employed. (And let's assume $100 per hour for our player.) Here the theoretical value to the player would be quite different. After all, with one confederate, the player would keep $76 – $25 = $51 of the extra profits. Paying two spotters would still leave $132 – $50 = $82 more for the player. Even three spotters, who would earn a total of $75, would leave an additional $172 – $75 = $97 for the player. And four (getting a little crowded now!), after their $100 salary, would nonetheless produce an additional $104 for the player. Not until the fifth spotter is hired would his presence be superfluous, as the player, after doling out $125 in salaries, would be left with "only" $103 extra for himself, a decrease of $1 compared to the four-spotter arrangement.

Here again, I believe that missed opportunities would preclude the use of four spotters, but a marginal case could be made for at least three. Obviously, there are several possible "variations on the theme," and this short piece is meant simply as a guideline to those who are contemplating the idea.

I have summarized the above findings in Table 1.1. I hope you have found this information useful, and I wish you success with your back-counting endeavors. Good luck, and ... good cards.

Postscript: *It has been 12 years since this article first appeared, yet, little has changed in my approach to the shoe game. I can't think of any advice that I would alter. Obviously, casino conditions are constantly changing, and many of the games described no longer exist in the same form as they were. The computer age has, however, made practicing a lot more fun than it was back then. Today, I believe the practice regimen, briefly described in the article's first paragraph, would be a lot more enjoyable, thanks to the genial software, such as Norm Wattenberger's* ***Casino Vérité,*** *that has been designed for such purposes. I'll be making further remarks on simulators and study aids, throughout the book.*

Finally, note the casual reference, in the next-to-last paragraph of the "Table-Hopper," to "when I write my first book." It may have taken 13 years, but I'm delighted to say that the time is now!

Table 1.1
Back-Counting with Spotters

No. of People	Hands Played per Hour[1]	Percent Increase in Profits	Extra Revenue (%) to Player (increase shared with spotters)	Hourly Profit for Player[2]	Extra Revenue ($) to Player (fixed wage, spotters)	Hourly Profit for Player[2]
player alone	25	—	—	100	—	100
one spotter	44	76	38	138	51	151
two spotters	58[3]	132[3]	44[3]	144[3]	82[3]	182[3]
three spotters	68[3]	172[3]	43[4]	143[4]	97[3]	197[3]
four spotters	76[3]	204[3]	—	—	104[3]	204[3]
five spotters	82[3]	228[3]	—	—	103[4]	203[4]

[1] Rounded to nearest whole number.
[2] Based on $100 per hour playing alone.
[3] Potentially lower because of missed opportunities (see text).
[4] Case of "diminishing returns."

June 1985 Cover Art by Paladin

Chapter 2

Betting Techniques and Win Rates

[Editor's note: Don wrote Part I of this article in the early 1980s, and included it in a newsletter he and his partner, Ken Feldman, sent out to their students. At the time, Schlesinger and Feldman held the New York franchise for Jerry L. Patterson's Blackjack Clinic. Don expanded the article considerably with Part II, which appeared in the June 1985 ***Blackjack Forum.*** *"The Ups and Downs of Your Bankroll," as it was initially titled, provides the most in-depth treatment of "bankroll fluctuations" ever presented for non-mathematicians. Though Schlesinger's work specifically analyzes the Hi-Lo counting system, his conclusions are applicable to all valid point-count systems played under similar conditions. — Arnold Snyder]*

Part I

On a recent trip to Las Vegas (one of my four losing ventures in 24 attempts), I blew a 65-unit lead and ended up a 45-unit loser — a swing of 110 units. Playing for the same stakes, a student of ours (one of our most successful ones) just encountered a similar catastrophe when, in one week in Las Vegas, his 57 units in profits evaporated and turned into 53 units' worth of red ink. No matter what stakes we play for, we have all experienced these incredible roller coaster-like fluctuations in our bankroll. I am writing this short essay for two reasons: 1) To assure you that such swings are, indeed, quite normal and to be expected, and 2) To explain to you why.

The culprit is something mathematicians call "standard deviation." Before we define the term, we need to consider another one: "expected value." Flip a coin 100 times. How many heads do we expect? Fifty, of course. Roll a die 600 times. How many sixes do we expect? One hundred. Bet $2 per hand for 100 hands at blackjack (a total of $200 wagered) at an average advantage of 1.5%. What profit do we *expect* to show? $200 x 1.5%, or $3. Thus, in layman's terms, "expected value" is nothing more than the "average" or "mean" outcome that we expect for a particular event, given a certain set of circumstances.

But rarely does the coin *actually* come up heads 50 times. Rare, indeed, would be *exactly* 100 sixes in 600 tosses of a die. And rarely do we win at blackjack for any given session the exact mathematical amount that we *expected* to. We are usually a certain amount above or below "expected value." The measure of how far we stray in either direction from "expected value" can be calculated very precisely, and it is this departure from "normal" that we label "standard deviation."

Now, for a blackjack game, standard deviation (s.d.) is approximately equal to 1.1 $\sqrt{N}$, where N = number of hands played. Take our prior example of 100 hands. S.d. = 1.1 $\sqrt{100}$ = 11. Thus, the s.d. for 100 hands is 11 units. If our units are $2 each, then one s.d. = 11 x $2, or $22. How do we *use* this figure? We take our "expected value" ($3, for the current example) and we both add and subtract one s.d. (in this case, $22) to the outcome. $3 + $22 = $25, and $3 – $22 = –$19. We then get a range (and a rather large one, at that) from –$19 to +$25. The actual outcome of our playing session will fall somewhere within this one-s.d. range 68.3% of the time (a little more than two-thirds). Thus, although we "expect" to win $3, we shouldn't be the least bit surprised if, instead, we were to win $25 or, alas, lose $19. The whims of s.d. assure us that such ups and downs will, indeed, occur with alarming regularity.

If we are willing to expand our range to two s.d.s in each direction, we will achieve an actual result that lies inside this widened range 95.4% of the time. And if we go up or down three s.d.s, we include 99.7% of the actual results (that is, practically *all* of them).

A good weekend's worth of play will yield, roughly, 1,000 hands. S.d. = 1.1 $\sqrt{1{,}000}$ = 35. (This is actually an underestimate of the true s.d., which will be determined with more precision, in Part II, below.) Thus, s.d. for 1,000 hands is 35 units. Now, let's get down to cases. If you spread from $5 to $40 (two hands of $30 would be better!), your average unit (or bet) will be about $10. So, for the weekend, you will wager 1,000 x $10, or $10,000 (more than you thought, huh?). At a 1.5% edge, your "expected" return is $10,000 x 1.5% = $150. But it rarely turns out that way, and you've been wondering why. Enter standard deviation (in this case, 35 units, or $350). Wonder no more! Adding one s.d. ($350) to $150, we get $500. Subtracting one s.d. from $150, we get $150 – $350 = –$200. You say you won $500? Enjoy it! Next time you're just as likely to *lose* $200! What's more, 95% of the time (19 times out of 20), your weekend's result will lie somewhere between –$550 and +$850. And you thought you were the only one riding the roller coaster! Multiply all of the above by a factor of 10 for the $50 to $400 player, and you begin to understand why high-stakes card counters often get ulcers!

Can you lose for an entire weekend? Week? Month? Year? (God forbid!) The answer is "yes" to all of the above. The reason? Standard deviation. But take heart. The greater the number of hands you play, the smaller the *percentage* of s.d. Thus, put in everyday terms, the more you play, the closer (on a % basis) your actual outcome will be to your "expected" outcome. And don't worry if it seems like just when you've reached the top, you slip back down again. Peter Griffin estimates that we're at an all-time high only 1% of the time. Thus, 99% of the time, we have the nagging feeling that we "once had more." The fluctuations will *always* be there. Play through the losing streaks, enjoy the winning sessions, and always remember — we have a partner in this exciting venture of ours: standard deviation.

Part II

Since I wrote the above article, about two years ago, several students have asked me to be more precise about standard deviation and win-rate calculations as they applied to our specific style of back-counting and betting (see Chapter 1). During the year 1983, Stanford Wong carried out the most complete set of computer simulations ever published for the Hi-Lo count and for the risk and return associated with 4-, 6-, and 8-deck games (See Wong's *BJ Newsletters,* Vol. 5). I have used Wong's data for much of the work in this article, but have expanded upon it and included some original material that, to my knowledge, has never before appeared in print.

When one scales his bets upward according to a rising true count, one not only increases expected return, but unavoidably, standard deviation as well. Indeed, larger risk (as measured by s.d.) is the price one pays for being able to achieve satisfactory rates of return in multiple-deck games. The purpose of this article is threefold: 1) to determine precise win rates for the back-counting and play-all-decks approaches to the game, 2) to determine accurate standard deviations for the same styles of play, and 3) to relate the two concepts by constructing tables that reflect the probability of being ahead after n hours of play in the various games. Along the way, I shall explain all the mathematical principles involved so as to facilitate the reading and to justify all of my contentions. Finally, some interesting conclusions will be drawn.

Here is a brief summary of my back-counting approach to the shoe game. I play only true counts of +1 and higher. I bet one unit at +1, two units at +2, four units at +3, six units at +4, and two hands of six units at +5 and higher. Thus, my top bet is two times six units no matter how high the true count goes. Determining how many hands per hour one actually gets to *play* with this technique is not a cut-and-dried objective calculation. But I believe the following to be a fair and accurate approach to the problem.

First, a few empirical facts gleaned from thousands of hours of experience and observations. Assuming four players and the dealer, it takes about 30 seconds for a competent dealer to deal one round. During that round, 5 hands x 2.7 cards per hand = 13.5 cards will be dealt, on average. Thus, it takes about two minutes to *deal* one deck. If we assume a 75% cut-card position at all times (for the 4-, 6-, or 8-deck game), we find that for the *dealing* alone, the 4-deck game takes 6 minutes (3 decks dealt), the 6-deck game, 9 minutes (4 1/2 decks dealt), and the 8-deck game, 12 minutes (6 decks dealt). Observation confirms the following shuffle times: 1 minute 20 seconds for 4 decks, 1 minute 40 seconds for 6 decks, and 2 minutes 20 seconds for 8 decks. Thus, it requires 7 1/3 minutes to deal and shuffle a 4-deck shoe, 10 2/3 minutes for a 6-deck variety, and 14 1/3 minutes for the 8-deck game. Dividing each into 60 minutes, we get the following *hourly* figures: 8.18 4-deck shoes, 5.62 6-deck shoes, and 4.18 8-deck shoes. Next, we must interpret how many *rounds* per hour this translates into. The results are as surprising as they are convenient.

When 3 decks are dealt, 156 cards are used. But when the cut card is reached, the hand must be completed; thus, on average, $\frac{13.5-1}{2} = 6.25$ extra cards will be seen. (We subtract one because if the cut card is poised to come out, the next hand is not dealt.) So, for the 4-deck game, 162 cards per shoe ÷ 13.5 cards per round = 12 rounds per shoe. Multiply this by 8.18 shoes per hour and we get 98 rounds per hour. Similar calculations for the 6-deck game yield 240 ÷ 13.5 = 17.78 x 5.62 = 100 rounds per hour. Finally, for 8 decks, 318 ÷ 13.5 = 23.56 x 4.18 = 98 rounds per hour. How very nice! Will anyone mind if I call all three exactly 100 rounds per hour? This is a great facilitator of later calculations.

Now, the question is, "Do I actually assume that, in back-counting, the number of rounds per hour actually *counted* is this 100 figure?" I say yes. Granted, by judiciously walking away from poor counts in progress, one ought to be able to do better, in theory. However, in practice, it is not that easy. Tables are crowded. What good is it to count if you can't play? Physically navigating the huge and unbelievably crowded Atlantic City casinos cuts down on mobility. Limited availability of higher-stakes tables invalidates the assumption that one can always find *instantly* another dealer ready to shuffle. For all these reasons, I believe it is realistic to approach the calculation in the following manner. Assume you actually have the right to sit at any given table and count all the rounds for the hour and that, furthermore, you may bet *only* the +1 and higher counts, betting, in effect, zero for the others. If you agree to accept this premise, then all that follows should be logical and easy to understand. *[Note: For the latest findings in this area, see Chapter 13, Part I. — D.S.]*

Wong's frequency distributions give us 29.18 hands per hour played for 4 decks, 26.77 for 6 decks, and 24.51 for 8 decks. Plug in the appropriate bets at the appropriate true counts, multiply horizontally by frequency and corresponding advantage, sum vertically, and you get the following hourly win rates (in units) for each game: 2.76 (4 decks), 1.68 (6 decks), and 1.12 (8 decks). I have prepared a table, for the reader's convenience, that summarizes not only these figures, but also other important data. (See Table 2.3.) It is interesting to note that if one decides to play all decks of the 4-deck game (and thus 100 rounds per hour, making minimum, one-unit bets 81.76% of the time), the win rate is nonetheless exactly 2.00 units per hour. (And this is for the slightly inferior Las Vegas game with no double after splits.)

One immediate practical application of this finding for the A.C. player is that it is more profitable to play the 4-deck Tropicana game in the play-all style than it is to back-count any other game (except the 4-deck Trop) in town. I am not sure that this is readily apparent to the average player. Please note that throughout this paper, I have purposely omitted any discussion of the 8-deck play-all approach. The win rate for this game is so poor as to make it totally unacceptable. Anyone considering playing blackjack in this style is, in my opinion, wasting his time.

As for the 6-deck play-all approach, the win rate (1.02 units per hour) is *less* than that of the 8-deck back-counting game while the s.d. (2.63 units per round, but 26.25 units per hour) is *greater.* Thus, for these reasons and others discussed later (camouflage), there is little to recommend for this style of play where 85% of the wagers placed would be at one unit and where the overall edge is a mere 0.65%.

The next part of our discussion continues the original topic of standard deviation. Once again, Wong's data are indispensable. (See, in particular, pp. 95–96 of the June 1983 *Blackjack World.*) I have included detailed charts of the necessary calculations that lead to the following sobering conclusions. For the 4-deck game, the one-round standard deviation for my back-counting approach is 5.50 units and the *hourly* s.d. is 29.70 units. (Remember, whereas the hourly expectation is the one-round expectation multiplied by the number of rounds per hour, the hourly s.d. is the one-round s.d. multiplied by the *square root* of the number of rounds played per hour.) Six-deck figures are 4.70 units for one-round s.d. and 24.32 units hourly s.d. For the 8-deck game, we have 4.15 units for one-round s.d. and 20.55 units for hourly s.d. (See Table 2.1.) Finally, for the play-all 4-deck approach, the one-round s.d. is 3.12 units whereas the hourly s.d. (for 100 rounds) is 31.23 units.

TABLE 2.1
Standard Deviation by Number of Decks

8-Deck Game, Deal 6

Count	Bet Squared (units)	Hands	Frequency	Var + (h−1)* Cov	Product
1–2	1	1	.1187	1.339	.1589
2–3	4	1	.0619	1.341	.3320
3–4	16	1	.0316	1.355	.6850
4–5	36	1	.0165	1.364	.8102
5–6	36	2	.0085	1.883	1.1524
6–7	36	2	.0043	1.897	.5873
7–8	36	2	.0020	1.888	.2719
8–9	36	2	.0009	1.886	.1222
9+	36	2	.0007	1.828	.0921
			.2451		4.2122

$\frac{4.2122}{.2451} = 17.1856.\ \sqrt{17.1856} = 4.15$ = one-round s.d. in units

6-Deck Game, Deal 4 1/2

Count	Bet Squared (units)	Hands	Frequency	Var + (h−1)* Cov	Product
1–2	1	1	.1177	1.341	.1578
2–3	4	1	.0653	1.343	.3508
3–4	16	1	.0371	1.357	.8055
4–5	36	1	.0213	1.366	1.0474
5–6	36	2	.0120	1.880	1.6243
6–7	36	2	.0067	1.892	.9127
7–8	36	2	.0036	1.881	.4876
8–9	36	2	.0020	1.860	.2678
9–10	36	2	.0010	1.838	.1323
10+	36	2	.0010	1.807	.1301
			.2677		5.9164

$\frac{5.9164}{.2677} = 22.1009.\ \sqrt{22.1009} = 4.70$ = one-round s.d. in units

4-Deck Game, Deal 3

Count	Bet Squared (units)	Hands	Frequency	Var + (h−1)* Cov	Product
1–2	1	1	.1094	1.314	.1438
2–3	4	1	.0684	1.322	.3617
3–4	16	1	.0421	1.332	.8972
4–5	36	1	.0275	1.338	1.3246
5–6	36	2	.0169	1.870	2.2754
6–7	36	2	.0107	1.916	1.4761
7–8	36	2	.0065	1.934	.9051
8–9	36	2	.0042	1.965	.5942
9–10	36	2	.0025	1.956	.3521
10+	36	2	.0036	1.931	.5005
			.2918		8.8308

$\frac{8.8308}{.2918} = 30.2632.\ \sqrt{30.2632} = 5.50$ = one-round s.d. in units

* Note: h = number of simultaneous hands played

For the play-all 4-deck game, the horizontal multiplication for counts below +1 adds only .922613 to the final "product" column. Thus, the new sum is 9.7534 for the full 100 rounds. Now, $\frac{9.7534}{1.00} = 9.7534$. $\sqrt{9.7534} = 3.12$ = one-round s.d. in units.

Why do I call the above "sobering conclusions"? Well, refer back to the original article on standard deviation (which supposes a *flat* bet equal to the *average* bet of the 1–12 spread scheme) and you will see how much greater the actual standard deviation really is. I don't suggest that there is any brilliant way to avoid the large s.d. — it is a way of life that comes with the territory — but I certainly think that it is incumbent upon all serious players and students of the game to understand the figures and their implications. And so, I have found a way to relate the two crucial concepts of expected win and standard deviation by studying the interaction of the two. This can be accomplished by exploring the probability of being ahead (winning $1 or more) after n hours of play for each of the four approaches discussed in this article. The only chart of this nature I have ever seen first appeared in Ken Uston's *Two Books on Blackjack* and was reprinted in his *Million Dollar Blackjack*. But Kenny's table was for a different bet scheme, a different count, and, most of all, a different style of play. For this reason, I believe that the present table provides insights into the back-counting approach that have never before appeared in print. For those who understand the cumulative normal distribution function, the calculations are straightforward. Simply determine, for a given number of hours, what percentage of the standard deviation the corresponding expected return represents. Then find, in this case, the area under the curve to the left of this figure. (See Table 8.10.) This is the probability of being ahead after the required number of hours. (See Table 2.2.)

Note that, in determining the probabilities for this array, an assumption is made that one's bank is sufficiently large so as to permit play to continue for the requisite number of hours, without regard for tapping out (losing one's entire stake) prematurely. See

TABLE 2.2
Probability of Being Ahead after n Hours of Play

Hours	8-deck	6-deck	4-deck	4-deck, play all
1	.52	.53	.54	.53
5	.55	.56	.58	.56
10	.58	.59	.62	.58
20	.60	.62	.66	.61
40	.64	.67	.72	.66
50	.65	.69	.74	.67
100	.71	.76	.82	.74
200	.78	.83	.90	.82
400	.86	.92	.97	.90
500	.89	.94	.98	.92

Chapter 8 for a complete discussion of this phenomenon.

Now, all of the mathematics in the world goes for naught, in my opinion, if it doesn't have some practical use at the tables. Much can be gleaned from a careful consideration of Table 2.3. And many of the unbelievable frustrations that are inherent in high-level blackjack play are more readily tolerated when one understands standard deviation. What, specifically, can we learn? For one thing, s.d. is a potent force to be respected greatly. In the 4-deck back-counting game, one is playing with a 2.47% advantage and one can win 2.76 units per hour. But even after 200 hours at the tables, one player in ten is still going to be losing. Will he understand why? After 500 hours, one unfortunate soul out of 50 will still be in the red. Try explaining to him that *someone* has to fall in the tail of the normal curve! What about the 20-hour weekend player? Well, he will lose once out of every three trips, while a ten-hour player loses three out of eight times. Finally, if you subdivide your records into daily or even hourly results, you're doing just fine if you win three out of five days and five out of nine hourly sessions. And these are all for the "best-case" 4-deck back-counting approach. Let's go on.

You're trying to make your fortune at the 8-deck game? Well, I hope you've got the patience (and the bankroll!) to prevail. God knows, it won't be easy. (For that matter, the Bishop knows it, too. He's been telling you for four years!) In fact, after 500 hours of play, one out of nine hardy souls will actually still be losing. Sad, huh? You're planning a one-week, 40-hour assault on the A.C. 6-deck game? Well, you should be a winner two out of three times, but don't expect any more. Finally, notice the similarity between the 6-deck back-counting figures and the 4-deck play-all numbers. They are almost identical. Given my choice, which would I prefer to play? It depends. Sedentary play is more comfortable, but it can be more risky to your cover. I don't think it looks good to play 82 hands out of 100 at one unit and then to play a few at 6 units or two hands of 6 units. Even with a good act, this is tough to pull off. On the other hand, 27 quality, back-counted hands per hour will produce 84% of the play-all approach profits and will, in my view, increase one's longevity. Never forget: You have to be welcome back the next time.

One thing is certain. To be successful at this game, one needs an inordinate amount of dedication. The ravages of standard deviation will test even the most hardened veteran. When this happens, take a break, reread this article, and go get 'em. Good luck, and … good cards!

P.S. to all of the above. I wonder how many readers would be interested in seeing the above materials expanded into a more complete work. One could, for example, consider single- and double-deck games, rules variations, different cut-card positions, different point counts, and, of course, different betting schemes. The combinations are almost endless. I have the energy to collate the material and to write the text. Perhaps someone has the interest, computer capability and know-how to generate the figures. Stanford? Peter? Any takers out there? Give me a call!

Postscript: *I'm pleased to say that the "P.S." that appeared at the end of the "Ups and Downs" article was directly responsible for the creation of Arnold Snyder's wonderful* **Beat the X-Deck Game** *series. I highly recommend it to all those who are interested in the statistical aspects of blackjack, as I am.*

What's more, I have vastly expanded upon the scope and content of this chapter with a brand-new analysis of win rates, effects of rules, penetration, and bet spreads, in this book's Chapters 9 and 10. For simulation "freaks," you might want to take a peek now!

TABLE 2.3
Analysis of Win Rates and Standard Deviation for Various Blackjack Games

Type of game and style of play	**8-deck, deal 6** AC rules, play TC +1 and higher	**6-deck, deal 4 1/2** AC rules, play TC +1 and higher	**4-deck, deal 3** LV Strip rules, play TC +1 and higher	**4-deck, deal 3** LV Strip rules, play all decks
1. Bankroll required for optimal bets (units)	400	400	360	360
2. Average bet size (units)	2.72	3.14	3.82	1.82
3. Actual no. of hands played per hour	24.51	26.77	29.18	100
4. Advantage (percent)	1.68	2.01	2.47	1.10
5. Win per hour (units) Rows 2 x 3 x 4	1.12	1.68	2.76	2.00
6. Divide your b.r. by this no. to get win per hour in units	357	238	130	180
7. Divide your b.r. by this no. to get avg. bet size in units	147	127	94	198
8. Standard deviation per round (units)	4.15	4.70	5.50	3.12
9. Standard deviation per hour (units)	20.55	24.32	29.70	31.23
10. Hours for expected win to equal s.d.	335	210	115	243
11. Hours to double bank*	357	238	130	180

*Note: bankroll requirements are not identical for all four games

From "The Gospel According to Don," Blackjack Forum, June 1989:

You never know where the idea for an article (or, at least, a mini-article) will pop up. In the December 1988 *Blackjack Forum,* a letter from a reader in Utah asked the following question: "My friend (who generally stays at the table unless the count gets very negative) argues that, when the count is high, he'll bet just one bet, his maximum bet of $500 a hand. I, however (back-counting), will play three or four spots at a time at $300 per spot, whereas my maximum normal single bet per hand would be $500. My friend states that it is more mathematically correct to play one spot against the dealer with a high count versus multiple hands against the dealer with a high count. I disagree. Could you please settle this argument."

Arnold Snyder's reply was adequate, given the shortage of space he had for his answer. "When the count is high, you'd theoretically like to have as much money on the table as your bankroll can afford. If you can get more on the table with multiple hands, then multiple hands are your best strategy, provided, of course, that multiple hands do not hurt you in some other way — notably, tipping off the casino that you may be a card counter."

With all due respect to the Bishop, I would like to expand upon his reply and examine, in greater detail, the question of when and why one ought to play multiple hands at the shoe game. Along the way, I trust that the reader from Utah will be satisfied with the result, which turns out to be a compromise between his approach and that of his friend.

Before considering the merits of multiple hands, it is important to understand the concept of optimal bet sizes. For a given bankroll, and a given advantage at a specific true count, an optimal bet is determined by multiplying the bankroll by the edge (in percent) and dividing the outcome by the variance (Griffin prefers the average squared result of a hand of blackjack, but the difference is slight) of hands played at the particular true. According to Stanford Wong, when one makes a bet at a true of +5 (level at which most pros "max out" and place their top bet), the wager should be 77% of the player's bankroll times the advantage. For the 6-deck A.C. game, given a $10,000 stake, for example, the correct bet is $196.

Now, when two or three hands are played, one must consider the covariance between these hands and wager on each one of them a percentage of the one-hand bet that creates risk and profit potential commensurate with the optimal one-hand play. For two hands, one divides by the sum of the variance and covariance. For n hands, the general formula is to divide by the sum of the variance and (n – 1) times the covariance. Thus, the "trade-off" for more money on the table is more risk, in the form of higher standard deviation. What are the optimal percentages? For two hands, 73% of the one-hand bet on each of the hands, and for three hands, 57% of the one-hand bet on each of the hands.

Our reader from Utah and his friend each make a maximum one-hand bet of $500. This means that, should they choose to play multiple hands, they are entitled to two hands of $500 x 73% = $365, or three hands of $500 x 57% = $285. Let us note, in passing, that Utah's idea to play "three or four hands of $300" is marginally acceptable for the three hands (a slight over-bet), but is definitely unacceptable for four hands, where he would be grossly overbetting his bankroll. Now the question arises: "Which is the superior approach? How do I know whether to bet the one hand, two, or three?" Here is how I suggest we figure it.

It is not sufficient to simply "get the greatest amount of money on the table" (obviously, the three hands) for any one given round, thereby assuming that this strategy will get the greatest amount on the table by the time one has finished playing. Each spot played uses up cards. We must consider, for any given shoe, how much money will actually be put on the table with each one of the approaches. It is only after considering the total "global" wager that one can determine the best approach. The "correct" answer is neither obvious nor intuitive.

Let us assume that, if we play long enough (just how long is easy to calculate), we will eventually have the opportunity to place 100 maximum bets (true of +5 or above). Furthermore, for the purposes of this demonstration, let head-up play against the dealer in a 6-deck, deal 4 1/2 (75%) game be the norm. Now, consider the effect of playing one, two, or three hands on the number of cards used per round. *[Note: 2.7 cards per completed hand is average. Therefore, head-on play uses 5.4 cards per round, and each additional hand at the table adds 2.7 cards to that total. — D.S.]* Consider, also, the effect of other players at the table. The following chart indicates exactly how many hands can be played under varying conditions, given our above scenario. In addition, I have multiplied hands played by dollar amount of wager to produce, in each box, a total amount of money wagered.

To determine, for any given box, how many hands (maximum = 100) can be played, use 5.4 as the numerator of a fraction. Next determine the total number of hands being played for the box in question. For example, playing with two other players and playing two hands equals five hands. Multiply the number of hands by 2.7 (in this case, 5 x 2.7 = 13.5). Use the result as the denominator of the fraction. Thus, 5.4/13.5 = 2/5 = 40%. Multiply 100 by this resulting fraction (for example, 40 hands).

On each horizontal line (Table 2.4), the maximum dollar amount is in **bold print,** indicating the optimal play.

The results are most interesting. If the goal is to get as much money on the table as possible, then, when playing alone, only one hand (of $500, in this illustration) should be used. When joined by one or two other players, we optimize potential with

TABLE 2.4
Optimal Number of Simultaneous Hands
Based on 100 Top-Bet Opportunities of One Hand, Playing Alone

Situation	Hands per round	Cards per round	1 hand of $500	2 hands of $365 = $730	3 hands of $285 = $855	Optimal play
Alone	2	5.4	100 x $500 = **$50,000**	67 x $730 = $48,910	50 x $855 = $42,750	1 hand
With 1 other player	3	8.1	67 x $500 = $33,500	50 x $730 = **$36,500**	40 x $855 = $34,200	2 hands
With 2 other players	4	10.8	50 x $500 = $25,000	40 x $730 = **$29,200**	33 x $855 = $28,215	2 hands
With 3 other players	5	13.5	40 x $500 = $20,000	33 x $730 = $24,090	29x $855 = **$24,795**	3 hands
With 4 other players	6	16.2	33 x $500 = $16,500	29 x $730 = $21,170	25 x $855 = **$21,375**	3 hands

two hands. Only when there is a total of four or five players at the table should three hands be played, and even here, I would strongly recommend against this. First of all, it isn't such a great idea to be playing with three or four other players. This cuts down too much on the hourly expectation. Secondly, the extra amount of money gotten on the table is only marginally greater than the two-hand approach, and the price paid in possible heat and pit attention for spreading to the third hand is most definitely not worth it.

Now, there is one instance, even when playing alone, when spreading to a second hand is a must. If you are proficient in eyeballing the discard tray and "know" when the last hand in the shoe is about to be dealt, then by all means, spread to the second hand. It doesn't matter any more that you are using up extra cards. You are guaranteed the completion of the round and, therefore, you would be foolish not to get the extra money on the table.

And there you have it. Thank you, reader from Utah, for a thought-provoking question that, I trust, I have answered to your satisfaction. To summarize: Play one hand if you are alone with the dealer; play two hands, each of which is 73% of your one-hand top wager, under all other circumstances. Finally, let's put the whole thing in perspective by considering the following: Whether you back-count or play all in any given hour, you play the same number of top-bet hands. Assuming 100 hands seen (or played) per hour, roughly 3% of them (three hands) are at a true of +5 or greater. Thus, it would take, on average, 33 hours to play the 100 top-bet hands in our illustration. If you never spread to two hands, you forfeit roughly $4,000 in action. Your edge for these hands is about 3%, so you lose about $120 over those 33 hours, or about $4 per hour. As is often the case with blackjack questions, sometimes we become so enamored of the theory that we lose sight of the practical, bottom-line implications.

From "The Gospel According to Don," *Blackjack Forum*, December 1990:

Q. *I have read several texts that have indicated that the formula for determining "expected value" is: Average bet x Number of hands per hour x 1.5% (approximate advantage) = Expected value per hour.*

It seems to me that the calculations do not consider the "spread" and, therefore, cannot be accurate. For instance, if I'm betting $5–$100 (1 to 20 spread), as opposed to $10–$100 (1 to 10 spread), my average bet would be lower for the first case, but my expected value will most certainly be higher. Can you shed some light on this topic?

A. This question, although relatively simple to answer, raises several interesting points that have caused a good deal of confusion for many readers. The first item is the misconception that a particular game has some "pre-ordained" or inherent advantage associated with it. In fact, it does not. What it does have is a frequency distribution of true counts, each of which represents a certain advantage or disadvantage. Until we specify a betting scheme, which associates an amount to be wagered with each true count, it is not possible to ascertain the overall advantage we will enjoy playing this game. And that is the problem here. The reader has assumed a fixed, 1.5% advantage for each of the two situations he describes. In reality, each spread has associated with it its own edge, and the two advantages are not the same.

Let's examine two cases, A and B. For each, to make matters simple, we'll assume that 1/3 of the time we labor under a 2% disadvantage, 1/3 of the time the game is neutral (edge = 0%), and 1/3 of the time we enjoy an edge of 1.5%. (I don't claim this represents reality — just humor me!) Also, to make the math come out even, we'll duplicate the 1–20 and 1–10 spreads, but with wagers of $6 to $120 and $12 to $120, respectively. We bet the lower amounts 2/3 of the time and the higher amount 1/3 of the time, when we have the edge. Now, consider the two cases, A and B, in Table 2.5.

TABLE 2.5
Expectation as a Function of "Spread"

	Frequency	x	Advantage	x	Bet	=	Win or Loss	Hourly Expectation
Case A	1/3		–2%		$6		–$.04	
	1/3		0%		$6		.00	
	1/3		1.5%		$120		.60	
							$.56	x 100 hands/hr = $56
Case B	1/3		–2%		$12		–$.08	
	1/3		0%		$12		.00	
	1/3		1.5%		$120		.60	
							$0.52	x 100 hands/hr = $52

Average bet for Case A is just (1/3) x ($6 + $6 + $120) = 1/3 x $132 = $44. For Case B: (1/3) x ($12 + $12 + $120) = 1/3 x $144 = $48. So far, our reader is correct in that the 1 to 20 spread (Case A) produces a lower average bet than the 1 to 10 spread (Case B) but, indeed, a higher expected win ($56 v. $52). How can this be? Because the advantages are not the same, as the reader had assumed. Divide the win by the total amount wagered to yield the hourly advantage. For Case A: 56/4,400 = 1.27%; for Case B: 52/4,800 = 1.08%.

It has always amused me when players compare the relative merits of different games by citing advantage alone. This reminds me of the old joke where the sportscaster has just given several final football scores and then says, "And now, a partial score: Notre Dame, 21 … " The point is, of course, that quoting advantage in a vacuum is as meaningless as judging a score when only one team is mentioned. If someone offered you the choice of playing a game with a 1.1% advantage or one with a 2% edge, which would you choose? Before you answer, look at Table 2.3. Notice the hourly win rate (2.00 units) of the play-all 4-deck game (1.10% edge). Now compare this to the 1.12 units won in the 8-deck back-counting game (1.68% edge), or to the 1.68 units won in the 6-deck back-counting game (2.01% edge).

What I would like you to retain from all of this is that the only thing that matters, when you evaluate a game solely for its profit potential, is how much money you will win per unit of time. Taken separately, the average bet (or the spread), the number of hands played per hour, and the overall percent advantage are relatively unimportant. It is the *product* that determines how much you will win, and it matters very little (except for the sake of cover) how that product is attained.

Of course, in this discussion, we are not considering the risk associated with the various approaches. It goes without saying that standard deviation would always play a major role in assessing the desirability of any game. More about this in Chapters 8, 9, and 10.

One final example. Suppose one game yields a 2% advantage, but the table is always full. Another has but a 1% edge, yet you can often play alone against the dealer. Which game should you play? I certainly hope you said the 1% one. For, at roughly 200 hands per hour, compared to about 50 for the full table, you will win, at the 1% game, about double the amount of the 2% game. Moral of the story: Quote *win rates,* not edges or spreads.

September 1992 Cover Art by Judy Robertson

Chapter 3

Evaluating the New Rules and Bonuses

In this chapter, I combine yet another feature-length article, "Lost in the Maze of New Rules," from the September 1992 ***Blackjack Forum,*** *with a follow-up, companion "Gospel" piece to analyze ten of the most popular "side bets," or new "gimmick" propositions, that have been introduced to blackjack in recent years.*

Passing fancy or wave of the future? Whereas it's too early to tell if the proliferation of exotic rules, bonuses, and options currently being offered at blackjack tables around the country is here to stay, one thing is certain: The "time traveler" who hasn't sat down to play for a year or two is in for culture shock! In an effort to stimulate what appears to be sagging public interest in table games, casinos are resorting to "gimmick" side bets and propositions to keep customers happy.

But will any of these variations actually enhance the player's expectation? Which, if any, lend themselves to a card-counting approach? What, exactly, are the changes in basic strategy that a player should employ while tackling these games? We shall attempt to answer all of these questions in the following article. In addition, where feasible, the mathematical calculations needed to derive the odds will be explained. Finally, where direct solutions prove to be cumbersome or unmanageable, we shall indicate the software (almost always Stanford Wong's *Blackjack Count Analyzer*) used to obtain the answers by simulation.

It all started, not quite four years ago, with the appearance of the over/under proposition. This side bet on the table layout permits the player to wager on the likelihood that the total points obtained with his initial two cards will be over or under 13 in value. (The player loses on a total of 13 exactly.) No need to analyze this option here. The over/under rule has been the subject of careful scrutiny by several blackjack experts including, of course, Arnold Snyder in his *Over/Under Report.* Count strategies were devised that actually offer greater profit potential than traditional blackjack approaches. For the sake of completeness, we include the numbers here, as a reference.

Basic strategy house advantage is 6.55% for the over and 10.07% for the under. Quite simply, one consults a frequency distribution for initial hands (Chambliss and Roginski, Braun, Imming, etc.) *[Note: These can now be found in Appendix A of this book. — D.S.]* and obtains the relevant numbers for starting totals of over 13 (46.72%), under 13 (44.97%), and 13 exactly — the "house" number (8.31%).

Of course, one must stipulate the number of decks being shuffled (I used six for the above percentages). Likewise, the edge attainable by card-counting techniques varies, as expected, according to decks used, spread employed, and depth of penetration (the latter, as always, being the most important). This advantage ranges from 0.1% all the way to greater than 4.0% (not readily attainable). On average, 1.5% to 2.0% should be achievable. Note, also, that at most casinos, there is a $100 maximum bet and, in all cases, the over/under bet may not exceed the wager on the original blackjack hand. Now, let's get on to the new proposition bets and rules.

Rule #1: Winning, Suited 6-7-8 Pays 2 to 1

Step #1: Determine the probability of obtaining a suited 6-7-8, given that you will automatically draw the third card regardless of the original situation (your hand v. the dealer's upcard). We'll assume the TropWorld 8-deck game in Atlantic City. Choose one specific order of one suited sequence (say, 6-7-8 spades). Probability of occurrence is 8/416 x 8/415 x 8/414 = .0000071. But we are not confined to 6-7-8 — any order will do (say, 7-8-6). There are six possibilities (or permutations). In addition, there are four suits. Thus, the desired probability is 6 x 4 x .0000071 = .0001704.

Step #2: Eliminate the percentage of starting hands that will sacrifice more than it's worth if we violate basic strategy. For example, we hold 7,8 of hearts; dealer shows a 5. The 2-to-1 bonus is not sufficient for us to draw in an attempt to find a 6 of hearts.

Wong's *Blackjack Count Analyzer* (henceforth, *BCA*) tells us that departing from basic to go for the 2-to-1 bonus is correct only for the case of 6,7 v. dealer's 2. *[Note: This is a change from the advice given in the March 1992* ***Blackjack Forum,*** *where the dealer's 3 was included. — D.S.]* So, all other holdings v. dealer's 2 and all starting two-card holdings v. 3–6 must be eliminated from consideration. A quick enumeration of these combinations leads us to the conclusion that roughly 14/39, or 35.90% of the original possibilities, must be cast aside. Thus, we are left with (1 – .3590) x .0001704 = .0001092 as our probability of occurrence.

Step #3: In their infinite greed and poor taste, the casinos actually require that our next-to-impossible-to-achieve suited 6-7-8 be a winning hand! Thus, should the dealer reach a final total of 21 (7.3% of the time) or, ultimately, reveal a natural (4.8% of the time), there is no bonus. Together, another 12.1% of the cases are eliminated. We're left with (1 – .121) = .879, which, multiplied by .0001092, equals .0000960.

Step #4: Determine our edge. Since the payoff is 2 to 1, we win one extra unit for each unit bet. Thus, our basic strategy advantage is simply our final probability, stated above, of .0000960, or roughly .01%. "Big deal," you say? Well, of course, you're absolutely right. Who can get excited over a bonus that offers us an extra penny for every $100 we bet? On to the next rule. (But it doesn't get any better!)

Rule #2: Winning 7-7-7 Pays 3 to 2

Step #1: Determine the probability of obtaining three consecutive 7s given, once again, the assumption that, with two 7s, we will draw automatically. We get 32/416 x 31/415 x 30/414 = .0004163.

Step #2: As for our 6-7-8, eliminate the percentage of starting hands (pair of 7s) that it would be foolish to sacrifice (fail to split) by trying for the bonus. In this instance, all correct basic strategy splits should be maintained. Since we split 7s in A.C. v. dealer's 2–7, roughly 6/13 or 46.15% of the hands are eliminated. We're left with (1 – .4615) x .0004163 = .0002241.

Step #3: Again, we get paid on only 87.9% of the 7-7-7s we achieve. Thus, .879 x .0002241 = .0001969.

Step #4: Finally, to determine our edge, note that, for each bonus of 3 to 2, we win only one-half unit for every unit bet. Our added advantage is, therefore, half of the number above, or .0000984. Our edge of .00984% (again, roughly .01%) is virtually identical to that of suited 6-7-8.

Certain shortcuts were used in the math for Rules #1 and #2. Small errors were introduced due to my neglecting the interaction of various upcards on final probabilities. Nonetheless, a more precise calculation, which accounted for these differences, was carried out by Peter Griffin, and I'm happy to say that it yielded almost identical results (.0000914 and .0000970, respectively).

Perspective is important. The $5 player, who averages perhaps $10 per hand, will wager about $600 per hour at a crowded table. Together, the above two bonuses will put exactly 12 extra cents per hour in the player's pocket. Enough said!

Rule #3: Red-Black

The Four Queens, in downtown Las Vegas, is experimenting with a side bet on the color of the dealer's upcard. Six decks are used and the player simply chooses, in a separate betting square, the "red" or "black" designation. The house advantage resides in the deuces. When the deuce of the color you bet on appears, your bet pushes instead of winning. Analysis of the casino's edge is straightforward. For simplicity, assume, for the moment, a one-deck game. Let's say we bet $1 on red for each of 52 hands. Red appears 26 times, but two are red deuces, which push. Therefore, we get back 24 x $2 = $48, plus 2 x $1 = $2, or $50. We've lost $2 out of $52 bet, or 1/26 of our wagers. This is the house edge: 3.85%.

It should also be obvious that this proposition lends itself admirably to card counting. An elementary plus-minus approach is all that's needed. Suppose we count each red card as +1 and each black as –1. For each deck, we need to overcome the 2-card (deuces) house edge. So, when the count per deck surpasses 2 (plus or minus), we have neutralized the house advantage. As that edge was 3.85%, we see that each true count (count per deck) is worth half of 3.85%, or roughly 1.92%. Compared to traditional card counting, where a plus-minus point is worth just over 0.50%, this proposition appears to be very interesting indeed! And that advantage can be increased.

As the deuces play an important role for the house, their removal from the deck is obviously beneficial to the card counter who is sufficiently motivated to keep a side count. We calculate an additional advantage of about 0.38% for each deuce above normal that is removed from the ranks of the color we are playing. For example, suppose one deck out of six has been dealt. Our running count is +10, indicating that 31 red and 21 black cards are gone. The true count is +10/5 = +2, so we have a dead-even situation … provided the "correct" number of black deuces (two, in this case) have been depleted.

But suppose our side count indicates that four black 2s are gone. Of the 260 cards remaining (of which 135 are black), there are eight black deuces left. So, if we bet $1 on each of the 260 cards, we'll win 135 – 8 = 127 times and push 8 times. Total return is 127 x $2 = $254 plus 8 x $1 = $8, or $262. We win $2 for a profit of 2/260 = 0.77%. Instead of being even with the house, our +2 true count with two deficient black deuces produced a slight edge. Since 0.77% divided by two is 0.385%, we estimate this latter number to represent the extra advantage provided by side-counting an imbalance of one extra deuce removed.

Finally, by simulating the game, to provide a frequency distribution of the red-black count, it is possible, as with traditional card counting, to devise a betting strategy to beat the house. Obviously, the bet spread employed will determine the ultimate advantage that can be gleaned by raising bets at trues of higher than +2 or lower than –2 (or even sooner, if the deuces permit). Without giving a precise figure, it is safe to say that a very substantial edge is attainable at this game. Ideally, one would assess the traditional count so that no money is lost while playing the actual blackjack portion. As with over/under, one would "switch," back and forth, when the red-black count indicated, so as to maximize potential profits. If forced to play alone, our advice would be to play basic strategy and count red-black, as the extra advantage gleaned from red-black play will more than offset the difference in edge that the traditional counter enjoys over the basic strategy player. The beauty of this proposition is, of course, that, unlike traditional card counting, we win on a count more extreme than 2 in either direction. Thus, "negative" counts aren't bad at all; they are, in fact, just as good as positive ones, provided they are large enough in absolute magnitude. Finally, we note that, in May, a $25 maximum bet limit was imposed. However, the Foxwoods (Connecticut) casino plans to offer the same proposition with a $50 maximum.

Rule #4: Winning, 5-Card 21 Pays 2 to 1

Let's return, momentarily, to A.C. for a look at another bonus rule currently offered at TropWorld. Here, a five-card winning hand totaling 21 receives a 2-to-1 payoff. Our analysis will be brief because the math needed for a direct calculation would be much too cumbersome. Stanford Wong's *BCA* to the rescue! Just input the new rule, run the simulations, and out pops the house edge. Compare this to the advantage without the bonus, and we get, approximately, an additional 0.2% for the player.

There are, obviously, several changes to basic strategy as we try for this five-card 21. Here, too, we get a hard-copy printout, instantly, from the *BCA*. There are some 24 deviations from the basic strategy for the standard 8-deck A.C. game. And whereas we will not catalog all of them, here are a few of the most important changes: hit four-card holdings of 14 v. 2; 13 v. 2, 3, 4, and 6; 12 v. 4, 5, and 6; soft 19 v. 10 and A; and soft 18 v. 2 through 8.

It would appear that this particular 8-deck TropWorld game, which offers the above bonus in conjunction with 6-7-8 suited, 7-7-7, and surrender, has an off-the-top edge of only about 0.185% for the casino — the best deal in town for the non-card-counting, basic strategy player.

Rule #5: Dealer 10-Up Natural Counts as Plain 21

In his May 1992 *Current Blackjack News,* Stanford Wong apprised us of an interesting opportunity at the Continental and Main Street Station casinos. (Unfortunately, in June, Main Street went bankrupt and closed its doors.) Of course, any discussion of advantage available from this rule must be kept in the perspective of the rather low limits tolerated in these casinos. Here is the proposition: When the dealer has a 10 up and, ultimately, reveals a natural, a) our natural 21 beats his, and b) our regular 21 ties his natural, but c) we lose all non-21 doubles and splits to his natural (with 10 up). Wong's *BCA* simulation showed a value to the player of about 0.25% for this attractive option. Let's attempt to replicate that figure by direct calculation.

Step #1: The dealer has a 10 up 4/13, or 30.77% of the time. He turns this into a natural 4/51 of the time (single-deck), or 7.84%. So, together, the dealer has a 10-up, ace in the hole natural 4/13 x 4/51 = 2.41% of the time.

Step #2: We find, from frequency distribution charts, that the player draws to 21 about 7.3% of the time (we assume knowledge of the dealer's 10) and has a natural (again, given the dealer's 10 up and ace in the hole) 2 x 15/50 x 3/49 = 3.67% of the time. Actually, the 7.3% is a conservative estimate based upon the dealer's probabilities. The player will achieve slightly more 21s by virtue of hitting soft 17 and 18, options not available to the dealer.

Step #3: So, when our natural gets paid 3 to 2 instead of pushing, we gain 1.5 units .0367 x .0241 = .088% of the time. And when our regular 21 pushes instead of losing one unit against the dealer's natural, we gain that one unit .073 x .0241 = .176% of the time.

Step #4: Together, we gain in expectation 1.5 x .088% = .132% from the first part and .176% from the second part. Our edge, therefore, is simply the sum of the two, or .308%.

Step #5: From this extra edge, we must subtract a loss in advantage from not splitting 8s or doubling 11 into the dealer's potential 10-up natural. The *BCA* gives us a quick answer to the two differences: We lose in expectation .045% by hitting the 8s and .037% by simply hitting the 11. Together, we forfeit .082%. *[Note: The minute extra loss from splitting aces and receiving non-tens on each was so negligible that it was ignored. — D.S.]*

Step #6: Subtracting this last number from the extra edge obtained in Step #4, we get the final player advantage from this attractive rule: 0.226%. This compares very nicely with the simulated answer of 0.25% (with a standard error of .02%) obtained by Wong.

Rule #6: Blackjack Jackpot

Although the "Blackjack Jackpot" at Merv Griffin's Resorts is currently in effect as we pen these lines (early June, 1992), we predict that this progressive jackpot will be gone by press time. In May, I described this promotion to Stanford Wong over the telephone and together we analyzed the payoffs. His June *Current Blackjack News* told the story. If you missed it, I'll summarize the math.

Resorts has designated one 12-table, $10-minimum pit as eligible for the jackpot, which starts at $250 and increases in value by one cent per second. To win, the player must receive a natural A,J of spades on the first hand of a freshly shuffled shoe. The probability of getting that hand is 2 x 8/416 x 8/415 = .0007414, or about 1 in 1,349. With 12 full tables of 7 spots and allowing for four shuffles per hour, there are 12 x 7 x 4 = 336 hands per hour eligible for the jackpot. With a mean of one jackpot every 1,349 hands, we can expect a "hit" once every 1,349/336 = 4.01 hours. We'll call it four. Now, at a penny per second, for four hours, the $250 will have grown to $250 + ($.01 x 4 x 60 x 60) = $394. Thus, the promotion is costing the casino $394/4 = $98.50 per hour, assuming maximum player participation. If all 84 spots are in use, the hourly expectation per player is simply $98.50/84 = $1.17.

Let's suppose that the average bet per player at these $10 minimum tables is $20. In addition, note that full tables average about 60 rounds per hour. Total hourly "action" per player is thus 60 x $20 = $1,200. If the casino retains about 2% of this amount, or $24, the jackpot is costing the house $1.17/$24 = 4.9% of its hourly profits in this pit. But, if we place only skilled basic strategy players at each spot (with about 0.40% negative expectation), the percentage loss to the casino is virtually quintupled! $1,200 x .40% = $4.80, and $1.17/$4.80 = 24.38%. Obviously, the ultimate tax on the establishment lies somewhere between these overly optimistic and pessimistic estimates. However, we predict that, by the time you read these lines, this potential 24%+ impost will have proven to be too steep, and the promotion will have been discontinued.

Rule #7: Super 7s

In the June 1992 issue of *Blackjack Forum,* Arnold Snyder's "Sermon" outlined, in great detail, the evolution of the "Super 7s" proposition. Allusion was made to the creator's (Ken Perrie's) desire to construct a payout on the various combinations such that the house would retain between 10 and 11 percent of the betting action. The Bishop and Perrie settled on the following: Three 7s suited on the first three cards, $5,000; three 7s (any suit), $500; two 7s suited on the first two cards, $100; two 7s (any suit), $50; and any 7 on the first card dealt, $3. Only one bet size is permissible: $1.

Other rules include: 1) If two 7s are split, the next card dealt is irrelevant and only the "two 7s" payout will apply; 2) Only the highest bonus achieved is paid; and 3) In case the dealer has blackjack, the player still receives a third card (and therefore a chance for the larger bonuses) if two 7s have been dealt.

With all of the above in mind, we set out to solve the mystery of the "Super 7s." Stay with me on this one — the math is a bit more complicated than in most of the previous illustrations.

Step #1: I chose to start from the top and work my way down. Three suited 7s are obtained by receiving any 7 for the first card (6-deck game), or 24/312, one of the remaining five matching 7s for the second, or 5/311, and, finally, one of the four matching 7s for the third, or 4/310. The probability is thus 24/312 x 5/311 x 4/310 = .0000159. The frequency of occurrence is once in every 62,893 attempts. (In fact, the more precise probability is once per 62,666 hands, but I did a little rounding above.)

Step #2: For any three 7s, we have 24/312 x 23/311 x 22/310 = .0004037, from which we must subtract the suited 7s calculated above, so as not to duplicate the payout: .0004037 – .0000159 = .0003878, or once in every 2,579 hands.

Step #3: Two suited 7s figures to be 24/312 x 5/311 = .0012367, or 1 out of 809.

Step #4: Any two 7s would be 24/312 x 23/311 = .0056888, from which we subtract the suited ones, therefore .0056888 – .0012367 = .0044521, or once in 225 hands. As an alternative calculation, we could have stipulated any 7 for the first card, or 24/312, and then insisted that the second 7 come from the 18 non-suited remaining ones, or 18/311. 24/312 x 18/311 = .0044521 — a more direct approach.

Step #5: Finally, any 7 on the first card alone is, of course, 24/312 (or 1/13) = .076923.

Step #6: Now comes the tricky part. For problems such as these, I like to assume enough $1 wagers to cover every possibility at least once, and then we see what happens. Well, we'll need 62,893 bets to get the big payout of $5,000 just that one time. At first glance, it would appear that we would simply divide each of the frequencies for the other bonuses into 62,893 to learn how many times we're entitled to each one. Unfortunately, it's not quite that simple. In three instances, we must eliminate the lower payout that is forgone when the fortuitous drawing of another 7 completes the bonus at the next level. For example, if my first two 7s are suited, I will try for the third one (more about why I might not try, later). I'll be successful four times out of 310, leaving 306/310 of the two, suited 7s to be paid. We'll use this fraction in a minute. Similarly, with two unsuited 7s, I'll go on to draw a third 7 22 times out of 310, leaving 288/310 unsuited double-7s to be paid. Finally, a singleton 7 will succeed in "moving up" 23 times out of 311, leaving 288/311 of the lone 7s to be cashed in. With these limitations on the payouts in mind, we proceed to:

Step #7: We need to establish the frequencies, per 62,893 deals, of each of the bonuses. We'll just divide and then adjust for the probabilities in Step #6. We get: 1) three suited 7s, once; 2) any three 7s, 62,893/2,579 = 24 times; 3) two suited 7s, 62,893/809 x 306/310 = 77 times; 4) any two 7s, 62,893/225 x 288/310 = 260 times; and 5) any 7, 62,893/13 x 288/311 = 4,480 times.

Step #8: We're almost done! Next we determine the total return for our $62,893 expenditure. Simply multiply the frequency of each bonus by its payoff, remembering to return the $1 wager in each instance. We obtain: (1 x $5,001) + (24 x $501) + (77 x $101) + (260 x $51) + (4,480 x $4) = $55,982. Our shortfall is $6,911, which, divided by $62,893, is 10.99% … the house edge at Super 7s! Note that an even more precise calculation, using the exact probabilities instead of the rounded frequencies obtained above, yields a house edge of 10.83%. This latter figure is more precise, but the rounded numbers were more convenient for our demonstration.

One last observation remains, and it is most interesting. As the maximum bet is $1, and as hitting 7,7 (both suited or otherwise) is clearly a serious violation of basic strategy versus the dealer's 2–7, there must be some initial bet on the blackjack hand that is large enough such that the loss from not splitting the 7s is greater than the positive expectation of the $1 wager. We need to know, for the two cases (suited and unsuited), what those expectations are. Two suited 7s pay $100. Once achieved, we get at least that much, no matter what our subsequent action is.

So the question is, "If I hit successfully, what extra amount do I win, and with what probability?" Well, I win $4,900 extra with 4/310 probability. Again, I can't lose on the proposition, because I'm already guaranteed the $100. The expectation is thus $4,900 x 4/310 = $63.23. Next, I consider the $400 more I win for three off-suited 7s. As this will occur with 18/310 probability, the additional expectation is $400 x 18/310 = $23.23. Together, I get $86.46 as my bonus expectation. Finally, I must determine my penalty percentage-wise, from drawing instead of splitting. But the answer is different for each dealer upcard! I must make six separate calculations, which I don't propose to do here.

Let's show one. Suppose the dealer has a 6 up. Wong's *BCA* tells us that, if we split, we have a positive .190 edge. But if we hit, we have a negative .304 expectation. The difference is a whopping .494. So the question is, "On what size bet will the .494 be larger than the $86.46 bonus expectation?" Answer: $86.46/.494 = $175. For smaller wagers, go for the third 7. For larger, it isn't worth it. We can use the same approach for each of the dealer's upcards as well as for the unsuited pair of 7s. Note that, in the latter instance, we're assured of $50, and we're contemplating going for $500, or an extra $450. The expectation is $450 x 22/310 = $31.94. So, for the same dealer 6, the break-even bet is $31.94/.494 = $65. As we progressively lower the value of the dealer's upcard, these break-even bet sizes become considerably larger.

Two valid points concerning all of the above were made by Ken Feldman and Peter Griffin when they read the original draft of this article. First, both pointed out that once a player is foolish enough to accept the premise of betting into a negative 11% edge, he is not going to be swayed by my "break-even" analysis. The player is obviously enticed by the big payoff, and logic is unlikely to prevail where greed rears its ugly head! Second, Feldman observed that players' continual betting of $1 chips or silver, dealer's constantly making change for the above, paying off winning bets and collecting losers, would all certainly tend to slow down the game. Thus, not only is the Super 7s a bad proposition, it is also a bad table at which to be sitting for the counter, who stands to lose volume and, therefore, profits.

Of all the exercises in probability theory contained in this article, surely the "Super 7s" was the most challenging!

Rule #8: Multi-Action Blackjack

It's being trumpeted as "Three times the action for your money" and, from the early returns, it looks as if "Multi-Action Blackjack" (also called "Multiple-Action" or "Triple Action") may be here to stay. The game has gained in popularity since its inception and is now offered at many of the plushest Strip casinos. In essence, all the rules of normal blackjack apply, except for the fact that, before the deal, we wager on a minimum of two spots and a maximum of three. (Some casinos require three wagers at all times.) The bets may be for the same or different amounts, and are placed vertically on a special layout. Players may bet on their "column" only. The catch is that one hand is dealt to each player, and the result of that single hand prevails for all of the player's wagers. The dealer, for his part, retains the same upcard against each player spot, but after each round (hand), he discards his hit cards only and plays out a new hand starting with the original upcard. Thus, the player's hand is the same for all two or three of his bets, and the dealer's first card is the same against each of the player's hands.

The casinos love the game because it seems that it has been doubling the "hold" (amount of money retained from the players' buy-ins) from the normal 16 to 18 percent to about 35%. Why? Many players, fearing that they will break, thereby losing all three hands at once, abandon basic strategy and refuse to hit stiff hands that are potential "busts." The result, of course, is even greater losses for the players who are, in effect, "punished" three times for violating correct basic strategy. Obviously, there is no mathematical justification for this kind of play. Basic strategy is still *de rigueur* for the non-counter.

The concern to the card counter lies elsewhere. He wants to know the proper method for bet-sizing on the multiple layout. (At least I know I wanted to know how to bet when I tackled this game for the first time in April at Caesars Palace!) It was clear to me that we obviously couldn't bet the optimal Kelly wager for one hand on each of the three. To do so would be to grossly overbet one's bankroll. It also didn't seem right to me to bet the traditional fraction of the ideal one-hand wager on each of the two or three hands because, intuitively, I sensed that the covariance between bets was not the same. Unfortunately, this did not seem to be the kind of problem I could solve by direct methods. Not to worry! By the end of June, Stanford Wong had simulated the game and published the results in the July 1992 *Current Blackjack News*.

Needless to say, if we have no advantage, but we are forced to wager, we bet as little as practical. This would call for two wagers, each of which is 70% of the amount we would like to have bet on just one hand. Of course, if our normal minimum bet exceeds the table minimum by at least double, and we can get away with it, why not just bet the table minimum on each of the two hands? When we have the edge, we switch to three hands and scale our bets in the following manner: Bet on each of the three hands 53% of what the optimal one-hand bet would be. Of course, the player should determine ahead of time a "tasteful" spread from the minimum two-hand bets to the maximum three-hand bets. My feeling is that much will be tolerated because of the very nature of the "Multi-Action" game, but, as always, it is unwise to make a complete pig of oneself!

By the time you read these lines, I am sure of two things: 1) Some of the rules and bonuses discussed in this article will have disappeared, and 2) new ones will have cropped up to take their place. So much the better! That's what we have "The Gospel" for! Keep writing in your questions. I enjoy hearing from you, and readers apparently enjoy the column, as well. As new rules develop, we'll attempt to analyze them in future issues of *Blackjack Forum.* Table 3.1 is a summary of the rules discussed.

Finally, my thanks to those who were kind enough to read the original draft of this article and to offer comments and suggestions: Stanford Wong, Peter Griffin, Arnold Snyder, Ken Feldman, Ron Wieck, and Joel Waller.

TABLE 3.1
Summary of New Rules

Rule	Basic Strategy Casino Edge*	Basic Strategy Changes Required	Edge to Counter
Over/Under	+6.55 (over) [d] +10.07 (under) [d]	None	Same or Superior to Traditional Counting
Winning 6-7-8 Suited Pays 2:1	–0.01 [d]	Hit 6,7 Suited v. Dealer 2	None
Winning 7-7-7 Pays 3:2	–0.01 [d]	None	None
Red-Black	+3.85 [d]	None	1.92% per Count per Deck
Winning, 5-Card 21 Pays 2:1	–0.20 [d]	Total of 24	None
10-Up Dealer BJ Is Plain 21	–0.25 [s] –0.226 [d]	Don't Dbl 11 v. 10 Don't Split 8s v. 10	None
Blackjack Jackpot	–24.38 to –4.9 [d]	None	None
Super 7s	+10.83 [d]	Depend on $ Bet on BJ Hand	Minimal Due to $1 Max Bet
Multi-Action	Same, but Hold Is Greater	None	Easier to Get Large Spread

* [d] = direct calculation; [s] = simulation. All figures in this column are percents.

From "The Gospel According to Don," *Blackjack Forum*, December 1992:

Q. *I enjoyed your "Lost in the Maze of New Rules" article very much (**BJF**, September 1992). In particular, you didn't just give the casino edge, but explained how to derive it. Some were easier to follow than others, but your explanations are always very clear.*

The "Royal Match" rule was probably established just as you were completing your article, so I can understand why it wasn't included. Could you cover this rule, as you did all the others?

A. Thanks for the kind words. "Lost in the Maze" was fun to write, and I was glad to learn from Arnold that *BJF* readers were pleased with the content.

I first learned of the "Royal Match" in the June 1992 *Blackjack Forum,* "Reno-Tahoe Insider," by John Gelt (p. 43). He described the single-deck side bet, as dealt by the Carson Nugget: "You bet that your first two cards will be of the same suit. If they are, you are paid 3 to 1. A suited King and Queen pay 10 to 1." Gelt went on to say that he calculated the house edge to be 3.8%, and he was correct.

A few days later, Stanford Wong's special June newsletter arrived, announcing the identical option at the Lady Luck, in downtown Las Vegas. Again, the correct 3.8% house advantage was stated. Wong's July 1992 issue brought news of four additional casinos that offered "Royal Match." Finally, his September 1992 edition of *Current Blackjack News* apprised us of the Reno Horseshoe's version of the bet: 2.5 to 1 for suited initial hands, and 25 to 1 for K,Q suited. These (inferior) payoffs were available at both the single- and six-deck games.

As you've asked for the math and, for the sake of completeness, as a follow-up to my article, I'll be happy to provide the analysis. But don't look for any good news; the Horseshoe's odds are horrendous, and the –3.8% is as good (?) as it gets. Let's start with single-deck at Lady Luck and the Nugget. There are 1,326 different two-card starting hands. We get this figure from the number of combinations of 52 cards taken two at a time: (52 x 51)/2 = 1,326. Now, how many "royal matches" are there? Obviously, only one for each suit, or four. In general, how many initial, two-card hands will be suited? Let's choose one suit, say, spades. There are (13 x 12)/2 = 78 possible two-spade initial hands, but one of these is the spade royal match, which we must eliminate so as not to double count it. So, 78 – 1 = 77. Multiply by four suits to obtain 77 x 4 = 308 suited hands.

In keeping with my normal methodology, let's bet $1 on each of the 1,326 possible combinations, see how much we get to "pick up" from winning bets, and calculate the shortfall as a percentage. For the four royal matches, we pick up (at 10-to-1 payoffs)

$11 four times, or $44. Similarly, we pick up $4 (3-to-1 odds) 308 times, for $1,232. Adding the two, we get to keep $44 + $1,232 = $1,276 of the $1,326 wagered. We're short $50, which, when expressed as a percentage of $1,326, equals 50/1,326, or 3.77%. Simple enough.

The Horseshoe is a … horse of a different color! The single-deck variety pays 25 to 1 for the royal match, so we keep $26 x 4 = $104. But we suffer from the skimpy 2.5-to-1 payout for the "common" matches. $3.50 x 308 = $1,078. Together, we get to keep only $104 + $1,078 = $1,182. The deficit is a whopping $144 out of $1,326, for a house edge of 10.86%.

Finally, let's look at 6-decks at Horseshoe. There are now 52 x 6 = 312 cards, so the calculations change. (312 x 311)/2 = 48,516 distinct opening hands. For each suit, we now have six kings that can match six queens, for 36 royals. For four suits, we get 4 x 36 = 144 royal matches. Again, first for one suit, we have (78 x 77)/2 = 3,003 common matches, or 4 x 3,003 = 12,012 in all. But we mustn't forget to remove the 144 royals: 12,012 – 144 = 11,868 non-royal matches. Now, for the payouts. At 25 to 1, we pick up $26 x 144 = $3,744 for the K,Q, and at 2.5 to 1, we keep $3.50 x 11,868 = $41,538 for the rest. Together, we keep $3,744 + $41,538 = $45,282. Unfortunately, this leaves us $3,234 short of our original investment. Now, 3,234/48,516 = 6.67%, the 6-deck house edge at the Horseshoe.

In closing, I might add that the Lady Luck odds were offered for Royal Match at their one-deck games only. Had the same deal been available at 6-deck, the edge would have been 1.1% … *for the player!* Now that I've shown you how to do the math, perhaps you'd like to verify this figure. I think someone should suggest that Lady Luck extend its Royal Match single-deck payoffs to 6-deck! But then, I think blackjacks should pay 2 to 1 everywhere!

As we put the finishing touches on this article, Suzanne LeCounte, an expert single-deck player, pointed out a potential benefit to the card counter sitting at a Royal Match table. It seems that, in the spirit of camaraderie, players who pick up their initial two-card hands often show them to one another while asking, "Did you get it?" So, the counter gets to see all of the usually concealed hands before making his play! In my December 1989 "Gospel," p. 21, I gave the value of seeing all the players' hole cards as at least 0.5% (see the following chapter, p. 47). Ironically, playing at a Royal Match table may, indeed, be profitable for the card counter, after all! *[Note: For an elegant analysis of the Royal Match proposition, see James Grosjean's* ***Beyond Counting,*** *pp. 36–38 — D.S.]*

Postscript: *I've always felt, perhaps erroneously, that everything that can be calculated analytically should be, and then, we should simulate! To my knowledge, the above article was then, and remains now, the only attempt to analyze the new rules* *from a purely theoretical point of view — which doesn't mean the math was perfect — it wasn't. But I have always considered it a challenge, over the years, to tackle blackjack problems head-on and to attempt to solve them mathematically, whenever I could. Many readers appreciate the explanations that accompany the analyses. Still, others "just want the facts" and the bottom line. For those who prefer computer-simulated results, I highly recommend Stanford Wong's* ***Basic Blackjack.*** *In it, you will find virtually every "variation on a theme" that you could ever hope to encounter at a blackjack table. Wong used his vaunted* ***Blackjack Count Analyzer*** *to develop the material. Check it out. You'll be glad you did.* [Note: Since I wrote the above lines, Mike Shackleford has performed extensive combinatorial analysis of virtually all of the side bets discussed here. Check out his wonderfully informative Web site: *www.wizardofodds.com* — D.S.]

Chapter 4

Some Statistical Insights

Although the original intent of the "Gospel" was to answer random questions on various topics, often, a single, more complex problem was analyzed. In this chapter, we group most of the multi-question "Gospels," and cover a potpourri of thorny blackjack posers, many of which had never before been answered in print.

From "The Gospel According to Don," *Blackjack Forum*, September 1989:

Q. *Page 134 of Ken Uston's* ***Million Dollar Blackjack*** *shows the following table for the allowable increase in total bet when playing multiple hands simultaneously:*

No. of Hands	Total Bet Increase
2	15%
3	22%
4	26%

These figures are significantly different from p. 68 of Snyder's ***Blackbelt in Blackjack*** *where he recommends a 50% increase for two hands and not more than 100% for three hands. On p. 100 of* ***Professional Blackjack,*** *Wong's optimal bet sizes are keyed to the player's percent advantage, but, in practice, it looks as if his figures are very close to Snyder's. Can Uston's figures be reconciled with Snyder/Wong?*

A. No, I can't reconcile Uston's figures with Snyder's or Wong's. (The latter two are not the same, by the way.) I believe Wong is right and Uston is much too conservative. Wong's numbers are based on equivalent risk of ruin to match higher win rates produced by getting more money on the table via multiple hands. Higher expectations are accompanied by higher standard deviations, so that the overall exposure to loss (as expressed by risk of ruin) remains identical. Note that, although Snyder agrees with Wong for two hands, his three-hand advice appears too aggressive. In last month's "Gospel," we gave a 57% figure for each of the three hands. The total increase would be 71%, compared to Snyder's 100% — a substantial difference.

Intuitively, Uston's numbers make very little sense to me. Why, for example, would you want to play *four* hands if you're going to get only an extra $26 per $100 on the table? Unfortunately, we can no longer ask Kenny what he had in mind. Use Wong's numbers with the confidence that they are accurate.

Q. *Many books carry the figure for the percentage of hands the dealer breaks (approximately 28%). But I have never seen a number for the percentage of time the* player *breaks using basic strategy. Can you tell me what it is?*

A. You're right. I've never seen this number in print anywhere. Strange. In any event, I am grateful to the ubiquitous Stanford Wong for providing the answer via a simulation of some 5 million hands. Here were the assumptions: 6 decks, Las Vegas Strip rules. For split pairs, we counted a break on one hand and a loss of any kind or a push on the second hand as a break. A break coupled with a win on the other hand was not counted. And now, the answer: The player using basic strategy breaks 15.9% of the time (standard error, .03%). For single-deck, Reno rules: 16.0%.

Q. *Now that surrender has returned to A.C. at the Claridge, I have a question regarding an apparent contradiction between the Hi-Lo index for surrendering 10,5 v. 10 and basic strategy for that play. Braun, on p. 135, gives zero for the index, but 10,5 v. 10 produces a –1 Hi-Lo running count (–.25 true) at the top of the four decks. Since I'm below zero, the count tells me to hit rather than surrender, but this goes against proper basic strategy. What do I do?*

A. You surrender. If a correct basic strategy play is apparently contradicted by the count, and it is the first hand of the shoe, basic strategy prevails. This is because the count provides one index number for the total (in this case, 15) without regard to the composition of the hand, that is, the individual cards that make up the total. In actuality, 8,7; 9,6; and 10,5 each have precise index numbers that, for expedience, are weighted for frequency and "blended" for all deck levels to produce the one index you use (zero, for this example). Note that Wong, in Table A4, p. 174 of *Professional Blackjack,* makes the necessary distinction and gives –1 for the 10,5 v. 10 play. Your true count of –1/4 = –.25 *exceeds* this index, so you surrender.

Several authors, notably Wong in his Appendixes, Revere, in his "Fine Points of Basic Strategy" discussion (pp. 68–69), and Griffin (several places, referring to "total dependent" v. "composition dependent" strategies) address this issue. Let me give you another hand that, I'll bet, many experienced count players would mishandle off the top of a shoe. What do you do with 10,2 v. 4? The Hi-Lo count is +1. The Revere Point Count is also +1. Respective true counts are +.25 and +.125. Both strategy matrices say "stand" at zero or higher. Basic strategy for the *total* of 12 v. 4 also says to stand, so, apparently, there is no contradiction. There's only one problem. The proper play is to *hit!* Here's why: Once again, the "blended" number rounds to zero. But the precise index for 10,2 v. 4 is about +.4. Note that the above true counts are *below* this number, and so you ought to hit. In fact, as a "fine point" (see Griffin, *The Theory of Blackjack,* pp. 20 and 176), this is correct basic strategy (up to, and including, seven decks).

But be careful not to apply the above basic strategy reasoning later in the shoe. This could get you into trouble. Suppose slightly more than two out of four decks have been dealt. You get 10,2 v. 4 and the Hi-Lo running count is +1. Your play? This time, it should be "stand." You now possess *more* information than the "fine point" basic strategist who will play his hand incorrectly. The true count is just above +.5. Therefore, a stand is indicated. Remember, basic strategy prevails over counting only at the top of the shoe. As the counter gathers information, through deck depletion, he has the upper hand.

As always, the practical side to all of this is humbling. Having "learned" all of the above, I doubt if the information will be worth $5 to you for the rest of your life!

From "The Gospel According to Don," *Blackjack Forum,* December 1989:

Q. *I'm interested in discovering how to figure out the advantage of seeing all the cards being dealt in a blackjack game. It is possible this issue has been addressed in some book or another and I just missed it.*

Basically, I wonder this: In Reno and Las Vegas, whenever I play a single-deck game, I have to try to peek at the other players' cards. I probably catch 40% of them.

Indeed, in using Snyder's Depth-Charging method, I deliberately seek out individuals where I catch a glimpse of their cards. (As a little side note: I've even become quite good at convincing other players to show me their hands by showing them my hand. It's amazing how that will stimulate others to show you their cards.)

On the other hand, whenever I play in multi-deck games in Reno, Las Vegas, and in Europe, I am able to see all the players' cards, with the exception of the dealer's hole card.

I know the formulas for calculating single-deck games v. multi-deck games, but how much of an advantage is it to see 100% of the players' cards, rather than the 40% or so I am able to glimpse in single-deck games?

A. The reader uses Snyder's "depth-charging" approach to the single-deck game. He plays at full tables, sits at third base and manages to see the two hands to his immediate right. The question is, therefore, if he were able to see all the cards (as in a face-up game), how much extra advantage would he glean from this valuable information? Finally, he plays Hi-Lo, which, for this particular approach, is rather ill suited given its 51% playing efficiency. When depth-charging, since one unit is always bet on the first hand, and two units maximum on the second (and last) hand before the shuffle, a non-ace-reckoned count would be more desirable. Here's how I would tackle this problem.

On p. 28 of *The Theory of Blackjack* (Griffin), there is a table, "How Much Can Be Gained by Perfect Play?" On p. 70, we find "Table of Exact Gain From Perfect Insurance." Together, these calculations can provide a precise answer for our questioner. We need to know, for each case, how many cards are seen before the first hand is played, and how many before the second. Next, we compute the strategy gains associated with each hand, sum up, average, convert to Hi-Lo advantage, and compare results. Betting advantage will not enter into the discussion as bets are placed prior to the deal, and the fact that all the cards are seen is irrelevant. Furthermore, we'll assume that the player is flat-betting.

A few pieces of information are necessary before we begin. On average, a completed hand contains 2.7 cards; thus, .7 hit cards per hand. Roughly 5% of all hands are blackjacks; 10% are doubled; 2% are split; and 16% of all hands are player breaks (a very useful statistic we learned in a previous "Gospel"). Now, let's put this all together. We'll call Case A the "see only two other hands" scenario, and Case B the "see all the cards" version. For Case A, before playing the first hand, our third baseman will see his two cards, the four cards to his right, the dealer's upcard, the 6 x .7 = 4.2 hit cards (assume double-down hit cards are visible), and the .33 x 4 x 2 = 2.67 hole cards of the four players farther to his right who expose their hand, before settlement, due to blackjacks (.05), doubles (.10), splits (.02), and breaks (.16) = .33.

Total number of cards seen: 13.87, or 14. Thus, the play of the hand takes place with 38 unseen cards. Strategy gain, in % (Griffin, p. 70, then, by interpolation, p. 28), is: .09 + .413 = .503.

For the second hand, 2.7 x 8 = 21.6 cards were seen from the first round, and an additional 13.87 will be seen during the second round, making 35.47; thus 16.5 unseen cards. Gain at this level is .34 + 2.46 = 2.80. We now add the results for the strategy gains from the two hands and average them:

$$\frac{.503 + 2.80}{2} = 1.65$$

Finally, the Hi-Lo captures only 51% of this edge, so 1.65 x .51 = .84, our final playing advantage for Case A. On to Case B.

For the play of the first hand, all 15 cards are seen plus the 4.2 hit cards; thus, 19.2 cards (33 unseen). The gain is .13 + .66 = .79. Next hand, 21.6 plus another 19.2 = 40.8 cards have been seen; therefore 11 unseen. Gain is *substantial:* .48 + 3.89 = 4.37. Total edge: (.79 + 4.37)/2 = 2.58. Hi-Lo gleans 2.58 x .51 = 1.32.

We now are ready to answer the question! Seeing all of the cards, instead of just two hands, produces 1.32 – .84 = .48 extra profit (an increase of nearly .5% in playing advantage). Translating to dollars and cents, let's figure 55 hands per hour and an average bet of 1.33 units. The nickel player will earn an additional $1.75 per hour; the quarter player, $8.75; and the black-chip player, $35.

Although he doesn't fully explain the methodology for the full-table situation, Arnold Snyder, in *Blackbelt in Blackjack,* pp. 80–81, reaches the same conclusion of .5% added advantage in his discussion of a virtually identical scenario. I am pleased that our results concur, as I worked out my answer prior to rereading his Chapter 10, "Unencountered Counter." See, also, the March 1984 *Nevada Blackjack* (Stanford Wong) for the description of a related, although not identical, topic, "Third Base at a Busy Table."

Obviously, the approach described above can be used for any game situation where not all the cards are seen. Use the charts from Griffin, plus the method I have outlined here, and you will know exactly how valuable knowledge of extra cards will be to you. Remember, also, that your count may have a different playing efficiency from the Hi-Lo 51% we used for this demonstration.

Q. *Risk of ruin is always a popular topic and one we are all concerned about. I'd like your opinion on a risk of ruin determination that uses the standard deviation formulas found in Arnold Snyder's* ***Beat the X-Deck Game*** *books.*

I have developed an 8-deck frequency distribution in which I enter the shoe at a 0% edge and exit with a –1/2% edge. The essential stats are $10 units; 10-to-1 bet *ratio; $34 average bet; 1.0% win rate; $48.20 = one-hand s.d.; 34.4¢ = one-hand gain. For 10,000 hands, my 67% (one s.d.) range is –$1,382 to +$8,258, and my 95% (two s.d.) range is –$6,202 to +$13,078.*

As you can see, there is a significant difference between the two left ends of each s.d., namely –$1,382 and –$6,202. Now, for my question. Is it safe to say that my risk of ruin at one s.d. (67% of range) is 17% if my bankroll is $1,382? Can I expect to double my bank 83% of the time and lose my bank 17% of the time if $1,382 equals my bank? If not, how can I calculate what bankroll I need for a riskier but more affordable 17% risk of ruin? [Note: For definitive answers to questions such as these, see the Risk of Ruin chapter. — D.S.]

A. First, although you mention a 1–10 spread, you don't say where your maximum 10-unit wager is made. I'm going to assume at +5 true (+2% advantage on Snyder's charts). If you're waiting any longer, in the 8-deck game, you're really wasting your time. Next, a few technical observations that don't really matter very much. The area under the normal curve plus or minus one s.d. is 68.28%, not 67%. Therefore, the area in each "tail" is 15.86%. For two s.d.s, the corresponding numbers are 95.44% and 2.28%.

Now, let's get down to cases. You probably play 25 hands per hour back-counting the 8-deck game. Thus, your arbitrary 10,000-hand (= 400 hours) level for analysis represents a lot of play. When you calculate your one s.d. downside for this as –$1,382, how do you know that this is as bad as one s.d. can get? Maybe a one-s.d. loss for fewer hours produces a more negative number. In fact, it does!

I don't know if you're familiar with "maximum-minimum" problems in calculus, but basically, if n = number of hours played, you are trying to maximize the loss represented by $241\sqrt{n} - 8.6n$, where, for the figures you supplied, $241 is the hourly s.d. and $8.60 is the hourly win. (34.4¢ x 25 hands = $8.60 and $48.20 x $\sqrt{25}$ = $241.) We take the first derivative of $241\sqrt{n} - 8.6n$ and set it equal to zero. Thus, $(1/2)(241)/\sqrt{n} - 8.6 = 0$. $120.5/\sqrt{n} = 8.6$. $120.5/8.6 = \sqrt{n}$. $14.01 = \sqrt{n}$, or n = 196. This means that, after 196 hours of play (compared to your 400), a one-s.d. loss is the most harmful. Now $\sqrt{196} = 14$ and 14 x $241 = $3,374. The win is 196 x $8.60 = $1,686, so the one-s.d. loss is $3,374 – $1,686 = $1,688. Note that this is $306 more than your figure. You can play around with different numbers of hours (do some below 196 and some above) to convince yourself that no one-s.d. loss can ever be more than $1,688.

But this is far from being your only problem! Whereas it is true that, at the end of 196 hours' worth of play, the probability of ruin would be 15.9%, this is true only if you are able to play through the entire 196 hours, no matter what happens along the way. It assumes that if you "bump" into the lower one-s.d. boundary in fewer hours, someone will come along and lend you money so that you can last through the 196 hours required for the analysis. The problem is, a) you have no such benefactor, and b) you wouldn't believe how often you smack into the barrier prematurely! Would you believe 46% of the time?! And you wanted to use $1,382, not $1,688, for a bankroll. That risk of ruin, for 196 hours of play, translates out to a whopping 54%. If you play "forever," the corresponding ultimate risks of ruin are, roughly, 61% and 66%, respectively. For a complete discussion of this concept, see Chapter 8.

These requirements are so much higher than you would like them to be because the 8-deck game has such a high s.d./win ratio. There is no way to make this go away. The game just flat out stinks! Most games do not have 675 units as the maximum 2-s.d. loss. In fact, for normal, decent games, I have always felt reasonably comfortable with 400 units of bank.

Finally, permit me to scold you just a little bit for not doing the following calculation. Your +5 true count edge is rarely more than 2.5%. According to the principles espoused by Wong and explained in the June 1989 "Gospel," roughly 75% of this edge, or 1.88%, is the optimal maximum bet. Now, $1,382 (your proposed bankroll) x 1.88% = about $26. To bet $100 would be to overbet by a multiple of four, virtually guaranteeing ruin. (Anything over a multiple of two is fatal!!) Note that, for your $100 top bet, optimal bankroll would be about 4 x $1,382 = $5,528. Earlier calculations suggest that, even at that level, your risk of ruin is too high.

I hope all of the above is helpful to you. This is an interesting topic about which there is a great deal of misunderstanding. More work needs to be done analyzing risk for different games and various betting styles. Indeed, George C. and I may pursue the idea and present our findings sometime in the not too distant future. *[See Chapter 10! — D.S.]*

Q. *What is a push on a single, tied hand worth to the house (for example, the house takes the bet only when the player and the house tie on 17)? How is this calculated? I tried using the tables in Braun's book, but because of double downs and multiple-card totals of 17, an exact answer does not seem possible from his data.*

A. Although I plan to do much more than simply answer your question, let me start with the information that you requested. Assuming 6-deck standard Strip rules, the basic strategy player ties the dealer, with a total of 17, 1.90% of the time. Thus, the loss is the same 1.90%. In order to answer this question, I once again called upon Stanford Wong to run a simulation — in this case, one that would reveal final-hand totals for both player and dealer. Wong provided the raw data that I, in turn, "massaged" a bit to produce Table 4.1.

A few explanations are necessary. My goal was to explain exactly what happens, from both the player's and dealer's perspectives, for every 100 hands dealt. There is a slight problem, because blackjacks interfere with the normal completion of a hand. It really doesn't matter at all what the player's two-card total is when the dealer reveals a natural unless, of course, the player has one, too. Similarly, the dealer doesn't play out his hand when the player breaks or has a blackjack (assuming head-on play) and, therefore, the ultimate total attained by the dealer is inapplicable.

Virtually everything else in the table should be self-explanatory. In an earlier "Gospel," when we revealed the 15.9% frequency for basic strategy player breaks, I pointed out that I believed this to be the first such mention of the figure in print. Likewise for the final player totals, I believe this is the first time such numbers have appeared in published form. There are a few interesting observations to be made upon examination of the data. Permit me to point them out.

Many "charity games" have a rule feature whereby the dealer wins all ties (except, perhaps, a blackjack). This would be worth 9.45% – 0.22% = 9.23% to the house. To compensate, partially, at these games, naturals often pay two to one instead of three to two. This extra half unit on 4.53% of the hands returns 2.27% to the player. But the impost is still a whopping 9.23% – 2.27% = 6.96% — much too much to overcome by card-counting techniques.

The doubling column reveals that 72.5% of the time, when doubling, we either "make it" (reach 20 or 21) or don't (achieve a total below 17). This may account for the frequent mood swings that accompany successful (euphoric!) or unsuccessful (inconsolable!) double downs. There isn't much "in between" remaining (2.62 hands per 100, or just 27.5% of all doubles).

TABLE 4.1
Final-Hand Probabilities (in %)

Outcome	Dealer's % Occurrence	Basic Strategy Player's % Occurrence: Untied	+	Tied	=	Total	(% via Doubling)
11–16	—	20.26		—		20.26	3.94
17	14.52	9.36		1.90		11.26	0.82
18	13.93	8.97		1.77		10.74	0.82
19	13.35	9.57		1.79		11.36	0.98
20	17.96	12.37		3.25		15.62	1.32
21 (non-BJ)	7.29	5.00		0.52		5.52	1.64
BJ	4.75	4.53		0.22		4.75	—
Break	28.20	15.94		—		15.94	—
Unplayed[1]	—	4.53		—		4.53	—
Totals	**100.00**	**90.53**	**+**	**9.45**	**=**	**99.98[2]**	**9.52**

[1] non-BJ, unplayed due to dealer's BJ

[2] less than 100% due to rounding

No attempt has been made, in this study, to catalog the extra hand(s) that may develop as the result of pair splitting. Where basic strategy called for a pair to be split, only the first hand was recorded. According to Julian Braun, p. 51 of *How to Play Winning Blackjack,* 21 hands per 1,000 dealt are splits. When doubling after splits is permitted, 23 per 1,000 are split (p. 53). Thus, 2.1 hands per 100 are split and, therefore, the results of those extra two hands have been omitted from the present discussion. Finally, for the sake of completeness, we note, again according to Braun (p. 155), that, on average, two hands per 55 dealt are surrendered, when this option is offered. This, in turn, would translate into 3.64 hands per 100 that the basic strategist would "toss in" rather than play to completion.

Obviously, many other questions concerning final-hand outcomes can be answered by consulting the table. We leave it to the reader to glean from the chart whatever information is needed.

From "The Gospel According to Don," *Blackjack Forum,* March 1991:

Q. *In the European no-hole-card game, the dealer "hits" his upcard only after all players have acted. Thus, the player at third base would seem to have much better information regarding the dealer's likely hand than is usually the case. Is this exploitable information?*

A. Not only is it non-exploitable information; it is no extra information at all! By the way, the same phenomenon exists in Puerto Rico, so the comments I'm about to make pertain to that game, as well.

It is a common error to think that the no-hole-card style of play for the dealer offers the player at third base some additional information. In fact, it does not. You may be surprised by this, but it is unquestionably true, and I will do my best to explain why.

Suppose, American style, the dealer "loads" his hole card and then deals all small cards to everyone at the table. His upcard is an ace, and you play your hand. Come the time for him to flip his hole card (we're in "no peek" Atlantic City), aren't you afraid, with all those small cards that just came out, that his chances of having the ten have increased? You should be! Now, you see the same upcard ace in Monte Carlo. No hole card. All the players get small hit cards. Dealer takes his hole card late, after all the players' hits. Is there a greater probability of a ten than there was in A.C.? Absolutely not! This may be hard for you to rationalize — but try. Perhaps the following "experiment" will help you to conceptualize the problem.

Suppose I take a 52-card pack, shuffle, and draw out a card face down. What is the probability it's a 10? Answer: 16 out of 52, or 4/13. Now, with that card lying on the table, I deal out, face up, ten small cards in plain sight. I ask you the question again:

"What is the probability that the face-down card is a 10?" Well, obviously, since we can see the ten small cards, there are only 42 cards left unknown, and 16 of them are 10s. Clearly, the answer is now 16 out of 42, or 8/21. Finally, if I deal out the ten small cards first, from the full pack of 52, and then deal a face down card, what is the probability now that it is a 10? Do you see that it is the exact same 16 out of 42?

Thus, you may be surprised and/or disappointed by the answer, but the third baseman in the no-hole-card variety game is no better off than if the dealer takes his hole card early. It makes no difference.

From "The Gospel According to Don," *Blackjack Forum*, June 1991:

Q. *My friend and I were informed of gaming promotions some Atlantic City casinos were thinking of using on selected tables. For one of the promotions, I couldn't find the percentage the player would be facing. Can you please help with the following: What is it worth if the casino offers a higher payout on blackjack? (Blackjack pays 3 to 1; 8-deck A.C. rules.)*

A. Your question is straightforward and lends itself to analysis with probability theory. I will do my best not only to answer, but also to explain along the way so that you may better understand the principles involved.

I will answer the question as you put it (3-to-1 payoff), but I will also respond for a 3-for-1 payoff (which is only 2-to-1 odds), because the advantage to the player is so ridiculously large for 3 to 1 that I don't believe any casino would be stupid enough to offer it for even one day. *[Note: In the summer of 1995, the Casino Morongo, in Cabazon, California, did, in fact, offer such odds. — D.S.]* You may know, for example, that once a year, at Christmas, Binion's Horseshoe in Las Vegas offers 2-to-1 bj payoffs and that: a) the promotion lasts only a few days, and b) the maximum bet to which the bonus applies is $5. *[Note: This annual promotion no longer exists. — D.S.]* You don't mention any restrictions in your question, but I assure you that a casino would be crazy not to have a maximum bet. It would also be crazy to offer 3 to 1. Be that as it may, here is the math involved.

There are 416 cards in 8 decks. 128 are tens and 32 are aces. As a blackjack consists of drawing a ten and then an ace, or vice versa, we have (128/416 x 32/415) + (32/416 x 128/415) = 4.75%. Now, I'll give you a great way to remember how frequently you get a blackjack in an 8-deck game. It's once in ... 21 hands! But we're not quite finished.

If you have a natural, but the dealer also has one, you won't get the bonus. So, we need to subtract from our 4.75% the percentage of the time the dealer ties the player with a blackjack of his own. This turns out to be 0.22%. We're left with 4.75%

– 0.22% = 4.53% uncontested player naturals. Since we're paid an extra half chip on each of these (2 instead of 1 1/2), our edge is .5 (.0453) = 2.27%. *[Note: A more precise value, with less rounding, is 2.263%. See Appendix C. — D.S.]*

If the casino were insane enough to offer 3 to 1, your blackjack would win an extra 1 1/2 chips, representing an edge of 1.5 (.0453) = 6.80%, and the casino is out of business! (Unless it limits the bets to, say, $5 and offers the promotion for one hour on alternate Wednesday mornings!)

Q. *I'm writing about an interesting surrender proposition. It involves early surrender v. 10 only. For example, a player has 17 against a dealer's 10. The bet is $100. The player wants to surrender his 17, so I buy the surrender bet by giving the player $50, which is one-half his original bet. I now own the hand. If the 17 wins, great! I get back $200 for my $50 investment. A push is also fine as it's like a win — an even-money payoff (I take the $100 back). This strategy can only be employed in a private game (this one's in Canada). In this instance, there are a couple of very amateur counters who tend to "over-surrender," especially 17 v. 10. I doubt that this "buying" procedure could be used in normal casino play, as the pit would probably not be too happy.*

I know I have an edge, but the question is, "How much?" Also, what about doing the same thing for his 14, 15, or 16 v. 10? Would that be good for me? Finally, suppose the true count is +5. Do I still have an advantage on the 17 v. 10 proposition? I would appreciate any advice you can give me.

A. This is an interesting question. Intuitively, without using any math, you should understand that, for the 16, 15, or 14 v. 10, if it is correct basic strategy for the player to early surrender and you volunteer to book ("buy") his bet, then obviously, it's a bad bet for you. It can't possibly be attractive for both of you simultaneously. That wouldn't make any sense. So, we need to know, for the hands we're discussing, if they are or are not net 50% losers (allowing for the dealer's 10 to turn into a natural).

The charts in Julian Braun's book (*How to Play Winning Blackjack,* Data House, 1980), pp. 82 and 83, are all we need to answer most of your questions. They give us, for 100,000 hands, how many times each situation will arise (for a 4-deck game) and what the net profit or loss is in each instance. What's more, and this is crucial, when the ten or ace is involved, they do not rule out the potential dealer blackjack (i.e., they are not referring to playable tens and aces, but rather to all tens and aces). This is precisely what we need to handle the early surrender aspect.

Here's what would happen for 16, 15, or 14, based on 100 decisions, if the player bet $100, you "bought" each hand for $50, and played out each one against the dealer's 10 upcard. For 16 v. 10: net loss = 56.8%. Since you pay, in effect, 50% for the hand, you lose .068/.50 = 13.6%. For 15 v. 10: net loss = 54.1%; you lose .041/.50 = 8.2%. And for 14 v. 10: net loss = 50.4% (close); so you lose .004/.50 = 0.8%.

You see, therefore, that in each instance, the correct basic strategy play is to early surrender (just marginally for 14 v. 10), and so, you would get the worst of it. Of course, if you were counting, there would come a time when you would have the edge.

As for the 17 v. 10, this is another story, and the hand would be a big winner for you if somehow you could book this bet alone, while finding a way to reject the aforementioned ones. In this case, the negative expectation for the player is just 46.2%. Since you buy the hand for 50% of the original bet, your gain is .038/.50 = 7.6%.

Obviously, this appears to be quite attractive ... until you calculate how frequently you get the opportunity to make the play! It looks like about once every 55 hands. So, depending on the speed of play and the number of players in the game, you'll earn the 7.6% on your $50 "buy" (= $3.80) about once or twice an hour. This is not something you'll get rich on, but obviously, it does have some value.

Finally, for 17 v. 10 at true count of +5, it would seem that the play is marginally with you. I didn't run a precise simulation, but in Table A4 of Wong's *Professional Blackjack,* p. 174, he gives the Hi-Lo indices for early surrender of 10,7 v. 10 and 9,8 v. 10. They are +5 and +6, respectively. As the 10,7 combination is four times more likely than the 9,8, I would say that, globally, for 17 v. 10, exactly +5 would be slightly disadvantageous for the player, therefore, advantageous for you. Note that Braun, on p. 135, gives just one number for 17 v. 10 and, predictably, it is +5. To be safe, +4 or below is clearly OK for you, and anything above +5 would appear to be a loser if you buy the bet for half. Hope this has been helpful.

Postscript: *Again, many of the above problems would probably be handled today, in a trice, by simulators. Progress is progress, and I applaud the technology; but I think it behooves all of us to reflect upon why some answers are the way they are. Analytical approaches permit the development of a "feel" and an "intuition" for the game, which can be stifled by simple acceptance of "whatever the computer says." I'd like to encourage readers to think about blackjack problems before they run to a simulator to determine an answer. Then, use the "sim" to verify your hypothesis. You'll feel a great sense of satisfaction, knowing that you were able to "dope out" the correct response.*

September 1986 Cover Art by Alison Finlay

Chapter 5

The "Illustrious 18"

"Don Schlesinger's article … is priceless."
— *George C., noted author of many fine blackjack reports*

"I would like to congratulate Don Schlesinger on his latest article, 'Attacking the Shoe.' A very lucid presentation of the facts that are extremely important to the shoe player. I view Don's qualification of playing strength versus betting strength … as a major contribution. …"
— *Jake Smallwood, frequent contributor to blackjack literature*

"Schlesinger's article … was very well done."
— *Marvin L. Master, frequent contributor to blackjack literature*

Perhaps, when all is said and done, and I've written my final lines on the game of blackjack, I will be best remembered for what you are about to read. "Attacking the Shoe," the feature article from ***Blackjack Forum,*** *September 1986, in which "The Illustrious 18" appeared for the first time, is, in many ways, my "legacy" to the game. I'm proud of how oft-quoted the concept has become, and I'm grateful for the impact* *that the discovery has had on an entire generation of card counters. Pay attention: What follows is important!*

The steely-nerved counter contemplates his big bet, checks the discard tray, does some lightning calculations, and stands on his 16 v. the dealer's 7. Dealer flips a 9 in the hole and breaks with a 10. Another large sum of money won thanks to the expert's knowledge of this seldom-used basic strategy departure. Right? Wrong!! In fact, the 16 v. 7 situation arises roughly once every 100 hands (actually .0096). The true count needed to stand (+9) occurs 0.61% of the time in a 4-deck, deal 3 game. Thus, the opportunity to *depart* from basic for this play presents itself .0061 x .0096 of the time, or about once in every 17,000 hands dealt! What's more, when it *does* occur, the player's "reward" for standing is, on average, $10 for every $500 bet. If our expert plays 100 hands per hour and 500 hours per year, this play comes up three times and nets the counter about $30 *annually!* And if your top bet is $100 and you play a more reasonable 200 hours per year, then knowing this number ($2.40 yearly gain) won't buy you two hot dogs a year on the Atlantic City boardwalk!

And so, this article — the purpose of which is twofold: 1) To explore the actual gain available from using the Hi-Lo count in a multiple-deck situation, and 2) To rank, in order of contribution to that gain, each individual departure from basic. Along the way, I will do my best to explain the methodology (a bit complex at times), and hope to demonstrate, at the conclusion, the adaptability of my approach for different counts, betting schemes, and rules.

More than ten years ago, when I taught myself to count cards using Lawrence Revere's text, I had no one to consult with for learning the index numbers. Over the years, as I grew more expert on the subject, I suspected that most of the numbers I had learned (some 165) contributed very little, if anything, to the overall gain available from the Point Count. Furthermore, despite hundreds of books and articles published on blackjack, I have not to this day seen a study that would tell a player *which* index numbers are most important to learn based on the *amount* of total gain that can be obtained by their use. In this respect, I believe that the chart presented near the end of this article contains information that has never before been published and that should provide some revealing facts for consideration by the blackjack-playing community.

In the third edition of Professor Peter Griffin's masterful work, *The Theory of Blackjack,* there is a new chart, p. 229, that indicates the gain (in 1000ths of a percent) achievable from perfect play with flat betting in a 4-deck game dealt to approximately the 80% level. Although the chart is extremely interesting, from the theoretical point of view, it has, unfortunately, little practical importance in that a) most 4-deck shoes aren't cut that deeply, b) no human plays "perfectly," and c) no serious player in his right mind flat-bets a 4-deck game. The problem, therefore, was to create a similar chart that would be more applicable to the style of play of most counters who come up against the 4-deck shoe game.

I received this admonition from Prof. Griffin in the first of two letters he wrote to me on the subject: "Certainly the direction of your thinking is valid, but it is difficult to write an article with *particular* recommendations for 'everybody' since 'everybody' has a slightly different count, wagering style, etc. If you can solve the problem for yourself, you're doing well." And later on: "Your proposed article will excite many, generate a tower of Babel. But there will be much misunderstanding." And so I have tried very hard to make the article pertinent to *most* counters and understandable to just about everyone. I will leave it to Prof. Griffin and the readers of *Blackjack Forum* to let me know if I have succeeded.

First, some generalities. For better or worse, most counters use the level one, Hi-Lo (plus-minus) system whose values are: 2–6, +1; 7–9, 0; 10s and As, –1. If you play a different count, you will receive plenty of information as to how to alter my numbers to fit your count. Next, I assumed that a 4-deck shoe was played to the 75% level. Perhaps the 6-deck game would have been a better choice (more of them in the world right now, I believe), but I wanted to use a compromise between the hand-dealt single- and double-deck Nevada games and the 6- and 8-deck shoe games. Finally — and this is very important — I assumed a large bet variation and parlaying approach that I have described in my two previous articles for *Blackjack Forum.* To wit: one unit below +2; two units from +2 to +3; four units from +3 to +4; six units from +4 to +5; and two hands of six units for +5 and above. Again, if your betting style differs from mine, a simple calculation and adjustment will permit you to use the final table for your personal method of wagering. Thus, in essence, I decided to solve the problem not for myself but rather for what I hope to be the *largest* segment of the *Blackjack Forum* readership. Frequenters of Atlantic City should not be put off by my choice of the 4-deck game as a model. The results of my work are entirely applicable to the A.C. games, and the plays that are recommended later on would be identical for 6- and 8-deck shoes. Only the magnitude of the gain would change; but more about that later.

Let's get on to the calculations. It is important to understand why the advantage *you* can get from playing the Hi-Lo numbers is different from what a computer can attain from "perfect" play.

Every count has a "playing efficiency" attached to it. In addition, for *each play* contemplated, the Hi-Lo (and every other count) has a "correlation coefficient" that measures how well (or poorly) the point count is able to convey information regarding a strategy departure for *that* particular play. The overall .51 playing efficiency of the Hi-Lo reveals that it does a mediocre job of capturing all the gain available from learning index numbers. What's more, because of certain *individual* correlations that are extremely low (14 v. 10 and our previous example of 16 v. 7 being two prime illustrations), the Hi-Lo (and many other counts) is ill suited to give any *worthwhile* information on how to play those hands. Note, for example, that in the case of 14 v. 10, there simply *is no number* for the play. Since the 7 is not counted in Hi-Lo, and since its removal from the deck plays such a crucial role in determining the departure from basic strategy for this play (see Griffin, bottom pp. 73 and 76), it is easy to understand why an imperfect knowledge (in fact, no knowledge at all) of the number of 7s remaining in the deck makes it impossible for the Hi-Lo to detect the correct moment for making such a strategic departure.

With these limitations in mind, our goal is to find those plays that *can* be sufficiently analyzed to produce some gain and to *quantify* that gain. The next problem stems from the fact that, with a large bet spread, not all departures, when made, will carry the same wager. Thus, it is necessary to calculate the average bet one has on the table *when* the variation is called into play. This is easily done. Wong's simulations (see footnote number 8 at the bottom of Table 5.1) provide the frequency with which each true count arises when playing the 4-deck, deal 3 game. If we assume (for convenience) that we will always have a mathematically correct bet out when called upon to make a departure, then we take a weighted average of how many units will be on the table when a certain count arises. For example, it is obvious that for counts below zero, hitting decisions will carry one-unit bets only. However, the moment we enter into non-negative territory, we must consider that if a departure to stand or double is made, we will have, at times, a multiple-unit bet out.

Let's go through a sample calculation. What will be the average bet, for my style, when a departure involving a +1 index number is used? From Wong's charts (bottom of p. 20), we see that 10.94% of the time, a one-unit bet will be involved. 6.84% of the time, we'll have two units riding. 4.21% of the time, we'll have four units out. Finally, since a number can apply to only one hand at a time, the fact that we may play *two* hands of six units is irrelevant for this discussion. Our top bet can be no more than six units and it will occur (2.75 + 1.69 +1.07 +.65 +.42 +.25 + .36) = 7.19% of the time. Our average bet is, therefore: (1 x 10.94) + (2 x 6.84) + (4 x 4.21) + (6 x 7.19) = 84.6/29.18 (the sum of the percentages) = 2.90 units. Similar calculations yield the average bet sizes at the other index number levels (see column 8 of Table 5.1).

The most challenging part still lies ahead. We must abandon the numbers provided in Griffin's p. 229 chart as being unsuitable for our use, as explained earlier. How, then, do we determine the gain available from a Hi-Lo number when one deals 75% of the 4-deck shoe, given a particular betting style? I am indebted to Prof. Griffin and to his second lengthy letter for the answer. Although the procedures for making the calculations, outlined on pp. 88–90, seem complex, one is able to follow, step by step, rather easily with a pocket calculator. One crucial additional piece of information, not available in this section of the text, was provided by Prof. Griffin's correspondence: To obtain the advantage at the 3/4 level of penetration, it is necessary to make two separate calculations. Use N = 208 and n = 52 for the first run, "a," and then N = 208 but n = 130 for the second set of calculations (call them "b"). Finally, weight the two results, a and b, by using $\frac{a+4b}{6}$ to obtain the results for the level desired (in this case, 75%). The weighting is an application of "Simpson's Rule" for approximating an integral and gives rather good results in these situations.

Let's "walk" through the 15 v. 10 calculation (a very important play) together. Keep in mind that before the task can be completed, we must already know the: 1) correlation coefficient, 2) sum of the squares, 3) full-deck favorability, 4) probability of occurrence, and 5) average bet for the play in question. These figures are either provided directly in Griffin (see references in the footnotes to the chart) or are easily obtained from Griffin (likewise, see footnotes).

Let's get back to 15 v. 10. As per bottom of p. 89, top of p. 90, we modify Step 1, top of p. 88 in Griffin, by multiplying the result by the correlation coefficient (which is .78 for 15 v. 10). Thus, for case "a," we get:

1) $b = .78 \times 51\sqrt{\dfrac{14.8(208 - 52)}{13 \times 207 \times 52}} = 5.11;$

2) $z = \dfrac{3.48}{5.11} = .68;$

3) UNLLI = .1478;

4) $.1478 \times 5.11 = .755;$

5) Probability of occurrence $= \dfrac{165}{1{,}326} \times \dfrac{188}{663} = .035.$

Therefore, gain at this level = .755 x .035 = .026. Repeating the entire process for case "b," where n = 130, we get a gain of .0022. Now, each separate gain must be multiplied by the average bet (6 units) associated with a +4 index number. Thus, "a" becomes 6 x .026 = .156, and "b" becomes 6 x .0022 = .013. Finally, the weighting:

$$\frac{a + 4b}{6} = \frac{.156 + 4(.013)}{6} = .035.$$

The entry of 37 (.037) in column 10 of Table 5.1 differs from this result by .002% and is due to certain rounding and estimation methods used to facilitate the numerous calculations that were necessary to generate the final numbers.

In a similar fashion, we carry out the identical process as above (tedious!) for each of the plays we are interested in. Along the way there are a few subjective decisions to be made. First, although I have included the gains available from splitting a pair of

10s against the dealer's 5 or 6, I don't recommend use of these departures. I believe that making these plays, along with jumping one's bet, is one of the biggest giveaways that you are a counter and thus should be avoided despite the decent amount of gain available from the splits. *[For more on this topic, see Chapter 14, p. 381. — D.S.]*

Other hands where deviation from basic strategy would occur with a six-unit bet are: 16 v. 7, 16 v. 8, 16 v. A, 15 v. 7, 15 v. 8, 15 v. 9, and 15 v. A. *None* of these plays produces more than a .002% gain, and it is doubtful that knowledge of *all* of the indices together contributes more than .01% to the overall gain attainable from using the Hi-Lo variations. With these "editorial" decisions explained, I offer the "Illustrious 18" — the top 18 departures from basic strategy that contribute the most gain to the Hi-Lo player's hands in a 4-deck, deal 3 situation.

Surely, there are some revealing facts that are readily apparent. Note that the insurance play *all by itself* is worth over one-third of all the I18 gain. The "Big 3" — insurance, 16 v. 10, and 15 v. 10 — account for nearly 60% of the total gain available from the 18 plays, while the top six plays contribute almost 75% of the advantage. The first half of the chart (nine plays) garners over 83% of the edge and the "even dozen" (top 12 plays) provides more than 90% of the total gain. I believe this is the first time that such information has been accurately quantified in a published article.

Of course, the ramifications of this study are quite clear. If you are a practicing Hi-Lo player and have diligently committed to memory 150 to 200 index numbers, you may be interested to learn that for this particular game and style of play, you might as well throw 90% of your numbers away and keep just the "Illustrious 18." On the other hand, if you have just mastered true count and were about to embark upon your study of the index-number matrix, I may have saved you a good deal of work. Learn the plays in the chart and forget about the rest. You can trust me that you won't be missing much!

Obviously, this study should not end here. At the moment, I don't have the time (or the energy!) to carry it any further, but the potential variations on the theme are endless. I, for one, would be interested in similar charts, ranking the order of dominance of the plays, for: 1) different numbers of decks, 2) different penetration levels, 3) different betting schemes, and 4) different point counts. Of course, plays that correlate poorly for one system might fare better in another. And plays that are important because of large bet variation might shrink in importance with a smaller spread. Finally, one glance at the gain from *single*-deck perfect play (Griffin, p. 30) will convince the reader of the obvious extra advantage to be gleaned from strategy variation in *this* game, regardless of the point count utilized.

On the contrary, one must expect a diminished gain for the 6-deck A.C. game, for example. Since the total gain from flat-bet, perfect play for all strategy variations is given by Griffin, pp. 228 and 230, as approximately .30%, compared to the 4-deck figure of .40%, pp. 228–29, it is reasonable to assume that only .30/40 = 75% of our 0.348% 4-deck gain, or 0.268%, would be available for the 6-deck variety. *[Note: Interestingly enough, it took almost ten years before any other blackjack researcher took me up on my above "challenge." In late 1995, Bryce Carlson published a most revealing study that both corroborated my findings and amplified and expanded on them. See* ***Blackjack Forum,*** *December 1995, pp. 88–90. Finally, see Chapter 13, p. 374, for the latest research in this area. — D.S.]*

Another observation comes to mind. In a previous article for *Blackjack Forum* (June 1985, p. 17), I indicated that the overall percentage gain available for the game studied in this article was 1.1%. On pp. 119–20, Griffin, there is a formula that permits us to estimate the betting gain available from wagering k units on each favorable hand, one otherwise. For the bet scheme in this article, k = 5.5. Thus, the "perfect" gain would be .32 (5.5 – 1) – .48 = .32 (4.5) – .48 = 1.44 – .48 = .96. Hi-Lo gain, with a betting efficiency of approximately .95 for the 4-deck game, would be .96 x .95 = .912. Now, our chart indicates that the Hi-Lo playing strategy gain is, roughly, .348. Together, .912 + .348 = 1.26. Not a bad comparison to the previously calculated 1.1%. The figures suggest that for this style of play, betting advantage is over 2.5 times as large as playing advantage.

There is an interesting corollary to this observation. As you know, I have advocated a back-counting approach to the shoe game in both of my previous *Blackjack Forum* articles. Indeed, in my June 1985 chart, the calculations show that whereas one wins 2.00 units per hour in the play-all 4-deck game, one ought to win 2.76 units per hour (a 38% increase) with the +1 and up "Wonging" style. Ironically, none of this extra advantage is the result of playing variations. In fact, since the five non-positive number "hits" that appear at the bottom of the chart are eliminated, there is actually a slight *decline* in the gain due to strategy departures (.027%, to be precise, if one were playing all the hands). In the 29 hands per hour that we *do* get to play, opportunities to vary from basic cannot increase from the 100 hands-per-hour level. Our extra gain comes strictly from avoiding all of the negative-expectation one-unit-bet hands that we are spared from playing at counts below +1 true. Originally, .55 units per hour were won due to playing departures:

$$\frac{.348}{1.26} \times 2.00 = .55.$$

Now, only .55 x $\frac{.321}{.348}$ = .507 units were won. Since the back-counting style produces 2.76 units of gain per hour, 2.76 – .507 = 2.253 of those units are the result of betting efficiency, while the remaining .507 are due to playing efficiency. This new ratio, $\frac{2.253}{.507}$ = 4.44, suggests that for the 4-deck back-counting approach, the betting contribution to gain is nearly *4.5* times as important as the playing strategy. Compare this observation to Griffin, pp. 47–48, where "Proper Balance between Betting and

TABLE 5.1

The "Illustrious 18" and the "Fab 4"

1 Hand	2 Inner Product	3 Correlation Coefficient	4 Sum of the Squares	5 Full-Deck Favorability	6 Prob. of Occurrence	7 Index Number	8 Avg. Bet (Units)	9 Hi-Lo Avg. Gain 1000ths of a pct.	10 Hi-Lo Actual Gain 1000ths of a pct.	11 Cumulative Gain 1000ths of a pct.	12 Cumulative % of Actual Gain
The "Illustrious 18"											
Insurance	23.52	0.76	95.7	7.200	0.0770	3	5.26	22.30	117	117	33.7
16 v. 10	7.73	0.56	19.1	0.067	0.0350	0	2.00	26.50	53	170	48.9
15 v. 10	9.47	0.78	14.8	3.480	0.0350	4	6.00	6.17	37	207	59.5
10,10 v. 5	33.55	0.96	121.5	15.530	0.0073	5	6.00	2.89	17	224	64.4
10,10 v. 6	30.81	0.92	111.7	12.715	0.0073	4	6.00	2.75	17	241	69.3
10 v. 10	10.10	0.84	14.6	3.280	0.0115	4	6.00	2.67	16	257	73.9
12 v. 3	13.90	0.70	39.0	2.040	0.0075	2	4.04	3.22	13	270	77.6
12 v. 2	12.60	0.63	39.9	4.080	0.0075	3	5.26	2.09	11	281	80.8
11 v. A	23.78	0.82	83.4	2.760	0.0025	1	2.90	3.45	10	291	83.6
9 v. 2	15.50	0.81	36.4	1.340	0.0028	1	2.90	3.10	9	300	86.2
10 v. A	26.53	0.94	78.9	8.850	0.0020	4	6.00	1.17	7	307	88.2
9 v. 7	20.17	0.86	55.3	6.700	0.0028	3	5.26	1.33	7	314	90.2
16 v. 9	8.14	0.53	23.7	3.280	0.0096	5	6.00	1.17	7	321	92.2
13 v. 2	14.90	0.67	50.1	1.460	0.0075	–1	1.00	7.00	7	328	94.3
12 v. 4	15.00	0.77	38.4	0.025	0.0075	0	1.00	6.00	6	334	96.0
12 v. 5	15.90	0.82	37.6	2.450	0.0075	–2	1.00	6.00	6	340	97.7
12 v. 6	13.30	0.65	41.4	1.450	0.0075	–1	1.00	4.00	4	344	98.9
13 v. 3	16.10	0.68	55.7	3.810	0.0075	–2	1.00	4.00	4	348	100.0
The "Fab 4" Surrenders											
14 v. 10	11.51	0.76	23.1	3.338	0.0202	3	5.26	12.74	67	67	55.4
15 v. 10	13.02	0.80	26.7	0.463	0.0202	0	1.00	22.60	23	90	74.4
15 v. 9	13.55	0.89	23.3	2.863	0.0055	2	4.04	4.95	20	110	90.9
15 v. A	11.41	0.71	25.9	1.907	0.0038	1	2.90	3.79	11	121	100.0

Footnotes: (Numbers refer to column headings):
2) From Griffin, 3rd Edition, pp. 74–85
3) Griffin, p. 71
4) Griffin, pp. 74–85 (12th columns)
5) Griffin, pp. 74–85 and 231–32
6) Griffin, p. 39
7) Wong, *Professional Blackjack*, p. 173
8) Wong, *Nevada Blackjack*, July '83, p. 110
9) Griffin, pp. 74–85 and correspondence
10) Columns 8 x 9
11) Vertical summation of Column 10
12) % of total gain of each entry in Column 11

Playing Strength" indicates a modest 1.5 to 1 ratio for the single-deck game. My suggestion is that when one is shopping around for a best single-parameter card-counting system, the paramount considerations are the *number of decks* one will be playing against and the style of play to be used. I am reasonably certain that, for my style of play, Wong's Halves and the Revere Point Count are the systems of choice. *[See the SCORE comparisons in Chapter 9 for more on this topic. — D.S.]*

And so, there you have it. I would hope that those who are reading these lines look upon the above study not at all as a "definitive work," but rather simply as a door-opener — a first attempt at viewing a rather complex problem from a fresh angle. Comments, both favorable or unfavorable, and suggestions for future articles are enthusiastically solicited.

As for me, I continue to play the Revere Point Count and, try as I may, I can't chase those 165 numbers I learned ten years ago from my head. Only now, I know once and for all that I am not very much better off for all this "knowledge" than if I had learned the "Illustrious 18" and left it at that! This has been a humbling (and, I hope, useful for you, dear readers) labor of love. I am indebted to Peter Griffin for his invaluable suggestions and explanations; to Stanford Wong; and finally, to a colleague at work, Derek Bandeen, who helped me with some computer programming and print-outs that greatly facilitated the preparation of this article.

From "The Gospel According to Don," *Blackjack Forum*, December 1995:

Q. *It's been quite some time since your "Attacking the Shoe" article, containing the now-famous "Illustrious 18" strategy index numbers, appeared in the September 1986* ***Blackjack Forum.*** *No surrender indices were included in the list, because Peter Griffin hadn't furnished the effects of removal for those plays in his book. Is there any way to obtain those numbers and to work out the surrender indices that are worthwhile to learn?*

A. Yes, and yes! I have wanted to do exactly what your are requesting for … well, nine years! Now, thanks to Steve Jacobs (more about him in a moment), I am able to supplement the original "Illustrious 18" with an additional four indices (the "Fab 4"?!) for surrender.

Readers will recall that, back in 1986, I set out to identify those strategy departures that contributed the most to our playing advantage. The idea was to make life simpler for newcomers and veterans alike; why bother to memorize 150–200 strategy departures when, at least for multi-deck, 15–20 would do? And so, the "Illustrious 18" were uncovered. The problem was, as mentioned above, there were no data for the surrender indices. Fast-forward nine years. Enter the Internet.

As many of you have read, in recent *Blackjack Forum* issues, there is an extremely lively and informative newsgroup, rec (for "recreational").gambling.blackjack, on the Internet. One of its most knowledgeable and prolific contributors is Steve Jacobs who, in my humble opinion (or, as they write in the group, IMHO!), is the "Godfather" of r.g.bj. It didn't take me long to strike up an ongoing e-mail correspondence with Steve, and I'd venture to say we've both learned a thing or two from each other. When a question arose on the Net asking for the importance of surrender index values, it led me to Steve: Maybe his program could furnish the desired effects of removal. I would do the rest, as per the 1986 article. Well, Steve came through in grand style! He is hereby publicly thanked and his contribution to what follows is duly and gratefully acknowledged. *[Note: With the addition of our new Appendix D, we now have complete sets of effects of removal for late surrender. — D.S.]*

First, a word about methodology. As a compromise between single-deck and 6- and 8-deck shoes, I calculated indices for the 4-deck, s17 game. While this may have the effect of leaving no one entirely happy (!), it does strive for some balance, given that I lacked the energy to grind out the numbers for five different games. We assumed the Hi-Lo count system, 75% penetration, and a 1–12 betting spread, according to the true count advantage.

Late surrender basic strategy is worth 0.069% for the 4-deck game. (For reference, we get 0.075% for 6-deck and only 0.022% for 1-deck.) The plays we make are initial 16 v. 9, 10, and A, and 15 (but not 8,7) v. dealer's 10. One might think that these are the indices we should learn. For the most part, one would be wrong! When determining which strategy departures are worthwhile, several factors come into play. Three are especially important here. First, the "full-deck favorability" must not be too large. This value tells us how much better it is to make the basic strategy play, as opposed to the next best choice, given a full deck. In essence it indicates how far from zero, in absolute value, our strategy index number lies. For example, what good is it to learn a departure with an index of +10 (or even worse, –10) if we never get to make the play? Such a play would have a very large "full-deck favorability."

A second important consideration, when we are "spreading big," is the average wager we will have on the table when a departure is called for. Well, for our basic strategy surrenders, the indices are all non-positive (i.e., zero or negative): 16 v. 9 = 0; 16 v. 10 = –3; 16 v. A = –2; and 15 v. 10 = 0. And, unlike all but the bottom five of the I18, we make a departure (in this case by *not* surrendering) only when the true count is *below* the index values in question. This ensures that, when we do depart from basic, we will always have a one-unit, minimum bet out for all plays. Finally, as conventional surrender occurs on the first two cards only, and not for multi-card holdings, the frequency with which we raise the white flag is rather low.

Not surprisingly, then, three of the four basic strategy surrender indices (those involving 16) are not worth learning. All three together glean only an extra 0.005% advantage. The lone exception is 15 v. 10, which soars to the second-highest surrender perch, with a 0.023% contribution to the counter's edge.

To find our additional three worthwhile plays, we need to look at indices that are reasonable enough to occur with sufficient frequency (relatively small "full-deck favorabilities") and where our average wager is a decent multiple of one unit, when the play is made. Now the idea will be to depart from basic by surrendering when, normally, we would not. We'll need to be at or *above* the critical index, instead of below it.

The envelope, please! And the winners are: 14 v. 10 (+3 index); 15 v. 9 (+2); and 15 v. A (+1). The extra gains are, respectively, a whopping 0.067%, 0.020%, and 0.011%. These three plays, coupled with 15 v. 10 (the "Fab 4"), add an additional 0.121% advantage to the basic strategy surrender value of 0.069%, making a grand total for the combination of 0.19%. (To be precise, the two gains cannot just be added, as there is some overlapping of the contribution of 15 v. 10 to both groups. Also, we must realize that when 15 and 16 are surrendered v. dealer 10, we do not get to use our valuable stand departures for these initial hands, at counts of +4 and 0, respectively.) You may learn more surrender indices, if you insist, but again, the idea is to capture the bulk of the attainable edge with the fewest departures possible. And, as was the case for the "Illustrious 18," the "Fab 4" get most of the surrender money.

Postscript: *As stated above, I was pleased, nine years after the appearance of the original article, to have "closed the circle," and delivered the missing surrender values. Subsequent studies to mine, notably by Bryce Carlson, and Olaf Vancura and Ken Fuchs, have confirmed the validity of my ground-breaking work. I take much satisfaction in having discovered a principle that has so greatly simplified the life of many a modern-day card counter.*

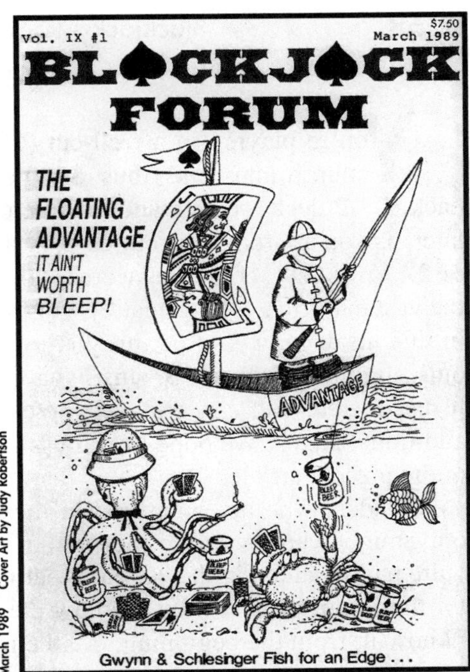

March 1989 Cover Art by Judy Robertson

Chapter 6

The "Floating Advantage"

As was the case for the "Illustrious 18," the "Floating Advantage," from the March 1989 ***Blackjack Forum,*** *contained another landmark discovery that was, right up to the date of its publication, misunderstood by virtually all of blackjack's leading authorities. But the concept never enjoyed the popularity of the "Illustrious 18," mainly, I feel, because the information could not be exploited as readily as its "illustrious" forerunner. Nonetheless, the project was noteworthy, if for no other reason, because it represented, at the time, the most ambitious simulation ever undertaken (by Dr. John Gwynn, Jr.) to study a single blackjack problem. I was honored to have collaborated with Professor Gwynn on this massive study, the entire text of which is provided below.*

Note: The original article that appeared in ***Blackjack Forum*** *was a condensed version of this longer paper, which sold separately, in the words of Snyder, "for serious blackjack fiends only."*

You're playing in a well-cut (75%) 2-deck game, betting $200 times the true count minus one (thus, $400 at a true of +3, for example). Early on in the pack, 1 1/2 decks remain and the true count is +2. Your bet? $200. A few moments later, as you approach the cut-card, at the 1 1/2-deck level, you once again have a true of 2. Your bet? If you answered, "the same $200 as before," chances are you're in the vast majority. Unfortunately, you're also unaware of the "floating advantage" that, in this instance, would permit you to make a $350 bet, as if the count were actually only slightly less than +3. Just what is this so-called "floating advantage," how was it discovered, and can a player increase his hourly win rate by capitalizing on this additional edge? We hope to address all of these questions and provide clear answers in the present article.

Readers who are interested in the bottom line and who are "numbers oriented" only, might want to skip my account of the cast of characters who participated, at times quite passionately, in the "Great Floating Advantage Debate." But I would be remiss, for those purists who want to know the whole story, if I didn't recount the episode, as I know it, from the beginning. Mathematicians, bear with me; I won't disappoint you later on!

Part I

For me, it started innocently enough in 1976. I had sent away for the $25 supplement to the Revere Point Count, offered in his book. The playing and betting strategies were geared to the multiple-deck game and, as such, contained information that Revere had withheld from *Playing Blackjack as a Business* (Lyle Stuart, 1975). As I studied the page entitled, "You Bet Your Money According to the Running Count," I noticed something odd. The author suggested, to cite just one example, that for the Las Vegas 4-deck Strip game, when the running count was +10 and 2 1/2 decks remained, a four-unit bet was called for. Later, when only one deck remained, a running count of +2 was apparently sufficient to make the same four-unit bet. Since the Revere Point Count adjusts running count to true count by dividing by the number of half-decks remaining to be played, the first bet represented a wager at a true count of +10/5 = +2. But the second bet, for the same amount of money, was being made at a true count of +2/2 = +1. I was confused. The immediate implication was that, at the 3-deck level, had I achieved a true of +2, I would have been entitled to bet more than four units. At the time, this didn't make sense to me. I believed that "true was true" no matter where in the deck it happened to appear. But, for the record, this was my first "brush" with the floating edge and, unless someone can prove otherwise, the first documented depiction of it in print (date of publication, 1969; revised, 1973).

In 1981, Ken Uston's *Million Dollar Blackjack* (Gambling Times) was released. On p. 128, a chart entitled, "Player Edge At Various True Counts" appeared. In it, in unmistakable fashion, the floating advantage surfaced (pun intended!) once again. The data were unambiguous: The same true count implied more advantage later in the deck than it did earlier. He may not, at the time, have understood completely why this was so, but none can deny that Uston, as Revere before him, recognized the concept of floating edge and made specific recommendations as to the betting technique necessary to exploit the concept.

Now the fun began! In the September 1982 *Blackjack Forum,* Arnold Snyder wrote an article entitled, "How True Is Your True Count?" In the article, whose computer simulations were done by Professor John Gwynn of California State University, Sacramento, Snyder's opening remarks were to the effect that, as the 4-deck shoe was depleted, a player would have the same advantage at a progressively lower true count. Snyder was astonished. "This was a revelation to me," he wrote. He advanced the "true is true" theory and further claimed that, "Expert opinion has always held that a running count of +6 with one deck dealt out of a 4-deck game indicates a player advantage equivalent to that of a running count of +4 with 2 decks dealt out or +1 with 3 1/2 decks dealt out. In all cases, the 'true' count is +2 (on a count-per-deck basis). John Gwynn's 23+ million-hand simulations of blackjack systems indicate otherwise."

There were two aspects of these remarks that troubled me greatly. First, by painting all "experts" with the same brush, Snyder did a disservice to Revere and Uston who, as I have shown, were aware of this "discovery" before Gwynn and Snyder. But the real controversy began when I dared to suggest, contrary to what Gwynn and Snyder believed, that even at a true count of zero, the disadvantage at the beginning of a 4-deck game was neutralized once the 3-deck level was reached. In other words, the advantage normally associated with a +1 true count earlier in the shoe actually "floated" down to the zero count once three decks were dealt. Little did I know, when I conveyed these beliefs in writing, not only to Snyder, but to Peter Griffin as well, what a brouhaha I would create. Here's what I wrote to Snyder in September 1982 (published in *Blackjack Forum,* December 1982): "That the same true count becomes worth more as the deck is depleted is a banality, known to me for seven years, known to Revere for longer, known to Uston. ..." I then made the arguments advanced above from Revere's pamphlet and Uston's p. 128. The plot thickened. The Reverend Bishop shot back: "Gwynn's data show a very slight increase in player advantage — at a true count of 0, as deck depletion occurs, in the 4-deck game, but nothing as radical as what is reported by Revere, and later, Uston. With all due respect to the late Lawrence Revere, this is one 'disclosure' he should have forgotten! It seems to me, Don, that the 'banality' which you have known for 7 years, which Revere had known for longer, is, alas, a mistake. This 'floating' off-the-top edge does not exist." Fighting words, those! I fired off an answer, published in the March 1983 *Blackjack Forum,* but the issue was far from resolved.

Undaunted, on December 19, 1982, moments after reading Snyder's retort, I wrote to Peter Griffin, hoping he would side with me on the zero-count business. At that time, the Las Vegas Hilton and Tropicana dealt a 4-deck game to the 3 1/2-deck level. I wrote, "I would anticipate an edge at these games even at a true of zero, with one deck (or less) remaining. With one-half deck left, the edge would, of course, be even greater." I also made several other mathematical observations in my letter. On December 27, Professor Griffin responded. While being most complimentary to me ("Based on your comments, I would guess your understanding of the game would outrank that of almost every published 'expert.'"), Griffin surprised me with his answer: "A running count of zero at the Tropicana would not, I believe, give you an edge with one deck left. Here's my intuitive explanation. Basic Strategy edge should still be –.5% (approximately). Hence, you require strategy variation to make up at least .5%. How can this be?" And, in a follow-up letter, dated January 10, 1983, Griffin added a few calculations that led to a similar conclusion.

What to do now? I had been turned down by the Supreme Court! There was no higher appeal. A year went by. I brooded and licked my wounds. And then, a double miracle occurred! Stanford Wong and Professor Griffin, working independently (much as Leibniz and Newton had, on the calculus — but of course with much less import than these studies on "floating advantage"!), had made a remarkable discovery: It wasn't just strategy variation that contributed to the increased edge for a true of zero at the one-deck level; *basic strategy itself* was worth .5% more at that level than at the start of the 4-deck pack. The bombshell first appeared in *Casino & Sports,* Vol. 24, p. 34. Peter Griffin had changed his mind! "Having spent much of the past eight years and almost half of this article (in which he quoted extensively from the two letters written to me), arguing that there is no reason to believe that a zero count occurring deep within a shoe (or deck) presages increased basic strategy advantage, I shall now do an about-face! I now suspect this increase does exist although it is neither of the magnitude nor origin previously proposed."

Eureka! Vindication was near. Seems that, at the one-deck level, extremely high counts produce less edge than expected for the basic strategist (many pushes) and the extreme negative counts were found to be even more unfavorable than previously thought (doubles, splits, and stands tend to be disastrous). But the overall basic strategy edge must remain constant, so the extra loss at the extremes must be compensated for somewhere in between. Guess where? Among other places, at zero, by God! These findings were reiterated by Griffin in the third edition of his master work, *The Theory of Blackjack,* 1986, pp. 208–14.

Almost immediately after the *Casino & Sports* article appeared, Wong published, in *Blackjack World,* February 1984, "Advantage Versus Penetration, Six Decks, A.C. Rules." On p. 23, we see how advantage, using basic strategy alone, increases, in general, for all counts, –4 to +4, as the pack is depleted, and, in particular, for the zero count. Wong's conclusion: "It appears reasonable to extrapolate this finding to apply to any number of decks shuffled together: After we have counted the pack down to where n decks are left, the edge with a count of zero is about the same as if we had started with n decks." Somewhere, I felt, Revere was smiling! But I will be the first to admit that I can't prove Revere actually knew all of this. Perhaps he was right by accident. But he was right. And so, I wrote once more to our illustrious Bishop. *Blackjack Forum,* March 1984, carried the following taunt (p. 38): "When you have a spare moment, I would like you to read Wong's February 1984 newsletter, pp. 22–25. Then I would like you to reread your December 1983 (sic! — it was 1982) *Blackjack Forum,* pp. 32–36. Next, I would like you to consider all your past 'sins' with regard to this topic." Snyder was gracious in defeat: "Revere was right! Uston was right *[perhaps for the wrong reason — D.S.].* I was wrong! AND DON SCHLESINGER WAS RIGHT!"

And so ended the "Great Floating Advantage Debate." I had taken on some heavyweights, had some fun, but most of all, stimulated some great minds to eventually seek out the truth and reach conclusions that, I hoped, would put extra money in your pockets and mine.

Postscript: *Four years have gone by. Arnold has published his series of booklets,* ***Beat the X-Deck Game,*** *in which he recommends betting according to the "floating advantage." You've come a long way, Bishop! But there was more work to be done. No one had yet quantified, in greater detail, the magnitude of the extra edge, on a true-count basis, for the single- and multiple-deck games. No one had suggested an accurate approach, based on these findings, for betting one's money in order to capitalize on the floating advantage, if, indeed, it was possible to do so. I turned to John Gwynn and proposed a joint effort. If he would provide the computer simulations, I would sift through the data, collate the material, calculate the improved win rates that, hypothetically, would result from adjustments to the "traditional" true-count betting approach, and publish my findings, for better or worse, in — where else? —* ***Blackjack Forum.*** *Professor Gwynn's monumental simulation is believed to be the most extensive ever run to study a blackjack-related problem. Nearly three-quarters of a billion hands were generated in order to produce data reliable enough for our paper.* [Note: Times change! Compare this "large" number, at the time, to our 20 trillion hands simulated for Chapter 10. — D.S.]

Part II

I won't tease you or beat around the bush. My initial intention in writing this article was to develop a betting strategy designed to capture the elusive, aforementioned "floating advantage" and put some extra money in everyone's pockets. I had even devised a clever acronym for my methods: "Playing SAFE" (**S**chlesinger's **A**djustments for **F**loating **E**dge). Eagerly, I awaited John Gwynn's data that would, I thought, confirm my hypothesis. It would be up to me to explain all the material to *Blackjack Forum* readers, and we would live happily (and richer!) ever after. Well, I've got some disappointing news. The "floating advantage" isn't what it's cracked up to be. Oh, it exists all right, just as it was presented in Part I. But, in simple terms, it's a question of "too little, too late." Permit me to explain.

The normal betting approach for most true-count players is to obtain a reliable frequency distribution chart for their systems; that is, a chart that breaks down each true count as to how often per 100 hands it occurs. In addition, a win rate per hand, expressed as a percent, is linked to each true count. Now, these win rates are "global," which means that they represent an average win rate across all portions of the deck (or shoe) for the particular true count in question. Whereas it is true that, at the start of a 2-deck game, for example, a true count of +2 is not as profitable as it is at the 50% (one-deck) level of the same game, and that this level, in turn, is less profitable than the +2 edge that the player has at the 75% (1 1/2-deck) level, a) the differences are not overly impressive, and b) none differs substantially from the "one true fits all" averaged result that most players use.

The direct (and sad) consequence of this phenomenon is that there simply aren't enough occasions with the sufficient extra advantage that present themselves so as to permit the bettor to exploit these discrepancies in systematic fashion. Briefly stated, the SAFE method appears to be a waste of time that would not put enough extra money in your pocket to be worth the effort. But John Gwynn, Arnold Snyder, and I felt this article should be written anyway, if for no other reason than to present the data and reach a definitive conclusion regarding the benefits (or lack thereof) of "floating advantage." I have extracted from Dr. Gwynn's massive simulations those results that best demonstrate the dilemma.

Before attempting to utilize the following tables, the reader is urged to refer to this chapter's Appendix for a detailed explanation of their construction. Through a gradual "fine-tuning" process, which consisted of several exchanges of phone calls

TABLE 6.1
.00 decks to .75 decks

Count	Hands	Net Win	Per Hand Expect.	Units Bet	Expect. Std. Err.	Per Hand Contr.	Units Won
≤ −1	47532028	−854550.5	−1.80	1	0.02	−0.708	−0.708
0	31067286	83665.5	0.27	1	0.02	0.069	0.069
1	6702201	58938.5	0.88	2	0.04	0.049	0.098
2	5910202	94372.5	1.60	4	0.05	0.078	0.312
3	4231085	70167.0	1.66	4	0.06	0.058	0.232
4	8924842	247146.0	2.77	4	0.04	0.205	0.820
≥ 5	16285066	989191.0	6.07	4	0.04	0.820	3.280
	120652710	688930.0	0.57		0.01	0.570	4.103

and letters, Professor Gwynn and I arrived at what we hoped would be the clearest and most practical presentation of the data. A full description of the methodology employed by Dr. Gwynn is furnished in his letters to me, reprinted in the Appendix of this chapter.

Table 6.1 presents a cumulative or "global" distribution for the one-deck game, dealt down to the 75% (generous!) level. Note that I have added a "Units Bet" column, not originally supplied by Dr. Gwynn, to represent my style of betting for such a game. If you bet less aggressively than this, the floating edge is useless to you, and if you bet more aggressively than this, blackjack, in general, is useless to you, because you're going to be kicked out of every casino you play in!

Now, take a look at Tables 6.2, 6.3, and 6.4. They present the same information, broken down by quarter-decks. Thus, Table 6.2 shows distribution of true counts and

TABLE 6.2
.00 decks to .25 decks

Count	Hands	Net Win	Per Hand Expect.	Units Bet	Expect. Std. Err.	Per Hand Contr.	Units Won
≤ −1	13421028	−157501.0	−1.17		0.03	−0.131	
0	21426454	11754.0	0.05		0.02	0.010	
1	4480394	33853.5	0.76		0.05	0.028	
2	1363291	18268.5	1.34		0.10	0.015	
3	2373672	35448.0	1.49		0.07	0.029	
4	1445455	32791.5	2.27		0.10	0.027	
≥ 5	1283626	44140.0	3.44		0.10	0.037	
	45793920	18754.5	0.04		0.02	0.016	

TABLE 6.3
.25 decks to .50 decks

Count	Hands	Net Win	Per Hand Expect.	Units Bet	Expect. Std. Err.	Per Hand Contr.	Units Won
≤ −1	17046131	−317615.0	−1.86		0.03	−0.263	
0	4621001	16479.0	0.36		0.05	0.014	
1	2221807	25085.0	1.13		0.08	0.021	
2	2045355	27872.5	1.36		0.08	0.023	
3	1857413	34719.0	1.87		0.09	0.029	
4	3139695	78716.0	2.15		0.07	0.065	
≥ 5	5564090	263587.0	4.74		0.05	0.218	
	36495492	128843.5	0.35		0.02	0.107	

TABLE 6.4
.50 decks to .75 decks

Count	Hands	Net Win	Per Hand Expect.	Units Bet	Expect. Std. Err.	Per Hand Contr.	Units Won
≤ −2	17064869	−379434.5	−2.22		0.03	−0.314	
0	5019831	55432.5	1.10		0.05	0.046	
2	2501556	48231.5	1.93		0.07	0.040	
4	4339692	135638.5	3.13		0.06	0.112	
≥ 6	9437350	681464.0	7.22		0.04	0.565	
	38363298	541332.0	1.41		0.02	0.449	

corresponding advantages for the first 13 cards dealt; Table 6.3, ditto for cards 14–26; and Table 6.4, cards 27–39 (all values rounded, as per Dr. Gwynn's description). I have purposely left the "Units Bet" column blank so that you may actively participate in the frustration! Take a pencil and, after referring back to Table 6.1, enter, for each quarter-deck segment, the wager that you feel should correspond to the given edge at each true count (3-unit bets permissible).

All done? Now, look at Tables 6.2A, 6.3A, and 6.4A. This is how I filled in mine. Note the arrows. They represent departures from the Table 6.1 global approach. Do you see the problem? That's right: too little, too late. The extra advantages pop up, all right. But by the time they appear, and by the nature of our betting spread, we just can't glean a lot of extra edge from this scheme. In fact, when all the smoke clears, we

TABLE 6.2A

.00 decks to .25 decks

Count	Hands	Net Win	Per Hand Expect.	Units Bet	Expect. Std. Err.	Per Hand Contr.	Units Won
≤ −1	13421028	−157501.0	−1.17	1	0.03	−0.131	−0.131
0	21426454	11754.0	0.05	1	0.02	0.010	0.010
1	4480394	33853.5	0.76	2	0.05	0.028	0.056
2	1363291	18268.5	1.34	4	0.10	0.015	0.060
3	2373672	35448.0	1.49	4	0.07	0.029	0.116
4	1445455	32791.5	2.27	4	0.10	0.027	0.108
≥ 5	1283626	44140.0	3.44	4	0.10	0.037	0.148
	45793920	18754.5	0.04		0.02	0.016	0.367

TABLE 6.3A

.25 decks to .50 decks

Count	Hands	Net Win	Per Hand Expect.	Units Bet	Expect. Std. Err.	Per Hand Contr.	Units Won
≤ −1	17046131	−317615.0	−1.86	1	0.03	−0.263	−0.263
0	4621001	16479.0	0.36	1	0.05	0.014	0.014
1	2221807	25085.0	1.13	→3*	0.08	0.021	0.063
2	2045355	27872.5	1.36	4	0.08	0.023	0.092
3	1857413	34719.0	1.87	4	0.09	0.029	0.116
4	3139695	78716.0	2.15	4	0.07	0.065	0.260
≥ 5	5564090	263587.0	4.74	4	0.05	0.218	0.872
	36495492	128843.5	0.35		0.02	0.107	1.154

TABLE 6.4A

.50 decks to .75 decks

Count	Hands	Net Win	Per Hand Expect.	Units Bet	Expect. Std. Err.	Per Hand Contr.	Units Won
≤ −2	17064869	−379434.5	−2.22	1	0.03	−0.314	−0.314
0	5019831	55432.5	1.10	→3†	0.05	0.046	0.138
2	2501556	48231.5	1.93	4	0.07	0.040	0.160
4	4339692	135638.5	3.13	4	0.06	0.112	0.448
≥ 6	9437350	681464.0	7.22	4	0.04	0.565	2.260
	38363298	541332.0	1.41		0.02	0.449	2.692
					Total (all segments)		**4.213**

* Note that, although this 3-unit bet is more than justified by the extra advantage available at a +1 true, there is a bet-jumping problem caused by the 1-unit to 3-unit sudden increase.

† Same comment as above, now applicable at a true of zero.

get a win rate of 4.103 units per 100 hands played for the un-SAFE (!) approach, and 4.213 units per 100 hands for the SAFE method. The increase, .11/4.103, expressed as a percent, is 2.681. And that's as good as it gets! Note, also, that the win rates here are important for comparative purposes only and should not be construed, at least for single-deck, as realistic. If a one-deck game is dealt to the 75% level, you're not on the Las Vegas Strip. You're in Reno, where starting disadvantage is worse. 75% penetration and Strip rules is Utopia — not the real world. Snyder's *Beat the 1-Deck Game,* for the above conditions, Reno style, gives 2.92 units won per 100 hands dealt (compared to our 4.10). *[See, also, Chapter 10, for more detail. — D.S.]*

Next, let us consider the charts for the two-deck game. Table 6.5 is the cumulative approach, dealt 75% to the 1 1/2-deck level. Again, this time, I feel a 1–6 spread is sufficiently bold. If you're in the "oink, oink" category, go figure your own results — I'm unsympathetic! Now, play with Tables 6.6, 6.7, 6.8, and 6.9. Then consult my adjustments in Tables 6.6A, 6.7A, 6.8A, and 6.9A.

Once more, the arrows represent the only opportunities I could find to capitalize on the floating edge. The result? 2.891 units won per 100 un-SAFE hands, and 2.945 units for the SAFE method. Big deal, you say? You're absolutely right! The .054/2.891 = 1.868% increase is negligible.

For the sake of completeness, I present the 6-deck game data, with the same parameters. But you know in advance that this is going to be a waste of our time. Suffice it to say that for the standard 75% (4 1/2-deck) cut, there is no extra advantage to be found. My traditional one to two hands of six units (1–12 spread) betting scheme, reported many times in prior *Blackjack Forum* articles, produces a cumulative win rate of 1.112 units per 100 hands, and the SAFE approach is identical. There is no increase. Some slight extra advantage (0.313%) can be found at the 5-deck level, but again, we're not talking about the real world here.

Stop the presses! No sooner had I penned these words, virtually ending the article, than I took a five-day vacation to Las Vegas. It is October 1988. And blackjack history is being made at the Dunes Hotel. Every 6-deck shoe in the place is being cut at the one-half-deck level! That's right — they're dealing 92% of the pack. What's more,

TABLE 6.5
.00 decks to 1.50 decks

Count	Hands	Net Win	Per Hand Expect.	Units Bet	Expect. Std. Err.	Per Hand Contr.	Units Won
≤−1	45605956	−780819.0	−1.71	1	0.02	−0.742	−0.742
0	17760340	−27713.0	−0.16	1	0.03	−0.026	−0.026
1	12724698	34982.0	0.27	1	0.03	0.033	0.033
2	8333774	69588.0	0.84	2	0.04	0.066	0.132
3	6015432	87393.0	1.45	4	0.05	0.083	0.332
4	4296808	87243.5	2.03	6	0.06	0.083	0.498
≥5	10542132	467825.5	4.44	6	0.04	0.444	2.664
	105279140	−61500.0	−0.06		0.01	−0.058	2.891

TABLE 6.6
.00 decks to .50 decks

Count	Hands	Net Win	Per Hand Expect.	Units Bet	Expect. Std. Err.	Per Hand Contr.	Units Won
≤ –1	13986768	–156501.5	–1.12		0.03	–0.149	
0	11390821	–30687.5	–0.27		0.03	–0.029	
1	6528992	9557.5	0.15		0.05	0.009	
2	2953702	20065.5	0.68		0.07	0.019	
3	1506325	21358.5	1.42		0.09	0.020	
4	329890	6370.5	1.93		0.20	0.006	
≥ 5	360605	9220.5	2.56		0.19	0.009	
	37057103	–120616.5	–0.33		0.02	–0.115	

TABLE 6.7
.50 decks to 1.00 decks

Count	Hands	Net Win	Per Hand Expect.	Units Bet	Expect. Std. Err.	Per Hand Contr.	Units Won
≤ –1	15930906	–267860.5	–1.68		0.03	–0.254	
0	3209928	–3785.0	–0.12		0.06	–0.004	
1	3892937	9275.0	0.24		0.06	0.009	
2	3908033	33112.5	0.85		0.06	0.031	
3	2467571	34115.5	1.38		0.07	0.032	
4	1523406	25979.0	1.71		0.09	0.025	
≥ 5	3256808	109866.5	3.37		0.07	0.104	
	34189589	–59297.0	–0.17		0.02	–0.056	

TABLE 6.8
1.00 decks to 1.25 decks

Count	Hands	Net Win	Per Hand Expect.	Units Bet	Expect. Std. Err.	Per Hand Contr.	Units Won
≤ –1	7904139	–163447.5	–2.07		0.04	–0.155	
0	1531151	1681.0	0.11		0.09	0.002	
1	1474364	9439.0	0.64		0.09	0.009	
2	726975	7246.5	1.00		0.13	0.007	
3	1277428	19298.0	1.51		0.10	0.018	
4	1096505	21744.0	1.98		0.11	0.021	
≥ 5	3013682	132827.0	4.41		0.07	0.126	
	17024244	28788.0	0.17		0.03	0.027	

TABLE 6.9
1.25 decks to 1.50 decks

Count	Hands	Net Win	Per Hand Expect.	Units Bet	Expect. Std. Err.	Per Hand Contr.	Units Won
≤ –1	7784143	–193009.5	–2.48		0.04	–0.183	
0	1628440	5078.0	0.31		0.09	0.005	
1	828405	6710.5	0.81		0.13	0.006	
2	745064	9163.5	1.23		0.13	0.009	
3	764108	12621.0	1.65		0.13	0.012	
4	1347007	33150.0	2.46		0.10	0.031	
≥ 5	3911037	215911.5	5.52		0.06	0.205	
	17008204	89625.5	0.53		0.03	0.085	

TABLE 6.6A
.00 decks to .50 decks

Count	Hands	Net Win	Per Hand Expect.	Units Bet	Expect. Std. Err.	Per Hand Contr.	Units Won
≤ –1	13986768	–156501.5	–1.12	1	0.03	–0.149	–0.149
0	11390821	–30687.5	–0.27	1	0.03	–0.029	–0.029
1	6528992	9557.5	0.15	1	0.05	0.009	0.009
2	2953702	20065.5	0.68	2	0.07	0.019	0.038
3	1506325	21358.5	1.42	4	0.09	0.020	0.080
4	329890	6370.5	1.93	6	0.20	0.006	0.036
≥ 5	360605	9220.5	2.56	6	0.19	0.009	0.054
	37057103	–120616.5	–0.33		0.02	–0.115	0.039

TABLE 6.7A
.50 decks to 1.00 decks

Count	Hands	Net Win	Per Hand Expect.	Units Bet	Expect. Std. Err.	Per Hand Contr.	Units Won
≤ –1	15930906	–267860.5	–1.68	1	0.03	–0.254	–0.254
0	3209928	–3785.0	–0.12	1	0.06	–0.004	–0.004
1	3892937	9275.0	0.24	1	0.06	0.009	0.009
2	3908033	33112.5	0.85	2	0.06	0.031	0.062
3	2467571	34115.5	1.38	4	0.07	0.032	0.128
4	1523406	25979.0	1.71	6*	0.09	0.025	0.150
≥ 5	3256808	109866.5	3.37	6	0.07	0.104	0.624
	34189589	–59297.0	–0.17		0.02	–0.056	0.715

TABLE 6.8A
1.00 decks to 1.25 decks

Count	Hands	Net Win	Per Hand Expect.	Units Bet	Expect. Std. Err.	Per Hand Contr.	Units Won
≤ –1	7904139	–163447.5	–2.07	1	0.04	–0.155	–0.155
0	1531151	1681.0	0.11	1	0.09	0.002	0.002
1	1474364	9439.0	0.64	→2	0.09	0.009	0.018
2	726975	7246.5	1.00	→3	0.13	0.007	0.021
3	1277428	19298.0	1.51	→4.5	0.10	0.018	0.081
4	1096505	21744.0	1.98	6	0.11	0.021	0.126
≥ 5	3013682	132827.0	4.41	6	0.07	0.126	0.756
	17024244	28788.0	0.17		0.03	0.027	0.849

TABLE 6.9A
1.25 decks to 1.50 decks

Count	Hands	Net Win	Per Hand Expect.	Units Bet	Expect. Std. Err.	Per Hand Contr.	Units Won
≤ –1	7784143	–193009.5	–2.48	1	0.04	–0.183	–0.183
0	1628440	5078.0	0.31	1	0.09	0.005	0.005
1	828405	6710.5	0.81	→2	0.13	0.006	0.012
2	745064	9163.5	1.23	→3.5	0.13	0.009	0.032
3	764108	12621.0	1.65	→5	0.13	0.012	0.060
4	1347007	33150.0	2.46	6	0.10	0.031	0.186
≥ 5	3911037	215911.5	5.52	6	0.06	0.205	1.230
	17008204	89625.5	0.53		0.03	0.085	1.342
						Total (all segments)	**2.945**

* Note that, technically, a 5-unit bet is called for here. However, the slight over-bet of 6 units, to be consistent with the "global approach," is more than compensated for by the restraint shown in the next category when no more than a 6-unit bet is ever placed.

they permit double after splits. Shoe heaven! Now, it just so happens that Professor Gwynn, in his original 6-deck runs, carried his calculations to the 5 1/2-deck level. The first incredible conclusion to be extracted from the data is that my 1–12 betting scheme earns 3.206 units per 100 hands with a play-all approach! Compare this to the 1.112 units earned at the 4 1/2-deck level and the 1.919 units won for 5 out of 6. Thus, seeing the one extra deck after the 75% level virtually triples your expectation, while seeing the extra one-half deck adds more than 50% to the 5-deck winnings. Such is the dramatic impact of deeper penetration.

But, more important still, for the purposes of this article, is the ultimate demonstration of the floating advantage. Table 6.10 presents the global statistics for 0–5 1/2-decks. Note the win rates for trues of +1 through +5. Now, take a look at the 5 1/4–5 1/2-deck segment (Table 6.11). Compare the two tables: +5 (embedded in the ≥ 5 line, 2.32) → +4 (2.34); +4 (1.72) → +3 (1.75); +3 (1.11) → +2 (1.10); +2 (.56) → +1 (.84); and +1 (.02) → 0 (.23). There couldn't be a clearer or more striking depiction of the "down-shifting" of advantage by one true count than in the Dunes game. And even though the final win rate, following SAFE, is only 3.088% (.099 units) better than the global rate of 3.206%, we have, at least, demonstrated without question a floating edge in an actual 6-deck game.

And there you have it. The debate has come full circle. I started the damn thing and I think I have ended it, also. I guess what the Messrs. Revere, Uston, Wong,

TABLE 6.10

.00 decks to 5.50 decks

Count	Hands	Net Win	Per Hand Expect.	Units Bet	Expect. Std. Err.	Per Hand Contr.	Units Won
≤ −1	189313871	−3289806.0	−1.74	1	0.01	−0.685	−0.685
0	108342451	−511185.5	−0.47	1	0.01	−0.106	−0.106
1	69352832	12562.0	0.02	1	0.01	0.003	0.003
2	39886514	224375.0	0.56	2	0.02	0.047	0.094
3	24717810	274662.5	1.11	4	0.02	0.057	0.228
4	15701358	270064.0	1.72	6	0.03	0.056	0.336
≥ 5	32755179	1335587.0	4.08	12	0.02	0.278	3.336
	480070015	−1683741.0	−0.35		0.01	−0.351	3.206

TABLE 6.11

5.25 decks to 5.50 decks

Count	Hands	Net Win	Per Hand Expect.	Units Bet	Expect. Std. Err.	Per Hand Contr.	Units Won
≤ −1	9903852	−263450.5	−2.66	1	0.04	−0.055	−0.055
0	1812968	4232.0	0.23	1	0.09	0.001	0.001
1	911412	7660.0	0.84	→3*	0.12	0.002	0.006
2	852042	9372.5	1.10	→4	0.13	0.002	0.008
3	863937	15129.5	1.75	→6	0.12	0.003	0.018
4	1576573	36940.5	2.34	→12	0.09	0.008	0.096
≥ 5	642899	338545.5	6.00	12	0.05	0.071	0.852
	21563683	148430.0	0.69		0.02	0.031	0.926
						Total (all segments)	**3.305**

* Note that, although this 3-unit bet is more than justified by the extra advantage available at a +1 true, there is a bet-jumping problem caused by the 1-unit to 3-unit sudden increase.

Griffin, Snyder, Gwynn, and yours truly have proven once and for all is that: a) As a pack is depleted, the same true count garners extra advantage, said advantage becoming especially interesting at those very levels when the shuffle is most likely to occur. This, in turn, leads to conclusion b) Because of shuffle-point and bet-spread limitations, the "floating advantage" is of more theoretical than practical import. I, for one, am disappointed by this, but that's the scientific approach. You formulate a hypothesis, gather the data, and let the chips fall where they may. Until next time, may all those chips be black ones!

Summary of Findings

No. of decks used	No. of decks dealt	Bet spread	Units won/ 100 hands played	Units won/ 100 SAFE hands	Win rate % enhancement
1	3/4	1–4	4.103	4.213	2.681
2	1 1/2	1–6	2.891	2.945	1.868
6	4 1/2	1–12	1.112	1.112	0.000
6	5	1–12	1.919	1.925	0.313
6	5 1/2	1–12	3.206	3.305	3.088

Appendix

Following is the edited text of John Gwynn's letters to me explaining his methodology. I thought that it should be presented in his own words, rather than paraphrased.

Dear Don,

I have used the Hi-Lo system for all variations in strategy; the strategy indices were taken from Wong's 1981 edition of ***Professional Blackjack.*** *I obtained the hit/stand, double, and split indices from Table A3 for 1-deck and from Table A4 for 2-deck and 6-deck (these tables show all indices from –20 to +20). Depending on the number of decks, insurance was taken according to Table 40.*

For decisions not governed by the tables, I used the basic strategy for the number of decks being simulated. I obtained this information from Peter Griffin's 1986 edition of ***The Theory of Blackjack.*** *All runs were made using these house rules:*

1. *Dealer stands on soft 17.*
2. *Double on any first two cards.*
3. *No doubling after splits.**
4. *No resplitting aces.*
5. *For non-aces, resplitting permitted up to a total of four hands.*

The true count was not computed exactly as the running count divided by the number of decks unseen. Instead, the number of decks unseen was approximated to the nearest quarter of a deck (with one exception). With 47 cards unseen, the number of

* Note: The Dunes game, described above, did permit das.

decks unseen would actually be 47/52; however, the program rounded this to 52/52, or one deck unseen. With 44 cards unseen, the number of decks unseen would actually be 44/52; here the program rounded to 39/52, or 3/4 of a deck unseen. The only exception was for six or fewer unseen cards; in this case, the program used 1/8 as the number of decks unseen.

Back in 1981, I made runs whose results indicated that using the exact fraction of decks unseen gave almost no significant improvement (at most 0.01% to 0.015% per hand played) over using this approximation. The approximate method is clearly a more realistic simulation for most players; I'll conjecture the great majority of them use some sort of approximation when they compute a true count. Each time a true count was computed to make a betting or strategy decision, it was computed as accurately as possible using the above-described approximate deck fraction. Except for insurance, this true count was then rounded to the nearest integer and used to make the decision. For example, if the running count was 4 and 76 cards were unseen, the true count would be computed using 78/52 = 6/4 = 3/2 as the number of decks unseen. This would be computed as 4/(3/2) = 2.67, which would then be rounded to 3 for use in making any decision other than insurance. For insurance, the unrounded value of 2.67 would be used to make the decision. [Note: Personally, I do not use rounded true count values to make strategy decisions. — D.S.]

This could clearly make a difference in some decisions. In the above example, suppose a system says the player should stand if the true count exceeds or equals 3. A player using the unrounded true count would hit; a player using the rounded true count would stand. It turns out not to make any significant difference whether the exact or the rounded true count is used. Based on additional runs I made in 1981, some systems did slightly better when the unrounded true count was used, whereas others did better when the rounded true count was used. The differences were less than 0.01% per hand played, if I recall correctly.

For insurance, the unrounded true count was used to make this decision since it is based on a non-integer index (1.4 for 1-deck, 2.4 for 2-deck, 3.0 for 6-deck).

In 1-deck and 2-deck play, shuffling took place when the number of unseen cards was 13 or fewer (for 1-deck) or 26 or fewer (for 2-deck) before a hand began. In 6-deck play, the shuffle point was 26 or fewer cards. Although the 6-deck shuffle point is unrealistic, the statistics are reported in such a way that you can consider any multiple of 13 cards as the shuffle point by ignoring the results from later quarters of the deck.

If a hand, once started, could not be completed without reshuffling, this hand was not counted in the total number of hands played, its outcome was not included in the net money won/lost, and the money bet on it was not included in the total amount of money bet. There were some such hands in 1-deck and 2-deck play, but there were none in 6-deck play.

Rather than just aborting these hands, I made the program finish each of them (without reshuffling but by re-dealing the cards first dealt off the top of this shuffled deck or shoe). The program kept a record of how many such hands were played and of the net win/loss of these hands. This provided an idea of how throwing out these hands affected the player's overall expectation. The answer is that the effect of aborting these hands is negligible. In no simulation would these hands have improved the player's expectation more than 0.0005% if they had been counted in the overall statistics.

The program gathers statistics by quarter of a deck; the number of cards unseen at the start of a hand determines the quarter. 0 to 0.25 decks means there were between 40 and 52 cards unseen for 1-deck play, between 92 and 104 cards unseen for 2-deck play, or between 300 and 312 cards unseen for 6-deck play. In each quarter, statistics are recorded for rounded true counts (before the hand was dealt) ranging from –20 to +20. Statistics for hands with rounded true counts less than –20 are lumped together with the –20s; statistics for hands with rounded true counts greater than 20 are included with the +20s.

How to Read the Do-It-Yourself Tables: *In each deck segment (and globally) the statistics for expectations as a function of the true count are reported in seven* [now eight — D.S.] *columns as follows:*

1. Specified true counts;

2. Hands played in this segment;

3. Net win for those hands, assuming the flat bet of 1 unit;

4. Expectation per hand played for these hands, in percent, computed as 100 times column 3 divided by column 2;

5. Number of units bet at the specified true count; [Not in Dr. Gwynn's printout. Added by me. — D.S.]

6. One standard error for the expectation in column 4, in percent, computed as 100 times a standard deviation divided by the square root of column 2;

7. Contribution per total hand played for these hands, in percent, computed as 100 times column 3 divided by the total number of hands played (throughout the deck for all true counts);

8. One standard error for the contribution in column 7, in percent, computed as column 6 times column 2 divided by the total number of hands played (throughout the deck for all true counts). [I have deleted this column and substituted a "Units Won" entry, along with subtotals for each segment. — D.S.]

Column 4 shows the expectation for hands played in this deck segment with a particular true count; from this value the player can tell what bet to make for the specified true count in this deck segment.

The standard deviation used to compute column 6 depends on the segments and the true count; it was obtained from sufficiently many hands played in this segment with the specified true count. However, this standard deviation may have been obtained from a different simulation run.

Column 7 shows how much the hands in this segment with the specified true count contribute to the overall expectation. This is what you get if you first compute the ratio of hands played in this segment with the specified true count (column 2) to the total number of hands played; then multiply this ratio by the expectation for hands in this segment (column 4). It is trivial to see that this computation yields column 7:

$$\frac{col.\,2}{total\ hands} \times col.\,4 = \frac{col.\,2}{total\ hands} \times \frac{col.\,3}{col.\,2} = \frac{col.\,3}{total\ hands} = col.\,7.$$

Summed over all segments and all true counts, the column 7 numbers add up to the expectation over all hands played (apart from minor rounding). The numbers from column 7 are used in applying a betting scheme in do-it-yourself fashion. One simply multiplies the value in column 7 times the bets placed in this segment for hands with the specified true count. [My column 5. — D.S.]

A segment summary follows the last line of 6-column statistics; there is no true count specified on this line. Below columns 2 and 3 are the respective totals for this segment, i.e., the number of hands played in this segment (for all true counts) and the net win for all hands in this segment. Below column 4 is the expectation per hand played for all hands in this segment, i.e., 100 times column 3 for this line divided by column 2 for this line. Below column 7 is the expectation per total hands played for these hands, i.e., 100 times column 3 for this line divided by the total number of hands played. Apart from minor rounding, column 7 of this line should be the total of the column 7 values in this segment and represents the contribution of this segment's expectation to the overall expectation (assuming a flat bet of 1 unit). [Column 8, "Units Won," is the product of columns 5 and 7. There is a subtotal for each column 8 in each deck segment. — D.S.]

I hope these data now provide you with what you need to finish the article. This has been an interesting study. Take care.

Sincerely,
John

Don Schlesinger's Comments: Many readers of *Blackjack Forum* are also familiar with Stanford Wong's *Blackjack Newsletters*. In Wong's many simulations published in 1983 and 1984, a slightly different method for generating true count intervals was used. Whereas Dr. Gwynn uses the rounding approach on each side of the count in question (True count = 2 means $1.5 \le TC < 2.5$), Wong uses the "at least" approach (True count = 2 means $2 \le TC < 3$), wherein each true count is the minimum for the interval considered. As a result, there are, obviously, two differences in the numbers associated with frequency of the count and advantage obtained with each count. For any given true, Wong's advantages are greater, but his frequencies are smaller. The two phenomena tend to offset each other, such that win rates generated by the two different methods produce reasonably similar results. Nonetheless, readers are cautioned not to compare directly win rates represented in this article with those outlined in my "Ups and Downs of Your Bankroll" article (June 1985 *Blackjack Forum,* pp. 8–17), where I used Wong's approach. *[See Chapter 2 — D.S.]*

Finally, one ought not to compare the unit bet sizes suggested for the various games (1, 2, or 6 decks) with those of another. For the sake of uniformity of comparison within a given game, I have assumed similar bet sizes at corresponding true counts for the "global" approach in all three games. In fact, to justify such bets, it appears that a slightly larger bankroll would be necessary for the 6-deck game where win rates at various trues compare unfavorably with the 1- and 2-deck versions. This is due largely to the differences in basic strategy starting disadvantages for the three games and to the greater playing-strategy-variation gains associated with the 1- and 2-deck varieties.

Postscript: *Well, it's now 1997, and the "Floating Advantage" has undergone a kind of renaissance, for two reasons. First, there remains, to this day, a profound misunderstanding of the concept. As a new generation of blackjack enthusiasts comes online, on the Internet, I'm surprised to see a renewed interest in the "floating advantage" principle. As questioners hypothesize on why, or why not, the edge actually exists — and to what degree — I am obliged, over and over, to point them in the direction of the above study. Perhaps, now, the idea will become more accessible and, hopefully, understandable, to newcomers and veterans alike.*

Finally, there currently exist at least two shoe games (in Las Vegas, at Harrah's and the Aladdin) where penetration is, once again, as deep as was the case at the Dunes, in October of 1988, and where the "floating advantage" is alive and well, and of more than just theoretical import! [Note: To no one's surprise, these games no longer exist! — D.S.]

TABLE 6.12

Summary of .00 decks to 4.50 decks

Count	Hands	Net Win	Per Hand Expect.	Units Bet	Expect. Std. Err.	Per Hand Contr.	Units Won
≤−1	149044570	−2320393.0	−1.56	1	0.01	−0.589	−0.589
0	102204912	−508786.5	−0.50	1	0.01	−0.129	−0.129
1	63530585	−15387.5	−0.20	1	0.01	−0.004	−0.004
2	34479737	173780.5	0.50	2*	0.02	0.044	0.088
3	19630167	201747.0	1.03	4*	0.03	0.051	0.204
4	10847091	170174.5	1.57	6*	0.04	0.043	0.258
≥5	13979096	422070.5	3.02	12	0.03	0.107	1.284
	393716158	−1876794.5	−0.48		0.01	−0.477	1.112

Per-Hand Expectations as a Function of True Count, Segment-by-Segment

TABLE 6.13

.00 decks to 4.00 decks

Count	Hands	Net Win	Per Hand Expect.	Units Bet	Expect. Std. Err.	Per Hand Contr.	Units Won
≤−1	128525539	−1905697.0	−1.48		0.01	−0.484	
0	99745008	−502294.0	−0.50		0.01	−0.128	
1	58757227	−17895.0	−0.03		0.02	−0.005	
2	30662659	152908.5	0.50		0.02	0.039	
3	15950605	156430.5	0.98		0.03	0.040	
4	8962407	138506.5	1.55		0.04	0.035	
≥5	7874641	216370.5	2.75		0.04	0.055	
	350478086	−1761670.0	−0.50		0.01	−0.447	

TABLE 6.14

4.00 decks to 4.25 decks

Count	Hands	Net Win	Per Hand Expect.	Units Bet	Expect. Std. Err.	Per Hand Contr.	Units Won
≤−1	10277445	−205161.5	−2.00		0.04	−0.052	
0	1205696	−2880.5	−0.24		0.10	−0.001	
1	2339886	−640.0	−0.03		0.08	−0.000	
2	2120738	10416.0	0.49		0.08	0.003	
3	1776728	22239.5	1.25		0.09	0.006	
4	1087545	18113.0	1.67		0.11	0.005	
≥5	2813790	90402.5	3.21		0.07	0.023	
	21621828	−67511.0	−0.31		0.02	−0.017	

TABLE 6.15

4.25 decks to 4.50 decks

Count	Hands	Net Win	Per Hand Expect.	Units Bet	Expect. Std. Err.	Per Hand Contr.	Units Won
≤−1	10241586	−209534.5	−2.05		0.04	−0.053	
0	1254208	−3612.0	−0.29		0.10	−0.001	
1	2433472	3147.5	0.13		0.07	0.001	
2	1696340	10456.0	0.62		0.09	0.003	
3	1902834	23077.0	1.21		0.08	0.006	
4	797139	13555.0	1.70		0.13	0.003	
≥5	3290665	115297.5	3.50		0.07	0.029	
	21616244	−47613.5	−0.22		0.02	−0.012	

* Again, technically, using the same bankroll as for the 1- and 2-deck games, the appropriate bets should be 1.5, 3 and 4.5 (or 5) units here. However, the constraint on the top bet (2 hands of 6 units) may justify the apparent slight over-betting done earlier. See, also, my comments at the end of the Appendix.

TABLE 6.13A
.00 decks to 4.00 decks

Count	Hands	Net Win	Per Hand Expect.	Units Bet	Expect. Std. Err.	Per Hand Contr.	Units Won
≤ –1	128525539	–1905697.0	–1.48	1	0.01	–0.484	–0.484
0	99745008	–502294.0	–0.50	1	0.01	–0.128	–0.128
1	58757227	–17895.0	–0.03	1	0.02	–0.005	–0.005
2	30662659	152908.5	0.50	2*	0.02	0.039	0.078
3	15950605	156430.5	0.98	4*	0.03	0.040	0.160
4	8962407	138506.5	1.55	6*	0.04	0.035	0.210
≥ 5	7874641	216370.5	2.75	12	0.04	0.055	0.660
	350478086	–1761670.0	–0.50		0.01	–0.447	0.491

TABLE 6.14A
4.00 decks to 4.25 decks

Count	Hands	Net Win	Per Hand Expect.	Units Bet	Expect. Std. Err.	Per Hand Contr.	Units Won
≤ –1	10277445	–205161.5	–2.00	1	0.04	–0.052	–0.052
0	1205696	–2880.5	–0.24	1	0.10	–0.001	–0.001
1	2339886	–640.0	–0.03	1	0.08	–0.000	–0.000
2	2120738	10416.0	0.49	2*	0.08	0.003	0.006
3	1776728	22239.5	1.25	4	0.09	0.006	0.024
4	1087545	18113.0	1.67	6*	0.11	0.005	0.030
≥ 5	2813790	90402.5	3.21	12	0.07	0.023	0.276
	21621828	–67511.0	–0.31		0.02	–0.017	0.283

TABLE 6.15A
4.25 decks to 4.50 decks

Count	Hands	Net Win	Per Hand Expect.	Units Bet	Expect. Std. Err.	Per Hand Contr.	Units Won
≤ –1	10241586	–209534.5	–2.05	1	0.04	–0.053	–0.053
0	1254208	–3612.0	–0.29	1	0.10	–0.001	–0.001
1	2433472	3147.5	0.13	1	0.07	0.001	0.001
2	1696340	10456.0	0.62	2	0.09	0.003	0.006
3	1902834	23077.0	1.21	4	0.08	0.006	0.024
4	797139	13555.0	1.70	6*	0.13	0.003	0.018
≥ 5	3290665	115297.5	3.50	12	0.07	0.029	0.348
	21616244	–47613.5	–0.22		0.02	–0.012	0.343
						Total (all segments)	**1.117**

* See footnote on p. 84.

TABLE 6.16
Summary of .00 decks to 5.00 decks

Count	Hands	Net Win	Per Hand Expect.	Units Bet	Expect. Std. Err.	Per Hand Contr.	Units Won
≤ –1	169369422	–2771602.5	–1.64	1	0.01	–0.634	–0.634
0	104953388	–515319.5	–0.49	1	0.01	–0.118	–0.118
1	66901449	–5143.5	–0.01	1	0.01	–0.001	–0.001
2	38264588	207237.0	0.54	2*	0.02	0.047	0.094
3	22443829	238602.0	1.06	4*	0.02	0.055	0.220
4	12847546	206255.0	1.61	6*	0.03	0.047	0.282
≥ 5	22143137	753947.0	3.40	12	0.03	0.173	2.076
	436923359	–1886024.5	–0.43		0.01	–0.432	1.919

Per-Hand Expectations as a Function of True Count, Segment-by-Segment

TABLE 6.17
.00 decks to 4.50 decks

Count	Hands	Net Win	Per Hand Expect.	Units Bet	Expect. Std. Err.	Per Hand Contr.	Units Won
≤ –1	149044570	–2320393.0	–1.56		0.01	–0.531	
0	102204912	–508786.5	–0.50		0.01	–0.116	
1	63530585	–15387.5	–0.02		0.01	–0.004	
2	34479737	173780.5	0.50		0.02	0.040	
3	19630167	201747.0	1.03		0.03	0.046	
4	10847091	170174.5	1.57		0.04	0.039	
≥ 5	13979096	422070.5	3.02		0.03	0.097	
	393716158	–1876794.5	–0.48		0.01	–0.430	

TABLE 6.18
4.50 decks to 4.75 decks

Count	Hands	Net Win	Per Hand Expect.	Units Bet	Expect. Std. Err.	Per Hand Contr.	Units Won
≤ –1	10195035	–221210.5	–2.17		0.04	–0.051	
0	1324454	–4132.5	–0.31		0.10	–0.001	
1	1971930	4703.5	0.24		0.08	0.001	
2	1776103	14034.5	0.79		0.09	0.003	
3	1615285	21730.0	1.35		0.09	0.005	
4	924099	17410.0	1.88		0.12	0.004	
≥ 5	3804282	145781.5	3.83		0.06	0.033	
	21611188	–21683.5	–0.10		0.02	–0.005	

TABLE 6.19
4.75 decks to 5.00 decks

Count	Hands	Net Win	Per Hand Expect.	Units Bet	Expect. Std. Err.	Per Hand Contr.	Units Won
≤ –1	10129817	–229999.9	–2.27		0.04	–0.053	
0	1424022	–2400.5	–0.17		0.10	–0.001	
1	1398934	5540.5	0.40		0.10	0.001	
2	2008748	19422.0	0.97		0.08	0.004	
3	1198377	15125.0	1.26		0.11	0.003	
4	1076356	18670.5	1.73		0.11	0.004	
≥ 5	4359759	186095.0	4.27		0.06	0.043	
	21596013	12453.5	0.06		0.02	0.003	

* See footnote on p. 84.

TABLE 6.17A

.00 decks to 4.50 decks

Count	Hands	Net Win	Per Hand Expect.	Units Bet	Expect. Std. Err.	Per Hand Contr.	Units Won
≤ −1	149044570	−2320393.0	−1.56	1	0.01	−0.531	−0.531
0	102204912	−508786.5	−0.50	1	0.01	−0.116	−0.116
1	63530585	−15387.5	−0.02	1	0.01	−0.004	−0.004
2	34479737	173780.5	0.50	2*	0.02	0.040	0.080
3	19630167	201747.0	1.03	4*	0.03	0.046	0.184
4	10847091	170174.5	1.57	6*	0.04	0.039	0.234
≥ 5	13979096	422070.5	3.02	12	0.03	0.097	1.164
	393716158	−1876794.5	−0.48		0.01	−0.430	1.011

TABLE 6.18A

4.50 decks to 4.75 decks

Count	Hands	Net Win	Per Hand Expect.	Units Bet	Expect. Std. Err.	Per Hand Contr.	Units Won
≤ −1	10195035	−221210.5	−2.17	1	0.04	−0.051	−0.051
0	1324454	−4132.5	−0.31	1	0.10	−0.001	−0.001
1	1971930	4703.5	0.24	1	0.08	0.001	0.001
2	1776103	14034.5	0.79	→3	0.09	0.003	0.009
3	1615285	21730.0	1.35	→4.5	0.09	0.005	0.0225
4	924099	17410.0	1.88	6	0.12	0.004	0.024
≥ 5	3804282	145781.5	3.83	12	0.06	0.033	0.396
	21611188	−21683.5	−0.10		0.02	−0.005	0.401

TABLE 6.19A

4.75 decks to 5.00 decks

Count	Hands	Net Win	Per Hand Expect.	Units Bet	Expect. Std. Err.	Per Hand Contr.	Units Won
≤ −1	10129817	−229999.9	−2.27	1	0.04	−0.053	−0.053
0	1424022	−2400.5	−0.17	1	0.10	−0.001	−0.001
1	1398934	5540.5	0.40	→1.5	0.10	0.001	0.0015
2	2008748	19422.0	0.97	→3	0.08	0.004	0.012
3	1198377	15125.0	1.26	→4.5	0.11	0.003	0.0135
4	1076356	18670.5	1.73	6	0.11	0.004	0.024
≥ 5	4359759	186095.0	4.27	12	0.06	0.043	0.516
	21596013	12453.5	0.06		0.02	0.003	0.513
						Total (all segments)	**1.925**

TABLE 6.20

Summary of .00 decks to 5.50 decks

Count	Hands	Net Win	Per Hand Expect.	Units Bet	Expect. Std. Err.	Per Hand Contr.	Units Won
≤ −1	189313871	−3289806.0	−1.74	1	0.01	−0.685	−0.685
0	108342451	−511185.5	−0.47	1	0.01	−0.106	−0.106
1	69352832	12562.0	0.02	1	0.01	0.003	0.003
2	39886514	224375.0	0.56	2*	0.02	0.047	0.094
3	24717810	274662.5	1.11	4*	0.02	0.057	0.228
4	15701358	270064.0	1.72	6*	0.03	0.056	0.336
≥ 5	32755179	1335587.0	4.08	12	0.02	0.278	3.336
	480070015	−1683741.0	−0.35		0.01	−0.351	3.206

* See footnote on p. 84.

Per-Hand Expectations as a Function of True Count, Segment-by-Segment

TABLE 6.21
.00 decks to 4.50 decks

Count	Hands	Net Win	Per Hand Expect.	Units Bet	Expect. Std. Err.	Per Hand Contr.	Units Won
≤–1	149044570	–2320393.0	–1.56		0.01	–0.483	
0	102204912	–508786.5	–0.50		0.01	–0.106	
1	63530585	–15387.5	–0.02		0.01	–0.003	
2	34479737	173780.5	0.50		0.02	0.036	
3	19630167	201747.0	1.03		0.03	0.042	
4	10847091	170174.5	1.57		0.04	0.035	
≥5	13979096	422070.5	3.02		0.03	0.088	
	393716158	–1876794.5	–0.48		0.01	–0.391	

TABLE 6.22
4.50 decks to 5.00 decks

Count	Hands	Net Win	Per Hand Expect.	Units Bet	Expect. Std. Err.	Per Hand Contr.	Units Won
≤–1	20324852	–451209.5	–2.22		0.02	–0.094	
0	2748476	–6533.0	–0.24		0.07	–0.001	
1	3370864	10244.0	0.30		0.06	0.002	
2	3784851	33456.5	0.88		0.06	0.007	
3	2813662	36855.0	1.31		0.07	0.008	
4	2000455	36080.5	1.80		0.08	0.008	
≥5	8164041	331876.5	4.07		0.04	0.069	
	43207201	–9230.0	–0.02		0.02	–0.002	

TABLE 6.23
5.00 decks to 5.25 decks

Count	Hands	Net Win	Per Hand Expect.	Units Bet	Expect. Std. Err.	Per Hand Contr.	Units Won
≤–1	10040597	–254753.0	–2.54		0.04	–0.053	
0	1576095	–98.0	–0.01		0.09	–0.000	
1	1539971	10045.0	0.65		0.09	0.002	
2	759884	7765.5	1.01		0.13	0.002	
3	1410044	20931.0	1.48		0.10	0.004	
4	1277239	26868.5	2.10		0.10	0.006	
≥5	4969143	243094.5	4.89		0.05	0.051	
	21582973	53853.5	0.25		0.02	0.011	

TABLE 6.24
5.25 decks to 5.50 decks

Count	Hands	Net Win	Per Hand Expect.	Units Bet	Expect. Std. Err.	Per Hand Contr.	Units Won
≤–1	9903852	–263450.5	–2.66		0.04	–0.055	
0	1812968	4232.0	0.23		0.09	0.001	
1	911412	7660.5	0.84		0.12	0.002	
2	852042	9372.5	1.10		0.13	0.002	
3	863937	15129.5	1.75		0.12	0.003	
4	1576573	36940.5	2.34		0.09	0.008	
≥5	5642899	338545.5	6.00		0.05	0.071	
	21563683	148430.0	0.69		0.02	0.031	

TABLE 6.21A
.00 decks to 4.50 decks

Count	Hands	Net Win	Per Hand Expect.	Units Bet	Expect. Std. Err.	Per Hand Contr.	Units Won
≤ −1	149044570	−2320393.0	−1.56	1	0.01	−0.483	−0.483
0	102204912	−508786.5	−0.50	1	0.01	−0.106	−0.106
1	63530585	−15387.5	−0.02	1	0.01	−0.003	−0.003
2	34479737	173780.5	0.50	2*	0.02	0.036	0.072
3	19630167	201747.0	1.03	4*	0.03	0.042	0.168
4	10847091	170174.5	1.57	6*	0.04	0.035	0.210
≥ 5	13979096	422070.5	3.02	12	0.03	0.088	1.056
	393716158	−1876794.5	−0.48		0.01	−0.391	0.914

TABLE 6.22A
4.50 decks to 5.00 decks

Count	Hands	Net Win	Per Hand Expect.	Units Bet	Expect. Std. Err.	Per Hand Contr.	Units Won
≤ −1	20324852	−451209.5	−2.22	1	0.02	−0.094	−0.094
0	2748476	−6533.0	−0.24	1	0.07	−0.001	−0.001
1	3370864	10244.0	0.30	1	0.06	0.002	0.002
2	3784851	33456.5	0.88	→3	0.06	0.007	0.021
3	2813662	36855.0	1.31	→4.5	0.07	0.008	0.036
4	2000455	36080.5	1.80	6	0.08	0.008	0.048
≥ 5	8164041	331876.5	4.07	12	0.04	0.069	0.828
	43207201	−9230.0	−0.02		0.02	−0.002	0.840

TABLE 6.23A
5.00 decks to 5.25 decks

Count	Hands	Net Win	Per Hand Expect.	Units Bet	Expect. Std. Err.	Per Hand Contr.	Units Won
≤ −1	10040597	−254753.0	−2.54	1	0.04	−0.053	−0.053
0	1576095	−98.0	−0.01	1	0.09	−0.000	−0.000
1	1539971	10045.0	0.65	→2	0.09	0.002	0.004
2	759884	7765.5	1.01	→3	0.13	0.002	0.006
3	1410044	20931.0	1.48	→5	0.10	0.004	0.020
4	1277239	26868.5	2.10	6	0.10	0.006	0.036
≥ 5	4969143	243094.5	4.89	12	0.05	0.051	0.612
	21582973	53853.5	0.25		0.02	0.011	0.625

TABLE 6.24A
5.25 decks to 5.50 decks

Count	Hands	Net Win	Per Hand Expect.	Units Bet	Expect. Std. Err.	Per Hand Contr.	Units Won
≤ −1	9903852	−263450.5	−2.66	1	0.04	−0.055	−0.055
0	1812968	4232.0	0.23	1	0.09	0.001	0.001
1	911412	7660.5	0.84	→3	0.12	0.002	0.006
2	852042	9372.5	1.10	→4	0.13	0.002	0.008
3	863937	15129.5	1.75	→6	0.12	0.003	0.018
4	1576573	36940.5	2.34	→12	0.09	0.008	0.096
≥ 5	5642899	338545.5	6.00	12	0.05	0.071	0.852
	21563683	148430.0	0.69		0.02	0.031	0.926
						Total (all segments)	**3.305**

* See footnote on p. 84.

September 1993 Cover Art by Judy Robertson

Chapter 7

Camouflage for the Basic Strategist and the Card Counter

It is a recognized fact that, to preserve longevity as a card counter, we cannot always bet "optimally." In this chapter, I examine, for the first time in print, what certain camouflage betting techniques would actually cost — first, for the basic strategist, and then, for the card counter.

From the September 1993 **Blackjack Forum,** *"How Dumb Can You Afford to Appear?"*

Part I

Q. *Being the blackjack fanatic that I am, there is a problem I hope you can help me with. Since I am a card counter, there are times that I have to make camouflage plays contrary to basic strategy. My friends, who are not counters, think that blackjack* *is a guessing game, and they also make plays contrary to basic strategy. I would like to know what these plays cost them and me, what edge the casino holds, and how much money and what percent it costs the player when making these plays. Enclosed is a copy of percentages it costs a player when deviating from basic strategy. I found most of the figures from Braun's book,* ***How to Play Winning Blackjack,*** *and from Wong's* ***Blackjack Count Analyzer.*** *If these numbers are correct, I would like to know how to use them.*

A. Your chart, which looks like it comes exclusively from Braun, is accurate if Braun is. It appears that he may have made a mistake for A,8 v. 2, but it really doesn't matter. The most accurate numbers come from pp. 231–32 of Griffin's third (elephant) edition. You can trust them 100%. (Note, however, the typo on p. 232 where the second A,9 should be A,8.) Another general problem concerns the treatment of the ace and ten. Percentages should always be against playable ones, as expectation doesn't matter if they ultimately turn into blackjacks.

My other comment is that, as a counter, I've never been a fan of making wrong plays for camouflage purposes. The pit thinks we make wrong plays every time we depart from strategy when the count tells us to. Throwing in truly wrong plays hardly seems necessary. But, if you insist, here's how to figure the cost.

First, at least promise me that you'll make "dumb" plays only with a minimum bet out. So, for all calculations, assume a one-unit wager. Next, you need the frequency with which each situation arises. You can't use Braun's p. 82 for all of the plays, because you may depart *after* drawing one or more cards; so the initial-hand frequency charts don't tell the full story. Instead, use Griffin's guidelines from p. 39 of his book. I used them for my "Illustrious 18" article. *[Note: The Appendixes to Wong's revised edition of* ***Professional Blackjack*** *now provide the most accurate information in print on this topic. — D.S.]*

Let me walk you through an example. Suppose you decide to stand on 15 against the dealer's 7. The chart says that the difference from correct hitting is 10%. Now, how frequently does the situation arise? For the upcard, 1/13 of the time. For the 15, 165 out of 1,326 times (see Griffin's explanations). So your 15 v. dealer's 7 occurs (165/1,326) x 1/13 = 0.96% of the time. Roughly speaking, you'll face the decision once an hour (one hand per hundred). And when you purposely play the hand incorrectly, you'll forfeit 10% of your wager. If you're a $10 player, it costs you a buck. If you're a black chip player, it will cost $10 an hour. So, if your win expectation is, say, two units per hour, this one departure would cost you 5% of your hourly expectation (a lot!) because: a) the hand occurs relatively frequently; and b) the "toll" for the mistake (10%) is rather high.

Coincidentally, I just got finished answering another letter where the writer wanted to know the cost of not doubling A,2 v. 5. The difference in expectation, compared to proper doubling, was 0.76% and the frequency of the hand was 0.092%. Do the math, and you'll find it costs the *black chip player* roughly seven cents per hour! Although this is a mistake, it would cost, on average, 1/140 the amount of our 15 v. 7 decision to stand.

Finally, there is one more subtle point to consider. Your friends, who guess and don't count cards, are not involved in this, but as a counter, you are. All of your figures apply when the count is more or less near the correct index number for the play in question. As soon as you consider making a wrong play at counts other than these, you must factor in how volatile (that is, sensitive) the move is with respect to the distance the count is from the critical index, or inflection point. Some plays are very sensitive to fluctuations in count, others not much at all. So it would behoove you to know how much extra damage you might be doing by varying at a time when the count causes your loss to be even greater than originally anticipated from your chart. For a discussion of all of this, see Griffin, pp. 28–31.

My conclusion is obvious. First, I wouldn't make mistakes at all. But if forced to, I'd damn well make an "error" where the percentage difference was rather small (say 0 to 4%) and the hand did not come up too frequently. Of course, if you're deciding that you must make the mistake *now,* as opposed to predetermining those hands for which you'll err and waiting for them to arise, the frequency really doesn't enter into the discussion.

Part II

Fine. That was then, this is now. Since responding to the above question, I've had somewhat of a change of heart. It occurred to me that there might be considerable interest in an article that would reflect the penalties, both conditional and absolute, associated with violating basic strategy *for the non-counter.* I spoke with Stanford Wong and Arnold Snyder, and they were both very enthusiastic, stressing the usefulness of this kind of information. That was all the encouragement I needed!

Once again, it is my hope that the following charts and explanations will provide data that were heretofore not cataloged (at least as readily!) in any other publication and, therefore, will add to readers' knowledge of this topic. Above all, as we develop a kind of "shopping list" for the camouflage-seeker, we shall try to define some practical limits as to just how "dumb" we can afford to appear.

As outlined above, there is a big difference between approaching this subject from the counter's perspective and that of the basic strategy player. The extra gains available to the counter, when he makes a departure, were analyzed (for 18 plays) in a previous chapter. Penalties for violating basic, when one is counting, become tricky to calculate, since there are so many variables to consider. I leave this whole area for a future study and assume that simulations would play a major role in the ultimate solutions.

When writing an article that, realistically, will be used by players of games ranging from single- to 8-deck, certain compromises must be struck. As was done previously, for studies involving "The Ups and Downs of Your Bankroll" and "Attacking the Shoe," I have chosen the 4-deck game as a surrogate for the entire spectrum. Now, it is true that 4-deck blackjack more resembles its 8-deck cousin than it does the one-deck variety, but it is also true that there are far more multiple-deck games (and players!) than single-deck. Hence, the 4-deck choice.

It is possible to obtain data regarding the frequency of player hands v. dealer upcards from many sources. *[If doing this study today, I would obtain the frequencies from simulation or from Wong's aforementioned charts. In addition, the* ***Professional Blackjack*** *Appendixes now provide the exact statistics first approximated in this article. — D.S.]* My point of departure was the p. 82 chart in Julian Braun's *How to Play Winning Blackjack.* But two modifications were made: one not very important, the other extremely so. By assuming that the dealer would check under his tens and aces, thus revealing a natural *before* the player could make a decision on his hand, the frequencies outlined below indicate "playable hands" only. Second, and much more important to the methodology, is the fact that a certain player total, say 15, is attainable in many ways other than just the initial two cards. Braun's frequencies do not reflect this. So, if we're studying the effect on expectation of purposely misplaying 15 v. 7, for example, we need an estimate of *all* situations in which the player might encounter this opportunity; that is, multi-card 15s must be added to the 2-card varieties.

Once the frequencies were obtained, the essence of the study — penalty in percent for making the second-best play — was researched. Here again, Braun's charts are useful, to a degree. We needed to decide what level of accuracy was required. After all, Griffin's charts, pp. 231–33 of the third edition of *The Theory of Blackjack,* provide all the information necessary to calculate exact penalties (derived from infinite-deck numbers) for any composition of the hand, any number of decks, most decisions. We decided not to use Griffin's numbers for several reasons: 1) laziness! It is cumbersome to carry out all of the aforementioned calculations; 2) lack of need for the kind of precision they afford as, ultimately, these small numbers are going to be multiplied by even smaller frequencies of occurrence; 3) the "raw," infinite-deck results are sufficiently close to Braun's charts so as to validate use of the latter; and finally, 4) Griffin doesn't furnish data for infinite-deck pair-splitting, so we would need to look elsewhere, anyway, for these.

Where we were unhappy with a Braun-rounded number of zero or one (more precision was desired), Wong's *Blackjack Count Analyzer* came to the rescue. Here, one can specify number of decks, the exact composition of the hand, the dealer upcard, and receive the gain or loss for each decision possible. Simple comparison provides the penalty for violating basic and going with choice number two instead.

Finally, the probabilities from the first two charts were multiplied so as to obtain the *absolute* penalty, in percent, for deviating from basic. The concept here is to express a value that reduces everything to cold dollars and cents.

What if a player decides, *before* sitting down, that he will *always* violate basic strategy for a particular play? For example, what if, upon seeing his neighbor insure a blackjack, the basic strategist boldly announces for all the pit to hear: "That's the best bet in the game; I *always* insure all of my blackjacks"? Well, if you're a black-chip ($100) player, averaging 100 hands per hour, you now know exactly what this "transgression" will cost you — $1.35 per hour. In fact, you know what *every* mistake will cost you!

To summarize: Table 7.2 penalties are *conditional.* They assume the condition that you already have the hand in front of you. Table 7.3 penalties are *absolute.* They take the viewpoint that you are about to play and are assessing the hourly "damage" that will be caused, on average, if you choose to misplay any of the hands listed.

Now, it is fair to ask, at this point, why a non-counting basic strategist would have any need to disguise his play at all. Why, indeed, should a player who does not have an edge over the house (at virtually all games) feel obliged to make an occasional "dumb" play? Well, several reasons have been pointed out to me. First, some high rollers who don't count have learned that they can play for comps and limit their losses (while having all expenses paid) simply by playing perfect basic strategy. Nonetheless, these players (because they play like machines and, therefore, provide only limited profit potential to the house) occasionally receive heat and, generally, are frowned upon. The ideal solution is to scan the charts and sprinkle in enough "mistakes" to placate the pit, all the while keeping the costs of camouflage to a minimum. In addition, I've been told, by a reputable source, that in some Far East locations, such as Macao, casinos have been known to actually bar perfect basic strategy players. Again, the advantages of camouflage are all too apparent.

Finally, many of us know that it is possible to win at blackjack by employing non-traditional methods that don't depend on counting. When we do this, it is obvious that, in order to gain an edge, we need to violate basic strategy from time to time. If we are being closely observed, wouldn't it be an effective use of misdirection to mix in a few truly "bad" plays so as to make it harder for pit personnel to ascertain exactly where our advantage resides? Clearly, the answer is yes.

In my conversation with Wong, we also discussed the fact that "announcing" to the table (and, especially, to any pit boss within earshot) that you "always" do something doesn't necessarily have to cost you the full penalty — or anything at all! Upon seeing your neighbor hit 12 v. 3, you might intone, "I would never hit in that situation." Later, if *you* get the hand, you should be "good to your word" and stand; but the point is, the situation may never materialize while you're at the table. In essence, you get the benefit of the camouflage without the cost that accompanies it. The other side of this coin is your on-the-spot, unplanned decision to make a "dumb" play in an attempt to shake a bothersome boss who can't make up his mind about you. Here, you make a departure *as soon as possible,* regardless of the hand (or almost!) and incur the instantaneous cost of the one-time maneuver. In this case, we throw away Tables 7.1 (frequency) and 7.3 (absolute penalty) to concentrate on conditional cost (Table 7.2) alone. That is, what will it cost me *right now* to do this?

Ultimately, each player will use these charts as best he sees fit. However, I thought it might be fun to scan Table 7.3 to create a few different "profile" players, while determining just how much their respective "acts" would cost. We can imagine, for example, the player who: splits too frequently; passes up good splits; doubles too often; fails to double when he should; hits stiffs that should be left alone; stand on stiffs that need to be hit; never hits 16 v. 10; doubles 11 v. A, and so on. In this regard, see Griffin, p. 150, where the author attempts to imagine just how bad the "world's worst player" might be by choosing from a "smorgasbord of prices paid for departure from basic strategy." Griffin's numbers are for single-deck and so differ, somewhat, from those found in my charts. In addition, the present charts afford the luxury of breaking down the broader categories (e.g., "never doubles") into their individual component plays so that the user may ascertain how the final, global number was reached.

Let's begin with a single play that can go a long way toward making you look like everyone else while, fortunately, costing very little: always insuring a blackjack. Price tag, as mentioned earlier, $1.35 per hour for the *black-chip* basic strategist and, therefore, mere cents (13.5¢) per 100 hands for the $10 player. Again, remember — if you aren't the first one at the table to get the natural against the dealer's ace, go ahead and expound your philosophy out loud, anyway. You're on record and, so far, it hasn't cost you a dime!

Interestingly enough, another extremely common play — not hitting 16 v. 10 — costs just about double that of insuring a blackjack, clocking in at 2.65¢/100 (or $2.65 per hour for the black-chip flat bettor). Note that the penalty, once you hold the hand, is a mere 0.75% of your wager (75¢ per black chip), but that you encounter 16 v. 10 more frequently than any other decision in the game, except 15 v. 10 — roughly 3.5 times per 100 hands played. On the contrary, you get to insure your blackjacks with only one-tenth the frequency of standing on 16 v. 10. Thus, recognize the importance of knowing the cost of the "spur of the moment decision," made once you have the hand already in front of you, as opposed to the "announce ahead of time" cost of camouflaging a play each time you are faced with the decision. Please remember that the above rate and all subsequent examples assume *hourly* expectations, based on 100 hands per hour.

TABLE 7.1

Hand Frequencies Based on 100,000 Playable Hands

Player hand v.:	2	3	4	5	6	7	8	9	10	A
8	179	179	185	179	179	185	185	185	681	124
9	271	271	271	271	271	271	277	277	1022	187
10	315	315	315	315	315	315	315	320	1181	215
11	364	364	364	364	364	364	364	364	1363	249
12	750	750	750	750	750	870	870	870	3210	600
13	750	750	750	750	750	900	900	900	3320	620
14	640	640	640	640	640	930	930	930	3430	640
15	640	640	640	640	640	960	960	960	3530	660
16	580	580	580	580	580	960	960	960	3530	660
A,2*	87	92	92	92	92	92	92	92	342	58
A,3*	92	87	92	92	92	92	92	92	342	58
A,4*	92	92	87	92	92	92	92	92	342	58
A,5*	92	92	92	87	92	92	92	92	342	58
A,6*	92	92	92	92	87	92	92	92	494	89
A,7*	92	92	92	92	92	87	134	133	550	99
A,8*	92	92	92	92	92	92	87	173	640	114
A,A	43	43	43	43	43	43	43	43	161	25
2,2	38	43	43	43	43	43	43	43	160	29
3,3	43	38	43	43	43	43	43	43	160	29
4,4	43	43	38	43	43	43	43	43	160	29
6,6	43	43	43	43	38	43	43	43	160	29
7,7	43	43	43	43	43	38	43	43	160	29
8,8	43	43	43	43	43	43	38	43	160	29
9,9	43	43	43	43	43	43	43	38	160	29
10,10	727	727	727	727	727	727	727	727	2598	497
Insurance										
w/10,10										727
w/A,10										346
w/other non-10,10										2952
w/non-10,non-10										3667
Total hands:										7692

* For A,2 through A,8, frequencies depend on whether or not doubling is involved as the best or second-best play. Thus, not all listings are two-card totals; some holdings may represent multi-card soft hands.

TABLE 7.2

Conditional Penalty, in Percent*, for Deviating from Basic Strategy

Player hand v.:	2	3	4	5	6	7	8	9	10	A
8	18.00	14.00	8.00	4.00	2.00	24.00	38.00	50.00	52.00	60.00
9	0.37	3.00	6.00	10.00	13.00	6.00	12.00	24.00	31.00	36.00
10	18.00	21.00	24.00	27.00	30.00	14.00	9.00	3.00	2.00	8.00
11	24.00	26.00	29.00	32.00	34.00	17.00	12.00	7.00	6.00	2.00
12	4.00	2.00	0.56	3.00	2.00	22.00	20.00	17.00	14.00	25.00
13	2.00	4.00	7.00	10.00	8.00	18.00	16.00	14.00	11.00	23.00
14	7.00	10.00	13.00	16.00	15.00	14.00	13.00	10.00	7.00	20.00
15	13.00	16.00	20.00	20.00	22.00	10.00	9.00	7.00	4.00	17.00
16	18.00	21.00	23.00	29.00	26.00	7.00	6.00	4.00	0.75	15.00
A,2	11.00	7.00	3.00	0.76	3.00	30.00	37.00	40.00	40.00	56.00
A,3	9.00	6.00	1.06	3.00	5.00	26.00	32.00	37.00	37.00	52.00
A,4	7.00	3.00	0.54	4.00	7.00	21.00	29.00	34.00	34.00	49.00
A,5	5.00	1.60	2.00	6.00	9.00	17.00	25.00	30.00	31.00	46.00
A,6	0.38	3.00	6.00	10.00	13.00	16.00	31.00	27.00	22.00	29.00
A,7	0.33	3.00	7.00	10.00	11.00	23.00	7.00	8.00	4.00	0.37
A,8	15.00	10.00	7.00	3.00	1.22	29.00	40.00	28.00	15.00	29.00
A,A	41.00	43.00	46.00	48.00	47.00	32.00	27.00	24.00	25.00	16.00
2,2	3.00	1.00	1.00	5.00	7.00	3.00	5.00	15.00	19.00	16.00
3,3	6.00	3.00	1.00	4.00	6.00	4.00	4.00	13.00	18.00	15.00
4,4	21.00	17.00	14.00	10.00	10.00	28.00	26.00	37.00	33.00	31.00
6,6	2.00	3.00	7.00	10.00	13.00	7.00	15.00	23.00	29.00	35.00
7,7	8.00	10.00	13.00	16.00	22.00	20.00	4.00	13.00	16.00	7.00
8,8	25.00	27.00	30.00	32.00	39.00	58.00	34.00	9.00	5.00	12.00
9,9	2.00	6.00	9.00	13.00	11.00	8.00	10.00	8.00	13.00	4.00
10,10	28.00	24.00	20.00	16.00	13.00	27.00	41.00	54.00	51.00	51.00
Insurance										3.60
w/10,10										4.60
w/A,10										3.90
w/other non-10,10										3.90
w/non-10,non-10										3.20

* All entries have been rounded. Rules assume no das.

TABLE 7.3

Absolute Penalty, in Cents per $100 Bet, for Deviating from Basic Strategy*

Player hand v.:	2	3	4	5	6	7	8	9	10	A
8	3.22	2.51	1.48	0.72	0.36	4.44	7.03	9.25	35.41	7.44
9	0.10	0.81	1.63	2.71	3.52	1.63	3.32	6.65	31.68	6.73
10	5.67	6.62	7.56	8.51	9.45	4.41	2.84	0.96	2.36	1.72
11	8.74	9.46	10.56	11.65	12.38	6.19	4.37	2.55	8.18	0.50
12	3.00	1.50	0.42	2.25	1.50	19.14	17.40	14.79	44.94	15.00
13	1.50	3.00	5.25	7.50	6.00	16.20	14.40	12.60	36.52	14.26
14	4.48	6.40	8.32	10.24	8.80	13.02	12.09	9.30	24.01	12.80
15	8.32	10.24	12.80	12.80	14.08	9.60	8.64	6.72	14.12	11.22
16	10.44	12.18	13.34	16.82	15.08	6.72	5.76	3.84	2.65	9.90
A,2	0.96	0.64	0.28	0.07	0.28	2.76	3.40	3.68	13.68	3.25
A,3	0.83	0.44	0.10	0.28	0.46	2.39	2.94	3.40	12.65	3.02
A,4	0.64	0.28	0.05	0.37	0.64	1.93	2.67	3.13	11.63	2.84
A,5	0.46	0.15	0.18	0.52	0.83	1.56	2.30	2.76	10.60	2.67
A,6	0.03	0.28	0.55	0.92	1.13	1.47	2.87	2.48	10.87	2.58
A,7	0.03	0.28	0.64	0.92	1.01	2.00	0.94	1.06	2.20	0.04
A,8	1.38	0.92	0.64	0.28	0.11	2.67	3.48	4.84	9.60	3.31
A,A	1.76	1.85	1.98	2.06	2.02	1.38	1.16	1.03	4.03	0.40
2,2	0.11	0.04	0.04	0.22	0.30	0.13	0.22	0.65	3.04	0.46
3,3	0.26	0.11	0.04	0.17	0.26	0.17	0.17	0.56	2.88	0.44
4,4	0.90	0.73	0.53	0.43	0.43	1.20	1.12	1.59	5.28	0.90
6,6	0.09	0.13	0.30	0.43	0.49	0.30	0.65	0.99	4.64	1.02
7,7	0.34	0.43	0.56	0.69	0.95	0.76	0.17	0.56	2.56	0.49
8,8	1.08	1.16	1.29	1.38	1.68	2.49	1.29	0.39	0.80	0.35
9,9	0.09	0.26	0.39	0.56	0.47	0.26	0.43	0.30	2.03	0.12
10,10	20.36	17.45	14.54	11.63	9.45	19.63	29.81	39.26	132.50	25.35
Insurance										27.93
w/10,10										3.34
w/A,10										1.35
w/other non-10,10										11.51
w/non-10,non-10										11.73

* Assuming $1 avg. bet and 100 hands/hr., loss is in cents. Assuming $100 avg. bet and 100 hands/hr., loss is in dollars.

Can you afford to insure your 10,10 twenties along with your naturals? In single-deck, Griffin tells us the cost is five times as great, but for 4-deck, it is roughly 2.5 times the impost, or 3.34¢ per $100. Insuring all the time becomes ridiculous, costing 28¢/$100. Nor could you possibly afford to "never hit a stiff" against the dealer's 7 through A, the penalty being a whopping $3.55/$100. Here, you might simply go for the more tasteful aforementioned 16 v. 10 and sprinkle in an occasional 16 v. 9 (3.84¢), 16 v. 8 (5.76¢), 16 v. 7, and 15 v. 9 (each 6.72¢).

Likewise, it would be far too costly to "hit stiffs v. dealer's small cards" all the time, even though the multi-deck $1.93 per $100 cost is far below the one-deck $3.20 charge. In this respect, it is interesting to note that hitting too frequently (as opposed to standing too much) seems to be a lesser crime in multi-deck than in single-deck. Here, the recommended plays would go along the lines of hitting 12 v. 4–6 (total cost only 4.17¢), 13 v. 2–4 (9.75¢), and an occasional 14 v. 2 (4.48¢) or 3 (6.40¢). Note, also, that if you are playing the "never hit a stiff v. big" game, you'll certainly also have to stand on 12 v. 2 (3.00¢) and 3 (1.50¢) for good measure!

In the doubling and splitting categories, we make the following observations: Doubling on all 10s and 11s (and, thereby, making three mistakes) costs a relatively small 4.58¢ for the trio of 10 v. 10, 10 v. A, and 11 v. A (the latter exacting a mere half a cent!).

And you certainly can afford the *tenth* of a cent it will cost per $100 to play 9 v. 2 wrong and double that, also. Come to think of it, there's nothing so bad, either, about doubling 9 v. 7 (1.63¢).

Never doubling at all takes a very heavy toll ($1.38), so we ought to confine our apparent "cowardice" to the cheaper plays, such as chickening out on 9 v. 3–6 (8.67¢ total) and, on the upper end, backing off on 10 v. 9 (less than a penny!), 10 v. 8 (2.84¢), 10 v. 7 (4.41¢), and 11 v. 9 (2.55¢) or 8 (4.37¢). Note that the *conditional* penalty for failing to double 11 v. 10 is actually *less* than that of passing up 11 v. 9, but we encounter the former play almost four times as frequently as the latter.

As for soft doubling, take a look at the charts and then go ahead and do anything you please! Not only are the frequencies of these plays extremely small, but the conditional penalties for playing incorrectly are, for the most part, relatively minor, as well. The result is that when the two are multiplied, they produce absolute results that are all almost entirely under a penny per $100. You can grab a whole bushel of "mistakes" here and not have them cost you even a dime per $100 bet.

An identical comment can be made for most pair-splitting. You just can't lose more than 3 cents per $100 on any split (too often or not enough) unless you are "dumb" enough to not split As v. the dealer's 10 (4.03¢), or you do split 2s (3.04¢), 4s (5.28¢) or 6s (4.64¢) against the ten. Here, some of the conditional penalties can be very steep (for example, failing to split 8s v. 7 is the second dumbest play, at 58%, on the entire conditional chart). But because the frequencies for most pair-splits are so low (0.043%), the *absolute* penalty, even for botching 8,8 v. 7, is only 2.49¢/$100, simply because it just isn't very likely that you will encounter the hand (once every 25 hours of play!). Oh yes, just in case you were too lazy to look — the "dumbest play on the chart" award goes to doubling 8 v. the dealer's A, instead of hitting. It would cost you 60% of your wager, assuming you already had the hand in front of you.

How, then, shall we interpret all of the above? Well, first, let's recall that, on average, the multi-deck basic strategist works against a 0.5% disadvantage. Thus, the black-chip player loses about $50 per hour, assuming 100 hands. If he's qualifying for RFB and airfare, he may have to play, say, 16 hours over four days. How much "extra" can he afford to lose by adding camouflage plays? It seems to me that the answer is entirely a function of the need, in the first place, to appear "dumb," and the willingness, on the player's part, to add possible losses to those already forecast ($800) by the above scenario. Certainly, the player needn't make all the same mistakes from one hour to the next. Appearing very "dumb" at the outset may remove any potential heat after just a few hours. Pit bosses often formulate "first impressions" that last for the duration of the trip. So you might not have to continue to make "mistakes" all the way through.

Finally, of the 254 plays on the chart, almost 100 of them carry penalties of less than one cent per $100. So, two distinct approaches are possible. You can arrive at a predetermined dollar amount of "dumbness" by heaping literally *dozens* of "miscues" upon one another, or you can play virtually error-free, saving your quality "eye-openers" for the five or six times that they will get the very best mileage. I've given you the numbers; now you go out and "pick your poison"!

And there you have it. Permit me a few final words before closing. This article can be improved in several ways. First, although I was very careful with the methodology, I am acutely aware of the fact that some values are only approximate, at best. Where frequencies involve only 2-card hands (e.g., soft doubles or pair splits), they are quite precise. But where the possibilities of multi-card totals exist (especially 12–16 v. all dealer upcards), I used Griffin's single-deck estimates (p. 39) for the multi-deck game. Extensive simulations might ultimately produce "better" numbers, but I had neither the time nor the inclination to do them. It is unlikely that any findings would be materially changed, and given the considerable amount of work that would be involved, I just don't think it would be worth the effort.

In addition, the conditional penalties for violating basic strategy change, sometimes drastically, from single- to multiple-deck. Using Wong's *Blackjack Count Analyzer* and Griffin's charts and methodology (pp. 74–86 and 231–33), it is possible to determine, precisely, the penalty for deviating from basic for any holding. But when one attempts to summarize that penalty across all ways of achieving the total, weighting the different outcomes properly, the task becomes overwhelming. We used, in Table 7.2, as conditional penalties, the 2-card values from Braun's charts (for the most part). But, for reasons explained earlier, we had to use Griffin's frequencies, especially for the "stiff" totals.

Obviously, there is an error in this approach. For example, when studying the penalty for standing on 16 v. 10, we used Braun's "global" 0.8%. That number averages the penalties for 10,6 and 9,7 by weighting the individual hands accordingly and summing up. This is fine — for the 2-card holding. But what about multi-card 16s? If I stand on 16 v. 10 when I have 2,3,5,4,2, there isn't any penalty at all! Obviously, I've made the right play, even though, as a basic strategist, I may not be aware of this. It was beyond the scope of this article to take these notions into consideration, but it is certain that ignoring them means the numbers in Table 7.2 and, therefore, in Table 7.3, as well, can be improved.

Interestingly, Braun states, on p. 27, "Please recognize that your total of 12 [the example he had been discussing involved basic strategy for 12 v. 2] may come from more than two cards. For example, your first two cards are 4,3. You hit and receive a 5. Your total is now 12 and the strategy above [hit, of course] now prevails." But Braun does not claim that he incorporated any of these possible multi-card holdings into his charts. In fact, he did not.

Lastly, as mentioned before, there was no attempt in this present work to catalog penalties from the card counter's perspective. Here, the task is gargantuan. Perhaps I will be equal to it one day, but it certainly will not be *this* day! And so, until next time, good luck, and ... good cards. *[Note: The deed is done! See Ian Andersen's* ***Burning the Tables in Las Vegas,*** *Chapter 7, "The Ultimate Gambit," for more information on the costs of cover plays to the counter. — D.S.]*

From "The Gospel According to Don," *Blackjack Forum*, March 1992:

Q. *I have been very successful using the betting methods and camouflage you have suggested over the years; but recently, while playing some of the Las Vegas double-deck games, I realized that I wasn't making my top bet (6 units) nearly as frequently as the frequency distribution would predict. I realize that because I never jump my bet (only parlay) nor ever raise a bet after a loss, there are times when the count justifies a top bet, but I can't put it out. Two questions: 1) Is there anything I can do to get out more top bets? and 2) Can you estimate how much your style of betting costs compared to the standard way of calculating win rates?*

A. Funny you should ask! Just last month, I proposed to Arnold a topic for a more substantial, "feature length" type of article such as I have done in the past for

Blackjack Forum. The suggested title: "The Effects of Camouflage Betting on Your Win Rate." I, too, have experienced what you described and, indeed, I have received many questions on this topic from friends, former students, and readers who use my betting approach.

George C. and I are going to try to get to the bottom of this question. I have some strong hunches on where the computer work will lead, but I'm going to keep them to myself for now — not to tease you, but simply because, if I'm wrong, there's no sense in misrepresenting the facts if simulations ultimately give the lie to my intuition. Very superficially, I'll answer each of your questions with the promise of a much more complete response, I hope, for the June issue.

I would definitely consider beginning with a two-unit wager off the top of a pack. Some 2-deck games offer surrender and/or double after splits. Consequently, the off-the-top house edge is minimal in these casinos (Hilton, Mirage, Flamingo, Riviera, for example), and it will cost you virtually nothing to routinely increase the first bet after a shuffle. But, on those occasions when you win and the count rises quickly, you will get to your top bet one hand sooner. The gains from this tactic will offset the slight aforementioned losses many times over.

I will leave the answer to your second question for the proposed article. One thing is clear, however. The cost of camouflage is greater as the number of decks decreases. This would appear to be common sense, as your ability to parlay to the top bet is more severely hampered as the number of rounds till a shuffle decreases. Again, I can guess as to what percentage of "ideal" you can glean, but I prefer to wait until we do our study.

From "The Gospel According to Don," *Blackjack Forum*, September 1994:

Q. *A while ago, you mentioned that you hoped to determine the ultimate cost to the player for following certain betting camouflage tactics that you have been associated with. I believe your style is to never increase the bet after a loss, never decrease after a win, and not touch the bet after a tie. Further, you have stated that "jumping" bets is the single most dangerous practice a counter can engage in, so you don't do that either. The most you bet after a win is a parlay, that is, twice the original bet. Finally, so as to "come down" gradually, you never decrease a bet (after a loss) by more than half. If, in fact, this is a correct portrayal of your betting style, have you determined what the cost is compared to unrestricted, "anything-goes" play?*

A. Yes … sort of! First, some background. Your depiction of my style of play is correct as far as you take it. Realizing (but never before having quantified it) that this set of camouflage "rules" could prove to be somewhat costly, I do find ways to compensate without necessarily violating the spirit of my own guidelines. More about that later. But I (and many others, apparently) have waited until now to actually be able to calculate what percentage of the total win one sacrifices by invoking the aforementioned ploys.

Enter John Imming and the magnificent *Real World Casino Universal Blackjack Engine* software he has developed, and the latest version of Stanford Wong's *Blackjack Count Analyzer.* Each of these spectacular programs now will provide answers to most of the questions we've been asking. Before I begin, I'd like to publicly thank Wong and especially John Imming for incorporating specific features into their software as a direct result of my wanting to research this thorny problem.

It is intuitive that the restrictions placed on bet-jumping are going to be damaging. For single- and double-deck games, in particular, the true count can change rapidly, necessitating correspondingly large jumps in bet sizes in order to make the correct Kelly wager. My 2-deck bet schedule is one unit until +2 true, then two units from 2 to 3 true, four units from 3 to 4 true, and six units for true 4 and above. Sometimes, it is "right" to go straight from one to six units, whereas I have to "pass through" two and four units along the way. Furthermore, I don't even permit myself to "advance" to two and then, eventually, four and six units unless I've won the previous hand. It's obvious that this "double whammy" on the upside is going to cost.

On the downside, if I win a hand, and the count deteriorates, I *don't* take the bet down. And when I do lose, I decrease in stages, backing off, as it were, gradually. The net result of all these maneuvers is that: 1) Sometimes I'm under-betting with respect to the count; 2) Sometimes I'm overbetting the count; 3) Sometimes I "steal" money, winning many big bets in a row that I'm not "entitled" to; and 4) Sometimes I save a fortune by avoiding a big losing streak right after the point where someone else would have jumped the bet and gotten clobbered. But (and don't think for one minute that I'm not aware of this) points (3) and (4) do not compensate for the "tax" imposed on me by points (1) and (2). I'm about to tell you how much that tax is, and how you can get at least a partial "refund" of the levy.

First, a disclaimer. This is a *preliminary* article. The date is July 19, and I need to meet Arnold's August 1 deadline for the September issue you're now reading. As I write, John Imming is feverishly modifying and debugging his version 7.0 — actually creating, changing, and rebuilding entire modules "on the fly." It's a massive undertaking. Wong, too, is incorporating new features that will eventually answer more and more of our questions. But for now, we need to draw the line somewhere and say, "Let's do what we can with what we have, and save the rest for later."

In the end, it is my plan to create a 3 x 4 matrix. Across the top: "2-deck, play all," "6-deck, play all," and "6-deck, back-counting." Down the side: "unrestricted betting," "total camouflage," "partial camouflage 1" (turn some on, some off), and "partial camouflage 2" (reverse of 1). For each scenario: an average bet size, win rate in %, and win per 100 hands played (or seen). Standard deviations for everything would be nice, too! For now, however, you're going to have to settle for much less: no back-counting (that is, no camouflage back-counting). We already know the old stuff. No "partial" camouflage. It's all or nothing. Wong can do the partial now, but time restrictions won't allow my simulations to be large enough for reliable data.

OK Enough suspense. Here's what I do have. We play Hi-Lo, Strip rules, double after split allowed. We use –10 to +10 strategy indices and head-on play. Penetration is to the 75% level. You have the bet spread for 2-deck; for 6-deck, go to 12 units at +5 and above. (Personally, I go to two hands of six, and one hand of 12 isn't the same thing; but the two hands caused programming problems. We'll straighten them out later.) Each simulation was a 20 million-hand run, so standard error is about 0.04%.

The 2-deck, unrestricted benchmark won 2.45 units per 100 hands played with an average bet of 1.79 units and, thus, a win rate of 1.36% of initial bets. The total camouflage package, unfortunately, gleaned only 51% of the win, checking in at 1.26 units per 100 hands. Average bet was 1.60 units (90% of unrestricted), and win rate was 0.78%. I knew the double-deck would be tough. We sacrifice a lot, but in a little while I'll tell you what you can do to get back some of the differential.

On to 6-deck. Here, the benchmark won 1.11 units per 100 hands, with an average bet of 1.74 units and a win rate of 0.64% of initial wagers. I expected the camouflage approach to do better here. I reasoned that there were many more hands dealt in the shoe game and that the count escalates more slowly. I figured eventually I would get to the big bets and I'd make a larger percentage of them here than in the 2-deck game. So, there wouldn't be as many "blown" opportunities. The good news is — I was right! The bad news is — not by as much as I thought. We managed to win nearly 70% of the unrestricted play, or 0.77 units per 100 hands. The average bet was 1.68 units (a full 97% of the non-camouflage total — good for junket players), but the win rate dropped off badly, from 0.64% to 0.46%, a decline of nearly 30% of the original rate. Frankly, I'm disappointed and a bit surprised. I didn't think the toll was this large. (Maybe Imming will find a last minute bug!) *[Editor's note: He didn't.]*

How, then, might we bridge some of the gaps caused by camouflage betting? First, for 2-deck, come "off the top" occasionally, if not all the time, with a two-unit bet. The idea is you're one parlay up the ladder without having to win a hand. And the off-the-top disadvantage is minimal (0.2%) for the 2-deck, das, Strip game. Second, cut back, on bad negative decks, to 3/4 of a unit. This will prevent some of the "drain."

Next, sit out some hands and/or take bathroom breaks (have to be quick for double-deck!). If you push with 20 (inevitably dropping the count), take the bet down, intoning, "What's the sense; I can't even win with a 20." Although pushes occur only once every 11 hands, the most common tie, by far, is 20-20. When this happens (perhaps once or twice per hour), the running count often drops by four (head-on). Halfway through a 6-deck shoe, the true drops by over one, and in the double-deckers, the true drops by four! So, taking down the bet is almost an imperative. Finally, for the die-hard impatient among you, "bend" the original camouflage guidelines (gently!). A good act and a lot of caution can go a long way.

If you must, I would violate the basic "dictates" in the following order: 1) Decrease after a win. What the heck, it's your money. You're not obligated to keep it out there; 2) Decrease by more than half after a loss. The "sequence" is over; back to square one. It can be justified; 3) Increase (tastefully!) after a loss. It's late; you're impatient. You "steam" a little. They might buy it; 4) The court of last resort — don't ever quote me as having advised you to jump a bet (that is, bet more than a parlay) after a win. This is anathema to me. But if you feel *compelled* to do it, follow Stanford Wong's advice and first take back all your money into your stack of chips. Then come out with a new bet that's one or more chips in excess of a parlay. Maybe they won't remember the previous bet (maybe they *will!*). But at least you don't stack up the parlay and then *add to it* from the chips in front of you. The latter is the *ultimate* in asking for trouble. But DON'T GET GREEDY! To quote the immortal Revere, "This is a discipline thing."

Of course, all of the 2-deck ploys can (and *should*) be used to recoup a good part of the sacrifice in 6-deck, too. And next time, I hope to report on back-counting — the only reasonable way to attack the multi-deck monsters. By the way, both Wong and Imming have installed a feature that will ignore the camouflage rules after a shuffle. For example, if you've just completed a shoe with a winning 12-unit bet, you may not want to start off the next shoe with such a large wager.

I've had feelings before about initial "breakthroughs" in articles I've written. For instance, I knew the "Illustrious 18" was destined to be a "biggie" and would generate a lot of interest. I think this topic may turn out to be the "hottest" of them all. Variations on the theme are virtually endless. We can change the count, rules, decks, penetration, and incorporate as much or as little camouflage as we choose. We're fortunate to have creative people like Imming and Wong, and we're equally fortunate to be living in a time when technological advances make programs and simulations such as these possible. *[Note: Since this article was written, advances in studying camouflage betting have been spectacular. See, in particular, Karel Janecek's* ***Statistical Blackjack Analyzer*** *and Norm Wattenberger's* ***Casino Vérité.*** *— D.S.]*

Ideas and suggestions for future studies are eagerly solicited. Please write to me at *AdvantagePlayer.com* with comments and questions. Better yet, get yourself a copy of the *Real World Casino UBE* and the *Blackjack Count Analyzer,* or the above-mentioned simulators. Then, you can answer your own questions! This has been fun for me; I hope you enjoyed, as well. Stay tuned, there's bound to be more!

From "The Gospel According to Don," *Blackjack Forum*, September 1995:

Q. *In the September 1994* ***BJF****, you published some initial findings on the cost of various camouflage methods you use. But you hadn't completed the study and you promised us more information. When do you think you'll have the rest of the numbers?*

A. Now! After all, a promise is a promise! Let me begin by acknowledging the work done by George C. in helping me to grind out the statistics you'll be reading shortly. One of my great frustrations, right now, is not having enough time to spend "playing" with the spectacular software that has been created in the past few years. I will give an unabashed plug (simply because they fully deserve it) to John Imming, Stanford Wong, Bryce Carlson, and Norm Wattenberger for, respectively, their *Universal Blackjack Engine*, *Blackjack Count Analyzer, Omega II Blackjack Casino,* and *Casino Vérité* programs. They have elevated simulation and practice routines to state-of-the-art. Unfortunately, with so little time, I don't derive the full benefit that I should from these magnificent research tools. But George keeps bailing me out, always at the ready to do some runs for me, and I want to publicly thank him for all of his contributions in this area.

Here was my intention, as promised on p. 44 of last September's issue *[Editor's note: pp. 104–5 of this book]:* "In the end, it is my plan to create a 3 x 4 matrix. Across the top, '2-deck, play all,' '6-deck, play all,' and '6-deck, back-counting.' Down the side: 'unrestricted betting' *[Editor's note: now called 'Case 1 – Benchmark'],* 'total camouflage,' 'partial camouflage 1' (turn some on, some off), and 'partial camouflage 2' (reverse of 1). For each scenario: an average bet size, win rate in %, and win per 100 hands played (or seen). Standard deviations for everything would be nice, too!" I'll leave it up to the reader to decide if we've delivered; but trust me, it's all here (and standard errors, too!).

Some quick background, in case you missed the September 1994 article (in fact, even if you didn't, now would be a good time to review the material, before continuing): We use the Hi-Lo, with –10 to +10 index numbers. Penetration is 75%. I've changed from head-on play to a more realistic two other players at the 2-deck game and three others at the 6-deck varieties. Obviously, this will affect win rates. Bet spread at the 2-deckers is: TC < 2, one unit; true of 2 to < 3, two units; true of 3 to < 4, four units; TC ≥ 4, six units. For 6-deck, ditto all of the above, then two hands of six units with TC ≥ 5. Finally, for back-counting, we don't enter the game until we have a true of +1 or higher.

Let's review the "total camouflage" scheme (Case 2, "Schlesinger's Total

Camouflage"): never increase your bet after a loss; never decrease your bet after a win; don't change your bet after a push; never bet more than a parlay after a win; never take down more than 50% of your bet after a loss. Some pretty serious restrictions — we all agree. Their costs are now fully documented, as Case 2, in each table. For Case 3, we "switched off" the bet-jumping restrictions. It was now permitted to vary your bet with mathematical conformity to your advantage, except that you still had to wait for a win to raise, and a loss to lower, the subsequent wagers. Finally, Case 4 restored the bet-jumping (up or down) prohibitions, but now permitted raising the bet after a loss (or push) and lowering it after a win (or push).

Findings: I hope that the tables on p. 109 are self-explanatory. I don't have to read them to you. But a few observations are in order. First, to reiterate my original advice: The total camouflage package is very costly. You need to follow my pp. 105–6 suggestions as to how to recapture some of the forgone profits. Next, it is clear that, of the two major categories of camouflage, failure to size bets correctly, by jumping (Case 4), is not as important as never raising at all after a loss or decreasing after a win. Needing "permission" to raise at all (even if it then can be "jumped"), as seen in Case 3, is, seemingly, the most damaging form of camouflage. Intuitively, this makes sense. When we *do* get to raise our bet, especially in the 6-deck games, we rarely get to "step it up" two or three categories at a time, anyway. To quote the famous Griffin *(Theory of Blackjack)* chapter subheading, "Opportunity Arises Slowly in Multiple Decks." Note, however, that the increase over total camouflage, when bet-jumping is allowed, is much greater in the double-deck games. Here, it doesn't much matter which of the two forms we switch back on; each restores a hefty portion of the total camouflage losses (52.1% and 55.7%, respectively).

As with many of my studies, the door is wide open for future enhancements. There still remain many variations on the themes. For example, we can incorporate other numbers of decks (one, eight), different bet spreads and camouflage methodologies, and, of course, different count systems. Better still — *you* can get some of the aforementioned software and customize a chart to fit your exact style of play. It will save me (and George!) some time, and you'll undoubtedly have some fun, in the bargain.

To recap: It is generally acknowledged that some form of betting camouflage is necessary if you play on a very frequent basis and/or for reasonably high stakes. Knowing the precise costs associated with the most common forms of camouflage is, obviously, essential if you are to make an intelligent decision as to which varieties to employ. I hope that the current article will help you to achieve this objective.

Postscript: *As mentioned in the chapter, I am indebted to Stanford Wong and John Imming for the enhancements to their simulators that permitted the foregoing studies* *to take place. Ironically, I believe that the capacity for blackjack software to handle betting camouflage has become more or less an "industry standard" today. I am glad to have been the one who provided the impetus for the inclusion of these features. We shall return to some of these concepts during our analysis of the data contained in Chapter 10.*

TABLE 7.4

2-Deck, Play All

	Hands (millions)	Advantage (%)	Units Won/ 100 hands	Avg. Bet (units)	S.D. (units)	SE (%)
Case 1	20.0	1.61	2.82	1.75	27.83	0.06
Case 2	25.8	0.93	1.42	1.52	21.85	0.04
Case 3	22.4	1.26	2.15	1.71	27.13	0.06
Case 4	26.2	1.34	2.20	1.64	24.35	0.05

TABLE 7.5

6-Deck, Play All

	Hands (millions)	Advantage (%)	Units Won/ 100 hands	Avg. Bet (units)	S.D. (units)	SE (%)
Case 1	23.7	0.63	0.91	1.44	23.95	0.05
Case 2	20.7	0.35	0.48	1.37	21.09	0.04
Case 3	18.0	0.39	0.56	1.44	23.05	0.05
Case 4	19.0	0.50	0.72	1.44	23.14	0.06

TABLE 7.6

6-Deck, Back-Counting*

	Hands (millions)	Advantage (%)	Units Won/ 100 hands	Avg. Bet (units)	S.D. (units)	SE (%)
Case 1	18.0	2.00	1.40	2.80	21.75	0.05
Case 2	19.8	1.46	1.10	3.01	19.54	0.04
Case 3	20.2	1.61	1.17	2.91	21.09	0.05
Case 4	20.6	1.90	1.40	2.95	21.14	0.05

* 25 rounds per hour actually played.

Explanation of Tables

Case 1: Benchmark (see p. 107)
Case 2: Schlesinger's Total Camouflage
Case 3: Same as Case 2, but bet-jumping OK
Case 4: Same as Case 2, but raising bets after loss or lowering after win OK

March 1993 Cover Art by Judy Robertson

Chapter 8

Risk of Ruin

"The best discussion I have seen of risk of ruin is an article by Don Schlesinger in the March 1993 issue of *Blackjack Forum.*"
— *Stanford Wong*

"Statistics have been used to prove and disprove anything and everything from the certainty of life on other planets to the existence of God. As blackjack players, we're looking for guidance on comparatively mundane matters. Should I hit? Should I stand? Should I split? But, the biggest question, the question that comes before all other questions, that precedes should I hit, should I stand, should I split … the question that Adam first asked himself that fateful day in Eden is … 'How much should I bet?'

"Alas, poor Adam understood nothing about his risk of ruin. Enter Don Schlesinger. From now on, when anyone asks me a question about risk of ruin, I'm referring him to this issue of *Blackjack Forum.* You want risk of ruin? We've got risk of ruin. … Read it. Study it. Use it. This one article contains everything you always wanted to know about risk of ruin. The answer to your Big Question: 'How much should I bet?' is right here in your hands!"
— *Arnold Snyder, excerpted from the "Sermon," March 1993* ***Blackjack Forum***

There is much to be excited about with the following material. For one thing, with the addition, at the conclusion of the chapter, of the never-before published "trip ruin formula," I feel I have provided the final piece of the "risk of ruin puzzle," which has plagued players and researchers alike for so many years. Simply stated, I believe the studies that follow represent the most complete and accurate information available anywhere on the subject of risk of ruin for the blackjack player. [Note: And with four brand-new formulas for this third edition, that information is now more complete than ever! — D.S.]

From the March 1993 *Blackjack Forum:*

I have wanted to write this article for a long time. I'll spare you the details as to why it's taken so long to get this information into print. Suffice it to say that I am fulfilling a promise I made to blackjack author George C. over three years ago to produce the numbers and accompanying analysis you're reading here today. Before we begin, a brief history of the evolution of this piece is appropriate.

Before 1989, every attempt I had ever seen to quantify risk of ruin for blackjack came with a qualifying statement to the effect that, because we weren't dealing with a simple plus or minus one situation for the payoffs, and especially because we varied our bet size along the way, classical "gambler's ruin" formulas did not apply perfectly to blackjack. In particular, Allan Wilson's and Ken Uston's descriptions of risk of ruin ran along those lines and, therefore, left us wanting more.

In 1989, in the first of two very fine papers, George C. published, on p. 8 of "How to Win $1 Million Playing Casino Blackjack," the "Ruin Formula":

$$\text{RUIN} = \left(\frac{1 - \frac{\text{w}}{\text{sd}}}{1 + \frac{\text{w}}{\text{sd}}} \right)^{\frac{\text{bank}}{\text{sd}}}$$

where *w* was the hourly win rate, *sd* the hourly standard deviation, and *bank* the player's bankroll in units. George also said, at the time, that he wouldn't give a lot of coverage to the topic, because he and I were going to "combine our efforts to produce what will be the most accurate coverage on the subject in the near future." Well, ... almost! Just a while later, that same year, George got tired of waiting for me and produced "Gambler's Ruin: The Most Accurate Numbers Ever Generated on This Subject." And, indeed, they were! The problem was, I don't think enough people got to read George's paper, which he marketed privately to, what one would assume was, a rather limited clientele.

Finally, in the summer of 1990, Patrick Sileo, of Carnegie Mellon University, presented, in London, at the 8th International Conference on Gambling and Risk Taking, a fascinating paper entitled, "The Evaluation of Blackjack Games Using a Combined Expectation and Risk Measure." In his work, reviewed by Carlson Chambliss in the December 1990 *Blackjack Forum,* Sileo presented the identical ruin formula that George C. had used, and he went on to give various simplified versions of the equation, once certain general assumptions were accepted by the reader. Two of those equations, below, were used to create the tables (8.1 and 8.2) that follow:

$$r \approx e^{\frac{-2E\beta}{\sigma^2}} \text{ and } \beta \approx -\frac{\sigma^2}{2E} \ln r.$$

So where do I come in, and how do the charts presented here advance our knowledge of this vitally important topic? A fair question! Both Sileo and George C. offered ruin figures for some of the "standard" approaches to various blackjack games. They picked a count (usually Hi-Lo), a set of rules, number of decks, level of penetration and, of course, a betting strategy. Then, they offered ruin figures for the games analyzed. George C. also included the very useful probabilities of being ahead after playing different numbers of hours at the games studied.

So what is left to be done? Simply this: The present tables permit you to determine for any approach you choose and for any conditions imaginable, the risk of ruin associated with the bankroll size of your choice! (See Table 8.1.) Alternatively, you may begin with a predetermined, desired risk of ruin and consult Table 8.2 to find out how many units of bank you will need for the safety you require. I have taken the already existing formulas and provided comprehensive spreadsheets that ought to cover just about any approach to any game you can think of. *[Note: See also the Chapter 10 tables. — D.S.]*

What I haven't done is to work out the expectations and standard deviations for you. Even back in 1989, George C. recognized the immensity of this endeavor when he wrote, "Donald Schlesinger and I are in the process of doing the numbers on what seems to be a giant task. There are so many games still being dealt today with different bet schemes and cut points, etc." So, I'm asking you to meet me halfway.

The two finest sources of information for determining the precise expectations and standard deviations per hand for the game you are currently playing are the newly developed *Blackjack Count Analyzer,* from Stanford Wong, and the five-book series, *Beat the X-Deck Game,* from Arnold Snyder. Although the latter presents literally hundreds of possible scenarios, with the win rates and s.d.s already worked out, the new *Blackjack Count Analyzer* is even more flexible in that it permits any count, any bet scheme, any set of rules, and any number of decks. You run the simulation and read the results. Of course, you need a computer. I assume that if you've invested in a computer, you'll automatically get Wong's software. A serious blackjack player would be crazy not to have it! Those without will have to rely on Snyder's numbers, and they are very well done.

For future editions, I am hereby offering to both Wong and Snyder that revised versions of their works may incorporate my charts directly, so that all the loose ends may be tied together. *[Again, since this article was written, Karel Janecek has developed his* ***Statistical Blackjack Analyzer (SBA)*** *software, which will also provide the data you need. And, don't forget the upcoming Chapter 10! Additionally, with* ***BJRM 2002,*** *John Auston has created the single most complete, easy-to-use tool for analyzing virtually any question involving risk of ruin. It's an absolutely must-have piece of software. Finally, see p. 145 for information about Norm Wattenberger's suite of risk-of-ruin formulas, now incorporated into his* ***Casino Vérité*** *software. — D.S.]*

How to Use the Tables

First, determine, for the game you're playing, what your per-hand expectation and standard deviation are. Now, you have a choice. If you already have an idea of what you're calling your bankroll (let's say 400 units), simply turn to the "Bankroll = __ Units" charts and select the appropriate one. Find the intersection of your expectation (horizontal numbers) and s.d. (vertical numbers) and, *voilà,* there's your risk of ruin! If your numbers fall between those listed, you may interpolate to get a reasonably close approximation to the exact values (or, use the equations directly). As the functions that define risk are not linear (that is, the graph would be a curve, and not a straight line), interpolation will not produce a precise answer.

Let's look at a quick example. Suppose expectation is .02 units per hand while s.d. is 2.50. For 400 units of bank, risk of ruin is given as 7.7%. Let's work this from another viewpoint. Say you decide to accept 8% (roughly one chance in 12) as your risk of ruin. You now turn to the "Risk of Ruin =" charts and find .08. At the intersection of .02 and 2.50, you'll find 395, which tells you that, for your desired level of risk, you'll need 395 units of bank. Neat, huh?!

A few observations are in order. First, there is a need to explain exactly what risk of ruin means as used throughout this article. As defined here, it means that you continue to play at the predetermined bet levels no matter where the fluctuations of your bankroll may take you. Theoretically, you continue to play until it's clear that you've built an insurmountable lead and will never lose your original bank, or ... you're wiped out. Now, I understand that most players don't play this way. Many, for example, if they lose half their bank, adjust their stakes downward (often to half) to re-establish the original bet-to-bankroll ratio. And certainly, this is a sound idea. If you play this way, then a simple adjustment to the tables will give you a fair estimate of your ultimate risk of ruin.

It is a well-known mathematical fact that, if you play perfect Kelly Criterion (and we never do, for a variety of reasons), the probability of doubling a bank before halving it is 67%. So, with perfect Kelly, you would halve your bank one-third of the time and then cut back your stakes to half. If you never cut back again despite possible future losses, your risk of ruin, from that point on, would be the same as that originally given by the tables; only, from the beginning, there was only a one-third chance that you would ever reach that level. So, your overall risk of ruin, using this approach, would be only one-third the value of the number in the table.

Now, here's a great trick. To get the exact risk of halving your bank for the particular parameters you're using, simply take the square root of your original risk! Thus, if you've determined that your acceptable ruin level is, say, .16 (16%), the square root of .16 is .40 (or 40%). This represents the probability of halving your bank. We now multiply one risk by the other (.16 x .40 = .064), and there you have it — the ultimate risk (6.4%) of tapping out if you cut back to half stakes after half your bank is lost. Note how we can use the charts in a variety of ways to check the validity of the square-root method. Look at the "Bankroll = 600 Units" chart, with expectation = .01 and s.d. = 2.0. Risk of ruin is .050. Now, to find the risk of losing half the bank (300 units), we can do one of three things: 1) Go to the "Bankroll = 300 Units" chart and, under .01, 2.0, find risk = .223 (which, of course, with a slight rounding error, is the square root of .050); 2) Stay on the 600-unit chart but halve your expectation, to .005, while keeping s.d. = 2.0. Read risk as .223; 3) Stay on the 600-unit chart and double both expectation and s.d., to .02 and 4.0, respectively. Again, read risk as .223.

All of this suggests that if you're comfortable with ultimate risk of, say, .05, you ought to look at the charts for risk of about .136 (use the "Risk of Ruin = .14" chart), because the square root of .136 = .369, and .136 x .369 = 0.05.

In closing, I wish to express my thanks to colleague Bill Margrabe who substantiated the validity of the "Ruin Formula" by comparing results to his own model developed for options pricing. And, finally, my public acknowledgment to George C., who started me on this journey and certainly shares credit for whatever good may come from this article.

TABLE 8.1
Risk of Ruin

Bankroll = 200 Units

Expectation per hand (in units)

SD/Hnd	.005	.010	.015	.020	.025	.030	.035	.040	.045	.050
1.50	0.411	0.169	0.069	0.029	0.012	0.005	0.002	0.001	0.000	0.000
1.75	0.520	0.271	0.141	0.073	0.038	0.020	0.010	0.005	0.003	0.001
2.00	0.607	0.368	0.223	0.135	0.082	0.050	0.030	0.018	0.011	0.007
2.25	0.674	0.454	0.306	0.206	0.139	0.093	0.063	0.042	0.029	0.019
2.50	0.726	0.527	0.383	0.278	0.202	0.147	0.106	0.077	0.056	0.041
2.75	0.768	0.589	0.452	0.347	0.267	0.205	0.157	0.121	0.093	0.071
3.00	0.801	0.641	0.513	0.411	0.329	0.264	0.211	0.169	0.135	0.108
3.25	0.828	0.685	0.567	0.469	0.388	0.321	0.266	0.220	0.182	0.151
3.50	0.849	0.721	0.613	0.520	0.442	0.375	0.319	0.271	0.230	0.195
3.75	0.867	0.752	0.653	0.566	0.491	0.426	0.370	0.321	0.278	0.241
4.00	0.883	0.779	0.687	0.607	0.535	0.472	0.417	0.368	0.325	0.287

Bankroll = 300 Units

Expectation per hand (in units)

SD/Hnd	.005	.010	.015	.020	.025	.030	.035	.040	.045	.050
1.50	0.264	0.069	0.018	0.005	0.001	0.000	0.000	0.000	0.000	0.000
1.75	0.375	0.141	0.053	0.020	0.007	0.003	0.001	0.000	0.000	0.000
2.00	0.472	0.223	0.105	0.050	0.024	0.011	0.005	0.002	0.001	0.001
2.25	0.553	0.306	0.169	0.093	0.052	0.029	0.016	0.009	0.005	0.003
2.50	0.619	0.383	0.237	0.147	0.091	0.056	0.035	0.021	0.013	0.008
2.75	0.673	0.452	0.304	0.205	0.138	0.093	0.062	0.042	0.028	0.019
3.00	0.717	0.513	0.368	0.264	0.189	0.135	0.097	0.069	0.050	0.036
3.25	0.753	0.567	0.427	0.321	0.242	0.182	0.137	0.103	0.078	0.058
3.50	0.783	0.613	0.480	0.375	0.294	0.230	0.180	0.141	0.110	0.086
3.75	0.808	0.653	0.527	0.426	0.344	0.278	0.225	0.181	0.147	0.118
4.00	0.829	0.687	0.570	0.472	0.392	0.325	0.269	0.223	0.185	0.153

Bankroll = 400 Units

Expectation per hand (in units)

SD/Hnd	.005	.010	.015	.020	.025	.030	.035	.040	.045	.050
1.50	0.169	0.029	0.005	0.001	0.000	0.000	0.000	0.000	0.000	0.000
1.75	0.271	0.073	0.020	0.005	0.001	0.000	0.000	0.000	0.000	0.000
2.00	0.368	0.135	0.050	0.018	0.007	0.002	0.001	0.000	0.000	0.000
2.25	0.454	0.206	0.093	0.042	0.019	0.009	0.004	0.002	0.001	0.000
2.50	0.527	0.278	0.147	0.077	0.041	0.021	0.011	0.006	0.003	0.002
2.75	0.589	0.347	0.205	0.121	0.071	0.042	0.025	0.015	0.009	0.005
3.00	0.641	0.411	0.264	0.169	0.108	0.069	0.045	0.029	0.018	0.012
3.25	0.685	0.469	0.321	0.220	0.151	0.103	0.071	0.048	0.033	0.023
3.50	0.721	0.520	0.375	0.271	0.195	0.141	0.102	0.073	0.053	0.038
3.75	0.752	0.566	0.426	0.321	0.241	0.181	0.137	0.103	0.077	0.058
4.00	0.779	0.607	0.472	0.368	0.287	0.223	0.174	0.135	0.105	0.082

TABLE 8.1
(continued)

Bankroll = 500 Units
Expectation per hand (in units)

SD/Hnd	.005	.010	.015	.020	.025	.030	.035	.040	.045	.050
1.50	0.108	0.012	0.001	0.000	0.000	0.000	0.000	0.000	0.000	0.000
1.75	0.195	0.038	0.007	0.001	0.000	0.000	0.000	0.000	0.000	0.000
2.00	0.287	0.082	0.024	0.007	0.002	0.001	0.000	0.000	0.000	0.000
2.25	0.372	0.139	0.052	0.019	0.007	0.003	0.001	0.000	0.000	0.000
2.50	0.449	0.202	0.091	0.041	0.018	0.008	0.004	0.002	0.001	0.000
2.75	0.516	0.267	0.138	0.071	0.037	0.019	0.010	0.005	0.003	0.001
3.00	0.574	0.329	0.189	0.108	0.062	0.036	0.020	0.012	0.007	0.004
3.25	0.623	0.388	0.242	0.151	0.094	0.058	0.036	0.023	0.014	0.009
3.50	0.665	0.442	0.294	0.195	0.130	0.086	0.057	0.038	0.025	0.017
3.75	0.701	0.491	0.344	0.241	0.169	0.118	0.083	0.058	0.041	0.029
4.00	0.732	0.535	0.392	0.287	0.210	0.153	0.112	0.082	0.060	0.044

Bankroll = 600 Units
Expectation per hand (in units)

SD/Hnd	.005	.010	.015	.020	.025	.030	.035	.040	.045	.050
1.50	0.069	0.005	0.000	0.000	0.000	0.000	0.000	0.000	0.000	0.000
1.75	0.141	0.020	0.003	0.000	0.000	0.000	0.000	0.000	0.000	0.000
2.00	0.223	0.050	0.011	0.002	0.001	0.000	0.000	0.000	0.000	0.000
2.25	0.306	0.093	0.029	0.009	0.003	0.001	0.000	0.000	0.000	0.000
2.50	0.383	0.147	0.056	0.021	0.008	0.003	0.001	0.000	0.000	0.000
2.75	0.452	0.205	0.093	0.042	0.019	0.009	0.004	0.002	0.001	0.000
3.00	0.513	0.264	0.135	0.069	0.036	0.018	0.009	0.005	0.002	0.001
3.25	0.567	0.321	0.182	0.103	0.058	0.033	0.019	0.011	0.006	0.003
3.50	0.613	0.375	0.230	0.141	0.086	0.053	0.032	0.020	0.012	0.007
3.75	0.653	0.426	0.278	0.181	0.118	0.077	0.050	0.033	0.021	0.014
4.00	0.687	0.472	0.325	0.223	0.153	0.105	0.072	0.050	0.034	0.024

Bankroll = 700 Units
Expectation per hand (in units)

SD/Hnd	.005	.010	.015	.020	.025	.030	.035	.040	.045	.050
1.50	0.045	0.002	0.000	0.000	0.000	0.000	0.000	0.000	0.000	0.000
1.75	0.102	0.010	0.001	0.000	0.000	0.000	0.000	0.000	0.000	0.000
2.00	0.174	0.030	0.005	0.001	0.000	0.000	0.000	0.000	0.000	0.000
2.25	0.251	0.063	0.016	0.004	0.001	0.000	0.000	0.000	0.000	0.000
2.50	0.326	0.106	0.035	0.011	0.004	0.001	0.000	0.000	0.000	0.000
2.75	0.396	0.157	0.062	0.025	0.010	0.004	0.002	0.001	0.000	0.000
3.00	0.459	0.211	0.097	0.045	0.020	0.009	0.004	0.002	0.001	0.000
3.25	0.515	0.266	0.137	0.071	0.036	0.019	0.010	0.005	0.003	0.001
3.50	0.565	0.319	0.180	0.102	0.057	0.032	0.018	0.010	0.006	0.003
3.75	0.608	0.370	0.225	0.137	0.083	0.050	0.031	0.019	0.011	0.007
4.00	0.646	0.417	0.269	0.174	0.112	0.072	0.047	0.030	0.019	0.013

TABLE 8.1
(continued)

Bankroll = 800 Units
Expectation per hand (in units)

SD/Hnd	.005	.010	.015	.020	.025	.030	.035	.040	.045	.050
1.50	0.029	0.001	0.000	0.000	0.000	0.000	0.000	0.000	0.000	0.000
1.75	0.073	0.005	0.000	0.000	0.000	0.000	0.000	0.000	0.000	0.000
2.00	0.135	0.018	0.002	0.000	0.000	0.000	0.000	0.000	0.000	0.000
2.25	0.206	0.042	0.009	0.002	0.000	0.000	0.000	0.000	0.000	0.000
2.50	0.278	0.077	0.021	0.006	0.002	0.000	0.000	0.000	0.000	0.000
2.75	0.347	0.121	0.042	0.015	0.005	0.002	0.001	0.000	0.000	0.000
3.00	0.411	0.169	0.069	0.029	0.012	0.005	0.002	0.001	0.000	0.000
3.25	0.469	0.220	0.103	0.048	0.023	0.011	0.005	0.002	0.001	0.001
3.50	0.520	0.271	0.141	0.073	0.038	0.020	0.010	0.005	0.003	0.001
3.75	0.566	0.321	0.181	0.103	0.058	0.033	0.019	0.011	0.006	0.003
4.00	0.607	0.368	0.223	0.135	0.082	0.050	0.030	0.018	0.011	0.007

Bankroll = 900 Units
Expectation per hand (in units)

SD/Hnd	.005	.010	.015	.020	.025	.030	.035	.040	.045	.050
1.50	0.018	0.000	0.000	0.000	0.000	0.000	0.000	0.000	0.000	0.000
1.75	0.053	0.003	0.000	0.000	0.000	0.000	0.000	0.000	0.000	0.000
2.00	0.105	0.011	0.001	0.000	0.000	0.000	0.000	0.000	0.000	0.000
2.25	0.169	0.029	0.005	0.001	0.000	0.000	0.000	0.000	0.000	0.000
2.50	0.237	0.056	0.013	0.003	0.001	0.000	0.000	0.000	0.000	0.000
2.75	0.304	0.093	0.028	0.009	0.003	0.001	0.000	0.000	0.000	0.000
3.00	0.368	0.135	0.050	0.018	0.007	0.002	0.001	0.000	0.000	0.000
3.25	0.427	0.182	0.078	0.033	0.014	0.006	0.003	0.001	0.000	0.000
3.50	0.480	0.230	0.110	0.053	0.025	0.012	0.006	0.003	0.001	0.001
3.75	0.527	0.278	0.147	0.077	0.041	0.021	0.011	0.006	0.003	0.002
4.00	0.570	0.325	0.185	0.105	0.060	0.034	0.019	0.011	0.006	0.004

Bankroll = 1000 Units
Expectation per hand (in units)

SD/Hnd	.005	.010	.015	.020	.025	.030	.035	.040	.045	.050
1.50	0.012	0.000	0.000	0.000	0.000	0.000	0.000	0.000	0.000	0.000
1.75	0.038	0.001	0.000	0.000	0.000	0.000	0.000	0.000	0.000	0.000
2.00	0.082	0.007	0.001	0.000	0.000	0.000	0.000	0.000	0.000	0.000
2.25	0.139	0.019	0.003	0.000	0.000	0.000	0.000	0.000	0.000	0.000
2.50	0.202	0.041	0.008	0.002	0.000	0.000	0.000	0.000	0.000	0.000
2.75	0.267	0.071	0.019	0.005	0.001	0.000	0.000	0.000	0.000	0.000
3.00	0.329	0.108	0.036	0.012	0.004	0.001	0.000	0.000	0.000	0.000
3.25	0.388	0.151	0.058	0.023	0.009	0.003	0.001	0.001	0.000	0.000
3.50	0.442	0.195	0.086	0.038	0.017	0.007	0.003	0.001	0.001	0.000
3.75	0.491	0.241	0.118	0.058	0.029	0.014	0.007	0.003	0.002	0.001
4.00	0.535	0.287	0.153	0.082	0.044	0.024	0.013	0.007	0.004	0.002

TABLE 8.2
Bankroll Requirements

Risk of Ruin = .03

Expectation per hand (in units)

SD/Hnd	.005	.010	.015	.020	.025	.030	.035	.040	.045	.050
1.50	789	394	263	197	158	131	113	99	88	79
1.75	1074	537	358	268	215	179	153	134	119	107
2.00	1403	701	468	351	281	234	200	175	156	140
2.25	1775	888	592	444	355	296	254	222	197	178
2.50	2192	1096	731	548	438	365	313	274	244	219
2.75	2652	1326	884	663	530	442	379	331	295	265
3.00	3156	1578	1052	789	631	526	451	394	351	316
3.25	3704	1852	1235	926	741	617	529	463	412	370
3.50	4296	2148	1432	1074	859	716	614	537	477	430
3.75	4931	2466	1644	1233	986	822	704	616	548	493
4.00	5610	2805	1870	1403	1122	935	801	701	623	561

Risk of Ruin = .04

Expectation per hand (in units)

SD/Hnd	.005	.010	.015	.020	.025	.030	.035	.040	.045	.050
1.50	724	362	241	181	145	121	103	91	80	72
1.75	986	493	329	246	197	164	141	123	110	99
2.00	1288	644	429	322	258	215	184	161	143	129
2.25	1630	815	543	407	326	272	233	204	181	163
2.50	2012	1006	671	503	402	335	287	251	224	201
2.75	2434	1217	811	609	487	406	348	304	270	243
3.00	2897	1448	966	724	579	483	414	362	322	290
3.25	3400	1700	1133	850	680	567	486	425	378	340
3.50	3943	1972	1314	986	789	657	563	493	438	394
3.75	4527	2263	1509	1132	905	754	647	566	503	453
4.00	5150	2575	1717	1288	1030	858	736	644	572	515

Risk of Ruin = .05

Expectation per hand (in units)

SD/Hnd	.005	.010	.015	.020	.025	.030	.035	.040	.045	.050
1.50	674	337	225	169	135	112	96	84	75	67
1.75	917	459	306	229	183	153	131	115	102	92
2.00	1198	599	399	300	240	200	171	150	133	120
2.25	1517	758	506	379	303	253	217	190	169	152
2.50	1872	936	624	468	374	312	267	234	208	187
2.75	2266	1133	755	566	453	378	324	283	252	227
3.00	2696	1348	899	674	539	449	385	337	300	270
3.25	3164	1582	1055	791	633	527	452	396	352	316
3.50	3670	1835	1223	917	734	612	524	459	408	367
3.75	4213	2106	1404	1053	843	702	602	527	468	421
4.00	4793	2397	1598	1198	959	799	685	599	533	479

TABLE 8.2
(continued)

Risk of Ruin = .06
Expectation per hand (in units)

SD/Hnd	.005	.010	.015	.020	.025	.030	.035	.040	.045	.050
1.50	633	317	211	158	127	106	90	79	70	63
1.75	862	431	287	215	172	144	123	108	96	86
2.00	1125	563	375	281	225	188	161	141	125	113
2.25	1424	712	475	356	285	237	203	178	158	142
2.50	1758	879	586	440	352	293	251	220	195	176
2.75	2128	1064	709	532	426	355	304	266	236	213
3.00	2532	1266	844	633	506	422	362	317	281	253
3.25	2972	1486	991	743	594	495	425	371	330	297
3.50	3446	1723	1149	862	689	574	492	431	383	345
3.75	3956	1978	1319	989	791	659	565	495	440	396
4.00	4501	2251	1500	1125	900	750	643	563	500	450

Risk of Ruin = .08
Expectation per hand (in units)

SD/Hnd	.005	.010	.015	.020	.025	.030	.035	.040	.045	.050
1.50	568	284	189	142	114	95	81	71	63	57
1.75	774	387	258	193	155	129	111	97	86	77
2.00	1010	505	337	253	202	168	144	126	112	101
2.25	1279	639	426	320	256	213	183	160	142	128
2.50	1579	789	526	395	316	263	226	197	175	158
2.75	1910	955	637	478	382	318	273	239	212	191
3.00	2273	1137	758	568	455	379	325	284	253	227
3.25	2668	1334	889	667	534	445	381	333	296	267
3.50	3094	1547	1031	774	619	516	442	387	344	309
3.75	3552	1776	1184	888	710	592	507	444	395	355
4.00	4041	2021	1347	1010	808	674	577	505	449	404

Risk of Ruin = .10
Expectation per hand (in units)

SD/Hnd	.005	.010	.015	.020	.025	.030	.035	.040	.045	.050
1.50	518	259	173	130	104	86	74	65	58	52
1.75	705	353	235	176	141	118	101	88	78	71
2.00	921	461	307	230	184	154	132	115	102	92
2.25	1166	583	389	291	233	194	167	146	130	117
2.50	1439	720	480	360	288	240	206	180	160	144
2.75	1741	871	580	435	348	290	249	218	193	174
3.00	2072	1036	691	518	414	345	296	259	230	207
3.25	2432	1216	811	608	486	405	347	304	270	243
3.50	2821	1410	940	705	564	470	403	353	313	282
3.75	3238	1619	1079	810	648	540	463	405	360	324
4.00	3684	1842	1228	921	737	614	526	461	409	368

TABLE 8.2
(continued)

Risk of Ruin = .12

Expectation per hand (in units)

SD/Hnd	.005	.010	.015	.020	.025	.030	.035	.040	.045	.050
1.50	477	239	159	119	95	80	68	60	53	48
1.75	649	325	216	162	130	108	93	81	72	65
2.00	848	424	283	212	170	141	121	106	94	85
2.25	1073	537	358	268	215	179	153	134	119	107
2.50	1325	663	442	331	265	221	189	166	147	133
2.75	1603	802	534	401	321	267	229	200	178	160
3.00	1908	954	636	477	382	318	273	239	212	191
3.25	2240	1120	747	560	448	373	320	280	249	224
3.50	2597	1299	866	649	519	433	371	325	289	260
3.75	2982	1491	994	745	596	497	426	373	331	298
4.00	3392	1696	1131	848	678	565	485	424	377	339

Risk of Ruin = .14

Expectation per hand (in units)

SD/Hnd	.005	.010	.015	.020	.025	.030	.035	.040	.045	.050
1.50	442	221	147	111	88	74	63	55	49	44
1.75	602	301	201	151	120	100	86	75	67	60
2.00	786	393	262	197	157	131	112	98	87	79
2.25	995	498	332	249	199	166	142	124	111	100
2.50	1229	614	410	307	246	205	176	154	137	123
2.75	1487	743	496	372	297	248	212	186	165	149
3.00	1770	885	590	442	354	295	253	221	197	177
3.25	2077	1038	692	519	415	346	297	260	231	208
3.50	2408	1204	803	602	482	401	344	301	268	241
3.75	2765	1382	922	691	553	461	395	346	307	276
4.00	3146	1573	1049	786	629	524	449	393	350	315

Risk of Ruin = .16

Expectation per hand (in units)

SD/Hnd	.005	.010	.015	.020	.025	.030	.035	.040	.045	.050
1.50	412	206	137	103	82	69	59	52	46	41
1.75	561	281	187	140	112	94	80	70	62	56
2.00	733	367	244	183	147	122	105	92	81	73
2.25	928	464	309	232	186	155	133	116	103	93
2.50	1145	573	382	286	229	191	164	143	127	115
2.75	1386	693	462	346	277	231	198	173	154	139
3.00	1649	825	550	412	330	275	236	206	183	165
3.25	1936	968	645	484	387	323	277	242	215	194
3.50	2245	1122	748	561	449	374	321	281	249	224
3.75	2577	1289	859	644	515	430	368	322	286	258
4.00	2932	1466	977	733	586	489	419	367	326	293

From "The Gospel According to Don," *Blackjack Forum*, March 1994:

Q. *What did you think of Idiot Savant's proposals regarding money management and trip bankroll requirements as presented in the December 1993* ***Blackjack Forum?***

A. "Idiot Savant" (IS, from here on) made an honest attempt to address an interesting topic (trip bankroll requirements and risk of losing the trip b.r. during a specified number of hours of play). Unfortunately, he fell victim to a mathematical trap that I alluded to in Chapter 4 (pp. 48–49), namely the "premature bumping into the barrier syndrome."

Because of this error in his methodology, I'm afraid most of the probabilities he presents are incorrect. In the pages that follow, I would like to propose an alternative solution to IS's problems while attempting to resolve some of the confusion that still may linger regarding proper bet-sizing and bankrolling.

To begin, let's mention that the concept of subdividing one's total blackjack-playing bankroll into smaller "trip" and even "session" bankrolls is certainly not original with IS's article. In 1973, in a privately sold $25 supplement to *Playing Blackjack as a Business,* Lawrence Revere proposed dividing the total playing b.r. into four parts, each of which would then be subdivided into 30 betting units. The problem here was that Revere was undoubtedly alluding to single-deck requirements, as 120 starting units is, unquestionably, insufficient capitalization for most of today's blackjack games.

In 1983, in the December *Blackjack World* (pp. 196–99), Stanford Wong discussed "Risk in Atlantic City Blackjack," and here, too, most of the data in the study were rendered useless, this time, by an unfortunate computer glitch. Recognizing the error (alas, several years later), Wong re-did the work and published the amended article in the March 1990 *BJF.* We shall return to some of his findings in a little while. Finally, when I was teaching the Blackjack Clinic for Jerry Patterson in the early 1980s (when early surrender was offered in A.C.), I introduced into the course materials the concept of trip and session bankrolls. Only here, recognizing the requirements as substantially larger than those suggested by Revere, I advocated, for most trips, dividing the total bankroll in half (IS would call this his "drawdown fraction," f = 1/2) and then subdividing that smaller amount into five session-bankrolls of 30 units each. Thus, I was recommending at least 300 units of bank, which I feel, is still inadequate for most of today's games.

Let's get down to some of the math so I can show you where IS went astray. Each game we play, coupled with our style of betting, produces a per-hand expected return and standard deviation. *BJF* readers will recall that the determination of these two parameters was both necessary and sufficient to produce the risk of ruin tables I generated for my March 1993 article. These calculations also played a major role in creating my Tables 2.2 and 2.3 *[Chapter 2]* for the "Ups and Downs of Your Bankroll" article in the June 1985 *Blackjack Forum,* where I documented probabilities of being ahead after n hours of play and analyzed win rates and standard deviations for various blackjack games.

Suppose we examine IS's numbers from his Chart 1, p. 35 (December 1993 *Blackjack Forum*). The per-hand expectation was .0385 units and the s.d. per hand was 3.5 units. As IS explains, if we play a certain number of hands on our trip (say 5,000), the expectation is a linear function (5,000 x .0385= 192.5), but the s.d. is a square-root function (therefore, $\sqrt{5,000}$ x 3.5 = 247.5). As it is also true that a one-s.d. or greater loss will occur roughly 16% of the time, had IS concluded that, *at the end of 5,000 hands of play* (if we are lucky enough to survive that long), we have a 16% chance of losing 247.5 – 192.5 = 55 or more units, he would have been correct. But IS claimed something very different. He concluded (erroneously, I'm afraid) that this 16% probability represented the chances that our maximum drawdown *for any time during* the entire 5,000-hand trip was 16%. It isn't! In fact, it's probably *double* that percentage. And IS makes this mistake throughout the entire article, applying the same logic to other probabilities of supposed intermediate losses. Each time, for example, that we see an entry of .16 in the charts, we are supposedly looking at a one-s.d. loss for the period. But what we might expect to lose *during* the period is the more salient number.

Enter the "premature bumping into the barrier syndrome"! Since many of you may not have been around in 1989 for my previous discussion, let me set the scene with the following analogy. Suppose I am standing just shy of ten paces to the right of the edge of the Grand Canyon. To my left, the abyss; to my right, plenty of good old terra firma! Now, suppose each time we toss a coin and it land heads, I move, with safety, one pace to the right. But for each tail, I stride one pace to the left, closer and closer to my final destiny.

I ask the following question: "If I flip the coin a maximum of 100 times, what is the probability that, at the end of the experiment, my mortal remains will be floating in the Colorado River?" Let's see how IS (and perhaps many of you) might have attacked the problem: "Since probability, p, of moving left or right = 1/2, my ultimate expected resting place is right where I'm standing now. But the standard deviation of the 100 tosses is $\sqrt{100}$ = 10; so a one-s.d. move to the left makes me part of the landscape! Since a one-s.d. move occurs roughly 16% of the time, that is how often I will meet my maker." Right? Wrong! In fact, I will meet my maker with exactly double the 16% frequency, or 32% of the time. Surprised? Permit me to explain.

It isn't enough to consider probabilities of "hitting the barrier" (or more specifically, the Canyon floor!) from an endpoint-only perspective. What if I take the fatal step on, say, the 70th toss? If there had been someone there to catch me and push me back, I might have followed with a series of heads, gotten my feet planted firmly on the ground, and never looked back (or down!). And, because of the $p = 1/2$ coin symmetry, I can expect to do just that exactly 50% of the times that I am "teetering on the brink." There's just one catch: No one is there to push me back.

Cut to blackjack. I have an expectation (slightly positive) for each hand. I have an s.d., as well. In addition, I have a limited amount of cash I have decided to take with me on this trip. If I calculate from an "endpoint" perspective only, I lose sight of all the times during the trip that I might have "taken the plunge" (read: "tapped out") only to bounce back and wind up solvent if some benefactor had come along when I was broke and lent me a few bucks for the next couple of hands.

I first pointed out this phenomenon to none other than Peter Griffin in September of 1987. In a letter to me, he wrote: "I doubted it at first, but not after I figured [it out]." Griffin went on to summarize: "Let E be the event that the player is at or below B (the "barrier") after exactly N trials, and F the event that the player has been at or below B at some stage up to and including the Nth trial. Then for large N, the following is approximately true: For $p > 1/2$, $P(E) < 1/2\ P(F)$, $P(F) > 2P(E)$; for $p = 1/2$, $P(E) = 1/2\ P(F)$, or $P(F) = 2P(E)$; and for $p < 1/2$, $P(E) > 1/2\ P(F)$, or $P(F) < 2P(E)$." Then, with characteristic candor and refreshing modesty, Griffin postscripted, "I learned something."

Here's what we can all learn from the above. The probability of being behind by *at least* a pre-specified amount at some time during a trip is a little more than double the probability of being behind by that same amount at exactly the end of that specified time. It is this premature bumping into the barrier that IS failed to reckon with in his analysis.

I am currently working with a colleague who models applications for calculating theoretical values of "exotic" options, including so-called "barrier" options. The principles involved are similar to those of our blackjack problem, although securities involve a continuous, "lognormal" distribution, whereas blackjack's distribution is discrete and more or less "normal." Let's hope that we'll be able to produce charts or, at the very least, a workable formula for analysis of short-term bankroll requirements. I am not a big fan of this way of thinking, but if the readers are interested, there's no harm in providing the data.

Apart from the central concept, there are many valid points made in the IS piece. Any time someone outlines how mentally grueling playing blackjack can be, I applaud him. It takes an inordinate amount of self-control to endure the ravages of standard deviation when we are losing. Betting wisely and conservatively is the cornerstone and hallmark of a successful approach to the game. My two all-time favorite books on blackjack are Revere's *Playing Blackjack as a Business* and Wong's *Professional Blackjack* (Pi Yee Press, 1980). *[Note: Since this writing, the new 1994 edition has appeared. — D.S.]* Each contains a statement I have often quoted while describing card counting to others. Coincidentally, each remark is found on p. 102 of the respective works. Wong writes, "You must be cold and calculating. Steely blue eyes will do." And later, "Know thyself — check your emotional readiness to lose before you set up a betting plan." Revere, less eloquent, is nonetheless right on the money when he intones, "This is a discipline thing."

Let's conclude with a computer simulation to verify our above-outlined thesis. In the March 1990 *BJF* piece, Wong describes the A.C. weekend player who bets $5 on true counts of zero to under one, $20 from one to under two, and $50 on two and above. The player walks on negative shoes. Wong simulates both 6- and 8-deck play for 1,000 and 4,000 hands (four simulations), and he is seeking the maximum drawdown experienced by the player under each scenario. I calculated, for the 6-deck game, the per-hand expectation and s.d. They were, respectively, $0.24 and $32.70. Thus, for 1,000 hands, expectation is $240 and s.d. is $1,034. At the *end* of the 1,000 hands, therefore, a one-s.d. loss would be $1,034 – $240 = $794. In the column labeled "Most dollars ever behind," next to "$800 or more," do we find 16%? Not at all! Instead, we read the now-expected 31.8%. For 4,000 hands, expectation is $960 and s.d. is $2,068. This time, a one-s.d. loss is $1,108. Not all too surprisingly, we find the "$1,100 or more" drawdown to be somewhat greater than twice 16% — a hefty 42.6%. What about two-s.d. drawdowns? They should occur not 2.3% of the time (area in the "tail" to the left of two s.d.s), but rather close to 4.6% of the time, in accordance with our tenets. Well, for case one, an extra s.d. loss would put us at –$1,828. Wong reckons the probability at about 3.5%. For case two, a two-s.d. loss is $3,176. Interpolating, we get a probability of about 4.3%. You get the idea.

To conclude: I liked Idiot Savant's article. I certainly do not agree with all of his observations and we know that the math is flawed; but he makes several valid points concerning psychological aspects of the game. Many of you may now wish to examine this approach using the corrected version of the math that has been provided in this piece. As mentioned above, I hope to have more data for you in the next issue.

Good luck, and … good cards.

From "The Gospel According to Don," *Blackjack Forum*, December 1994:

Q. *In your March 1994 "Gospel," when you discussed trip bankroll requirements and short-term risk of ruin calculation, you promised, for a future issue, detailed charts and more data on the topic. Have you done further work in this area?*

A. Yes, I have. You'll recall that I had hoped for a closed-form mathematical solution that would apply for any general case, but I still don't have that. *[Note: Patience! It's coming after all! — D.S.]* What I do have, however, is a rather detailed analysis of short-term (trip) bankroll requirements depending on one's tolerance for losing the entire stake and the number of hours (or hands) one intends to play on the trip. I analyze two different games, for four different trip lengths and five levels of probability of losing the trip bankroll. In all, there are 40 separate trip stakes that are calculated. What's more, I'll explain how to generalize the process so that you may do your own calculation for any style of play, any game, any number of hours of play. I'll also supply some guidelines based on my own style of play and risk tolerance.

Let's get started. I'm proposing that we study a two-deck game, Strip rules, Hi-Lo count, 75% penetration, and a 1–6 spread (max bet at true count ≥ 4). Also, we'll consider a six-decker, back-counting, 1–12 spread (max bet at true ≥ 5), Strip rules, Hi-Lo count, 75% penetration. The play-all double-deck will yield 100 hands per hour with expectation (μ) of about 2.50 units and standard deviation (σ) of 27 units, while the back-counting game (27 hands *played* per hour) has μ = 1.68 units and σ = 24.32 units.

Next, let's consider some typical trip lengths (I'm purposely choosing lengths that are "perfect squares" so that you'll follow the s.d. calculations more easily): 1) 16 hours of play (call it a weekend); 2) 25 hours (three to five days); 3) 36 hours (five to seven days); and 4) 49 hours (seven to ten days).

The final consideration, before we get to the math, will be a range of "acceptable probabilities of losing the entire trip stake." In other words, we're going to ask you for your personal level of risk aversion; that is, "I wouldn't want to lose all this money I'm taking with me more than once in every ___ trips." If, for example, you fill in the blank with a 10, then that's a 10% probability of losing all the money *at some time* during the trip. If you'll go for one in five, that's the 20% probability level, etc.

Methodology: For each of the two games described, I calculate the mean and standard deviation of each trip length. For example, 2-deck, 16 hours, yields μ = 16 x 2.50 = 40 units, and $\sigma = \sqrt{16}$ x 27 = 4 x 27 = 108 units, respectively. Now comes the tricky part (directly from the March article). The cumulative normal distribution tells us the probability associated with the "tails" of the aforementioned percentages. But this is *not* what we want. For example, if I accept a 20% probability of loss of trip stake, I need to consult the standard normal entry for a "tail" of 10%, in accordance with the hypothesis that the probability of losing a specified amount at any time *during* the trip is approximately double the "end" probability. (In fact, it is actually slightly *more* than double, for our positive-expectation blackjack games; but I'm going to ignore that fact in the hopes of not making this any more complicated than need be.) The resulting entry in Table 8.3, below, is 1.28 standard deviations (to the left — a loss). In a similar fashion, I associate the appropriate s.d. losses for each probability level. They are, respectively: 1.96, 1.65, 1.28, 0.84, and 0.67.

The final calculation of the required trip bankroll is simply the appropriate number of s.d.s, multiplied by the trip-length σ, reduced by the trip expectation. Staying with our same example, for 20% risk of ruin, 1.28 (s.d.s) x 108 (the 16-hour σ) – 40 (the 16-hour μ) = 138.24 – 40 $\approx$ 98 units. Note, also, that I have chosen to express all stakes in terms of *starting* units. Others may prefer to express these guidelines as multiples of the *maximum,* rather than the minimum, bet. If this is more convenient for you, simply divide all "Required Trip Bankroll" numbers in the 2-deck charts by 6 and all 6-deck numbers by 12.

Conclusion: If I go away for a long weekend (16 hours) of double-deck play, and

TABLE 8.3

Trip Bankroll Requirements (in units)

(Preliminary Findings. See Table 8.4 for more accurate results)

			Probability of Ruin				
playing hours	**μ**	**σ**	**5%**	**10%**	**20%**	**40%**	**50%**
2-Deck[1]							
16	40	108	172	138	98	51	33
25	63	135	202	160	110	51	33
36	90	162	228	177	117	51	33
49	123	189	248	189	119	51	33
Number of hours until max. loss:			112	79	48	21	13
6-Deck[2]							
16	27	97	164	134	98	55	38
25	42	122	196	159	114	60	40
36	60	146	226	180	126	62	40
49	82	170	251	199	136	62	40
Number of hours until max. loss:			201	143	86	37	24

For bankroll requirements in shaded areas, maximum loss is expected to occur prior to end of trip.

μ = expectation in units (table entries rounded)

σ = standard deviation in units (table entries rounded)

[1] 2-Deck Assumptions: 75% dealt; Hi-Lo count; 1–6 spread; 100 hands/hr; μ = 2.50 units/hr; σ = 27.00 units/hr

[2] 6-Deck Assumptions: 75% dealt; Hi-Lo count; 1–12 spread; 27 hands/hr; μ = 1.68 units/hr; σ = 24.32 units/hr

5% probability of ruin = 1.96 s.d.s
10% probability of ruin = 1.65 s.d.s
20% probability of ruin = 1.28 s.d.s
40% probability of ruin = 0.84 s.d.s
50% probability of ruin = 0.67 s.d.s

I accept that once every five such trips I'll lose all the money I bring with me, I should bring roughly 100 starting bets (units). Need more of a comfort zone, tapping out just once every ten trips? Then you'd better bring along at least 138 units.

But there's more to consider. As you can see from the charts, as trip hours increase, for the low probability levels (5% and 10%), bankroll requirements continue to increase as well. That is, for our relatively short (49 hours) longest trip, we have not reached, for either game, the maximum bankroll requirement for the respective probability level. Now, look at the 20%, 40% and 50% entries, as we progress from 16 hours up to 49 hours. Notice what happens? In most cases, the b.r. requirements eventually cap out; that is, they reach a maximum below which it makes no sense to descend. Here's the explanation. In these instances, using the original methodology yields an erroneous (and sometimes ridiculously low) answer.

For example, in the 2-deck game, accepting 40% risk of trip ruin for 49 hours would lead to a b.r. determination of 36 units. But that's impossible, since for a *smaller* number of hours (for example, 16), I already required 51 units of bank for the same 40% level. Since I have to "survive" the 16- (as well as the 25- and 36-) hour levels to be able to reach 49, I'm subjected to the *maximum* requirement for the level, wherever it occurs, from that moment on.

Basically, what's happening here is, if they don't get you early, they're not likely to get you at all. With a limited bank, survive the early going and they probably won't touch you as the hours begin to mount. To calculate these hour levels at which maximum losses occur for the various numbers of standard deviations, we'll need a little calculus. I'll include the derivation below, for the more mathematically minded among you.

If n = the required number of s.d.s (e.g., 1.96, 1.65, etc.), σ = the hourly standard deviation for the game in question, μ = the hourly expectation for the game, and x = the number of hours at which the maximum loss occurs, then we are looking to *minimize* the following expression: $x\mu - \sqrt{x}\, n\sigma$. To do so, we need to take the first derivative of the expression with respect to x, set the result equal to zero, and solve for x. We get: $\mu - n\sigma/ 2\sqrt{x} = 0$, which means: $\mu = n\sigma/2\sqrt{x}$. Squaring both sides yields: $\mu^2 = n^2\sigma^2/4x$, or $x = n^2\sigma^2/ 4\mu^2 = (n\sigma/2\mu)^2$.

This last expression tells us the number of hours required for a maximum loss to occur at any standard-deviation level. Let's try an example. For the 6-deck game, at the 40% (0.84 s.d.) level, why do we max out at 62 units (36-hour entry) and not go any higher at 49 hours? In this case, n = 0.84, σ = 24.32, and μ = 1.68. So x, the number of hours at which a maximum 0.84-s.d. loss occurs, equals $[(0.84 \cdot 24.32)/(2 \cdot 1.68)]^2$ = 37 hours. This tells us not to go below the 62-unit entry (at 36 hours) for any higher number of hours to be played. Such entries are designated by shading in the tables.

Another very important idea to retain is that these trip bankroll requirements are not to be confused with *overall,* total b.r. requirements for playing blackjack on an ongoing basis. I mentioned, in the March article, that I wasn't fond of these artificial "trip" calculations, but that I would provide them as a guideline for readers who find them useful. Although you may decide that taking a 10% risk of losing, say, 160 units is acceptable for a 25-hour trip, should this be *all* the money you have available for playing blackjack in this manner? No, it shouldn't. To ascertain that (larger) number, we need to consult my March 1993 "Risk of Ruin" article *[now p. 112 — D.S.]* where the formula tells us we need about $-\frac{\sigma^2}{2E}$ ln r units of bank to play at the risk level r. For r = .10, solving yields 336 units as the appropriate bankroll. The trip b.r., therefore, was about *half* of what the total stake should be. Why? Because sooner or later (we hope for "later," but what if it's "sooner"?!), even with just a 1 in 10 chance of tapping out, we *will* lose the entire trip stake. And when that happens, we don't want to be out of business. The "reserve" permits us to jump back on the horse and continue playing.

Remember, also, that just because an event has a 10% probability of occurring, this doesn't mean that the average number of trips you'll take to experience your first wipeout will be ten. It isn't. In fact, in this instance, it's between six and seven (6.6). So, half the time, if you play over and over, you'll have your first disaster by the sixth or seventh outing. Ten is the *mean* of the experiment, but 6.6 is the median. You might want to store in your memory, for future reference, that as the number of trials gets larger and the probability of an event's happening is rather small (say 1 in 100), the median is very well approximated by taking .69 and dividing it by the relevant probability. Thus, for an event that happens, on average, once every hundred trials, we expect the "over/under" to be 69 (i.e., .69/.01 = 69). The moral here is that, often, events that are "unlikely" to happen actually *do* happen earlier than we might imagine they will.

With this final thought in mind, let me offer my habitual warning: All the numbers generated here assumed you used a proportional, Kelly betting scheme to achieve the original expectations and standard deviations. Also, no provision was made for cutting back the unit size midway through the trip if you should lose, say, half the stake. I assume you're willing to lose what you brought. Now, let's look at the charts. Consider, for example, the 36-hour, 6-deck entries. Say I accept 20% risk of losing a 126-unit trip bankroll. Now suppose I do something dumb and decide, from the outset, to *double* my original bet size, effectively cutting my stake to only 63 units of this higher, base wager. Immediately, I have *doubled* my risk of tapping out on this trip from 20% to 40% (62 units). *[Note: See p. 131; actual risk is even greater than these preliminary findings indicated. — D.S.]* I shouldn't have to drum this into you anymore: DON'T OVERBET! Don't listen to anyone who tells you to bet more than 100% Kelly. You should probably bet much *less* than 100% Kelly! Remember: Play perfectly, but bet stupidly (and double Kelly is just plain dumb), and you have *absolutely* nothing!

Interestingly enough (at least for me), the figures confirm levels of bankroll that I have instinctively used for many years. For a typical junket-type trip, where I expect to play 20 to 25 hours, I will usually bring 150 units of bank with me. Notice that, regardless of the game (2-deck or 6-deck), I am playing at about the 10% level of losing the entire stake. That is my comfort zone. You're entitled to a different one, but I haven't lost too many trip bankrolls these past twenty years! In fact, it's probably been somewhat fewer than 1 in 10, because I play in a more conservative manner than the assumptions of "pure" Kelly (see the camouflage discussion). And, of course, in keeping with all of the above, I consider my total playing bankroll to be at least 400 units. Ask any savvy Wall Street pro, whose trading has stood the test of time, what the secret to success is and, invariably, he will tell you: "For every trade, don't worry about how much you can make; just consider how much you might lose." You could do worse than to apply these words of wisdom to your blackjack money management.

From "The Gospel According to Don," *Blackjack Forum*, March 1995:

Q. *I think your "Trip Bankroll Requirements" article in the last* ***BJF*** *broke new ground while shedding light on an important topic. Have you refined the technique in any way since December?*

A. Have I ever! In fact, I have obtained the long-sought-after formula that will yield the risk of ruin for any trip length and trip bankroll size. I am indebted to a former colleague, who wishes to remain anonymous (we'll call him c.c.), for this historic breakthrough. After a few discussions with c.c., it was clear to me that we were on the right track. I am confident that we have found the correct approach.

I am amending and expanding the chart that appears on p. 127 (from the December 1994 *BJF*). A few comments are in order. I did state, in both the September and December articles, that, because of the upward "drift" due to the blackjack player's positive expectation, the actual ruin probabilities were somewhat *more* than double the "final" probabilities. But (a) I purposely ignored this effect, and (b) for some cases, I underestimated how much more than double these numbers could be. It turns out that for the case of, say, 49 hours, the risk of tapping out prior to the end of the trip can be as much as *triple* the final probabilities.

Another revelation of the formula is that, as hours increase, the "final" (or endpoint perspective) risk of ruin decreases, after having peaked, but the ultimate risk of losing one's bank must keep on increasing until, in the limit, it approaches the risk of ruin numbers associated with "playing forever" that were generated in my March 1993 article. As a result, the risk of ever losing one's bank on the way to a 100- or 400-hour play can easily be upward of five to ten times the "final" probabilities, depending on the desired level of risk.

Table 8.4 can serve as a sample of much more that will come in the near future. The entries are self-explanatory. Again, one final cautionary word. Note how, in each chart, the b.r. requirements for the 40% ruin level for moderate trip lengths are just slightly less than half those of the 10% level. Stated another way, if one doubles his customary bet size, thereby effectively halving his trip stake, the risk of tapping out nearly quadruples! If this isn't your idea of fun, don't even think of betting any multiple of Kelly. *[Note: The effect on ruin of doubling the bet, in these final findings, is even more severe than in our preliminary estimate, on p. 129. — D.S.]*

Stay tuned. There's definitely going to be more on this topic.

TABLE 8.4
Trip Bankroll Requirements (in units)
(Final Results)

playing hours	μ	σ	Probability of Ruin 5%	10%	20%	40%	50%
2-Deck[1]							
1	2.5	27	51	42	33	21	17
4	10	54	98	81	62	40	33
9	23	81	140	116	88	56	44
16	40	108	179	147	111	69	54
25	63	135	214	175	131	81	63
36	90	162	246	200	149	91	70
49	123	189	274	222	164	99	76
100	250	270	340	271	196	116	89
400	1000	540	427	330	231	132	100
6-Deck[2]							
1	1.68	24.32	46	39	30	19	16
4	7	49	95	80	62	41	33
9	15	73	131	108	83	53	42
16	27	97	169	139	106	67	53
25	42	122	204	168	127	80	62
36	60	146	237	194	146	59	71
49	82	170	268	218	163	100	78
100	168	243	346	278	205	123	95
400	672	486	485	379	268	155	118

μ = expectation in units (table entries rounded)

σ = standard deviation in units (table entries rounded)

[1] 2-Deck Assumptions: 75% dealt; Hi-Lo count; 1–6 spread; 100 hands/hr

[2] 6-Deck Assumptions: 75% dealt; Hi-Lo count; 1–12 spread; 27 hands/hr

The Short-Term, or Trip, Ruin Formula

Well, enough beating around the bush. It has been nearly two years since the charts on the previous page were generated using my short-term, or "trip," risk of ruin formula. It is now time to share that formula with the blackjack-playing community. I saved it, if you will, as a "bonus" for readers of the second edition of my book.

Before I reveal the formula, there is an immediate caveat: Although extremely accurate, it is not 100% "correct," in a rigorously mathematical way. You can't imagine how many highly qualified "quants" I approached with this problem, in an attempt to find what is referred to as a "closed-form" solution. If you have not seen one, all of these years, it's probably because none exists! The difficulty lies in the fact that the formula is intended to be applied to "continuous" processes, whereas playing blackjack involves a series of "discrete" plays. By assuming, nonetheless, that play takes place in a somewhat continuous, rather than discrete, manner, we obtain altogether satisfactory, if not "perfect" results. *[Note: For much more on this concept, and for four new formulas, see "The Formulas Are Put to the Test," later in this chapter. — D.S.]*

I might note, in passing, that, as an analogous situation, in the financial world, for nearly 25 years, investors have been valuing stock options using the famous "Black-Scholes" formula, which assumes, among other things, that stocks trade in a continuous process. In point of fact, they don't; but everyone uses the formula anyway, and it has become a universally accepted "standard" in the industry and the literature. Such, I believe, should be the case with the following short-term risk of ruin formula, which I now present for the first time. Let:

N ( ) = the cumulative normal distribution function of ( ),
e = the base of the natural logarithm system (≈ 2.7183),
B = the short-term bankroll, expressed as a number of starting units,
ev = expected value, or win, for the trip, in units,
sd = standard deviation of the win, in units, and
var = variance of the win (the square of the sd), in units squared.

Then, the risk, r, of losing one's entire bankroll at any time during the trip is given by:

$$r = N((-B - ev)/sd) + e^{((-2 \times ev \times B)/var)} \times N((-B + ev)/sd).$$

Note that the quantities inside the parentheses will usually be negative numbers, for relatively short trips, since the bankroll, B, that we take with us will normally be considerably larger than our expected win, ev. Note, as well, that we should be familiar with the first two "pieces" of the formula. The first simply expresses the probability of losing the entire bankroll at the *end* of the required number of hours of play, while the second is our old friend, the long-term risk of ruin formula, from earlier in this chapter. (See top of p. 113.)

A few observations are in order, before we proceed to a practical example. We have been stating, all along, that previous attempts to quantify analytically this short-term risk of ruin have been erroneous, because they usually consider just the "end result" mentality while failing to take into account the interim period, as well. The formula shows us, quite clearly, that the actual "trip" risk is equal to the "final" risk *increased* by some percentage of the ultimate, long-term risk. Intuitively, this makes good sense.

In addition, let us observe that, if the trip length (expressed as number of hours, or hands) gets very large, our e.v. will increase much faster than our s.d., causing the first term, N ((–B – ev)/sd) to tend toward zero, while N ((–B + ev)/sd) will tend toward one. This leaves us with $r = e^{((-2 \times ev \times B)/var)}$, as we would expect.

Finally, as one more "reasonableness" check, note that when ev = 0, as would be the case for our coin-tossing "Grand Canyon experiment" (see pp. 123–24), the final risk, r, becomes N (–B/sd) + e^0 x N (–B/sd) = N (–B/sd) + N (–B/sd) = 2N (–B/sd). This result is in keeping with our reflection principle, which states that the risk of "hitting the barrier" at any time during the period is *precisely* double the risk of being there at the *end* of the period.

Any weakness in the accuracy of the formula is probably to the conservative side; that is, it may tend to slightly overstate the true risk of ruin. Again, this appears to be because we are assuming that play is continuous when, in fact, it occurs in discrete "coups" that we call "hands." *[Note: Again, see confirmation of this hypothesis, later in the chapter, when the formulas are "put to the test." — D.S.]*

A single application of the formula, by way of illustration, should suffice. In our next two chapters, literally dozens of scenarios will provide the reader with an almost endless stream of data to supply to the formula, if such short-term ruin information should be desired. For now, let's consider a four-day junket to our favorite Caribbean resort. Las Vegas Strip rules will be in effect, with s17 and das. What's more, we'll face a nicely cut 6-deck shoe, where five decks are dealt. A quick glance ahead to Table 10.44, p. 237, tells us that, with our customary 1-to-12 spread, we can expect to win 2.33 units (of $14.99 each) per hour (100 hands played), with a corresponding s.d. of 40.13 units. We'll be playing four hours a day, for four days, on this 16-hour junket. So, the question is, how many units of bank should we bring along?

In keeping with our previous approach (pp. 126–27), we need to answer the "comfort zone" question. Do we mind losing one out of every five such trip stakes? Too upsetting? Will tapping out once out of every ten junkets be acceptable? Probably, it ought to be. I've always made it a rule of thumb, as stated earlier, to take along about half of my normal "total" playing bankroll of 400 units on such trips. Sometimes, instead of 200 units, I'll get a little bold and take only 150. So, now, the question is: With these two sizes of trip bankroll, what risk of ruin am I facing? Let's apply the formula to each case.

For the 150 units, B = 150; ev = 16 x 2.33 = 37.28; sd = $\sqrt{16}$ x 40.13 = 4 x 40.13 = 160.52; var = $(160.52)^2$ = 25,767. For your convenience, I've provided, at the end of the chapter, tables of the cumulative normal probability distribution and the exponential function. So, let's proceed with the calculation.

$$
\begin{aligned}
r &= N((-150 - 37.28)/160.52) + e^{((-2 \times 37.28 \times 150)/25{,}767)} \times N((-150 + 37.28)/160.52) \\
&= N(-1.17) + e^{-.434} \times N(-.702) \\
&= .1210 + .6479 \times .2358 \\
&= .2738
\end{aligned}
$$

Thus, the risk of ruin is just slightly greater than 27% (which, in turn, is, predictably, just over double our "end" ruin value of 12.10%). Note that a calculator was used to find $e^{-.434}$ precisely. Our tables give $e^{-.40}$ and $e^{-.45}$, from which one could come close, by straight interpolation, to the exact value. Alternatively, one could use the original version of the long-term ruin formula (see p. 112), where powers of e do not appear.

For the 200-unit bank, we find:

$$
\begin{aligned}
r &= N((-200 - 37.28)/160.52) + e^{((-2 \times 37.28 \times 200)/25{,}767)} \times N((-200 + 37.28)/160.52) \\
&= N(-1.48) + e^{-.579} \times N(-1.01) \\
&= .0694 + .5605 \times .1562 \\
&= .1569
\end{aligned}
$$

So by bringing an extra 50 units of bank, we've succeeded in almost halving our trip risk of ruin, from 27% to just under 16%. Many would consider the peace of mind obtained from this extra "cushion" well worth the effort. And calculations show that bringing another 35 units of bank would reduce trip risk even further, to the 10% level.

We hope to have provided the reader with a powerful, never-before published, new tool: the trip ruin formula. Just follow the above steps, by plugging in the variables for your favorite game, and you will know just how many units of bank you'll need to feel "comfortable" on your next blackjack-playing excursion.

***Postscript**: Once again, at my request, Stanford Wong modified his **BCA** to be able to handle inquiries into "trip ruin" calculations. His "nadir" entry was created to furnish trip ruin information in simulation mode. I am, nonetheless, excited over the prospect that my analytic formula will become generally accepted and, eventually, widely used by the blackjack-playing community. Finally, there is an entire "Trip Risk of Ruin" module, with comprehensive data, in **BJRM 2002** and an impressive array of ruin and goal-reaching calculators in Norm Wattenberger's **CVCX** (more about this later).*

Expanding Our Horizons: Four New "Double-Barrier" Formulas

You may have noticed that, for each of the two ROR formulas presented on the preceding pages, there was no upper goal, which, if reached, would have caused the player to either stop playing or, in some manner, to reassess his current situation. For example, had the goal been attained, a team's joint bankroll might have been "broken," or, upon reaching a goal, an individual player might have decided to change his betting stakes. As soon as we introduce the notion of an upper boundary, in addition to our lower boundary (in each case, zero, which means we have lost all of our bank and are ruined), the formulas for calculating not only risk of ruin, but also for successfully reaching the goal before being ruined, become extremely complex. Indeed, I was in possession of such a "double-barrier" formula before the second edition of *BJA* went to print, but I purposely chose not to include the math in that revised edition, for fear that it was too intimidating and complex. Well, I've changed my mind!

Not only do I now plan to furnish the aforementioned formula, but I will also throw in, free of charge (!), three related formulas that can be derived from the new one. Thus, I'll present four new "double-barrier" formulas that will permit you to ascertain answers to ruin and goal-reaching problems that have never before been considered in any work on blackjack.

A bit of background is in order, before we begin. As I mentioned earlier, in the field of finance, there is an area of options theory that deals with the pricing of so-called "exotic" options. The term refers to options that have some form of non-standard features or payoffs that differentiate them from "plain-vanilla" options. One such group of exotic options is called "double-barrier," because the instruments involve a payoff that takes on one value should the underlying security (on which the option's price is based) reach a certain upper limit, and another payoff (often zero) should the security touch a second, lower boundary. How does all of this relate to blackjack? Well, stocks have some expected return, or "drift," just as we have a positive e.v. for the blackjack game we're playing. And stocks have a standard deviation, just as our games have s.d.s and variances. So, is it not reasonable to assume, when investigating probabilities of reaching an upper goal before going broke, or, conversely, of losing our entire bank before achieving some goal, that these options formulas might apply to the game of blackjack? It turns out that the answer is yes.

Next comes the notion of time. Do we have a time limit or not? Should the formulas specify that we must achieve our upper goal (or possibly go broke trying) within some finite time limit (expressed as number of hands or hours played), or should we assume that we're going to play "forever"? The answer to that question will determine which one of the four formulas we will use. For, it is clear that, in keeping with the above scheme, we have four variations on the theme. What are the chances that I: 1) reach an upper goal before going broke, given a certain time constraint; 2) go broke, before reaching an upper goal, given a time constraint; 3) reach an upper goal before going broke, with no time limit; or 4) go broke, before reaching an upper goal, with no time limit? (For the sake of completeness, I reiterate that both of the formulas presented earlier in this chapter explore ROR with and without a time constraint, but with no upper goal.) Here, then, are the formulas that provide answers to the four questions above. (I am indebted to two friends, both brilliant financial engineers, Professors Peter Carr and Kim Lee, for their valuable assistance not only in locating the formulas but also in helping me to understand how they are applied.)

Now, I already told you that I withheld the following, because, frankly, I was afraid that it would scare you to death. The math faint of heart might want to take a deep breath before continuing to read. Suppose I possess some bankroll, B, and have some goal, G (G > B). What's more, I wish to attain that goal by time T (expressed in number of hands played). In addition, the blackjack game I am playing has positive per-hand e.v. (expectation) μ and per-hand variance (the square of standard deviation) σ^2. Bankroll, goal, expectation, and standard deviation are all expressed in starting units of bank. Then the probability of reaching goal G by time T, before going bankrupt at B = 0, is:

Formula 1

$$e^{\frac{\mu(G-B)}{\sigma^2}}\left\{\left[\frac{\sinh\left(\frac{\mu}{\sigma^2}B\right)}{\sinh\left(\frac{\mu}{\sigma^2}G\right)}\right]-\frac{\sigma^2}{G^2}\sum_{k=1}^{\infty}\frac{e^{-\delta_k T}}{\delta_k}\left[k\pi\sin\left(\frac{k\pi(G-B)}{G}\right)\right]\right\}$$

$$\text{where } \sinh(x)=\frac{e^x-e^{-x}}{2}, \text{ and } \delta_k=\frac{1}{2}\left(\frac{\mu^2}{\sigma^2}+\frac{k^2\pi^2\sigma^2}{G^2}\right)$$

There, I've said it! You may now breathe again! Obviously, it is beyond the scope of this text to go into some of the math contained in the above formula. I'm not going to get into infinite series and hyperbolic sines, all of which you may find in any standard math texts, if you have the courage to do the research. So, for most of you, you're either going to have to take my word for figuring out some of the RORs and goal-reaching probabilities I'm about to explore, or you can go one step further and obtain some blackjack software products that now contain all of this chapter's formulas as an extra added feature. More about these genial applications in a little while.

Suppose, then, that I'm playing a 4.5/6, s17, das, ls game, with a 1–16 spread, and a play-all approach. From the Chapter 10 Table 10.51, p. 240, I learn that my per-hand e.v. is .0342 units, while my per-hand s.d. is 5.60 units. I have a $10,000 bank (the optimal unit size is $10.91), and I'm wondering what the chances are that I might hit a goal of $15,000 (I win $5,000) in my next 50 hours of play, without, of course, tapping out in the process. So, in the above formula, B is \$10,000/\$10.91 = 916.59 units, G is \$15,000/\$10.91 = 1,374.89 units, μ is .0342, σ^2 is 5.60^2, or 31.36, and e and π are the well-known mathematical constants, equal to approximately 2.7183 and 3.1416, respectively. Plugging in the values, I learn that my chances of obtaining the goal are 38.62%; that is, if I play this game for 50 hours, I have an almost 39% probability of being ahead by \$5,000 *at some point* up to and including the final hour.

The ability to perform such calculations can be of great relevance to teams, who often wonder what their chances are of doubling a joint bankroll, for example, without tapping out, and within a pre-specified number of hours. Here's an example: The team has a \$100,000 bank (unit is now \$109.10) and wonders what the chances are that the bank will be doubled within the next 100 hours. The answer? 24.33%. Obviously, there are numerous variations on the theme, and, a little later, we'll explore several additional scenarios.

But, what about the dark side of the above? Suppose we're interested not in our likelihood of achieving an upper goal, but rather in the probability of going bankrupt while attempting to reach that goal. Well, that requires the second of our formulas, which, as it turns out, is simply a modification of the above monstrosity. Actually, the derivation of this second formula is rather elegant, as we invoke a form of symmetry to turn the first equation into the second. (Thanks are due to Kim Lee for pointing this out to me.) Consider that, in the above equation, I have edge μ, and am attempting to win G – B dollars before losing B dollars. Well, then, what would my adversary be attempting? With edge now $-\mu$, he would be attempting to win my bank, B, from me, before he would "lose," when I would reach my upper goal of G – B dollars. So, all I need to do to turn the above goal-reaching formula into the required ROR formula is to interchange the expressions G – B and B (that is, wherever I see one expression I replace it with the other), and change μ to $-\mu$, in the exponent of e, at the very beginning of the formula. The result is the second of our double-barrier formulas, namely, the probability of going broke, within a specified period of time, before reaching a goal:

Formula 2

$$e^{\frac{-\mu B}{\sigma^2}}\left\{\left[\frac{\sinh\left(\frac{\mu}{\sigma^2}(G-B)\right)}{\sinh\left(\frac{\mu}{\sigma^2}G\right)}\right]-\frac{\sigma^2}{G^2}\sum_{k=1}^{\infty}\frac{e^{-\delta_k T}}{\delta_k}\left[k\pi\sin\left(\frac{k\pi B}{G}\right)\right]\right\}$$

Again, let's try an example. As our team attempts to double its \$100,000 bank in the above-mentioned 100 hours of play, what is the likelihood that it will tap out before reaching the goal? The formula advises that ROR is 3.30%. Note that, as with the goal-reaching scenario, we stop playing when any one of three criteria are met: 1) We reach our goal, 2) We reach the end of our time limit, or 3) We lose the entire bank.

Until now, our new formulas have concentrated on reaching a goal or going broke, within the confines of a certain time limit. But, what if we relax that time constraint? What if we do not specify an amount of time that we are interested in playing? Suppose, instead, that we wish to know if we might reach a certain goal, or tap out trying, regardless of how long either might take. Well, in those instances, the T value in our equation tends to infinity; that is, it has no upper bound. And, conveniently, when we permit T to become infinitely large, the value of the entire infinite series part of our equations tends to zero and, therefore, drops out. How very fortuitous! So, we're left with two modifications of the original formula, and this time, it is actually possible to use a calculator to work through some of the math.

Let's begin with a goal-reaching example. Suppose we return to our original scenario of the 4.5/6, s17, das, ls game, with a 1–16 spread, and a play-all approach. Suppose, again, that we wish to turn our $10,000 bank into $15,000; however, this time, I do not specify 50 hours of play, or any time limit at all. I simply would like to know what my chances are of turning a 50% profit on my bank ($10,000 becomes $15,000), no matter how long it may take to do so. Intuitively, I know that my answer will be considerably greater than the previously obtained value of 38.62%, because I'm not required to reach my goal within the 50-hour time period designated in our first example. Here is the new, streamlined formula for reaching my goal, G – B, before going bankrupt (B = 0), with no time constraint:

Formula 3

$$e^{\frac{\mu(G-B)}{\sigma^2}}\left[\frac{\sinh\left(\frac{\mu}{\sigma^2}B\right)}{\sinh\left(\frac{\mu}{\sigma^2}G\right)}\right]$$

This time, let's actually try to work out the math together. Beginning to the right, we evaluate $(\mu/\sigma^2)B$ and $(\mu/\sigma^2)G$. The first expression is (.0342/31.36) x 916.59028 = .9995978, while the second becomes (.0342/31.36) x 1,374.88543 = 1.4993967. (Please pardon the seven-decimal-place accuracy. I need it to demonstrate a principle a little later on.) On to the hyperbolic sine functions. Sinh (.9995978) = $(e^{.9995978} - e^{-.9995978})/2$ = 1.1745807, while sinh (1.4993967) = $(e^{1.4993967} - e^{-1.4993967})/2$ = 2.1278606. So, the quotient of the two hyperbolic sines becomes 1.1745807/2.1278606 = .5520008. Now, for the first part of the formula. In units, our goal, G – B, is 5,000/10.91 = 458.2951, while μ/σ^2 is, once again, 0.0342/31.36 = .0010905. The product is 458.2951 x .0010905 = .4997708, and $e^{.4997708}$ = 1.6483434. We're almost done! Multiplying 0.5520008 by 1.6483434, we get 0.9098869 = 90.98869%. And that's our final answer: Given that we may play "forever," we have almost a 91% chance of turning our original stake of $10,000 into $15,000.

Well, then, what's the flip side to the above? If we can play for as long as we want, what is the probability that we'll go broke as we attempt to reach the $15,000 level? We turn to our second original equation, above, the one that deals with risk of ruin, and, once again, we simply ignore the infinite series that deals with the time constraint. So, the modified formula becomes:

Formula 4

$$e^{\frac{-\mu B}{\sigma^2}}\left[\frac{\sinh\left(\frac{\mu}{\sigma^2}(G-B)\right)}{\sinh\left(\frac{\mu}{\sigma^2}G\right)}\right]$$

Thus, with no time limit, what is the chance that we'll be unsuccessful in reaching our $15,000 goal and that we'll go broke, instead? Well, you know the drill by now. Once again, we'll start with the hyperbolic sines. Note that the denominator (2.1278606) is identical to our previous one, so there's no extra work to do. The numerator has changed, however we have already worked out the value: $(\mu/\sigma^2)(G - B) = .0010905 \times 458.2951 = .4997708$. Sinh (.4997708) = $(e^{.4997708} - e^{-.4997708})/2 = .5208369$. So, the quotient of the two hyperbolic sines is .5208369 /2.1278606 = .2447702. Finally, the coefficient of this expression is now $e^{-\mu B/\sigma^2} = e^{-.9995978} = .3680274$. And, .3680274 x .2447702 = .0900821 = 9.00821%, which is our desired ROR.

There's something interesting about this value. It would appear that we might have ascertained the ROR simply by subtracting the likelihood of reaching our goal (90.98869%) from 100%. That value of 9.01131% is (save for a negligible rounding error of .0031%) virtually identical to our formula result of 9.00821%. Do you see what's going on? Given that we have, in essence, forever to either achieve our goal or go broke trying, those two probabilities should add, in the limit, to 100%. After all, if I have no time constraint, it seems reasonable that I will, with probability 100% (certainty), either reach my goal or go bankrupt in the process of trying.

Note that the concept of adding the goal-reaching probability and the ROR before reaching the goal probability, to obtain 100%, does not work when we have a time constraint. For the team example illustrating our first two new formulas, those probabilities (24.33% and 3.30%) did not sum to 100%. Do you understand why? Given a time limit of 100 hours, the shortfall of 100% – (24.33% + 3.30%) = 72.37% represents the quite substantial probability that, within the given time period, we will neither achieve our goal nor go broke, but rather linger at some intermediate point. And, as we have seen, as we increase the designated time period, this "lingering" probability gets smaller and smaller, tending to zero, in the limit. Indeed, were we permitted to play forever in our quest to double the bank, we learn that we have precisely an 88.07% probability of attaining our goal, while maintaining an 11.93% chance of tapping out.

The Formulas Are Put to the Test

No sooner had I decided that these four new formulas would be a welcome addition to this ROR chapter (already quite complete in its treatment of the topic) than I wondered if their accuracy could be verified by simulation. I turned to Norm Wattenberger (much more about him in Chapter 10) to see if his *Casino Vérité* software would be up to the task. At the time, it wasn't. But, Norm had already incorporated the new formulas into his suite of ROR and goal-reaching calculators, so it was a logical extension of these applications to consider building new ROR simulation modules for his vaunted software.

I was intrigued! Here was the perfect way to put not only our four new formulas, but also the two original no-goal formulas, to the test. Norm would use all six formulas, and we would agree on a slew of scenarios involving many different-sized bankrolls, a wide range of number of hours played, and an assortment of goals and time limits. Norm would build the necessary software and would provide side-by-side comparisons of the simulator's and the formulas' results. Thus, we would put the formulas to the test to see what, if any, shortcomings or deficiencies any of them might have. Of course, my hope was that the applicability of the formulas to the game of blackjack and, therefore, their robustness and accuracy, would be confirmed. What follows, then, are the fruits of our collaborative labor.

Methodology: It is obvious that, to test the accuracy of the formulas, we needed to begin by creating, in *CVData,* a set of base simulations of various games that would serve as the reference sims upon which the subsequent risk-of-ruin or goal-reaching probabilities would be based. We decided on the following four game scenarios, all of which presumed a $10,000 bank and use of the "Illustrious 18" and "Fab 4" indices (other non-counting players at the table used the appropriate basic strategy):

1) Single-deck, dealt to the 35/52 level, h17, one player to our right. A 1–3 bet spread was accomplished using a $25 unit and the following ramp: At TCs (floored) of zero and below, one unit; at +1, two units; and at +2 and higher, 3 units. The two vital pieces of information needed for the formulas are, of course, the win rate, μ, and the standard deviation, σ. They turned out to be 1.675 and 20.658, respectively. *[Note: As Norm made certain assumptions for these special sims, results obtained here should be expected to differ slightly from similar game simulations found in Chapters 9 and 10. However, as the same methodology was used throughout this study, results are internally consistent, and that is of paramount importance. In addition, as we moved from the testing of one formula to the next, because of some of the constraints put on the player's bankroll or number of hands ultimately played, some of*

the expectations and standard deviations, for the same base game, may have changed slightly. — D.S.]

2) Double-deck, dealt to the 70/104 level, s17, das, two players to our right. A 1–6 bet spread, with $25 unit. We bet one unit at TC = 0 or lower; two units at +1; four units at +2; five units at +3; and six units at +4 and higher. $\mu = 2.476$ and $\sigma = 31.125$.

3) Six-deck, dealt to the 4.5/6 level, s17, das, ls, three players to our right. We're back-counting, using a 1–4 spread, and we enter the game as soon as the TC reaches +2. The unit size is $100, and we bet $100 at +2; $125 at +3; $200 at +4 and +5; $300 at +6 and +7; and $400 at +8 and higher. $\mu = .4765$ and $\sigma = 7.052$.

4) Eight-deck, dealt to the 6/8 level, s17, das, three players to our right. We're back-counting, using a 1–4 spread, and we enter the game as soon as the TC reaches +2. The unit size is $75, and we bet $75 at +2; $100 at +3; $150 at +4; $200 at +5 and +6; and $300 at +7 and higher. $\mu = .2969$ and $\sigma = 6.385$.

It was felt that these games corresponded reasonably closely to conditions that are available in the Northern Nevada (1), Las Vegas Strip (2) and (3), and Atlantic City (4) areas, respectively. (In all, 80 billion hands were dealt to generate the above data, while 445 billion hands were simulated to complete the entire study that follows.)

Next would come the actual testing of the six formulas. For the first ROR formula, found on p. 112, we decided simply to ascertain the long-term risk of ruin of each of the four base games. As a reminder, the term "lifetime" refers to the fact that we have no predetermined stopping point on the upside. We play until we win, ostensibly, large amounts of money, or until we lose our entire stake. For this long-term ROR formula, some players often make the erroneous assumption that, should we, for example, double our bank, we stop and begin the process again. There is no such stopping point associated with this first formula. Table 8.5 summarizes our findings.

TABLE 8.5
Lifetime ROR Comparisons

Game	Sim ROR	Eqn 1 ROR (p. 112)	Eqn 2 ROR (p. 113)	% Difference (Eqn 1)
#1	4.21%	4.30%	4.33%	2.09%
#2	12.76%	12.89%	12.94%	1.01%
#3	14.49%	14.67%	14.69%	1.23%
#4	14.20%	14.32%	14.34%	0.85%

Note that, in each instance, two equations are referenced. This is because p. 112 contains a more precise formula than the excellent approximation to that formula, found on p. 113. Whereas I've been using the second, p. 113, risk formula in my ROR discussions and tables, we thought, for accuracy's sake, that the values ascertained from the first, p. 112, formula would be more pertinent. The "% Difference (Eqn 1)" shaded area subtracts the simulation values from the equation 1 (p. 112) values and then forms a percentage of difference by dividing the result by the equation 1 value. Thus, we see that, in all four cases, the sim values are quite close to the predicted, formula outputs. There is little doubt as to the accuracy of the long-term ROR formula.

For the trip ROR formula, found on p. 132, simulations were more extensive. We wanted to test the accuracy of the formula using several different combinations of trip lengths and bankroll sizes. Our game #2 was used to carry out all of the simulations. Table 8.6 furnishes the relevant data.

Across the top, bankroll sizes of 20, 50, 100, 200, 400, and 800 units were assumed. Down the left-hand column, trip lengths of 1, 5, 10, 20, 50, 100, and 400 hours are provided. Of course, the higher hour values might apply more logically to a team

TABLE 8.6

Trip ROR Comparisons

Bankroll (units)

	20			50			100		
Hours	sim	eqn	% diff	sim	eqn	% diff	sim	eqn	% diff
1	44.91%	49.02%	8.38%	8.40%	9.45%	11.11%	0.16%	0.10%	-60.00%
5	71.49%	73.48%	2.71%	39.99%	41.56%	3.78%	11.14%	11.74%	5.11%
10	77.75%	79.30%	1.95%	51.95%	53.27%	2.48%	23.06%	23.90%	3.51%
20	82.62%	83.84%	1.46%	61.77%	62.81%	1.66%	35.14%	36.01%	2.42%
50	86.52%	87.49%	1.11%	70.18%	70.98%	1.13%	48.15%	48.81%	1.35%
100	88.28%	89.16%	0.99%	73.80%	74.60%	1.07%	53.84%	54.52%	1.25%
400	89.63%	90.44%	0.90%	77.34%	77.78%	0.57%	60.14%	60.56%	0.69%

	200			400			800		
Hours	sim	eqn	% diff	sim	eqn	% diff	sim	eqn	% diff
1	0.00%	0.00%	0.00%	0.00%	0.00%	0.00%	0.00%	0.00%	0.00%
5	0.26%	0.25%	-4.00%	0.00%	0.00%	0.00%	0.00%	0.00%	0.00%
10	2.44%	2.56%	4.69%	0.00%	0.00%	0.00%	0.00%	0.00%	0.00%
20	8.68%	9.00%	3.56%	0.16%	0.15%	-4.03%	0.00%	0.00%	0.00%
50	20.47%	20.69%	1.06%	2.28%	2.33%	2.15%	0.00%	0.00%	0.00%
100	27.82%	28.42%	2.11%	6.06%	6.24%	2.88%	0.11%	0.12%	5.17%
400	36.07%	36.61%	1.48%	12.90%	12.97%	0.54%	1.37%	1.42%	3.52%

trip rather than to that of an individual. A session was aborted when the time limit was reached, or when the player went bankrupt. In all, 35 billion hands were dealt to produce the data you are reading. Juxtaposed, are the sim and equation results for risk of ruin, with the percentage of deviation between the two once again highlighted in the shaded area. In virtually all cases, we note that the sims reported slightly lower values than those of the formula, in keeping with the comment I made on p. 133, namely that the formula might tend to slightly overstate the true risk of ruin. And although the comparisons bear out that hypothesis, the results are quite satisfactory, with many scenarios producing extremely close ROR values.

Now the fun would really begin. It was time to put the new formulas up against *CVData,* to see how they fared. I am pleased to report that they acquitted themselves quite admirably! It's a bit tricky to construct a table when three variables are involved, as the printed page is two-dimensional, but we hope that the following format will be satisfactory. First up were the goal-reaching and ROR formulas, where a time constraint was imposed. We examined many different scenarios from a couple of different viewpoints. Table 8.7 summarizes the findings.

Game #3 was used to produce the data. A total of 115 billion hands were simulated for this table. The first two scenarios represented modest banks (perhaps for playing a session or two) with equally modest goals. Note that, in all cases where goals are mentioned, the starting number of units of bank is included in the goal. So, our first row depicts the likelihood that a player with a 20-unit bank, who plays for ten hours, will reach a *total* of 30 units (that is, a 10-unit profit) at some point within the allotted time frame.

The simulation provided information not only on the goal-reaching probabilities but also on those for ROR. In addition, since it was rarely if ever certain that one or the other barrier would be attained, a third category, "PrNeither," documented those instances when we failed to achieve our goal but did not lose our entire stake in the process (that is, we still had part of our bank left at the conclusion of the time period).

For the last five rows, we supposed a near-Kelly-optimal bank of 100 units (remember, this is a back-counting scenario), and attempted to double that bank, given a variety of time constraints that ranged from the rather challenging 30 hours (not much chance to reach the goal — or to go broke, for that matter) to the rather generous 200 and then 2,000 hours, when the probability of success became quite large. Note that ROR also increases as the number of hours grows larger, and that it is the "PrNeither" that shrinks, tending to zero, in the long run. The interpretation is that, given sufficient time to achieve our goal, in the limit, we will either do just that or go broke trying.

Once again, the shaded "% Differences" are well within the range of acceptability, with, as expected, the larger number of hours of play (more like the continuous, rather than the discrete, case) producing the tightest comparisons.

TABLE 8.7

Reach Goal or ROR, with Time Constraints Comparisons

			Simulation			Goal Eqn.			% Differences		
Bankroll	**Hours**	**Goal**	PrReach	ROR	PrNeither	PrReach	ROR	PrNeither	PrReach	ROR	PrNeither
20	10	30	58.90%	19.25%	21.85%	63.90%	20.65%	15.45%	7.82%	6.78%	-41.42%
50	20	80	31.30%	3.67%	65.03%	33.32%	3.53%	63.15%	6.06%	-3.97%	-2.98%
100	30	200	0.66%	0.12%	99.22%	0.56%	0.09%	99.36%	-17.86%	-36.47%	0.13%
100	50	200	4.22%	0.67%	95.11%	4.39%	0.66%	94.95%	3.87%	-1.52%	-0.17%
100	100	200	20.91%	3.17%	75.92%	21.92%	3.31%	74.77%	4.61%	4.23%	-1.54%
100	200	200	49.92%	7.41%	42.67%	50.82%	7.67%	41.51%	1.77%	3.39%	-2.79%
100	2,000	200	86.96%	13.04%	0.00%	86.85%	13.15%	0.00%	-0.13%	0.84%	0.00%

TABLE 8.8

Reach Goal, with No Time Constraints Comparisons

Bankroll (units)

		10			20			50			100			150	
Goal	sim	eqn	% diff	sim	eqn	% diff	sim	eqn	% diff	sim	eqn	% diff	sim	eqn	% diff
50	27.26%	26.24%	-3.89%	49.10%	48.83%	-0.55%	–	–	–	–	–	–	–	–	–
100	18.71%	17.73%	-5.53%	33.65%	32.87%	-2.37%	67.61%	67.47%	-0.21%	–	–	–	–	–	–
200	15.32%	14.35%	-6.76%	27.63%	26.81%	-3.06%	55.44%	54.76%	-1.24%	81.30%	81.17%	-0.16%	93.85%	93.88%	0.03%
400	14.60%	13.62%	-7.20%	26.16%	25.35%	-3.20%	52.32%	51.84%	-0.93%	77.01%	76.68%	-0.43%	89.17%	89.01%	-0.18%

Finally, where we maintained a goal but permitted play to continue with no time limit, we used game #4 to simulate the following, as reported in Table 8.8.

Here, the idea was to furnish different-sized bankrolls, ranging from 10 to 150 units, and to give our player all the time he needed to reach the four goals listed down the left-hand column. Thus, the simulation used its "unlimited number of rounds" mode to model the no-time-constraint feature. In all, some 215 billion rounds were dealt to produce the data.

Once again, as bankroll size increases, the tightness of the fit between simulation and formula becomes readily apparent. And several interesting observations can be made as we peruse the data. First, with only 10 units of bank, we have, nonetheless, a reasonable shot (better than one chance in four) at winning four times our stake and reaching 50 units. Incredibly, we have one chance in four of turning a mere 20 units of bank into 400 units, without going broke. And, at the other end of the spectrum, with a decent stake of 150 units, we learn that reaching 400 units (or, all the money in the world, for that matter) occurs with almost 90% likelihood. We didn't insist on simulating past the 400-unit goal (probability of 89.17% to attain), as it became apparent that, if they didn't get us early, they weren't going to get us at all, as evidenced by the formula's lower limit of 88.74% probability of achieving even the loftiest of goals.

It wasn't necessary to carry out ROR sims for the no-time-constraint scenarios above, because of an interesting phenomenon, discussed before. If you have all the time in the world to reach a goal, you're either going to reach it, or you're going to go broke trying: there is no in between. And so, for all cases documented in Table 8.8, if you'd like to know your risk of ruin, simply subtract the probability of reaching your goal from 100%, and *voilà,* that's your ROR. In each case, a goal of 800 units was more than sufficient to produce an asymptotic result; that is, the percentages for goal-reaching or ROR did not change if we made the goal any higher.

The interrelatedness of the six formulas should also be apparent. For example, it is clear that, for the ROR formula with both goal and time constraint (Table 8.7), if we were to make the goal extremely large, the formula would return the same ROR values as our trip ROR formula, which, in essence, has infinite goal. And, for the ROR formula with a goal but no time constraint, once again, making the goal quite large yields RORs altogether consistent with the first ROR formula, where both time and goal are assumed to be infinite. So, the first of our new formulas presented here can, in a manner of speaking, be considered the "Holy Grail," from which all five of the other formulas can be derived (some, it is true, more simply than others, I'm afraid). And, thanks to Norm Wattenberger, you are now able to access calculators, as part of his software *(CVCX* and *CVData),* that will deliver to your fingertips the end results of scenarios utilizing all six formulas. If you don't already have them, I urge you to explore these incredibly imaginative and useful applications.

I'm delighted to have now presented the full complement of ROR formulas, whose accuracy, as demonstrated by the above simulations, is quite reliable. Years ago, as I worked with options strategies, I was aware of the similarities between the "barrier" formulas, employed to value certain derivatives products, and the risk-of-ruin and goal-reaching problems I was studying for blackjack. It may have taken awhile, but I have finally pulled together all the pieces and confirmed the validity of applying the options equations to our blackjack scenarios. I hope our findings will prove useful to both players and researchers alike. This has been a fascinating study for me, and I once again thank Norm Wattenberger for his invaluable collaboration on this project.

TABLE 8.9

Cumulative Normal Probability Distribution (–3.09 to 0)

	.00	.01	.02	.03	.04	.05	.06	.07	.08	.09
-3.00	0.0013	0.0013	0.0013	0.0012	0.0012	0.0011	0.0011	0.0011	0.0010	0.0010
-2.90	0.0019	0.0018	0.0018	0.0017	0.0016	0.0016	0.0015	0.0015	0.0014	0.0014
-2.80	0.0026	0.0025	0.0024	0.0023	0.0023	0.0022	0.0021	0.0021	0.0020	0.0019
-2.70	0.0035	0.0034	0.0033	0.0032	0.0031	0.0030	0.0029	0.0028	0.0027	0.0026
-2.60	0.0047	0.0045	0.0044	0.0043	0.0041	0.0040	0.0039	0.0038	0.0037	0.0036
-2.50	0.0062	0.0060	0.0059	0.0057	0.0055	0.0054	0.0052	0.0051	0.0049	0.0048
-2.40	0.0082	0.0080	0.0078	0.0075	0.0073	0.0071	0.0069	0.0068	0.0066	0.0064
-2.30	0.0107	0.0104	0.0102	0.0099	0.0096	0.0094	0.0091	0.0089	0.0087	0.0084
-2.20	0.0139	0.0136	0.0132	0.0129	0.0125	0.0122	0.0119	0.0116	0.0113	0.0110
-2.10	0.0179	0.0174	0.0170	0.0166	0.0162	0.0158	0.0154	0.0150	0.0146	0.0143
-2.00	0.0228	0.0222	0.0217	0.0212	0.0207	0.0202	0.0197	0.0192	0.0188	0.0183
-1.90	0.0287	0.0281	0.0274	0.0268	0.0262	0.0256	0.0250	0.0244	0.0239	0.0233
-1.80	0.0359	0.0351	0.0344	0.0336	0.0329	0.0322	0.0314	0.0307	0.0301	0.0294
-1.70	0.0446	0.0436	0.0427	0.0418	0.0409	0.0401	0.0392	0.0384	0.0375	0.0367
-1.60	0.0548	0.0537	0.0526	0.0516	0.0505	0.0495	0.0485	0.0475	0.0465	0.0455
-1.50	0.0668	0.0655	0.0643	0.0630	0.0618	0.0606	0.0594	0.0582	0.0571	0.0559
-1.40	0.0808	0.0793	0.0778	0.0764	0.0749	0.0735	0.0721	0.0708	0.0694	0.0681
-1.30	0.0968	0.0951	0.0934	0.0918	0.0901	0.0885	0.0869	0.0853	0.0838	0.0823
-1.20	0.1151	0.1131	0.1112	0.1093	0.1075	0.1056	0.1038	0.1020	0.1003	0.0985
-1.10	0.1357	0.1335	0.1314	0.1292	0.1271	0.1251	0.1230	0.1210	0.1190	0.1170
-1.00	0.1587	0.1562	0.1539	0.1515	0.1492	0.1469	0.1446	0.1423	0.1401	0.1379
-0.90	0.1841	0.1814	0.1788	0.1762	0.1736	0.1711	0.1685	0.1660	0.1635	0.1611
-0.80	0.2119	0.2090	0.2061	0.2033	0.2005	0.1977	0.1949	0.1922	0.1894	0.1867
-0.70	0.2420	0.2389	0.2358	0.2327	0.2296	0.2266	0.2236	0.2206	0.2177	0.2148
-0.60	0.2743	0.2709	0.2676	0.2643	0.2611	0.2578	0.2546	0.2514	0.2483	0.2451
-0.50	0.3085	0.3050	0.3015	0.2981	0.2946	0.2912	0.2877	0.2843	0.2810	0.2776
-0.40	0.3446	0.3409	0.3372	0.3336	0.3300	0.3264	0.3228	0.3192	0.3156	0.3121
-0.30	0.3821	0.3783	0.3745	0.3707	0.3669	0.3632	0.3594	0.3557	0.3520	0.3483
-0.20	0.4207	0.4168	0.4129	0.4090	0.4052	0.4013	0.3974	0.3936	0.3897	0.3859
-0.10	0.4602	0.4562	0.4522	0.4483	0.4443	0.4404	0.4364	0.4325	0.4286	0.4247
-0.00	0.5000	0.4960	0.4920	0.4880	0.4840	0.4801	0.4761	0.4721	0.4681	0.4641

TABLE 8.10
Cumulative Normal Probability Distribution (0 to 3.09)

	.00	.01	.02	.03	.04	.05	.06	.07	.08	.09
0.00	0.5000	0.5040	0.5080	0.5120	0.5160	0.5199	0.5239	0.5279	0.5319	0.5359
0.10	0.5398	0.5438	0.5478	0.5517	0.5557	0.5596	0.5636	0.5675	0.5714	0.5753
0.20	0.5793	0.5832	0.5871	0.5910	0.5948	0.5987	0.6026	0.6064	0.6103	0.6141
0.30	0.6179	0.6217	0.6255	0.6293	0.6331	0.6368	0.6406	0.6443	0.6480	0.6517
0.40	0.6554	0.6591	0.6628	0.6664	0.6700	0.6736	0.6772	0.6808	0.6844	0.6879
0.50	0.6915	0.6950	0.6985	0.7019	0.7054	0.7088	0.7123	0.7157	0.7190	0.7224
0.60	0.7257	0.7291	0.7324	0.7357	0.7389	0.7422	0.7454	0.7486	0.7517	0.7549
0.70	0.7580	0.7611	0.7642	0.7673	0.7704	0.7734	0.7764	0.7794	0.7823	0.7852
0.80	0.7881	0.7910	0.7939	0.7967	0.7995	0.8023	0.8051	0.8078	0.8106	0.8133
0.90	0.8159	0.8186	0.8212	0.8238	0.8264	0.8289	0.8315	0.8340	0.8365	0.8389
1.00	0.8413	0.8438	0.8461	0.8485	0.8508	0.8531	0.8554	0.8577	0.8599	0.8621
1.10	0.8643	0.8665	0.8686	0.8708	0.8729	0.8749	0.8770	0.8790	0.8810	0.8830
1.20	0.8849	0.8869	0.8888	0.8907	0.8925	0.8944	0.8962	0.8980	0.8997	0.9015
1.30	0.9032	0.9049	0.9066	0.9082	0.9099	0.9115	0.9131	0.9147	0.9162	0.9177
1.40	0.9192	0.9207	0.9222	0.9236	0.9251	0.9265	0.9279	0.9292	0.9306	0.9319
1.50	0.9332	0.9345	0.9357	0.9370	0.9382	0.9394	0.9406	0.9418	0.9429	0.9441
1.60	0.9452	0.9463	0.9474	0.9484	0.9495	0.9505	0.9515	0.9525	0.9535	0.9545
1.70	0.9554	0.9564	0.9573	0.9582	0.9591	0.9599	0.9608	0.9616	0.9625	0.9633
1.80	0.9641	0.9649	0.9656	0.9664	0.9671	0.9678	0.9686	0.9693	0.9699	0.9706
1.90	0.9713	0.9719	0.9726	0.9732	0.9738	0.9744	0.9750	0.9756	0.9761	0.9767
2.00	0.9772	0.9778	0.9783	0.9788	0.9793	0.9798	0.9803	0.9808	0.9812	0.9817
2.10	0.9821	0.9826	0.9830	0.9834	0.9838	0.9842	0.9846	0.9850	0.9854	0.9857
2.20	0.9861	0.9864	0.9868	0.9871	0.9875	0.9878	0.9881	0.9884	0.9887	0.9890
2.30	0.9893	0.9896	0.9898	0.9901	0.9904	0.9906	0.9909	0.9911	0.9913	0.9916
2.40	0.9918	0.9920	0.9922	0.9925	0.9927	0.9929	0.9931	0.9932	0.9934	0.9936
2.50	0.9938	0.9940	0.9941	0.9943	0.9945	0.9946	0.9948	0.9949	0.9951	0.9952
2.60	0.9953	0.9955	0.9956	0.9957	0.9959	0.9960	0.9961	0.9962	0.9963	0.9964
2.70	0.9965	0.9966	0.9967	0.9968	0.9969	0.9970	0.9971	0.9972	0.9973	0.9974
2.80	0.9974	0.9975	0.9976	0.9977	0.9977	0.9978	0.9979	0.9979	0.9980	0.9981
2.90	0.9981	0.9982	0.9982	0.9983	0.9984	0.9984	0.9985	0.9985	0.9986	0.9986
3.00	0.9987	0.9987	0.9987	0.9988	0.9988	0.9989	0.9989	0.9989	0.9990	0.9990

TABLE 8.11
Exponential Functions

x	e^x	e^{-x}	x	e^x	e^{-x}
0.00	1.000	1.000	2.20	9.025	0.111
0.05	1.051	0.951	2.30	9.974	0.100
0.10	1.105	0.905	2.40	11.023	0.091
0.15	1.162	0.861	2.50	12.183	0.082
0.20	1.221	0.819	2.60	13.464	0.074
0.25	1.284	0.779	2.70	14.880	0.067
0.30	1.350	0.741	2.80	16.445	0.061
0.35	1.419	0.705	2.90	18.174	0.055
0.40	1.492	0.670	3.00	20.086	0.050
0.45	1.568	0.638	3.10	22.198	0.045
0.50	1.649	0.607	3.20	24.533	0.041
0.55	1.733	0.577	3.30	27.113	0.037
0.60	1.822	0.549	3.40	29.964	0.033
0.65	1.916	0.522	3.50	33.116	0.030
0.70	2.014	0.497	3.60	36.598	0.027
0.75	2.117	0.472	3.70	40.447	0.025
0.80	2.226	0.449	3.80	44.701	0.022
0.85	2.340	0.427	3.90	49.402	0.020
0.90	2.460	0.407	4.00	54.598	0.018
0.95	2.586	0.387	4.10	60.340	0.017
1.00	2.718	0.368	4.20	66.686	0.015
1.10	3.004	0.333	4.30	73.700	0.014
1.20	3.320	0.301	4.40	81.451	0.012
1.30	3.669	0.273	4.50	90.017	0.011
1.40	4.055	0.247	4.60	99.484	0.010
1.50	4.482	0.223	4.70	109.947	0.009
1.60	4.953	0.202	4.80	121.510	0.008
1.70	5.474	0.183	4.90	134.290	0.007
1.80	6.050	0.165	5.00	148.413	0.007
1.90	6.686	0.150	5.10	164.022	0.006
2.00	7.389	0.135	5.20	181.272	0.006
2.10	8.166	0.123	5.30	200.337	0.005

July 1999 Cover Art by Corey Singleton

Chapter 9

Before You Play, Know the SCORE!

The following material first appeared on the **rge21.com** *Web site* [now *AdvantagePlayer.com* — D.S.], *originally as just Part I. Reaction to the study was so positive that John Auston and I decided to expand the concept to counts other than just Hi-Lo. Part II was added, and the article you're about to read then was published, in its entirety, in* ***Blackjack Forum.***

I'm proud to report that the SCORE methodology has become the gold standard for comparing the relative attractiveness of various blackjack games. I am gratified by the overwhelmingly enthusiastic response the piece has generated.

Part I

Consider the following scenario: A group of four friends, with an hour to spare, enters a mega-casino where every conceivable form of blackjack game is offered. We have some one-deck games, a few double-deckers, a high-rollers' pit with 6-deck games, and a large number of 8-deck shoes on the floor (suitable for back-counting).

Having diligently studied the principles of fixed-Kelly betting, optimal-bet spreading, and risk of ruin (much more on all of this later), the players decide to split up and to attack each of these games in the most "intelligent" manner possible. They each have a $10,000 total bankroll, and they're convinced that a 1–3 spread is all the single-deck game will allow (h17, two players at the table); 1–6 will be tolerated at double-deck (s17, das, three players); 1–12 is acceptable at the 6-deck game (s17, das, ls, play all, four players); and the 8-deck game will be back-counted only, with a 1–8 spread (s17, das, ls, four players).

Furthermore, the SD (single-deck) game will yield about 31–32 cards of penetration, the DD (double-deck) about 70, the 6-deck shoe is dealt to the 5-deck level, and the 8-deck games are generously cut at 6.5 decks played. Which of our players will have the most productive hour? You're not sure? Well, read on; you're about to find out why, before you play, you need to know the SCORE ("**S**tandardized **C**omparison **O**f **R**isk and **E**xpectation").

Some background: Many of you, in attempting to answer the above question, might go running to Chapter 10, and I'd be flattered if you did. The win rates of various games, and the DIs ("Desirability Indexes") that John Auston, Norm Wattenberger, and I provided have become widely accepted in the blackjack-playing community and are often quoted in discussions about the relative merits of different game conditions. But there are some drawbacks to the charts in their current form. *[Note: No longer! See below. — D.S.]*

First, in an effort to simplify matters somewhat, a rather rigid and standard betting pattern was used in playing the games. Careful perusal of the charts will indicate many situations where a one-unit bet may have been made when a two-unit bet would have been permissible, or where a 12-unit bet was made rather early, indicating an overbet, at that particular count. What's more, because of the vast differences in game conditions, the risks of ruin associated with those games fluctuated somewhat wildly, from some very low figures to some unacceptably high ones. Perhaps there was a better way. *[Note: Indeed, there was a better way. That new and improved methodology was used not only to create the studies in this chapter but also was adapted for the following chapter, as well. For those of you who are already familiar with the now-famous Chapter 10 charts, this represents a major change in their derivation and presentation. — D.S.]*

There exists today a wide body of literature on the theory of optimal betting. These studies focus especially on optimal betting within the constraints of an upper and lower bound, which is almost always the case when playing blackjack. Significant contributions to our knowledge of this vitally important topic have been made, in various venues, by Winston Yamashita, Ed Thorp, Karel Janecek, Brett Harris, Michael Lea, and David D'Aquin, to mention only a few of the most notable participants. What's more, the germ of the idea was actually presented in a fascinating paper, written by Patrick Sileo, entitled "The Evaluation of Blackjack Games Using a Combined Expectation and Risk Measure," first presented at the 1990 8th International Conference on Risk and Gambling, in London (See Eadington, W. R. and J. A. Cornelius, eds. (1992) *Gambling and Commercial Gaming: Essays in Business, Economics, Philosophy and Science.* Reno: Institute for the Study of Gambling & Commercial Gaming, University of Nevada, Reno. See, also, Richard Reid's "Mathematics of Blackjack" Web site, at *bjmath.com*). We'll get back to this seminal work in a little while.

Armed with all the pertinent formulas, John Auston set out to create software that would permit any player to determine, virtually instantly, the precise optimal betting pattern to maximize the logarithmic growth of his bankroll, given a certain game, and a bet spread, ranging from one to x units. The "One Second Sim," just part of the genial *Blackjack Risk Manager (BJRM)* that Auston designed, is one of the most valuable tools a player could ever own. It was used to create all of the charts and SCOREs that, we hope, you will be consulting for a long time to come.

As I was lecturing one day on applying risk-management techniques to equity portfolios, I was explaining the "Sharpe Ratio," and, of course, I always think of the DI when Sharpe is mentioned. *[Note: See the next chapter for a full explanation of these concepts. — D.S.]* I note, on p. 203, how the two concepts are, indeed, quite similar. But, in that same lecture, I explained yet another risk-adjusted-return measure, much less well known, but not without considerable merit: the Modigliani-Modigliani ("M-squared") model.

Briefly, the model attempts to give units to an otherwise unitless Sharpe Ratio. In a variation on Sharpe, the Modiglianis assume that a benchmark portfolio is chosen. One notes its volatility (annualized standard deviation of continuously compounded returns) and its mean return. Now, consider several different portfolio managers, each with different volatilities and returns for their portfolios. Instead of calculating their Sharpe Ratios, we do the following: Each gets to either lever (make more risky) or delever (make less risky) the portfolio by adding, in the first instance, futures contracts (or buying on margin), or, in the second case, Treasury bills, to the stock holdings. They do this until each portfolio's volatility has been adjusted to match exactly the volatility of the original, benchmark portfolio. But, in applying leverage, returns are increased (risk too!); and in delevering, returns are diminished. And so is risk. So now, we have a more meaningful basis for comparing the returns of the various portfolios, given that all of their risks have been made identical. The concept is quite compelling.

And then, I thought of blackjack. Why not apply this same principle to the comparison of blackjack games? Why not indeed! Suppose we gave a player a standard (benchmark) bankroll of, say, $10,000 and insisted that, regardless of the game he chose to play, he would have to structure his approach in such a manner as to have one single risk of ruin. Well, which one? The one associated with applying the Kelly Criterion, subject to a fixed betting schedule (where predetermined bet sizes don't change as bank fluctuates), and where an optimal bet "ramp" would be applied, once the bet spread was enunciated. Some of you know that this risk of ruin turns out to be, roughly, 13.5%.

No sooner had I begun to feverishly lay the groundwork for this article in my mind (I was in Tokyo at the time) than I received an e-mail from John Auston proposing virtually the same approach! I will resist the temptation to observe, immodestly, that "great minds think alike" (!) and simply state that I was convinced, then more than ever, that a revolutionary idea was in the making.

And then, I reread Patrick Sileo's aforementioned paper, only to discover what a true visionary he had been, back in 1990. For, although he had chosen a different level of risk, the index he proposed in his work was unmistakably the one that I have re-baptized SCORE. Judge for yourself:

> *"This index may be interpreted as the expected per hand dollar win for a player with a $10,000 bankroll who will tolerate a 5% chance of ruin. It is assumed that this hypothetical player uses the same betting scheme as was used to calculate E* [author's note: expected return] *and s* [note: standard deviation] *(although he will in general have a different size unit bet). This index will be used to evaluate various games and playing styles in what follows. It will henceforth be referred to as the Total Return Index, or more simply as TR."*

What makes this measure so appealing is twofold: First, it puts an absolute dollar value on an hour's worth of play. We see, immediately, from the charts exactly what our hourly "wages" will be. Second, and this is crucial, it relates more realistically the comparisons we are trying to establish between games and conditions. If two DIs are given as, say, 7.07 and 5.00, the point has been made that we really don't grasp from these numbers, whose ratio is 7.07/5.00 = 1.41, that the first game is actually worth twice as much as the second, on a risk-adjusted basis. This is because the SCORE squares the DI values, permitting us to understand that $(7.07/5.00)^2 = 2.00$, and that we will win precisely double by choosing the first game over the second.

What follows, then, is a resurrection of this variation on the DI or Sharpe Ratio of a blackjack game. This square of DI is the return on investment, which, for the record, has been touted for a long time by experts such as Abdul Jalib, to name just one of its major proponents. In fairness to Sileo, he did not have, in 1990, the amazing technology and simulation capabilities that we have today. This study would not have been possible without the previously mentioned *BJRM* software, nor without the optimal betting spread theories and formulas that have since been developed.

Methodology: To interpret and comprehend the charts on pp. 156–58, let's get back to our four players to make sure that their approaches to the games are completely understood by the reader. All have the same bankroll and all use the "Illustrious 18" and "Fab 4" surrender indices. All play Hi-Lo (See Part II, which starts on p. 164, for SCORE comparisons of several other counts.) After each decides on the maximum tolerable bet spread for the game he is playing, he enters the game conditions into the "One Second Sim." Instantly, he receives the size of his betting unit (this changes according to the game!), his bet schedule, and all the pertinent statistics that readers of previous editions of this work have come to know and love from the "World's Greatest Blackjack Simulation" charts (see next chapter). A few new ones, such as Brett Harris's "N0," "number of hands to double the bank," and "exponential growth rate" are supplied, as well. Of course, we know the risk of ruin ahead of time: It will always be 13.5%. *[Note: Since the publication of this material, John Auston has released a new version of* ***BJRM.*** *Because of slight improvements to the software code, users may find SCOREs that differ slightly, in some cases, from those reported here. — D.S.]*

Do not let that number "spook" you. Although the entire study was conducted using the fixed-Kelly principle, which maximizes log-growth of the bankroll and leads to the 13.5% ROR number, once you obtain the software, you are altogether free to increase your bank size, thereby lowering risk of ruin to any level you desire. The keystrokes take but a few seconds.

A few explanations are in order. First, although we imposed a specific level of penetration for each of the games in our original scenario, John has supplied four different depths for each game. So, in wandering from table to table, if one of our players encounters better (or worse) penetration, he will immediately be able to assess its impact on his hourly bottom line. Of course, only one set of rules for each game will apply to our current example, but the full range of combinations (eight or nine different columns) is offered in the charts.

Certain decisions had to be made. We realized that betting odd amounts such as $37 or $242 is simply not done in actual casino practice. But, we did it anyway! (We rounded to the nearest whole dollar.) We wanted the comparisons to be as objective as possible, devoid of any subjective adjustments to unit size or rounding. However, if you use the *BJRM* software, you may round all bets to the nearest, say, $5, or $25, increments.

A compromise as to the upper limit of a top bet was also reached. A true count of +9 would be the ultimate upper boundary. Nonetheless, virtually all the sims had us reaching the top bet well before that level.

Next — and we hope this will not prove to be too problematic — we assumed, for the sake of easy comparison, that all games would be played at the rate of 100 hands played or seen per hour. The problem here is that, again, to be as realistic as possible,

TABLE 9.1
Hi-Lo 1-Deck ($ Won/100)

	S17	S17 DAS	S17 LS	S17 DAS LS	H17	H17 DAS	H17 LS	H17 DAS LS	H17 D10
26	(1.15)								
1–2	18.08	29.31	24.51	37.60	5.90	16.22	10.48	21.51	0.02
1–3	38.39	52.80	51.15	67.83	22.83	36.65	33.64	47.63	4.02
1–4	54.41	69.72	71.86	89.75	36.69	52.18	52.22	68.21	13.32
31	(1.20)								
1–2	41.13	57.28	54.76	72.83	23.32	36.46	34.80	50.68	5.13
1–3	75.54	95.24	100.00	122.79	53.79	70.49	75.45	95.12	22.76
1–4	100.84	121.77	133.85	158.16	77.49	95.48	107.42	128.92	41.46
35	(1.20)								
1–2	49.12	66.67	65.76	86.42	29.54	43.59	43.37	60.31	8.47
1–3	93.24	114.44	122.73	147.53	68.59	87.09	95.00	117.04	34.14
1–4	125.58	148.50	165.38	191.06	98.32	119.17	134.86	159.25	57.28
39	(1.25)								
1–2	79.31	101.28	104.73	128.88	54.02	72.83	76.87	98.71	23.15
1–3	139.88	166.24	184.77	214.73	109.79	133.13	150.98	178.16	63.95
1–4	182.33	210.41	242.02	273.30	151.03	176.55	206.01	236.16	98.26

TABLE 9.2
Hi-Lo 2-Deck ($ Won/100)

	S17	S17 DAS	S17 LS	S17 DAS LS	H17	H17 DAS	H17 LS	H17 DAS LS
52	(0.95)							
1–4	10.10	17.73	18.90	28.93	1.55	8.19	7.91	16.67
1–6	18.48	27.63	31.45	42.65	9.41	16.17	19.40	28.95
1–8	24.74	34.08	39.85	51.74	15.11	22.60	27.50	37.36
62	(0.95)							
1–4	20.12	30.23	33.75	46.53	9.22	17.32	19.42	30.60
1–6	33.53	44.89	52.58	66.76	20.91	30.36	37.12	49.18
1–8	42.43	54.18	65.13	79.44	29.39	39.38	49.09	61.83
70	(1.00)							
1–4	30.40	42.35	48.30	63.51	16.63	26.44	31.53	43.90
1–6	48.64	62.18	73.43	90.20	32.89	44.13	54.60	68.95
1–8	60.58	74.47	90.15	107.41	44.52	56.20	70.64	85.80
78	(1.00)							
1–4	44.38	58.79	67.27	84.99	27.91	39.51	47.39	62.49
1–6	68.68	84.64	100.70	120.01	50.13	63.80	79.26	95.71
1–8	84.75	101.14	122.55	142.15	65.77	79.81	100.30	117.68

TABLE 9.3

Hi-Lo 6-Deck Play-All ($ Won/100)

	S17	S17 DAS	S17 LS	S17 DAS LS	H17	H17 DAS	H17 LS	H17 DAS LS
4/6	(0.90)							
1–8	3.55	7.80	8.31	14.88	0.04	1.69	3.04	7.24
1–10	5.34	9.91	11.93	18.11	0.99	2.97	5.58	10.37
1–12	6.65	11.75	14.25	20.76	1.87	4.70	7.99	13.13
4.5/6	(0.95)							
1–8	8.40	14.03	16.67	24.71	2.17	6.60	9.06	15.28
1–10	11.41	17.27	21.18	29.43	4.72	9.20	13.12	19.91
1–12	13.76	19.79	24.52	32.91	7.23	11.90	16.44	23.24
5/6	(0.95)							
1–8	17.30	24.93	30.27	40.45	9.16	15.03	19.92	28.62
1–10	22.04	29.92	36.65	47.26	13.29	19.69	26.02	35.05
1–12	25.71	33.62	41.64	52.04	17.01	23.43	30.77	40.19
5.5/6	(0.95)							
1–8	37.48	47.98	59.81	73.74	25.15	34.18	44.83	56.85
1–10	45.64	56.40	70.80	85.26	32.81	42.31	55.56	67.92
1–12	51.85	62.75	79.29	93.62	39.09	48.56	64.09	76.51

TABLE 9.4

Hi-Lo 6-Deck Back-Count ($ Won/100 observed)

(Integer to right of $ amount is TC of 1st bet)

	S17	S17 DAS	S17 LS	S17 DAS LS	H17	H17 DAS	H17 LS	H17 DAS LS
4/6	(0.90)							
1–1	16.83 2	20.43 2	25.32 2	31.08 2	14.10 3	16.22 3	21.67 3	25.93 2
1–2	20.16 2	24.35 2	30.93 2	35.87 2	16.84 3	19.53 2	26.42 2	31.00 2
1–4	21.83 2	25.40 2	33.04 2	38.71 1	18.29 2	20.95 2	28.10 2	32.84 2
1–8	22.49 1	26.71 1	33.94 2	39.76 1	19.05 2	21.76 2	29.38 2	33.14 2
1–12	22.69 1	26.85 1	34.82 1	40.72 1	18.77 2	21.81 2	29.20 1	34.42 1
4.5/6	(0.95)							
1–1	26.56 3	30.49 2	39.22 3	43.94 2	23.19 3	25.76 3	35.19 3	38.46 3
1–2	30.98 2	35.87 2	46.16 2	52.70 2	26.58 3	30.26 3	40.63 2	45.65 2
1–4	34.40 2	38.99 2	51.52 2	56.84 2	29.50 2	33.59 2	44.95 2	50.96 2
1–8	35.43 2	40.00 1	51.85 2	58.76 1	30.63 2	34.59 2	46.55 2	51.50 2
1–12	35.54 1	41.13 1	52.86 1	59.91 1	30.55 2	34.69 1	46.28 2	52.78 1
5/6	(0.95)							
1–1	42.86 3	46.30 3	60.87 3	65.91 3	38.78 4	41.67 4	54.17 3	60.87 3
1–2	49.42 2	55.42 2	70.43 2	79.41 2	44.44 3	50.00 3	65.52 3	71.43 3
1–4	54.70 2	61.07 2	77.89 2	85.87 2	48.06 2	54.16 2	71.00 2	79.17 2
1–8	56.15 2	62.43 1	80.13 2	89.04 1	50.71 2	55.67 2	73.37 2	80.12 2
1–12	56.80 1	63.84 1	81.23 1	90.21 1	50.37 2	55.95 2	72.80 3	81.55 1
5.5/6	(0.95)							
1–1	76.92 4	79.49 4	106.25 4	112.50 4	70.73 4	75.00 4	97.06 4	106.06 4
1–2	89.09 3	94.45 3	124.44 3	131.82 3	82.46 3	87.75 4	117.39 3	122.50 4
1–4	96.59 2	105.86 2	138.03 2	147.89 2	88.37 3	95.66 2	126.32 2	136.49 2
1–8	100.00 2	108.14 1	140.87 2	151.43 1	91.17 2	98.84 2	130.60 2	139.41 2
1–12	100.38 1	109.42 1	141.46 1	153.26 1	92.03 2	99.20 2	130.30 2	141.36 1

TABLE 9.5

Hi-Lo 8-Deck Play-All ($ Won/100)

	S17	S17 DAS	S17 LS	S17 DAS LS	H17	H17 DAS	H17 LS	H17 DAS LS
5.5/8	(0.90)							
1–8	0.99	3.19	3.86	8.51	—	0.15	0.52	2.62
1–10	2.24	5.38	6.54	10.96	0.04	1.06	1.87	4.69
1–12	3.01	6.74	8.31	13.03	0.08	1.67	3.28	6.79
6/8	(0.95)							
1–8	2.01	6.72	7.47	14.03	0.00	1.62	2.61	6.84
1–10	4.03	9.50	10.59	17.16	0.71	2.92	5.02	10.42
1–12	5.32	11.40	13.25	19.74	1.46	4.90	7.23	12.73
6.5/8	(0.95)							
1–8	5.25	11.18	13.78	21.11	1.16	4.70	6.44	12.40
1–10	8.90	14.63	18.01	25.54	2.85	8.04	10.15	16.76
1–12	10.91	16.98	21.18	28.77	5.18	10.17	13.13	19.98
7/8	(0.95)							
1–8	12.12	18.54	23.16	31.98	5.32	11.22	14.12	21.15
1–10	16.17	23.01	28.84	38.04	9.36	15.31	19.41	27.02
1–12	19.40	26.48	33.37	42.65	12.54	18.53	23.65	31.48

TABLE 9.6

Hi-Lo 8-Deck Back-Count ($ Won/100 observed)

(Integer to right of $ amount is TC of 1st bet)

	S17	S17 DAS	S17 LS	S17 DAS LS	H17	H17 DAS	H17 LS	H17 DAS LS
5.5/8	(0.90)							
1–1	11.93 2	14.85 2	18.39 2	22.22 2	9.41 3	11.54 3	15.15 3	17.74 3
1–2	14.61 2	17.21 2	22.33 2	25.51 2	10.78 3	13.97 2	18.42 2	21.30 2
1–4	15.70 2	18.10 1	23.28 2	27.27 1	11.79 2	15.17 2	19.69 2	22.53 2
1–8	16.09 1	19.20 1	24.72 1	28.76 1	12.40 2	15.52 2	20.22 2	23.30 1
1–12	16.29 1	19.27 1	25.15 1	28.92 1	12.25 2	15.57 2	20.13 2	23.75 1
6/8	(0.95)							
1–1	16.90 3	20.65 2	25.32 2	30.14 2	14.10 3	16.90 3	21.67 3	25.00 2
1–2	19.20 2	24.56 2	30.93 2	35.16 2	15.71 2	20.00 2	25.71 2	30.30 2
1–4	21.38 2	25.51 1	32.71 2	37.82 1	17.50 2	22.02 2	28.69 2	33.04 2
1–8	21.96 2	26.54 1	33.33 2	40.00 1	18.30 2	22.86 2	29.61 2	33.53 2
1–12	21.85 1	27.24 1	34.43 1	40.49 1	18.07 2	22.68 1	29.31 2	34.50 1
6.5/8	(0.95)							
1–1	24.62 3	27.87 3	37.26 3	40.82 3	21.43 3	24.62 3	30.91 3	35.85 3
1–2	29.29 2	32.98 2	43.04 2	48.00 2	25.00 3	29.33 3	36.73 2	41.98 2
1–4	31.78 2	35.83 2	46.60 2	52.09 2	26.42 2	32.06 2	41.03 2	46.43 2
1–8	32.54 2	37.02 1	48.49 2	54.22 1	28.12 2	32.86 2	42.38 2	48.19 2
1–12	32.56 2	37.63 1	49.00 1	55.56 1	28.23 2	33.01 2	42.20 2	48.82 1
7/8	(0.95)							
1–1	36.21 3	40.00 3	52.63 4	57.78 3	32.79 3	37.93 3	47.50 4	52.08 3
1–2	42.86 3	47.76 3	61.11 2	68.12 2	39.73 3	44.29 3	55.93 3	61.40 2
1–4	46.66 2	52.63 2	67.71 2	74.20 2	41.24 3	47.10 2	60.78 2	69.07 2
1–8	48.27 2	54.05 1	70.06 2	77.06 1	43.46 2	48.98 2	63.74 2	70.31 2
1–12	48.15 2	55.07 1	71.12 1	78.51 1	43.68 2	49.33 2	63.60 2	70.90 1

we put two players at the single-deck games, three at double-deck, and four at shoes. Well, it simply isn't true that the two-player games and the four-player games will be played at the same speeds. Borrowing from the charts supplied by Stanford Wong, in *Professional Blackjack* (see, specifically, pp. 234 and 237), we have attempted to suggest certain "factor" adjustments to the SCOREs provided in the charts. Next to each level of penetration is a number, such as 1.15, or 0.90, that represents our best estimate as to how the SCOREs might best be adjusted, by multiplying. See, also, in this book, pp. 18–19, for more on the kind of calculations and methodology involved.

Thus, the SD, s17, 26/52, 1–3 SCORE of $38.39 might best be multiplied by 1.15 ($44.15) to better reflect the approximately 115 hands per hour one might expect at such a game. Similarly, the 6-deck back-counting, s17, das, ls, 4/6, 1–12 SCORE of $40.72 should probably be scaled down to 90% of that win (only 90 hands per hour), or $36.65. Notice, then, that the two SCOREs have "flip-flopped," with the readjustments causing the SD game to be more appealing, after all, than its 6-deck counterpart, despite the apparent superiority of the 6-deck SCORE. We agonized over whether to incorporate the factors into the "raw" SCOREs, so as to save you the multiplication. But, because not all dealers deal at the same speed, nor are some shuffles as elaborate as others, we thought it best to present the uniform 100 hand-per-hour numbers and let you use, or not use, the factors, at your discretion.

Interpretation of the data: Needless to say, much can be gleaned from these charts, which, we feel, ought to be the final arbiters insofar as game desirability is concerned (not to mention system power, since SCOREs of other counts are consulted, in Part II). Nonetheless, there are some startling observations to be made using just the Hi-Lo figures provided in this Part I.

Do not lose sight of the fact that every single cell represents the return from a game where the players have the exact same bank and exact same risk of ruin. This is a crucial element in our dissertation. As with the Modigliani-squared model, the playing field has been completely leveled. The comparisons are as pure, and as "apples-to-apples," as you could hope to achieve. And, therein lies the beauty of the SCOREs. Please note that the following examples, for simplicity, do not make use of the factors.

With a 1–4 spread, adding just five cards of penetration (from 26 to 31) to a SD, s17 game will almost double your SCORE, while, with respect to this same benchmark game, moving up to Northern Nevada (h17, d10) will, depending on penetration, deprive you of anywhere from roughly 45 to 75% of your profits!

Want to virtually double your SCORE at double-deck? Well, for the same penetration and spread (70/104, 1–6), you could switch from h17 ($32.89) to s17, das ($62.18); or, alternatively, for the same rules (s17) and spread (1–6),

TABLE 9.7
Hi-Lo 1-Deck

	S17	H17D10
26	(1.15)	
1–4	54.41	13.32
31	(1.20)	
1–4	100.84	41.46

TABLE 9.8
Hi-Lo 2-Deck

	S17	S17DAS	H17
62 1–6	(0.95) 33.53	44.89	20.91
70 1–6	(1.00) 48.64	62.18	32.89
78 1–6	(1.00) 68.68	84.64	50.13

TABLE 9.9
Hi-Lo 2-Deck

	S17DAS
52	(0.95)
1–4	17.73
1–6	27.63
1–8	34.08

seek out better penetration (62/104 = $33.53, while 78/104 = $68.68). (See Table 9.8.)

Finally, for the same rules (s17, das) and penetration (52/104), spreading 1–8 ($34.08) just about doubles your 1–4 ($17.73) SCORE. (See Table 9.9.)

In an effort reminiscent of the Chapter 10 explanations, it is tempting to recite more of the entries, to elucidate how very damaging (h17) or rewarding (ls) some of the rules can be. We shall resist said temptation, but we invite the reader to scrutinize carefully the SCOREs, in order to best appreciate the full effects of these rules variations.

Moving on to the shoe games presents some eye-opening opportunities and will lead to what I expect to be an incredibly revealing and appealing fresh approach to back-counting. But let's start with play all. Or, should I say *don't* play all! By now, your eye has adjusted to what should be some "acceptable" SCOREs in the $50 range. For normal levels of penetration, do you see any such $50 entries in these play-all charts? Zilch! Except for the $52.04 (the SCORE for one of our four original friends!) to be had from the s17, das, ls, 5/6, 1–12 game, there's not a single situation that will provide a $50 per hour return for a $10,000 bank. Sad. On the other hand, if you are ever fortunate enough to find the rarer-than-hens'-teeth, deeply dealt 5.5/6 game, virtually all returns are instantly doubled with respect to the 5/6 variety. Again, such is the power of penetration. Needless to say, playing all, at the 8-deck game, is an utter waste of time and effort unless, once again, you are lucky enough to find an extraordinary cut with extremely liberal rules.

But, it was when "playing" with the 6-deck back-counting SCOREs and bet spreads that John and I found some astonishing results, with what we believe to be some very far-reaching applications. One of the important features of the optimal-betting theory is that, given a bank and a spread, the size of our unit may very well vary depending on the approach to the game employed. Specifically, to be successful, a play-all approach will normally require a wide spread, with a typically small starting unit, which is wagered in all negative-advantage situations. On the contrary, since no bets are made without a player edge, back-counting affords the opportunity for a larger betting unit, with a correspondingly smaller spread. However, until I began examining the SCOREs, I never realized the full implications of the theory. Stay with me: There's some good stuff coming!

I won't harp on the fact that, for shoes, back-counting is *de rigueur.* You've heard that from me many times before. Nor will I read any specific SCOREs to you, to hammer this point home. Suffice it to say that, with good rules and good penetration, back-counting shoes can provide some of the best returns in the game. But, I think it will come as somewhat of an eye-opener when I describe to you just how some of those returns can be achieved.

Let's go to the s17, das, ls, 4.5/6 game, which, by the way, can currently be found all over the Las Vegas Strip — or, at the very least, at several well-known, upper-end establishments. The SCOREs for these charts have an added number after them. This figure represents the true count at which it is optimal to enter the game. Now, whereas most of these counts are the expected "one" or "two" that you would assume represent the first instance of positive edge (not always!), there are a few cryptic "threes" and — how can this be? — even an occasional "four"! Surely, these are misprints.

Well, no, they are not. Permit me to explain.

When the optimal bet spread is ascertained, subject to the constraints of an upper and lower bound, we don't always enter the game at the first appearance of a positive edge. Without getting too involved, this is true simply because we are constantly studying the tradeoff between risk and reward and, although betting as early as possible will ultimately put more dollars on the table, it may do so at the expense of also increasing our hourly standard deviation disproportionately with respect to the dollars gained at the earlier count. In other words, we're not getting enough reward (or "bang for the buck") for the extra risk we're taking on. And so, we wait for the next higher true count before entering the game.

But, there is much more. Depending on the spread chosen, look at the SCOREs for the aforementioned game. While we were playing with the array, I suggested to John that we explore some spreads that no one would have thought possible — levels such as only 1–4, 1–2, or, blasphemy of blasphemies, flat-betting! The results were as surprising as they were, well, downright spectacular! With my traditional approach to this game, I would be spreading 1–12, entering at TC = +1, and winning $59.91 per hour, for a $10,000 bank. Although you can't see it on the charts, it also would be true that my unit size would be $43. Now, what if I did the unthinkable and decided to spread only 1–2 in such a game? Well, three things change: First, I wait until a true of +2 to enter; second, my unit size increases, understandably, to a much larger $116; and third, my win rate is lower, but not by all that much. It drops to $52.70, fully 88% of the 1–12 amount. The start of a plan? Maybe. Read on.

TABLE 9.10
Hi-Lo 6-Deck

	S17DASLS	
4.5/6	(0.95)	
1–1	43.94	2
1–2	52.70	2
1–4	56.84	2
1–8	58.76	1
1–12	59.91	1

Perhaps some of you remember the first time Stanford Wong described back-counting in the original edition of *Professional Blackjack*. Although he may not have mentioned the betting units ($500), Wong's concept was to flat-bet shoe games, making the one-size-fits-all wager whenever he had the edge. What if we tried that approach in our 4.5/6 game? The entry point remains +2 (note that, for the 5/6 game, it would actually gravitate to +3; Wong probably didn't know this back then!), while the unit size is bigger still, at $152, and the win is reduced to $43.94. We drop to 73% of the 1–12 SCORE, but think of the camouflage implications. We never vary our bet size! Now, increase your bank to $33,000 and *voilà,* you're flat-betting purple ($500) chips, winning $145 per hour and living the life of the prototypical Wonger! Note that, back with your $10,000 bank, if you were to take the same approach at 5/6 games, it might have never occurred to you that the "right" way to flat-bet this game is with $227 units, waiting until a true of +3 to enter. And you will win the very respectable sum of $65.91 per hour, with no bet variation. Again, and I promise I'll say this for the last time, all this is occurring with exactly the same risk of ruin as all the other approaches.

Implications: Clearly, for John and me, the above applications to the back-counting approach were the most revealing findings of our SCORE study. I think we're going to change the way many players Wong games! I'm going to change my own approach, so don't think this is a "do as I say, not as I do"-type article. Hey, you're never too old to learn!

Next, I think the SCOREs will provide, for thousands of players, the quintessential guide to system comparisons, as well as the fairest, most comprehensive method for determining the "fair value" of any given blackjack game. As such, I think this latter feature may have a useful application to team play. Often, team members are compensated according to an hourly stipend that can depend on the quality of the game being played. For example, if the team manager deems that game "A" has a greater hourly expected return than game "B," he might take that into account when determining the size of player shares of the team win. Perhaps managers will now have a better handle, risk-adjusted, as to what those relative stipends ought to be.

John sees yet another important application. Sometimes, when we are in an area where casinos are somewhat spaced out from one another, the question may arise as to whether or not it is worth it to leave our present location to "seek greener (literally!) pastures." For example, suppose one Mississippi casino offers a game with a $45 SCORE, while 30 minutes away, another offers a $60 game. Suppose, as well, that I have four hours to devote to playing before catching my plane and that the two casinos in question are equidistant from the airport. Well, if I stay where I am, I can expect to earn $180 for the evening. But, if I make the half-hour drive, I can earn 3.5 x $60 = $210, and that extra $30 (minus the cost of the gas!) may make the ride worth the effort.

John even has a great quote, attributed to Abraham Lincoln, that summarizes the principle: "Given one hour to chop wood, spend 45 minutes sharpening the axe"! The point is, doing your homework and "shopping" SCOREs, as well as not being afraid to get in your car in search of the superior game, can surely "sharpen the axe" so that it will, eventually, "cut more wood" (win more money) when the time comes to play. The best return, for the moment, may be no return at all if you're on your way to a game with a SCORE superior to the present one.

Conclusions and recommendations: As with any study of this type (and I've done a few!), we always leave the door open for further research. As mentioned, SCOREs were also prepared for the Unbalanced Zen 2 count and the K-O Preferred. And more are on the way (see Part II). But no methodology will entirely satisfy everyone. Some will not be happy with the count chosen to explain the concepts; others will protest the non-rounded bet sizes that were used; still others may find that the use of factors can further complicate the issue. No one promised you a rose garden! Studies of this type can become extremely tedious if too many variables are permitted to roam free. Entries multiply to the point of unwieldiness. But, there is a solution: Get John Auston's magnificent *BJRM* and customize your SCOREs to your heart's content. Vary the bank, change the spreads, use a different risk of ruin, or compare counts. Knock yourself out! If nothing else, I hope this article has given you the impetus to think along the lines of the importance of risk-adjusted returns and return on investment.

Of course, even the SCORE charts cannot factor in every aspect of the complicated activity we engage in. If we play for comps — and many of us do — it certainly is true that the play-all approach is more conducive to building a rating and accumulating hours than is back-counting. In fact, for some, the whole idea of back-counting is to "hit and run" and to maintain anonymity. Of course, this approach gets the most money; but, if we do receive some comps from playing all, then is it not fair to suggest that their value be considered when comparing the play-all SCOREs to the back-counting ones?

Yet another consideration is the playing of two hands, each at roughly 75% of the correct one-hand wager. On p. 205, I explain how this approach can increase profits while holding risk at equivalent levels to the one-hand scenario. In an e-mail to me, a friend, who posts frequently to blackjack Web sites under the pseudonym "MathProf," observed the following: "If the play-all shoe player always played two spots, the raw SCORE would be improved by a factor of 1.46. We have to deflate this due to slower play. From Wong's data, the number of rounds falls to 83.5% (from 91 rounds per hour down to 76, using a 4/6 game as an example). Hence, net effect is to raise the SCORE by 1.46 x .835 = 1.22, or a 22% increase. Going to your next paragraph, this improves the play-all earnings from $49.44 to $60.32 and moves the 6-decker into third place, much closer to the leaders." I might add that even playing two hands at just very favorable counts will probably increase the SCORE by at least 10%.

As for our four friends, well, you should know by now who won the most money in that hour they had to play. Here are the SCOREs: SD, the winner, with 1.20 x $53.79 = $64.55, just edged out DD, the runner-up, who clocked in at 1.00 x $62.18 = $62.18.

In third place was our 8-deck back-counter, who earned 0.95 x \$54.22 = \$51.51; and bringing up the rear was the 6-deck "player of all," who, despite his 1–12 spread, garnered only 0.95 x \$52.04 = \$49.44 (but remember MathProf's suggestion above, which can raise this figure to \$60.32). If you got the right order before consulting the SCOREs, congratulations!

Now, on to …

Part II

It's been roughly four months since we penned the above lines. The reaction to the posting of Part I of this article on the RGE Publishing Web site was, frankly, quite positive. We are convinced that the SCORE methodology is already widely accepted as the most efficient and meaningful way to judge the true value of a blackjack game and a player's approach to that game. But, as we hinted in that same article, there was more work to be done. Clearly, the exhaustive Hi-Lo charts produced by John provide a wealth of information *for that particular count.* And, whereas Hi-Lo remains the most widely used count system in existence today, others — notably the unbalanced K-O — are rapidly catching up in popularity.

And so, this Part II, the aim of which is to expand our analyses to other count systems and to furnish the basis for the most reliable "apples-to-apples" comparisons of count systems that we know of today. Of course, as always, there are concessions to be made. It just isn't possible, in a single *BJF* issue, to produce the equivalent in number of the Hi-Lo charts for each of the many counts used by our readers. So, compromises were the order of the day. Although we know we won't please everyone, here is the structure we decided upon.

As the aforementioned K-O is perhaps now the second most popular system employed by today's "new breed" of card counter, it is only natural to include this level-one unbalanced count as our entry to test the mettle of the level-one balanced standard-bearer, Hi-Lo. Thus, we begin with a complete set of K-O charts. There may be some surprises along the way, as readers wonder if K-O can "hold its own" or, perhaps, even outperform its worthy predecessor. If not, then for "unbalanced lovers," might we step up one level, to the Unbalanced Zen (level two) to see just how much more power can be obtained from the slightly greater complexity of this excellent count?

What's that? While we've gravitated to level two, shouldn't we give the balanced counts another go and see how one of their representatives might stack up against the Unbalanced Zen? What more likely candidate than the (balanced) Zen? Indeed. But, now the plot thickens. We anticipate your needs; we sympathize with your concerns! You don't use any of these counts; instead, you play Hi-Opt I, Hi-Opt II, Advanced Omega II (AOII), the Revere Point Count (RPC), Halves, the Uston Advanced Point

Count (UAPC), or the Revere Advanced Point Count, 1973 edition (RAPC). It's only fair, therefore, that we examine: a level-one balanced count that neutralizes the ace; two more level twos that do the same thing; a level-two balanced, ace-reckoned count; a level three that does the same; a level-three balanced, ace-neutral count; and, finally, a level-four, balanced, ace-neutral system.

But, please don't expect miracles. As I said before, complete charts for all of the above are well beyond the scope of the present article. For this last set of counts (all but K-O), you'll have to settle for six representative games for which we do the comparisons: SD, 31/52, s17; SD, 39/52, h17, d10; DD, 62/104, s17, das; the same game, with h17; and 6-deck, 5/6, s17, das, ls, for both play all and back-counting.

Finally — and this is very important — having already completed the "World's Greatest Blackjack Simulation," John and I do not want this article to turn into the "World's Greatest Pissing Contest"! Neither of us is a system seller or developer; rather, we are blackjack researchers. As such, we have no "stake" in the results we present, no preconceived bias or prejudice for any particular count. John just cranks up his computer, grinds out the results, in his usual impeccable manner, and I do my best to analyze the data and point out a few interesting items along the way. I suspect this article will continue to generate interest and debate long after its publication. Nothing would please me more. With that in mind, let's take a look at the SCOREs.

Interpretation of the data: We'll follow our above-stated game plan and begin with the great Hi-Lo–K-O debate. We must mention, up front, that although the K-O Preferred Count was used in all the simulations, no attempt was made to incorporate a true count, which has come to be known as "TKO." Not only was K-O using the Preferred version (which means that only a few index values were used from the "Illustrious 18" and "Fab 4"), but the values that were picked were not necessarily optimized for the penetration levels we used. This, along with the location of K-O's pivot (equivalent to +4 Hi-Lo), is what helps to cause the back and forth results that Hi-Lo and K-O show when compared across a wide variety of games. The allure of K-O is its simplicity, and the question is: Can an unbalanced, running-count system hold its own against the true-counted Hi-Lo? The answer, quite simply, is: Yes it can.

This is not to say, however, that K-O is always the equal of Hi-Lo. A quick glance at the charts will instantly confirm that, often, K-O has SCOREs somewhat lower. This is true across virtually the entire spectrum of hand-held games to shoes. "Close, but no cigar" comes to mind as a possible slogan for K-O. But then, we look at

TABLE 9.11

4.5/6 Back-Count

	K-O				Hi-Lo			
	S17		S17DAS		S17		S17DAS	
1–1	28.57	2	31.67	2	26.56	3	30.49	2
1–2	31.65	2	37.62	-1	30.98	2	35.87	2
1–4	35.12	-1	39.45	-1	34.40	2	38.99	2
1–8	36.06	-1	40.22	-4	35.43	2	40.00	1
1–12	35.72	2	41.06	-4	35.54	1	41.13	1

a few special situations (for example, 6-deck back-count, 4.5/6, s17; or s17 with das) and there is K-O nudging its nose in front of Hi-Lo.

Fame is fleeting, and the superiority doesn't last long, but there is a point to be made: Strictly speaking, it would appear that, in most instances, Hi-Lo SCOREs are better than K-O's. But, throughout this part of the article, we will continually emphasize the tradeoff between achieving a slightly higher rating and the ease of use of the systems being employed. It is undeniably true that using K-O is simpler than using Hi-Lo. And, if your game of choice happens to be one of those where K-O already outperforms Hi-Lo, well, the decision seems like a no-brainer. When such is not the case, only you can decide if the extra effort to true-count is worth the extra dollars indicated by the SCORE differentials.

As for those differentials, now would be a good time to point out the following: Because of cover and camouflage, they probably are not attainable in the "real world." And so, when we compare a SCORE of, say, $50 with one of, say, $55, it may very well be that only 60% of these values is truly achievable. So, in actuality, you may be deciding between two SCOREs that are even closer, at first glance, than you may imagine (in this case, $30 and $33).

Let us note, in passing, that Hi-Opt I has little to recommend for itself in that it requires both true counting and a side-count of aces for betting purposes. For the extra aggravation of having to ace-adjust on every bet we make, not to mention computing the true count, one would certainly demand that this level-one count produce SCOREs clearly superior to those of Hi-Lo and K-O. Unfortunately, much of the time, it doesn't. Whereas, if you are already proficient in Hi-Opt I, there is little reason to abandon it for Hi-Lo or K-O, there is no sense for a new player to choose Humble's "first born" as an introduction to card counting.

As an illustration, let's use a "benchmark" game: 5/6, play all, s17, das, ls, with a 1–12 spread. Our level-one representatives earned SCOREs of $52.04, $50.50, and $53.13, respectively, with Hi-Opt I edging out Hi-Lo and K-O. Instead of agonizing over which, if any, to use, suppose we are game for a higher, level-two count. We still are looking for the simplicity of an unbalanced system, but this time, we're willing to expend a little more effort, without resorting to true counting. Then take a look at George C.'s impressive Unbalanced Zen 2. Here, the SCORE of $55.96 is almost 11% higher than K-O's, is 7.5% superior to Hi-Lo, and is 5.3% better than Hi-Opt I.

We expect more from level-two counts and, clearly, in this instance, UBZ2 delivers. As I scan over the charts (not all of which are reproduced here), it isn't surprising that, virtually across the board, UBZ2 outperforms our level-one entrants; what is interesting is

TABLE 9.12
5/6 S17 DAS LS

	Hi-Lo	K-O	Hi-Opt I	UBZ2
1–12	52.04	50.50	53.13	55.96

that, depending on the particular game, that superiority can be quite substantial. Thus, we would heartily endorse UBZ2 for the hand-held games, where its outperformance is significant. On the contrary, there is little to recommend once we step up to the 6- and 8-deck shoes, where UBZ2 is very close, if not inferior, to Hi-Lo and K-O.

Well, what if we remain on level two, but are willing to take on true count? Will we be further compensated for our efforts? Specifically, will the Zen SCORE be better than its unbalanced cousin? Although our answer seems to depend on the specific game studied, one thing appears clear: In all but one situation, Zen does not outperform UBZ2 by a margin sufficient enough to make true counting seem worthwhile. The exception seems to be the 6-deck back-counting game, where Zen's 7–8% superiority, when using a spread, may be worthwhile for some players. If we're going to use a level-two count that employs true-count conversion, we probably should look elsewhere. Which leads us to the Revere Point Count.

Here, the comparison shifts back to the UBZ2. If Zen isn't preferable to George C.'s count, might we find extra value in the RPC? Somewhat disappointingly (at least for me), the answer is no. It was thought that, for the multi-deck game, which emphasizes betting efficiency, especially with a large spread, Revere might prove more desirable than Zen. The SCOREs tell a different story. Although some of the results are "too close to call," once again, it is clear that, for all games other than the aforementioned 6-decker with back-counting, the RPC will not outperform the UBZ2, nor is it any stronger, in general, than the Zen.

TABLE 9.13
5/6 S17 DAS LS

	Hi-Lo		K-O		Hi-Opt I		UBZ2		Zen		RPC	
	Back-Count											
1–8	89.04	1	86.66	−1	89.24	1	88.50	−6	95.20	3	95.18	1
1–12	90.21	1	86.47	−1	89.91	1	89.10	−6	96.57	1	97.08	1
	Play All											
1–8	40.45		38.40		41.18		44.17		44.41		43.51	
1–10	47.26		45.36		48.02		50.98		51.55		50.93	
1–12	52.04		50.50		53.13		55.96		56.95		56.34	

I told you before that John and I had no axe to grind for any particular system or method of counting. Which is not to say that we don't both have our likes and dislikes when it comes to certain features of card counting. Specifically, we both have an aversion to side-counting aces. Laziness would come to mind (!), but now we have some data to help us decide if this extra mental exertion is really worth the effort. Again, it all depends.

Remember, the SCOREs in this article are for the player with a $10,000 bank. Many of you have considerably less than that and, yes, I know, many have more. The point is, as you examine the Hi-Opt II and Advanced Omega II SCOREs, you need to once again decide, along with John and me, if the higher numbers justify both side-counting aces and using true-count adjustments. Whereas some of the highest SCOREs in blackjack can be attained using these two counts, one has to wonder if "simpler" still isn't better.

For example, is the mental effort of side-counting aces, for level two, greater or less than having no side-count at all, but designating one single card, the five, as level three? For me, this is not an equal tradeoff: Count the five a little higher and drop the ace side-count. And so, our level-three entrant, the august Halves, arrives on the scene. It is an elegant, powerful count. To be sure, true counting is required. But, we are definitely getting some bang for our buck. Our benchmark game returns $59.67, within pennies of AOII, albeit almost $2 below Hi-Opt II.

TABLE 9.14

5/6 S17 DAS LS Play All

	AOII	Hi-Opt II	Halves	UAPC	RAPC
1–8	47.27	48.50	46.72	46.32	47.08
1–10	54.61	55.96	54.08	53.79	54.59
1–12	60.21	61.61	59.67	59.43	60.13

TABLE 9.15

31/52 S17

	UBZ2	Zen	RPC	AOII	Hi-Opt II	Halves	UAPC	RAPC
1–2	42.15	43.46	38.89	44.25	46.09	41.06	45.76	43.80
1–3	78.07	79.10	74.68	82.11	84.04	77.53	84.35	80.71
1–4	104.39	105.26	101.29	109.47	111.97	103.87	112.63	108.27

However, as can be expected, because of playing efficiency, Halves yields to our ace-neutral, level-two counts when SD and DD games are examined. Compromises, compromises! It should be obvious that no one count can be "the best" for all occasions. "Different horses for different courses" comes to mind.

One thing is abundantly clear: The use of the formerly high-priced, "advanced" level-three and -four "designer" counts of Uston and Revere cannot possibly be justified in today's environment. The UAPC — requiring level three, true counting, and ace-adjustment — will not outperform Halves (no side-count) in our benchmark game and actually underperforms in 6-deck back-counting, as well. Nor does it equal the level-two Hi-Opt II or AOII in those same games. In fact, even the RPC, level two and with no side-count, holds its own in these situations versus the UAPC. And although

Uston's count has SCOREs slightly better than Halves and the RPC, in double-deck games, and considerably better than Revere in SD, it fails to out-SCORE Humble's Hi-Opt II or Carlson's AOII by a margin sufficient, in our opinion, to justify the added complexity. In short, no matter what game we play, the UAPC should not be our first choice.

Finally, imagine stepping up to level *four,* true counting, and keeping a side of aces! We should expect the moon from Revere's famous 1973 Advanced Point Count. But, we really don't get it. It is never the "best" count in our array, losing consistently to level-two Hi-Opt II, for all the games we considered, and losing, more often than not, depending on the game, to Halves in 6-deck and to AOII in SD and DD. For me, the simple conclusion is: The most advanced counts really are not worth the time, effort, or money spent acquiring, learning, and using them. To see this, just look at the SCOREs.

In fairness, I would like to make a point here. For uniformity and ease of comparison, we used a limited set of strategy departures for all our SCOREs. Systems that are ace-neutral *depend* on playing departures for a good part of their power, especially in the hand-held games. In a certain sense, to "rob" them of, say, 20–30% of this power is unfair. To be sure, all systems gain in value if 100 additional indices are employed. But, that gain is not constant across all systems. And so, one of our entries, which may underperform a given "rival" on these pages, may very well outperform that same count, when both are fully parameterized.

Of course, there is a different argument, which may cut the other way. In their zeal to promote the efficiency of their counts, some authors quote results for games that, quite frankly, just don't exist in the real world. If you find a count that is capable of out-SCORE-ing another one only for DD, 78/104, s17, das, ls, you may as well move on — you're unlikely to encounter such conditions any day soon!

TABLE 9.16
39/52 H17 D10

	Hi-Lo	K-O	UBZ2	Zen	RPC	AOII	Hi-Opt II	Halves	UAPC	RAPC
1–2	23.15	20.24	27.08	26.73	23.05	30.51	31.27	22.76	31.48	27.49
1–3	63.95	59.67	71.55	70.59	66.23	77.98	79.82	65.80	78.88	72.13

It is also interesting to note, across the board, how extremely sensitive the SD games are to bet spreads. While this is no new revelation, one glance at the northern Nevada 39/52, h17, d10 SCOREs indicates, for all counts, the enormous jump in values from the unacceptable 1–2 spreads to the highly attractive SCOREs a 1–3 spread can garner.

Again, we shall refrain from over-analyzing the charts and will leave further observations of the type above to the reader. Part of the allure of presenting such a wide variety of "canned" SCOREs is that you may peruse the data to your heart's content and draw all the inferences you may choose. *[Note: And wait till you see what's coming in the next chapter! — D.S.]*

Conclusions: Blackjack technology has come a long way since the days of our illustrious predecessors, Braun, Revere, Humble, Uston, et al. We now have the ability, with genial software, to test the power of the systems these pioneers gave us, and we have the methodology, thanks to developments in optimal-betting formulas, risk of ruin calculations, and the present SCORE technique, to assess the relative merits of these systems.

As is often the case, the present article should be a beginning, not an end. I mentioned earlier that we suspect there will be more written on this subject. John and I have already discussed many variations on the theme and several possible "spinoffs" from the current work. It may be somewhat premature to hint at those topics right now; suffice it to say that, as has happened numerous times in the past, others will take up where we have left off, and we, ourselves, will endeavor to contribute more.

Until that next time, good luck, and … good cards.

[Editor's note: On the following pages, we have reproduced selected data from John Auston's computer simulations, showing comparisons among eleven different card-counting systems in various single-, double-, and six-deck games, with various bet spreads. Following these system-comparison pages, we are presenting a full set of K-O charts for single-, double-, and six-deck games exactly like the charts for the Hi-Lo count presented in Part I of this article.

*Some of you may be wondering: "What happened to the Red Seven Count?" I know a lot of **BJF** subscribers use the Red Seven. Unfortunately, John was unable to get the Red Seven SCORE charts done in time for this issue of **Blackjack Forum,** and we felt it was more important to compare counts that differed substantially from one another — balanced and unbalanced, multiple levels, with and without ace side-counts. We suspect Red Seven will perform similarly to K-O and Hi-Lo and other ace-reckoned, level-one counts. But, John has promised me that he will be getting us some Red Seven SCOREs for the next issue, and we'll have a look at how Red Seven stacks up against both K-O and Hi-Lo in various games. So, stay tuned! — Arnold Snyder]*

***Postscript:** Indeed, John prepared the Red Seven SCORE data for the subsequent issue of **BJF.** Some of the relevant Red Seven/K-O/Hi-Lo comparisons follow the complete K-O SCORE charts. As can clearly be seen, Red Seven acquits itself quite admirably, besting both K-O and Hi-Lo in many situations.*

TABLE 9.17
31/52 S17 ($ Won/100 observed, 1.20 factor)

	Hi-Lo	K-O	Hi-Opt I	UBZ2	Zen	RPC	AOII	Hi-Opt II	Halves	UAPC	RAPC
1–2	41.13	38.27	39.11	42.15	43.46	38.89	44.25	46.09	41.06	45.76	43.80
1–3	75.54	72.93	74.00	78.07	79.10	74.68	82.11	84.04	77.53	84.35	80.71
1–4	100.84	98.70	99.16	104.39	105.26	101.29	109.47	111.97	103.87	112.63	108.27

TABLE 9.18
39/52 H17 D10 ($ Won/100 observed, 1.25 factor)

	Hi-Lo	K-O	Hi-Opt I	UBZ2	Zen	RPC	AOII	Hi-Opt II	Halves	UAPC	RAPC
1–2	23.15	20.24	27.12	27.08	26.73	23.05	30.51	31.27	22.76	31.48	27.49
1–3	63.95	59.67	70.33	71.55	70.59	66.23	77.98	79.82	65.80	78.88	72.13
1–4	98.26	94.08	106.44	108.21	107.05	103.25	117.05	120.00	102.72	118.28	109.76

TABLE 9.19
62/104 S17 DAS ($ Won/100 observed, 0.95 factor)

	Hi-Lo	K-O	Hi-Opt I	UBZ2	Zen	RPC	AOII	Hi-Opt II	Halves	UAPC	RAPC
1–4	30.23	28.92	30.28	33.25	31.77	32.44	35.15	36.58	32.23	35.05	35.10
1–6	44.89	44.02	44.78	48.75	47.15	48.48	51.64	53.19	48.31	51.03	51.52
1–8	54.18	53.59	54.27	58.46	57.18	58.81	62.05	63.63	58.45	61.52	61.95

TABLE 9.20
62/104 H17 DAS ($ Won/100 observed, 0.95 factor)

	Hi-Lo	K-O	Hi-Opt I	UBZ2	Zen	RPC	AOII	Hi-Opt II	Halves	UAPC	RAPC
1–4	17.32	16.38	17.08	20.56	19.36	18.30	20.64	22.40	19.06	20.37	21.28
1–6	30.36	30.04	29.98	35.11	34.15	32.25	35.30	37.57	33.28	35.59	36.64
1–8	39.38	39.20	39.12	44.90	44.10	41.95	45.30	47.96	43.31	45.67	46.83

TABLE 9.21

5/6 S17 DAS LS ($ Won/100 observed, 0.95 factor)

Integer to right of $ amount is TC/RC of 1st bet (KO IRC = –20, UBZ2 = –24)

	Hi-Lo	K-O	Hi-Opt I	UBZ2	Zen	RPC	AOII	Hi-Opt II	Halves	UAPC	RAPC
Back-Count											
1–1	65.91 3	68.89 2	66.67 3	71.43 2	72.73 5	71.43 3	73.91 5	76.00 2	74.42 3	72.55 3	73.68 5
1–2	79.41 2	77.19 2	78.46 2	81.43 –1	83.82 3	84.38 2	87.14 3	89.65 2	88.06 2	86.88 3	87.88 3
1–4	85.87 2	85.10 –1	85.88 2	86.84 –4	93.02 3	91.77 2	94.55 3	97.03 1	96.63 2	94.84 2	95.92 2
1–8	89.04 1	86.66 –1	89.24 1	88.50 –6	95.20 3	95.18 1	96.77 2	102.45 1	98.80 1	97.55 2	99.41 2
1–12	90.21 1	86.47 –1	89.91 1	89.10 –6	96.57 1	97.08 1	97.43 1	102.97 1	100.89 1	97.46 2	100.39 1
Play All											
1–8	40.45	38.40	41.18	44.17	44.41	43.51	47.27	48.50	46.72	46.32	47.08
1–10	47.26	45.36	48.02	50.98	51.55	50.93	54.61	55.96	54.08	53.79	54.59
1–12	52.04	50.50	53.13	55.96	56.95	56.34	60.21	61.61	59.67	59.43	60.13

TABLE 9.22
K-O (P) 1-Deck ($ Won/100)

	S17	S17 DAS	S17 LS	S17 DAS LS	H17	H17 DAS	H17 LS	H17 DAS LS	H17 D10
26	(1.15)								
1–2	15.09	26.38	21.45	32.47	5.17	11.24	8.03	15.70	—
1–3	36.80	49.46	46.90	61.89	21.24	31.33	29.22	40.80	3.36
1–4	52.46	65.76	66.78	83.07	35.01	47.11	47.45	60.41	12.08
31	(1.20)								
1–2	38.27	53.37	49.05	65.22	21.60	33.46	29.52	44.19	4.00
1–3	72.93	91.31	92.94	113.26	51.28	67.21	68.40	87.37	21.09
1–4	98.70	118.23	127.13	147.84	74.90	92.56	99.55	120.29	39.55
35	(1.20)								
1–2	44.98	60.00	58.97	75.86	27.12	39.84	42.61	53.37	6.94
1–3	87.56	106.53	114.62	135.21	64.94	81.42	92.63	107.87	30.58
1–4	119.63	140.20	157.39	179.04	95.52	113.65	132.80	150.56	54.31
39	(1.25)								
1–2	73.22	92.13	93.67	114.49	49.32	66.50	68.64	86.75	20.24
1–3	133.92	156.04	172.21	197.06	103.67	125.00	140.63	163.75	59.67
1–4	177.37	201.25	230.07	256.20	145.17	168.39	196.14	220.28	94.08

TABLE 9.23
K-O(P) 2-Deck ($ Won/100)

	S17	S17 DAS	S17 LS	S17 DAS LS	H17	H17 DAS	H17 LS	H17 DAS LS
52	(0.95)							
1–4	10.62	17.05	17.55	26.26	2.50	6.81	6.77	12.79
1–6	19.70	26.92	29.89	40.04	9.54	15.96	17.64	25.74
1–8	26.14	33.49	38.41	48.88	15.06	22.18	25.54	33.98
62	(0.95)							
1–4	18.97	28.92	31.66	43.20	8.94	16.38	17.47	25.70
1–6	32.51	44.02	50.88	63.14	20.51	30.04	33.86	44.31
1–8	41.81	53.59	63.54	75.82	29.07	39.20	46.00	56.63
70	(1.00)							
1–4	29.71	41.48	44.94	57.99	15.45	24.95	28.26	39.67
1–6	48.26	61.19	70.21	84.35	31.45	42.97	51.36	64.37
1–8	60.87	73.97	86.64	101.17	42.83	55.53	67.62	81.09
78	(1.00)							
1–4	42.78	55.80	61.70	78.92	25.40	36.90	41.69	57.24
1–6	67.75	81.78	95.47	114.40	48.28	61.33	72.57	90.28
1–8	83.85	98.39	117.10	136.45	63.99	77.71	94.36	112.01

TABLE 9.24

K-O(P) 6-Deck Play-All ($ Won/100)

	S17	S17 DAS	S17 LS	S17 DAS LS	H17	H17 DAS	H17 LS	H17 DAS LS
4/6	(0.90)							
1–8	2.88	7.73	7.49	12.57	0.14	2.45	2.45	5.82
1–10	5.20	10.57	10.49	16.07	0.73	4.44	4.67	9.01
1–12	7.08	12.36	12.96	18.48	2.06	6.45	6.86	11.72
4.5/6	(0.95)							
1–8	8.07	13.52	15.84	22.74	1.92	6.66	7.77	13.09
1–10	11.29	17.13	20.23	27.36	4.18	9.95	11.77	17.48
1–12	13.81	19.79	23.57	30.85	6.78	12.54	15.02	20.83
5/6	(0.95)							
1–8	17.16	24.43	29.45	38.40	7.96	15.10	18.14	26.68
1–10	21.97	29.62	36.02	45.36	12.59	19.82	24.20	33.38
1–12	25.80	33.59	41.31	50.50	16.89	23.74	29.08	38.75
5.5/6	(0.95)							
1–8	36.86	47.54	57.89	71.54	24.44	34.16	42.05	52.58
1–10	45.06	56.27	68.84	82.83	32.59	42.54	53.26	63.80
1–12	51.26	62.45	77.06	91.07	38.81	49.07	61.53	71.87

TABLE 9.25

K-O(P) 6-Deck Back-Count ($ Won/100 observed)

Integer to right of $ amount is RC of 1st bet (IRC = −20)

	S17	S17 DAS	S17 LS	S17 DAS LS	H17	H17 DAS	H17 LS	H17 DAS LS
4/6	(0.90)							
1–1	17.39 −1	21.95 −1	25.68 −1	28.57 −1	14.67 2	17.39 2	22.03 2	25.00 −1
1–2	20.69 −1	25.71 −1	30.11 −1	33.71 −1	17.32 −1	20.69 −1	26.03 2	29.47 −1
1–4	21.83 −4	26.82 −4	31.85 −4	36.91 −4	18.30 −1	22.22 −1	27.74 −1	31.68 −4
1–8	22.71 −4	28.40 −4	32.93 −4	37.58 −4	19.16 −1	23.31 −1	29.07 −1	32.35 −4
1–12	22.90 −4	28.20 −4	33.47 −4	38.43 −4	19.09 −1	23.08 −1	29.12 −4	33.08 −4
4.5/6	(0.95)							
1–1	28.57 2	31.67 2	41.18 2	44.00 2	23.19 2	28.13 2	34.55 2	39.62 2
1–2	31.65 2	37.62 −1	45.98 −1	51.81 −1	27.06 2	31.82 −1	39.71 2	44.44 −1
1–4	35.12 −1	39.45 −1	50.48 −1	54.44 −1	29.16 −1	34.59 −1	42.99 −1	48.62 −1
1–8	36.06 −1	40.22 −4	51.67 −1	56.67 −4	29.33 −1	35.25 −1	44.64 −1	50.00 −1
1–12	35.72 2	41.06 −4	51.75 −4	57.51 −4	30.37 −1	35.68 −1	44.80 −1	50.00 −4
5/6	(0.95)							
1–1	45.61 2	49.09 2	61.70 2	68.89 2	39.13 5	44.07 2	54.05 5	62.50 2
1–2	52.11 2	57.35 2	72.41 2	77.19 2	46.05 2	51.39 2	64.52 2	72.88 2
1–4	55.00 −1	60.87 −1	78.35 2	85.10 −1	47.67 2	54.10 −1	69.07 2	76.77 −1
1–8	55.20 −1	61.66 −1	79.21 −1	86.66 −1	49.60 2	54.69 −1	70.59 −1	80.69 −1
1–12	57.07 −1	63.54 −1	81.30 −1	86.47 −1	50.00 −1	55.81 −1	71.50 −1	80.80 −1
5.5/6	(0.95)							
1–1	77.78 5	85.71 5	106.67 5	117.24 5	71.05 5	78.38 5	100.00 5	103.23 5
1–2	88.52 2	98.30 2	124.00 2	132.65 2	81.25 2	90.16 2	113.46 2	121.57 2
1–4	94.79 2	104.35 2	133.34 2	142.11 2	87.00 2	95.84 2	118.64 2	128.75 2
1–8	95.59 2	104.95 −1	133.54 −1	144.17 −1	86.70 2	95.46 2	122.89 2	128.98 2
1–12	95.54 −1	104.90 −1	134.97 −1	144.58 −1	87.02 2	96.50 2	124.40 2	130.17 −1

TABLE 9.26

K-O(P) 8-Deck Play-All ($ Won/100)

	S17	S17 DAS	S17 LS	S17 DAS LS	H17	H17 DAS	H17 LS	H17 DAS LS
5.5/8	(0.90)							
1-8	0.78	3.69	3.83	7.68	—	0.18	0.30	2.21
1-10	2.32	5.09	5.52	9.82	0.02	1.06	1.60	4.44
1-12	3.31	6.92	7.81	12.06	0.19	2.00	3.04	5.96
6/8	(0.95)							
1-8	2.53	5.48	6.66	11.87	0.06	1.47	1.43	4.79
1-10	4.66	8.02	9.80	14.94	0.56	3.23	3.76	7.80
1-12	6.19	9.86	12.38	17.42	1.79	4.79	5.85	9.90
6.5/8	(0.95)							
1-8	5.65	10.77	11.93	18.51	1.00	3.63	5.10	9.83
1-10	8.56	14.24	16.10	22.77	3.25	5.87	8.65	13.99
1-12	10.95	16.69	19.34	25.99	4.98	8.93	11.47	17.06
7/8	(0.95)							
1-8	12.72	18.40	22.12	30.24	4.27	9.75	12.35	18.85
1-10	16.97	22.96	27.99	36.29	8.47	13.88	17.57	24.43
1-12	20.41	26.45	32.57	41.13	11.63	17.15	21.87	29.06

TABLE 9.27

K-O(P) 8-Deck Back-Count ($ Won/100 observed)

Integer to right of $ amount is RC of 1st bet (IRC = -28)

	S17	S17 DAS	S17 LS	S17 DAS LS	H17	H17 DAS	H17 LS	H17 DAS LS
5.5/8	(0.90)							
1-1	12.09 -1	14.46 -1	18.31 -1	20.90 -1	10.39 2	11.83 -1	15.19 -1	17.57 -1
1-2	14.18 -4	17.21 -4	21.70 -4	25.25 -4	11.45 -1	14.05 -1	17.82 -1	21.10 -4
1-4	15.61 -4	18.63 -4	23.36 -4	27.13 -4	12.05 -4	14.95 -4	18.90 -4	22.86 -4
1-8	15.59 -7	19.03 -7	23.81 -7	28.04 -7	12.55 -4	15.17 -4	19.35 -4	23.26 -4
1-12	15.93 -4	19.33 -7	24.10 -7	28.11 -7	12.57 -4	15.36 -4	20.00 -4	23.29 -4
6/8	(0.95)							
1-1	17.44 -1	19.75 -1	26.47 -1	28.79 -1	15.28 2	16.85 -1	20.78 -1	23.61 -1
1-2	20.36 -1	22.58 -1	30.00 -1	33.33 -1	17.39 2	19.66 -1	24.24 -1	28.05 -1
1-4	21.31 -4	24.53 -4	32.24 -4	35.94 -4	18.24 -1	20.93 -1	26.36 -1	30.00 -4
1-8	21.91 -4	25.34 -4	32.96 -4	36.40 -7	18.92 -1	21.20 -4	27.32 -4	31.21 -4
1-12	22.38 -4	25.40 -4	33.57 -4	37.14 -4	10.05 -4	21.48 -4	27.63 -4	31.67 -4
6.5/8	(0.95)							
1-1	25.40 2	28.81 2	35.29 2	38.78 2	22.39 2	24.62 2	30.91 2	34.62 2
1-2	29.03 -1	33.33 -1	41.86 -1	45.83 -1	25.88 2	27.71 2	35.80 -1	40.45 -1
1-4	31.36 -1	35.47 -4	44.66 -1	49.59 -4	27.08 -1	30.43 -1	38.46 -1	42.48 -1
1-8	32.46 -1	36.67 -1	46.41 -1	51.32 -4	28.37 -1	30.96 -1	39.52 -1	44.02 -1
1-12	32.36 -4	36.99 -4	46.52 -4	51.59 -4	28.51 -1	31.17 -4	39.89 -1	44.12 -1
7/8	(0.95)							
1-1	39.29 2	41.82 2	54.35 2	60.00 2	34.78 5	37.29 2	47.37 5	52.08 2
1-2	45.83 2	47.89 2	62.71 2	67.24 2	38.46 2	42.67 2	55.56 2	60.65 2
1-4	47.54 -1	52.55 -1	67.00 -1	73.19 -2	40.60 -1	45.67 -1	57.95 -1	64.08 -1
1-8	49.71 -1	53.29 -1	67.92 -1	75.00 -1	41.38 -1	46.84 -1	60.23 -1	66.24 -1
1-12	49.41 -1	53.44 -4	69.14 -1	74.75 -1	42.69 -1	46.55 -1	60.83 -1	66.31 -1

TABLE 9.28

Hi-Lo v. Red Seven v. K-O

Spread	Hi-Lo	Red Seven	K-O
1-Deck: 39/52 H17 DAS (Play All)			
1–2	72.83	72.43	66.50
1–3	133.13	131.95	125.00
1–4	176.55	174.72	168.39
2-Deck: 78/104 S17 DAS (Play All)			
1–4	58.79	58.50	55.80
1–6	84.64	83.78	81.78
1–8	101.14	99.75	98.39
6-Deck: 4.5/6 S17 DAS LS (Play All)			
1–8	24.71	26.39	22.74
1–10	29.43	30.99	27.36
1–12	32.91	34.53	30.85
6-Deck: 5/6 S17 DAS LS (Play All)			
1–8	40.45	40.72	38.40
1–10	47.26	47.00	45.36
1–12	52.04	51.89	50.50
6-Deck: 5.5/6 S17 DAS LS (Play All)			
1–8	73.74	68.24	71.54
1–10	85.26	77.57	82.83
1–12	93.62	84.46	91.07
8-Deck: 6/8 S17 DAS LS (Back-Count)			
1–1	30.14	31.34	28.79
1–4	37.82	38.40	35.94
1–12	40.49	39.94	37.14
8-Deck: 6.5/8 S17 DAS LS (Back-Count)			
1–1	40.82	40.98	38.78
1–4	52.09	51.52	49.59
1–12	55.56	52.64	51.59
8-Deck: 7/8 S17 DAS LS (Back-Count)			
1–1	57.78	56.36	60.00
1–4	74.20	70.94	73.19
1–12	78.51	72.29	74.75

More on SCORE

It has been about two years since I first presented the notion of SCORE, in an article for the Summer 1999 *Blackjack Forum.* Since then, I am pleased to report that the concept of using a **S**tandardized **C**omparison **O**f **R**isk and **E**xpectation has become, more or less, the "industry norm," either when players attempt to describe the relative strengths of different card-counting systems, or when they wish to assess the desirability of games with varying rules, all played using the same system. As was the case when my "Illustrious 18" achieved such widespread acceptability, my delight in having provided something useful for the blackjack-playing community is often tempered by the inevitable misunderstandings that sometimes accompany new ideas. And so, this essay, the sole purpose of which is to clear up some of the occasional misuses and abuses of the SCORE name. Along the way, we'll have a little fun with some tangential ideas.

One of the first annoyances to surround SCORE was that players became so comfortable with the notion that they decided to render generic the acronym. Sloppy in their usage, they began to refer to the "Score" of a particular game, or, even worse, the "score." Ever vigilant, I began to chastise the writers, and, much in the manner of corporations who doggedly protect their registered trademarks and service marks, I insisted that SCORE be given its proper acronymous, all-caps form (although I stopped short of appending a ® to the term!).

While it's flattering for a product or, in this instance, a concept, to become so universal that the "brand name" itself becomes the everyday description of the item, companies spend fortunes protecting their trademarks. Just for fun, let's review a few of the most well-known products. When you stick those two pieces of torn paper back together again, are you sure you're using Scotch tape (short for "Scotch brand cellophane tape"), or, perhaps, could you be using a different brand? We may all clean our ears with Q-tips, or do we? You may use a cheaper brand of … that's right: cotton swab. Sick with a bad cold? You may blow your nose so many times that you run out of Kleenex. Or, were those "tissues" really Kleenex *brand?* And, if that nose gets sore from all the blowing, you'll want to put some Vaseline on it, right? Well, at the very least, you'll be applying *petroleum jelly;* it may or may not be the Vaseline kind! And, while we're poking around the medicine cabinet, what do you pull out to put on that cut on your finger? A Band-Aid, of course. Well, that is, an "adhesive bandage." It could be another brand. Finally, let's consider three examples from the food area: What's America's all-time favorite "chocolate sandwich cookie"? The Oreo, without a doubt. Dare we mention that there could be other brands of this delight? Upon seeing a non-Nabisco one, would anyone call the cookie something other than an Oreo? I doubt it. And when we ordered a light dessert, were we able to verify that the package actually said "Jell-O," or might we have been served an alternative brand of "gelatin dessert"? To conclude, for an after-dinner sweet, we might decide to suck on a couple of LifeSavers. I'll bet you don't know what the generic description of this food is. Well, it can be found on the package, just above the name: We're actually eating "roll candy"!

So, why this long digression? Simply because I'd like everyone to understand that, if you're using the original, "trademarked" article, you're quoting a SCORE, and not a "Score," a "score," or any other form of "standardized hourly win rate." Otherwise, I'm afraid SCORE is going to go the way of "thermos," "aspirin," "xerox," "escalator," and "ping-pong," all of which were once brand names but are now lowercase entries in most dictionaries (but not all), because they replaced the generic words ("vacuum bottle," "acetylsalicylic acid (!)," "xerography," "moving stairway," "table tennis," etc.) they described, and thereby lost their trademarks in the process.

But, there is much more. Were it simply a matter of forgetting a few capital letters here and there, I wouldn't have bothered penning these lines. No, the abuses are considerably more serious when it comes to actually *defining* the term. First used exclusively for Hi-Lo, I then expanded SCORE's horizons to include many of the other most popularly used counting systems. But, *in all instances,* and I can't stress this enough, the SCORE of a game is a precisely defined concept. By *definition* (mine!), SCORE is the expected dollar amount won, after playing (or watching) 100 rounds, by a Kelly-constrained, optimally-betting counter in possession of a $10,000 bank. By convention, we agree to use the "Illustrious 18" and "Fab 4," while a further constraint defines, depending on the number of decks being shuffled together, how many players will be at the table. Finally, we actually create two SCOREs for each situation, one for the person who plays all and another for the back-counter.

Now, here's where the fun begins! Despite the precision employed to define the term, players started using SCORE (even if it *was* capitalized), to describe *any kind* of hourly win rate, no matter which of the original parameters were no longer invoked. Got $50,000 to play the game? No problem. Your "SCORE" will be five times as great as it normally is. Or will it? Well, I contend that your "hourly win rate" will be five times as great as your SCORE, but that's not quite the same thing, now, is it? Or, suppose you decide that strict Kelly betting is too rich for your blood, and you're going to bet some fraction of Kelly, such as one-half. Simple. Just fire up your trusty simulator *(BJRM? SBA? CV?)* and find what some are defining as the new "SCORE," ostensibly, half of the original. But, do you still have the right to call this result the SCORE? No, you don't. Once again, it's the hourly win rate that has been altered, not the original SCORE.

Suppose, although I decreed that shoe games would have four players, you decide to play with only one other person? If you now get about 160–170 rounds per hour, will your "SCORE" be 1.6 or 1.7 times the original SCORE? Not if I can help it! Now, interestingly enough, the above variations on three of the original premises of SCORE — bankroll size, Kelly fraction, and rounds per hour — will not affect the absolute *rankings* of games that are SCOREd; rather, they will preserve that same order while in some cases altering the *magnitude* of the values. On the other hand, all the while invoking the SCORE appellation, players have gone much further indeed. And these modifications to the spirit of the original acronym change not only a single SCORE, but also, potentially, the rankings of several SCOREs. Let's take a look.

Often, players will write, "How much will the SCORE suffer if I employ a certain form of cover, such as Ian Andersen's 'Ultimate Gambit'?" By now, you know my somewhat sarcastic answer: "The SCORE won't suffer at all, but you'll probably lose some of your hourly expectation." And so it goes with a host of other possible variations on what I had expected to be an invariable theme. What else can be changed? The rather unrealistic assumption of fractional-unit bet sizes, such as 1.7 or 5.3. We really don't bet that way, and software such as *BJRM* and *CV* permits us to specify "integer bets only." Such a modification will change the hourly win rate slightly and might, under certain circumstances, even change the rankings of SCOREs.

I don't intend an exhaustive list, but two other common alterations to the original premise are widely discussed. The first is a bet scheme other than the Kelly-optimal one. I'm sure you can think of several reasons why one might be forced to deviate from a strictly optimal betting approach. In so doing, we encounter the same "What does this do to my SCORE?" query. Of course, I'd like you to ask a different question! Finally, for systems where playing efficiency is very high, and where the bet spread may not be too wide, it is quite possible — in fact, advisable — to use many more indices than just the 18 "Illustrious" ones. Once again, the original SCORE and the new win rate will not be the same.

So, here's the big question: What should we call all of these *adjusted* SCOREs? Please permit me one more digression. When, several years ago, I introduced readers to a SCORE-related term, the "Desirability Index," or "DI," I mentioned that the concept was inspired by the famous "Sharpe Ratio," whose originator, William Sharpe, won the Nobel Prize in Economics for his efforts. Well, imitation is the sincerest form of flattery, and it wasn't long before other economists and financial researchers began to propose other risk-adjusted-return measures that resembled, but were not identical to, the Sharpe Ratio. I described one such metric, the Modigliani-squared value, above. This measure, although different from Sharpe, when applied to a list of individual investment opportunities, will rank those choices in the same order as Sharpe. On the other hand, at least two other risk-adjusted-return measures, those suggested by Sortino and Treynor, not only alter the Sharpe *value,* but almost always lead to different *rankings* of the investment choices, as well. I won't bore you with the details here. Suffice it to say that, once again, there is an analogy to the SCORE discussion.

For me, SCORE (the square of the very *special* DI obtained by following the original guidelines) is analogous to the Sharpe Ratio. We may choose to alter that original value in many different ways, and I have suggested several above. But, when we do this, we have an *adjusted* SCORE, or a modified version; we no longer have *the* SCORE, of which there is only one, per set of game conditions and count system. Alas, I may be fighting a losing battle (more on this below). In language, usage often gains the upper hand over strict logic. Players may simply not have the patience to refer to a "modified-bank-SCORE," a "cover-SCORE," or a "fractional-Kelly-SCORE." They may balk at specifying a "non-optimal-bet-SCORE," or an "expanded-index-array-SCORE." You get the idea.

I've been accused, more than once in my lifetime, of being too much of a purist or perfectionist, and of being too rigid at times. Perhaps this is one such occasion; perhaps it isn't. I'd really like to preserve the sanctity of the meaning of SCORE, just as I described it two years ago. I wish players would stop bandying around the acronym, using it for every "hourly win rate" (risk-adjusted or otherwise) that comes their way. But, in the final analysis, perhaps I should be flattered that SCORE has taken root in the blackjack player's vernacular, and I should be happy, rather than annoyed, that the term has achieved, at least in our small circle, equal footing with Oreos, Q-tips, and Scotch tape! I need *you* to be the judge. Here, for example, are the (opposing) opinions of two people for whom I have the utmost respect:

Kim Lee writes: "You're being a bit narrow about the meaning of SCORE." And later on, "Give the people what they want!" MathProf intones: "I really am not sure why you are so upset. Sometimes I think you are like Columbus, who really didn't understand the significance of what he had discovered!" Fighting words, those. (For the record, I really didn't "discover" SCORE, as I documented above, but I did coin the term, standardize its parameters, and I tried like the dickens to explain the concept and popularize its usage. I guess that should count for something!)

I need to invoke one final analogy. Back to our Oreos! Consider Double Stuf Oreos. They *are* Oreos of a sort (Oreos on steroids!), but they're a special kind of Oreo, different from the original, one-and-only Oreo. So, we can call them Double Stuf Oreos (continue to use the Oreo name, but now with a modifier), or, eventually, evolve to just Double Stufs (the new name — the modifier — supplants the old, and "Oreo" vanishes altogether), but we can't call them just plain Oreos, because that appellation invokes a singular thought. Nor can we call just any chocolate sandwich cookie an Oreo, lest we allow that registered trademark of Nabisco to go the way of the thermos and the escalator.

So, I see three possibilities: 1) Allow SCORE to "roam free," unencumbered by the original "trademarked" constraints I have imposed on it, and let it become somewhat of a generic term (this isn't necessarily a bad thing); 2) Go the "Double Stuf Oreos" route, preserving SCORE in various forms of modified designations (see above), where the term has some description in front of it; or 3) Create an entirely new term to describe a SCORE-like value, which follows the $(ev/sd)^2$, or DI^2 dictates, but falls short of being the "genuine article." I'd really like to avoid that, if possible.

Wait! Just as I was getting ready to wrap this up, I got what I think is an inspirational idea (better late than never!) that may just solve the entire problem. But, you're going to have to bear with me for one more important digression.

Our age of technology has created so many new products at such a dizzying pace that, from time to time, we actually need to reconsider what we used to call certain things, before the advent of their newer incarnations made a *retronym,* or retrofitted word, necessary. For example, in my day, a "watch" was a "watch," or, alternatively, a "wristwatch." Enter the *digital* watch. So, to distinguish the more modern version of this timepiece from its predecessor, linguists were forced to come up with "*analog* watch." Similarly, the standard cradle telephone of yesteryear, which used to have a dial, was just a plain "telephone" until the touch-tone (pushbutton) phone came along. Thus was born the "*rotary* (or *rotary-dial*) phone," so that we could differentiate between the two. And although there really isn't a single term we can all agree upon, most now call "*snail* mail" what used to be simply "mail," due to the arrival upon the scene of the ubiquitous "*electronic* mail," or "e-mail."

Finally, how can we overlook a retronym that popped up right in our own backyard? The surrender option traditionally was offered to the player only *after* it was verified that the dealer did not have a natural. We didn't get to surrender against the dealer's blackjack. And then came a new form of the option, where we *did* get to wave the white flag, even in the face of the dealer's mighty natural. And, this highly beneficial rule was called "*early* surrender." Well, now, what to do about the *original* surrender, so as to avoid confusion with the new term? Enter "*late* surrender"! (For the record, you may be interested to learn that it was Stanford Wong who coined the term.)

What does all of this have to do with finding the best solution to our SCORE predicament? Well, in considering the above examples, I suddenly remembered what had to be the single greatest marketing disaster in history — the most monumental corporate miscalculation since Ford said to the public, "You'll love these Edsels"! I'm thinking of Coca-Cola's catastrophic decision to scrap the world's most preferred soft drink in favor of the "New Coke." When the clamor died down, two things were crystal clear: 1) the new product was a resounding failure that had to be taken off the market, and 2) the "old" Coca-Cola had to be brought back. But now, there was a marketing dilemma: what to call the returning product, which was, of course, nothing more than the original "Coca-Cola." Ah, but with the creation of the ill-fated new version, there was now a need to distinguish the genuine article from its intended, but subsequently aborted, replacement. And so, "Coca-Cola *Classic*" was born!

Thank you for your patience. We now return to the SCORE discussion! I'm going to propose something bold. In so doing, I may very well be shooting myself in the foot, as it were, reversing direction in mid-field, to suggest a compromise. I think what is called for here is a *retronym* for my original concept of SCORE, to leave the scene free and clear for all the other SCOREs you wanted to refer to before I got on your case. So, what should we call the prototype? *Classic* SCORE? Too pretentious. *Original* SCORE? Accurate, but a bit dull. *Article* SCORE? Too specific a reference. *Baseline* SCORE? Hmm. We seem to be on the right track, but that's not the exact word I'm looking for. I've got it! *Benchmark* SCORE. That's precisely what it is. The SCOREs in the original article above are all *benchmarks.*

So, what do you think? You want to tinker around with bankroll, cover, indices, Kelly fractions, number of players at the table, or all of the above at the same time? Tinker away! Go ahead and create all the SCOREs your hearts desire. You have my permission. But, let's agree, when we do have those special, prototypical SCOREs in mind, to pay them some respect and call them "benchmark SCOREs," OK? This may take some getting used to, because, until now, it was usually to these benchmarks that most players were referring when they did mention SCORE in the first place. Will you guys have the patience to append the extra word, for clarity, so we may all understand that there are now garden-variety, run-of-the-mill SCOREs and then the *special* ones I created so that we could compare games and systems? Please give this your consideration.

The ball is in your court. I eagerly await your comments, reactions, and suggestions.

Postscript 1: *Several months have passed since I made the above remarks readily available on the most popular blackjack Web sites. The verdict is in, and my proposal for a retronym seems to have been widely disregarded by the blackjack community. Instead, Richard Reid has suggested keeping SCORE in its pure form and designating any SCORE other than an "original" as a* **custom-***SCORE, or "c-SCORE," for short. The idea seems to be catching on, and I couldn't be more pleased. We get to preserve the "purity" of the SCORE term, while avoiding confusion by prefixing a "c-" to any SCORE other than a benchmark variety. This compromise has my enthusiastic endorsement, and I hope that it will become the "industry norm" in the years to come.*

Postcript 2: *Nope. Didn't happen. I've lost the battle. It is now the summer of 2003, and one thing is perfectly clear: Every darn hourly win rate has become, in the players' eyes, a SCORE, whether I like it or not. Although I persevere, trying desperately to maintain the aforementioned distinctions, I am a minority of one. The blackjack community has spoken, and SCORE has become a generic term depicting* *the hourly win rate of any game and approach to it under discussion. Win some, lose some. I'm less upset than you might imagine, as the SCORE concept has become so widely accepted as to have taken root as the gold standard for game comparisons. I'm grateful for that acceptance, and I bow to the public's desire to render SCORE generic. Now, if you'll excuse me, all of this has given me a gigantic headache, so I think I'm going to open my thermos and take a couple of aspirin!*

Chapter 10

The "World's Greatest Blackjack Simulation"

Preface for the third edition: *In the 1950s, one of my favorite television shows, which aired on Sunday evenings, was "You Asked For It!" The host was Art Baker, and the program's producers traveled around the world to film the most entertaining and unusual events and stunts that viewers had requested by letter or postcard. There were some utterly amazing features, and at the conclusion of each segment, Baker would intone, in staccato fashion, "You-Asked-For-It," as the music played four chords in the background. Well, cut to 2003, and strike up the music, because, with the all-new, entirely revised Chapter 10 charts that follow, I can truly affirm that ... "You Asked For It!" Permit me to explain.*

Since the first appearance of the charts, in 1997, players have been asking for several critical changes in their scope and content. For one, they wanted the newer SCORE methodology to be reflected both in the reporting of the win rates, as well as in the betting schemes devised for each game. In short, players wanted optimal betting ramps to be used, and they wanted the varying unit sizes for those bet schemes to be furnished, as well. Other enhancements, such as the standard deviations for each true count and the flooring (as opposed to truncating) method for calculating those counts, were also requested. Finally, the most accurate index numbers, corresponding to the flooring technique, would also be required. Enter Norm Wattenberger.

To any serious student of the game, Norm can hardly be a stranger. His suite of ***Casino Vérité*** *products has become the gold standard in blackjack simulation software. Sometime after the appearance of the second edition of* ***BJA,*** *Norm decided to create a special product,* ***CVCX*** *(****C****asino* ***V****érité* ***C****hapter* ***X****), that would enable the user to reproduce not only the charts provided in the book, but also thousands of others, varying penetration by one card at a time (!), for a wide range of pack depths. The software would also employ the SCORE methodology and would, therefore, address all of the items on readers' "wish lists," some of which were mentioned above.*

As the date drew near to revise ***BJA*** *once again and to contemplate additions and enhancements for the third edition, I approached Norm and asked him if he would consider redoing all of the Chapter 10 charts, utilizing the* ***CVCX*** *approach. I was delighted when he responded, "When do we start?" What followed was a complete revision of the charts that you have grown so fond of, and I am pleased to present them as one of the highlights of the many added features in this third edition. You will be interested to learn that, for each chart,* 20 billion *rounds of blackjack were simulated! And, given the various rules sets, different numbers of decks employed, and the play-all and back-counting approaches, over* 20 trillion *hands were generated for this monumental study. A quick calculation on my part ascertained that this represented more hands of blackjack than have been dealt in every casino, worldwide, since the advent of the game!*

As excited as I was to have the new charts, a big problem loomed insofar as the prefatory remarks to the tables were concerned. Acknowledgment of the landmark, pioneering work, done by John Auston, for the first two editions, simply could not be omitted. There was, after all, a story to tell — the genesis and evolution of the charts. And so, I have decided to preserve much of the explanatory text — especially the earlier material — for reference purposes. However, if there is mention of a number that is still relevant to a chart, I have included the updated values, pertaining to the present charts, in square brackets and bold, italic font ***[like this].*** *Furthermore, where a specific aspect of the methodology is discussed, an update will be provided, again, in brackets, indicating a change. It is hoped that, by using this approach, the new material will virtually leap off the page and that all of the revised data will thus be readily apparent. Finally, to simplify the reader's task of assimilating all of this information, a new "Summary of the Simulation Methodology" is included, just before the presentation of the charts. For those who are already familiar with the text of this chapter, you might want to cut directly to the chase, beginning on p. 210. I trust that you will enjoy the brand-new* ***Casino Vérité*** *Chapter 10 charts.*

It all started so innocently. A questioner in Stanford Wong's "BJ21" Web group had asked whether a certain 2-deck game, with various rules furnished, would be superior to another 2-deck game where, once again, all rules, penetration levels, etc., were stipulated. Because of the well-known and oft-quoted values of rules such as "das" (double after splits permitted) and "ls" (late surrender), several readers responded by simply tallying up the starting disadvantage to the *basic strategist* and offering the more favorable number as if it were also, logically, the superior game for the counter, as well. But, I knew otherwise.

I knew, for example, that, despite the 0.14% afforded the basic strategist who may double after splitting, the rule didn't "inflate" much for the card counter; that is, despite a hefty bet spread and the use of strategy index numbers, the extra gain provided to the counter, by invoking das, just wasn't very much more than 0.14% of initial money wagered. On the other hand, I also knew that the late surrender rule worked very differently, once in the arsenal of the card counter. For the 6-deck game, the basic strategy extra advantage for conventional surrender is 0.073%. What's more, as the number of decks decreases, so too does the surrender edge, shrinking all the way to 0.023% for single-deck. But what about the card counter? What could he expect to glean from surrender in, say, a well-cut 6-deck game in which a decent (1–12) bet spread was employed? Well, the answer is a very different 0.20% (nearly triple the basic strategy number) for the play-all approach, and an extremely impressive 0.32% (more than four times the basic strategy gain!) for the back-counter who employs a 1–4 spread. ***[Whereas comparisons such as these are valid if one keeps the same betting pattern and unit size for the various rules sets, they are less meaningful when, as is the case with the SCORE methodology, the bet ramps and unit sizes change, as the rules change. Thus, we learn that, for the play-all game, the unit size was 11.22 for no surrender, but 14.31 when surrender was permitted. Similarly, when surrender was allowed, and back-counting was employed, the unit size grew to 104.93 from a previous, non-surrender 87.69.]***

I was reasonably certain that many readers did not fully understand the difference in the values of some of blackjack's fundamental rules as they pertained not to the basic strategist, but rather to the card counter. I posted the findings, and then, the wheels started turning. Here was the subject for a study that, on the scale I envisioned it, had never been done before. I would consider rules such as double after splits, surrender, hit soft seventeen, and double on ten and eleven only. I would choose a count system (invariably, Hi-Lo, for my studies), and would simulate a "base" game, followed by the same game with different rules. Then, I would determine the value of those rules, separately, and in combination, to the card counter. Already a big project, little did I realize, at the time, what an incredible journey I and my partner in this mission, John

Auston, were about to embark on. Before I continue, a word about the indefatigable Mr. Auston.

John was a relative newcomer to the blackjack scene, an honest researcher who showed an inquisitive mind, a sharp wit, a good writing ability, and a great proficiency with computers and simulations. He climbed the learning curve very quickly, and several of his posts to various Internet blackjack groups began to impress me. We started an e-mail correspondence and, some months later, when my thoughts of the aforementioned project began to crystallize, I turned to John, coincidentally, as I had to another John (Gwynn), eight years prior, for the now-famous "Floating Advantage" study. Here is the partial text of an e-mail I sent to John Auston, enlisting his help in a piece of research that I knew I did not have the time to carry out alone:

"How long do you think it would take to do 250, 100-million-hand simulations ***[now 20 billion hands, as previously mentioned]?*** Basically, that's what I think it would take to make this 'masterful.' Base games will be 1-, 2-, and 6-deck ***[8-deck was added for the second edition],*** Strip rules (i.e., s17, doa, no rsa, split to 4 hands, peek, ins). For each game, we add, separately (a different sim), to the base rules, the following: 1) das; 2) h17; 3) ls; 4) das and h17; 5) das and ls; 6) h17 and ls; and 7) das, h17, and ls. This would make eight sims for each, except that, for one-deck, I would add h17 and d10 (one sim) for Northern Nevada players.

"Next, we define the levels of penetration. ..."

And so it went. I continued to specify all of the criteria for this proposed study, and ended by asking John if he wanted to join me in this exciting endeavor. I'll leave it up to him to decide if he rues the day that he said "yes"!

It is now four and one-half months later. Literally *hundreds* of e-mails have been exchanged between the two of us. Three different simulators were used, before we agreed upon the "ideal" approach. Untold hours of computer time and programming have gone into producing the *fifty-six pages* ***[now seventy-three!]*** of charts, analyzing over 500 distinct blackjack games, that you are about to read. For, you see, once begun, the project began to take on a life all its own! Not content with the primary study and its completely fascinating results, I kept pushing for more. And Auston never quit. The French have a saying: *L'homme propose, et Dieu dispose* — "Man proposes, and God disposes." Here, it was "Schlesinger proposes and Auston disposes"! Quite simply, as more and more data became available, I decided that this was going to be, immodest as I know it must sound, the greatest blackjack simulation ever done. I'll let you be the judge.

Methodology

After several "false starts" with other simulators (due mainly to problems with calculating the precise standard deviations, properly handling the rounding versus truncating question, speed, and a few other minor details), we decided on Karel Janecek's "Statistical Blackjack Analyzer" (henceforth referred to as "SBA") as the vehicle of choice. Janecek's brilliant software simply met more of all the criteria we had imposed for the manner in which we wanted to carry out the study. ***[For this third edition,* Casino Vérité *was used.]*** As stated above, the count system to be used was Hi-Lo. I have had to make this decision many times before, with other research I have performed. Hi-Lo becomes the system of choice for the simplest of reasons: I am quite certain that it is employed by more card counters than any other method. Since I could not hope to appeal to all players at once, it certainly made sense to appeal to the largest segment of the card-counting population.

Next, a betting strategy, or, more correctly, an entire slew of betting schemes, would be determined. For the play-all games, I decided upon spreads of 1–2, 1–3, and 1–4, for single-deck; 1–4, 1–6, and 1–8, for 2-deck; and 1–8, 1–10, and 1–12, for 6-deck. ***[For both 6- and 8-deck, spreads of 1–8, 1–12, and 1–16 are now used. In addition, a new array of spreads for multi-deck back-counting, namely 1–2, 1–3, and 1–4, is now employed.]*** Obviously, once again, I knew I couldn't please everyone here. But, I think the vast majority of card counters will identify with one of the three choices for the game they play most frequently.

Of course, the scaling of the bets would also have to be determined. Here, it is obvious that, depending on the game studied, and the advantages ascertained at each true count, different bet schedules could be employed. But, keep in mind that, before the project "sprouted wings," the original objective was to simply choose a method of play and to determine how much additional advantage could accrue to the counter, as rules variations were introduced. I didn't want to get too fancy with a multitude of bet variations. So, I settled upon raising the one-deck bet to two units at a true of one and to reaching the max bet of three or four units (if we permitted such a spread) at a true of two. For double-deck, we bet two units at TC = +2, four units at +3, four or six units at +4, and again, depending on the spread permitted, either four, six, or eight units at TC = +5 or higher. Finally, for 6-deck, the same as above, through +3, then six units at +4, and either eight, ten, or twelve units at TC = +5 or higher. ***[These schemes were abandoned in favor of the Kelly-optimal approach used in the SCORE methodology. In addition, for each game, a second bet ramp, utilizing integral, casino-chip unit sizes, where little or no bet-jumping was permitted, is also furnished, to provide a customized, practical, "real-world" approach.]*** And, yes, I know, you may bet differently. But, as you will learn shortly, nearly one *trillion* hands ***[make that 20 trillion!]*** of blackjack were simulated for this study! There just had to be some limits to the ranges of the various parameters that were input.

Playing variations were the next roadblock to consider. Oh, how we agonized over these! In the end, I opted for the "Illustrious 18," not because they were, after all, my "baby," but rather because, once again, they represented a compromise between those players who use more strategy indices and those who may use fewer, or none at all. Too, I wanted the value of the rules (let's not lose sight of the primary objective) to be truly representative of the way most of us play the game under contemporary conditions; and if I'm reading today's messages correctly, many go the route of the "Illustrious 18."

But how to determine them? Brilliant as it was, when we began, Janecek's *SBA* did not devise playing indices (it does now, by the way). We might have used the published indices from Wong's *Professional Blackjack,* but they are for 1-deck and 4-deck, while we were studying other games. So, we returned to John Imming's vaunted *UBE,* whose methodology for determining correct indices for playing departures is, to my knowledge, the most accurate approach that exists. ***[The CV index generator, more accurate still, was used to produce the new indices.]*** You will find the precise indices for each game, including separate values for the h17 games, in Table 10.1, on p. 213. Note, as well, that the "Fab 4" surrender values were also used, when the rule was offered, and that, for the one-deck game, which I never studied when generating the "Illustrious 18," I deemed it appropriate to include the doubling indices for 8 v. 5 and 6, because I feel that they are important enough to "make the cut," so to speak.

Next came the eternal "truncate versus round" debate for the true count and how it is ultimately used by the player. I have already discussed this phenomenon somewhat in detail in the Appendix to the "Floating Advantage" chapter, on pp. 82–83. Simply stated, the concepts of "bet four units at a true count of +3," or "stand on 12 v. 3 at a true count of +2 or higher," can actually be interpreted differently by today's players. To many researchers and players, "a true count of +3" means "a true count between 2.5 and just below 3.5." That is, many players *round* the true count, at least for the purposes of betting their money. Inherently, there is absolutely nothing wrong or incorrect with this approach. I simply don't do it that way, and neither has Stanford Wong, for all the years he has been simulating and producing blackjack data. We prefer to *truncate,* which is to say that, for us, when we bet, "a true count of +3" means that the TC is *at least* precisely +3, and not yet +4. ***[Times change. The methodology of choice for both reckoning the true count and for subsequent use of indices is*** **flooring.** ***For positive counts, flooring and truncating produce identical results. However, for negative counts, flooring rounds to the left, or downwards, whereas truncating rounds to the right, or upwards. So, if a count is, say, –2.6, truncating produces –2, whereas flooring yields –3.]***

Now, this approach guarantees that, if you simulate in the rounding manner, your numbers are going to look different from mine. For any specified interval, logically enough, a positive rounded true count will appear more frequently than the truncated ***[or floored]*** true count. As compensation, the advantage associated with the positive rounded true count will, of necessity, be smaller than its truncated ***[or floored]*** counterpart. And so, when all the smoke clears, and we get around to determining *win rates,* there is little difference in the two approaches, as noted in the "Floating Advantage" remarks.

As for the playing indices, it is generally accepted practice that we do not make a departure until we have fully attained the index number in question. That is, in the aforementioned scenario of 12 v. 3, it is understood that, if we were able to determine a *precise* true count of, say, 2.8, and recognizing that we "need" 3.0 to stand, we would, indeed, *hit* the hand in question. Interestingly enough, recent studies performed by Ken Fuchs, and published in his excellent joint venture with Olaf Vancura, *Knock-Out Blackjack,* as well as in the Internet groups, provide compelling data as to the relative unimportance of being "off by one" when we use strategy departure numbers. Which leads to the next question: How, exactly, did we program the simulator to compute (or rather, to estimate) the true count, in the first place?

Another area for compromise! Devotees of *UBE* or Carlson's *Omega II Blackjack Casino* simulators are usually quick to put the "true count precision" setting on the adjustment that is most accurate — namely, to one card. When this is done, throughout the simulations, the true count is calculated by considering the *precise* number of cards remaining to be dealt, and by utilizing that exact number in the determination of the true count. The resulting values, furnished by the sim, are, therefore, quite accurate. There's only one problem: Human beings are incapable of such precision in their own play. So, what's the sense of reporting frequencies and data that do not correspond to anyone's reality in a casino? Still, researchers may yearn for "perfection." They may want to know what is actually happening at the table, whether or not the counter is capable of discerning these results. What to do? Well, "real world" calculations, for most players, are probably along the lines of true-count adjustment to the nearest deck, or, more ambitiously, to the nearest half-deck. As a compromise, we chose the "quarter-deck" setting. As such, the values you will read are near enough to what a human can achieve to satisfy the practitioners, but are also close enough to "perfect" to allay the fears of the most meticulous theoreticians. ***[This time, we used quarter-deck estimates for the one- and two-deck games, but half-deck precision for 6- and 8-deck.]***

It is interesting to note, parenthetically, when discussing the above topic, that setting the true count precision to "one card" does not imply that the simulation results, which are, by definition, more exact, lead to higher win rates. Often, they do not. This is true because a more crude approximation of the true count may, indeed, cause us to evaluate some lower counts as one number higher, thereby attributing a greater frequency to the count in question. As a result, we are, technically, overbetting somewhat. The consequences can be slightly larger win rates, to be sure, but somewhere, there must be a price to pay, and it is in the form of correspondingly higher risks of ruin (higher standard deviations) for the "cruder" approach. ***[The above no longer holds if optimal betting is employed. Quite simply, greater precision in estimating true counts leads to slightly higher win rates than if quarter-deck or half-deck estimates are employed. But the effects are minimal.]***

What else was left to agree upon, before the simulations could begin? Well, penetration levels, number of players at the table, and sizes of the sims themselves would probably be the final considerations. These were relatively simple to establish. Realizing that there had to be some practical limits to the amount of data generated, we settled on four levels of penetration for each game. For single-deck, 50% (26 cards dealt), 60% (31 cards), 67% (35 cards), and 75% (39 cards) were chosen. I must say that, in this area, there was some discussion as to whether to introduce, by the above methodology, the dreaded "cut-card effect," or to opt, rather, for a fixed number of rounds to be dealt. Here, I deferred to Arnold Snyder, who has more experience in playing single-deck than I do. As we aimed for realism, Snyder assured me that, much more often than not, single-deck dealers go by the "feel" of the remaining cards in their hand to tell them when to shuffle, and not by some predetermined number of rounds that they would be willing to deal, based on the number of players at the table. To be sure, there can be some casinos that have just such "rounds" guidelines, but we decided that they were the exception, rather than the rule.

For 2-deck, penetration percentage levels were identical to single-deck, producing depths of 52, 62, 70, and 78 cards, respectively. Finally, for the 6-deck shoe, we settled upon 4, 4.5, 5, and 5.5 decks dealt, corresponding to my evaluation of poor, average, good, and exceptional levels of penetration for these games. ***[For 8-deck, penetration levels were 5.5, 6, 6.5, and 7.]*** After all the sims were done (months of work!), we learned that, due to a quirk in the *SBA*'s methodology for defining penetration (since corrected), the shuffle card actually appeared *before* the card mentioned in the depth to which we dealt. Thus, 78/104 meant that, at the end of a round, if the stop card was poised to come out after the 77th card was dealt, a shuffle occurred. Faced with re-running all the sims (a nightmare!), we decided to simply explain the phenomenon, whose effects are entirely minimal for shoe games and are very small even for 1- and 2-deck. In fact, we may easily turn the feature into a more realistic interpretation of penetration by noting that, as presently constructed, our charts furnish an automatic "burn card" for every game studied! ***[This feature was retained for the new sims.]***

How many players to put at each table was, again, a purely subjective decision. I hope you will be comfortable with our choices of two players at the single-deck game, three at the table for 2-deck, and four players attacking the 6-deck ***[and 8-deck]*** shoe. At all such multi-player games, we got to play last, thus gleaning a little extra advantage from cards glimpsed to our right, before the play of our own hand. Note that, while the 1- and 2-deck games were dealt face down, the 6-deck ***[and 8-deck]*** shoes were dealt "open," as is customary in most casinos.

Finally, as the last item in our simulation specifications, we needed to agree upon how many rounds would be dealt for each and every one of our 500+ different game scenarios. John and I felt that it made no sense to be so meticulous about everything else and then to skimp on the accuracy of the findings we were about to generate. Only a number large enough to produce relatively small standard errors for the frequencies and win rates would be acceptable. Given the incredible speed of the *SBA* and Mr. Auston's computers, we settled upon the norm for each line of data produced: *400 million rounds* ***[now 20 billion]!*** And since, depending on the game studied, there were an average of either three, four, or five hands dealt per round (including the dealer's), a little multiplication leads to the rather mind-boggling conclusion that, to produce the data you are about to read, in excess of 800 *billion* ***[20 trillion]*** hands of blackjack were dealt by computer simulation!

If the adjective "greatest" can be used to depict not only the concept of "most extraordinary" or "most accomplished," but also one of sheer size, as in "biggest," "largest," or "hugest," then, shamelessly and unapologetically, I proclaim that what you are about to read is The "World's Greatest Blackjack Simulation"!

How to Interpret the Data

Again, let us not lose sight of the original mission — to determine the value, to the card counter, of various rules, as they are encountered under different game conditions. Before examining, in detail, the simulations for each broad category of game, we summarize our findings in a one-page chart. Here, depending on the number of decks being dealt, we are afforded eight or nine different game rules scenarios, each with four levels of penetration and three bet spreads. So, for our first summary page, the 8-deck play-all approach, 96 different games are examined. What's that? You always back-count the shoe game and never play all? Oh, perhaps I forgot to tell you: For the ***[8-]***, 6-, and 2-deck games, we ran all the sims for the back-counting approach, as well! (More about this in a little while.)

The "base" game is always the one with the dealer standing on soft 17 ("S17," in the charts) and the poorest level of penetration. The idea is to introduce a new rule and to summarize the effect of that rule by indicating, in the appropriate column, the new win rate ***[now SCORE]*** (in large, **bold** numbers) and, to its right, the differential, expressed as extra units ***[dollars]*** won or lost, per 100 hands played (or seen, for back-counting), *in comparison to the base game*. Careful distinction must be made not only between the extra *units* ***[dollars]*** provided by a favorable rule and the *differential in the win percentages* for the games being compared, but also between that very same differential in win percentages and the *increase in winnings* ***[as expressed by SCORE],*** percentagewise, that the favorable rule provides. The latter two, in particular, can be a source of great confusion. A single example will suffice, to illustrate the above points.

Let's consult the 6-deck, play-all summary page (p. 232) and consider the base, 6-deck, 5.0/6, 1–12 spread game. We see a SCORE of 25.88. (Ignore, for the moment, the small "18.31" below the bold **25.88.**) Now, move one column to the right, where das is permitted. We learn that the new win rate is **34.61** per 100 hands played, and that, furthermore, this represents (small number to the right) an 8.73 increase over the **25.88** figure for our base game. Well, how much of a *percentage* increase, over the 25.88 SCORE, is a 34.61 win? Clearly, it is (34.61 – 25.88)/25.88 = 33.7%. Later, when we look at the individual charts, we will learn that the "global" percentage win rate of initial bets, for the base game, is 0.95%, while that of the das game is 1.08%. It is this differential of 0.13% that the rule of das adds to the percentage win rate of the base game. ***[Once again, we must advise caution in making this kind of comparison, as the percentage win rates are operating on* different *unit sizes.]*** As such, this value can *not* be gleaned from the summary pages, but must be ascertained by judicious comparisons of the individual charts, on a line-by-line basis.

It is important, therefore, to understand that, in this specific instance, das adds to the card counter's win percentage just about the same as it does to the basic strategist's advantage (or, often, disadvantage); but, in terms of actual increase in *earnings,* the rule augments our hourly stipend ***[as determined by SCORE]*** by almost 34% over the base game's hourly win, expressed in dollars.

Returning to the summary pages, this time for back-counting (p. 233), let's stay in our original column, for the base game, but travel vertically this time, through different penetrations and bet spreads. Here, the incredible importance of spread and, especially, of penetration becomes evident. Again, a few examples will prove helpful. The 4.0/6, 1–2 spread game wins **20.83** per 100 hands played. (If we agree to the standard of 100 hands played per hour, we may shorten further comments by referring to "win per hour.") Subsequent spreading to 1–3 or 1–4 produces relatively small increases in wins, to **22.13** and **22.62** dollars per hour, respectively. These become our base figures, or "norms." Now, let's move down the column.

When we consider the 4.5/6 game, two items become immediately apparent. First, all the previous SCOREs are improved substantially, with the extra dollars won per hour cataloged *underneath* each new win rate, and always with reference to the *original* 4.0/6 game. Thus, our 1–4 spread now wins **35.02** dollars per hour, a full 54.8% increase in earnings over the **22.62** rate associated with the 4.0/6 game. What's more, within the *same* 4.5/6 game, the value of spreading (2.30 more dollars won, per extra unit added to the 1–2 spread) is considerably more than it was for the 4.0/6 game, where the increase in spread from 1–2 to 1–3 added only 1.30 dollars per hour.

And, of course, the trend continues. By the time we reach the exceptional level of 5.5/6 penetration, our 1–4 spread wins **97.56** dollars, or almost *four and one-half* times the base figure of **22.62,** while each unit of extra spread that we can get away with (from 1–2 to 1–4) adds an additional 5.14 or 2.20 dollars of win — a far cry from the 1.30 or 0.49 of the 4.0/6 game.

But it is the charts themselves, and not the summary pages, that represent the incredible compendium of information to be found on the following pages. Permit me to explain their evolution and the meaning of each column of data provided. Originally, in keeping with our guiding premise of importance of rules variations, we intended to report, for each game, the following only: "Bet Spread," "%W/L" (that is, the hourly win rate, expressed as a percentage of the *initial* bets), and "Win/100" (or, the number of units won per 100 hands played or observed). Indeed, reporting just these three items, for each game and set of rules, would have represented, in my humble opinion, a monumental achievement and would have broken considerable, new ground in the area of blackjack simulations. Day by day, as the data came pouring in to me, from Auston's computers, I sifted through the output, looking for bugs, or anomalous findings. Inevitably, there were lines that looked "funny." Painstakingly, John and I "debugged" any errors, often resorting to lines of information not originally intended to be seen by the reader.

And then, one day, as if by revelation, it came to me. Why withhold any of the data that had been generated in the course of calculating the rules' values we had set out to find? Why not provide *everything?* The information was there, after all. And so, we decided to furnish, as a start: the frequencies of all true counts, expressed as percents ("TC Freq. (%)"), the standard errors of those frequencies ("Freq. SE %") ***[we no longer report the SEs, because with the huge sample size of each sim, SE is so small as to be, for all intents and purposes, zero],*** the win rate, or "expected value" associated with each true count interval ("EV W/L (%)"), the standard errors of those win rates ("EV SE %") ***[gone!],*** and the size, in units, of bets made at each true-count level ("Sim Bets") ***[now "Play-All Bets" and "Back-Count," for both the "Optimal" and "Practical" approaches].***

It must be acknowledged, at this time, that the pioneer for all such simulation studies remains Stanford Wong. His 1983 simulations, published in volume 5 of his newsletters, are the models from which all such work has evolved and flowed, in the past thirteen years. In particular, the July 1983 issue of *Nevada Blackjack* carried, on pp. 114–16, an enlightening study that very well could have served as the title for this chapter: "Importance of Rules Variations to Card Counters." But, immodestly, I must say that the present work, both in its depth and sheer size, has "pushed the envelope" and treads on some ground that even the inimitable Wong did not walk.

Inspired, as it were, by the memory of the aforementioned landmark simulations, I decided that we could do more. Surely, the average bet size could be provided ("Avg. Bet ($)"); so, too, could the standard deviations, both for a single hand, and for 100 hands played, or seen (the back-counting approach) ***[now just "SD/100 ($)"].*** We also thought it proper to furnish the standard errors associated with the percentage win rates. They appear to the right of the win rates, designated as "SE %." ***[Still no need for these, as the sample sizes were huge.]***

[The next three paragraphs are rendered obsolete by the SCORE methodology. You may skip them.]

I had begun to exhaust the possibilities for reporting data when I decided to really "take a chance" and test John's patience! Once a bankroll size is chosen, the risk of ruin for any given game is fully described by that game's per-hand win rate and per-hand standard deviation (recall the Ruin Formula, from p. 112). But, we had decided to report that information. Why not designate a bankroll level (or two!), in units, and provide a unique service to the reader by calculating, for each of the games studied, the risk of ruin for each? Why not, indeed.

No one bankroll size can be "optimal," or ideal, for the range of games we studied. Fixing the "b.r." (bankroll) size, in units, would guarantee wide disparities in the risks of ruin reported for each game. No matter. The alternative was to designate one or two fixed rates of risk of ruin (along the lines of our Chapter 8), and to furnish the bankrolls necessary to stay within the risk boundaries. I decided against this rather subjective approach, and in favor of the former scheme. And so, to the extreme right of the charts, for 2-deck and 6-deck, I have furnished the long-term risks of ruin associated with total bankrolls of 400 and 800 units, for every game we studied. For single-deck, where 400 units may actually be considered too *much* bank by some, ruin figures for an alternative 300-unit bank are supplied. Consider these an extra bonus. I hope you find them useful.

Two columns remain to be explained, before I move on to some analysis of the massive quantity of data simulated. The first is found to the far right of the first group of information, after the "Sim Bets" sizes, and is entitled "Optm Bet 400 Bank." By now, the reader should be fully familiar with the concept of Kelly-Criterion wagering and the determination of the optimal bet size for a given true count and set of playing conditions. Recall that, for a given bank of, say, n units, the proper Kelly bet to place at any particular true count is simply the product of the advantage associated with that count and the bank size, n, divided by the variance of outcomes of a hand of blackjack played at the particular true. Well, although we had to draw the line somewhere, and we do not report those variances in our charts, we do the next best thing (and, perhaps, a much better thing), and that is to furnish, directly, for the 400-unit bankroll, the optimal bets for each true count and for each game.

Obviously, once the 400-unit optimal wagers are established, optimal bets for any other chosen bank size are readily attainable through a direct proportion. What's more, the line-by-line variances can be "backed out," quite easily, from the tables, by simple multiplication and division. ***[No need. Standard deviations are now provided for each true count. Squaring those values gives variances.]*** Here's an example. Suppose I'm interested in the per-hand variance of the 5.0/6 base game, at a true count of +5. Simply square the indicated standard deviation of 1.140, yielding 1.300 as the variance. Similar calculation of the variance, at the same true count level, for the 5.0/6 game, but where das is permitted, yields a predictably higher value of 1.313 (1.146 squared), whereas, for the late surrender game, the salutary effect of this rule as a variance reducer is clearly discernible: variance = 1.245 (1.116 squared). We shall return to this very beneficial "fringe benefit" of late surrender in later comments.

Finally, were one to bet according to the optimal bets suggested, but nonetheless respecting the restrictions of the bet spreads allowable (more about this in a moment), a win rate associated with this optimal betting style can be determined. This is the figure furnished to the extreme right of the charts, entitled "Optm Win/100" ***[now reported under the "Results" section as "W/100 ($)" and to the far right, risk-adjusted, as SCORE].*** Note that, unless one is back-counting, the zero optimal wagers associated with true counts of zero and below must become one-unit wagers, while optimal bets that exceed the prescribed guidelines for bet spread must be reduced to the maximum wager permitted. Thus, the "Optimal Win" ***[SCORE]*** columns report, more accurately, the win associated with bets scaled as precisely as possible, given the above two constraints.

As an aside, now may be the appropriate time to mention that, for the frequency and win rate of the zero true-count lines, the interval is actually a double one, ranging from just above –1 to just below +1. At the time we utilized it, the *SBA* did not break apart this interval into two (or, perhaps, three — with "exactly zero" a possibility) separate categories (subsequent versions of the simulator may improve upon this minor "flaw"). Whereas we would have preferred the greater precision, it is to be noted that, because the advantage is almost always not great enough to warrant raising one's bet in this territory, nothing is lost by lumping the "0–" and "0+" counts together, on one line. ***[The flooring methodology reports the interval consisting of a true count of exactly zero and all trues up until, but not including, +1 in the TC = 0 line. However, true counts below zero, but greater than –1 are now lumped in with*** **all** ***negative true counts, and are reported in the TC < 0 line, as per the flooring algorithm.]***

Having now explained the complete genesis and significance of the features of the charts, it behooves us to provide further interpretation and analysis of the results obtained.

An Appreciation of the Data

Many observations "jump" off the pages of data provided by the charts. For me, one of the most striking is the incredible value of the surrender option. Because of its deceptively small value to the basic strategy player, many counters undoubtedly underestimate the advantage-enhancing properties of conventional, or late, surrender. In multi-deck, from the weakest 8-deck, play-all base game (5.5/8, s17, 1–8 spread), where surrender adds 3.36 dollars to the "plain" game, to the 2-deck, s17, back-counting game, dealt 78/104, with a 1–4 spread, where surrender adds almost 15 times (14.79) the number of dollars (49.71) to the basic win, the gains are impressive. Furthermore, as stated earlier, the "inflation" of the value of the rule, from its various basic strategy levels, is quite astounding. Thus, we see that, under the proper circumstances (in this case, the 5.5/6 back-counted game, with a 1–4 spread), the increase to the percentage win rate, attributable to surrender, can mount to as high as 0.36% over the base game (3.21% v. 2.85%) ***[again, remember that the percentages operate on different unit sizes].*** This gain represents a multiple in value of 4.9 over the basic strategy figure of 0.073%.

But the allure of surrender doesn't end with its expectation-enhancing properties. As previously mentioned, surrender has the highly desirable quality of increasing gain while simultaneously *decreasing* risk (as measured by variance). In the world of finance, such a feat is, indeed, revered by the average investor, who has been taught that, in general, to increase returns, one needs to be willing to take on more risk. Rare, indeed, is that judicious mix of assets that permits one to increase returns over some measured benchmark, while managing to produce *lower* risk than that of the same benchmark.

Well, in blackjack circles, surrender represents this "Holy Grail." I am altogether convinced that most card counters, even the most experienced, do not fully appreciate the "therapeutic" value of the surrender feature. And fortunately, at least in Las Vegas, many such games still exist. Ironically, savvy managers, in whose casinos surrender is offered, realize that the amount lost to this rule by the uninitiated, who abuse the "privilege," thereby routinely leaving valuable money on the table, is far more than the amount taken from the casino by enlightened basic strategists and card counters. Indeed, where comparative studies of "hold" have been conducted by casinos that introduce surrender, invariably, profits increase over the non-surrender, "year over year" figures. We have, in essence, a "win-win" situation, whereby surrender is not only good for the card counter, but seemingly, enhances casinos' profits at the same time!

By rather stark contrast, it would appear that the "elasticity" of the double-after-split rule cannot hold a candle to its surrender cousin. In fact, little improvement, from a win-percentage standpoint, can be achieved by the counter, despite liberal spreads and the deepest of penetration. No matter what the game, for any 6-deck variety, the 0.14% basic strategy value simply cannot be enhanced by more than a couple of hundredths of one percent. Indeed, for most games, the increase is virtually nonexistent. Once again, we mustn't confuse this percent increase of the win rate with the increase, percentagewise, in dollars won ***[SCORE],*** due to the das rule. It is altogether possible to nearly quadruple the amount of dollars won per hour, simply by seeking das. But to claim this, one needs to begin with an entirely unacceptable 4.0/6, h17, 1–8, play-all game, whose SCORE is a puny 0.86. Adding das almost quadruples the win, to 3.14 dollars per hour. Clearly, neither provides a playable game.

In a similar vein, the dealer who hits soft 17 actually takes *less* money, on a percent basis, from the counter than he does from the basic strategist. This is not a new discovery, and has been documented, most recently, by Bryce Carlson, in his excellent *Blackjack for Blood* (see, in particular, the final paragraph of p. 145). Our own research confirmed Bryce's findings, whereby the penalty to the basic strategist for this rule, which ranges from 0.195%, for 1-deck, to 0.217%, for 8-deck, is actually *reduced* for the card counter. For the back-counted 8-deck games, this reduced advantage averaged only 0.15%, while the 2-deck variety cost, on average, about the same. On the other hand, whether we played all at single-deck, 2-deck, or 6-deck, the percentage penalty approached the full basic strategy values (around 0.20%) for games where small spreads were employed. As the spread ranges increased, the imposts of h17 "retreated" back to the 0.16% to 0.18% ranges — slightly lower than the basic strategy penalties.

These findings are somewhat ironic in that, if one were asked to state a preference for, say, back-counting a 2-deck game offering surrender, but where the dealer hit soft 17, or playing a double-decker with no surrender, but featuring the s17 rule, many an experienced player might be tempted to analyze as follows: "Let's see, for 2-deck, where the dealer hits soft 17, surrender is worth only 0.067% over the "base" game *[Note: It is worth only 0.052% for the 2-deck, s17 variety — D.S.]*, whereas the penalty for h17, versus s17, is 0.204%. I'll take the s17 game, with no surrender." Wrong!! In fact, it was this very question, which I alluded to in my introductory remarks, that was the catalyst and driving force behind this entire study.

The correct answer, as we now know, is, Run to the surrender game, despite h17. For, in this version, the player advantage will range anywhere from 2.15% to 3.56%, depending, as always, on bet spread and penetration. The best the s17, no surrender games can offer are 2.00% to 3.29%, respectively. So, what looked like about a 0.14% superiority for the latter game turned out, in reality, to be anywhere from a 0.15% to 0.27% inferiority. ***[And, in terms of SCORE, there is really no contest. The s17 game wins from 51.32 to 156.68, while the h17, ls game garners from 61.05 to 190.14, representing increases of from 19.0% to 21.4% in the actual dollars won.]***

Finally, still on the topic of rules variation, we note that, for single-deck, Northern

Nevada rules, where the player may double on 10 or 11 only (d10), and where, in addition, the dealer hits on soft 17, the former rule alone, for the basic strategist, carries a hefty 0.277% penalty. If, temporarily, we consider h17 to be our "norm" (as it is in Northern Nevada, as well as in downtown Las Vegas), then the card counter suffers, by comparison to h17, with all doubling allowed, to the augmented tune of anywhere from 0.31% (26/52 game, 1–2 spread) to 0.42% (39/52 game, 1–4 spread). Thus d10 falls into the category of rules that are "inflated" (like surrender) for card counters, with respect to basic strategists, rather than those that are "deflated," like h17, or those that remain relatively stable, such as das.

Since you are looking at the one-deck charts, now may be as good a time as any to allay your fears that there may be some kind of misprint with respect to the true count frequencies. Perhaps you've noticed the absence of any data for the +3 and +7 lines and wonder what the heck is going on. No, there's no mistake. Permit me to explain.

You may recall that part of our assumptions for the simulations had us estimating true count to the nearest quarter-deck. In addition, we supposed two players against the dealer, for the single-deck sims. Finally, we *truncated* ***[floored]*** our trues for betting purposes. These three provisos conspired to produce the seemingly strange absence of any true counts of +3 or +7. Here's why. After the first hand is dealt to two players, on average, about eight cards will have been used. Since 13 cards constitute a quarter-deck, when we go to estimate the true count at the start of the second hand, we'll be dividing our running count by three-quarters. Suppose the running count is +2. Well, 2 divided by 3/4 is 8/3, or 2 2/3. Truncating ***[Flooring],*** we drop the 2/3 and catalog the count as +2. If the running count is one higher, at +3, division by 3/4 produces a true of exactly +4. Do you see what happened? We skipped over any possibility of a +3 true count!

The same kind of reasoning obviates the chance for a true count of +7, as a running count of +5 yields a true of 6 2/3, which truncates ***[floors]*** to 6, while a running +6 jumps our true to +8. I hear some objections. What if the first hand uses only six cards, such that we are dividing by a whole deck? If the running count is then +3, won't the true count also be +3? Well, yes it would, except for one thing — if only six cards are used on the first round, then everyone stood pat with a two-card hand. Under these circumstances, it just isn't possible for the running count to be +3; in fact, it's likely to be quite negative. Finally, since further estimates of the true count, deeper into the pack, cause us to divide either by a half-deck (effectively multiplying by two), or by a quarter-deck (multiplying by four), *no* odd true counts can be obtained, let alone counts of +3 or +7.

Although we set out to study the aforementioned effects of rules variations for the card counter, our simulations led to many other areas of interest. Here, too, there were no earth-shattering surprises, but we did confirm, in a more precise manner, concepts that were well known for a long time, to wit: Back-counting the shoe game is an incredibly powerful approach and is vastly superior to play all, not only because of enhanced earnings, but because, here again, as was the case for late surrender, our hourly standard deviation actually increases at a much slower rate than our SCORE. ***[With optimal betting, we now standardize risk of ruin, but surrender permits larger unit sizes and, therefore, greater wins, for the same ROR, than we can achieve in non-surrender games.]***

Consider, for example, the 5.0/6 game, s17, with das. A 1–12 spread wins 34.64 dollars per hour, for the play-all approach, with a standard deviation of 46.14 units (of $12.75 each), for a total of $588.27 per hour. The same game, back-counted instead, with a 1–4 spread, produces a win per 100 hands *seen* (but only 18 actually played), of 63.30 dollars, with an hourly standard deviation of 8.49 units (now $93.74 each), for a total of $795.53. As a result, with identical risks of ruin, SCORE is increased 82.7%, by back-counting, while standard deviation increases by only 35.2%. Such is the power of the back-counting approach.

Note, by the way, in this regard, that for all the back-counting data, the numbers of hands actually played per hour are indicated to the right of the word "Back-Count," where the percentages of true counts greater than or equal to +1, +2, and +3 are clearly labeled. All back-counting simulations were predicated on the notion that the player plays only when he has the advantage, said edge arriving virtually all the time when the true count reaches +1. ***[Note, however, that with the small spreads of 1–2 to 1–4 that were used in our new sims, most entries into the game took place at +2, the optimal entry point for this style of back-counting.]***

It must be pointed out here that, if you have run simulations for back-counting, you may be surprised that the instances of true counts greater than or equal to one ***[or two]*** seem lower, for the present charts, than for what you determined. You may, for example, have ascertained that one plays, say, 37 or 38 hands per 100 seen, rather than 27 or 28 for entry at +1, and correspondingly lower values for +2 entry. What's going on? Well, you're rounding; that is, you're calling any true count between 0.5 and 1.5 a true of 1. I'm not. What's more, I would caution you to be careful, because I wonder if you truly do back-count by entering a game not when you've attained a full true count of +1, but rather when you have a true of just 0.5. What's that? You don't play that way? You enter the game when you have a *bona fide* true count of *at least* +1 or higher? Then you shouldn't be simulating by rounding!

The next "truism," graphically depicted by the data, is the power of increased spread. Within a given category of game, running one's eyes down the column, through the three spreads offered, is, well, an eye-opener! Most card counters realize how important it is to obtain the largest spread consistent with being welcome back in the casino the next time. But not every player realizes just how many more units can be obtained by a judicious increase in one's spread. Hopefully, perusing the charts will help counters to home in on the dramatic extra advantages available when spreads are increased even minimally.

Perhaps the best illustrations of this phenomenon can be found in one-deck games where, often, adding just a single unit to the spread can almost double the hourly win rate of the game in question. Consider, for example, the Reno game (h17, d10). For 39 cards dealt, a 1–3 spread wins 60.54 dollars per hour, while 1–2 garners only 21.12. The percentage improvement in earnings is, therefore, 187%! And, what's more, adding a fourth unit to the spread makes the 1–4 game, with its 94.15 win per hour, almost 56% more desirable than "one rung below," with our 1–3 spread. For shallower levels of penetration, these improvements can be even more dramatic.

In real estate, we are told that there are three important factors to consider: location, location, and location! What study of simulations could ever be complete without reference to the three most vital factors in determining the desirability of any counting game: penetration, penetration, and penetration?! Already, I had alerted you to this phenomenon in our "Floating Advantage" and SCORE chapters. In particular, the summaries, at the conclusion of the "Floating Advantage" study, portrayed the incredible extra earnings that accrued to the player who could find, for example, a very deeply cut 6-deck shoe game, such as our 5.5/6 variety. The present study confirms the prior findings and alerts us to the general fact that, for the 6-deck game, the lowest level of spread, but in the next-higher category of penetration, is always superior to the highest spread in the next-lower category of penetration.

For example, should I prefer the 5.0/6, s17 game, where I can spread 1–16, or a 5.5/6 game, where only a 1–8 spread is attainable? No contest — the latter out-SCOREs the former, 37.51 to 31.25 dollars per hour, a 20% increase. So powerful is the multiplying effect of penetration in the shoe game that, for the poorly cut, 1–8 spread, 4.0/6, s17 base game, one hardly wins at all, garnering an infinitesimal 4.17 dollars per hour. Keep the same rules and spread, but play 5.5/6 decks, and the win soars to the aforementioned 37.51 dollars per hour — a nine-fold increase!

The summary pages become a gold mine of information for "instant comparisons." For example, it can't be intuitive or apparent if I should prefer to play a 2-deck game, with a 1–8 spread, but where only 50% of the cards are dealt, or another double-decker, where I have 60% penetration, but feel only a 1–4 spread is prudent. It turns out that the win rate for the 50% game is 24.36 dollars per hour, while the SCORE for the 60% game is only 19.05 dollars. Dozens of such "tradeoffs" can be studied in a trice, using the summary tables. In each case, we judiciously measure how much gain is available, for identical levels of risks of ruin, and we opt for the game with the higher SCORE.

In fact, as it turns out, risk (as expressed by standard deviation), reward, and risk of ruin are inextricably linked, by virtue of the mathematical formulas enunciated in our previous Chapter 8. All of this got me to thinking that I could do one more service for you with our charts. Permit me to explain.

In the world of finance, risk and return are often compared through a measure, named after Nobel Prize winner William Sharpe. The "Sharpe Ratio" attempts to define the desirability of a given investment by comparing the incremental return it affords the investor over some riskless choice (such as a Treasury bill), to the extra risk taken on by the alternative investment. Suppose T-bills yield 5%. Suppose, further, that Investment A garners a 15% return with a 10% risk, as measured by its "volatility," or annualized standard deviation of the continuously compounded returns of the asset. Investment A's Sharpe Ratio is, therefore, calculated to be (15% – 5%)/10% = 1.00. Now, suppose another investment choice, B, could yield, say, 20%, but with a volatility of 17%. Which investment do I prefer? Well, although many might run to the higher return (B), Sharpe's ratio ranks A's 1.00 over B's (20% – 5%)/17% = 0.88.

Now, in the world of blackjack, for a given bankroll, such as the $10,000 proposed in our charts, it is possible to create a measure, entirely analogous to the Sharpe Ratio, simply by forming, for any game, the ratio of "win rate/standard deviation." Thus, we should be able to compare the desirability of one given set of rules, conditions, bet spread, etc., to any other such scenario in an attempt to ascertain the "superior" game. ***[As it turns out, this DI (see below) was the precursor of SCORE, for SCORE is simply the square of the DI. In the present charts, we furnish both of these risk-adjusted-return measures.]***

I have christened this ratio the "Desirability Index" (DI) and have defined it, for any game, to be equal to one thousand times (for convenience of expression) the ratio of that game's per-hand-seen win rate to the per-hand-seen standard deviation. The DI can be found to the extreme right of the charts, as the first of two entries under the heading, "Rankings." The second entry is the all-important SCORE. Although DIs and SCOREs have been furnished for the 2-deck back-counting games, as well, we must admit that, in today's casino environment, it is rarely practical or wise to use the back-counting approach against double-deckers.

Now, when you're strolling around, looking for the game that offers the best bang for your buck, you need only to consult the relevant DIs ***[SCOREs]*** to be able to determine the most desirable games. Of course, we recognize the fact that some games are more a pipe dream than a reality. Whereas it's interesting to note that the highest SCORE ($258.05!) is attributed to the single-deck, s17 game, with das and ls, dealt to the 75% level, at which we manage a 1–4 spread, it is also quite apparent that the probability of finding such utopian conditions is virtually nil!

I've just e-mailed John Auston to tell him that there is no end to the richness of the "pearls" contained in the charts, and that I could go on making observations forever. But that won't do; besides, I think part of the allure of these tables is that you can now "lose yourself" in them and compare and contrast to your heart's content. Perhaps you will make discoveries that, ultimately, you will convey to me and that I might document in some future article. I'd be delighted to hear from you. There is, however, one more very important area to be considered. I'll address it in the next and final section.

Can You Win at These Rates?

I'm tempted to be somewhat flippant and reply, "It really doesn't matter, because this was just a comparative study, intended to determine the value of rules variations to the card counter. Even if the win rates are not entirely realistic, they do, nevertheless, portray the *relative* merits of the different rules." But such an answer would be a cop-out. It isn't fair to produce all these data, warn you that, in a real casino, you may not actually achieve these results, and then leave it at that. I'm going to be much more specific in pointing out to you why you could actually win *more* than the rates cataloged here, and why, in all likelihood, you'll probably win somewhat *less*.

How you can win more: I've identified at least four primary reasons why your play could, conceivably, produce wins in excess of those documented in the following charts. Permit me to explain.

First, you may not use the Hi-Lo, but rather a count whose win rate is superior. For obvious reasons, it was beyond the scope of this study to replicate the data for an entire gamut of count systems. You wouldn't be reading this book today, if we had undertaken such a project, and I'm afraid poor Mr. Auston ***[and later, Mr. Wattenberger]*** would have wound up in the loony bin! Many such comparisons have already been published. Indeed, it is common practice for the purveyors of count systems to extol the virtues of their particular counts by making just such side-by-side comparisons to the eternal reference point — the Hi-Lo.

Although I've just "begged off," by telling you I wouldn't engage in such an endeavor, I think it is only fair, nonetheless, to give you some idea of how much higher-level, more sophisticated approaches and counts could conceivably add to your win rates. Five to ten percent. There, I've said it! Can we move on? What's that? You want proof? *More* simulations? Well, maybe just a couple. ***[The aforementioned "couple" turned into hundreds, for Part II of the SCORE chapter that you've just finished reading.]***

As for the "better" counts? Well, how about a level two, such as Carlson's Omega II (with a side count of aces); a level three — Wong's Halves; and, for good measure, an unbalanced count — George C.'s Unbalanced Zen 2. Here, we relied upon the published indices of each author rather than to custom generate the values for the I18. The respective win rates (with percentage improvements over optimal Hi-Lo indicated in parentheses) are: 2.63 (+8.2%), 2.56 (+5.3%), and 2.60 (+7.0%). ***[See the previous SCORE chapter for more detailed and accurate information.]*** Of course, more complexity can lead to player error. So, it makes no sense to step up to a higher level if you don't intend to play perfectly. Dissipating your newfound gains, through increased error rates, would be a total waste of time, don't you think?

How else might you win more than our simulations? Well, perhaps the easiest method of all is to play with fewer players at the tables than those assumed for our models. If you're playing one-deck, try to get a table for yourself. Playing alone, instead of with a companion, can increase your hourly haul by a full 50%, simply by virtue of the extra hands played per unit of time. Similarly, for 2-deck, try to play with only one other person at the table (or head-on), instead of with two partners. Finally, for the shoe game, whereas I don't recommend solo play (you can't sit out hands, and negative shoes can prove interminable), playing with fewer than three buddies can certainly enhance profits.

Incidentally, we performed a quick sim of the 5/6, 6-deck, play-all, das game, spread 1–12, to see how win rates might be affected by playing alone. Here the idea was to test whether or not head-on play, in addition to producing the obviously greater number of hands per hour, could actually generate a slightly higher win rate. It didn't. The differences in results (if, indeed, there were any at all) were totally insignificant, and they made it clear that the player's edge, as a function of numbers of players at the table, is virtually invariant, for the shoe game.

In what other ways could we win more? Well, larger spreads would do the trick. They might also get you thrown out of all the casinos you play in, and that's why I don't recommend them, but the mathematics can be compelling. It is an incontestable fact that, for one-deck, spreading higher than 1–4 can produce extraordinary returns. Do you dare? Ditto for 2-deck spreads greater than 1–8 or 6-deck with more than 1–16. Here, I don't believe extra sims are required. If you let your eyes scroll down, for any given column, there is a very orderly progression of win rates ***[SCOREs]*** as spread is increased by, say, an extra two units ***[now four units for play all, or one unit for back-counting].*** You can easily determine, by this simple inspection, what a larger spread could do for your hourly win.

I will, at this point, mention another way to achieve a greater win. When the maximum bet is contemplated (such as eight units), spread to two hands (in this case, of six units) instead. The resulting win is always superior to the one-hand approach, and, you maintain your ultimate risk of ruin. The win is increased, logically enough, because more money is put on the table. There is a slight penalty because of the extra cards used in playing the second hand, but unless you are playing alone, this slight reduction in effective penetration is more than offset by the larger total wager placed in these situations. Standard deviation is also increased somewhat by the two-handed approach, but this is to be expected. Thus, the two-handed ploy is an elegant tactic that I heartily recommend to all players.

Finally, I can think of yet another way to increase the total win: Learn more index numbers, beyond the "Illustrious 18." Here again, I believe ample studies have been done on this issue, so I won't belabor the point. In multi-deck games, I have stated that the "Illustrious 18" will garner almost 80% of all the gain available from learning the full set of indices associated with your system. But that could entail learning more than 100 additional values. Are you up to the task? Suppose you decide to add just another, say, 30 indices. What will that extra work provide? Well, of course, it matters *which* indices you choose, but if you select wisely, you will probably add another 5% to 10% to your win. A quick sim using indices from –1 to +6, in 2-deck, yielded a 5% improvement, for example. For single-deck, because of the increased importance of strategy variation, relative to bet spread, the I18 glean perhaps only 75% of the total win available, so here, there can be extra incentive to "widen your horizons" somewhat.

Well, that's the good news. I hate to end this narrative on a negative note, but it would be irresponsible of me not to report to you why you may *not* win as much as the charts suggest. Here are some of the reasons.

Why you may win less: To begin with, each of the four aforementioned reasons for winning more has a "flip side," that is, a corresponding scenario that could cause you to win less than the published rates. Let's review them quickly.

1) Your count may be ill suited for the game you have chosen and might provide results inferior to Hi-Lo. For example, it is common knowledge that, in the shoe game, because of the increased importance of betting accuracy and the correspondingly decreased emphasis on playing departures, counts that neutralize the ace are somewhat "handicapped" when employed in shoe games. So, if you use, say, Hi-Opt I, without a side count of aces, in a shoe game, where you spread 1–12, you are going to underperform Hi-Lo, without a doubt. By how much? Well, a quick sim shows the SCORE to be 28.12, instead of the 34.61 dollars won with Hi-Lo, at the 5.0/6, das game. This sacrifice of almost 19% of profits is much too great, in my opinion, given the slight differences between the two counts. Hi-Opt I shoe players take note: Perhaps it's time to switch to Hi-Lo.

2) You may be forced to play at crowded tables. The effect is just the reverse of seeking solitude. Whereas your actual win rate, per 100 hands played, is virtually unaffected, at a full table, unquestionably, your hourly *play* rate is curtailed sharply, from 100 hands per hour, to about only 60. It is undeniable that playing at crowded tables can be detrimental to your wealth.

3) You may feel uncomfortable with the spreads outlined. For example, at one deck, no spread may be tolerated, and you may be forced to flat bet. Winning is still possible, but only under very special conditions. For the shoe, if you still believe that 1–8 draws too much heat and that you are incapable of "pulling it off," your win rate

is going to suffer dramatically. Instead of wondering by how much, I strongly suggest that you cultivate a better "act" so that you don't have to scrounge for pennies!

4) You might not use any strategy indices at all, at the moment, or perhaps only, say, insurance, 16 v. 10, and 15 v. 10 (already an excellent start, I might add!). In this case, your game will certainly suffer, as the "Illustrious 18" surely provide increased advantage over a game where only betting variation is relied upon to "get the money." How much are you leaving on the table? With our 2-deck reference game, and optimal bets, the "fall-off" is considerable: Using no strategy indices at all, we win only 65% of the amount garnered when the I18 are employed. Another way to interpret the same data is as follows: If you're currently using no indices at all, adding just the "Illustrious 18" will *increase* your profits by almost 54%! If that isn't sufficient motivation to get cracking, I don't know what is!

Were the aforementioned items the only impediments to winning at the prescribed rates, I'd tell you not to be too concerned. Unfortunately, they aren't. Here are some additional reasons to worry.

5) You are not a computer! Therefore, it is safe to say that you aren't capable of winning with the accuracy that the computer does. Just how much this "human element" subtracts from your win rate is a highly individualistic assessment. Fortunately, we have on the scene today some extraordinary software that makes practicing and honing your counting skills a real pleasure, rather than a chore. At the top of the list is Norm Wattenberger's incomparable *Casino Vérité.* I urge you to give it a serious look if you are intent upon testing your skills in the casinos. Errors can manifest themselves in many ways: You may lose the count, bet an inappropriate amount, miscalculate the true count, employ an improper index and therefore play a hand incorrectly, or all of the above! The computer makes no such errors. It doesn't get tired. It doesn't act on impulse (you shouldn't either, but sometimes, the flesh is weak!). And so, the first additional warning: You, mere mortal that you are, will probably win less than the computer, all else being equal.

6) In all likelihood, you don't split tens. I wrestled a long time with this one. I almost reduced the I18 to the "Sweet 16," and didn't include ten-splitting. In the end, as documented earlier, and at Peter Griffin's urging, I left the plays in. But even I have eschewed the "pleasure" of destroying my twenty, for a long time already. I used to do it, and then, I stopped. If you're with me on this one, then the charts all overstate, to some degree, your ultimate edge. By how much? Well, obviously, that depends on the game and the style of play, but, in ballpark terms, it's safe to say that refusing to split tens probably costs about 4–6% of total SCORE per hour. Just to be sure, we ran a sim! Once again using the optimal bets, our 2-deck, s17, das, 70/104, 1–6 spread, 60.28-dollar hourly-win game dropped off to 57.19 dollars per hour, for a loss of 3.09 dollars (or 5.13%) — smack in the middle of our above estimate.

But, there's good news. John and I figured out how to get back nearly half of the loss from refusing to split tens. We ran a sim ***[a new one was done by Norm]*** that replaced the two discarded indices with two new ones, namely doubling a total of 8 v. the dealer's 5 or 6, at true counts of +4 and +2, respectively, for the double-deck game. Almost 40% of the lost advantage was restored! Splitting the difference between 60.28 dollars won with the I18, and the reduced 57.19 figure, without ten-splitting, the newly reconstituted I18 managed to win 58.37 dollars per hour. So, if you're afraid to split tens, and are still willing to learn 18 indices, the above should prove quite helpful. I'll even go so far as to recommend, formally, that you make the switchover.

7) In single-deck, you may encounter the dreaded "preferential shuffle." To me, this should not be a big concern, for this simple reason: They're only going to do it to me once or twice before a) I detect the maneuver, and b) I get the hell out of the casino (or, at the very least, I change dealers). Much too much has been written about this devious casino maneuver, which is, unquestionably, beyond a shadow of a doubt, blatant, flagrant, outright cheating. Do you detect a bit of vehemence on my part over this issue?! For the uninitiated, permit me to digress.

In single-deck, in particular, a dealer has the right, more or less, to "shuffle at will." Unlike the shoe game, or any game at which there is a shuffle-card, to which level the dealer almost invariably deals, in single-deck, for want of a better description, the dealer shuffles whenever he feels like it. Of course, he has guidelines, but, in general, he has great latitude in deciding whether he will or won't deal "one more round." Now, suppose the dealer is capable of counting cards or, at the very least, can spot a rash of small or large cards when they appear. The dealer may be perfectly willing to continue dealing more deeply into the deck, provided he knows the count is negative and, therefore, the house has the edge. When the count is positive, in lieu of continuing the deal, the dealer shuffles, thereby destroying the players' advantage. This ploy is all the more insidious in that it cheats card counter and "average Joe" alike. *Everyone* at such a table is condemned to fight insurmountable odds in attempting to win money.

Players have a right to think at the table; in essence, dealers don't. The latter play by fixed, immutable laws. Proponents (poor misguided souls!) of preferential shuffling opine that "what's sauce for the goose is sauce for the gander," and if counters can back-count, thus avoiding negative decks while playing only positive ones, dealers should have the same prerogative. Nonsense! Players may double, split, and stand on any total they please. Dealers may not double, split, or stand on any total lower than 17. The rules of the game are not symmetrical; they don't apply equally on either side of the table. The laws of all "mainstream" casino establishments insist that, to ensure fairness, the cards be dealt in "random" fashion. Without belaboring the semantics of the term, it is clear that deciding to continue to deal when the player is "handicapped," while refusing to deal when that same player is "advantaged," destroys the randomness of the dealing process. To claim otherwise is absurd. End of pontificating.

Moral of the story: Preferential shuffling is a cheating practice employed by some casinos. The act is inimical to your bankroll. Avoid it like the plague.

8) Some casinos might cheat in a more conventional manner! Whereas I am firmly entrenched in the camp of those who believe that, for the very large majority of cases, you are going to be on the receiving end of a fair game in the casino of your choice, unless you are in on the scam (and are therefore a criminal), cheating, no matter how rare it may be, can only hurt you. It certainly can't help! And so, you must be ever vigilant. Unfortunately, it is also no doubt the case that if you *are* ever cheated, you probably will not be aware of the action, and even if you are, you likely won't be able to do much about it other than to discontinue your play at the offensive casino.

9) I've saved what I consider to be the worst for last. And it brings us full circle, back to our Chapter 7, on camouflage tactics. The computer always made the mathematically correct bet, no matter what that entailed: raise the wager after a loss? — no problem; jump the bet to four times the previous amount? — go right ahead. The computer knows no fear of reprisals. Unfortunately, we do. And so, we compromise. We camouflage our play. We don't jump our bets; rather, we escalate slowly. We don't raise our bet after a loss; rather, we wait to win and then parlay. Occasionally, to throw a pit boss off our scent, we purposely play a hand or two incorrectly, under the guise of smart "cover" play. Some of us may also tip, from time to time.

All of the above costs us, sometimes quite dearly. I tried to catalog the damage, in Chapter 7. More work needs to be done in this area, but one thing is clear: If you employ any camouflage whatsoever in your overall approach to the game (and rare is the individual who doesn't), you cannot possibly win at 100% of the published rates. I did try to suggest ways, at the end of Chapter 7, to cushion the blow, but you're going to take some amount of a "hit," no matter what it might be. I wish it weren't the case. I wish they'd leave us alone and let us spread and play to our heart's content. But they won't. And we have to do what we have to do.

So there you have it. Writing this chapter, and planning, with John Auston ***[and Norm Wattenberger],*** the format of the simulations it contains, has been an exciting, if somewhat exhausting, adventure. I have visions of players everywhere, carrying around either the book itself, or copies of this individual chapter, as they go off to the far corners of the world to play this crazy game we all love. If that should be the case, then surely John, Norm, and I shall have received all the satisfaction we could have possibly hoped for when we undertook this challenging endeavor. I hope you enjoy the "World's Greatest Blackjack Simulation."

Postscript 1: *Well it's now three years later, the returns are in, and quite frankly, the response to Chapter 10 has been overwhelmingly positive. Players are, indeed, carrying the book with them, wherever they go, and have written to say how grateful* *they are for the information contained in the following pages. Needless to say, John and I couldn't be more delighted.*

Postscript 2: *Despite the prodigious amount of information contained in this chapter, there were no charts in the original study for the 8-deck game, alas, so prevalent in Atlantic City and Connecticut (and now, as we write these lines, rearing its ugly head in Las Vegas, as well).*

It occurred to me that I might want to provide some data for our east-coast brethren who have no choice but to tackle these multi-deck monsters. And so, once again, I turned to the ever-obliging John Auston and asked him to grind out more tables, this time for the 8-deck game.

A few observations are in order. First, it should be obvious to all but the most masochistic among you that the play-all approach to these games is an utter waste of time. Assuming no better than 6/8 penetration (which is quite the norm in A.C.), the non-surrender game, even allowing a generous 1–16 spread, garners a pathetic $14.25 per hour! If you're really timid, and laying down 1–8 is all you can handle, you may as well stay home and save your time and effort, as the paltry 7.57 SCORE clearly indicates. With late surrender, a 1–16 spread, and 6.5/8 dealt, we finally crack the $30-per-hour barrier, with a SCORE of 33.01; but the 5.75 DI (implying an hourly s.d. of more than 17 times expectation) is still quite mediocre.

No — if you're planning to find profit potential in these 8-deck horrors, you're either going to have to find a cut-card placement of 7 decks or deeper (good luck!), or you'll have to back-count. The latter approach can yield respectable win rates, DIs, and SCOREs, under certain conditions. Here, I would look for games with surrender ***[not currently available in A.C., but offered in the Las Vegas Strip 8-deck games]*** *and, at the worst, 6/8 cuts. With a 1–4 spread (you'll play 25 hands), a DI of 6.11, and, therefore, a SCORE of 37.33, these are, in my view, marginally playable games.*

Note, again, the power of surrender, where the SCOREs for the 6/8 back-count game are virtually identical to those of the 6.5/8 back-count game that doesn't offer surrender. And, of course, simply by back-counting in general, we are better off at the 6/8 game, without surrender (SCORE = 26.60, for a 1–4 spread) than we are at some 6.5/8 play-all games, with surrender (for example, SCORE = 21.00, with a 1–8 spread). Of course, many other observations are possible, but you can make them on your own, I'm sure. Enjoy the 8-deck tables, which were added to the revised, second edition.

Summary of the Simulation Methodology

For those who have skipped some or all of the preceding text and who want a summary of the methodology relating to the following simulation tables, this outline is for you!

Creating the Simulations:

1) Norm Wattenberger's *Casino Vérité CVCX* software was used to create the tables.
2) Twenty billion rounds were dealt to produce *each* table.
3) The count system used was Hi-Lo.
4) The "Illustrious 18" and "Fab 4" indices were used for multi-deck. Two extra indices (for doubling 8 v. 5 and 6) were used for single-deck.
5) True count "bins" were created using the flooring technique, and flooring was used when invoking any index to make a strategy departure.
6) True count estimation precision was to the nearest quarter-deck, for single-deck and double-deck, and to the nearest half-deck, for 6- and 8-deck games.
7) Levels of penetration for single-deck were: 26, 31, 35, and 39 cards; for double-deck: 52, 62, 70, and 78 cards; for 6-deck: 4/6, 4.5/6, 5/6, and 5.5/6; and for 8-deck: 5.5/8, 6/8, 6.5/8, and 7/8.
8) One burn card was used for all numbers of decks before a fresh deal began.
9) For single-deck, there was one player to our right; for double-deck, two players to our right; and for 6- and 8-deck, three players to our right. The other players played the basic strategy that was appropriate for the rules set they were facing.
10) Bet spreads of 1–2, 1–3, and 1–4 were used for single-deck; 1–4, 1–6, and 1–8 for 2-deck; 1–8, 1–12, and 1–16 for 6- and 8-deck, play all; and 1–2, 1–3, and 1–4 for 2-, 6-, and 8-deck back-counting.
11) Two types of bet schemes were employed. The first is optimal betting. It permits fractional units, and the unit size itself was calculated to two decimal places. No attempt at cover betting is made, and, therefore, bets will occasionally "jump," beyond a parlay. The second, "practical" bet scheme uses integral-chip-size wagers and is designed to be more suitable for actual casino play. It permits only modest bet-jumping, if any at all. However, it should be noted that, if the TC jumps by more than one, from the start of one hand to the start of the next, it is inevitable that bet-jumping will occur, as, again, no attempt is made to engage in cover betting in these charts.
12) Rules vary, but, for all games, splitting was permitted to a total of four hands and the dealer peeked under tens and aces. Aces could be split only once.

Reading the Charts:

1) The first column, "TC," lists the floored true counts. All counts below zero ("< 0") and above +9 ("> 9") are lumped together, in a single entry.
2) "TC Freq. (%)" lists the frequency, in percent, of each true count.
3) "EV W/L (%)" furnishes the player advantage or disadvantage (negative sign) for the particular true count.

4) "Std. Dev." is the standard deviation, expressed in initial-bet units, of a hand of blackjack, played at the given true count.
5) "Play-All Bets" has two sections: "Optimal ($)" and "Practical ($)." In the "Optimal $" section, for each true count, there are three values, separated by commas. They represent the bet sizes, in dollars (rounded to the nearest whole integer) corresponding to the three suggested bet spreads (in units) of 1–8, 1–12, and 1–16, respectively. The "Practical $" section lists wagers in the same manner, only this time, integral chip multiples are used, to make the bets conform to real-world play.
6) The "Back-Count" title is followed by three percentages, rounded to the nearest integer. They represent the frequencies of hands played, per 100 observed, for the back-counter who enters the game at true counts of +1, +2, or +3, respectively. Optimal and Practical wagers are listed in similar fashion to those provided for Play All, only this time, spreads (in units) are 1–2, 1–3, and 1–4. A zero entry for the wager indicates that we do not place a wager at that true count. In all instances of wagering, for both Optimal and Practical, and for Play All or Back-Counting, the first bet indicated in any column represents the size in dollars of the betting unit for that scheme.
7) "Bet Spread" gives line-by-line descriptions that are to be used for the data that follow to the right. "PA Opt.," followed by a bet spread, means "Play All, Optimal." "PA Pract." refers to "Play All, Practical." "BC Opt." is "Back-Count, Optimal," and "BC Pract." is "Back-Count, Practical."
8) "Kelly Bank (units)" is the size, in units, of the bank required to make Kelly-optimal wagers, for the various bet schemes presented. Since, in all cases, the bankroll is presumed to be $10,000, more precise unit sizes than those listed can be ascertained by dividing $10,000 by the number of units in the Kelly bank.
9) "Avg. Bet ($)" is the average bet size, in dollars.
10) Under "Results," "W/L (%)" is the overall percentage advantage that the player enjoys, and "W/100 ($)" is the dollar amount won, per 100 hands played (play all) or observed (if back-counting).
11) Under "Risk Measures," "SD/100 ($)" is the standard deviation, expressed in dollars, per 100 hands played (play all) or observed (if back-counting). "ROR (%)" is the risk of ruin, expressed as a percent. Note that, for optimal wagering, except for a few rounding entries, ROR should always be equal to 13.5%. For the practical bet schemes, the hourly win rate may be higher than the optimal win rate, but when this is the case, the corresponding ROR for the practical scheme will also be higher, resulting in a lower DI and SCORE than for the optimal case. Finally, "N0 (hands)" (N-zero) is a term coined by the

blackjack researcher and system developer Brett Harris. It is the number of hands one must play so that expectation is equal to a one-standard-deviation result; that is, e.v. – s.d. = 0. Some may consider this number of hands to be a fair approximation of the "long run," while others may prefer to use the number of hands for e.v. to equal a 2-s.d. result to be an indicator of the long run.

12) Under "Rankings," "DI" is the "Desirability Index," as explained on p. 203, and "SCORE" is, well, … the SCORE, as explained in the previous chapter.

TABLE 10.1
The Simulation Indices

	6- & 8-Deck	2-Deck	1-Deck
Insurance	+3	+3	+2
16 v. 9	+5	+5	+5
16 v. 10	0	0	+1
15 v. 10	+4	+4	+4
13 v. 2	−1	0	+1
13 v. 3	−2	−2	0
12 v. 2	+4	+4	+5
12 v. 3	+2	+2	+3
12 v. 4	0	+1	+1
12 v. 5	−1	−1	0
12 v. 6	−1/−3*	0/−3	+1/−2
11 v. Ace	+1/−1	0/−2	−2/−3
10 v. 10	+4	+4	+2
10 v. Ace	+4/+3	+3/+2	+2/+1
9 v. 2	+1	+1	+1
9 v. 7	+4	+3	+3
10,10 v. 5	+5	+5	+5
10,10 v. 6	+5	+5/+4	+5
Surrender			
15 v. 9	+2	+2	+2
15 v. 10	0	0	0
15 v. Ace	+2/−1	+1/−2	0/−2
14 v. 10	+3	+3	+4
1-Deck Extras			
8 v. 5	N/A	N/A	+3
8 v. 6	N/A	N/A	+3

* Where two indices are shown, the second is for h17.

TABLE 10.2
8–Deck Play-All SCORE Summary

		S17	S17 DAS		S17 LS		S17 DAS LS		H17		H17 DAS		H17 LS		H17 DAS LS	
5.5/8	**1–8**	**1.63**	**4.67**	3.04	**4.99**	3.36	**9.71**	8.08	**0.02**	–1.61	**0.95**	–0.68	**1.24**	–0.39	**4.02**	2.39
	1–12	**3.94**	**7.66**	3.72	**8.54**	4.60	**13.83**	9.89	**1.06**	–2.88	**3.09**	–0.85	**3.88**	–0.06	**7.54**	3.60
	1–16	**5.61**	**9.65**	4.04	**10.97**	5.36	**16.52**	10.91	**2.25**	–3.36	**4.72**	–0.89	**5.90**	0.29	**9.97**	4.36
6.0/8	**1–8**	**3.51**	**7.57**	4.06	**8.33**	4.82	**14.19**	10.68	**0.58**	–2.93	**2.56**	–0.95	**3.23**	–0.28	**7.13**	3.62
		1.88	2.90		3.34		4.48		0.56		1.61		1.99		3.11	
	1–12	**6.80**	**11.58**	4.78	**13.16**	6.36	**19.69**	12.89	**2.73**	–4.07	**5.73**	–1.07	**7.14**	0.34	**11.95**	5.15
		2.86	3.92		4.62		5.86		1.67		2.64		3.26		4.41	
	1–16	**9.20**	**14.25**	5.05	**16.42**	7.22	**23.13**	13.93	**4.64**	–4.56	**8.14**	–1.06	**10.13**	0.93	**15.29**	6.09
		3.59	4.60		5.45		6.61		2.39		3.42		4.23		5.32	
6.5/8	**1–8**	**6.75**	**12.05**	5.30	**13.64**	6.89	**21.00**	14.25	**2.26**	–4.49	**5.47**	–1.28	**6.78**	0.03	**12.09**	5.34
		5.12	7.38		8.65		11.29		2.24		4.52		5.54		8.07	
	1–12	**11.65**	**17.64**	5.99	**20.46**	8.81	**28.39**	16.74	**5.97**	–5.68	**10.27**	–1.38	**12.73**	1.08	**18.96**	7.31
		7.71	9.98		11.92		14.56		4.91		7.18		8.85		11.42	
	1–16	**14.98**	**21.23**	6.25	**24.89**	9.91	**33.01**	18.03	**8.94**	–6.04	**13.65**	–1.33	**16.92**	1.94	**23.48**	8.50
		9.37	11.58		13.92		16.49		6.69		8.93		11.02		13.51	
7.0/8	**1–8**	**12.80**	**19.84**	7.04	**22.83**	10.03	**32.04**	19.24	**6.09**	–6.71	**11.13**	–1.67	**13.66**	0.86	**20.90**	8.10
		11.17	15.17		17.84		22.33		6.07		10.18		12.42		16.88	
	1–12	**20.10**	**27.81**	7.71	**32.59**	12.49	**42.34**	22.24	**12.39**	–7.71	**18.49**	–1.61	**22.71**	2.61	**30.80**	10.70
		16.16	20.15		24.05		28.51		11.33		15.40		18.83		23.26	
	1–16	**24.93**	**32.83**	7.90	**38.79**	13.86	**48.68**	23.75	**17.01**	–7.92	**23.47**	–1.46	**28.83**	3.90	**37.16**	12.23
		19.32	23.18		27.82		32.16		14.76		18.75		22.93		27.19	

Legend: **X** y
z

X = SCORE; y = $ gain from rules; z = $ gain from penetration

TABLE 10.3

8-Deck Back-Count SCORE Summary

		S17	S17 DAS		S17 LS		S17 DAS LS		H17		H17 DAS		H17 LS		H17 DAS LS	
5.5/8	**1-2**	**14.57**	**18.17**	3.60	**21.70**	7.13	**26.03**	11.46	**11.12**	–3.45	**14.20**	–0.37	**17.60**	3.03	**21.47**	6.90
	1-3	**15.44**	**18.96**	3.52	**22.78**	7.34	**27.02**	11.58	**11.98**	–3.46	**15.09**	–0.35	**18.76**	3.32	**22.58**	7.14
	1-4	**15.75**	**19.20**	3.45	**23.13**	7.38	**28.06**	12.31	**12.39**	–3.36	**15.43**	–0.32	**19.21**	3.46	**22.96**	7.21
6.0/8	**1-2**	**20.35**	**24.84**	4.49	**29.67**	9.32	**35.04**	14.69	**16.63**	–3.72	**20.00**	–0.35	**24.70**	4.35	**29.46**	9.11
		5.78	6.67		7.97		9.01		5.51		5.80		7.10		7.99	
	1-3	**21.73**	**26.14**	4.41	**31.40**	9.67	**36.68**	14.95	**17.44**	–4.29	**21.43**	–0.30	**26.55**	4.82	**31.24**	9.51
		6.29	7.18		8.62		9.66		5.46		6.34		7.79		8.66	
	1-4	**22.28**	**26.60**	4.32	**32.03**	9.75	**37.33**	15.05	**18.10**	–4.18	**22.02**	–0.26	**27.33**	5.05	**31.92**	9.64
		6.53	7.40		8.90		9.27		5.71		6.59		8.12		8.96	
6.5/8	**1-2**	**28.93**	**34.42**	5.49	**41.34**	12.41	**47.91**	18.98	**25.09**	–3.84	**28.86**	–0.07	**35.83**	6.90	**41.03**	12.10
		14.36	16.25		19.64		21.88		13.97		14.66		18.23		19.56	
	1-3	**31.16**	**36.57**	5.41	**44.15**	12.99	**50.66**	19.50	**26.26**	–4.90	**30.83**	–0.33	**38.02**	6.86	**43.92**	12.76
		15.72	17.61		21.37		23.64		14.28		15.74		19.26		21.34	
	1-4	**32.13**	**37.42**	5.29	**45.31**	13.18	**51.70**	19.57	**26.91**	–5.22	**31.85**	–0.28	**39.34**	7.21	**45.13**	13.00
		16.38	18.22		22.18		23.64		14.52		16.42		20.13		22.17	
7.0/8	**1-2**	**43.94**	**49.52**	5.58	**59.87**	15.93	**67.49**	23.55	**38.99**	–4.95	**44.19**	0.25	**54.33**	10.39	**60.13**	16.19
		29.37	31.35		38.17		41.46		27.87		29.99		36.73		38.66	
	1-3	**46.33**	**53.17**	6.84	**64.10**	17.77	**72.12**	25.79	**41.18**	–5.15	**46.30**	–0.03	**57.01**	10.68	**63.97**	17.64
		30.89	34.21		41.32		45.10		29.20		31.21		38.25		41.39	
	1-4	**48.02**	**54.72**	6.70	**66.14**	18.12	**74.02**	26.00	**42.00**	–6.02	**48.01**	–0.01	**58.81**	10.79	**66.09**	18.07
		32.27	35.52		43.01		45.96		29.61		32.58		39.60		43.13	

Legend: **X** y
z

X = SCORE; y = $ gain from rules; z = $ gain from penetration

TABLE 10.4
5.5/8 S17

TC	TC Freq. (%)	EV W/L (%)	Std. Dev.	Play-All Bets Optimal ($)	Play-All Bets Practical ($)	Back-Count 24% 12% 5% Optimal ($)	Back-Count 24% 12% 5% Practical ($)	Bet Spread	Kelly Bank (units)	Avg. Bet ($)	Results W/L (%)	Results W/100 ($)	Risk Measures SD/100 ($)	Risk Measures ROR (%)	Risk Measures N0 (hands)	Rankings DI	Rankings SCORE
<0	45.76	−1.26	1.138	4,4,4	5	0	0	PA Opt. (1–8)	2,817	7.67	0.21	1.63	127.54	13.50	615,540	1.28	1.63
0	30.50	−0.38	1.133	4,4,4	5	0	0	PA Pract. (1–8)	2,000	8.78	0.17	1.46	142.45	23.80	958,448	1.02	1.04
1	11.85	0.17	1.133	13,13,13	10	0	0	PA Opt. (1–12)	2,500	10.35	0.38	3.94	198.60	13.50	253,511	1.99	3.94
2	6.49	0.67	1.130	28,48,53	25	75,66,60	75,75,50	PA Pract. (1–12)	2,000	9.86	0.33	3.26	184.93	14.80	320,810	1.77	3.12
3	2.76	1.19	1.131	28,48,64	40,60,50	93,93,93	100	PA Opt. (1–16)	2,487	11.55	0.49	5.62	236.85	13.60	178,259	2.37	5.61
4	1.53	1.73	1.135	28,48,64	40,60,80	135,135,135	125	PA Pract. (1–16)	2,000	10.12	0.40	4.09	199.73	12.90	238,461	2.05	4.19
5	0.60	2.35	1.142	28,48,64	40,60,80	150,180,180	150,200,200	BC Opt. (1–2)	133	93.93	1.31	14.57	381.74	13.50	68,460	3.82	14.57
6	0.32	2.99	1.146	28,48,64	40,60,80	150,199,228	150,225,200	BC Pract. (1–2)	133	94.20	1.30	14.54	381.30	13.60	68,676	3.81	14.54
7	0.11	3.59	1.145	28,48,64	40,60,80	150,199,240	150,225,200	BC Opt. (1–3)	151	92.68	1.40	15.44	392.88	13.50	64,512	3.93	15.44
8	0.05	4.17	1.141	28,48,64	40,60,80	150,199,240	150,225,200	BC Pract. (1–3)	133	99.98	1.39	16.52	421.58	15.60	65,280	3.92	15.35
9	0.02	4.86	1.137	28,48,64	40,60,80	150,199,240	150,225,200	BC Opt. (1–4)	167	90.70	1.46	15.75	396.93	13.50	63,301	3.97	15.75
>9	0.01	5.53	1.131	28,48,64	40,60,80	150,199,240	150,225,200	BC Pract. (1–4)	200	85.25	1.48	15.00	380.45	12.60	64,330	3.94	15.54

TABLE 10.5
6.0/8 S17

TC	TC Freq. (%)	EV W/L (%)	Std. Dev.	Play-All Bets Optimal ($)	Play-All Bets Practical ($)	Back-Count 25% 13% 7% Optimal ($)	Back-Count 25% 13% 7% Practical ($)	Bet Spread	Kelly Bank (units)	Avg. Bet ($)	Results W/L (%)	Results W/100 ($)	Risk Measures SD/100 ($)	Risk Measures ROR (%)	Risk Measures N0 (hands)	Rankings DI	Rankings SCORE
<0	45.92	−1.32	1.138	5,5,5	5	0	0	PA Opt. (1–8)	1,931	11.06	0.32	3.51	187.29	13.60	285,270	1.87	3.51
0	29.02	−0.37	1.133	5,5,5	5	0	0	PA Pract. (1–8)	2,000	9.27	0.28	2.62	151.93	10.30	334,965	1.73	2.98
1	11.62	0.18	1.133	14,14,14	10	0	0	PA Opt. (1–12)	1,818	13.75	0.49	6.77	260.75	13.40	147,137	2.61	6.80
2	6.77	0.68	1.131	42,53,53	25	84,73,66	75	PA Pract. (1–12)	2,000	10.61	0.48	5.04	200.70	8.20	158,890	2.51	6.29
3	3.06	1.20	1.131	42,66,84	40,60,50	93,93,94	100	PA Opt. (1–16)	1,889	14.84	0.62	9.18	303.36	13.50	108,695	3.03	9.20
4	1.88	1.75	1.136	42,66,84	40,60,80	136,136,136	150	PA Pract. (1–16)	2,000	11.02	0.58	6.35	221.59	7.50	121,774	2.87	8.21
5	0.82	2.36	1.142	42,66,84	40,60,80	168,181,181	150,200,200	BC Opt. (1–2)	119	104.17	1.45	20.35	451.08	13.60	48,981	4.51	20.35
6	0.50	2.99	1.145	42,66,84	40,60,80	168,219,228	150,225,200	BC Pract. (1–2)	133	100.88	1.46	19.74	439.65	13.00	49,680	4.49	20.15
7	0.21	3.57	1.144	42,66,84	40,60,80	168,219,263	150,225,300	BC Opt. (1–3)	137	102.97	1.57	21.73	466.16	13.50	46,131	4.66	21.73
8	0.12	4.17	1.141	42,66,84	40,60,80	168,219,263	150,225,300	BC Pract. (1–3)	133	108.98	1.58	23.08	495.38	15.30	45,988	4.66	21.71
9	0.05	4.73	1.137	42,66,84	40,60,80	168,219,263	150,225,300	BC Opt. (1–4)	152	101.03	1.64	22.28	472.00	13.50	44,747	4.72	22.28
>9	0.04	5.67	1.131	42,66,84	40,60,80	168,219,263	150,225,300	BC Pract. (1–4)	133	110.33	1.62	23.97	509.55	15.80	45,077	4.70	22.14

TABLE 10.6
6.5/8 S17

TC	TC Freq. (%)	EV W/L (%)	Std. Dev.	Play-All Bets Optimal ($)	Play-All Bets Practical ($)	Back-Count 26% 15% 8% Optimal ($)	Back-Count 26% 15% 8% Practical ($)	Bet Spread	Kelly Bank (units)	Avg. Bet ($)	Results W/L (%)	Results W/100 ($)	Risk Measures SD/100 ($)	Risk Measures ROR (%)	Risk Measures N0 (hands)	Rankings DI	Rankings SCORE
<0	46.03	–1.39	1.139	7,7,7	5	0	0	PA Opt. (1–8)	1,413	15.20	0.44	6.79	259.88	13.70	148,080	2.60	6.75
0	27.63	–0.37	1.133	7,7,7	5	0	0	PA Pract. (1–8)	2,000	9.78	0.42	4.11	161.32	4.20	153,687	2.55	6.50
1	11.33	0.19	1.133	15,14,14	10	0	0	PA Opt. (1–12)	1,352	18.01	0.65	11.62	341.27	13.50	85,877	3.41	11.65
2	6.95	0.70	1.131	55,54,54	25	95,83,74	100,75,75	PA Pract. (1–12)	2,000	11.39	0.64	7.31	216.37	4.40	87,727	3.38	11.41
3	3.27	1.22	1.132	57,89,95	40,60,50	95,95,95	100	PA Opt. (1–16)	1,428	19.05	0.79	14.97	387.07	13.50	66,728	3.87	14.98
4	2.23	1.77	1.136	57,89,112	40,60,80	137,137,137	150	PA Pract. (1–16)	2,000	12.02	0.78	9.31	244.07	4.40	68,798	3.81	14.54
5	1.03	2.39	1.141	57,89,112	40,60,80	183,183,183	200	BC Opt. (1–2)	105	117.07	1.65	28.93	537.90	13.50	34,616	5.38	28.93
6	0.73	2.98	1.144	57,89,112	40,60,80	190,228,228	200,225,200	BC Pract. (1–2)	100	124.40	1.65	30.83	573.30	15.30	34,647	5.38	28.93
7	0.32	3.61	1.143	57,89,112	40,60,80	190,249,277	200,225,300	BC Opt. (1–3)	120	116.36	1.79	31.16	558.22	13.60	32,170	5.58	31.16
8	0.24	4.15	1.140	57,89,112	40,60,80	190,249,296	200,225,300	BC Pract. (1–3)	133	115.43	1.79	31.02	557.55	13.60	32,244	5.56	30.95
9	0.10	4.80	1.137	57,89,112	40,60,80	190,249,296	200,225,300	BC Opt. (1–4)	135	114.34	1.87	32.13	566.83	13.50	31,062	5.67	32.13
>9	0.14	5.81	1.130	57,89,112	40,60,80	190,249,296	200,225,300	BC Pract. (1–4)	133	118.13	1.86	33.03	584.10	14.50	31,329	5.65	31.97

TABLE 10.7
7.0/8 S17

TC	TC Freq. (%)	EV W/L (%)	Std. Dev.	Play-All Bets Optimal ($)	Play-All Bets Practical ($)	Back-Count 28% 17% 10% Optimal ($)	Back-Count 28% 17% 10% Practical ($)	Bet Spread	Kelly Bank (units)	Avg. Bet ($)	Results W/L (%)	Results W/100 ($)	Risk Measures SD/100 ($)	Risk Measures ROR (%)	Risk Measures N0 (hands)	Rankings DI	Rankings SCORE
<0	46.11	–1.47	1.140	10,10,9	10	0	0	PA Opt. (1–8)	958	21.16	0.61	12.80	357.80	13.50	78,103	3.58	12.80
0	26.33	–0.36	1.133	10,10,9	10	0	0	PA Pract. (1–8)	1,000	19.36	0.60	11.58	325.70	11.30	79,108	3.56	12.64
1	10.95	0.19	1.133	15,15,15	15	0	0	PA Opt. (1–12)	982	23.78	0.85	20.12	448.38	13.50	49,721	4.48	20.10
2	7.05	0.72	1.132	56,56,56	40	0,97,86	0,100,75	PA Pract. (1–12)	1,000	22.51	0.85	19.22	431.07	12.60	50,302	4.46	19.88
3	3.39	1.23	1.133	84,96,96	80,100,100	147,97,96	150,100,100	PA Opt. (1–16)	1,072	24.59	1.01	24.94	499.33	13.50	40,123	4.99	24.93
4	2.52	1.81	1.137	84,122,140	80,120,150	147,140,140	150	PA Pract. (1–16)	1,000	24.73	1.03	25.57	515.19	14.50	40,595	4.96	24.64
5	1.22	2.39	1.141	84,122,149	80,120,160	183,183,184	200	BC Opt. (1–2)	68	181.20	2.53	43.94	662.94	13.60	22,721	6.63	43.94
6	0.98	2.99	1.143	84,122,149	80,120,160	229,229,229	200	BC Pract. (1–2)	67	184.35	2.53	44.64	675.00	13.90	22,803	6.61	43.75
7	0.47	3.56	1.142	84,122,149	80,120,160	273,273,273	300	BC Opt. (1–3)	103	134.07	2.08	46.33	680.70	13.50	21,581	6.81	46.33
8	0.40	4.16	1.139	84,122,149	80,120,160	293,291,321	300	BC Pract. (1–3)	100	138.40	2.07	47.61	701.00	14.40	21,688	6.79	46.13
9	0.19	4.64	1.137	84,122,149	80,120,160	293,291,344	300	BC Opt. (1–4)	116	131.99	2.19	48.02	692.98	13.50	20,833	6.93	48.02
>9	0.40	5.98	1.128	84,122,149	80,120,160	293,291,344	300	BC Pract. (1–4)	133	127.80	2.18	46.35	672.30	12.90	21,039	6.89	47.53

TABLE 10.8
5.5/8 S17 DAS

TC	TC Freq. (%)	EV W/L (%)	Std. Dev.	Play-All Bets Optimal ($)	Play-All Bets Practical ($)	Back-Count 24% 12% 5% Optimal ($)	Back-Count 24% 12% 5% Practical ($)	Bet Spread	Kelly Bank (units)	Avg. Bet ($)	Results W/L (%)	Results W/100 ($)	Risk Measures SD/100 ($)	Risk Measures ROR (%)	Risk Measures N0 (hands)	Rankings DI	Rankings SCORE
<0	45.76	–1.12	1.147	6,5,5	5	0	0	PA Opt. (1–8)	1,702	12.96	0.36	4.66	216.10	13.50	214,154	2.16	4.67
0	30.50	–0.24	1.141	6,5,5	5	0	0	PA Pract. (1–8)	2,000	8.78	0.31	2.76	143.33	6.80	269,666	1.92	3.70
1	11.84	0.32	1.140	24,24,24	10	0	0	PA Opt. (1–12)	1,848	14.71	0.52	7.62	276.86	13.40	130,491	2.77	7.67
2	6.49	0.83	1.138	47,64,65	25	84,74,67	75	PA Pract. (1–12)	2,000	9.86	0.48	4.74	186.05	6.50	154,056	2.55	6.49
3	2.76	1.35	1.138	47,65,84	40,60,50	104,104,105	100	PA Opt. (1–16)	1,903	15.62	0.62	9.68	310.70	13.60	103,605	3.11	9.65
4	1.53	1.89	1.142	47,65,84	40,60,80	145,145,147	150	PA Pract. (1–16)	2,000	10.12	0.55	5.60	200.91	6.20	128,478	2.79	7.78
5	0.60	2.50	1.148	47,65,84	40,60,80	168,190,191	150,200,200	BC Opt. (1–2)	119	104.46	1.46	18.17	426.26	13.50	55,138	4.26	18.17
6	0.32	3.14	1.151	47,65,84	40,60,80	168,223,239	150,225,200	BC Pract. (1–2)	133	97.43	1.47	17.06	401.03	12.00	55,484	4.25	18.10
7	0.11	3.77	1.150	47,65,84	40,60,80	168,223,270	150,225,300	BC Opt. (1–3)	134	102.67	1.55	18.96	435.40	13.60	52,594	4.35	18.96
8	0.05	4.26	1.146	47,65,84	40,60,80	168,223,270	150,225,300	BC Pract. (1–3)	133	103.20	1.56	19.14	439.88	13.90	52,900	4.35	18.94
9	0.02	4.99	1.142	47,65,84	40,60,80	168,223,270	150,225,300	BC Opt. (1–4)	148	100.48	1.61	19.19	438.10	13.60	52,254	4.38	19.20
>9	0.01	5.61	1.136	47,65,84	40,60,80	168,223,270	150,225,300	BC Pract. (1–4)	133	103.73	1.58	19.49	446.03	14.20	52,318	4.37	19.09

TABLE 10.9
6.0/8 S17 DAS

TC	TC Freq. (%)	EV W/L (%)	Std. Dev.	Play-All Bets Optimal ($)	Play-All Bets Practical ($)	Back-Count 25% 13% 7% Optimal ($)	Back-Count 25% 13% 7% Practical ($)	Bet Spread	Kelly Bank (units)	Avg. Bet ($)	Results W/L (%)	Results W/100 ($)	Risk Measures SD/100 ($)	Risk Measures ROR (%)	Risk Measures N0 (hands)	Rankings DI	Rankings SCORE
<0	45.92	–1.18	1.147	7,7,6	5	0	0	PA Opt. (1–8)	1,354	16.37	0.46	7.56	275.09	13.50	132,042	2.75	7.57
0	29.02	–0.23	1.141	7,7,6	5	0	0	PA Pract. (1–8)	2,000	9.27	0.43	4.01	152.85	3.20	145,292	2.62	6.88
1	11.62	0.32	1.140	25,25,25	10	0	0	PA Opt. (1–12)	1,397	18.40	0.63	11.57	340.33	13.50	86,326	3.40	11.58
2	6.77	0.84	1.138	59,65,65	25	92,80,73	100,75,75	PA Pract. (1–12)	2,000	10.61	0.63	6.64	201.89	3.80	92,447	3.29	10.81
3	3.06	1.36	1.138	59,86,103	40,60,50	105,105,106	100	PA Opt. (1–16)	1,557	18.97	0.75	14.24	377.47	13.50	70,164	3.77	14.25
4	1.88	1.92	1.143	59,86,103	40,60,80	147,147,148	150	PA Pract. (1–16)	2,000	11.02	0.73	8.02	222.88	4.00	77,228	3.60	12.95
5	0.82	2.51	1.148	59,86,103	40,60,80	184,190,191	200	BC Opt. (1–2)	109	114.54	1.61	24.84	498.35	13.40	40,252	4.98	24.84
6	0.50	3.15	1.151	59,86,103	40,60,80	184,238,238	200,225,200	BC Pract. (1–2)	100	119.90	1.62	26.04	523.20	14.90	40,494	4.98	24.78
7	0.21	3.73	1.149	59,86,103	40,60,80	184,241,283	200,225,300	BC Opt. (1–3)	125	112.81	1.73	26.14	511.31	13.40	38,157	5.11	26.14
8	0.12	4.31	1.146	59,86,103	40,60,80	184,241,292	200,225,300	BC Pract. (1–3)	133	108.98	1.74	25.44	498.15	12.90	38,388	5.11	26.09
9	0.05	4.88	1.142	59,86,103	40,60,80	184,241,292	200,225,300	BC Opt. (1–4)	137	110.75	1.79	26.60	515.81	13.50	37,630	5.16	26.60
>9	0.04	5.81	1.135	59,86,103	40,60,80	184,241,292	200,225,300	BC Pract. (1–4)	133	110.33	1.78	26.37	512.40	13.50	37,671	5.15	26.48

TABLE 10.10
6.5/8 S17 DAS

TC	TC Freq. (%)	EV W/L (%)	Std. Dev.	Play-All Bets Optimal ($)	Play-All Bets Practical ($)	Back-Count 26% 15% 8% Optimal ($)	Back-Count 26% 15% 8% Practical ($)	Bet Spread	Kelly Bank (units)	Avg. Bet ($)	Results W/L (%)	Results W/100 ($)	Risk Measures SD/100 ($)	Risk Measures ROR (%)	Risk Measures N0 (hands)	Rankings DI	Rankings SCORE
<0	46.04	–1.25	1.148	10,9,8	10	0	0	PA Opt. (1–8)	1,037	20.80	0.58	12.09	347.11	13.60	82,980	3.47	12.05
0	27.63	–0.23	1.141	10,9,8	10	0	0	PA Pract. (1–8)	1,000	20.12	0.56	11.34	329.65	12.40	84,505	3.44	11.83
1	11.33	0.33	1.141	26,26,26	25	0	0	PA Opt. (1–12)	1,123	22.64	0.78	17.63	420.01	13.50	56,700	4.2	17.64
2	6.95	0.86	1.138	66,66,66	50	103,90,80	100,100,75	PA Pract. (1–12)	1,000	22.68	0.76	17.34	417.24	13.60	57,899	4.16	17.27
3	3.27	1.38	1.139	77,107,107	80,100,100	107,107,107	100	PA Opt. (1–16)	1,216	23.33	0.91	21.24	460.77	13.50	47,101	4.61	21.23
4	2.23	1.93	1.143	77,107,132	80,120,150	148,148,148	150	PA Pract. (1–16)	1,000	24.37	0.90	22.01	485.06	15.40	48,568	4.54	20.58
5	1.03	2.55	1.148	77,107,132	80,120,160	193,194,194	200	BC Opt. (1–2)	97	127.36	1.80	34.42	586.76	13.50	28,971	5.87	34.42
6	0.73	3.14	1.149	77,107,132	80,120,160	206,237,238	200	BC Pract. (1–2)	100	124.40	1.81	33.81	576.50	13.00	29,091	5.86	34.39
7	0.32	3.75	1.148	77,107,132	80,120,160	206,270,285	200,300,300	BC Opt. (1–3)	111	126.03	1.93	36.57	604.73	13.60	27,318	6.05	36.57
8	0.24	4.29	1.145	77,107,132	80,120,160	206,270,322	200,300,300	BC Pract. (1–3)	100	129.70	1.92	37.32	619.90	14.30	27,620	6.02	36.26
9	0.10	4.91	1.141	77,107,132	80,120,160	206,270,322	200,300,300	BC Opt. (1–4)	124	123.70	2.02	37.42	611.72	13.60	26,720	6.12	37.42
>9	0.14	5.94	1.134	77,107,132	80,120,160	206,270,322	200,300,300	BC Pract. (1–4)	133	118.13	2.02	35.84	587.25	12.50	26,833	6.1	37.24

TABLE 10.11
7.0/8 S17 DAS

TC	TC Freq. (%)	EV W/L (%)	Std. Dev.	Play-All Bets Optimal ($)	Play-All Bets Practical ($)	Back-Count 28% 17% 10% Optimal ($)	Back-Count 28% 17% 10% Practical ($)	Bet Spread	Kelly Bank (units)	Avg. Bet ($)	Results W/L (%)	Results W/100 ($)	Risk Measures SD/100 ($)	Risk Measures ROR (%)	Risk Measures N0 (hands)	Rankings DI	Rankings SCORE
<0	46.11	–1.34	1.149	13,12,11	15,10,10	0	0	PA Opt. (1–8)	781	26.72	0.74	19.84	445.39	13.50	50,439	4.45	19.84
0	26.33	–0.22	1.141	13,12,11	15,10,10	0	0	PA Pract. (1–8)	667	27.93	0.73	20.27	462.32	15.00	52,045	4.38	19.22
1	10.95	0.34	1.141	26,26,26	25	0	0	PA Opt. (1–12)	848	28.58	0.97	27.87	527.32	13.60	35,952	5.27	27.81
2	7.04	0.88	1.139	68,68,68	50	117,103,92	125,100,100	PA Pract. (1–12)	1,000	24.31	0.97	23.64	449.26	9.60	36,116	5.26	27.68
3	3.39	1.40	1.139	102,108,108	100	117,108,108	125,100,100	PA Opt. (1–16)	937	29.07	1.13	32.84	572.99	13.50	30,456	5.73	32.83
4	2.52	1.97	1.143	102,142,151	120,120,150	151,151,151	150	PA Pract. (1–16)	1,000	26.53	1.14	30.34	531.28	11.60	30,663	5.71	32.62
5	1.22	2.56	1.147	102,142,171	120,120,160	194,194,194	200	BC Opt. (1–2)	86	144.90	2.06	49.51	703.69	13.30	20,234	7.04	49.52
6	0.98	3.17	1.148	102,142,171	120,120,160	234,241,241	200	BC Pract. (1–2)	80	149.75	2.03	50.59	721.38	14.30	20,304	7.01	49.18
7	0.47	3.70	1.147	102,142,171	120,120,160	234,281,282	250,300,300	BC Opt. (1–3)	97	143.55	2.23	53.17	729.14	13.40	18,813	7.29	53.17
8	0.40	4.30	1.144	102,142,171	120,120,160	234,310,329	250,300,300	BC Pract. (1–3)	100	138.40	2.23	51.27	704.70	12.60	18,870	7.28	52.93
9	0.19	4.76	1.141	102,142,171	120,120,160	234,310,366	250,300,400	BC Opt. (1–4)	109	141.23	2.33	54.72	739.70	13.40	18,304	7.40	54.72
>9	0.40	6.11	1.133	102,142,171	120,120,160	234,310,368	250,300,400	BC Pract. (1–4)	100	142.00	2.32	54.65	741.80	13.70	18,458	7.37	54.27

TABLE 10.12
5.5/8 S17 LS

TC	TC Freq. (%)	EV W/L (%)	Std. Dev.	Play-All Bets Optimal ($)	Play-All Bets Practical ($)	Back-Count 24% 12% 5% Optimal ($)	Back-Count 24% 12% 5% Practical ($)	Bet Spread	Kelly Bank (units)	Avg. Bet ($)	Results W/L (%)	Results W/100 ($)	Risk Measures SD/100 ($)	Risk Measures ROR (%)	Risk Measures N0 (hands)	Rankings DI	Rankings SCORE
<0	45.75	−1.23	1.130	6,6,6	5	0	0	PA Opt. (1–8)	1,597	13.69	0.37	4.98	223.34	13.50	200,338	2.23	4.99
0	30.51	−0.28	1.121	6,6,6	5	0	0	PA Pract. (1–8)	2,000	8.79	0.32	2.78	140.03	5.90	253,719	1.99	3.95
1	11.84	0.31	1.118	25,25,25	10	0	0	PA Opt. (1–12)	1,681	15.86	0.54	8.57	292.28	13.60	117,025	2.92	8.54
2	6.49	0.86	1.112	50,70,70	25	94,83,75	100,75,75	PA Pract. (1–12)	2,000	9.87	0.50	4.91	181.58	5.10	136,486	2.71	7.32
3	2.77	1.43	1.109	50,72,94	40,60,50	116,116,117	125	PA Opt. (1–16)	1,711	16.97	0.65	11.02	331.18	13.60	91,135	3.31	10.97
4	1.53	2.04	1.111	50,72,94	40,60,80	166,166,166	200	PA Pract. (1–16)	2,000	10.12	0.58	5.86	195.97	4.70	111,837	2.99	8.94
5	0.60	2.72	1.117	50,72,94	40,60,80	188,218,219	200	BC Opt. (1–2)	107	116.92	1.56	21.70	465.90	13.40	46,272	4.66	21.70
6	0.32	3.40	1.120	50,72,94	40,60,80	188,249,272	200,225,300	BC Pract. (1–2)	100	128.00	1.57	23.93	514.40	16.40	46,324	4.65	21.64
7	0.11	4.09	1.119	50,72,94	40,60,80	188,249,302	200,225,300	BC Opt. (1–3)	121	115.17	1.66	22.78	477.26	13.40	43,871	4.77	22.78
8	0.05	4.74	1.115	50,72,94	40,60,80	188,249,302	200,225,300	BC Pract. (1–3)	133	115.43	1.68	23.02	486.00	14.30	44,553	4.74	22.43
9	0.02	5.39	1.110	50,72,94	40,60,80	188,249,302	200,225,300	BC Opt. (1–4)	132	112.84	1.72	23.14	481.01	13.60	43,093	4.81	23.13
>9	0.01	6.21	1.105	50,72,94	40,60,80	188,249,302	200,225,300	BC Pract. (1–4)	133	118.65	1.73	24.48	511.20	15.40	43,714	4.79	22.93

TABLE 10.13
6.0/8 S17 LS

TC	TC Freq. (%)	EV W/L (%)	Std. Dev.	Play-All Bets Optimal ($)	Play-All Bets Practical ($)	Back-Count 25% 13% 7% Optimal ($)	Back-Count 25% 13% 7% Practical ($)	Bet Spread	Kelly Bank (units)	Avg. Bet ($)	Results W/L (%)	Results W/100 ($)	Risk Measures SD/100 ($)	Risk Measures ROR (%)	Risk Measures N0 (hands)	Rankings DI	Rankings SCORE
<0	45.91	−1.29	1.131	8,8,7	10,10,5	0	0	PA Opt. (1–8)	1,252	17.51	0.48	8.32	288.63	13.50	120,056	2.89	8.33
0	29.02	−0.28	1.121	8,8,7	10,10,5	0	0	PA Pract. (1–8)	1,000	19.13	0.44	8.42	304.04	16.20	130,388	2.77	7.67
1	11.61	0.32	1.118	25,25,25	25,25,10	0	0	PA Opt. (1–12)	1,260	20.01	0.66	13.14	362.81	13.50	75,985	3.63	13.16
2	6.77	0.87	1.113	64,70,70	50,50,25	103,90,81	100,100,75	PA Pract. (1–12)	1,000	21.80	0.65	14.06	398.22	17.00	80,219	3.53	12.47
3	3.06	1.44	1.109	64,95,115	80,120,50	117,117,117	125	PA Opt. (1–16)	1,396	20.73	0.79	16.41	405.25	13.50	60,907	4.05	16.42
4	1.88	2.06	1.112	64,95,115	80,120,80	166,166,167	200	PA Pract. (1–16)	2,000	11.03	0.77	8.44	217.41	2.80	66,355	3.88	15.07
5	0.82	2.70	1.117	64,95,115	80,120,80	206,217,217	200	BC Opt. (1–2)	97	128.33	1.72	29.67	544.72	13.50	33,726	5.45	29.67
6	0.50	3.40	1.119	64,95,115	80,120,80	206,269,272	200,300,300	BC Pract. (1–2)	100	132.60	1.73	30.81	567.90	14.80	33,997	5.43	29.43
7	0.21	4.05	1.118	64,95,115	80,120,80	206,269,324	200,300,300	BC Opt. (1–3)	112	126.53	1.85	31.40	560.35	13.40	31,918	5.60	31.40
8	0.12	4.69	1.115	64,95,115	80,120,80	206,269,325	200,300,300	BC Pract. (1–3)	100	139.40	1.84	34.42	616.20	16.30	32,087	5.59	31.21
9	0.05	5.28	1.111	64,95,115	80,120,80	206,269,325	200,300,300	BC Opt. (1–4)	123	124.18	1.92	32.03	565.95	13.60	31,163	5.66	32.03
>9	0.04	6.28	1.104	64,95,115	80,120,80	206,269,325	200,300,300	BC Pract. (1–4)	133	126.83	1.93	32.95	585.68	14.70	31,642	5.63	31.65

TABLE 10.14
6.5/8 S17 LS

TC	TC Freq. (%)	EV W/L (%)	Std. Dev.	Play-All Bets Optimal ($)	Play-All Bets Practical ($)	Back-Count 26% 15% 8% Optimal ($)	Back-Count 26% 15% 8% Practical ($)	Bet Spread	Kelly Bank (units)	Avg. Bet ($)	Results W/L (%)	Results W/100 ($)	Risk Measures SD/100 ($)	Risk Measures ROR (%)	Risk Measures N0 (hands)	Rankings DI	Rankings SCORE
<0	46.03	–1.36	1.131	11,10,9	10	0	0	PA Opt. (1–8)	938	22.61	0.60	13.65	369.36	13.50	73,301	3.69	13.64
0	27.63	–0.27	1.121	11,10,9	10	0	0	PA Pract. (1–8)	1,000	20.13	0.59	11.80	322.04	10.30	74,483	3.66	13.42
1	11.32	0.32	1.118	26,26,26	25	0	0	PA Opt. (1–12)	1,001	24.86	0.82	20.49	452.31	13.60	48,876	4.52	20.46
2	6.95	0.88	1.113	71,71,71	50	116,101,90	125,100,100	PA Pract. (1–12)	1,000	23.36	0.82	19.22	428.57	12.30	49,721	4.48	20.11
3	3.28	1.46	1.110	85,118,118	80,120,125	118,118,118	125	PA Opt. (1–16)	1,077	25.71	0.97	24.89	498.94	13.50	40,179	4.99	24.89
4	2.23	2.07	1.112	85,120,148	80,120,160	167,167,168	200	PA Pract. (1–16)	1,000	25.44	0.98	24.97	505.30	14.10	40,951	4.94	24.41
5	1.03	2.73	1.116	85,120,148	80,120,160	219,219,219	200	BC Opt. (1–2)	86	142.93	1.93	41.34	642.94	13.60	24,177	6.43	41.34
6	0.74	3.38	1.118	85,120,148	80,120,160	232,270,271	250,300,300	BC Pract. (1–2)	80	154.13	1.92	44.37	692.13	15.60	24,327	6.41	41.10
7	0.32	4.07	1.117	85,120,148	80,120,160	232,303,326	250,300,300	BC Opt. (1–3)	99	141.78	2.07	44.15	664.45	13.50	22,672	6.64	44.15
8	0.24	4.66	1.114	85,120,148	80,120,160	232,303,361	250,300,400	BC Pract. (1–3)	100	147.60	2.08	46.02	695.10	14.80	22,834	6.62	43.82
9	0.10	5.33	1.110	85,120,148	80,120,160	232,303,361	250,300,400	BC Opt. (1–4)	111	139.24	2.17	45.31	673.08	13.40	22,048	6.73	45.31
>9	0.14	6.44	1.103	85,120,148	80,120,160	232,303,361	250,300,400	BC Pract. (1–4)	100	150.80	2.14	48.57	724.30	15.70	22,211	6.71	44.97

TABLE 10.15
7.0/8 S17 LS

TC	TC Freq. (%)	EV W/L (%)	Std. Dev.	Play-All Bets Optimal ($)	Play-All Bets Practical ($)	Back-Count 28% 17% 10% Optimal ($)	Back-Count 28% 17% 10% Practical ($)	Bet Spread	Kelly Bank (units)	Avg. Bet ($)	Results W/L (%)	Results W/100 ($)	Risk Measures SD/100 ($)	Risk Measures ROR (%)	Risk Measures N0 (hands)	Rankings DI	Rankings SCORE
<0	46.10	–1.45	1.132	14,13,12	15,15,10	0	0	PA Opt. (1–8)	703	29.24	0.78	22.80	477.83	13.50	43,774	4.78	22.83
0	26.33	–0.27	1.121	14,13,12	15,15,10	0	0	PA Pract. (1–8)	667	28.64	0.78	22.29	471.57	13.40	44,758	4.73	22.33
1	10.94	0.33	1.118	26,26,26	25	0	0	PA Opt. (1–12)	754	31.51	1.03	32.57	570.91	13.50	30,691	5.71	32.59
2	7.04	0.90	1.114	72,73,73	50	0,116,103	0,125,100	PA Pract. (1–12)	667	32.54	1.05	34.19	605.07	15.40	31,329	5.65	31.93
3	3.40	1.47	1.111	114,119,119	120,125,125	175,119,119	200,125,125	PA Opt. (1–16)	830	32.15	1.21	38.81	622.84	13.50	25,785	6.23	38.79
4	2.52	2.10	1.113	114,159,169	120,180,160	175,169,169	200	PA Pract. (1–16)	1,000	27.67	1.23	34.00	549.75	10.50	26,144	6.18	38.24
5	1.22	2.72	1.116	114,159,193	120,180,160	219,219,219	200	BC Opt. (1–2)	57	215.88	2.89	59.87	773.68	13.60	16,651	7.74	59.87
6	0.98	3.39	1.117	114,159,193	120,180,160	272,272,272	300	BC Pract. (1–2)	50	236.20	2.88	65.22	845.80	16.10	16,828	7.71	59.47
7	0.47	3.99	1.116	114,159,193	120,180,160	320,320,321	300	BC Opt. (1–3)	86	161.42	2.39	64.10	800.70	13.50	15,580	8.01	64.10
8	0.40	4.64	1.113	114,159,193	120,180,160	350,349,374	400,375,400	BC Pract. (1–3)	80	172.38	2.37	68.00	851.75	15.30	15,689	7.98	63.75
9	0.19	5.12	1.111	114,159,193	120,180,160	350,349,412	400,375,400	BC Opt. (1–4)	97	158.80	2.50	66.14	813.28	13.50	15,111	8.13	66.14
>9	0.42	6.61	1.102	114,159,193	120,180,160	350,349,412	400,375,400	BC Pract. (1–4)	100	163.30	2.50	67.82	836.80	14.40	15,233	8.10	65.67

TABLE 10.16

5.5/8 S17 DAS LS

TC	TC Freq. (%)	EV W/L (%)	Std. Dev.	Play-All Bets Optimal ($)	Play-All Bets Practical ($)	Back-Count 24% 12% 5% Optimal ($)	Back-Count 24% 12% 5% Practical ($)	Bet Spread	Kelly Bank (units)	Avg. Bet ($)	Results W/L (%)	Results W/100 ($)	Risk Measures SD/100 ($)	Risk Measures ROR (%)	Risk Measures N0 (hands)	Rankings DI	Rankings SCORE
<0	45.76	-1.10	1.139	9,8,7	10,10,5	0	0	PA Opt. (1–8)	1,164	19.00	0.51	9.71	311.65	13.50	102,984	3.12	9.71
0	30.51	-0.14	1.129	9,8,7	10,10,5	0	0	PA Pract. (1–8)	1,000	18.16	0.46	8.42	287.75	13.10	116,790	2.93	8.56
1	11.83	0.46	1.126	36,36,36	25,25,10	0,0,55	0,50,50	PA Opt. (1–12)	1,275	20.57	0.67	13.83	371.95	13.50	72,246	3.72	13.83
2	6.49	1.01	1.120	69,81,81	50,50,25	102,91,81	100,75,75	PA Pract. (1–12)	1,000	20.32	0.64	13.04	369.99	14.90	80,505	3.52	12.42
3	2.77	1.60	1.116	69,94,114	80,120,50	128,128,128	125	PA Opt. (1–16)	1,404	21.09	0.78	16.51	406.43	13.50	60,541	4.06	16.52
4	1.53	2.20	1.118	69,94,114	80,120,80	176,176,176	200,150,200	PA Pract. (1–16)	2,000	10.12	0.73	7.37	197.15	2.30	71,655	3.74	13.96
5	0.60	2.89	1.123	69,94,114	80,120,80	205,229,218	200,150,200	BC Opt. (1–2)	98	127.47	1.72	26.03	510.17	13.40	38,487	5.10	26.03
6	0.32	3.55	1.126	69,94,114	80,120,80	205,272,218	200,150,200	BC Pract. (1–2)	100	128.00	1.73	26.34	517.50	14.00	38,718	5.09	25.91
7	0.11	4.25	1.124	69,94,114	80,120,80	205,272,218	200,150,200	BC Opt. (1–3)	110	125.09	1.82	27.02	519.82	13.60	36,892	5.20	27.02
8	0.05	4.88	1.120	69,94,114	80,120,80	205,272,218	200,150,200	BC Pract. (1–3)	200	76.70	1.31	23.84	460.05	10.50	37,208	5.18	26.85
9	0.02	5.46	1.115	69,94,114	80,120,80	205,272,218	200,150,200	BC Opt. (1–4)	183	85.80	1.38	28.06	529.70	13.60	35,695	5.30	28.06
>9	0.01	6.19	1.109	69,94,114	80,120,80	205,272,218	200,150,200	BC Pract. (1–4)	200	82.30	1.40	27.40	519.30	13.10	35,920	5.28	27.85

TABLE 10.17

6.0/8 S17 DAS LS

TC	TC Freq. (%)	EV W/L (%)	Std. Dev.	Play-All Bets Optimal ($)	Play-All Bets Practical ($)	Back-Count 25% 13% 7% Optimal ($)	Back-Count 25% 13% 7% Practical ($)	Bet Spread	Kelly Bank (units)	Avg. Bet ($)	Results W/L (%)	Results W/100 ($)	Risk Measures SD/100 ($)	Risk Measures ROR (%)	Risk Measures N0 (hands)	Rankings DI	Rankings SCORE
<0	45.92	-1.16	1.139	10,10,9	10	0	0	PA Opt. (1–8)	979	22.87	0.62	14.16	376.75	13.50	70,494	3.77	14.19
0	29.02	-0.14	1.129	10,10,9	10	0	0	PA Pract. (1–8)	1,000	19.13	0.59	11.26	305.95	9.00	73,829	3.68	13.54
1	11.61	0.46	1.126	36,36,37	25	0,0,61	0	PA Opt. (1–12)	1,037	24.71	0.80	19.65	443.79	13.40	50,756	4.44	19.69
2	6.77	1.02	1.120	81,81,81	50	111,98,82	100	PA Pract. (1–12)	1,000	21.80	0.80	17.35	400.65	11.50	53,325	4.33	18.75
3	3.06	1.61	1.116	82,115,129	80,120,125	129,129,129	125	PA Opt. (1–16)	1,171	25.05	0.92	23.12	480.92	13.50	43,228	4.81	23.13
4	1.88	2.22	1.118	82,115,137	80,120,160	178,178,178	200	PA Pract. (1–16)	1,000	23.40	0.93	21.71	464.88	13.40	45,852	4.67	21.82
5	0.82	2.88	1.123	82,115,137	80,120,160	222,228,228	200	BC Opt. (1–2)	90	138.78	1.88	35.04	592.00	13.50	28,471	5.92	35.04
6	0.50	3.54	1.125	82,115,137	80,120,160	222,280,246	200,300,300	BC Pract. (1–2)	100	132.60	1.89	33.67	571.30	12.70	28,739	5.89	34.74
7	0.21	4.20	1.124	82,115,137	80,120,160	222,293,246	200,300,300	BC Opt. (1–3)	103	136.74	2.00	36.68	605.59	13.40	27,269	6.06	36.68
8	0.12	4.86	1.120	82,115,137	80,120,160	222,293,246	200,300,400	BC Pract. (1–3)	100	139.40	2.00	37.43	619.80	14.20	27,464	6.04	36.47
9	0.05	5.46	1.116	82,115,137	80,120,160	222,293,246	200,300,400	BC Opt. (1–4)	163	96.00	1.55	37.34	611.02	13.50	26,809	6.11	37.33
>9	0.04	6.50	1.108	82,115,137	80,120,160	222,293,246	200,300,400	BC Pract. (1–4)	100	141.00	2.03	38.54	634.40	14.70	27,152	6.08	36.92

TABLE 10.18
6.5/8 S17 DAS LS

TC	TC Freq. (%)	EV W/L (%)	Std. Dev.	Play-All Bets Optimal ($)	Play-All Bets Practical ($)	Back-Count 26% 15% 8% Optimal ($)	Back-Count 26% 15% 8% Practical ($)	Bet Spread	Kelly Bank (units)	Avg. Bet ($)	Results W/L (%)	Results W/100 ($)	Risk Measures SD/100 ($)	Risk Measures ROR (%)	Risk Measures N0 (hands)	Rankings DI	Rankings SCORE
<0	46.03	–1.23	1.140	13,12,11	15,10,10	0	0	PA Opt. (1–8)	752	28.30	0.74	20.99	458.27	13.50	47,626	4.58	21.00
0	27.63	–0.13	1.129	13,12,11	15,10,10	0	0	PA Pract. (1–8)	667	27.05	0.73	19.72	444.72	13.60	50,832	4.43	19.66
1	11.32	0.47	1.126	37,37,37	25	0	0	PA Opt. (1–12)	841	29.78	0.95	28.45	532.87	13.50	35,205	5.33	28.39
2	6.95	1.03	1.120	82,83,82	50	124,108,97	125,100,100	PA Pract. (1–12)	1,000	23.36	0.97	22.74	431.15	8.60	35,948	5.27	27.81
3	3.28	1.62	1.117	106,130,130	120,120,125	130,130,130	125	PA Opt. (1–16)	948	30.04	1.10	33.04	574.52	13.50	30,298	5.75	33.01
4	2.23	2.23	1.119	106,143,169	120,120,160	178,178,179	200	PA Pract. (1–16)	1,000	25.44	1.13	28.83	508.27	10.70	31,081	5.67	32.17
5	1.03	2.91	1.123	106,143,169	120,120,160	231,231,231	200	BC Opt. (1–2)	81	153.26	2.08	47.92	692.16	13.40	20,872	6.92	47.91
6	0.74	3.52	1.124	106,143,169	120,120,160	248,279,279	250,300,300	BC Pract. (1–2)	80	154.13	2.08	48.05	696.13	13.70	21,033	6.90	47.65
7	0.32	4.24	1.123	106,143,169	120,120,160	248,324,337	250,300,300	BC Opt. (1–3)	93	151.56	2.23	50.65	711.77	13.40	19,738	7.12	50.66
8	0.24	4.82	1.119	106,143,169	120,120,160	248,324,385	250,300,400	BC Pract. (1–3)	100	147.60	2.23	49.55	699.10	13.10	19,866	7.09	50.25
9	0.10	5.47	1.115	106,143,169	120,120,160	248,324,387	250,300,400	BC Opt. (1–4)	103	148.72	2.31	51.70	718.96	13.60	19,313	7.19	51.70
>9	0.14	6.61	1.107	106,143,169	120,120,160	248,324,387	250,300,400	BC Pract. (1–4)	100	150.80	2.30	52.18	728.30	13.90	19,466	7.16	51.33

TABLE 10.19
7.0/8 S17 DAS LS

TC	TC Freq. (%)	EV W/L (%)	Std. Dev.	Play-All Bets Optimal ($)	Play-All Bets Practical ($)	Back-Count 28% 17% 10% Optimal ($)	Back-Count 28% 17% 10% Practical ($)	Bet Spread	Kelly Bank (units)	Avg. Bet ($)	Results W/L (%)	Results W/100 ($)	Risk Measures SD/100 ($)	Risk Measures ROR (%)	Risk Measures N0 (hands)	Rankings DI	Rankings SCORE
<0	46.10	–1.32	1.141	17,15,13	15	0	0	PA Opt. (1–8)	597	34.91	0.92	32.03	566.01	13.50	31,230	5.66	32.04
0	26.33	–0.13	1.129	17,15,13	15	0	0	PA Pract. (1–8)	667	32.04	0.91	29.17	515.96	11.10	31,275	5.65	31.97
1	10.93	0.48	1.126	38,38,38	40	0	0	PA Opt. (1–12)	672	36.32	1.17	42.35	650.73	13.50	23,622	6.51	42.34
2	7.04	1.05	1.121	84,84,84	75	139,123,109	150,125,100	PA Pract. (1–12)	667	35.93	1.16	41.70	641.46	13.10	23,663	6.50	42.26
3	3.40	1.63	1.118	131,131,131	120,125,125	139,131,131	150,125,125	PA Opt. (1–16)	745	36.63	1.33	48.65	697.74	13.50	20,544	6.98	48.68
4	2.52	2.26	1.120	134,179,180	120,180,200	180,180,180	200	PA Pract. (1–16)	667	38.16	1.32	50.26	724.52	14.70	20,776	6.94	48.12
5	1.22	2.90	1.123	134,179,215	120,180,200	230,230,231	200	BC Opt. (1–2)	72	172.48	2.35	67.49	821.52	13.50	14,793	8.22	67.49
6	0.98	3.54	1.123	134,179,215	120,180,240	278,281,281	300	BC Pract. (1–2)	67	183.45	2.34	71.46	871.80	15.00	14,909	8.20	67.19
7	0.47	4.14	1.122	134,179,215	120,180,240	278,329,329	300	BC Opt. (1–3)	81	170.98	2.54	72.12	849.24	13.60	13,846	8.49	72.12
8	0.40	4.80	1.119	134,179,215	120,180,240	278,368,384	300,375,400	BC Pract. (1–3)	80	172.38	2.53	72.53	856.38	13.80	13,952	8.47	71.74
9	0.19	5.29	1.116	134,179,215	120,180,240	278,368,426	300,375,400	BC Opt. (1–4)	91	168.28	2.64	74.02	860.41	13.60	13,493	8.60	74.02
>9	0.42	6.76	1.106	134,179,215	120,180,240	278,368,438	300,375,400	BC Pract. (1–4)	100	163.30	2.65	72.11	841.30	13.00	13,615	8.57	73.48

TABLE 10.20
5.5/8 H17

TC	TC Freq. (%)	EV W/L (%)	Std. Dev.	Play-All Bets Optimal ($)	Play-All Bets Practical ($)	Back-Count 24% 12% 5% Optimal ($)	Back-Count 24% 12% 5% Practical ($)	Bet Spread	Kelly Bank (units)	Avg. Bet ($)	Results W/L (%)	Results W/100 ($)	Risk Measures SD/100 ($)	Risk Measures ROR (%)	Risk Measures N0 (hands)	Rankings DI	Rankings SCORE
<0	45.76	–1.50	1.142	0,2,2	1,2,2	0	0	PA Opt. (1–8)	Expectations for all of these approaches are either negative or so negligible as to render the games unplayable.								
0	30.51	–0.59	1.138	0,2,2	1,2,2	0	0	PA Pract. (1–8)									
1	11.84	–0.03	1.136	0,2,2	1,2,2	0	0	PA Opt. (1–12)									
2	6.49	0.49	1.134	3,26,38	8,5,5	0,56,51	0,50,50	PA Pract. (1–12)	Data are meaningless in these circumstances, and so they are not reported.								
3	2.76	1.03	1.135	3,26,38	8,10,10	98,80,80	100,75,75	PA Opt. (1–16)									
4	1.53	1.58	1.139	3,26,38	8,24,25	121,121,121	125	PA Pract. (1–16)									
5	0.60	2.21	1.146	3,26,38	8,24,32	169,169,169	200,150,200	BC Opt. (1–2)	102	121.52	1.70	11.12	333.48	13.60	89,527	3.34	11.12
6	0.32	2.81	1.149	3,26,38	8,24,32	196,169,204	200,150,200	BC Pract. (1–2)	100	127.60	1.71	11.76	353.40	15.20	89,695	3.33	11.08
7	0.11	3.51	1.148	3,26,38	8,24,32	196,169,204	200,150,200	BC Opt. (1–3)	177	80.58	1.25	11.98	346.15	13.60	83,177	3.46	11.98
8	0.05	4.08	1.145	3,26,38	8,24,32	196,169,204	200,150,200	BC Pract. (1–3)	200	74.80	1.26	11.18	323.55	11.80	83,454	3.45	11.93
9	0.02	4.77	1.140	3,26,38	8,24,32	196,169,204	200,150,200	BC Opt. (1–4)	196	79.23	1.32	12.39	351.91	13.50	80,464	3.52	12.39
>9	0.01	5.43	1.134	3,26,38	8,24,32	196,169,204	200,150,200	BC Pract. (1–4)	200	79.45	1.34	12.66	360.75	14.30	81,326	3.51	12.31

TABLE 10.21
6.0/8 H17

TC	TC Freq. (%)	EV W/L (%)	Std. Dev.	Play-All Bets Optimal ($)	Play-All Bets Practical ($)	Back-Count 25% 13% 7% Optimal ($)	Back-Count 25% 13% 7% Practical ($)	Bet Spread	Kelly Bank (units)	Avg. Bet ($)	Results W/L (%)	Results W/100 ($)	Risk Measures SD/100 ($)	Risk Measures ROR (%)	Risk Measures N0 (hands)	Rankings DI	Rankings SCORE
<0	45.92	–1.57	1.143	2,3,4	2,3,5	0	0	PA Opt. (1–8)	4,599	4.22	0.14	0.58	76.06	13.50	1,729,719	0.76	0.58
0	29.02	–0.58	1.138	2,3,4	2,3,5	0	0	PA Pract. (1–8)	5,000	2.95	–0.01	–0.03	47.64	100.00		–0.07	0.00
1	11.62	–0.02	1.136	2,3,4	2,3,5	0	0	PA Opt. (1–12)	3,093	8.01	0.34	2.73	165.24	13.50	365,961	1.65	2.73
2	6.77	0.50	1.134	17,39,39	5,10,10	0,65,58	0,75,50	PA Pract. (1–12)	3,333	5.34	0.24	1.30	102.28	8.40	622,857	1.27	1.61
3	3.06	1.04	1.135	17,39,60	10,25,25	106,80,80	100,75,75	PA Opt. (1–16)	2,658	9.93	0.47	4.65	215.39	13.60	215,566	2.15	4.64
4	1.88	1.60	1.140	17,39,60	16,36,50	123,123,123	125	PA Pract. (1–16)	2,000	8.10	0.26	2.08	163.72	21.20	619,513	1.27	1.61
5	0.82	2.21	1.146	17,39,60	16,36,80	168,168,169	200	BC Opt. (1–2)	94	133.20	1.87	16.63	407.80	13.50	60,371	4.08	16.63
6	0.50	2.84	1.149	17,39,60	16,36,80	213,194,215	200	BC Pract. (1–2)	100	133.00	1.88	16.67	410.50	13.80	60,422	4.06	16.49
7	0.21	3.46	1.148	17,39,60	16,36,80	213,194,230	200,225,200	BC Opt. (1–3)	155	91.50	1.42	17.44	417.66	13.50	57,227	4.18	17.44
8	0.12	4.11	1.145	17,39,60	16,36,80	213,194,230	200,225,200	BC Pract. (1–3)	133	98.85	1.42	18.85	453.83	16.10	58,117	4.15	17.25
9	0.05	4.62	1.141	17,39,60	16,36,80	213,194,230	200,225,200	BC Opt. (1–4)	174	89.88	1.50	18.10	425.52	13.50	55,390	4.25	18.10
>9	0.04	5.55	1.134	17,39,60	16,36,80	213,194,230	200,225,200	BC Pract. (1–4)	200	85.50	1.53	17.59	415.95	13.10	55,854	4.23	17.87

TABLE 10.22
6.5/8 H17

TC	TC Freq. (%)	EV W/L (%)	Std. Dev.	Play-All Bets Optimal ($)	Play-All Bets Practical ($)	Back-Count 26% 15% 8% Optimal ($)	Back-Count 26% 15% 8% Practical ($)	Bet Spread	Kelly Bank (units)	Avg. Bet ($)	Results W/L (%)	Results W/100 ($)	Risk Measures SD/100 ($)	Risk Measures ROR (%)	Risk Measures N0 (hands)	Rankings DI	Rankings SCORE
<0	46.04	–1.64	1.143	4,5,5	5	0	0	PA Opt. (1–8)	2,446	8.38	0.27	2.26	150.37	13.50	442,405	1.50	2.26
0	27.63	–0.58	1.138	4,5,5	5	0	0	PA Pract. (1–8)	2,000	7.68	0.15	1.17	127.62	23.80	1,189,684	0.92	0.84
1	11.33	–0.01	1.136	4,5,5	5	0	0	PA Opt. (1–12)	1,892	12.40	0.48	5.97	244.32	13.60	167,614	2.44	5.97
2	6.95	0.52	1.134	33,40,40	10	0,0,66	0,0,75	PA Pract. (1–12)	2,000	8.63	0.40	3.43	169.62	9.20	244,549	2.02	4.09
3	3.27	1.05	1.136	33,63,82	25	121,103,82	125,100,75	PA Opt. (1–16)	1,912	13.92	0.64	8.95	298.93	13.60	111,882	2.99	8.94
4	2.23	1.61	1.140	33,63,84	40,60,50	124,124,124	125	PA Pract. (1–16)	2,000	8.92	0.52	4.61	186.97	7.10	164,491	2.47	6.08
5	1.03	2.24	1.145	33,63,84	40,60,80	171,171,171	200	BC Opt. (1–2)	83	148.63	2.10	25.09	500.87	13.40	39,712	5.01	25.09
6	0.73	2.84	1.147	33,63,84	40,60,80	216,216,216	200	BC Pract. (1–2)	80	153.75	2.09	25.89	518.00	14.50	40,077	5.00	24.97
7	0.32	3.50	1.147	33,63,84	40,60,80	242,267,265	250,300,300	BC Opt. (1–3)	97	146.50	2.23	26.26	512.48	13.50	38,194	5.12	26.26
8	0.24	4.06	1.144	33,63,84	40,60,80	242,310,265	250,300,300	BC Pract. (1–3)	100	148.50	2.23	26.69	522.60	14.10	38,310	5.11	26.08
9	0.10	4.67	1.140	33,63,84	40,60,80	242,310,265	250,300,300	BC Opt. (1–4)	151	103.18	1.74	26.91	518.75	13.50	37,127	5.19	26.91
>9	0.14	5.72	1.133	33,63,84	40,60,80	242,310,265	250,300,300	BC Pract. (1–4)	133	108.98	1.75	28.53	552.23	15.40	37,544	5.17	26.69

TABLE 10.23
7.0/8 H17

TC	TC Freq. (%)	EV W/L (%)	Std. Dev.	Play-All Bets Optimal ($)	Play-All Bets Practical ($)	Back-Count 28% 17% 10% Optimal ($)	Back-Count 28% 17% 10% Practical ($)	Bet Spread	Kelly Bank (units)	Avg. Bet ($)	Results W/L (%)	Results W/100 ($)	Risk Measures SD/100 ($)	Risk Measures ROR (%)	Risk Measures N0 (hands)	Rankings DI	Rankings SCORE
<0	46.11	–1.72	1.144	7,8,8	5,10,10	0	0	PA Opt. (1–8)	1,417	14.23	0.43	6.06	246.76	13.40	164,114	2.47	6.09
0	26.33	–0.57	1.139	7,8,8	5,10,10	0	0	PA Pract. (1–8)	2,000	8.20	0.37	3.03	138.78	4.30	209,767	2.18	4.76
1	10.95	0.00	1.137	7,8,8	5,10,10	0	0	PA Opt. (1–12)	1,275	18.11	0.68	12.42	352.03	13.60	80,656	3.52	12.39
2	7.04	0.54	1.135	42,42,42	10,25,25	0	0	PA Pract. (1–12)	1,000	19.21	0.66	12.72	378.89	17.00	88,726	3.36	11.27
3	3.39	1.07	1.137	56,83,83	25,50,50	137,119,103	150,125,100	PA Opt. (1–16)	1,322	19.53	0.87	17.02	412.40	13.50	58,821	4.12	17.01
4	2.52	1.65	1.141	56,94,121	40,120,125	137,127,127	150,125,125	PA Pract. (1–16)	1,000	20.80	0.85	17.73	448.45	17.10	63,975	3.95	15.63
5	1.22	2.25	1.145	56,94,121	40,120,160	172,172,172	200	BC Opt. (1–2)	73	169.75	2.40	38.99	624.34	13.60	25,690	6.24	38.99
6	0.98	2.87	1.146	56,94,121	40,120,160	218,218,219	200	BC Pract. (1–2)	67	184.35	2.39	42.18	677.25	15.70	25,817	6.23	38.80
7	0.47	3.45	1.145	56,94,121	40,120,160	263,263,263	300	BC Opt. (1–3)	84	167.52	2.57	41.18	641.68	13.50	24,331	6.42	41.18
8	0.40	4.05	1.143	56,94,121	40,120,160	273,310,311	300	BC Pract. (1–3)	80	172.13	2.55	42.02	657.38	14.30	24,498	6.39	40.86
9	0.19	4.54	1.140	56,94,121	40,120,160	273,349,350	300	BC Opt. (1–4)	97	164.53	2.67	42.00	648.09	13.40	23,796	6.48	42.00
>9	0.40	5.89	1.132	56,94,121	40,120,160	273,357,414	300,375,400	BC Pract. (1–4)	100	164.30	2.65	41.71	646.20	13.50	24,014	6.45	41.66

TABLE 10.24
5.5/8 H17 DAS

TC	TC Freq. (%)	EV W/L (%)	Std. Dev.	Play-All Bets Optimal ($)	Play-All Bets Practical ($)	Back-Count 24% 12% 5% Optimal ($)	Back-Count 24% 12% 5% Practical ($)	Bet Spread	Kelly Bank (units)	Avg. Bet ($)	Results W/L (%)	Results W/100 ($)	Risk Measures SD/100 ($)	Risk Measures ROR (%)	Risk Measures N0 (hands)	Rankings DI	Rankings SCORE
<0	45.76	−1.37	1.151	3,4,4	3,5,5	0	0	PA Opt. (1–8)	3,662	5.79	0.17	0.96	97.62	13.60	1,049,480	0.98	0.95
0	30.51	−0.45	1.147	3,4,4	3,5,5	0	0	PA Pract. (1–8)	3,333	6.33	0.17	1.04	106.78	16.10	1,052,191	0.98	0.95
1	11.84	0.12	1.144	9,9,9	10	0	0	PA Opt. (1–12)	2,813	8.89	0.35	3.10	175.81	13.60	323,851	1.76	3.09
2	6.49	0.65	1.141	22,43,50	24,25,25	73,64,58	75,75,50	PA Pract. (1–12)	2,000	9.86	0.29	2.82	186.69	19.80	436,722	1.51	2.29
3	2.76	1.19	1.142	22,43,58	24,60,50	91,91,91	100	PA Opt. (1–16)	2,781	10.20	0.46	4.74	217.17	13.60	211,951	2.17	4.72
4	1.53	1.75	1.146	22,43,58	24,60,80	133,133,133	125	PA Pract. (1–16)	2,000	10.11	0.36	3.66	201.60	16.50	303,387	1.81	3.29
5	0.60	2.37	1.152	22,43,58	24,60,80	146,179,179	150,200,200	BC Opt. (1–2)	137	91.75	1.30	14.20	376.89	13.50	70,608	3.77	14.20
6	0.32	2.98	1.155	22,43,58	24,60,80	146,193,224	150,200,200	BC Pract. (1–2)	133	94.20	1.29	14.48	384.75	14.20	70,651	3.76	14.16
7	0.11	3.63	1.154	22,43,58	24,60,80	146,193,232	150,225,200	BC Opt. (1–3)	155	90.64	1.40	15.09	388.47	13.60	66,038	3.88	15.09
8	0.05	4.23	1.150	22,43,58	24,60,80	146,193,232	150,225,200	BC Pract. (1–3)	133	99.30	1.38	16.24	420.08	15.90	67,239	3.87	14.94
9	0.02	4.92	1.145	22,43,58	24,60,80	146,193,232	150,225,200	BC Opt. (1–4)	172	88.71	1.46	15.43	392.80	13.60	64,699	3.93	15.43
>9	0.01	5.59	1.139	22,43,58	24,60,80	146,193,232	150,225,200	BC Pract. (1–4)	200	85.25	1.48	14.99	383.90	13.10	65,502	3.90	15.25

TABLE 10.25
6.0/8 H17 DAS

TC	TC Freq. (%)	EV W/L (%)	Std. Dev.	Play-All Bets Optimal ($)	Play-All Bets Practical ($)	Back-Count 25% 13% 7% Optimal ($)	Back-Count 25% 13% 7% Practical ($)	Bet Spread	Kelly Bank (units)	Avg. Bet ($)	Results W/L (%)	Results W/100 ($)	Risk Measures SD/100 ($)	Risk Measures ROR (%)	Risk Measures N0 (hands)	Rankings DI	Rankings SCORE
<0	45.92	−1.44	1.152	4,5,5	5	0	0	PA Opt. (1–8)	2,253	9.25	0.28	2.56	160.00	13.50	390,246	1.60	2.56
0	29.02	−0.44	1.147	4,5,5	5	0	0	PA Pract. (1–8)	2,000	9.27	0.24	2.19	153.37	15.50	490,448	1.43	2.04
1	11.62	0.13	1.144	10,10,10	10	0	0	PA Opt. (1–12)	2,032	12.22	0.47	5.72	239.44	13.50	174,473	2.39	5.73
2	6.77	0.66	1.141	36,51,51	25	82,72,65	75	PA Pract. (1–12)	2,000	10.60	0.44	4.62	202.59	10.50	192,288	2.28	5.20
3	3.06	1.20	1.142	36,59,79	40,60,50	92,92,92	100	PA Opt. (1–16)	2,027	13.56	0.60	8.12	285.26	13.40	122,823	2.85	8.14
4	1.88	1.77	1.147	36,59,79	40,60,80	134,134,135	125	PA Pract. (1–16)	2,000	11.02	0.54	5.95	223.64	9.30	141,269	2.66	7.08
5	0.82	2.38	1.152	36,59,79	40,60,80	164,179,179	150,200,200	BC Opt. (1–2)	122	102.29	1.46	20.00	447.23	13.40	50,099	4.47	20.00
6	0.50	3.00	1.154	36,59,79	40,60,80	164,215,225	150,225,200	BC Pract. (1–2)	133	97.35	1.45	18.92	424.05	12.20	50,340	4.46	19.91
7	0.21	3.62	1.153	36,59,79	40,60,80	164,215,258	150,225,300	BC Opt. (1–3)	140	101.31	1.58	21.43	462.92	13.40	46,694	4.63	21.43
8	0.12	4.26	1.150	36,59,79	40,60,80	164,215,258	150,225,300	BC Pract. (1–3)	133	105.45	1.57	22.30	482.48	14.80	46,915	4.62	21.36
9	0.05	4.80	1.146	36,59,79	40,60,80	164,215,258	150,225,300	BC Opt. (1–4)	155	99.50	1.65	22.02	469.28	13.50	45,365	4.69	22.02
>9	0.04	5.76	1.138	36,59,79	40,60,80	164,215,258	150,225,300	BC Pract. (1–4)	133	106.88	1.62	23.21	497.25	15.30	45,741	4.67	21.79

TABLE 10.26
6.5/8 H17 DAS

TC	TC Freq. (%)	EV W/L (%)	Std. Dev.	Play-All Bets Optimal ($)	Play-All Bets Practical ($)	Back-Count 26% 15% 8% Optimal ($)	Back-Count 26% 15% 8% Practical ($)	Bet Spread	Kelly Bank (units)	Avg. Bet ($)	Results W/L (%)	Results W/100 ($)	Risk Measures SD/100 ($)	Risk Measures ROR (%)	Risk Measures N0 (hands)	Rankings DI	Rankings SCORE
<0	46.04	–1.51	1.152	6,7,7	5	0	0	PA Opt. (1–8)	1,597	13.32	0.41	5.48	233.98	13.60	182,461	2.34	5.47
0	27.63	–0.43	1.147	6,7,7	5	0	0	PA Pract. (1–8)	2,000	9.78	0.38	3.69	162.84	6.20	194,734	2.27	5.14
1	11.33	0.14	1.144	11,11,11	10	0	0	PA Opt. (1–12)	1,454	16.52	0.62	10.25	320.51	13.50	97,332	3.21	10.27
2	6.95	0.68	1.142	50,52,52	25	0,82,73	0,75,75	PA Pract. (1–12)	2,000	11.39	0.61	6.91	218.36	5.50	100,004	3.16	10.01
3	3.27	1.22	1.143	50,82,93	40,60,50	129,93,93	125,100,100	PA Opt. (1–16)	1,527	17.68	0.77	13.64	369.50	13.50	73,273	3.70	13.65
4	2.23	1.79	1.147	50,82,105	40,60,80	136,136,136	150	PA Pract. (1–16)	2,000	12.02	0.74	8.93	246.27	5.30	76,054	3.63	13.14
5	1.03	2.40	1.151	50,82,105	40,60,80	181,181,182	200	BC Opt. (1–2)	78	158.88	2.26	28.86	537.22	13.40	34,639	5.37	28.86
6	0.73	2.99	1.153	50,82,105	40,60,80	225,225,225	200,225,200	BC Pract. (1–2)	80	160.63	2.24	28.91	539.88	13.70	34,958	5.36	28.68
7	0.32	3.66	1.152	50,82,105	40,60,80	258,245,276	250,225,300	BC Opt. (1–3)	123	114.66	1.79	30.83	555.27	13.40	32,400	5.55	30.83
8	0.24	4.21	1.149	50,82,105	40,60,80	258,245,291	250,225,300	BC Pract. (1–3)	133	115.43	1.80	31.13	562.43	14.00	32,652	5.53	30.63
9	0.10	4.86	1.145	50,82,105	40,60,80	258,245,291	250,225,300	BC Opt. (1–4)	137	112.77	1.88	31.85	564.35	13.60	31,427	5.64	31.85
>9	0.14	5.89	1.138	50,82,105	40,60,80	258,245,291	250,225,300	BC Pract. (1–4)	133	118.13	1.87	33.17	589.13	14.80	31,583	5.63	31.70

TABLE 10.27
7.0/8 H17 DAS

TC	TC Freq. (%)	EV W/L (%)	Std. Dev.	Play-All Bets Optimal ($)	Play-All Bets Practical ($)	Back-Count 28% 17% 10% Optimal ($)	Back-Count 28% 17% 10% Practical ($)	Bet Spread	Kelly Bank (units)	Avg. Bet ($)	Results W/L (%)	Results W/100 ($)	Risk Measures SD/100 ($)	Risk Measures ROR (%)	Risk Measures N0 (hands)	Rankings DI	Rankings SCORE
<0	46.11	–1.59	1.153	10,10,9	10	0	0	PA Opt. (1–8)	1,041	19.31	0.58	11.10	333.55	13.50	89,860	3.34	11.13
0	26.33	–0.43	1.147	10,10,9	10	0	0	PA Pract. (1–8)	1,000	16.74	0.53	8.79	282.83	11.10	103,532	3.11	9.66
1	10.94	0.15	1.144	11,11,11	10	0	0	PA Opt. (1–12)	1,040	22.29	0.83	18.48	429.99	13.50	54,062	4.30	18.49
2	7.04	0.70	1.142	53,53,53	25	0,0,85	0,100,75	PA Pract. (1–12)	1,000	19.21	0.82	15.68	380.99	11.50	59,038	4.12	16.95
3	3.39	1.23	1.143	77,94,94	50	146,126,94	150,100,100	PA Opt. (1–16)	1,136	23.22	1.01	23.46	484.47	13.50	42,620	4.84	23.47
4	2.52	1.83	1.147	77,115,139	80,120,125	146,139,139	150	PA Pract. (1–16)	1,000	20.80	1.01	20.96	450.87	12.70	46,272	4.65	21.62
5	1.22	2.42	1.151	77,115,141	80,120,160	182,182,182	200	BC Opt. (1–2)	69	180.11	2.56	44.19	664.69	13.40	22,698	6.65	44.19
6	0.98	3.04	1.152	77,115,141	80,120,160	229,229,229	200	BC Pract. (1–2)	67	184.35	2.56	45.14	680.70	14.10	22,730	6.63	43.98
7	0.47	3.60	1.151	77,115,141	80,120,160	272,272,272	300	BC Opt. (1–3)	79	177.71	2.72	46.30	680.45	13.60	21,642	6.80	46.30
8	0.40	4.23	1.148	77,115,141	80,120,160	291,321,321	300	BC Pract. (1–3)	100	138.40	2.09	47.96	707.00	14.60	21,695	6.78	46.03
9	0.19	4.70	1.145	77,115,141	80,120,160	291,359,339	300	BC Opt. (1–4)	118	130.69	2.21	48.01	692.87	13.50	20,836	6.93	48.01
>9	0.40	6.05	1.136	77,115,141	80,120,160	291,378,339	300	BC Pract. (1–4)	133	127.80	2.20	46.73	677.93	13.10	21,051	6.89	47.52

TABLE 10.28
5.5/8 H17 LS

TC	TC Freq. (%)	EV W/L (%)	Std. Dev.	Play-All Bets Optimal ($)	Play-All Bets Practical ($)	Back-Count 24% 12% 5% Optimal ($)	Back-Count 24% 12% 5% Practical ($)	Bet Spread	Kelly Bank (units)	Avg. Bet ($)	Results W/L (%)	Results W/100 ($)	Risk Measures SD/100 ($)	Risk Measures ROR (%)	Risk Measures N0 (hands)	Rankings DI	Rankings SCORE
<0	45.75	−1.46	1.133	3,4,4	3,5,5	0	0	PA Opt. (1–8)	3,121	6.75	0.18	1.23	111.19	13.50	808,043	1.11	1.24
0	30.52	−0.49	1.124	3,4,4	3,5,5	0	0	PA Pract. (1–8)	3,333	6.33	0.18	1.16	104.20	11.80	809,627	1.11	1.24
1	11.84	0.13	1.119	11,11,11	10	0	0	PA Opt. (1–12)	2,445	10.21	0.38	3.88	197.07	13.50	257,241	1.97	3.88
2	6.48	0.70	1.114	26,49,56	24,25,25	84,73,66	75	PA Pract. (1–12)	2,000	9.87	0.32	3.12	181.93	15.10	338,910	1.72	2.95
3	2.77	1.27	1.111	26,49,67	24,60,50	103,103,103	100	PA Opt. (1–16)	2,381	11.73	0.50	5.88	242.82	13.50	169,593	2.43	5.90
4	1.53	1.90	1.113	26,49,67	24,60,80	153,153,153	150	PA Pract. (1–16)	2,000	10.12	0.40	4.04	196.34	12.30	236,174	2.06	4.23
5	0.60	2.59	1.119	26,49,67	24,60,80	167,207,207	150,200,200	BC Opt. (1–2)	120	104.90	1.41	17.60	419.51	13.40	56,581	4.19	17.60
6	0.32	3.26	1.122	26,49,67	24,60,80	167,220,259	150,225,300	BC Pract. (1–2)	133	97.50	1.42	16.42	391.80	11.80	56,901	4.19	17.57
7	0.11	3.97	1.121	26,49,67	24,60,80	167,220,265	150,225,300	BC Opt. (1–3)	136	103.69	1.52	18.76	433.14	13.50	53,424	4.33	18.76
8	0.05	4.68	1.117	26,49,67	24,60,80	167,220,265	150,225,300	BC Pract. (1–3)	133	103.20	1.52	18.61	429.68	13.40	53,365	4.33	18.75
9	0.02	5.26	1.113	26,49,67	24,60,80	167,220,265	150,225,300	BC Opt. (1–4)	151	101.49	1.59	19.21	438.31	13.50	52,189	4.38	19.21
>9	0.01	6.07	1.106	26,49,67	24,60,80	167,220,265	150,225,300	BC Pract. (1–4)	133	106.43	1.58	20.01	458.03	14.90	52,316	4.37	19.09

TABLE 10.29
6.0/8 H17 LS

TC	TC Freq. (%)	EV W/L (%)	Std. Dev.	Play-All Bets Optimal ($)	Play-All Bets Practical ($)	Back-Count 25% 13% 7% Optimal ($)	Back-Count 25% 13% 7% Practical ($)	Bet Spread	Kelly Bank (units)	Avg. Bet ($)	Results W/L (%)	Results W/100 ($)	Risk Measures SD/100 ($)	Risk Measures ROR (%)	Risk Measures N0 (hands)	Rankings DI	Rankings SCORE
<0	45.91	−1.52	1.133	5,6,6	5	0	0	PA Opt. (1–8)	1,954	10.62	0.30	3.22	179.61	13.50	310,249	1.80	3.23
0	29.03	−0.48	1.124	5,6,6	5	0	0	PA Pract. (1–8)	2,000	9.27	0.26	2.43	149.58	11.40	380,473	1.62	2.63
1	11.61	0.14	1.119	11,11,11	10	0	0	PA Opt. (1–12)	1,739	14.03	0.51	7.15	267.25	13.60	140,010	2.67	7.14
2	6.76	0.70	1.114	41,57,57	25	94,82,73	100,75,75	PA Pract. (1–12)	2,000	10.61	0.48	5.06	197.42	7.40	152,223	2.56	6.57
3	3.06	1.28	1.112	41,69,91	40,60,50	104,104,104	100	PA Opt. (1–16)	1,749	15.50	0.65	10.10	318.30	13.40	98,717	3.18	10.13
4	1.88	1.92	1.114	41,69,91	40,60,80	154,154,155	150	PA Pract. (1–16)	2,000	11.03	0.59	6.52	217.83	6.40	111,614	2.99	8.96
5	0.82	2.59	1.119	41,69,91	40,60,80	188,207,207	200	BC Opt. (1–2)	107	116.68	1.57	24.70	496.92	13.40	40,580	4.97	24.70
6	0.50	3.27	1.122	41,69,91	40,60,80	188,245,260	200,225,300	BC Pract. (1–2)	100	119.90	1.57	25.38	511.30	14.30	40,521	4.96	24.64
7	0.21	3.94	1.120	41,69,91	40,60,80	188,245,294	200,225,300	BC Opt. (1–3)	122	115.68	1.71	26.55	515.28	13.60	37,612	5.15	26.55
8	0.12	4.63	1.117	41,69,91	40,60,80	188,245,294	200,225,300	BC Pract. (1–3)	133	109.05	1.71	25.07	486.75	12.10	37,757	5.15	26.53
9	0.05	5.24	1.113	41,69,91	40,60,80	188,245,294	200,225,300	BC Opt. (1–4)	136	113.62	1.79	27.33	522.72	13.50	36,602	5.23	27.33
>9	0.04	6.23	1.106	41,69,91	40,60,80	188,245,294	200,225,300	BC Pract. (1–4)	133	114.15	1.81	27.71	531.15	14.10	36,835	5.22	27.22

TABLE 10.30
6.5/8 H17 LS

TC	TC Freq. (%)	EV W/L (%)	Std. Dev.	Play-All Bets Optimal ($)	Play-All Bets Practical ($)	Back-Count 26% 15% 8% Optimal ($)	Back-Count 26% 15% 8% Practical ($)	Bet Spread	Kelly Bank (units)	Avg. Bet ($)	Results W/L (%)	Results W/100 ($)	Risk Measures SD/100 ($)	Risk Measures ROR (%)	Risk Measures N0 (hands)	Rankings DI	Rankings SCORE
<0	46.03	–1.60	1.134	7,8,8	5,10,10	0	0	PA Opt. (1–8)	1,398	15.17	0.45	6.78	260.41	13.50	147,462	2.60	6.78
0	27.64	–0.47	1.124	7,8,8	5,10,10	0	0	PA Pract. (1–8)	2,000	9.78	0.41	4.03	158.81	4.10	155,291	2.54	6.44
1	11.32	0.14	1.119	11,12,12	10	0	0	PA Opt. (1–12)	1,260	18.82	0.68	12.75	356.79	13.60	78,589	3.57	12.73
2	6.94	0.71	1.115	57,57,57	25	0,93,83	0,100,75	PA Pract. (1–12)	1,000	17.63	0.60	10.64	335.35	15.10	99,338	3.17	10.07
3	3.28	1.30	1.113	57,95,105	40,50,50	147,105,105	150,100,100	PA Opt. (1–16)	1,313	20.18	0.84	16.92	411.30	13.50	59,088	4.11	16.92
4	2.23	1.92	1.115	57,95,122	40,120,125	155,155,155	150	PA Pract. (1–16)	1,000	18.77	0.77	14.41	389.63	15.00	73,110	3.70	13.67
5	1.03	2.60	1.119	57,95,122	40,120,160	208,208,208	200	BC Opt. (1–2)	68	181.86	2.44	35.83	598.64	13.50	27,851	5.99	35.83
6	0.74	3.26	1.120	57,95,122	40,120,160	260,260,260	300	BC Pract. (1–2)	67	184.95	2.46	36.74	615.00	14.20	28,005	5.97	35.69
7	0.32	3.96	1.119	57,95,122	40,120,160	294,278,316	300	BC Opt. (1–3)	108	130.56	1.94	38.02	616.56	13.50	26,339	6.17	38.02
8	0.24	4.57	1.117	57,95,122	40,120,160	294,278,330	300	BC Pract. (1–3)	100	134.80	1.96	39.56	643.20	14.70	26,382	6.15	37.83
9	0.10	5.25	1.113	57,95,122	40,120,160	294,278,330	300	BC Opt. (1–4)	121	128.46	2.04	39.34	627.25	13.50	25,446	6.27	39.34
>9	0.14	6.35	1.105	57,95,122	40,120,160	294,278,330	300	BC Pract. (1–4)	133	123.23	2.07	38.33	613.20	13.10	25,600	6.25	39.07

TABLE 10.31
7.0/8 H17 LS

TC	TC Freq. (%)	EV W/L (%)	Std. Dev.	Play-All Bets Optimal ($)	Play-All Bets Practical ($)	Back-Count 28% 17% 10% Optimal ($)	Back-Count 28% 17% 10% Practical ($)	Bet Spread	Kelly Bank (units)	Avg. Bet ($)	Results W/L (%)	Results W/100 ($)	Risk Measures SD/100 ($)	Risk Measures ROR (%)	Risk Measures N0 (hands)	Rankings DI	Rankings SCORE
<0	46.10	–1.69	1.135	11,11,10	10	0	0	PA Opt. (1–8)	906	21.92	0.62	13.68	369.58	13.60	73,166	3.70	13.66
0	26.33	–0.47	1.124	11,11,10	10	0	0	PA Pract. (1–8)	1,000	16.76	0.57	9.57	275.86	8.10	83,091	3.47	12.04
1	10.94	0.15	1.119	12,12,12	10	0	0	PA Opt. (1–12)	903	25.31	0.90	22.72	476.54	13.50	44,043	4.77	22.71
2	7.03	0.73	1.115	59,59,59	25	0,0,96	0,0,100	PA Pract. (1–12)	1,000	19.24	0.89	17.03	371.40	8.50	47,561	4.59	21.02
3	3.40	1.31	1.113	88,106,105	50	165,143,106	200,150,100	PA Opt. (1–16)	981	26.39	1.09	28.78	536.94	13.50	34,687	5.37	28.83
4	2.52	1.95	1.116	88,133,157	80,120,125	165,157,157	200,150,150	PA Pract. (1–16)	1,000	20.84	1.09	22.76	439.57	9.50	37,300	5.18	26.81
5	1.22	2.59	1.119	88,133,163	80,120,160	207,207,207	200	BC Opt. (1–2)	60	204.96	2.76	54.32	737.04	13.70	18,369	7.37	54.33
6	0.98	3.28	1.120	88,133,163	80,120,160	261,261,261	300	BC Pract. (1–2)	50	236.20	2.74	62.23	847.40	17.60	18,561	7.34	53.92
7	0.47	3.88	1.118	88,133,163	80,120,160	310,310,310	300	BC Opt. (1–3)	70	202.29	2.94	57.01	755.06	13.40	17,560	7.55	57.01
8	0.40	4.54	1.116	88,133,163	80,120,160	331,365,365	400	BC Pract. (1–3)	67	208.50	2.97	59.39	788.25	14.60	17,610	7.53	56.77
9	0.19	5.05	1.113	88,133,163	80,120,160	331,408,384	400,450,400	BC Opt. (1–4)	104	148.30	2.39	58.81	766.84	13.50	17,019	7.67	58.81
>9	0.42	6.52	1.104	88,133,163	80,120,160	331,430,384	400,450,400	BC Pract. (1–4)	100	150.60	2.41	60.47	790.20	14.40	17,059	7.65	58.55

TABLE 10.32
5.5/8 H17 DAS LS

	TC	EV		Play-All Bets		Back-Count 24% 12% 5%			Kelly	Avg.	Results		Risk Measures			Rankings	
TC	Freq. (%)	W/L (%)	Std. Dev.	Optimal ($)	Practical ($)	Optimal ($)	Practical ($)	Bet Spread	Bank (units)	Bet ($)	W/L (%)	W/100 ($)	SD/100 ($)	ROR (%)	N0 (hands)	DI	SCORE
<0	45.76	−1.32	1.142	6,6,6	5	0	0	PA Opt. (1–8)	1,788	12.17	0.33	4.00	200.41	13.40	249,082	2.00	4.02
0	30.52	−0.34	1.132	6,6,6	5	0	0	PA Pract. (1–8)	2,000	8.78	0.28	2.45	141.18	8.50	330,684	1.74	3.02
1	11.83	0.28	1.127	22,22,22	10	0	0	PA Opt. (1–12)	1,816	14.64	0.52	7.53	274.65	13.50	132,485	2.75	7.54
2	6.48	0.85	1.121	45,66,67	25	92,81,74	100,75,75	PA Pract. (1–12)	2,000	9.87	0.47	4.60	183.07	6.40	158,034	2.51	6.32
3	2.77	1.44	1.118	45,66,88	40,60,50	115,115,116	125	PA Opt. (1–16)	1,818	15.90	0.63	9.97	315.77	13.50	100,307	3.16	9.97
4	1.53	2.06	1.120	45,66,88	40,60,80	164,164,165	200	PA Pract. (1–16)	2,000	10.12	0.55	5.56	197.54	5.80	126,457	2.81	7.91
5	0.60	2.74	1.125	45,66,88	40,60,80	184,216,217	200	BC Opt. (1–2)	108	115.34	1.57	21.47	463.40	13.60	46,530	4.63	21.47
6	0.32	3.42	1.128	45,66,88	40,60,80	184,244,270	200,225,300	BC Pract. (1–2)	100	128.00	1.58	23.99	518.40	16.80	46,656	4.63	21.41
7	0.11	4.15	1.127	45,66,88	40,60,80	184,244,297	200,225,300	BC Opt. (1–3)	123	113.67	1.67	22.58	475.19	13.50	44,389	4.75	22.58
8	0.05	4.78	1.122	45,66,88	40,60,80	184,244,297	200,225,300	BC Pract. (1–3)	133	115.50	1.68	23.11	489.83	14.60	44,963	4.72	22.25
9	0.02	5.51	1.118	45,66,88	40,60,80	184,244,297	200,225,300	BC Opt. (1–4)	135	111.44	1.73	22.96	479.12	13.50	43,452	4.79	22.96
>9	0.01	6.23	1.111	45,66,88	40,60,80	184,244,297	200,225,300	BC Pract. (1–4)	133	118.65	1.74	24.58	515.10	15.70	43,844	4.77	22.76

TABLE 10.33
6.0/8 H17 DAS LS

	TC	EV		Play-All Bets		Back-Count 25% 13% 7%			Kelly	Avg.	Results		Risk Measures			Rankings	
TC	Freq. (%)	W/L (%)	Std. Dev.	Optimal ($)	Practical ($)	Optimal ($)	Practical ($)	Bet Spread	Bank (units)	Bet ($)	W/L (%)	W/100 ($)	SD/100 ($)	ROR (%)	N0 (hands)	DI	SCORE
<0	45.92	−1.39	1.142	7,7,7	5	0	0	PA Opt. (1–8)	1,358	16.04	0.45	7.14	267.08	13.50	140,315	2.67	7.13
0	29.03	−0.33	1.132	7,7,7	5	0	0	PA Pract. (1–8)	2,000	9.27	0.41	3.80	150.53	3.50	156,498	2.53	6.38
1	11.61	0.28	1.127	22,22,22	10	0	0	PA Opt. (1–12)	1,341	18.77	0.64	11.92	345.70	13.50	83,724	3.46	11.95
2	6.76	0.86	1.122	59,68,68	25	102,88,80	100,100,75	PA Pract. (1–12)	2,000	10.61	0.63	6.65	198.64	3.40	89,226	3.35	11.21
3	3.06	1.45	1.119	59,89,110	40,60,50	116,116,116	125	PA Opt. (1–16)	1,456	19.70	0.78	15.27	391.04	13.50	65,381	3.91	15.29
4	1.88	2.07	1.121	59,89,110	40,60,80	165,165,165	200	PA Pract. (1–16)	2,000	11.03	0.74	8.17	219.14	3.30	72,030	3.73	13.89
5	0.82	2.74	1.125	59,89,110	40,60,80	203,217,217	200	BC Opt. (1–2)	98	126.85	1.73	29.46	542.80	13.60	33,856	5.43	29.46
6	0.50	3.43	1.128	59,89,110	40,60,80	203,265,270	200,300,300	BC Pract. (1–2)	100	132.60	1.74	30.94	572.30	15.10	34,303	5.41	29.23
7	0.21	4.09	1.126	59,89,110	40,60,80	203,265,320	200,300,300	BC Opt. (1–3)	113	125.23	1.86	31.25	558.94	13.50	32,054	5.59	31.24
8	0.12	4.74	1.122	59,89,110	40,60,80	203,265,320	200,300,300	BC Pract. (1–3)	100	139.40	1.85	34.60	620.90	16.60	32,203	5.57	31.04
9	0.05	5.35	1.118	59,89,110	40,60,80	203,265,320	200,300,300	BC Opt. (1–4)	125	122.95	1.93	31.92	564.95	13.50	31,291	5.65	31.92
>9	0.04	6.45	1.111	59,89,110	40,60,80	203,265,320	200,300,300	BC Pract. (1–4)	133	126.83	1.94	33.15	590.18	15.00	31,695	5.62	31.54

TABLE 10.34
6.5/8 H17 DAS LS

				Play-All Bets		Back-Count 26% 15% 8%					Results		Risk Measures			Rankings	
TC	TC Freq. (%)	EV W/L (%)	Std. Dev.	Optimal ($)	Practical ($)	Optimal ($)	Practical ($)	Bet Spread	Kelly Bank (units)	Avg. Bet ($)	W/L (%)	W/100 ($)	SD/100 ($)	ROR (%)	N0 (hands)	DI	SCORE
<0	46.03	−1.47	1.143	10,10,9	10	0	0	PA Opt. (1–8)	1,018	20.95	0.58	12.10	347.66	13.60	82,683	3.48	12.09
0	27.64	−0.33	1.132	10,10,9	10	0	0	PA Pract. (1–8)	1,000	20.13	0.55	11.15	324.62	12.00	84,762	3.44	11.80
1	11.32	0.29	1.127	23,23,23	25	0	0	PA Opt. (1–12)	1,052	23.58	0.80	19.00	435.47	13.60	52,759	4.35	18.96
2	6.94	0.87	1.122	69,69,69	50	114,100,89	125,100,100	PA Pract. (1–12)	1,000	23.36	0.80	18.64	431.99	13.60	53,710	4.31	18.61
3	3.28	1.47	1.120	79,114,117	80,120,125	117,117,117	125	PA Opt. (1–16)	1,126	24.56	0.96	23.43	484.56	13.40	42,578	4.85	23.48
4	2.23	2.08	1.121	79,114,142	80,120,160	165,165,166	200	PA Pract. (1–16)	1,000	25.44	0.96	24.43	509.27	15.20	43,456	4.80	23.02
5	1.03	2.76	1.125	79,114,142	80,120,160	218,218,218	200	BC Opt. (1–2)	88	141.30	1.93	41.03	640.58	13.40	24,402	6.41	41.03
6	0.74	3.41	1.126	79,114,142	80,120,160	228,269,269	250,300,300	BC Pract. (1–2)	80	154.13	1.93	44.53	697.38	16.00	24,559	6.39	40.78
7	0.32	4.12	1.125	79,114,142	80,120,160	228,299,326	250,300,300	BC Opt. (1–3)	100	140.25	2.09	43.92	662.71	13.60	22,759	6.63	43.92
8	0.24	4.69	1.122	79,114,142	80,120,160	228,299,356	250,300,400	BC Pract. (1–3)	100	147.60	2.09	46.24	700.40	15.10	22,983	6.60	43.59
9	0.10	5.41	1.117	79,114,142	80,120,160	228,299,356	250,300,400	BC Opt. (1–4)	112	137.79	2.18	45.13	671.77	13.60	22,187	6.72	45.13
>9	0.14	6.51	1.110	79,114,142	80,120,160	228,299,356	250,300,400	BC Pract. (1–4)	100	150.80	2.16	48.82	729.70	15.90	22,359	6.69	44.76

TABLE 10.35
7.0/8 H17 DAS LS

				Play-All Bets		Back-Count 28% 17% 10%					Results		Risk Measures			Rankings	
TC	TC Freq. (%)	EV W/L (%)	Std. Dev.	Optimal ($)	Practical ($)	Optimal ($)	Practical ($)	Bet Spread	Kelly Bank (units)	Avg. Bet ($)	W/L (%)	W/100 ($)	SD/100 ($)	ROR (%)	N0 (hands)	DI	SCORE
<0	46.10	−1.56	1.144	13,13,12	15,15,10	0	0	PA Opt. (1–8)	741	27.63	0.76	20.90	457.23	13.50	47,811	4.57	20.90
0	26.33	−0.32	1.132	13,13,12	15,15,10	0	0	PA Pract. (1–8)	667	28.64	0.75	21.47	475.34	14.90	49,039	4.52	20.40
1	10.93	0.30	1.127	24,23,23	25	0	0	PA Opt. (1–12)	786	30.20	1.02	30.77	554.94	13.50	32,468	5.55	30.80
2	7.03	0.88	1.123	70,70,70	50	0,115,102	0,125,100	PA Pract. (1–12)	667	32.54	1.03	33.48	609.78	16.50	33,172	5.49	30.15
3	3.40	1.48	1.120	108,118,118	120,125,125	174,118,118	200,125,125	PA Opt. (1–16)	864	30.94	1.20	37.10	609.58	13.50	26,913	6.10	37.16
4	2.52	2.11	1.122	108,153,167	120,180,160	174,168,168	200	PA Pract. (1–16)	1,000	27.67	1.21	33.57	554.01	11.20	27,235	6.06	36.71
5	1.22	2.75	1.125	108,153,185	120,180,160	218,218,218	200	BC Opt. (1–2)	58	214.64	2.92	60.13	775.47	13.30	16,631	7.75	60.13
6	0.98	3.43	1.126	108,153,185	120,180,160	271,271,271	300	BC Pract. (1–2)	50	236.20	2.90	65.85	852.00	16.20	16,766	7.73	59.73
7	0.47	4.03	1.124	108,153,185	120,180,160	319,319,319	300	BC Opt. (1–3)	87	159.96	2.40	63.97	799.78	13.50	15,614	8.00	63.97
8	0.40	4.68	1.121	108,153,185	120,180,160	348,345,373	400,375,400	BC Pract. (1–3)	80	172.38	2.39	68.43	858.00	15.50	15,746	7.97	63.60
9	0.19	5.21	1.118	108,153,185	120,180,160	348,345,408	400,375,400	BC Opt. (1–4)	98	157.57	2.52	66.09	812.93	13.50	15,150	8.13	66.09
>9	0.42	6.67	1.109	108,153,185	120,180,160	348,345,408	400,375,400	BC Pract. (1–4)	100	163.30	2.51	68.29	843.00	14.60	15,234	8.10	65.61

TABLE 10.36
6–Deck Play-All SCORE Summary

		S17	S17 DAS		S17 LS		S17 DAS LS		H17		H17 DAS		H17 LS		H17 DAS LS	
4.0/6	**1-8**	**4.17**	**8.48**	4.31	**9.35**	5.18	**15.46**	11.29	**0.86**	–3.31	**3.14**	–1.03	**3.85**	–0.32	**8.17**	4.00
	1-12	**7.57**	**12.56**	4.99	**14.29**	6.72	**20.98**	13.41	**3.20**	–4.37	**6.44**	–1.13	**7.91**	0.34	**13.12**	5.55
	1-16	**9.99**	**15.22**	5.23	**17.57**	7.58	**24.38**	14.39	**5.18**	–4.81	**8.88**	–1.11	**10.93**	0.94	**16.49**	6.50
4.5/6	**1-8**	**8.69**	**14.64**	5.95	**16.63**	7.94	**24.66**	15.97	**3.44**	–5.25	**7.29**	–1.40	**8.92**	0.23	**15.10**	6.41
		4.52	6.16		7.28		9.20		2.58		4.15		5.07		6.93	
	1-12	**14.06**	**20.64**	6.58	**23.99**	9.93	**32.50**	18.44	**7.76**	–6.30	**12.64**	–1.42	**15.55**	1.49	**22.58**	8.52
		6.49	8.08		9.70		11.52		4.56		6.20		7.64		9.46	
	1-16	**17.62**	**24.44**	6.82	**28.70**	11.08	**37.32**	19.70	**11.00**	–6.62	**16.28**	–1.34	**20.05**	2.43	**27.40**	9.78
		7.63	9.22		11.13		12.94		5.82		7.40		9.12		10.91	
5.0/6	**1-8**	**17.50**	**25.61**	8.11	**29.51**	12.01	**40.03**	22.53	**9.46**	–8.04	**15.64**	–1.86	**19.07**	1.57	**27.58**	10.08
		13.33	17.13		20.16		24.57		8.60		12.50		15.22		19.41	
	1-12	**25.88**	**34.61**	8.73	**40.55**	14.67	**51.53**	25.65	**17.00**	–8.88	**24.16**	–1.72	**29.54**	3.66	**38.84**	12.96
		18.31	22.05		26.26		30.55		13.80		17.72		21.63		25.72	
	1-16	**31.25**	**40.17**	8.92	**47.45**	16.20	**58.52**	27.27	**22.20**	–9.05	**29.71**	–1.54	**36.36**	5.11	**45.90**	14.65
		21.26	24.95		29.88		34.14		17.02		20.83		25.43		29.41	
5.5/6	**1-8**	**37.51**	**48.96**	11.45	**57.04**	19.53	**71.39**	33.88	**25.14**	–12.37	**34.92**	–2.59	**42.21**	4.70	**54.39**	16.88
		33.34	40.48		47.69		55.93		24.28		31.78		38.36		46.22	
	1-12	**51.86**	**63.89**	12.03	**75.34**	23.48	**90.12**	38.26	**38.87**	–12.99	**49.61**	–2.25	**60.14**	8.28	**73.03**	21.17
		44.29	51.33		61.05		69.14		35.67		43.17		52.23		59.91	
	1-16	**60.88**	**73.01**	12.13	**86.57**	25.69	**101.36**	40.48	**47.97**	–12.91	**58.97**	–1.91	**71.58**	10.70	**84.57**	23.69
		50.89	57.79		69.00		76.98		42.79		50.09		60.65		68.08	

Legend: **X** y
z

X = SCORE; y = $ gain from rules; z = $ gain from penetration

TABLE 10.37

6–Deck Back-Count SCORE Summary

		S17	S17 DAS		S17 LS		S17 DAS LS		H17		H17 DAS		H17 LS		H17 DAS LS	
4.0/6	**1–2**	**20.83**	**25.42**	4.59	**30.48**	9.65	**35.87**	15.04	**16.69**	–4.14	**20.50**	–0.33	**25.17**	4.34	**30.34**	9.51
	1–3	**22.13**	**26.62**	4.49	**32.11**	9.98	**37.39**	15.26	**17.73**	–4.40	**21.86**	–0.27	**26.92**	4.79	**32.02**	9.89
	1–4	**22.62**	**27.02**	4.40	**32.69**	10.07	**38.31**	15.69	**18.34**	–4.28	**22.39**	–0.23	**27.62**	5.00	**32.63**	10.01
4.5/6	**1–2**	**31.74**	**37.75**	6.01	**45.39**	13.65	**52.37**	20.63	**27.33**	–4.41	**31.47**	–0.27	**38.99**	7.25	**45.27**	13.53
		10.91	12.33		14.91		16.50		10.64		10.97		13.82		14.93	
	1–3	**34.04**	**39.97**	5.93	**48.32**	14.28	**55.18**	21.14	**28.48**	–5.56	**33.81**	–0.23	**41.55**	7.51	**48.24**	14.20
		11.91	13.35		16.21		17.79		10.75		11.95		14.63		16.22	
	1–4	**35.02**	**40.82**	5.80	**49.49**	14.47	**56.20**	21.18	**29.38**	–5.64	**34.83**	–0.19	**42.88**	7.86	**49.45**	14.43
		12.40	13.80		16.80		17.89		11.04		12.44		15.26		16.82	
5.0/6	**1–2**	**50.97**	**57.57**	6.60	**69.40**	18.43	**78.16**	27.19	**45.31**	–5.66	**51.23**	0.26	**63.00**	12.03	**69.69**	18.72
		30.14	32.15		38.92		42.29		28.62		30.73		37.83		39.35	
	1–3	**53.79**	**61.63**	7.84	**74.20**	20.41	**83.30**	29.51	**47.66**	–6.13	**53.75**	–0.04	**65.92**	12.13	**74.25**	20.46
		31.66	35.01		42.09		45.91		29.93		31.89		39.00		42.23	
	1–4	**55.61**	**63.29**	7.68	**76.42**	20.81	**85.36**	29.75	**48.50**	–7.11	**55.63**	0.02	**68.13**	12.52	**76.52**	20.91
		32.99	36.27		43.73		47.05		30.16		33.24		40.51		43.89	
5.5/6	**1–2**	**90.22**	**98.89**	8.67	**119.57**	29.35	**129.56**	39.34	**82.56**	–7.66	**90.81**	0.59	**110.53**	20.31	**120.08**	29.86
		69.39	73.47		89.09		93.69		65.87		70.31		85.36		89.74	
	1–3	**95.36**	**104.38**	9.02	**125.83**	30.47	**137.65**	42.29	**87.47**	–7.89	**96.06**	0.70	**117.15**	21.79	**126.38**	31.02
		73.23	77.76		93.72		100.26		69.74		74.20		90.23		94.36	
	1–4	**97.56**	**107.96**	10.40	**129.85**	32.29	**142.02**	44.46	**89.03**	–8.53	**97.91**	0.35	**118.89**	21.33	**130.17**	32.61
		74.94	80.94		97.16		103.71		70.69		75.52		91.27		97.54	

Legend: **X** y
z

X = SCORE; y = $ gain from rules; z = $ gain from penetration

TABLE 10.38
4.0/6 S17

TC	TC Freq. (%)	EV W/L (%)	Std. Dev.	Play-All Bets Optimal ($)	Play-All Bets Practical ($)	Back-Count 26% 14% 7% Optimal ($)	Back-Count 26% 14% 7% Practical ($)	Bet Spread	Kelly Bank (units)	Avg. Bet ($)	Results W/L (%)	Results W/100 ($)	Risk Measures SD/100 ($)	Risk Measures ROR (%)	Risk Measures N0 (hands)	Rankings DI	Rankings SCORE
<0	44.71	−1.32	1.139	6,6,5	5	0	0	PA Opt. (1–8)	1,804	12.10	0.35	4.17	204.26	13.50	239,437	2.04	4.17
0	29.69	−0.36	1.133	6,6,5	5	0	0	PA Pract. (1–8)	2,000	9.40	0.31	2.92	153.99	8.50	277,144	1.90	3.61
1	11.63	0.20	1.133	15,15,15	10	0	0	PA Opt. (1–12)	1,750	14.64	0.52	7.59	275.12	13.60	132,029	2.75	7.57
2	7.12	0.70	1.131	44,55,55	25	83,72,65	75	PA Pract. (1–12)	2,000	10.77	0.50	5.39	203.38	7.40	142,641	2.65	7.02
3	3.18	1.22	1.132	44,69,86	40,60,50	95,95,95	100	PA Opt. (1–16)	1,849	15.64	0.64	9.98	316.09	13.50	100,072	3.16	9.99
4	1.97	1.77	1.136	44,69,86	40,60,80	137,137,138	150	PA Pract. (1–16)	2,000	11.19	0.60	6.69	224.06	7.00	112,333	2.98	8.90
5	0.83	2.36	1.141	44,69,86	40,60,80	166,181,181	150,200,200	BC Opt. (1–2)	121	103.49	1.44	20.83	456.47	13.40	48,172	4.56	20.83
6	0.49	2.96	1.144	44,69,86	40,60,80	166,217,227	150,225,200	BC Pract. (1–2)	133	100.35	1.45	20.28	446.03	13.10	48,514	4.55	20.68
7	0.20	3.59	1.144	44,69,86	40,60,80	166,217,261	150,225,300	BC Opt. (1–3)	138	102.10	1.55	22.13	470.40	13.60	45,260	4.70	22.13
8	0.11	4.17	1.141	44,69,86	40,60,80	166,217,261	150,225,300	BC Pract. (1–3)	133	108.00	1.56	23.50	499.73	15.30	45,316	4.70	22.11
9	0.04	4.77	1.137	44,69,86	40,60,80	166,217,261	150,225,300	BC Opt. (1–4)	153	100.20	1.62	22.62	475.60	13.50	44,221	4.76	22.62
>9	0.03	5.56	1.131	44,69,86	40,60,80	166,217,261	150,225,300	BC Pract. (1–4)	133	109.13	1.59	24.28	512.10	15.70	44,412	4.74	22.48

TABLE 10.39
4.5/6 S17

TC	TC Freq. (%)	EV W/L (%)	Std. Dev.	Play-All Bets Optimal ($)	Play-All Bets Practical ($)	Back-Count 27% 16% 9% Optimal ($)	Back-Count 27% 16% 9% Practical ($)	Bet Spread	Kelly Bank (units)	Avg. Bet ($)	Results W/L (%)	Results W/100 ($)	Risk Measures SD/100 ($)	Risk Measures ROR (%)	Risk Measures N0 (hands)	Rankings DI	Rankings SCORE
<0	45.00	−1.41	1.139	8,8,7	10,10,5	0	0	PA Opt. (1–8)	1,233	17.48	0.50	8.70	294.83	13.50	115,131	2.95	8.69
0	27.68	−0.35	1.134	8,8,7	10,10,5	0	0	PA Pract. (1–8)	1,000	18.85	0.48	9.05	315.42	16.20	121,473	2.87	8.24
1	11.25	0.21	1.133	16,16,16	15,15,10	0	0	PA Opt. (1–12)	1,263	19.96	0.71	14.06	374.98	13.50	71,090	3.75	14.06
2	7.38	0.72	1.131	56,56,56	40,40,25	96,83,74	100,75,75	PA Pract. (1–12)	1,000	21.63	0.70	15.13	411.68	16.80	74,036	3.67	13.50
3	3.48	1.25	1.133	65,95,97	80,100,50	97,97,97	100	PA Opt. (1–16)	1,345	20.92	0.84	17.61	419.75	13.50	56,742	4.20	17.62
4	2.45	1.79	1.136	65,95,119	80,120,80	139,139,139	150	PA Pract. (1–16)	2,000	12.51	0.84	10.49	252.99	3.80	58,162	4.15	17.19
5	1.11	2.40	1.141	65,95,119	80,120,80	184,184,184	200	BC Opt. (1–2)	104	118.60	1.67	31.74	563.38	13.60	31,514	5.63	31.74
6	0.81	2.98	1.143	65,95,119	80,120,80	192,229,229	200,225,200	BC Pract. (1–2)	100	124.70	1.67	33.49	594.50	15.00	31,493	5.63	31.73
7	0.34	3.63	1.143	65,95,119	80,120,80	192,250,278	200,225,300	BC Opt. (1–3)	120	117.68	1.80	34.04	583.46	13.50	29,402	5.83	34.04
8	0.26	4.13	1.140	65,95,119	80,120,80	192,250,297	200,225,300	BC Pract. (1–3)	133	115.80	1.81	33.65	578.55	13.40	29,517	5.82	33.83
9	0.10	4.80	1.137	65,95,119	80,120,80	192,250,297	200,225,300	BC Opt. (1–4)	135	115.62	1.89	35.01	591.75	13.40	28,590	5.92	35.02
>9	0.14	5.79	1.130	65,95,119	80,120,80	192,250,297	200,225,300	BC Pract. (1–4)	133	118.43	1.88	35.72	605.03	14.20	28,722	5.90	34.85

TABLE 10.40
5.0/6 S17

TC	TC Freq. (%)	EV W/L (%)	Std. Dev.	Play-All Bets Optimal ($)	Play-All Bets Practical ($)	Back-Count 29% 18% 11% Optimal ($)	Back-Count 29% 18% 11% Practical ($)	Bet Spread	Kelly Bank (units)	Avg. Bet ($)	Results W/L (%)	Results W/100 ($)	Risk Measures SD/100 ($)	Risk Measures ROR (%)	Risk Measures N0 (hands)	Rankings DI	Rankings SCORE
<0	45.20	–1.51	1.141	12,11,10	10	0	0	PA Opt. (1–8)	838	24.83	0.71	17.48	418.32	13.50	57,117	4.18	17.50
0	25.90	–0.34	1.134	12,11,10	10	0	0	PA Pract. (1–8)	1,000	20.25	0.70	14.20	340.34	8.60	57,445	4.17	17.41
1	10.74	0.22	1.133	17,17,17	15	0	0	PA Opt. (1–12)	890	27.21	0.95	25.86	508.75	13.50	38,649	5.09	25.88
2	7.49	0.75	1.132	58,58,58	40	0,99,88	0,100,100	PA Pract. (1–12)	1,000	23.79	0.97	22.95	453.27	10.70	39,008	5.06	25.63
3	3.64	1.27	1.133	95,99,99	80,100,100	149,99,99	150,100,100	PA Opt. (1–16)	974	27.98	1.12	31.24	559.03	13.50	31,999	5.59	31.25
4	2.84	1.84	1.137	95,135,142	80,120,150	149,142,143	150	PA Pract. (1–16)	1,000	26.31	1.15	30.25	544.04	12.90	32,345	5.56	30.91
5	1.36	2.41	1.140	95,135,164	80,120,160	185,185,185	200	BC Opt. (1–2)	67	184.87	2.59	50.97	713.96	13.50	19,575	7.14	50.97
6	1.13	3.02	1.142	95,135,164	80,120,160	232,232,232	200	BC Pract. (1–2)	67	185.40	2.58	51.05	716.70	13.50	19,749	7.12	50.74
7	0.54	3.57	1.141	95,135,164	80,120,160	274,274,274	300	BC Opt. (1–3)	101	138.39	2.14	53.79	733.39	13.50	18,562	7.33	53.79
8	0.46	4.15	1.139	95,135,164	80,120,160	298,297,320	300	BC Pract. (1–3)	100	140.10	2.14	54.32	742.40	13.90	18,693	7.32	53.54
9	0.22	4.63	1.137	95,135,164	80,120,160	298,297,351	300,300,400	BC Opt. (1–4)	114	136.17	2.25	55.61	745.68	13.50	17,970	7.46	55.61
>9	0.46	5.97	1.129	95,135,164	80,120,160	298,297,351	300,300,400	BC Pract. (1–4)	100	143.90	2.23	58.11	782.60	14.90	18,144	7.43	55.14

TABLE 10.41
5.5/6 S17

TC	TC Freq. (%)	EV W/L (%)	Std. Dev.	Play-All Bets Optimal ($)	Play-All Bets Practical ($)	Back-Count 30% 20% 13% Optimal ($)	Back-Count 30% 20% 13% Practical ($)	Bet Spread	Kelly Bank (units)	Avg. Bet ($)	Results W/L (%)	Results W/100 ($)	Risk Measures SD/100 ($)	Risk Measures ROR (%)	Risk Measures N0 (hands)	Rankings DI	Rankings SCORE
<0	45.29	–1.66	1.142	18,17,15	20,15,15	0	0	PA Opt. (1–8)	545	36.40	1.03	37.52	612.49	13.50	26,657	6.12	37.51
0	24.28	–0.32	1.134	18,17,15	20,15,15	0	0	PA Pract. (1–8)	500	37.72	1.03	38.80	635.42	14.60	26,820	6.11	37.28
1	10.12	0.23	1.133	18,18,18	20	0	0	PA Opt. (1–12)	589	38.72	1.34	51.85	720.13	13.50	19,277	7.20	51.86
2	7.48	0.79	1.133	61,61,61	50	0,0,108	0,0,100	PA Pract. (1–12)	667	35.46	1.35	47.72	664.11	11.40	19,372	7.18	51.62
3	3.61	1.31	1.134	102,102,102	100	180,151,108	200,150,100	PA Opt. (1–16)	658	39.15	1.56	60.87	780.29	13.50	16,421	7.80	60.88
4	3.17	1.90	1.137	147,147,147	150	180,151,147	200,150,150	PA Pract. (1–16)	667	37.91	1.55	58.74	754.80	12.70	16,512	7.78	60.56
5	1.47	2.50	1.140	147,192,192	160,180,200	192,192,192	200	BC Opt. (1–2)	56	224.61	3.13	90.23	949.89	13.30	11,071	9.50	90.22
6	1.50	3.05	1.140	147,204,234	160,180,200	234,234,234	200	BC Pract. (1–2)	50	236.80	3.12	94.88	1002.60	15.10	11,185	9.46	89.56
7	0.65	3.74	1.140	147,204,243	160,180,240	288,288,288	300	BC Opt. (1–3)	66	219.57	3.39	95.36	976.46	13.60	10,497	9.77	95.36
8	0.78	4.09	1.136	147,204,243	160,180,240	317,317,317	300	BC Pract. (1–3)	67	216.90	3.40	94.50	970.05	13.20	10,537	9.74	94.90
9	0.29	4.92	1.135	147,204,243	160,180,240	360,382,382	400,450,400	BC Opt. (1–4)	93	168.63	2.85	97.55	987.68	13.40	10,249	9.88	97.56
>9	1.35	6.31	1.124	147,204,243	160,180,240	360,452,431	400,450,400	BC Pract. (1–4)	100	160.90	2.85	93.01	943.40	12.30	10,290	9.86	97.19

TABLE 10.42
4.0/6 S17 DAS

TC	TC Freq. (%)	EV W/L (%)	Std. Dev.	Play-All Bets Optimal ($)	Play-All Bets Practical ($)	Back-Count 26% 14% 7% Optimal ($)	Back-Count 26% 14% 7% Practical ($)	Bet Spread	Kelly Bank (units)	Avg. Bet ($)	Results W/L (%)	Results W/100 ($)	Risk Measures SD/100 ($)	Risk Measures ROR (%)	Risk Measures N0 (hands)	Rankings DI	Rankings SCORE
<0	44.72	−1.19	1.147	8,7,7	10,5,5	0	0	PA Opt. (1–8)	1,302	17.37	0.49	8.50	291.29	13.60	117,826	2.91	8.48
0	29.69	−0.22	1.141	8,7,7	10,5,5	0	0	PA Pract. (1–8)	1,000	19.38	0.46	8.83	315.24	16.90	127,456	2.80	7.84
1	11.63	0.34	1.140	27,27,27	25,10,10	0	0	PA Opt. (1–12)	1,365	19.28	0.65	12.60	354.40	13.60	79,613	3.54	12.56
2	7.12	0.86	1.138	62,67,67	50,25,25	91,79,72	100,75,75	PA Pract. (1–12)	2,000	10.77	0.65	7.00	204.55	3.50	85,389	3.42	11.70
3	3.18	1.39	1.139	62,88,104	80,60,50	107,107,107	100	PA Opt. (1–16)	1,537	19.77	0.77	15.26	390.15	13.60	65,707	3.90	15.22
4	1.97	1.94	1.143	62,88,104	80,60,80	149,149,150	150	PA Pract. (1–16)	2,000	11.19	0.75	8.36	225.32	3.70	72,642	3.71	13.77
5	0.83	2.50	1.147	62,88,104	80,60,80	182,190,191	200	BC Opt. (1–2)	110	113.81	1.60	25.42	504.15	13.50	39,232	5.04	25.42
6	0.49	3.13	1.150	62,88,104	80,60,80	182,237,238	200,225,200	BC Pract. (1–2)	100	119.20	1.60	26.66	529.80	14.90	39,373	5.03	25.32
7	0.20	3.71	1.149	62,88,104	80,60,80	182,238,282	200,225,300	BC Opt. (1–3)	126	111.88	1.71	26.62	516.02	13.50	37,636	5.16	26.62
8	0.11	4.34	1.146	62,88,104	80,60,80	182,238,290	200,225,300	BC Pract. (1–3)	133	108.00	1.72	25.90	502.50	12.90	37,715	5.15	26.57
9	0.04	4.76	1.142	62,88,104	80,60,80	182,238,290	200,225,300	BC Opt. (1–4)	138	109.88	1.76	27.03	519.88	13.50	37,074	5.20	27.02
>9	0.03	5.85	1.135	62,88,104	80,60,80	182,238,290	200,225,300	BC Pract. (1–4)	133	109.13	1.75	26.70	514.88	13.40	37,186	5.19	26.89

TABLE 10.43
4.5/6 S17 DAS

TC	TC Freq. (%)	EV W/L (%)	Std. Dev.	Play-All Bets Optimal ($)	Play-All Bets Practical ($)	Back-Count 27% 16% 9% Optimal ($)	Back-Count 27% 16% 9% Practical ($)	Bet Spread	Kelly Bank (units)	Avg. Bet ($)	Results W/L (%)	Results W/100 ($)	Risk Measures SD/100 ($)	Risk Measures ROR (%)	Risk Measures N0 (hands)	Rankings DI	Rankings SCORE
<0	45.01	−1.28	1.148	11,10,9	10	0	0	PA Opt. (1–8)	947	23.08	0.63	14.65	382.61	13.50	68,315	3.83	14.64
0	27.68	−0.21	1.141	11,10,9	10	0	0	PA Pract. (1–8)	1,000	20.71	0.62	12.91	339.28	10.60	69,066	3.80	14.47
1	11.25	0.35	1.140	27,27,27	25	0	0	PA Opt. (1–12)	1,046	24.76	0.83	20.64	454.33	13.50	48,439	4.54	20.64
2	7.38	0.88	1.139	68,68,68	50	104,90,81	100,100,75	PA Pract. (1–12)	1,000	23.49	0.83	19.43	431.11	12.30	49,230	4.51	20.32
3	3.48	1.41	1.139	85,108,108	80,100,100	108,108,109	100	PA Opt. (1–16)	1,161	25.26	0.97	24.41	494.35	13.50	40,921	4.94	24.44
4	2.45	1.97	1.143	85,115,138	80,120,150	150,150,151	150	PA Pract. (1–16)	1,000	25.32	0.97	24.50	502.22	14.30	42,020	4.88	23.79
5	1.11	2.56	1.147	85,115,138	80,120,160	194,194,195	200	BC Opt. (1–2)	96	128.92	1.82	37.75	614.43	13.50	26,462	6.14	37.75
6	0.81	3.15	1.148	85,115,138	80,120,160	208,239,239	200	BC Pract. (1–2)	100	124.70	1.83	36.71	597.80	12.80	26,533	6.14	37.71
7	0.34	3.76	1.148	85,115,138	80,120,160	208,271,286	200,300,300	BC Opt. (1–3)	111	127.50	1.95	39.97	632.18	13.40	25,017	6.32	39.97
8	0.26	4.32	1.145	85,115,138	80,120,160	208,271,324	200,300,300	BC Pract. (1–3)	100	129.90	1.94	40.39	641.70	14.00	25,229	6.29	39.61
9	0.10	4.90	1.141	85,115,138	80,120,160	208,271,324	200,300,300	BC Opt. (1–4)	124	125.13	2.03	40.82	638.85	13.40	24,460	6.39	40.82
>9	0.14	5.90	1.134	85,115,138	80,120,160	208,271,324	200,300,300	BC Pract. (1–4)	133	118.43	2.04	38.77	608.25	12.30	24,607	6.37	40.62

TABLE 10.44
5.0/6 S17 DAS

				Play-All Bets		Back-Count 29% 18% 11%					Results		Risk Measures			Rankings	
TC	TC Freq. (%)	EV W/L (%)	Std. Dev.	Optimal ($)	Practical ($)	Optimal ($)	Practical ($)	Bet Spread	Kelly Bank (units)	Avg. Bet ($)	W/L (%)	W/100 ($)	SD/100 ($)	ROR (%)	N0 (hands)	DI	SCORE
<0	45.21	−1.39	1.149	14,13,12	15,15,10	0	0	PA Opt. (1–8)	697	30.53	0.84	25.60	506.03	13.50	39,058	5.06	25.61
0	25.90	−0.20	1.142	14,13,12	15,15,10	0	0	PA Pract. (1–8)	667	29.16	0.83	24.25	483.47	12.50	39,731	5.02	25.17
1	10.74	0.37	1.141	28,28,28	25	0	0	PA Opt. (1–12)	784	32.07	1.08	34.64	588.27	13.50	28,891	5.88	34.61
2	7.49	0.91	1.139	70,70,70	50	120,106,94	125,100,100	PA Pract. (1–12)	667	32.52	1.08	34.97	602.03	14.50	29,646	5.81	33.74
3	3.64	1.43	1.140	110,110,110	100	120,110,110	125,100,100	PA Opt. (1–16)	869	32.52	1.24	40.13	633.79	13.50	24,889	6.34	40.17
4	2.84	2.01	1.143	115,153,154	120,150,150	154,154,154	150	PA Pract. (1–16)	1,000	28.13	1.26	35.36	559.90	10.40	25,072	6.32	39.89
5	1.36	2.57	1.146	115,153,184	120,180,160	196,196,196	200	BC Opt. (1–2)	83	149.48	2.12	57.57	758.80	13.60	17,391	7.59	57.57
6	1.13	3.19	1.147	115,153,184	120,180,160	240,242,243	200	BC Pract. (1–2)	80	150.88	2.10	57.50	760.13	13.60	17,476	7.56	57.23
7	0.54	3.73	1.146	115,153,184	120,180,160	240,284,284	250,300,300	BC Opt. (1–3)	95	148.06	2.29	61.63	785.02	13.40	16,211	7.85	61.63
8	0.46	4.34	1.144	115,153,184	120,180,160	240,317,332	250,300,300	BC Pract. (1–3)	100	140.10	2.30	58.45	746.20	12.20	16,326	7.83	61.35
9	0.22	4.76	1.141	115,153,184	120,180,160	240,317,366	250,300,400	BC Opt. (1–4)	107	145.57	2.40	63.30	795.53	13.40	15,809	7.96	63.29
>9	0.46	6.10	1.133	115,153,184	120,180,160	240,317,375	250,300,400	BC Pract. (1–4)	100	143.90	2.39	62.32	786.50	13.30	15,938	7.92	62.79

TABLE 10.45
5.5/6 S17 DAS

				Play-All Bets		Back-Count 30% 20% 13%					Results		Risk Measures			Rankings	
TC	TC Freq. (%)	EV W/L (%)	Std. Dev.	Optimal ($)	Practical ($)	Optimal ($)	Practical ($)	Bet Spread	Kelly Bank (units)	Avg. Bet ($)	W/L (%)	W/100 ($)	SD/100 ($)	ROR (%)	N0 (hands)	DI	SCORE
<0	45.30	−1.53	1.151	21,19,17	20,20,15	0	0	PA Opt. (1–8)	476	42.30	1.16	48.91	699.73	13.50	20,432	7.00	48.96
0	24.28	−0.18	1.142	21,19,17	20,20,15	0	0	PA Pract. (1–8)	500	39.14	1.18	46.01	661.38	12.20	20,672	6.96	48.39
1	10.12	0.38	1.141	29,29,29	25	0	0	PA Opt. (1–12)	534	43.83	1.46	63.92	799.32	13.50	15,649	7.99	63.89
2	7.47	0.95	1.140	73,73,73	50	0,129,113	0,125,125	PA Pract. (1–12)	500	43.38	1.48	64.33	809.64	14.00	15,845	7.95	63.13
3	3.61	1.47	1.141	113,113,113	125	187,129,113	200,125,125	PA Opt. (1–16)	603	43.90	1.66	73.08	854.43	13.50	13,697	8.54	73.01
4	3.17	2.07	1.143	158,158,158	150	187,158,158	200,150,150	PA Pract. (1–16)	667	39.90	1.71	68.02	799.67	11.80	13,819	8.51	72.35
5	1.47	2.66	1.146	168,203,203	160,200,200	203,203,203	200	BC Opt. (1–2)	53	234.21	3.29	98.89	994.35	13.60	10,103	9.94	98.89
6	1.50	3.23	1.146	168,225,246	160,240,240	246,246,246	200	BC Pract. (1–2)	50	236.80	3.28	99.78	1007.20	13.90	10,185	9.91	98.14
7	0.65	3.89	1.145	168,225,265	160,240,240	297,297,297	300	BC Opt. (1–3)	77	181.71	2.83	104.38	1021.63	13.60	9,572	10.22	104.38
8	0.78	4.28	1.141	168,225,265	160,240,240	328,328,328	300	BC Pract. (1–3)	80	172.50	2.82	98.82	968.75	12.10	9,600	10.20	104.06
9	0.29	5.07	1.139	168,225,265	160,240,240	375,388,391	400,375,500	BC Opt. (1–4)	89	177.09	3.00	107.96	1039.07	13.30	9,266	10.39	107.96
>9	1.35	6.45	1.128	168,225,265	160,240,240	375,388,451	400,375,500	BC Pract. (1–4)	80	182.63	3.01	111.61	1081.00	14.70	9,378	10.32	106.59

TABLE 10.46
4.0/6 S17 LS

TC	TC Freq. (%)	EV W/L (%)	Std. Dev.	Play-All Bets Optimal ($)	Play-All Bets Practical ($)	Back-Count 26% 14% 7% Optimal ($)	Back-Count 26% 14% 7% Practical ($)	Bet Spread	Kelly Bank (units)	Avg. Bet ($)	Results W/L (%)	Results W/100 ($)	Risk Measures SD/100 ($)	Risk Measures ROR (%)	Risk Measures N0 (hands)	Rankings DI	Rankings SCORE
<0	44.71	−1.30	1.131	8,8,7	10,10,5	0	0	PA Opt. (1–8)	1,202	18.59	0.50	9.35	305.73	13.50	106,977	3.06	9.35
0	29.70	−0.27	1.122	8,8,7	10,10,5	0	0	PA Pract. (1–8)	1,000	19.39	0.47	9.07	307.91	14.80	115,248	2.94	8.67
1	11.62	0.33	1.118	27,27,27	25,25,10	0	0	PA Opt. (1–12)	1,228	20.97	0.68	14.28	378.02	13.50	69,999	3.78	14.29
2	7.12	0.89	1.113	67,72,72	50,50,25	102,89,81	100,100,75	PA Pract. (1–12)	1,000	22.13	0.67	14.84	403.22	16.10	73,827	3.68	13.54
3	3.18	1.47	1.110	67,98,117	80,120,50	119,119,120	125	PA Opt. (1–16)	1,373	21.60	0.81	17.61	419.15	13.60	56,906	4.19	17.57
4	1.97	2.08	1.112	67,98,117	80,120,80	168,168,168	200	PA Pract. (1–16)	2,000	11.19	0.79	8.81	219.68	2.60	62,245	4.01	16.07
5	0.83	2.73	1.116	67,98,117	80,120,80	204,220,220	200	BC Opt. (1–2)	98	127.73	1.71	30.48	552.05	13.50	32,922	5.52	30.48
6	0.49	3.41	1.119	67,98,117	80,120,80	204,267,273	200,300,300	BC Pract. (1–2)	100	131.90	1.72	31.68	575.80	14.80	32,993	5.50	30.27
7	0.20	4.06	1.118	67,98,117	80,120,80	204,267,323	200,300,300	BC Opt. (1–3)	112	125.79	1.83	32.11	566.70	13.60	31,185	5.67	32.11
8	0.11	4.69	1.115	67,98,117	80,120,80	204,267,323	200,300,300	BC Pract. (1–3)	100	138.20	1.82	35.08	621.10	16.20	31,312	5.65	31.90
9	0.04	5.33	1.111	67,98,117	80,120,80	204,267,323	200,300,300	BC Opt. (1–4)	124	123.45	1.90	32.69	571.78	13.50	30,655	5.72	32.69
>9	0.03	6.33	1.105	67,98,117	80,120,80	204,267,323	200,300,300	BC Pract. (1–4)	133	125.40	1.91	33.50	589.20	14.60	30,888	5.69	32.33

TABLE 10.47
4.5/6 S17 LS

TC	TC Freq. (%)	EV W/L (%)	Std. Dev.	Play-All Bets Optimal ($)	Play-All Bets Practical ($)	Back-Count 27% 16% 9% Optimal ($)	Back-Count 27% 16% 9% Practical ($)	Bet Spread	Kelly Bank (units)	Avg. Bet ($)	Results W/L (%)	Results W/100 ($)	Risk Measures SD/100 ($)	Risk Measures ROR (%)	Risk Measures N0 (hands)	Rankings DI	Rankings SCORE
<0	45.00	−1.39	1.132	12,11,10	10	0	0	PA Opt. (1–8)	854	25.12	0.66	16.66	407.77	13.60	60,165	4.08	16.63
0	27.69	−0.27	1.122	12,11,10	10	0	0	PA Pract. (1–8)	1,000	20.72	0.65	13.45	331.34	8.60	60,688	4.06	16.47
1	11.24	0.34	1.118	27,27,27	25	0	0	PA Opt. (1–12)	931	27.18	0.88	24.03	489.82	13.60	41,689	4.90	23.99
2	7.38	0.90	1.113	73,73,73	50	117,101,91	125,100,100	PA Pract. (1–12)	1,000	24.20	0.89	21.52	442.60	11.10	42,300	4.86	23.64
3	3.49	1.49	1.111	94,121,121	80,120,125	121,121,121	125	PA Opt. (1–16)	1,028	27.83	1.03	28.72	535.74	13.50	34,849	5.36	28.70
4	2.45	2.10	1.112	94,129,156	80,120,160	169,169,170	200	PA Pract. (1–16)	1,000	26.46	1.05	27.77	522.97	13.10	35,465	5.31	28.20
5	1.11	2.77	1.116	94,129,156	80,120,160	222,222,223	200	BC Opt. (1–2)	86	144.84	1.95	45.39	673.72	13.40	22,055	6.74	45.39
6	0.81	3.40	1.117	94,129,156	80,120,160	233,273,273	250,300,300	BC Pract. (1–2)	80	154.38	1.94	48.18	717.38	15.30	22,220	6.72	45.12
7	0.34	4.09	1.117	94,129,156	80,120,160	233,304,329	250,300,300	BC Opt. (1–3)	99	143.55	2.10	48.32	695.13	13.40	20,665	6.95	48.32
8	0.26	4.66	1.114	94,129,156	80,120,160	233,304,362	250,300,400	BC Pract. (1–3)	100	148.00	2.10	49.91	720.70	14.60	20,860	6.93	47.96
9	0.10	5.39	1.110	94,129,156	80,120,160	233,304,362	250,300,400	BC Opt. (1–4)	110	140.94	2.19	49.49	703.43	13.60	20,183	7.03	49.49
>9	0.14	6.43	1.104	94,129,156	80,120,160	233,304,362	250,300,400	BC Pract. (1–4)	100	151.10	2.16	52.53	749.70	15.40	20,392	7.01	49.10

TABLE 10.48
5.0/6 S17 LS

TC	TC Freq. (%)	EV W/L (%)	Std. Dev.	Play-All Bets Optimal ($)	Play-All Bets Practical ($)	Back-Count 29% 18% 11% Optimal ($)	Back-Count 29% 18% 11% Practical ($)	Bet Spread	Kelly Bank (units)	Avg. Bet ($)	Results W/L (%)	Results W/100 ($)	Risk Measures SD/100 ($)	Risk Measures ROR (%)	Risk Measures N0 (hands)	Rankings DI	Rankings SCORE
<0	45.19	–1.50	1.133	16,14,13	15	0	0	PA Opt. (1–8)	626	33.41	0.88	29.53	543.20	13.50	33,883	5.43	29.51
0	25.90	–0.26	1.122	16,14,13	15	0	0	PA Pract. (1–8)	667	29.91	0.89	26.56	492.56	11.20	34,379	5.39	29.08
1	10.73	0.35	1.118	28,28,28	25	0	0	PA Opt. (1–12)	698	35.32	1.15	40.57	636.80	13.50	24,660	6.37	40.55
2	7.49	0.93	1.114	75,75,75	50	0,119,105	0,125,100	PA Pract. (1–12)	667	34.32	1.17	40.22	636.62	13.70	25,060	6.32	39.91
3	3.65	1.50	1.112	122,122,122	120,125,125	178,122,122	200,125,125	PA Opt. (1–16)	768	35.92	1.32	47.48	688.82	13.50	21,075	6.89	47.45
4	2.84	2.13	1.114	128,172,172	120,180,200	178,172,172	200	PA Pract. (1–16)	667	36.86	1.35	49.67	730.34	15.50	21,624	6.80	46.25
5	1.36	2.76	1.116	128,172,208	120,180,200	222,222,222	200	BC Opt. (1–2)	56	220.39	2.95	69.40	833.00	13.60	14,418	8.33	69.40
6	1.14	3.42	1.117	128,172,208	120,180,240	275,275,275	300	BC Pract. (1–2)	50	237.40	2.94	74.48	896.80	15.60	14,529	8.30	68.96
7	0.54	4.00	1.116	128,172,208	120,180,240	322,322,322	300	BC Opt. (1–3)	84	166.30	2.46	74.20	861.43	13.50	13,482	8.61	74.20
8	0.46	4.67	1.113	128,172,208	120,180,240	356,356,377	400,375,400	BC Pract. (1–3)	80	174.38	2.44	77.42	901.25	14.80	13,567	8.59	73.80
9	0.22	5.17	1.110	128,172,208	120,180,240	356,356,420	400,375,400	BC Opt. (1–4)	95	163.62	2.57	76.42	874.21	13.60	13,072	8.74	76.42
>9	0.47	6.60	1.102	128,172,208	120,180,240	356,356,420	400,375,400	BC Pract. (1–4)	100	165.70	2.57	77.30	887.40	14.00	13,179	8.71	75.87

TABLE 10.49
5.5/6 S17 LS

TC	TC Freq. (%)	EV W/L (%)	Std. Dev.	Play-All Bets Optimal ($)	Play-All Bets Practical ($)	Back-Count 30% 20% 13% Optimal ($)	Back-Count 30% 20% 13% Practical ($)	Bet Spread	Kelly Bank (units)	Avg. Bet ($)	Results W/L (%)	Results W/100 ($)	Risk Measures SD/100 ($)	Risk Measures ROR (%)	Risk Measures N0 (hands)	Rankings DI	Rankings SCORE
<0	45.29	–1.66	1.134	24,21,19	25,20,20	0	0	PA Opt. (1–8)	426	46.56	1.23	57.05	755.22	13.50	17,530	7.55	57.04
0	24.27	–0.24	1.122	24,21,19	25,20,20	0	0	PA Pract. (1–8)	400	46.68	1.26	58.68	783.63	14.70	17,837	7.49	56.07
1	10.11	0.37	1.118	29,29,29	25	0	0	PA Opt. (1–12)	475	48.43	1.56	75.35	867.99	13.50	13,276	8.68	75.34
2	7.47	0.96	1.115	77,77,77	50	0,0,127	0,150,125	PA Pract. (1–12)	500	45.04	1.59	71.77	833.06	12.60	13,477	8.61	74.22
3	3.62	1.54	1.113	125,125,125	125	211,176,127	200,150,125	PA Opt. (1–16)	534	48.62	1.78	86.63	930.44	13.50	11,549	9.30	86.57
4	3.16	2.17	1.114	175,175,175	200	211,176,175	200	PA Pract. (1–16)	500	48.44	1.83	88.84	961.84	14.60	11,722	9.24	85.32
5	1.47	2.85	1.116	188,229,229	200	229,229,229	200	BC Opt. (1–2)	47	263.98	3.52	119.57	1093.50	13.60	8,369	10.93	119.57
6	1.50	3.42	1.116	188,253,275	200,240,300	274,274,274	300	BC Pract. (1–2)	50	255.00	3.53	115.96	1063.20	12.70	8,401	10.91	118.97
7	0.65	4.16	1.114	188,253,300	200,240,320	335,335,335	300	BC Opt. (1–3)	57	257.50	3.80	125.83	1121.72	13.30	7,957	11.22	125.83
8	0.79	4.55	1.111	188,253,300	200,240,320	368,368,368	400	BC Pract. (1–3)	67	211.65	3.01	129.53	1160.55	14.40	8,019	11.16	124.55
9	0.30	5.46	1.108	188,253,300	200,240,320	422,444,445	400,450,500	BC Opt. (1–4)	79	198.89	3.21	129.86	1139.51	13.40	7,696	11.40	129.85
>9	1.38	6.95	1.098	188,253,300	200,240,320	422,529,506	400,450,500	BC Pract. (1–4)	80	202.13	3.21	131.94	1161.38	14.00	7,741	11.36	129.08

TABLE 10.50
4.0/6 S17 DAS LS

TC	TC Freq. (%)	EV W/L (%)	Std. Dev.	Play-All Bets Optimal ($)	Play-All Bets Practical ($)	Back-Count 26% 14% 7% Optimal ($)	Back-Count 26% 14% 7% Practical ($)	Bet Spread	Kelly Bank (units)	Avg. Bet ($)	Results W/L (%)	Results W/100 ($)	Risk Measures SD/100 ($)	Risk Measures ROR (%)	Risk Measures N0 (hands)	Rankings DI	Rankings SCORE
<0	44.71	−1.17	1.140	11,10,9	10	0	0	PA Opt. (1–8)	950	23.94	0.65	15.48	393.22	13.50	64,652	3.93	15.46
0	29.70	−0.13	1.130	11,10,9	10	0	0	PA Pract. (1–8)	1,000	19.39	0.61	11.90	309.85	8.40	67,797	3.84	14.76
1	11.62	0.48	1.125	38,38,38	25	0,0,61	0,0,50	PA Opt. (1–12)	1,025	25.61	0.82	21.01	458.07	13.50	47,642	4.58	20.98
2	7.12	1.05	1.120	83,83,84	50	110,97,83	100,100,75	PA Pract. (1–12)	1,000	22.13	0.82	18.10	405.71	11.10	50,243	4.46	19.90
3	3.18	1.62	1.117	84,117,130	80,120,125	130,130,130	125	PA Opt. (1–16)	1,166	25.87	0.94	24.43	493.79	13.60	41,012	4.94	24.38
4	1.97	2.24	1.119	84,117,137	80,120,160	179,179,179	200	PA Pract. (1–16)	1,000	23.75	0.95	22.49	470.12	13.00	43,696	4.78	22.88
5	0.83	2.89	1.122	84,117,137	80,120,160	221,230,230	200	BC Opt. (1–2)	91	137.80	1.86	35.86	598.88	13.40	27,894	5.99	35.87
6	0.49	3.56	1.125	84,117,137	80,120,160	221,282,244	200,300,200	BC Pract. (1–2)	100	131.90	1.88	34.55	579.20	12.70	28,022	5.97	35.59
7	0.20	4.21	1.123	84,117,137	80,120,160	221,290,244	200,300,200	BC Opt. (1–3)	103	135.56	1.98	37.39	611.46	13.60	26,765	6.11	37.39
8	0.11	4.87	1.120	84,117,137	80,120,160	221,290,244	200,300,200	BC Pract. (1–3)	100	138.20	1.97	38.09	624.80	14.20	26,893	6.10	37.17
9	0.04	5.41	1.116	84,117,137	80,120,160	221,290,244	200,300,200	BC Opt. (1–4)	164	96.55	1.55	38.31	618.96	13.50	26,081	6.19	38.31
>9	0.03	6.45	1.109	84,117,137	80,120,160	221,290,244	200,300,200	BC Pract. (1–4)	200	87.80	1.58	35.48	577.25	11.90	26,441	6.15	37.78

TABLE 10.51
4.5/6 S17 DAS LS

TC	TC Freq. (%)	EV W/L (%)	Std. Dev.	Play-All Bets Optimal ($)	Play-All Bets Practical ($)	Back-Count 27% 16% 9% Optimal ($)	Back-Count 27% 16% 9% Practical ($)	Bet Spread	Kelly Bank (units)	Avg. Bet ($)	Results W/L (%)	Results W/100 ($)	Risk Measures SD/100 ($)	Risk Measures ROR (%)	Risk Measures N0 (hands)	Rankings DI	Rankings SCORE
<0	45.00	−1.26	1.141	14,13,11	15,15,10	0	0	PA Opt. (1–8)	705	30.75	0.80	24.70	496.57	13.60	40,568	4.97	24.66
0	27.69	−0.12	1.130	14,13,11	15,15,10	0	0	PA Pract. (1–8)	667	31.37	0.79	24.90	502.85	13.90	40,782	4.95	24.52
1	11.24	0.49	1.125	39,39,39	40,40,25	0	0	PA Opt. (1–12)	799	32.09	1.01	32.52	570.12	13.50	30,763	5.70	32.50
2	7.38	1.06	1.121	85,85,85	75,75,50	125,109,97	125,100,100	PA Pract. (1–12)	667	34.67	1.00	34.67	612.98	15.80	31,268	5.66	31.98
3	3.49	1.64	1.118	114,132,132	120,125,125	132,132,132	125	PA Opt. (1–16)	917	32.19	1.16	37.29	610.89	13.50	26,803	6.11	37.32
4	2.45	2.26	1.119	114,150,174	120,180,160	180,180,181	200	PA Pract. (1–16)	1,000	26.46	1.20	31.73	526.08	10.10	27,489	6.03	36.37
5	1.11	2.92	1.122	114,150,174	120,180,160	232,232,233	200	BC Opt. (1–2)	80	154.94	2.10	52.37	723.65	13.50	19,135	7.24	52.37
6	0.81	3.55	1.123	114,150,174	120,180,160	250,282,282	250,300,300	BC Pract. (1–2)	80	154.38	2.10	52.08	721.50	13.50	19,159	7.22	52.09
7	0.34	4.26	1.122	114,150,174	120,180,160	250,326,339	250,300,300	BC Opt. (1–3)	92	153.07	2.24	55.18	742.86	13.50	18,119	7.43	55.18
8	0.26	4.83	1.119	114,150,174	120,180,160	250,326,386	250,300,400	BC Pract. (1–3)	100	148.00	2.25	53.63	724.80	12.90	18,285	7.40	54.75
9	0.10	5.51	1.115	114,150,174	120,180,160	250,326,389	250,300,400	BC Opt. (1–4)	103	150.16	2.33	56.20	749.72	13.50	17,812	7.50	56.20
>9	0.14	6.57	1.108	114,150,174	120,180,160	250,326,389	250,300,400	BC Pract. (1–4)	100	151.10	2.32	56.33	753.90	13.70	17,931	7.47	55.81

TABLE 10.52
5.0/6 S17 DAS LS

				Play-All Bets		Back-Count 29% 18% 11%					Results		Risk Measures			Rankings	
TC	TC Freq. (%)	EV W/L (%)	Std. Dev.	Optimal ($)	Practical ($)	Optimal ($)	Practical ($)	Bet Spread	Kelly Bank (units)	Avg. Bet ($)	W/L (%)	W/100 ($)	SD/100 ($)	ROR (%)	N0 (hands)	DI	SCORE
<0	45.20	−1.38	1.142	19,16,14	20,15,15	0	0	PA Opt. (1–8)	536	39.32	1.02	40.00	632.72	13.50	24,981	6.33	40.03
0	25.90	−0.11	1.130	19,16,14	20,15,15	0	0	PA Pract. (1–8)	500	41.02	1.00	40.86	648.82	14.30	25,215	6.30	39.65
1	10.73	0.50	1.126	40,40,40	50,40,40	0	0	PA Opt. (1–12)	616	40.48	1.27	51.50	717.88	13.40	19,403	7.18	51.53
2	7.49	1.09	1.121	86,86,86	75	142,125,112	150,125,100	PA Pract. (1–12)	667	37.80	1.28	48.26	672.78	11.80	19,438	7.17	51.45
3	3.64	1.66	1.119	133,133,133	125	142,133,133	150,125,125	PA Opt. (1–16)	700	40.52	1.44	58.49	764.96	13.50	17,090	7.65	58.52
4	2.84	2.30	1.120	149,183,183	160,180,200	183,183,183	200	PA Pract. (1–16)	667	40.34	1.44	58.12	762.95	13.50	17,229	7.62	58.03
5	1.36	2.92	1.122	149,195,228	160,180,200	232,232,232	200	BC Opt. (1–2)	70	177.65	2.42	78.16	884.10	13.60	12,799	8.84	78.16
6	1.14	3.58	1.122	149,195,228	160,180,240	284,284,284	300	BC Pract. (1–2)	67	184.95	2.41	81.09	919.05	14.40	12,826	8.82	77.85
7	0.54	4.18	1.121	149,195,228	160,180,240	285,332,333	300	BC Opt. (1–3)	80	175.92	2.61	83.30	912.67	13.40	11,998	9.13	83.30
8	0.46	4.82	1.118	149,195,228	160,180,240	285,376,386	300,375,400	BC Pract. (1–3)	80	174.38	2.60	82.50	906.13	13.30	12,063	9.10	82.89
9	0.22	5.31	1.115	149,195,228	160,180,240	285,376,427	300,375,400	BC Opt. (1–4)	90	173.11	2.71	85.37	923.87	13.30	11,723	9.24	85.36
>9	0.47	6.77	1.106	149,195,228	160,180,240	285,376,447	300,375,400	BC Pract. (1–4)	100	165.70	2.73	82.12	892.20	12.60	11,810	9.20	84.72

TABLE 10.53
5.5/6 S17 DAS LS

				Play-All Bets		Back-Count 30% 20% 13%					Results		Risk Measures			Rankings	
TC	TC Freq. (%)	EV W/L (%)	Std. Dev.	Optimal ($)	Practical ($)	Optimal ($)	Practical ($)	Bet Spread	Kelly Bank (units)	Avg. Bet ($)	W/L (%)	W/100 ($)	SD/100 ($)	ROR (%)	N0 (hands)	DI	SCORE
<0	45.29	−1.53	1.144	26,23,20	25,25,20	0	0	PA Opt. (1–8)	381	52.62	1.36	71.38	844.94	13.50	14,013	8.45	71.39
0	24.27	−0.09	1.130	26,23,20	25,25,20	0	0	PA Pract. (1–8)	400	53.85	1.35	72.79	864.13	14.20	14,089	8.42	70.97
1	10.11	0.52	1.126	41,41,41	50	0	0	PA Opt. (1–12)	438	53.64	1.68	90.10	949.29	13.40	11,095	9.49	90.12
2	7.47	1.12	1.122	89,89,89	100	0,152,132	0,150,125	PA Pract. (1–12)	400	58.45	1.65	96.18	1016.48	15.50	11,170	9.46	89.52
3	3.62	1.70	1.120	136,136,136	150	219,152,136	200,150,150	PA Opt. (1–16)	495	53.47	1.90	101.43	1006.76	13.50	9,867	10.07	101.36
4	3.16	2.34	1.121	187,186,187	200	219,187,187	200	PA Pract. (1–16)	500	55.60	1.86	103.35	1029.80	14.10	9,930	10.04	100.72
5	1.47	3.00	1.122	210,239,239	200	239,239,239	200	BC Opt. (1–2)	46	273.54	3.68	129.56	1138.20	13.20	7,707	11.38	129.56
6	1.50	3.58	1.121	210,274,285	200,300,300	284,284,284	300	BC Pract. (1–2)	50	255.00	3.70	121.29	1068.60	11.80	7,774	11.35	128.86
7	0.65	4.34	1.120	210,274,323	200,300,320	346,346,346	300	BC Opt. (1–3)	66	213.15	3.18	137.65	1173.26	13.40	7,264	11.73	137.65
8	0.79	4.70	1.117	210,274,323	200,300,320	377,377,377	400	BC Pract. (1–3)	67	211.65	3.17	136.55	1166.55	13.20	7,304	11.71	137.03
9	0.30	5.63	1.113	210,274,323	200,300,320	438,454,454	400,450,500	BC Opt. (1–4)	76	207.76	3.36	142.02	1191.68	13.40	7,046	11.92	142.02
>9	1.38	7.11	1.102	210,274,323	200,300,320	438,455,526	400,450,500	BC Pract. (1–4)	80	206.63	3.34	140.20	1180.38	13.20	7,083	11.88	141.07

TABLE 10.54

4.0/6 H17

TC	TC Freq. (%)	EV W/L (%)	Std. Dev.	Play-All Bets Optimal ($)	Play-All Bets Practical ($)	Back-Count 26% 14% 7% Optimal ($)	Back-Count 26% 14% 7% Practical ($)	Bet Spread	Kelly Bank (units)	Avg. Bet ($)	Results W/L (%)	Results W/100 ($)	Risk Measures SD/100 ($)	Risk Measures ROR (%)	Risk Measures N0 (hands)	Rankings DI	Rankings SCORE
<0	44.72	−1.57	1.143	3,3,4	3,3,5	0	0	PA Opt. (1–8)	3,838	5.15	0.17	0.86	92.74	13.60	1,163,391	0.93	0.86
0	29.69	−0.57	1.139	3,3,4	3,3,5	0	0	PA Pract. (1–8)	3,333	4.94	0.09	0.47	83.90	26.60	3,255,813	0.56	0.31
1	11.63	0.00	1.136	3,3,4	3,3,5	0	0	PA Opt. (1–12)	2,889	8.73	0.37	3.21	178.83	13.70	312,571	1.79	3.20
2	7.12	0.53	1.134	21,41,41	10	0,64,56	0,75,50	PA Pract. (1–12)	3,333	5.41	0.27	1.45	103.38	6.70	508,992	1.40	1.96
3	3.17	1.06	1.136	21,42,63	24,25,25	105,82,82	100,75,75	PA Opt. (1–16)	2,558	10.55	0.49	5.19	227.52	13.60	193,261	2.28	5.18
4	1.97	1.63	1.140	21,42,63	24,36,50	125,125,125	125	PA Pract. (1–16)	2,000	8.15	0.27	2.22	164.10	19.20	546,401	1.35	1.83
5	0.83	2.21	1.145	21,42,63	24,36,80	169,169,169	200	BC Opt. (1–2)	95	132.02	1.85	16.69	408.49	13.50	59,548	4.09	16.69
6	0.49	2.85	1.148	21,42,63	24,36,80	211,191,216	200	BC Pract. (1–2)	100	132.00	1.86	16.78	412.40	13.90	60,259	4.07	16.56
7	0.20	3.43	1.148	21,42,63	24,36,80	211,191,226	200,225,200	BC Opt. (1–3)	157	90.71	1.40	17.73	421.11	13.60	56,300	4.21	17.73
8	0.11	4.04	1.145	21,42,63	24,36,80	211,191,226	200,225,200	BC Pract. (1–3)	133	97.95	1.40	19.12	457.05	16.10	57,112	4.18	17.49
9	0.04	4.58	1.141	21,42,63	24,36,80	211,191,226	200,225,200	BC Opt. (1–4)	177	88.86	1.48	18.33	428.23	13.60	54,612	4.28	18.34
>9	0.03	5.53	1.135	21,42,63	24,36,80	211,191,226	200,225,200	BC Pract. (1–4)	200	84.50	1.51	17.81	418.40	13.10	55,251	4.26	18.12

TABLE 10.55

4.5/6 H17

TC	TC Freq. (%)	EV W/L (%)	Std. Dev.	Play-All Bets Optimal ($)	Play-All Bets Practical ($)	Back-Count 27% 16% 9% Optimal ($)	Back-Count 27% 16% 9% Practical ($)	Bet Spread	Kelly Bank (units)	Avg. Bet ($)	Results W/L (%)	Results W/100 ($)	Risk Measures SD/100 ($)	Risk Measures ROR (%)	Risk Measures N0 (hands)	Rankings DI	Rankings SCORE
<0	45.01	−1.66	1.143	5,6,6	5	0	0	PA Opt. (1–8)	2,047	10.38	0.33	3.44	185.34	13.60	291,262	1.85	3.44
0	27.69	−0.56	1.139	5,6,6	5	0	0	PA Pract. (1–8)	2,000	7.89	0.22	1.75	131.70	13.30	566,364	1.33	1.76
1	11.25	0.01	1.136	5,6,6	5	0	0	PA Opt. (1–12)	1,688	14.24	0.55	7.78	278.58	13.60	128,919	2.79	7.76
2	7.38	0.54	1.135	39,42,42	10	0,0,66	0,0,75	PA Pract. (1–12)	2,000	8.93	0.47	4.20	175.82	6.60	174,815	2.39	5.72
3	3.48	1.09	1.137	39,71,84	25	121,104,84	125,100,75	PA Opt. (1–16)	1,737	15.66	0.70	10.99	331.68	13.50	90,891	3.32	11.00
4	2.45	1.65	1.140	39,71,92	40,60,50	127,127,127	125	PA Pract. (1–16)	2,000	9.23	0.59	5.45	193.63	5.50	126,227	2.81	7.92
5	1.11	2.25	1.145	39,71,92	40,60,80	172,172,172	200	BC Opt. (1–2)	83	149.69	2.11	27.33	522.77	13.40	36,538	5.23	27.33
6	0.81	2.85	1.146	39,71,92	40,60,80	217,217,217	200	BC Pract. (1–2)	80	153.50	2.10	28.01	537.13	14.30	36,799	5.22	27.20
7	0.34	3.48	1.146	39,71,92	40,60,80	242,265,265	250,300,300	BC Opt. (1–3)	97	147.42	2.23	28.48	533.69	13.40	35,139	5.34	28.48
8	0.26	4.03	1.144	39,71,92	40,60,80	242,308,265	250,300,300	BC Pract. (1–3)	100	148.30	2.24	28.79	541.30	14.00	35,326	5.32	28.28
9	0.10	4.70	1.140	39,71,92	40,60,80	242,311,265	250,300,300	BC Opt. (1–4)	151	104.52	1.75	29.38	542.01	13.50	34,009	5.42	29.38
>9	0.14	5.63	1.134	39,71,92	40,60,80	242,311,265	250,300,300	BC Pract. (1–4)	133	109.20	1.76	30.84	571.73	15.20	34,401	5.39	29.09

TABLE 10.56
5.0/6 H17

				Play-All Bets		Back-Count 29% 18% 11%					Results		Risk Measures			Rankings	
TC	TC Freq. (%)	EV W/L (%)	Std. Dev.	Optimal ($)	Practical ($)	Optimal ($)	Practical ($)	Bet Spread	Kelly Bank (units)	Avg. Bet ($)	W/L (%)	W/100 ($)	SD/100 ($)	ROR (%)	N0 (hands)	DI	SCORE
<0	45.20	−1.77	1.144	9,9,8	10	0	0	PA Opt. (1–8)	1,144	17.91	0.53	9.47	307.58	13.50	105,721	3.08	9.46
0	25.91	−0.55	1.139	9,9,8	10	0	0	PA Pract. (1–8)	1,000	17.49	0.49	8.54	295.08	14.10	119,389	2.89	8.37
1	10.74	0.02	1.137	9,9,8	10	0	0	PA Opt. (1–12)	1,106	21.44	0.79	16.99	412.28	13.50	58,853	4.12	17.00
2	7.49	0.57	1.136	44,44,44	25	0	0	PA Pract. (1–12)	1,000	20.30	0.79	15.97	400.42	13.60	62,867	3.99	15.91
3	3.64	1.11	1.137	70,86,86	50	140,121,104	150,125,100	PA Opt. (1–16)	1,191	22.60	0.98	22.21	471.20	13.50	45,029	4.71	22.20
4	2.84	1.69	1.141	70,108,130	80,120,125	140,130,130	150,125,125	PA Pract. (1–16)	1,000	22.12	0.98	21.73	475.43	14.60	47,869	4.57	20.89
5	1.36	2.26	1.144	70,108,134	80,120,160	173,173,173	200	BC Opt. (1–2)	72	173.55	2.45	45.31	673.11	13.40	22,140	6.73	45.31
6	1.13	2.89	1.145	70,108,134	80,120,160	221,221,221	200	BC Pract. (1–2)	67	185.40	2.44	48.27	718.95	15.30	22,157	6.71	45.07
7	0.54	3.42	1.145	70,108,134	80,120,160	261,261,262	300	BC Opt. (1–3)	83	171.13	2.61	47.66	690.31	13.40	20,988	6.90	47.66
8	0.46	4.06	1.143	70,108,134	80,120,160	279,311,311	300	BC Pract. (1–3)	80	173.50	2.60	48.09	699.50	14.00	21,127	6.88	47.27
9	0.22	4.54	1.140	70,108,134	80,120,160	279,349,350	300	BC Opt. (1–4)	96	168.02	2.71	48.50	696.51	13.40	20,575	6.96	48.50
>9	0.46	5.84	1.132	70,108,134	80,120,160	279,363,418	300,375,400	BC Pract. (1–4)	100	166.00	2.70	47.76	688.70	13.30	20,759	6.93	48.09

TABLE 10.57
5.5/6 H17

				Play-All Bets		Back-Count 30% 20% 13%					Results		Risk Measures			Rankings	
TC	TC Freq. (%)	EV W/L (%)	Std. Dev.	Optimal ($)	Practical ($)	Optimal ($)	Practical ($)	Bet Spread	Kelly Bank (units)	Avg. Bet ($)	W/L (%)	W/100 ($)	SD/100 ($)	ROR (%)	N0 (hands)	DI	SCORE
<0	45.30	−1.92	1.146	15,14,13	15	0	0	PA Opt. (1–8)	672	29.58	0.85	25.14	501.38	13.50	39,782	5.01	25.14
0	24.28	−0.53	1.139	15,14,13	15	0	0	PA Pract. (1–8)	667	29.61	0.86	25.42	507.99	13.90	39,920	5.00	25.04
1	10.12	0.04	1.137	15,14,13	15	0	0	PA Opt. (1–12)	695	32.90	1.18	38.87	623.44	13.50	25,726	6.23	38.87
2	7.47	0.61	1.136	47,47,47	40	0	0	PA Pract. (1–12)	667	33.41	1.18	39.57	636.03	14.10	25,836	6.22	38.70
3	3.61	1.15	1.138	89,89,89	100	171,144,117	200,150,125	PA Opt. (1–16)	755	33.89	1.42	47.97	692.57	13.50	20,846	6.93	47.97
4	3.17	1.75	1.141	119,134,135	120,125,125	171,144,134	200,150,125	PA Pract. (1–16)	667	35.85	1.40	50.30	730.77	15.10	21,111	6.88	47.38
5	1.47	2.35	1.144	119,173,180	120,180,200	180,180,180	200	BC Opt. (1–2)	59	213.31	3.00	82.06	905.81	13.30	12,178	9.06	82.06
6	1.50	2.92	1.143	119,173,212	120,180,200	223,223,223	200	BC Pract. (1–2)	50	236.80	2.99	90.73	1005.60	16.50	12,265	9.02	81.40
7	0.65	3.59	1.144	119,173,212	120,180,240	275,275,275	300	BC Opt. (1–3)	70	209.38	3.26	87.48	935.33	13.30	11,443	9.35	87.47
8	0.78	3.98	1.140	119,173,212	120,180,240	307,307,307	300	BC Pract. (1–3)	67	215.70	3.26	90.08	964.65	14.20	11,450	9.34	87.20
9	0.29	4.83	1.138	119,173,212	120,180,240	341,373,373	400	BC Opt. (1–4)	85	203.65	3.41	89.03	943.51	13.50	11,229	9.44	89.03
>9	1.35	6.21	1.127	119,173,212	120,180,240	341,431,469	400,450,500	BC Pract. (1–4)	80	207.75	3.45	91.86	975.63	14.40	11,276	9.42	88.65

TABLE 10.58

4.0/6 H17 DAS

TC	TC Freq. (%)	EV W/L (%)	Std. Dev.	Play-All Bets Optimal ($)	Play-All Bets Practical ($)	Back-Count 26% 14% 7% Optimal ($)	Back-Count 26% 14% 7% Practical ($)	Bet Spread	Kelly Bank (units)	Avg. Bet ($)	Results W/L (%)	Results W/100 ($)	Risk Measures SD/100 ($)	Risk Measures ROR (%)	Risk Measures N0 (hands)	Rankings DI	Rankings SCORE
<0	44.72	–1.44	1.152	5,5,5	5	0	0	PA Opt. (1–8)	2,079	10.33	0.31	3.15	177.32	13.60	317,775	1.77	3.14
0	29.69	–0.43	1.147	5,5,5	5	0	0	PA Pract. (1–8)	2,000	9.40	0.27	2.50	155.42	12.60	386,486	1.61	2.59
1	11.63	0.15	1.144	12,12,12	10	0	0	PA Opt. (1–12)	1,943	13.15	0.49	6.45	253.78	13.50	155,199	2.54	6.44
2	7.12	0.68	1.142	39,52,52	25	82,71,64	75	PA Pract. (1–12)	2,000	10.77	0.46	4.98	205.24	9.40	170,191	2.42	5.88
3	3.17	1.21	1.143	39,62,81	40,60,50	93,93,93	100	PA Opt. (1–16)	1,979	14.39	0.62	8.86	297.93	13.50	112,662	2.98	8.88
4	1.97	1.78	1.147	39,62,81	40,60,80	136,136,136	150	PA Pract. (1–16)	2,000	11.19	0.56	6.30	226.08	8.50	128,778	2.79	7.76
5	0.83	2.42	1.151	39,62,81	40,60,80	163,182,183	150,200,200	BC Opt. (1–2)	123	101.63	1.44	20.50	452.68	13.40	48,892	4.53	20.50
6	0.49	3.02	1.154	39,62,81	40,60,80	163,213,227	150,225,200	BC Pract. (1–2)	133	100.35	1.45	20.30	450.00	13.50	49,019	4.51	20.34
7	0.20	3.62	1.153	39,62,81	40,60,80	163,213,255	150,225,300	BC Opt. (1–3)	141	100.41	1.56	21.86	467.56	13.50	45,891	4.68	21.86
8	0.11	4.23	1.150	39,62,81	40,60,80	163,213,255	150,225,300	BC Pract. (1–3)	133	108.00	1.56	23.57	504.15	15.70	45,829	4.67	21.85
9	0.04	4.81	1.146	39,62,81	40,60,80	163,213,255	150,225,300	BC Opt. (1–4)	157	98.46	1.63	22.39	473.17	13.50	44,749	4.73	22.39
>9	0.03	5.58	1.139	39,62,81	40,60,80	163,213,255	150,225,300	BC Pract. (1–4)	133	109.13	1.60	24.35	516.53	16.20	44,905	4.71	22.22

TABLE 10.59

4.5/6 H17 DAS

TC	TC Freq. (%)	EV W/L (%)	Std. Dev.	Play-All Bets Optimal ($)	Play-All Bets Practical ($)	Back-Count 27% 16% 9% Optimal ($)	Back-Count 27% 16% 9% Practical ($)	Bet Spread	Kelly Bank (units)	Avg. Bet ($)	Results W/L (%)	Results W/100 ($)	Risk Measures SD/100 ($)	Risk Measures ROR (%)	Risk Measures N0 (hands)	Rankings DI	Rankings SCORE
<0	45.01	–1.53	1.152	7,7,7	5	0	0	PA Opt. (1–8)	1,390	15.61	0.47	7.30	269.95	13.60	137,251	2.70	7.29
0	27.69	–0.42	1.147	7,7,7	5	0	0	PA Pract. (1–8)	2,000	10.08	0.44	4.45	167.75	4.20	142,104	2.65	7.04
1	11.25	0.17	1.144	13,13,13	10	0	0	PA Opt. (1–12)	1,344	18.54	0.68	12.64	355.53	13.50	79,074	3.56	12.64
2	7.38	0.70	1.142	54,54,54	25	0,82,73	100,75,75	PA Pract. (1–12)	2,000	11.81	0.67	7.95	225.72	4.40	80,715	3.52	12.40
3	3.48	1.24	1.143	58,89,95	40,60,50	129,95,95	100	PA Opt. (1–16)	1,420	19.66	0.83	16.29	403.44	13.60	61,449	4.03	16.28
4	2.45	1.81	1.147	58,89,113	40,60,80	138,138,138	150	PA Pract. (1–16)	2,000	12.50	0.81	10.15	255.24	4.40	63,234	3.98	15.82
5	1.11	2.44	1.151	58,89,113	40,60,80	185,185,185	200	BC Opt. (1–2)	78	159.92	2.27	31.47	560.94	13.40	31,827	5.61	31.47
6	0.81	3.01	1.152	58,89,113	40,60,80	227,227,227	200,225,200	BC Pract. (1–2)	100	124.70	1.68	33.62	599.70	15.40	31,856	5.61	31.43
7	0.34	3.68	1.152	58,89,113	40,60,80	258,246,277	200,225,300	BC Opt. (1–3)	122	116.12	1.81	33.81	581.40	13.50	29,512	5.81	33.81
8	0.26	4.22	1.149	58,89,113	40,60,80	258,246,292	200,225,300	BC Pract. (1–3)	133	115.80	1.82	33.84	583.58	13.70	29,766	5.80	33.62
9	0.10	4.84	1.145	58,89,113	40,60,80	258,246,292	200,225,300	BC Opt. (1–4)	137	114.09	1.90	34.83	590.20	13.50	28,772	5.90	34.83
>9	0.14	5.79	1.138	58,89,113	40,60,80	258,246,292	200,225,300	BC Pract. (1–4)	133	118.43	1.89	35.93	610.20	14.60	28,850	5.89	34.67

TABLE 10.60
5.0/6 H17 DAS

TC	TC Freq. (%)	EV W/L (%)	Std. Dev.	Play-All Bets Optimal ($)	Play-All Bets Practical ($)	Back-Count 29% 18% 11% Optimal ($)	Back-Count 29% 18% 11% Practical ($)	Bet Spread	Kelly Bank (units)	Avg. Bet ($)	Results W/L (%)	Results W/100 ($)	Risk Measures SD/100 ($)	Risk Measures ROR (%)	Risk Measures N0 (hands)	Rankings DI	Rankings SCORE
<0	45.21	−1.64	1.154	11,11,10	10	0	0	PA Opt. (1–8)	897	23.09	0.68	15.65	395.43	13.50	63,945	3.95	15.64
0	25.91	−0.41	1.147	11,11,10	10	0	0	PA Pract. (1–8)	1,000	20.24	0.67	13.53	343.40	10.10	64,418	3.94	15.51
1	10.74	0.18	1.144	14,14,14	15	0	0	PA Opt. (1–12)	933	25.82	0.94	24.18	491.55	13.50	41,393	4.92	24.16
2	7.49	0.73	1.143	56,56,56	40	0,98,86	0,100,75	PA Pract. (1–12)	1,000	23.78	0.94	22.37	457.26	11.70	41,783	4.89	23.94
3	3.64	1.27	1.144	89,97,97	80,100,100	148,98,97	150,100,100	PA Opt. (1–16)	1,020	26.71	1.11	29.69	545.08	13.50	33,655	5.45	29.71
4	2.84	1.86	1.147	89,129,141	80,120,150	148,141,141	150	PA Pract. (1–16)	1,000	26.31	1.13	29.77	548.76	13.80	33,979	5.42	29.43
5	1.36	2.45	1.150	89,129,157	80,120,160	185,185,185	200	BC Opt. (1–2)	68	183.75	2.62	51.23	715.78	13.30	19,551	7.16	51.23
6	1.13	3.06	1.151	89,129,157	80,120,160	231,231,231	200	BC Pract. (1–2)	67	185.40	2.61	51.61	722.70	13.70	19,616	7.14	51.00
7	0.54	3.61	1.150	89,129,157	80,120,160	273,273,273	300	BC Opt. (1–3)	102	137.02	2.16	53.75	733.18	13.60	18,605	7.33	53.75
8	0.46	4.23	1.148	89,129,157	80,120,160	296,293,321	300	BC Pract. (1–3)	100	140.10	2.15	54.76	748.60	14.10	18,661	7.31	53.50
9	0.22	4.69	1.145	89,129,157	80,120,160	296,293,345	300	BC Opt. (1–4)	116	134.79	2.27	55.63	745.90	13.40	17,991	7.46	55.63
>9	0.46	6.01	1.136	89,129,157	80,120,160	296,293,345	300	BC Pract. (1–4)	133	129.83	2.27	53.40	719.48	12.70	18,153	7.42	55.07

TABLE 10.61
5.5/6 H17 DAS

TC	TC Freq. (%)	EV W/L (%)	Std. Dev.	Play-All Bets Optimal ($)	Play-All Bets Practical ($)	Back-Count 30% 20% 13% Optimal ($)	Back-Count 30% 20% 13% Practical ($)	Bet Spread	Kelly Bank (units)	Avg. Bet ($)	Results W/L (%)	Results W/100 ($)	Risk Measures SD/100 ($)	Risk Measures ROR (%)	Risk Measures N0 (hands)	Rankings DI	Rankings SCORE
<0	45.30	−1.79	1.155	17,16,15	15	0	0	PA Opt. (1–8)	573	34.79	1.00	34.89	590.97	13.50	28,622	5.91	34.92
0	24.28	−0.39	1.147	17,16,15	15	0	0	PA Pract. (1–8)	667	29.61	1.02	30.07	510.77	9.90	28,871	5.89	34.65
1	10.12	0.19	1.145	17,16,15	15	0	0	PA Opt. (1–12)	617	37.30	1.33	49.66	704.37	13.50	20,161	7.04	49.61
2	7.47	0.77	1.143	59,59,59	40	0,0,107	0,0,100	PA Pract. (1–12)	667	34.20	1.36	46.41	661.34	11.90	20,306	7.02	49.25
3	3.61	1.31	1.145	100,100,100	100	179,150,107	200,150,100	PA Opt. (1–16)	686	37.78	1.56	58.95	767.95	13.50	16,956	7.68	58.97
4	3.17	1.92	1.147	140,146,146	120,150,150	179,150,146	200,150,150	PA Pract. (1–16)	667	36.65	1.57	57.57	753.63	13.10	17,137	7.64	58.35
5	1.47	2.54	1.150	140,192,192	120,180,200	192,192,192	200	BC Opt. (1–2)	56	223.56	3.17	90.81	952.98	13.50	10,994	9.53	90.81
6	1.50	3.10	1.149	140,195,233	120,180,200	235,235,235	200	BC Pract. (1–2)	50	236.80	3.16	95.91	1010.40	15.20	11,078	9.49	90.09
7	0.65	3.78	1.149	140,195,233	120,180,240	286,286,287	300	BC Opt. (1–3)	67	218.67	3.42	96.06	980.17	13.30	10,410	9.80	96.06
8	0.78	4.15	1.145	140,195,233	120,180,240	317,317,317	300	BC Pract. (1–3)	67	216.90	3.44	95.56	977.40	13.30	10,464	9.78	95.58
9	0.29	4.99	1.143	140,195,233	120,180,240	357,382,382	400,450,400	BC Opt. (1–4)	94	167.60	2.88	97.91	989.50	13.40	10,221	9.89	97.91
>9	1.35	6.38	1.131	140,195,233	120,180,240	357,450,426	400,450,400	BC Pract. (1–4)	100	160.90	2.88	93.90	950.70	12.40	10,251	9.88	97.55

TABLE 10.62
4.0/6 H17 LS

TC	TC Freq. (%)	EV W/L (%)	Std. Dev.	Play-All Bets Optimal ($)	Play-All Bets Practical ($)	Back-Count 26% 14% 7% Optimal ($)	Back-Count 26% 14% 7% Practical ($)	Bet Spread	Kelly Bank (units)	Avg. Bet ($)	Results W/L (%)	Results W/100 ($)	Risk Measures SD/100 ($)	Risk Measures ROR (%)	Risk Measures N0 (hands)	Rankings DI	Rankings SCORE
<0	44.71	−1.53	1.133	6,6,6	5	0	0	PA Opt. (1–8)	1,821	11.64	0.33	3.86	196.31	13.60	259,243	1.96	3.85
0	29.70	−0.47	1.125	6,6,6	5	0	0	PA Pract. (1–8)	2,000	9.40	0.29	2.72	151.54	9.40	311,541	1.79	3.22
1	11.62	0.15	1.119	12,12,12	10	0	0	PA Opt. (1–12)	1,675	14.91	0.53	7.89	281.19	13.50	126,546	2.81	7.91
2	7.12	0.72	1.115	44,58,58	25	92,80,72	100,75,75	PA Pract. (1–12)	2,000	10.77	0.50	5.40	199.94	6.70	137,092	2.70	7.30
3	3.18	1.31	1.113	44,72,93	40,60,50	106,106,106	100	PA Opt. (1–16)	1,715	16.28	0.67	10.91	330.56	13.50	91,527	3.31	10.93
4	1.97	1.94	1.114	44,72,93	40,60,80	157,157,157	150	PA Pract. (1–16)	2,000	11.19	0.61	6.84	220.08	5.90	103,526	3.11	9.66
5	0.83	2.60	1.118	44,72,93	40,60,80	185,208,208	200	BC Opt. (1–2)	108	115.64	1.56	25.17	501.74	13.60	39,591	5.02	25.17
6	0.49	3.27	1.121	44,72,93	40,60,80	185,241,261	200,225,300	BC Pract. (1–2)	100	119.20	1.56	25.91	517.50	14.40	39,923	5.01	25.08
7	0.20	3.93	1.120	44,72,93	40,60,80	185,241,289	200,225,300	BC Opt. (1–3)	124	114.50	1.68	26.92	518.90	13.60	37,059	5.19	26.92
8	0.11	4.58	1.117	44,72,93	40,60,80	185,241,289	200,225,300	BC Pract. (1–3)	133	108.00	1.69	25.45	490.58	12.10	37,230	5.19	26.92
9	0.04	5.19	1.113	44,72,93	40,60,80	185,241,289	200,225,300	BC Opt. (1–4)	139	112.32	1.76	27.62	525.52	13.40	36,130	5.26	27.62
>9	0.03	6.18	1.107	44,72,93	40,60,80	185,241,289	200,225,300	BC Pract. (1–4)	133	112.73	1.77	27.92	532.50	14.00	36,428	5.24	27.49

TABLE 10.63
4.5/6 H17 LS

TC	TC Freq. (%)	EV W/L (%)	Std. Dev.	Play-All Bets Optimal ($)	Play-All Bets Practical ($)	Back-Count 27% 16% 9% Optimal ($)	Back-Count 27% 16% 9% Practical ($)	Bet Spread	Kelly Bank (units)	Avg. Bet ($)	Results W/L (%)	Results W/100 ($)	Risk Measures SD/100 ($)	Risk Measures ROR (%)	Risk Measures N0 (hands)	Rankings DI	Rankings SCORE
<0	45.00	−1.63	1.134	8,9,8	10	0	0	PA Opt. (1–8)	1,197	17.70	0.50	8.93	298.75	13.50	112,093	2.99	8.92
0	27.70	−0.46	1.125	8,9,8	10	0	0	PA Pract. (1–8)	1,000	18.86	0.48	9.00	310.30	15.40	118,872	2.90	8.41
1	11.24	0.16	1.119	13,13,13	15	0	0	PA Opt. (1–12)	1,167	20.98	0.74	15.56	394.33	13.50	64,315	3.94	15.55
2	7.37	0.74	1.115	59,59,59	40	0,93,83	0,100,75	PA Pract. (1–12)	1,000	21.63	0.72	15.64	404.61	14.80	66,927	3.87	14.94
3	3.49	1.33	1.113	67,103,107	80,100,100	147,108,108	150,100,100	PA Opt. (1–16)	1,229	22.25	0.90	20.03	447.78	13.50	49,877	4.48	20.05
4	2.45	1.96	1.115	67,103,130	80,120,150	158,158,158	150	PA Pract. (1–16)	1,000	23.47	0.89	20.87	476.34	15.90	52,094	4.38	19.20
5	1.11	2.63	1.118	67,103,130	80,120,160	211,211,211	200	BC Opt. (1–2)	68	183.02	2.45	38.99	624.49	13.40	25,576	6.24	38.99
6	0.81	3.26	1.119	67,103,130	80,120,160	260,260,260	300	BC Pract. (1–2)	67	184.80	2.47	39.70	637.20	14.00	25,697	6.23	38.81
7	0.34	3.98	1.119	67,103,130	80,120,160	295,279,318	300	BC Opt. (1–3)	108	132.11	1.96	41.54	644.60	13.40	24,126	6.45	41.55
8	0.26	4.56	1.116	67,103,130	80,120,160	295,279,330	300	BC Pract. (1–3)	100	135.00	1.97	42.80	666.20	14.50	24,228	6.43	41.29
9	0.10	5.22	1.113	67,103,130	80,120,160	295,279,330	300	BC Opt. (1–4)	121	129.93	2.05	42.88	654.83	13.50	23,334	6.55	42.88
>9	0.14	6.32	1.106	67,103,130	80,120,160	295,279,330	300	BC Pract. (1–4)	133	123.53	2.09	41.45	635.33	12.90	23,465	6.52	42.56

TABLE 10.64
5.0/6 H17 LS

TC	TC Freq. (%)	EV W/L (%)	Std. Dev.	Play-All Bets Optimal ($)	Play-All Bets Practical ($)	Back-Count 29% 18% 11% Optimal ($)	Back-Count 29% 18% 11% Practical ($)	Bet Spread	Kelly Bank (units)	Avg. Bet ($)	Results W/L (%)	Results W/100 ($)	Risk Measures SD/100 ($)	Risk Measures ROR (%)	Risk Measures N0 (hands)	Rankings DI	Rankings SCORE
<0	45.19	−1.74	1.135	13,12,11	15,10,10	0	0	PA Opt. (1–8)	784	26.03	0.73	19.07	436.71	13.50	52,415	4.37	19.07
0	25.91	−0.45	1.125	13,12,11	15,10,10	0	0	PA Pract. (1–8)	667	27.36	0.72	19.59	458.10	15.40	54,683	4.28	18.29
1	10.73	0.17	1.119	14,14,14	15	0	0	PA Opt. (1–12)	813	29.15	1.01	29.56	543.53	13.50	33,853	5.44	29.54
2	7.48	0.76	1.116	61,61,61	40	0,0,98	0,0,100	PA Pract. (1–12)	1,000	23.80	1.01	24.11	445.51	8.80	34,144	5.41	29.29
3	3.64	1.35	1.114	102,109,109	100	169,146,109	200,150,100	PA Opt. (1–16)	886	30.17	1.21	36.38	603.02	13.50	27,497	6.03	36.36
4	2.84	1.99	1.116	102,148,160	120,120,160	169,160,160	200	PA Pract. (1–16)	1,000	26.62	1.23	32.67	544.65	11.00	27,793	6.00	35.98
5	1.36	2.62	1.118	102,148,181	120,120,160	210,210,210	200	BC Opt. (1–2)	59	209.36	2.82	63.00	793.77	13.60	15,853	7.94	63.00
6	1.14	3.29	1.119	102,148,181	120,120,160	263,263,263	300	BC Pract. (1–2)	50	237.40	2.80	71.10	898.80	17.10	16,025	7.91	62.58
7	0.54	3.89	1.118	102,148,181	120,120,160	311,311,311	300	BC Opt. (1–3)	69	206.61	2.99	65.92	811.88	13.30	15,197	8.12	65.92
8	0.46	4.56	1.115	102,148,181	120,120,160	337,367,367	400	BC Pract. (1–3)	67	223.65	2.96	70.74	874.50	15.50	15,256	8.09	65.43
9	0.22	5.04	1.113	102,148,181	120,120,160	337,408,392	400,450,400	BC Opt. (1–4)	102	153.08	2.45	68.13	825.39	13.50	14,667	8.25	68.13
>9	0.47	6.52	1.104	102,148,181	120,120,160	337,437,392	400,450,400	BC Pract. (1–4)	100	160.70	2.46	71.88	874.80	15.20	14,803	8.22	67.52

TABLE 10.65
5.5/6 H17 LS

TC	TC Freq. (%)	EV W/L (%)	Std. Dev.	Play-All Bets Optimal ($)	Play-All Bets Practical ($)	Back-Count 30% 20% 13% Optimal ($)	Back-Count 30% 20% 13% Practical ($)	Bet Spread	Kelly Bank (units)	Avg. Bet ($)	Results W/L (%)	Results W/100 ($)	Risk Measures SD/100 ($)	Risk Measures ROR (%)	Risk Measures N0 (hands)	Rankings DI	Rankings SCORE
<0	45.29	−1.90	1.137	20,19,17	20,20,15	0	0	PA Opt. (1–8)	505	39.23	1.08	42.16	649.71	13.40	23,679	6.50	42.21
0	24.27	−0.43	1.125	20,19,17	20,20,15	0	0	PA Pract. (1–8)	500	38.08	1.08	41.13	634.74	12.90	23,828	6.48	41.99
1	10.11	0.19	1.120	20,19,17	20,20,15	0	0	PA Opt. (1–12)	540	42.15	1.43	60.14	775.54	13.50	16,630	7.76	60.14
2	7.46	0.80	1.117	64,64,64	50,50,40	0	0,0,125	PA Pract. (1–12)	500	43.62	1.44	62.99	817.70	15.10	16,857	7.70	59.34
3	3.61	1.39	1.115	111,112,112	100	202,170,137	200,200,125	PA Opt. (1–16)	599	42.70	1.68	71.59	846.04	13.50	13,968	8.46	71.58
4	3.16	2.03	1.117	158,163,163	160,200,200	202,170,163	200	PA Pract. (1–16)	667	38.90	1.73	67.12	801.51	12.30	14,258	8.37	70.12
5	1.47	2.71	1.118	158,217,217	160,200,200	217,217,217	200	BC Opt. (1–2)	49	253.14	3.39	110.53	1051.41	13.70	9,056	10.51	110.53
6	1.50	3.28	1.118	158,222,263	160,240,240	263,263,263	300	BC Pract. (1–2)	50	255.00	3.41	111.74	1065.40	13.90	9,081	10.49	110.00
7	0.65	4.06	1.117	158,222,267	160,240,240	325,325,325	300	BC Opt. (1–3)	59	247.79	3.68	117.15	1082.31	13.40	8,542	10.82	117.15
8	0.79	4.43	1.114	158,222,267	160,240,240	357,357,357	400	BC Pract. (1–3)	50	268.20	3.56	122.80	1148.20	15.40	8,743	10.69	114.37
9	0.30	5.35	1.111	158,222,267	160,240,240	405,434,434	400,500,500	BC Opt. (1–4)	73	241.47	3.83	118.90	1090.39	13.40	8,413	10.90	118.89
>9	1.38	6.85	1.100	158,222,267	160,240,240	405,509,549	400,500,500	BC Pract. (1–4)	80	202.25	3.08	126.48	1163.75	15.30	8,463	10.87	118.11

TABLE 10.66

4.0/6 H17 DAS LS

TC	TC Freq. (%)	EV W/L (%)	Std. Dev.	Play-All Bets Optimal ($)	Play-All Bets Practical ($)	Back-Count 26% 14% 7% Optimal ($)	Back-Count 26% 14% 7% Practical ($)	Bet Spread	Kelly Bank (units)	Avg. Bet ($)	Results W/L (%)	Results W/100 ($)	Risk Measures SD/100 ($)	Risk Measures ROR (%)	Risk Measures N0 (hands)	Rankings DI	Rankings SCORE
<0	44.72	−1.40	1.142	8,8,7	10,10,5	0	0	PA Opt. (1–8)	1,291	17.20	0.48	8.16	285.77	13.50	122,361	2.86	8.17
0	29.71	−0.32	1.133	8,8,7	10,10,5	0	0	PA Pract. (1–8)	1,000	19.38	0.44	8.46	310.37	17.30	134,592	2.72	7.42
1	11.62	0.30	1.127	24,24,24	25,25,10	0	0	PA Opt. (1–12)	1,301	19.80	0.66	13.12	362.17	13.50	76,258	3.62	13.12
2	7.11	0.88	1.122	62,70,70	50,50,25	101,88,80	100,100,75	PA Pract. (1–12)	1,000	22.12	0.65	14.28	406.42	17.70	81,002	3.51	12.35
3	3.18	1.48	1.120	62,92,112	80,120,50	118,118,118	125	PA Opt. (1–16)	1,429	20.63	0.80	16.49	406.04	13.50	60,638	4.06	16.49
4	1.97	2.10	1.121	62,92,112	80,120,80	167,167,168	200	PA Pract. (1–16)	2,000	11.19	0.77	8.55	221.39	3.00	66,969	3.86	14.93
5	0.83	2.76	1.124	62,92,112	80,120,80	202,218,219	200	BC Opt. (1–2)	99	126.45	1.72	30.34	550.83	13.50	32,880	5.51	30.34
6	0.49	3.44	1.127	62,92,112	80,120,80	202,264,271	200,300,300	BC Pract. (1–2)	100	131.90	1.73	31.86	580.20	15.00	33,081	5.49	30.15
7	0.20	4.10	1.126	62,92,112	80,120,80	202,264,318	200,300,300	BC Opt. (1–3)	114	124.58	1.84	32.02	565.89	13.50	31,311	5.66	32.02
8	0.11	4.76	1.122	62,92,112	80,120,80	202,264,318	200,300,300	BC Pract. (1–3)	100	138.20	1.83	35.29	625.80	16.50	31,428	5.64	31.81
9	0.04	5.38	1.118	62,92,112	80,120,80	202,264,318	200,300,300	BC Opt. (1–4)	126	122.28	1.91	32.63	571.21	13.40	30,625	5.71	32.63
>9	0.03	6.35	1.111	62,92,112	80,120,80	202,264,318	200,300,300	BC Pract. (1–4)	133	125.40	1.93	33.73	593.70	14.80	30,945	5.68	32.28

TABLE 10.67

4.5/6 H17 DAS LS

TC	TC Freq. (%)	EV W/L (%)	Std. Dev.	Play-All Bets Optimal ($)	Play-All Bets Practical ($)	Back-Count 27% 16% 9% Optimal ($)	Back-Count 27% 16% 9% Practical ($)	Bet Spread	Kelly Bank (units)	Avg. Bet ($)	Results W/L (%)	Results W/100 ($)	Risk Measures SD/100 ($)	Risk Measures ROR (%)	Risk Measures N0 (hands)	Rankings DI	Rankings SCORE
<0	45.00	−1.49	1.143	11,10,9	10	0	0	PA Opt. (1–8)	913	23.60	0.64	15.08	388.54	13.50	66,278	3.89	15.10
0	27.70	−0.31	1.133	11,10,9	10	0	0	PA Pract. (1–8)	1,000	20.72	0.62	12.88	333.96	9.90	67,229	3.86	14.88
1	11.24	0.31	1.127	25,25,25	25	0	0	PA Opt. (1–12)	984	25.94	0.87	22.59	475.15	13.50	44,299	4.75	22.58
2	7.37	0.90	1.122	71,71,71	50	115,100,90	125,100,100	PA Pract. (1–12)	1,000	24.19	0.87	21.03	446.06	12.10	44,989	4.71	22.22
3	3.48	1.50	1.120	88,120,120	80,120,125	120,120,120	125	PA Opt. (1–16)	1,067	26.81	1.02	27.41	523.49	13.50	36,500	5.23	27.40
4	2.45	2.12	1.122	88,122,150	80,120,160	169,169,169	200	PA Pract. (1–16)	1,000	26.45	1.03	27.34	527.03	13.90	37,160	5.19	26.91
5	1.11	2.79	1.124	88,122,150	80,120,160	221,221,221	200	BC Opt. (1–2)	87	143.58	1.96	45.26	672.79	13.40	22,037	6.73	45.27
6	0.81	3.43	1.125	88,122,150	80,120,160	231,271,271	250,300,300	BC Pract. (1–2)	80	154.38	1.96	48.48	722.75	15.60	22,207	6.71	44.99
7	0.34	4.13	1.124	88,122,150	80,120,160	231,301,327	250,300,300	BC Opt. (1–3)	100	142.31	2.11	48.24	694.59	13.40	20,734	6.95	48.24
8	0.26	4.72	1.122	88,122,150	80,120,160	231,301,358	250,300,400	BC Pract. (1–3)	100	148.10	2.11	50.25	726.10	14.80	20,921	6.92	47.89
9	0.10	5.44	1.117	88,122,150	80,120,160	231,301,358	250,300,400	BC Opt. (1–4)	112	139.73	2.20	49.45	703.22	13.40	20,255	7.03	49.45
>9	0.14	6.47	1.110	88,122,150	80,120,160	231,301,358	250,300,400	BC Pract. (1–4)	100	151.20	2.18	52.90	755.30	15.60	20,386	7.00	49.05

TABLE 10.68
5.0/6 H17 DAS LS

TC	TC Freq. (%)	EV W/L (%)	Std. Dev.	Play-All Bets Optimal ($)	Play-All Bets Practical ($)	Back-Count 29% 18% 11% Optimal ($)	Back-Count 29% 18% 11% Practical ($)	Bet Spread	Kelly Bank (units)	Avg. Bet ($)	Results W/L (%)	Results W/100 ($)	Risk Measures SD/100 ($)	Risk Measures ROR (%)	Risk Measures N0 (hands)	Rankings DI	Rankings SCORE
<0	45.20	–1.61	1.144	15,14,13	15	0	0	PA Opt. (1–8)	663	31.81	0.87	27.56	525.21	13.50	36,243	5.25	27.59
0	25.91	–0.30	1.133	15,14,13	15	0	0	PA Pract. (1–8)	667	29.91	0.87	25.86	496.41	12.20	36,849	5.21	27.15
1	10.73	0.33	1.127	26,26,26	25	0	0	PA Opt. (1–12)	722	34.14	1.14	38.82	623.18	13.50	25,752	6.23	38.84
2	7.48	0.92	1.123	73,73,73	50	0,118,104	0,125,100	PA Pract. (1–12)	667	34.31	1.16	39.65	641.49	14.50	26,182	6.18	38.20
3	3.64	1.52	1.121	121,121,121	120,125,125	177,121,121	200,125,125	PA Opt. (1–16)	795	34.84	1.32	45.89	677.50	13.50	21,780	6.78	45.90
4	2.84	2.15	1.122	121,166,171	120,180,200	177,171,171	200	PA Pract. (1–16)	667	36.86	1.34	49.19	735.86	16.20	22,383	6.68	44.69
5	1.36	2.79	1.124	121,166,201	120,180,200	221,221,221	200	BC Opt. (1–2)	57	219.20	2.98	69.69	834.71	13.20	14,339	8.35	69.69
6	1.14	3.45	1.125	121,166,201	120,180,240	273,273,273	300	BC Pract. (1–2)	50	237.40	2.97	75.19	903.60	15.80	14,438	8.32	69.25
7	0.54	4.05	1.123	121,166,201	120,180,240	321,321,321	300	BC Opt. (1–3)	85	165.10	2.48	74.24	861.61	13.50	13,483	8.62	74.25
8	0.46	4.71	1.120	121,166,201	120,180,240	354,353,375	400,375,400	BC Pract. (1–3)	80	174.50	2.46	78.01	907.88	15.00	13,548	8.59	73.84
9	0.22	5.21	1.117	121,166,201	120,180,240	354,353,415	400,375,400	BC Opt. (1–4)	96	162.55	2.59	76.52	874.75	13.50	13,059	8.75	76.52
>9	0.47	6.66	1.108	121,166,201	120,180,240	354,353,415	400,375,400	BC Pract. (1–4)	100	165.80	2.59	77.92	893.90	14.20	13,167	8.72	75.97

TABLE 10.69
5.5/6 H17 DAS LS

TC	TC Freq. (%)	EV W/L (%)	Std. Dev.	Play-All Bets Optimal ($)	Play-All Bets Practical ($)	Back-Count 30% 20% 13% Optimal ($)	Back-Count 30% 20% 13% Practical ($)	Bet Spread	Kelly Bank (units)	Avg. Bet ($)	Results W/L (%)	Results W/100 ($)	Risk Measures SD/100 ($)	Risk Measures ROR (%)	Risk Measures N0 (hands)	Rankings DI	Rankings SCORE
<0	45.29	–1.77	1.146	22,20,18	20	0	0	PA Opt. (1–8)	446	44.83	1.21	54.29	737.51	13.40	18,385	7.38	54.39
0	24.28	–0.29	1.133	22,20,18	20	0	0	PA Pract. (1–8)	500	39.48	1.23	48.47	660.24	10.80	18,547	7.34	53.90
1	10.11	0.34	1.127	27,27,27	25	0	0	PA Opt. (1–12)	492	47.10	1.55	73.09	854.61	13.50	13,689	8.55	73.03
2	7.46	0.96	1.124	76,76,76	50	0,0,126	0,150,125	PA Pract. (1–12)	500	45.04	1.58	71.21	839.20	13.20	13,892	8.49	72.00
3	3.61	1.56	1.122	124,124,124	125	210,175,126	200,150,125	PA Opt. (1–16)	552	47.42	1.78	84.55	919.64	13.40	11,823	9.20	84.58
4	3.16	2.20	1.123	174,174,174	160,200,200	210,175,174	200	PA Pract. (1–16)	500	48.44	1.83	88.44	968.78	15.10	11,999	9.13	83.34
5	1.47	2.88	1.124	179,228,227	160,200,200	228,228,228	200	BC Opt. (1–2)	48	262.76	3.55	120.08	1095.80	13.20	8,319	10.96	120.08
6	1.50	3.45	1.124	179,244,273	160,240,300	273,273,273	300	BC Pract. (1–2)	50	255.00	3.57	117.03	1070.60	12.90	8,373	10.93	119.48
7	0.65	4.21	1.122	179,244,290	160,240,320	334,334,334	300	BC Opt. (1–3)	57	256.48	3.83	126.38	1124.16	13.40	7,921	11.24	126.38
8	0.79	4.59	1.119	179,244,290	160,240,320	367,367,367	400	BC Pract. (1–3)	67	211.80	3.03	130.57	1168.80	14.50	8,022	11.17	124.80
9	0.30	5.52	1.115	179,244,290	160,240,320	420,443,443	400,450,500	BC Opt. (1–4)	80	197.79	3.24	130.17	1140.98	13.30	7,682	11.41	130.17
>9	1.38	6.99	1.104	179,244,290	160,240,320	420,526,503	400,450,500	BC Pract. (1–4)	80	202.25	3.24	133.04	1169.50	14.20	7,732	11.38	129.41

TABLE 10.70
2-Deck Play-All SCORE Summary

		S17	S17 DAS		S17 LS		S17 DAS LS		H17		H17 DAS		H17 LS		H17 DAS LS	
52/104	1-4	**9.83**	**17.21**	7.38	**17.55**	7.72	**26.98**	17.15	**3.12**	-6.71	**7.89**	-1.94	**8.21**	-1.62	**15.36**	5.53
	1-6	**18.30**	**27.03**	8.73	**29.24**	10.94	**39.96**	21.66	**9.53**	-8.77	**16.15**	-2.15	**17.94**	-0.36	**26.92**	8.62
	1-8	**24.36**	**33.57**	9.21	**37.27**	12.91	**48.39**	24.03	**14.79**	-9.57	**22.25**	-2.11	**25.32**	0.96	**35.00**	10.64
62/104	1-4	**19.05**	**28.85**	9.80	**30.78**	11.73	**43.11**	24.06	**9.05**	-10.00	**16.38**	-2.67	**18.00**	-1.05	**27.86**	8.81
		9.22	11.64		13.23		16.13		5.93		8.49		9.79		12.50	
	1-6	**32.25**	**43.51**	11.26	**48.52**	16.27	**62.31**	30.06	**20.13**	-12.12	**29.48**	-2.77	**33.72**	1.47	**45.50**	13.25
		13.95	16.48		19.28		22.35		10.60		13.33		15.78		18.58	
	1-8	**41.30**	**53.03**	11.73	**60.34**	19.04	**74.51**	33.21	**28.52**	-12.78	**38.66**	-2.64	**44.94**	3.64	**57.44**	16.14
		16.94	19.46		23.07		26.12		13.73		16.41		19.62		22.44	
70/104	1-4	**28.91**	**40.74**	11.83	**44.54**	15.63	**59.15**	30.24	**16.02**	-12.89	**25.42**	-3.49	**28.45**	-0.46	**40.67**	11.76
		19.08	23.53		26.99		32.17		12.90		17.53		20.24		25.31	
	1-6	**46.94**	**60.28**	13.34	**68.37**	21.43	**84.51**	37.57	**31.78**	-15.16	**43.34**	-3.60	**50.09**	3.15	**64.39**	17.45
		28.64	33.25		39.13		44.55		22.25		27.19		32.15		37.47	
	1-8	**58.98**	**72.80**	13.82	**83.94**	24.96	**100.54**	41.56	**43.26**	-15.72	**55.59**	-3.39	**65.08**	6.10	**80.11**	21.13
		34.62	39.23		46.67		52.15		28.47		33.34		39.76		45.11	
78/104	1-4	**42.74**	**56.95**	14.21	**63.25**	20.51	**80.55**	37.81	**26.73**	-16.01	**38.59**	-4.15	**43.70**	0.96	**58.65**	15.91
		32.91	39.74		45.70		53.57		23.61		30.70		35.49		43.29	
	1-6	**66.99**	**82.82**	15.83	**94.91**	27.92	**113.85**	46.86	**48.86**	-18.13	**62.91**	-4.08	**73.12**	6.13	**90.28**	23.29
		48.69	55.79		65.67		73.89		39.33		46.76		55.18		63.36	
	1-8	**83.14**	**99.47**	16.33	**115.68**	32.54	**135.06**	51.92	**64.43**	-18.71	**79.38**	-3.76	**93.33**	10.19	**111.30**	28.16
		58.78	65.90		78.41		86.67		49.64		57.13		68.01		76.30	

Legend: **X** y
z

X = SCORE; y = $ gain from rules; z = $ gain from penetration

TABLE 10.71
2-Deck Back-Count SCORE Summary

		S17	S17 DAS		S17 LS		S17 DAS LS		H17		H17 DAS		H17 LS		H17 DAS LS	
52/104	1-2	**51.32**	**58.81**	7.49	**70.07**	18.75	**78.65**	27.33	**43.66**	–7.66	**50.91**	–0.41	**61.05**	9.73	**69.59**	18.27
	1-3	**54.04**	**61.40**	7.36	**73.60**	19.56	**82.39**	28.35	**46.60**	–7.44	**53.76**	–0.28	**64.80**	10.76	**73.19**	19.15
	1-4	**55.05**	**63.04**	7.99	**74.88**	19.83	**84.73**	29.68	**47.81**	–7.24	**54.85**	–0.20	**66.28**	11.23	**74.50**	19.45
62/104	1-2	**78.21**	**87.95**	9.74	**105.26**	27.05	**116.74**	38.53	**70.22**	–7.99	**77.99**	–0.22	**95.02**	16.81	**104.74**	26.53
		26.89	29.14		35.19		38.09		26.56		27.08		33.97		35.15	
	1-3	**83.15**	**92.66**	9.51	**111.69**	28.54	**122.93**	39.78	**73.55**	-9.60	**83.06**	-0.09	**100.51**	17.36	**111.27**	28.12
		29.11	31.26		38.09		40.54		26.95		29.30		35.71		38.08	
	1-4	**85.05**	**94.39**	9.34	**114.15**	29.10	**125.45**	40.40	**75.72**	-9.33	**85.07**	0.02	**103.22**	18.17	**113.77**	28.72
		30.00	31.35		39.27		40.72		27.91		30.22		36.94		39.27	
70/104	1-2	**106.35**	**117.32**	10.97	**140.97**	34.62	**154.46**	48.11	**96.84**	-9.51	**106.48**	0.13	**129.55**	23.20	**140.21**	33.86
		55.03	58.51		70.90		75.81		53.18		55.57		68.50		70.62	
	1-3	**113.19**	**124.44**	11.25	**150.51**	37.32	**163.80**	50.61	**101.48**	-11.71	**112.98**	-0.21	**136.53**	23.34	**149.63**	36.44
		59.15	63.04		76.91		81.41		54.88		59.22		71.73		76.44	
	1-4	**116.16**	**127.13**	10.97	**154.34**	38.18	**167.32**	51.16	**104.67**	-11.49	**116.05**	-0.11	**140.69**	24.53	**153.51**	37.35
		61.11	64.09		79.46		82.59		56.86		61.20		74.41		79.01	
78/104	1-2	**144.17**	**155.62**	11.45	**189.65**	45.48	**203.00**	58.83	**132.91**	-11.26	**144.64**	0.47	**175.64**	31.47	**188.88**	44.71
		92.85	96.81		119.58		124.35		89.25		93.73		114.59		119.29	
	1-3	**152.40**	**166.06**	13.66	**200.87**	48.47	**216.69**	64.29	**139.84**	-12.56	**152.63**	0.23	**184.47**	32.07	**200.01**	47.61
		98.36	104.66		127.27		134.30		93.24		98.87		119.67		126.82	
	1-4	**156.68**	**169.94**	13.26	**206.39**	49.71	**221.72**	65.04	**143.43**	-13.25	**157.08**	0.40	**190.14**	33.46	**205.63**	48.95
		101.63	106.90		131.51		136.99		95.62		102.23		123.86		131.13	

Legend: **X** y
z

X = SCORE; y = $ gain from rules; z = $ gain from penetration

TABLE 10.72
52/104 S17

TC	TC Freq. (%)	EV W/L (%)	Std. Dev.	Play-All Bets Optimal ($)	Play-All Bets Practical ($)	Back-Count 27% 18% 11% Optimal ($)	Back-Count 27% 18% 11% Practical ($)	Bet Spread	Kelly Bank (units)	Avg. Bet ($)	Results W/L (%)	Results W/100 ($)	Risk Measures SD/100 ($)	Risk Measures ROR (%)	Risk Measures N0 (hands)	Rankings DI	Rankings SCORE
<0	37.29	−1.46	1.142	13,14,13	15	0	0	PA Opt. (1–4)	744	22.55	0.44	9.82	313.48	13.50	101,649	3.13	9.83
0	35.97	−0.22	1.137	13,14,13	15	0	0	PA Pract. (1–4)	667	23.99	0.44	10.44	335.67	15.70	103,377	3.11	9.67
1	8.70	0.45	1.136	35,35,35	25	0	0	PA Opt. (1–6)	707	27.80	0.66	18.30	427.80	13.50	54,653	4.28	18.30
2	7.27	0.94	1.137	54,73,73	60,50,50	113,98,87	125,100,75	PA Pract. (1–6)	667	26.49	0.65	17.10	406.70	12.60	56,565	4.21	17.69
3	3.38	1.42	1.139	54,85,108	60,90,100	113,109,110	125,100,100	PA Opt. (1–8)	741	29.81	0.82	24.32	493.52	13.50	41,048	4.94	24.36
4	3.84	2.00	1.141	54,85,108	60,90,120	153,153,154	150	PA Pract. (1–8)	667	29.04	0.81	23.50	484.16	13.40	42,482	4.85	23.55
5	1.29	2.60	1.141	54,85,108	60,90,120	200,200,200	200	BC Opt. (1–2)	88	142.25	2.00	51.32	716.36	13.60	19,464	7.16	51.32
6	0.93	3.11	1.145	54,85,108	60,90,120	227,237,238	200	BC Pract. (1–2)	80	148.75	1.98	53.00	742.88	14.60	19,646	7.13	50.90
7	0.53	3.64	1.144	54,85,108	60,90,120	227,278,279	250,300,300	BC Opt. (1–3)	102	140.55	2.13	54.04	735.09	13.40	18,491	7.35	54.04
8	0.42	4.29	1.140	54,85,108	60,90,120	227,295,331	250,300,300	BC Pract. (1–3)	100	137.70	2.13	53.00	722.50	13.10	18,583	7.34	53.80
9	0.16	4.76	1.137	54,85,108	60,90,120	227,295,347	250,300,300	BC Opt. (1–4)	115	138.01	2.21	55.05	741.94	13.60	18,189	7.42	55.05
>9	0.22	5.86	1.129	54,85,108	60,90,120	227,295,347	250,300,300	BC Pract. (1–4)	133	127.65	2.23	51.29	693.53	11.90	18,276	7.40	54.70

TABLE 10.73
62/104 S17

TC	TC Freq. (%)	EV W/L (%)	Std. Dev.	Play-All Bets Optimal ($)	Play-All Bets Practical ($)	Back-Count 30% 21% 13% Optimal ($)	Back-Count 30% 21% 13% Practical ($)	Bet Spread	Kelly Bank (units)	Avg. Bet ($)	Results W/L (%)	Results W/100 ($)	Risk Measures SD/100 ($)	Risk Measures ROR (%)	Risk Measures N0 (hands)	Rankings DI	Rankings SCORE
<0	38.70	−1.60	1.143	18,19,18	20	0	0	PA Opt. (1–4)	546	31.29	0.61	19.05	436.44	13.50	52,463	4.36	19.05
0	31.74	−0.21	1.137	18,19,18	20	0	0	PA Pract. (1–4)	500	30.74	0.59	18.11	423.98	13.30	54,749	4.27	18.25
1	8.70	0.46	1.136	36,36,36	25	0	0	PA Opt. (1–6)	532	37.04	0.87	32.23	567.89	13.50	31,018	5.68	32.25
2	7.39	0.96	1.137	73,74,74	50	130,113,99	125,125,100	PA Pract. (1–6)	500	36.12	0.88	31.73	565.72	13.70	31,768	5.61	31.46
3	3.78	1.46	1.140	73,113,113	80,120,125	130,113,113	125	PA Opt. (1–8)	563	39.13	1.06	41.27	642.64	13.50	24,217	6.43	41.30
4	4.18	2.02	1.142	73,113,142	80,120,150	155,155,155	150	PA Pract. (1–8)	500	39.76	1.07	42.57	671.20	15.10	24,871	6.34	40.22
5	1.85	2.64	1.144	73,113,142	80,120,160	202,202,202	200	BC Opt. (1–2)	77	163.22	2.30	78.22	884.39	13.40	12,802	8.84	78.21
6	1.36	3.21	1.145	73,113,142	80,120,160	245,245,245	200	BC Pract. (1–2)	80	155.25	2.28	74.00	838.00	12.10	12,824	8.83	77.97
7	0.77	3.83	1.143	73,113,142	80,120,160	260,293,294	250,300,300	BC Opt. (1–3)	89	160.92	2.48	83.15	911.83	13.30	12,020	9.12	83.15
8	0.67	4.37	1.139	73,113,142	80,120,160	260,337,337	250,300,300	BC Pract. (1–3)	80	163.88	2.43	83.11	916.13	13.70	12,146	9.07	82.30
9	0.35	4.99	1.135	73,113,142	80,120,160	260,339,388	250,375,400	BC Opt. (1–4)	101	158.62	2.57	85.05	922.19	13.30	11,763	9.22	85.05
>9	0.51	6.30	1.126	73,113,142	80,120,160	260,339,398	250,375,400	BC Pract. (1–4)	100	156.10	2.54	82.58	898.20	12.80	11,825	9.19	84.51

TABLE 10.74
70/104 S17

TC	TC Freq. (%)	EV W/L (%)	Std. Dev.	Play-All Bets Optimal ($)	Play-All Bets Practical ($)	Back-Count 31% 22% 15% Optimal ($)	Back-Count 31% 22% 15% Practical ($)	Bet Spread	Kelly Bank (units)	Avg. Bet ($)	Results W/L (%)	Results W/100 ($)	Risk Measures SD/100 ($)	Risk Measures ROR (%)	Risk Measures N0 (hands)	Rankings DI	Rankings SCORE
<0	39.70	–1.73	1.144	23,23,21	25,25,20	0	0	PA Opt. (1–4)	435	38.84	0.74	28.91	537.70	13.50	34,595	5.38	28.91
0	29.20	–0.20	1.137	23,23,21	25,25,20	0	0	PA Pract. (1–4)	400	42.10	0.74	30.93	576.88	15.60	34,797	5.36	28.76
1	8.72	0.50	1.137	39,39,39	50	0	0	PA Opt. (1–6)	433	44.97	1.04	46.90	685.10	13.50	21,306	6.85	46.94
2	7.47	1.02	1.138	79,79,79	75	0,127,113	0,125,125	PA Pract. (1–6)	400	48.73	1.04	50.51	738.58	15.60	21,369	6.84	46.78
3	3.36	1.46	1.140	92,113,113	100,125,125	180,127,113	200,125,125	PA Opt. (1–8)	466	47.03	1.25	58.94	768.01	13.40	16,957	7.68	58.98
4	4.50	2.05	1.142	92,138,157	100,150,150	180,157,158	200,150,150	PA Pract. (1–8)	500	45.98	1.24	56.98	744.16	12.70	17,056	7.66	58.63
5	2.28	2.72	1.144	92,138,171	100,150,160	207,207,208	200	BC Opt. (1–2)	56	225.59	3.16	106.35	1031.26	13.20	9,394	10.31	106.35
6	1.71	3.35	1.145	92,138,171	100,150,160	255,255,256	300	BC Pract. (1–2)	50	241.60	3.14	113.15	1101.40	15.40	9,467	10.27	105.53
7	0.68	3.83	1.143	92,138,171	100,150,160	293,293,294	300	BC Opt. (1–3)	79	181.62	2.78	113.19	1063.94	13.30	8,839	10.64	113.19
8	0.96	4.43	1.139	92,138,171	100,150,160	341,341,342	300	BC Pract. (1–3)	80	179.75	2.79	112.24	1058.38	13.40	8,890	10.61	112.47
9	0.57	5.15	1.134	92,138,171	100,150,160	360,381,400	400,375,500	BC Opt. (1–4)	88	178.16	2.91	116.16	1077.80	13.60	8,601	10.78	116.16
>9	0.86	6.69	1.124	92,138,171	100,150,160	360,381,452	400,375,500	BC Pract. (1–4)	80	187.75	2.93	123.07	1148.25	15.40	8,697	10.72	114.87

TABLE 10.75
78/104 S17

TC	TC Freq. (%)	EV W/L (%)	Std. Dev.	Play-All Bets Optimal ($)	Play-All Bets Practical ($)	Back-Count 32% 24% 16% Optimal ($)	Back-Count 32% 24% 16% Practical ($)	Bet Spread	Kelly Bank (units)	Avg. Bet ($)	Results W/L (%)	Results W/100 ($)	Risk Measures SD/100 ($)	Risk Measures ROR (%)	Risk Measures N0 (hands)	Rankings DI	Rankings SCORE
<0	40.34	–1.87	1.146	28,28,26	25	0	0	PA Opt. (1–4)	354	47.29	0.90	42.75	653.76	13.50	23,395	6.54	42.74
0	27.39	–0.18	1.137	28,28,26	25	0	0	PA Pract. (1–4)	400	43.20	0.89	38.61	592.43	11.00	23,555	6.52	42.48
1	8.20	0.52	1.137	40,40,40	50	0	0	PA Opt. (1–6)	362	53.68	1.25	67.02	818.49	13.50	14,926	8.18	66.99
2	7.61	1.07	1.138	83,83,83	75	0,141,121	0,150,125	PA Pract. (1–6)	400	50.68	1.24	62.73	768.73	11.90	15,020	8.16	66.58
3	3.05	1.46	1.140	113,113,113	100,125,125	199,141,121	200,150,125	PA Opt. (1–8)	387	56.00	1.49	83.16	911.80	13.50	12,028	9.12	83.14
4	4.83	2.09	1.142	113,161,161	100,150,200	199,161,161	200	PA Pract. (1–8)	400	57.38	1.49	85.73	946.43	14.70	12,189	9.06	82.05
5	2.28	2.78	1.144	113,166,207	100,150,200	212,212,212	200	BC Opt. (1–2)	50	250.26	3.50	144.17	1200.68	13.50	6,928	12.01	144.17
6	2.13	3.37	1.143	113,166,207	100,150,200	258,258,258	300	BC Pract. (1–2)	50	251.80	3.49	144.65	1208.00	13.60	6,979	11.97	143.38
7	0.62	3.83	1.143	113,166,207	100,150,200	293,293,293	300	BC Opt. (1–3)	71	202.87	3.12	152.41	1234.46	13.30	6,558	12.35	152.40
8	1.33	4.40	1.140	113,166,207	100,150,200	339,339,339	300	BC Pract. (1–3)	67	217.95	3.08	161.72	1315.95	15.20	6,623	12.29	151.03
9	0.60	5.29	1.133	113,166,207	100,150,200	399,412,412	400,450,500	BC Opt. (1–4)	82	198.14	3.29	156.68	1251.74	13.50	6,388	12.52	156.68
>9	1.62	6.77	1.123	113,166,207	100,150,200	399,424,486	400,450,500	BC Pract. (1–4)	80	211.50	3.26	165.64	1330.63	15.20	6,454	12.45	154.98

TABLE 10.76
52/104 S17 DAS

				Play-All Bets		Back-Count 27% 18% 11%					Results		Risk Measures			Rankings	
TC	TC Freq. (%)	EV W/L (%)	Std. Dev.	Optimal ($)	Practical ($)	Optimal ($)	Practical ($)	Bet Spread	Kelly Bank (units)	Avg. Bet ($)	W/L (%)	W/100 ($)	SD/100 ($)	ROR (%)	N0 (hands)	DI	SCORE
<0	37.29	−1.33	1.150	18,17,16	20,15,15	0	0	PA Opt. (1–4)	566	29.68	0.58	17.21	414.89	13.50	58,079	4.15	17.21
0	35.98	−0.08	1.144	18,17,16	20,15,15	0	0	PA Pract. (1–4)	500	33.42	0.58	19.38	467.22	16.90	58,121	4.15	17.21
1	8.70	0.60	1.143	46,46,46	50,40,40	0,0,75	0,0,75	PA Opt. (1–6)	579	33.85	0.80	27.03	519.94	13.50	37,009	5.20	27.03
2	7.27	1.09	1.143	71,83,83	80,75,75	121,105,83	125,100,75	PA Pract. (1–6)	667	29.61	0.80	23.70	455.97	10.20	37,015	5.20	27.03
3	3.37	1.58	1.145	71,104,121	80,90,120	121,120,120	125	PA Opt. (1–8)	628	35.16	0.96	33.63	579.39	13.50	29,797	5.79	33.57
4	3.84	2.16	1.147	71,104,128	80,90,120	164,164,164	200	PA Pract. (1–8)	667	32.84	0.96	31.56	545.04	11.90	29,825	5.79	33.52
5	1.29	2.77	1.147	71,104,128	80,90,120	210,210,211	200	BC Opt. (1–2)	83	151.66	2.15	58.81	766.90	13.40	16,980	7.67	58.81
6	0.93	3.25	1.150	71,104,128	80,90,120	242,246,246	200	BC Pract. (1–2)	80	159.50	2.13	61.21	803.63	15.00	17,215	7.62	58.01
7	0.53	3.77	1.149	71,104,128	80,90,120	242,286,286	250,300,300	BC Opt. (1–3)	96	149.74	2.27	61.40	783.63	13.40	16,277	7.84	61.40
8	0.42	4.42	1.144	71,104,128	80,90,120	242,314,299	250,300,300	BC Pract. (1–3)	100	153.10	2.25	62.22	800.30	14.30	16,555	7.78	60.46
9	0.16	4.90	1.141	71,104,128	80,90,120	242,314,299	250,300,300	BC Opt. (1–4)	134	119.18	1.98	63.04	793.96	13.40	15,851	7.94	63.04
>9	0.22	5.98	1.133	71,104,128	80,90,120	242,314,299	250,300,300	BC Pract. (1–4)	133	120.90	1.99	64.15	814.05	14.40	16,116	7.88	62.09

TABLE 10.77
62/104 S17 DAS

				Play-All Bets		Back-Count 30% 21% 13%					Results		Risk Measures			Rankings	
TC	TC Freq. (%)	EV W/L (%)	Std. Dev.	Optimal ($)	Practical ($)	Optimal ($)	Practical ($)	Bet Spread	Kelly Bank (units)	Avg. Bet ($)	W/L (%)	W/100 ($)	SD/100 ($)	ROR (%)	N0 (hands)	DI	SCORE
<0	38.71	−1.47	1.152	23,22,20	25,20,20	0	0	PA Opt. (1–4)	440	38.57	0.75	28.88	537.08	13.60	34,671	5.37	28.85
0	31.75	−0.07	1.144	23,22,20	25,20,20	0	0	PA Pract. (1–4)	400	40.98	0.74	30.16	563.73	15.00	34,901	5.35	28.63
1	8.69	0.61	1.143	47,47,47	50	0	0	PA Opt. (1–6)	456	43.19	1.01	43.49	659.62	13.50	22,979	6.60	43.51
2	7.39	1.11	1.144	85,85,85	75	138,119,105	150,125,100	PA Pract. (1–6)	500	40.14	1.01	40.34	612.18	11.60	23,030	6.59	43.41
3	3.78	1.62	1.146	91,123,123	100,120,125	138,124,124	150,125,125	PA Opt. (1–8)	502	44.47	1.19	53.02	728.20	13.50	18,851	7.28	53.03
4	4.18	2.18	1.148	91,132,159	100,120,160	166,166,166	200	PA Pract. (1–8)	500	44.20	1.19	52.64	723.52	13.30	18,892	7.27	52.93
5	1.85	2.80	1.149	91,132,159	100,120,160	212,212,213	200	BC Opt. (1–2)	73	172.47	2.45	87.94	937.85	13.30	11,388	9.38	87.95
6	1.36	3.35	1.150	91,132,159	100,120,160	253,254,254	300	BC Pract. (1–2)	67	190.80	2.44	97.13	1038.30	16.30	11,411	9.36	87.52
7	0.77	3.98	1.148	91,132,159	100,120,160	275,302,303	300	BC Opt. (1–3)	84	170.09	2.61	92.67	962.64	13.50	10,793	9.63	92.66
8	0.67	4.51	1.144	91,132,159	100,120,160	275,345,346	300	BC Pract. (1–3)	80	180.50	2.59	97.34	1016.50	15.10	10,897	9.58	91.69
9	0.35	5.11	1.139	91,132,159	100,120,160	275,357,395	300,375,400	BC Opt. (1–4)	95	167.53	2.70	94.39	971.56	13.60	10,602	9.72	94.39
>9	0.51	6.41	1.129	91,132,159	100,120,160	275,357,419	300,375,400	BC Pract. (1–4)	100	172.60	2.68	96.55	1000.20	14.40	10,721	9.65	93.19

TABLE 10.78
70/104 S17 DAS

TC	TC Freq. (%)	EV W/L (%)	Std. Dev.	Play-All Bets Optimal ($)	Play-All Bets Practical ($)	Back-Count 31% 22% 15% Optimal ($)	Back-Count 31% 22% 15% Practical ($)	Bet Spread	Kelly Bank (units)	Avg. Bet ($)	Results W/L (%)	Results W/100 ($)	Risk Measures SD/100 ($)	Risk Measures ROR (%)	Risk Measures N0 (hands)	Rankings DI	Rankings SCORE
<0	39.71	-1.60	1.153	27,26,24	25	0	0	PA Opt. (1-4)	367	46.08	0.88	40.73	638.28	13.50	24,551	6.38	40.74
0	29.20	-0.05	1.145	27,26,24	25	0	0	PA Pract. (1-4)	400	43.95	0.89	39.16	615.63	12.60	24,695	6.36	40.47
1	8.71	0.65	1.144	50,50,50	50	0	0	PA Opt. (1-6)	386	51.02	1.18	60.31	776.43	13.50	16,592	7.76	60.28
2	7.47	1.17	1.144	90,90,90	100	153,133,118	150,125,125	PA Pract. (1-6)	400	50.58	1.18	59.74	770.43	13.30	16,626	7.75	60.12
3	3.35	1.62	1.146	109,124,123	100,125,125	153,133,124	150,125,125	PA Opt. (1-8)	420	52.55	1.39	72.77	853.20	13.40	13,739	8.53	72.80
4	4.50	2.21	1.148	109,156,168	100,150,200	168,168,168	200	PA Pract. (1-8)	400	56.35	1.40	78.95	926.95	15.80	13,785	8.52	72.55
5	2.28	2.88	1.150	109,156,190	100,150,200	218,218,218	200	BC Opt. (1-2)	65	192.14	2.73	117.32	1083.08	13.50	8,524	10.83	117.32
6	1.71	3.50	1.150	109,156,190	100,150,200	264,264,265	300	BC Pract. (1-2)	67	197.10	2.72	120.06	1112.70	14.10	8,598	10.79	116.42
7	0.68	3.98	1.148	109,156,190	100,150,200	302,302,302	300	BC Opt. (1-3)	75	190.01	2.93	124.44	1115.48	13.50	8,027	11.16	124.44
8	0.96	4.58	1.144	109,156,190	100,150,200	307,350,351	300,375,400	BC Pract. (1-3)	80	193.13	2.93	126.44	1138.75	14.10	8,104	11.10	123.27
9	0.57	5.27	1.138	109,156,190	100,150,200	307,399,407	300,375,500	BC Opt. (1-4)	85	186.61	3.04	127.13	1127.56	13.40	7,864	11.28	127.13
>9	0.86	6.80	1.127	109,156,190	100,150,200	307,399,470	300,375,500	BC Pract. (1-4)	80	202.13	3.06	138.56	1233.13	16.00	7,913	11.24	126.25

TABLE 10.79
78/104 S17 DAS

TC	TC Freq. (%)	EV W/L (%)	Std. Dev.	Play-All Bets Optimal ($)	Play-All Bets Practical ($)	Back-Count 32% 24% 16% Optimal ($)	Back-Count 32% 24% 16% Practical ($)	Bet Spread	Kelly Bank (units)	Avg. Bet ($)	Results W/L (%)	Results W/100 ($)	Risk Measures SD/100 ($)	Risk Measures ROR (%)	Risk Measures N0 (hands)	Rankings DI	Rankings SCORE
<0	40.35	-1.75	1.155	33,31,28	25	0	0	PA Opt. (1-4)	307	54.59	1.04	56.90	754.67	13.50	17,566	7.55	56.95
0	27.40	-0.03	1.145	33,31,28	25	0	0	PA Pract. (1-4)	400	45.10	1.05	47.18	631.00	9.30	17,891	7.48	55.91
1	8.20	0.67	1.144	51,51,51	50	0	0	PA Opt. (1-6)	325	59.96	1.38	82.87	910.06	13.50	12,074	9.10	82.82
2	7.61	1.22	1.145	93,93,93	100	0,147,126	0,150,125	PA Pract. (1-6)	400	52.55	1.38	72.44	800.13	10.30	12,197	9.05	81.97
3	3.05	1.62	1.146	123,124,123	100,125,125	206,147,126	200,150,125	PA Opt. (1-8)	356	61.52	1.62	99.43	997.33	13.40	10,053	9.97	99.47
4	4.82	2.26	1.147	130,172,171	100,150,200	206,171,171	200	PA Pract. (1-8)	400	59.25	1.63	96.47	973.43	13.00	10,181	9.91	98.21
5	2.28	2.94	1.150	130,185,222	100,150,200	222,222,222	200	BC Opt. (1-2)	49	259.10	3.65	155.62	1247.45	13.10	6,425	12.47	155.62
6	2.13	3.52	1.148	130,185,224	100,150,200	267,267,267	300	BC Pract. (1-2)	50	251.80	3.64	150.88	1212.80	12.70	6,468	12.44	154.77
7	0.62	3.98	1.148	130,185,224	100,150,200	302,302,302	300	BC Opt. (1-3)	68	211.41	3.26	166.07	1288.59	13.40	6,019	12.89	166.06
8	1.33	4.56	1.144	130,185,224	100,150,200	348,348,348	300	BC Pract. (1-3)	67	217.95	3.23	169.57	1321.20	14.00	6,076	12.83	164.72
9	0.60	5.42	1.137	130,185,224	100,150,200	412,419,419	400,450,500	BC Opt. (1-4)	79	206.18	3.43	169.93	1303.61	13.50	5,884	13.04	169.94
>9	1.62	6.90	1.126	130,185,224	100,150,200	412,441,503	400,450,500	BC Pract. (1-4)	80	211.50	3.41	173.23	1335.75	14.20	5,944	12.97	168.19

TABLE 10.80
52/104 S17 LS

TC	TC Freq. (%)	EV W/L (%)	Std. Dev.	Play-All Bets Optimal ($)	Play-All Bets Practical ($)	Back-Count 27% 18% 11% Optimal ($)	Back-Count 27% 18% 11% Practical ($)	Bet Spread	Kelly Bank (units)	Avg. Bet ($)	Results W/L (%)	Results W/100 ($)	Risk Measures SD/100 ($)	Risk Measures ROR (%)	Risk Measures N0 (hands)	Rankings DI	Rankings SCORE
<0	37.29	–1.48	1.134	18,19,17	20,20,15	0	0	PA Opt. (1–4)	545	30.67	0.57	17.56	418.97	13.50	56,950	4.19	17.55
0	35.99	–0.15	1.125	18,19,17	20,20,15	0	0	PA Pract. (1–4)	500	33.42	0.57	19.13	456.56	15.90	57,019	4.19	17.55
1	8.70	0.58	1.119	46,46,46	50,50,40	0	0	PA Opt. (1–6)	539	36.01	0.81	29.21	540.74	13.50	34,200	5.41	29.24
2	7.27	1.11	1.117	73,89,89	80,100,100	135,117,104	125,125,100	PA Pract. (1–6)	500	39.18	0.81	31.90	590.12	16.00	34,222	5.41	29.23
3	3.37	1.64	1.117	73,111,132	80,120,120	135,132,132	125	PA Opt. (1–8)	576	37.84	0.99	37.22	610.46	13.50	26,825	6.10	37.27
4	3.84	2.29	1.117	73,111,139	80,120,120	184,184,184	200	PA Pract. (1–8)	667	34.64	0.99	34.44	567.51	11.70	27,153	6.07	36.83
5	1.29	2.98	1.115	73,111,139	80,120,120	240,240,240	200	BC Opt. (1–2)	74	169.91	2.29	70.07	837.12	13.50	14,294	8.37	70.07
6	0.92	3.52	1.119	73,111,139	80,120,120	270,282,282	250,300,300	BC Pract. (1–2)	80	162.00	2.28	66.61	798.75	12.30	14,373	8.34	69.56
7	0.52	4.09	1.118	73,111,139	80,120,120	270,328,328	250,300,300	BC Opt. (1–3)	85	167.93	2.43	73.60	857.94	13.50	13,596	8.58	73.60
8	0.42	4.82	1.113	73,111,139	80,120,120	270,352,390	250,375,400	BC Pract. (1–3)	80	171.50	2.42	74.75	874.50	14.10	13,687	8.55	73.05
9	0.16	5.40	1.110	73,111,139	80,120,120	270,352,415	250,375,400	BC Opt. (1–4)	96	164.93	2.52	74.88	865.33	13.60	13,371	8.65	74.88
>9	0.22	6.61	1.102	73,111,139	80,120,120	270,352,415	250,375,400	BC Pract. (1–4)	100	162.50	2.52	73.82	856.10	13.30	13,457	8.62	74.35

TABLE 10.81
62/104 S17 LS

TC	TC Freq. (%)	EV W/L (%)	Std. Dev.	Play-All Bets Optimal ($)	Play-All Bets Practical ($)	Back-Count 30% 21% 13% Optimal ($)	Back-Count 30% 21% 13% Practical ($)	Bet Spread	Kelly Bank (units)	Avg. Bet ($)	Results W/L (%)	Results W/100 ($)	Risk Measures SD/100 ($)	Risk Measures ROR (%)	Risk Measures N0 (hands)	Rankings DI	Rankings SCORE
<0	38.69	–1.63	1.135	24,24,22	25,25,20	0	0	PA Opt. (1–4)	414	40.73	0.76	30.79	554.80	13.50	32,462	5.55	30.78
0	31.76	–0.14	1.125	24,24,22	25,25,20	0	0	PA Pract. (1–4)	400	42.83	0.76	32.51	586.30	15.10	32,544	5.55	30.75
1	8.70	0.58	1.120	46,46,46	50	0	0	PA Opt. (1–6)	417	46.60	1.04	48.54	696.59	13.50	20,613	6.97	48.53
2	7.39	1.12	1.118	90,90,90	100	154,134,118	150,125,125	PA Pract. (1–6)	400	48.60	1.03	50.21	721.78	14.50	20,673	6.96	48.39
3	3.79	1.68	1.118	97,135,134	100,125,125	154,135,135	150,125,125	PA Opt. (1–8)	453	48.42	1.25	60.29	776.78	13.40	16,574	7.77	60.34
4	4.18	2.31	1.117	97,144,176	100,150,160	185,185,185	200	PA Pract. (1–8)	500	46.04	1.23	56.81	733.50	12.10	16,665	7.75	59.99
5	1.84	3.00	1.118	97,144,176	100,150,160	240,240,240	200	BC Opt. (1–2)	65	193.46	2.61	105.26	1025.96	13.40	9,511	10.26	105.26
6	1.36	3.63	1.119	97,144,176	100,150,160	290,290,290	300	BC Pract. (1–2)	67	190.65	2.60	103.48	1011.15	13.00	9,544	10.23	104.74
7	0.77	4.29	1.117	97,144,176	100,150,160	308,344,345	300	BC Opt. (1–3)	75	190.97	2.81	111.69	1056.89	13.30	8,955	10.57	111.69
8	0.67	4.92	1.113	97,144,176	100,150,160	308,397,398	300,375,500	BC Pract. (1–3)	80	182.75	2.79	106.44	1010.50	12.30	9,024	10.53	110.95
9	0.35	5.63	1.108	97,144,176	100,150,160	308,402,459	300,375,500	BC Opt. (1–4)	85	188.14	2.91	114.15	1068.39	13.30	8,763	10.68	114.15
>9	0.51	7.06	1.099	97,144,176	100,150,160	308,402,472	300,375,500	BC Pract. (1–4)	80	191.88	2.94	117.47	1107.00	14.60	8,876	10.61	112.61

TABLE 10.82
70/104 S17 LS

TC	TC Freq. (%)	EV W/L (%)	Std. Dev.	Play-All Bets Optimal ($)	Play-All Bets Practical ($)	Back-Count 31% 22% 15% Optimal ($)	Back-Count 31% 22% 15% Practical ($)	Bet Spread	Kelly Bank (units)	Avg. Bet ($)	Results W/L (%)	Results W/100 ($)	Risk Measures SD/100 ($)	Risk Measures ROR (%)	Risk Measures N0 (hands)	Rankings DI	Rankings SCORE
<0	39.69	−1.77	1.137	29,29,27	25	0	0	PA Opt. (1–4)	340	49.26	0.90	44.51	667.39	13.50	22,443	6.67	44.54
0	29.20	−0.13	1.126	29,29,27	25	0	0	PA Pract. (1–4)	400	43.98	0.91	39.97	601.55	10.90	22,645	6.64	44.15
1	8.72	0.62	1.121	49,49,49	50	0	0	PA Opt. (1–6)	349	55.50	1.23	68.34	826.88	13.40	14,623	8.27	68.37
2	7.47	1.19	1.118	95,95,95	100	0,150,133	0,150,125	PA Pract. (1–6)	400	50.60	1.23	62.03	751.90	11.10	14,696	8.25	68.05
3	3.36	1.68	1.118	118,134,135	100,125,125	212,150,135	200,150,125	PA Opt. (1–8)	377	57.57	1.46	83.94	916.20	13.40	11,911	9.16	83.94
4	4.50	2.33	1.118	118,172,186	100,150,200	212,186,186	200	PA Pract. (1–8)	400	56.38	1.47	82.67	904.00	13.10	11,956	9.15	83.63
5	2.28	3.06	1.119	118,172,212	100,150,200	245,245,245	200	BC Opt. (1–2)	47	265.75	3.56	140.97	1187.38	13.50	7,098	11.87	140.97
6	1.71	3.76	1.119	118,172,212	100,150,200	300,301,301	300	BC Pract. (1–2)	50	248.00	3.57	132.16	1115.00	11.80	7,114	11.85	140.48
7	0.68	4.29	1.117	118,172,212	100,150,200	344,344,345	300	BC Opt. (1–3)	67	214.10	3.14	150.51	1226.89	13.30	6,643	12.27	150.51
8	0.96	4.96	1.113	118,172,212	100,150,200	401,401,401	400,450,500	BC Pract. (1–3)	67	213.15	3.14	149.91	1226.55	13.40	6,700	12.22	149.38
9	0.57	5.78	1.108	118,172,212	100,150,200	425,450,472	400,450,500	BC Opt. (1–4)	75	210.17	3.28	154.33	1242.36	13.40	6,480	12.42	154.34
>9	0.86	7.48	1.096	118,172,212	100,150,200	425,450,533	400,450,500	BC Pract. (1–4)	80	206.38	3.32	153.52	1245.00	13.70	6,578	12.33	152.04

TABLE 10.83
78/104 S17 LS

TC	TC Freq. (%)	EV W/L (%)	Std. Dev.	Play-All Bets Optimal ($)	Play-All Bets Practical ($)	Back-Count 32% 24% 16% Optimal ($)	Back-Count 32% 24% 16% Practical ($)	Bet Spread	Kelly Bank (units)	Avg. Bet ($)	Results W/L (%)	Results W/100 ($)	Risk Measures SD/100 ($)	Risk Measures ROR (%)	Risk Measures N0 (hands)	Rankings DI	Rankings SCORE
<0	40.34	−1.92	1.138	35,34,31	25	0	0	PA Opt. (1–4)	282	58.81	1.08	63.25	795.28	13.50	15,812	7.95	63.25
0	27.39	−0.11	1.126	35,34,31	25	0	0	PA Pract. (1–4)	400	45.10	1.07	48.45	616.40	7.80	16,186	7.86	61.78
1	8.20	0.63	1.121	51,51,51	50	0	0	PA Opt. (1–6)	293	65.54	1.45	94.95	974.19	13.40	10,535	9.74	94.91
2	7.61	1.23	1.119	98,98,98	100	0,166,143	0,200,150	PA Pract. (1–6)	400	52.58	1.44	75.49	780.68	8.30	10,692	9.67	93.51
3	3.05	1.68	1.118	135,135,135	100,125,125	234,166,143	200,200,150	PA Opt. (1–8)	318	67.67	1.71	115.73	1075.56	13.40	8,643	10.76	115.68
4	4.83	2.35	1.118	142,189,189	100,150,200	234,189,189	200	PA Pract. (1–8)	400	59.28	1.71	101.26	949.10	10.50	8,787	10.67	113.83
5	2.28	3.12	1.119	142,205,250	100,150,200	249,249,249	200	BC Opt. (1–2)	43	293.83	3.92	189.65	1377.22	13.20	5,269	13.77	189.65
6	2.13	3.77	1.118	142,205,252	100,150,200	302,302,302	300	BC Pract. (1–2)	50	260.00	3.94	168.73	1229.00	10.60	5,301	13.73	188.50
7	0.62	4.29	1.117	142,205,252	100,150,200	344,344,344	300	BC Opt. (1–3)	60	238.05	3.51	200.87	1417.34	13.50	4,979	14.17	200.87
8	1.33	4.91	1.113	142,205,252	100,150,200	396,396,396	400,500,500	BC Pract. (1–3)	50	262.60	3.52	222.10	1580.00	16.70	5,056	14.06	197.62
9	0.60	5.94	1.106	142,205,252	100,150,200	468,485,485	400,500,500	BC Opt. (1–4)	70	232.66	3.69	206.40	1436.59	13.40	4,844	14.37	206.39
>9	1.62	7.55	1.096	142,205,252	100,150,200	468,498,571	400,600,600	BC Pract. (1–4)	67	240.30	3.71	214.86	1504.35	14.60	4,905	14.28	204.00

TABLE 10.84
52/104 S17 DAS LS

TC	TC Freq. (%)	EV W/L (%)	Std. Dev.	Play-All Bets Optimal ($)	Play-All Bets Practical ($)	Back-Count 27% 18% 11% Optimal ($)	Back-Count 27% 18% 11% Practical ($)	Bet Spread	Kelly Bank (units)	Avg. Bet ($)	Results W/L (%)	Results W/100 ($)	Risk Measures SD/100 ($)	Risk Measures ROR (%)	Risk Measures N0 (hands)	Rankings DI	Rankings SCORE
<0	37.29	−1.36	1.142	23,22,20	25,20,20	0	0	PA Opt. (1–4)	442	37.80	0.71	26.97	519.38	13.50	37,084	5.19	26.98
0	35.99	−0.01	1.133	23,22,20	25,20,20	0	0	PA Pract. (1–4)	400	40.68	0.71	29.03	560.35	15.70	37,271	5.18	26.84
1	8.70	0.72	1.126	57,57,57	50	0,101,89	0,100,100	PA Opt. (1–6)	462	42.01	0.95	39.99	632.14	13.50	25,034	6.32	39.96
2	7.27	1.26	1.124	90,100,100	100	143,101,100	150,100,100	PA Pract. (1–6)	500	39.18	0.96	37.48	593.44	11.90	25,070	6.32	39.89
3	3.37	1.79	1.123	90,130,142	100,120,150	143,142,142	150	PA Opt. (1–8)	504	43.22	1.12	48.45	695.61	13.50	20,672	6.96	48.39
4	3.83	2.44	1.123	90,130,159	100,120,160	194,194,194	200	PA Pract. (1–8)	500	43.14	1.13	48.68	700.38	13.70	20,700	6.95	48.31
5	1.29	3.11	1.121	90,130,159	100,120,160	247,247,248	200	BC Opt. (1–2)	70	179.23	2.44	78.65	886.84	13.40	12,707	8.87	78.65
6	0.93	3.68	1.124	90,130,159	100,120,160	286,292,292	300	BC Pract. (1–2)	67	183.00	2.42	79.81	902.55	13.90	12,792	8.84	78.18
7	0.53	4.25	1.123	90,130,159	100,120,160	286,303,338	300	BC Opt. (1–3)	99	143.18	2.15	82.39	907.73	13.50	12,138	9.08	82.39
8	0.42	4.98	1.118	90,130,159	100,120,160	286,303,356	300,300,400	BC Pract. (1–3)	100	142.40	2.14	81.50	900.10	13.30	12,197	9.05	81.98
9	0.16	5.52	1.114	90,130,159	100,120,160	286,303,356	300,300,400	BC Opt. (1–4)	112	141.28	2.24	84.73	920.52	13.60	11,802	9.20	84.73
>9	0.22	6.71	1.105	90,130,159	100,120,160	286,303,356	300,300,400	BC Pract. (1–4)	100	145.40	2.21	85.95	938.00	14.10	11,896	9.16	83.96

TABLE 10.85
62/104 S17 DAS LS

TC	TC Freq. (%)	EV W/L (%)	Std. Dev.	Play-All Bets Optimal ($)	Play-All Bets Practical ($)	Back-Count 30% 21% 13% Optimal ($)	Back-Count 30% 21% 13% Practical ($)	Bet Spread	Kelly Bank (units)	Avg. Bet ($)	Results W/L (%)	Results W/100 ($)	Risk Measures SD/100 ($)	Risk Measures ROR (%)	Risk Measures N0 (hands)	Rankings DI	Rankings SCORE
<0	38.70	−1.51	1.144	29,27,24	25	0	0	PA Opt. (1–4)	348	48.17	0.90	43.09	656.59	13.50	23,193	6.57	43.11
0	31.76	0.00	1.133	29,27,24	25	0	0	PA Pract. (1–4)	400	42.80	0.90	38.65	589.60	10.80	23,271	6.56	42.98
1	8.70	0.73	1.127	57,57,57	50	0,0,103	0,0,100	PA Opt. (1–6)	368	52.85	1.18	62.29	789.38	13.50	16,052	7.89	62.31
2	7.39	1.28	1.124	101,101,101	100	162,140,103	200,150,100	PA Pract. (1–6)	400	49.53	1.19	58.97	747.95	12.10	16,085	7.88	62.16
3	3.78	1.84	1.124	115,145,145	100,150,150	162,145,145	200,150,150	PA Opt. (1–8)	412	53.83	1.38	74.51	863.19	13.50	13,421	8.63	74.51
4	4.18	2.46	1.123	115,163,194	100,150,200	195,195,195	200	PA Pract. (1–8)	400	54.38	1.39	75.81	878.80	14.00	13,441	8.63	74.42
5	1.84	3.15	1.124	115,163,194	100,150,200	249,249,249	200	BC Opt. (1–2)	62	202.93	2.76	116.74	1080.52	13.30	8,576	10.80	116.74
6	1.36	3.78	1.124	115,163,194	100,150,200	299,299,299	300	BC Pract. (1–2)	50	228.60	2.73	130.18	1216.20	17.10	8,725	10.70	114.57
7	0.77	4.46	1.122	115,163,194	100,150,200	324,354,354	400	BC Opt. (1–3)	71	200.28	2.95	122.93	1108.77	13.50	8,133	11.09	122.93
8	0.67	5.06	1.117	115,163,194	100,150,200	324,405,405	400,450,400	BC Pract. (1–3)	67	205.35	2.96	126.50	1145.40	14.30	8,205	11.04	121.97
9	0.35	5.74	1.113	115,163,194	100,150,200	324,421,412	400,450,400	BC Opt. (1–4)	97	162.09	2.62	125.45	1120.08	13.40	7,967	11.20	125.45
>9	0.51	7.19	1.102	115,163,194	100,150,200	324,421,412	400,450,400	BC Pract. (1–4)	100	159.30	2.63	123.59	1107.10	13.20	8,023	11.16	124.63

TABLE 10.86
70/104 S17 DAS LS

TC	TC Freq. (%)	EV W/L (%)	Std. Dev.	Play-All Bets Optimal ($)	Play-All Bets Practical ($)	Back-Count 31% 22% 15% Optimal ($)	Back-Count 31% 22% 15% Practical ($)	Bet Spread	Kelly Bank (units)	Avg. Bet ($)	Results W/L (%)	Results W/100 ($)	Risk Measures SD/100 ($)	Risk Measures ROR (%)	Risk Measures N0 (hands)	Rankings DI	Rankings SCORE
<0	39.70	−1.65	1.146	34,32,29	25	0	0	PA Opt. (1–4)	296	56.60	1.05	59.17	769.09	13.50	16,899	7.69	59.15
0	29.21	0.01	1.133	34,32,29	25	0	0	PA Pract. (1–4)	400	43.98	1.05	46.29	604.85	7.90	17,085	7.65	58.57
1	8.72	0.77	1.128	60,60,60	50	0	0	PA Opt. (1–6)	317	61.65	1.37	84.52	919.31	13.50	11,831	9.19	84.51
2	7.47	1.34	1.125	106,106,106	100	180,156,138	200,150,150	PA Pract. (1–6)	400	51.43	1.38	70.89	774.80	9.40	11,951	9.15	83.70
3	3.36	1.84	1.124	135,145,145	100,150,150	180,156,146	200,150,150	PA Opt. (1–8)	346	63.20	1.59	100.48	1002.69	13.40	9,949	10.03	100.54
4	4.50	2.49	1.124	135,189,197	100,150,200	197,197,197	200	PA Pract. (1–8)	400	57.20	1.62	92.40	924.35	11.40	10,008	10.00	99.93
5	2.28	3.21	1.125	135,189,231	100,150,200	254,254,254	300	BC Opt. (1–2)	56	225.34	3.06	154.46	1242.88	13.20	6,473	12.43	154.46
6	1.71	3.92	1.124	135,189,231	100,150,200	310,310,310	300	BC Pract. (1–2)	50	245.20	3.06	168.12	1355.20	15.90	6,492	12.41	153.92
7	0.68	4.46	1.122	135,189,231	100,150,200	354,354,355	400	BC Opt. (1–3)	64	222.80	3.28	163.80	1279.82	13.40	6,106	12.80	163.80
8	0.96	5.11	1.117	135,189,231	100,150,200	360,409,410	400,450,500	BC Pract. (1–3)	67	226.35	3.30	167.43	1312.20	14.00	6,145	12.76	162.80
9	0.57	5.90	1.112	135,189,231	100,150,200	360,468,478	400,450,500	BC Opt. (1–4)	72	218.99	3.41	167.31	1293.48	13.50	5,977	12.94	167.32
>9	0.86	7.61	1.100	135,189,231	100,150,200	360,468,553	400,450,600	BC Pract. (1–4)	67	235.50	3.44	181.36	1405.65	15.60	6,008	12.90	166.46

TABLE 10.87
78/104 S17 DAS LS

TC	TC Freq. (%)	EV W/L (%)	Std. Dev.	Play-All Bets Optimal ($)	Play-All Bets Practical ($)	Back-Count 32% 24% 16% Optimal ($)	Back-Count 32% 24% 16% Practical ($)	Bet Spread	Kelly Bank (units)	Avg. Bet ($)	Results W/L (%)	Results W/100 ($)	Risk Measures SD/100 ($)	Risk Measures ROR (%)	Risk Measures N0 (hands)	Rankings DI	Rankings SCORE
<0	40.34	−1.80	1.147	40,37,34	50,25,25	0	0	PA Opt. (1–4)	251	66.24	1.22	80.60	897.50	13.50	12,420	8.97	80.55
0	27.40	0.03	1.133	40,37,34	50,25,25	0	0	PA Pract. (1–4)	200	76.95	1.21	93.42	1049.55	18.30	12,627	8.90	79.22
1	8.20	0.78	1.128	62,62,62	50	0	0	PA Opt. (1–6)	268	71.94	1.58	113.75	1067.01	13.40	8,785	10.67	113.85
2	7.61	1.39	1.125	110,110,110	100	200,172,147	200,200,150	PA Pract. (1–6)	400	53.33	1.59	84.58	801.35	7.10	8,978	10.55	111.40
3	3.05	1.84	1.124	145,145,145	150	200,172,147	200,200,150	PA Opt. (1–8)	295	73.46	1.84	135.06	1162.17	13.40	7,403	11.62	135.06
4	4.82	2.52	1.124	160,199,199	200,150,200	200,199,199	200	PA Pract. (1–8)	400	60.03	1.86	111.36	967.53	9.20	7,550	11.51	132.48
5	2.28	3.28	1.124	160,224,259	200,150,200	259,259,259	300	BC Opt. (1–2)	50	248.74	3.39	203.00	1424.70	13.40	4,926	14.25	203.00
6	2.13	3.93	1.123	160,224,271	200,150,200	312,312,312	300	BC Pract. (1–2)	50	253.00	3.39	206.50	1451.40	13.90	4,945	14.23	202.40
7	0.62	4.46	1.122	160,224,271	200,150,200	354,354,354	400	BC Opt. (1–3)	58	246.83	3.65	216.69	1472.01	13.40	4,613	14.72	216.69
8	1.33	5.06	1.118	160,224,271	200,150,200	400,405,405	400,500,500	BC Pract. (1–3)	50	274.60	3.66	241.81	1647.80	16.60	4,644	14.67	215.33
9	0.60	6.06	1.110	160,224,271	200,150,200	400,491,491	400,500,500	BC Opt. (1–4)	68	240.72	3.83	221.72	1489.04	13.30	4,510	14.89	221.72
>9	1.62	7.68	1.099	160,224,271	200,150,200	400,516,589	400,600,600	BC Pract. (1–4)	67	252.45	3.85	233.72	1574.55	14.80	4,539	14.84	220.32

TABLE 10.88
52/104 H17

TC	TC Freq. (%)	EV W/L (%)	Std. Dev.	Play-All Bets Optimal ($)	Play-All Bets Practical ($)	Back-Count 27% 18% 11% Optimal ($)	Back-Count 27% 18% 11% Practical ($)	Bet Spread	Kelly Bank (units)	Avg. Bet ($)	Results W/L (%)	Results W/100 ($)	Risk Measures SD/100 ($)	Risk Measures ROR (%)	Risk Measures N0 (hands)	Rankings DI	Rankings SCORE
<0	37.27	−1.72	1.146	7,10,10	5,10,10	0	0	PA Opt. (1–4)	1,343	12.65	0.25	3.12	176.68	13.60	320,778	1.77	3.12
0	35.98	−0.42	1.140	7,10,10	5,10,10	0	0	PA Pract. (1–4)	2,000	8.15	0.25	2.00	113.94	4.60	324,530	1.76	3.08
1	8.69	0.27	1.139	21,21,21	10,20,20	0	0	PA Opt. (1–6)	1,024	19.55	0.49	9.54	308.63	13.60	105,069	3.09	9.53
2	7.27	0.76	1.141	30,59,59	20,50,50	104,90,79	100,100,75	PA Pract. (1–6)	1,000	19.17	0.48	9.17	298.03	12.70	105,629	3.08	9.46
3	3.38	1.27	1.144	30,59,84	20,60,80	104,97,97	100	PA Opt. (1–8)	954	22.82	0.65	14.78	384.57	13.50	67,567	3.85	14.79
4	3.84	1.85	1.149	30,59,84	20,60,80	140,140,140	150	PA Pract. (1–8)	1,000	21.33	0.65	13.78	358.57	11.70	67,709	3.84	14.76
5	1.30	2.47	1.150	30,59,84	20,60,80	187,187,187	200	BC Opt. (1–2)	96	130.43	1.85	43.66	660.76	13.60	22,915	6.61	43.66
6	0.93	2.98	1.150	30,59,84	20,60,80	207,225,225	200	BC Pract. (1–2)	100	130.40	1.86	43.82	664.10	13.70	22,989	6.60	43.55
7	0.53	3.53	1.148	30,59,84	20,60,80	207,268,268	200,300,300	BC Opt. (1–3)	111	129.01	2.00	46.60	682.67	13.60	21,466	6.83	46.60
8	0.43	4.19	1.144	30,59,84	20,60,80	207,269,316	200,300,300	BC Pract. (1–3)	100	137.90	1.99	49.52	727.20	15.30	21,582	6.81	46.38
9	0.17	4.64	1.140	30,59,84	20,60,80	207,269,316	200,300,300	BC Opt. (1–4)	127	126.87	2.09	47.81	691.50	13.40	20,908	6.91	47.81
>9	0.22	5.77	1.132	30,59,84	20,60,80	207,269,316	200,300,300	BC Pract. (1–4)	133	127.80	2.09	48.14	698.18	13.90	21,025	6.89	47.53

TABLE 10.89
62/104 H17

TC	TC Freq. (%)	EV W/L (%)	Std. Dev.	Play-All Bets Optimal ($)	Play-All Bets Practical ($)	Back-Count 30% 21% 13% Optimal ($)	Back-Count 30% 21% 13% Practical ($)	Bet Spread	Kelly Bank (units)	Avg. Bet ($)	Results W/L (%)	Results W/100 ($)	Risk Measures SD/100 ($)	Risk Measures ROR (%)	Risk Measures N0 (hands)	Rankings DI	Rankings SCORE
<0	38.69	−1.86	1.147	13,15,15	15	0	0	PA Opt. (1–4)	789	21.39	0.42	9.04	300.83	13.50	110,375	3.01	9.05
0	31.75	−0.41	1.140	13,15,15	15	0	0	PA Pract. (1–4)	667	24.09	0.42	9.99	336.36	17.10	113,365	2.97	8.82
1	8.69	0.28	1.140	22,22,22	20	0	0	PA Opt. (1–6)	673	28.81	0.70	20.12	448.71	13.50	49,661	4.49	20.13
2	7.38	0.79	1.141	51,60,60	50	0,105,93	0,100,100	PA Pract. (1–6)	667	28.13	0.70	19.66	439.23	13.00	49,888	4.48	20.03
3	3.78	1.31	1.144	51,89,100	60,90,100	154,105,100	150,100,100	PA Opt. (1–8)	682	31.82	0.90	28.52	534.06	13.50	35,070	5.34	28.52
4	4.18	1.87	1.150	51,89,117	60,90,120	154,141,141	150	PA Pract. (1–8)	667	31.41	0.90	28.35	531.83	13.40	35,191	5.33	28.41
5	1.85	2.51	1.151	51,89,117	60,90,120	189,189,189	200	BC Opt. (1–2)	65	191.55	2.72	70.22	838.07	13.50	14,253	8.38	70.22
6	1.36	3.09	1.150	51,89,117	60,90,120	234,234,234	200	BC Pract. (1–2)	67	187.50	2.72	68.76	822.15	12.90	14,322	8.36	69.96
7	0.77	3.73	1.147	51,89,117	60,90,120	283,283,284	300	BC Opt. (1–3)	95	150.15	2.35	73.55	857.62	13.50	13,577	8.58	73.55
8	0.67	4.28	1.143	51,89,117	60,90,120	307,315,328	300	BC Pract. (1–3)	100	147.50	2.34	72.09	842.70	13.10	13,661	8.55	73.19
9	0.35	4.89	1.139	51,89,117	60,90,120	307,315,371	300,300,400	BC Opt. (1–4)	108	147.62	2.46	75.72	870.14	13.40	13,210	8.70	75.72
>9	0.51	6.22	1.130	51,89,117	60,90,120	307,315,371	300,300,400	BC Pract. (1–4)	100	151.60	2.43	76.99	887.50	14.10	13,285	8.67	75.25

TABLE 10.90
70/104 H17

TC	TC Freq. (%)	EV W/L (%)	Std. Dev.	Play-All Bets Optimal ($)	Play-All Bets Practical ($)	Back-Count 31% 22% 15% Optimal ($)	Back-Count 31% 22% 15% Practical ($)	Bet Spread	Kelly Bank (units)	Avg. Bet ($)	Results W/L (%)	Results W/100 ($)	Risk Measures SD/100 ($)	Risk Measures ROR (%)	Risk Measures N0 (hands)	Rankings DI	Rankings SCORE
<0	39.71	–2.00	1.148	17,19,18	15,20,20	0	0	PA Opt. (1–4)	601	28.40	0.56	16.03	400.25	13.50	62,469	4.00	16.02
0	29.21	–0.40	1.141	17,19,18	15,20,20	0	0	PA Pract. (1–4)	667	25.94	0.56	14.59	364.68	11.10	62,433	4.00	16.01
1	8.71	0.32	1.141	25,25,25	25	0	0	PA Opt. (1–6)	532	36.37	0.87	31.77	563.75	13.50	31,459	5.64	31.78
2	7.46	0.85	1.142	65,65,65	60,50,50	0,0,106	0,125,100	PA Pract. (1–6)	500	36.90	0.87	32.25	574.84	14.20	31,752	5.61	31.48
3	3.36	1.31	1.144	67,100,100	60,100,100	170,148,106	200,125,100	PA Opt. (1–8)	559	39.23	1.10	43.23	657.70	13.50	23,118	6.58	43.26
4	4.49	1.89	1.150	67,113,143	60,120,150	170,148,143	200,150,150	PA Pract. (1–8)	500	41.08	1.11	45.46	694.48	15.10	23,338	6.55	42.85
5	2.28	2.58	1.152	67,113,143	60,120,160	194,194,194	200	BC Opt. (1–2)	59	214.05	3.04	96.84	984.12	13.40	10,329	9.84	96.84
6	1.71	3.23	1.150	67,113,143	60,120,160	244,244,244	200	BC Pract. (1–2)	50	230.20	3.00	102.84	1053.80	15.60	10,508	9.76	95.24
7	0.68	3.73	1.147	67,113,143	60,120,160	283,283,284	300	BC Opt. (1–3)	67	210.64	3.23	101.48	1007.38	13.60	9,857	10.07	101.48
8	0.96	4.34	1.143	67,113,143	60,120,160	332,332,332	300	BC Pract. (1–3)	80	172.13	2.63	101.12	1008.00	13.60	9,936	10.03	100.63
9	0.57	5.06	1.138	67,113,143	60,120,160	340,390,390	400,375,400	BC Opt. (1–4)	95	167.46	2.80	104.67	1023.05	13.30	9,555	10.23	104.67
>9	0.86	6.60	1.127	67,113,143	60,120,160	340,445,423	400,375,400	BC Pract. (1–4)	100	161.70	2.78	100.56	985.90	12.50	9,604	10.20	104.04

TABLE 10.91
78/104 H17

TC	TC Freq. (%)	EV W/L (%)	Std. Dev.	Play-All Bets Optimal ($)	Play-All Bets Practical ($)	Back-Count 32% 24% 16% Optimal ($)	Back-Count 32% 24% 16% Practical ($)	Bet Spread	Kelly Bank (units)	Avg. Bet ($)	Results W/L (%)	Results W/100 ($)	Risk Measures SD/100 ($)	Risk Measures ROR (%)	Risk Measures N0 (hands)	Rankings DI	Rankings SCORE
<0	40.34	–2.14	1.150	22,23,23	20,25,25	0	0	PA Opt. (1–4)	452	36.96	0.72	26.74	517.03	13.50	37,422	5.17	26.73
0	27.41	–0.38	1.141	22,23,23	20,25,25	0	0	PA Pract. (1–4)	500	32.56	0.72	23.35	452.96	10.30	37,663	5.15	26.57
1	8.20	0.34	1.141	26,26,26	25	0	0	PA Opt. (1–6)	428	45.04	1.09	48.83	698.97	13.50	20,475	6.99	48.86
2	7.60	0.90	1.142	69,69,69	50	0,0,116	0,0,125	PA Pract. (1–6)	400	45.93	1.10	50.29	722.98	14.60	20,659	6.96	48.39
3	3.05	1.31	1.144	89,100,100	80,100,100	190,159,116	200,150,125	PA Opt. (1–8)	442	48.34	1.33	64.42	802.69	13.50	15,518	8.03	64.43
4	4.82	1.94	1.149	89,140,147	80,150,150	190,159,147	200,150,150	PA Pract. (1–8)	400	50.23	1.35	67.76	847.98	15.10	15,666	7.99	63.85
5	2.28	2.64	1.152	89,140,181	80,150,200	199,199,199	200	BC Opt. (1–2)	53	239.18	3.38	132.88	1152.69	13.20	7,533	11.53	132.89
6	2.13	3.25	1.149	89,140,181	80,150,200	246,246,246	200	BC Pract. (1–2)	50	238.80	3.37	132.47	1153.40	13.50	7,589	11.49	131.91
7	0.62	3.73	1.147	89,140,181	80,150,200	283,283,283	300	BC Opt. (1–3)	63	234.49	3.63	139.84	1182.48	13.40	7,144	11.83	139.84
8	1.33	4.31	1.144	89,140,181	80,150,200	329,329,329	300	BC Pract. (1–3)	67	221.70	3.64	132.80	1126.65	12.10	7,203	11.79	138.94
9	0.61	5.21	1.137	89,140,181	80,150,200	381,403,403	400,450,500	BC Opt. (1–4)	86	188.06	3.17	143.43	1197.64	13.50	6,970	11.98	143.43
>9	1.61	6.71	1.127	89,140,181	80,150,200	381,477,463	400,450,500	BC Pract. (1–4)	80	192.50	3.18	147.33	1237.50	14.50	7,051	11.91	141.75

TABLE 10.92
52/104 H17 DAS

TC	TC Freq. (%)	EV W/L (%)	Std. Dev.	Play-All Bets Optimal ($)	Play-All Bets Practical ($)	Back-Count 27% 18% 11% Optimal ($)	Back-Count 27% 18% 11% Practical ($)	Bet Spread	Kelly Bank (units)	Avg. Bet ($)	Results W/L (%)	Results W/100 ($)	Risk Measures SD/100 ($)	Risk Measures ROR (%)	Risk Measures N0 (hands)	Rankings DI	Rankings SCORE
<0	37.27	-1.59	1.154	12,13,13	10,15,15	0	0	PA Opt. (1-4)	846	20.00	0.39	7.91	280.89	13.60	126,674	2.81	7.89
0	35.98	-0.28	1.148	12,13,13	10,15,15	0	0	PA Pract. (1-4)	1,000	16.72	0.39	6.59	234.69	9.10	126,829	2.81	7.88
1	8.69	0.43	1.146	32,32,32	25	0	0	PA Opt. (1-6)	771	25.77	0.63	16.15	401.93	13.50	61,908	4.02	16.15
2	7.26	0.92	1.147	47,70,70	40,50,50	112,97,85	100,100,75	PA Pract. (1-6)	667	26.51	0.61	16.18	411.12	14.70	64,523	3.94	15.50
3	3.38	1.42	1.150	47,78,102	40,90,100	112,108,108	100	PA Opt. (1-8)	786	28.18	0.79	22.28	471.75	13.50	44,942	4.72	22.25
4	3.84	2.00	1.155	47,78,102	40,90,120	150,150,151	150	PA Pract. (1-8)	667	29.07	0.78	22.65	489.59	15.10	46,722	4.63	21.41
5	1.30	2.62	1.155	47,78,102	40,90,120	196,196,197	200	BC Opt. (1-2)	89	140.22	2.01	50.91	713.50	13.60	19,637	7.13	50.91
6	0.93	3.16	1.155	47,78,102	40,90,120	224,237,237	200	BC Pract. (1-2)	100	130.40	2.02	47.54	667.40	11.80	19,742	7.12	50.74
7	0.53	3.69	1.153	47,78,102	40,90,120	224,278,278	200,300,300	BC Opt. (1-3)	103	138.61	2.15	53.76	733.21	13.60	18,629	7.33	53.76
8	0.43	4.35	1.148	47,78,102	40,90,120	224,290,330	200,300,300	BC Pract. (1-3)	100	137.90	2.15	53.46	730.70	13.50	18,654	7.32	53.52
9	0.17	4.82	1.145	47,78,102	40,90,120	224,290,341	200,300,300	BC Opt. (1-4)	117	136.13	2.23	54.84	740.57	13.60	18,246	7.41	54.85
>9	0.22	5.92	1.136	47,78,102	40,90,120	224,290,341	200,300,300	BC Pract. (1-4)	133	127.80	2.24	51.79	701.55	12.20	18,325	7.38	54.50

TABLE 10.93
62/104 H17 DAS

TC	TC Freq. (%)	EV W/L (%)	Std. Dev.	Play-All Bets Optimal ($)	Play-All Bets Practical ($)	Back-Count 30% 21% 13% Optimal ($)	Back-Count 30% 21% 13% Practical ($)	Bet Spread	Kelly Bank (units)	Avg. Bet ($)	Results W/L (%)	Results W/100 ($)	Risk Measures SD/100 ($)	Risk Measures ROR (%)	Risk Measures N0 (hands)	Rankings DI	Rankings SCORE
<0	38.70	-1.73	1.156	17,18,17	15,20,15	0	0	PA Opt. (1-4)	595	28.76	0.57	16.38	404.77	13.50	61,088	4.05	16.38
0	31.75	-0.27	1.148	17,18,17	15,20,15	0	0	PA Pract. (1-4)	667	25.26	0.57	14.45	357.26	10.40	61,168	4.04	16.35
1	8.69	0.44	1.147	33,33,33	25	0	0	PA Opt. (1-6)	564	35.03	0.84	29.44	542.95	13.50	33,933	5.43	29.48
2	7.38	0.94	1.148	67,71,72	60,50,50	0,112,98	125,100,100	PA Pract. (1-6)	500	35.36	0.83	29.51	551.92	14.40	35,003	5.35	28.59
3	3.78	1.47	1.151	67,106,111	60,100,100	162,112,111	125,100,100	PA Opt. (1-8)	592	37.39	1.03	38.69	621.76	13.50	25,864	6.22	38.66
4	4.18	2.03	1.155	67,106,135	60,120,120	162,152,152	150	PA Pract. (1-8)	667	31.85	1.05	33.28	536.88	9.90	26,017	6.20	38.43
5	1.85	2.66	1.157	67,106,135	60,120,120	199,199,199	200	BC Opt. (1-2)	62	200.98	2.88	77.99	883.17	13.40	12,841	8.83	77.99
6	1.36	3.26	1.155	67,106,135	60,120,120	244,244,244	200	BC Pract. (1-2)	80	155.38	2.30	74.59	846.38	12.40	12,863	8.81	77.66
7	0.77	3.88	1.152	67,106,135	60,120,120	293,293,293	250,300,300	BC Opt. (1-3)	90	159.26	2.50	83.06	911.37	13.30	12,047	9.11	83.06
8	0.67	4.43	1.148	67,106,135	60,120,120	323,335,337	250,300,300	BC Pract. (1-3)	100	147.50	2.50	76.96	846.50	11.60	12,086	9.09	82.65
9	0.35	5.06	1.143	67,106,135	60,120,120	323,335,388	250,300,400	BC Opt. (1-4)	102	156.89	2.60	85.07	922.32	13.40	11,767	9.22	85.07
>9	0.51	6.37	1.133	67,106,135	60,120,120	323,335,393	250,300,400	BC Pract. (1-4)	100	151.70	2.59	82.00	891.50	12.60	11,820	9.20	84.60

TABLE 10.94
70/104 H17 DAS

				Play-All Bets		Back-Count 31% 22% 15%					Results		Risk Measures			Rankings	
TC	TC Freq. (%)	EV W/L (%)	Std. Dev.	Optimal ($)	Practical ($)	Optimal ($)	Practical ($)	Bet Spread	Kelly Bank (units)	Avg. Bet ($)	W/L (%)	W/100 ($)	SD/100 ($)	ROR (%)	N0 (hands)	DI	SCORE
<0	39.72	−1.87	1.157	21,22,20	20	0	0	PA Opt. (1–4)	475	35.97	0.71	25.37	504.14	13.50	39,346	5.04	25.42
0	29.22	−0.25	1.148	21,22,20	20	0	0	PA Pract. (1–4)	500	31.62	0.70	22.01	441.60	10.40	40,218	4.98	24.85
1	8.71	0.48	1.148	36,36,36	25	0	0	PA Opt. (1–6)	459	42.72	1.01	43.37	658.31	13.50	23,082	6.58	43.34
2	7.46	1.01	1.148	76,77,77	50	0,126,112	0,125,100	PA Pract. (1–6)	500	36.90	1.02	37.80	577.64	10.30	23,352	6.54	42.82
3	3.35	1.47	1.151	84,111,111	80,100,100	178,126,112	200,125,100	PA Opt. (1–8)	491	45.08	1.23	55.57	745.60	13.50	17,986	7.46	55.59
4	4.49	2.05	1.156	84,131,154	80,120,150	178,154,154	200,150,150	PA Pract. (1–8)	500	41.06	1.26	51.68	697.72	11.90	18,227	7.41	54.86
5	2.28	2.74	1.157	84,131,163	80,120,160	205,205,205	200	BC Opt. (1–2)	56	223.77	3.19	106.48	1031.79	13.50	9,394	10.32	106.48
6	1.71	3.40	1.155	84,131,163	80,120,160	255,255,255	300	BC Pract. (1–2)	50	241.60	3.17	114.14	1110.60	15.60	9,458	10.28	105.61
7	0.68	3.88	1.152	84,131,163	80,120,160	293,293,293	300	BC Opt. (1–3)	80	179.63	2.81	112.98	1062.92	13.30	8,847	10.63	112.98
8	0.96	4.50	1.148	84,131,163	80,120,160	342,342,342	300	BC Pract. (1–3)	80	179.88	2.81	113.06	1067.00	13.60	8,896	10.60	112.28
9	0.57	5.23	1.142	84,131,163	80,120,160	356,377,401	400,375,400	BC Opt. (1–4)	89	176.33	2.95	116.05	1077.29	13.60	8,610	10.77	116.05
>9	0.86	6.75	1.131	84,131,163	80,120,160	356,377,447	400,375,400	BC Pract. (1–4)	100	169.30	2.96	112.13	1045.80	12.80	8,703	10.72	114.96

TABLE 10.95
78/104 H17 DAS

				Play-All Bets		Back-Count 32% 24% 16%					Results		Risk Measures			Rankings	
TC	TC Freq. (%)	EV W/L (%)	Std. Dev.	Optimal ($)	Practical ($)	Optimal ($)	Practical ($)	Bet Spread	Kelly Bank (units)	Avg. Bet ($)	W/L (%)	W/100 ($)	SD/100 ($)	ROR (%)	N0 (hands)	DI	SCORE
<0	40.34	−2.02	1.159	26,26,25	25	0	0	PA Opt. (1–4)	379	44.46	0.87	38.59	621.25	13.50	25,907	6.21	38.59
0	27.42	−0.23	1.148	26,26,25	25	0	0	PA Pract. (1–4)	400	43.18	0.86	37.04	597.98	12.60	26,049	6.19	38.37
1	8.20	0.50	1.148	38,38,38	50	0	0	PA Opt. (1–6)	382	51.35	1.23	62.91	793.20	13.50	15,891	7.93	62.91
2	7.60	1.06	1.149	80,80,80	75	0,140,121	0,150,125	PA Pract. (1–6)	400	49.88	1.21	60.25	761.10	12.40	15,958	7.92	62.67
3	3.05	1.47	1.151	106,111,111	100	198,140,121	200,150,125	PA Opt. (1–8)	401	54.13	1.47	79.35	890.97	13.40	12,594	8.91	79.38
4	4.81	2.10	1.155	106,157,157	100,150,150	198,157,157	200,150,150	PA Pract. (1–8)	400	54.15	1.45	78.42	881.60	13.20	12,637	8.89	79.12
5	2.28	2.81	1.157	106,157,199	100,150,200	210,210,210	200	BC Opt. (1–2)	51	248.68	3.54	144.64	1202.68	13.20	6,908	12.03	144.64
6	2.13	3.41	1.154	106,157,199	100,150,200	256,256,256	300	BC Pract. (1–2)	50	251.80	3.53	146.04	1217.80	13.80	6,957	11.99	143.82
7	0.62	3.88	1.152	106,157,199	100,150,200	293,293,293	300	BC Opt. (1–3)	71	201.03	3.16	152.63	1235.50	13.50	6,549	12.35	152.63
8	1.33	4.47	1.148	106,157,199	100,150,200	339,339,339	300	BC Pract. (1–3)	67	207.90	3.16	158.00	1283.25	14.40	6,601	12.31	151.60
9	0.61	5.38	1.141	106,157,199	100,150,200	395,413,413	400,450,500	BC Opt. (1–4)	83	196.64	3.32	157.09	1253.26	13.30	6,368	12.53	157.08
>9	1.61	6.86	1.130	106,157,199	100,150,200	395,421,483	400,450,500	BC Pract. (1–4)	80	201.38	3.35	162.02	1298.13	14.50	6,421	12.48	155.77

TABLE 10.96

52/104 H17 LS

				Play-All Bets		Back-Count 27% 18% 11%					Results		Risk Measures			Rankings	
TC	TC Freq. (%)	EV W/L (%)	Std. Dev.	Optimal ($)	Practical ($)	Optimal ($)	Practical ($)	Bet Spread	Kelly Bank (units)	Avg. Bet ($)	W/L (%)	W/100 ($)	SD/100 ($)	ROR (%)	N0 (hands)	DI	SCORE
<0	37.28	–1.72	1.137	12,14,14	10,15,15	0	0	PA Opt. (1–4)	803	20.91	0.39	8.21	286.54	13.50	121,916	2.87	8.21
0	36.01	–0.34	1.127	12,14,14	10,15,15	0	0	PA Pract. (1–4)	1,000	16.71	0.39	6.56	228.97	8.20	121,829	2.87	8.21
1	8.69	0.41	1.121	32,32,32	25	0	0	PA Opt. (1–6)	707	27.78	0.65	17.96	423.60	13.50	55,747	4.24	17.94
2	7.26	0.95	1.120	50,76,76	40,50,50	126,109,96	125,100,100	PA Pract. (1–6)	667	26.48	0.63	16.64	400.35	12.50	57,921	4.16	17.27
3	3.37	1.49	1.120	50,85,112	40,90,120	126,119,119	125	PA Opt. (1–8)	711	30.70	0.82	25.31	503.15	13.50	39,499	5.03	25.32
4	3.83	2.13	1.123	50,85,112	40,90,120	169,169,169	200	PA Pract. (1–8)	667	29.72	0.83	24.56	495.48	13.50	40,667	4.96	24.58
5	1.29	2.82	1.123	50,85,112	40,90,120	224,224,224	200	BC Opt. (1–2)	79	157.91	2.15	61.05	781.37	13.60	16,395	7.81	61.05
6	0.93	3.41	1.123	50,85,112	40,90,120	252,270,271	250,300,300	BC Pract. (1–2)	80	162.00	2.14	62.44	802.38	14.30	16,481	7.78	60.55
7	0.53	3.98	1.120	50,85,112	40,90,120	252,318,318	250,300,300	BC Opt. (1–3)	92	156.18	2.30	64.79	804.93	13.40	15,415	8.05	64.80
8	0.42	4.72	1.116	50,85,112	40,90,120	252,327,380	250,300,400	BC Pract. (1–3)	100	158.20	2.30	65.50	818.80	14.10	15,627	8.00	63.99
9	0.16	5.31	1.112	50,85,112	40,90,120	252,327,384	250,300,400	BC Opt. (1–4)	104	153.40	2.40	66.28	814.13	13.50	15,108	8.14	66.28
>9	0.22	6.48	1.104	50,85,112	40,90,120	252,327,384	250,300,400	BC Pract. (1–4)	100	162.70	2.38	69.79	860.40	15.10	15,195	8.11	65.79

TABLE 10.97

62/104 H17 LS

				Play-All Bets		Back-Count 30% 21% 13%					Results		Risk Measures			Rankings	
TC	TC Freq. (%)	EV W/L (%)	Std. Dev.	Optimal ($)	Practical ($)	Optimal ($)	Practical ($)	Bet Spread	Kelly Bank (units)	Avg. Bet ($)	W/L (%)	W/100 ($)	SD/100 ($)	ROR (%)	N0 (hands)	DI	SCORE
<0	38.69	–1.87	1.138	18,20,19	20	0	0	PA Opt. (1–4)	550	30.80	0.58	18.00	424.21	13.50	55,610	4.24	18.00
0	31.77	–0.33	1.127	18,20,19	20	0	0	PA Pract. (1–4)	500	30.72	0.56	17.27	417.72	13.80	58,436	4.14	17.10
1	8.69	0.41	1.122	33,33,33	25	0	0	PA Opt. (1–6)	508	38.27	0.88	33.74	580.65	13.50	29,671	5.81	33.72
2	7.38	0.97	1.120	73,77,77	50	0,126,111	0,125,100	PA Pract. (1–6)	500	36.12	0.89	31.98	556.66	12.70	30,299	5.74	33.00
3	3.78	1.54	1.121	73,118,122	80,120,125	183,126,122	200,125,125	PA Opt. (1–8)	528	41.16	1.09	44.94	670.39	13.50	22,242	6.70	44.94
4	4.18	2.15	1.124	73,118,152	80,120,160	183,170,170	200	PA Pract. (1–8)	500	40.18	1.11	44.65	672.30	13.80	22,662	6.64	44.11
5	1.85	2.86	1.125	73,118,152	80,120,160	226,226,226	200	BC Opt. (1–2)	55	228.34	3.09	95.02	974.79	13.20	10,528	9.75	95.02
6	1.36	3.51	1.123	73,118,152	80,120,160	278,278,279	300	BC Pract. (1–2)	50	238.60	3.08	98.94	1018.20	14.70	10,578	9.72	94.41
7	0.77	4.19	1.119	73,118,152	80,120,160	334,334,334	300	BC Opt. (1–3)	79	179.95	2.68	100.51	1002.53	13.50	9,950	10.03	100.51
8	0.67	4.81	1.115	73,118,152	80,120,160	367,378,387	400,375,400	BC Pract. (1–3)	80	182.88	2.66	101.31	1014.50	13.90	10,040	9.99	99.72
9	0.35	5.53	1.111	73,118,152	80,120,160	367,378,445	400,375,400	BC Opt. (1–4)	90	177.08	2.80	103.22	1016.01	13.40	9,688	10.16	103.22
>9	0.51	6.96	1.101	73,118,152	80,120,160	367,378,445	400,375,400	BC Pract. (1–4)	100	175.90	2.77	101.71	1006.80	13.40	9,800	10.10	102.05

TABLE 10.98
70/104 H17 LS

TC	TC Freq. (%)	EV W/L (%)	Std. Dev.	Play-All Bets Optimal ($)	Play-All Bets Practical ($)	Back-Count 31% 22% 15% Optimal ($)	Back-Count 31% 22% 15% Practical ($)	Bet Spread	Kelly Bank (units)	Avg. Bet ($)	Results W/L (%)	Results W/100 ($)	Risk Measures SD/100 ($)	Risk Measures ROR (%)	Risk Measures N0 (hands)	Rankings DI	Rankings SCORE
<0	39.71	–2.02	1.139	23,24,23	25	0	0	PA Opt. (1–4)	432	38.95	0.73	28.40	533.39	13.50	35,155	5.33	28.45
0	29.22	–0.32	1.127	23,24,23	25	0	0	PA Pract. (1–4)	400	38.05	0.72	27.26	522.38	13.50	36,748	5.22	27.23
1	8.71	0.45	1.122	36,36,36	25	0	0	PA Opt. (1–6)	411	46.93	1.07	50.08	707.75	13.50	19,965	7.08	50.09
2	7.46	1.03	1.121	82,82,82	50	0,142,126	0,150,125	PA Pract. (1–6)	400	44.65	1.08	48.38	691.28	13.20	20,420	7.00	48.99
3	3.36	1.54	1.121	92,122,122	100,125,125	202,142,126	200,150,125	PA Opt. (1–8)	436	49.78	1.31	65.12	806.75	13.50	15,367	8.07	65.08
4	4.49	2.17	1.124	92,146,172	100,150,200	202,172,172	200	PA Pract. (1–8)	400	50.43	1.35	68.21	855.33	15.40	15,717	7.98	63.60
5	2.28	2.92	1.125	92,146,184	100,150,200	231,231,231	200	BC Opt. (1–2)	50	253.70	3.43	129.55	1138.11	13.20	7,715	11.38	129.55
6	1.71	3.64	1.123	92,146,184	100,150,200	289,289,289	300	BC Pract. (1–2)	50	248.00	3.44	127.09	1118.20	13.00	7,752	11.37	129.17
7	0.68	4.19	1.119	92,146,184	100,150,200	334,334,334	300	BC Opt. (1–3)	70	202.66	3.01	136.53	1168.43	13.60	7,324	11.68	136.53
8	0.96	4.85	1.115	92,146,184	100,150,200	390,390,390	400,450,500	BC Pract. (1–3)	67	213.15	3.00	143.14	1229.70	14.80	7,384	11.64	135.49
9	0.57	5.67	1.110	92,146,184	100,150,200	404,426,460	400,450,500	BC Opt. (1–4)	79	199.13	3.16	140.69	1186.19	13.50	7,111	11.86	140.69
>9	0.86	7.37	1.099	92,146,184	100,150,200	404,426,505	400,450,500	BC Pract. (1–4)	80	206.38	3.19	147.03	1248.25	15.00	7,211	11.78	138.76

TABLE 10.99
78/104 H17 LS

TC	TC Freq. (%)	EV W/L (%)	Std. Dev.	Play-All Bets Optimal ($)	Play-All Bets Practical ($)	Back-Count 32% 24% 16% Optimal ($)	Back-Count 32% 24% 16% Practical ($)	Bet Spread	Kelly Bank (units)	Avg. Bet ($)	Results W/L (%)	Results W/100 ($)	Risk Measures SD/100 ($)	Risk Measures ROR (%)	Risk Measures N0 (hands)	Rankings DI	Rankings SCORE
<0	40.35	–2.17	1.141	29,29,28	25	0	0	PA Opt. (1–4)	343	48.39	0.90	43.71	661.07	13.50	22,877	6.61	43.70
0	27.42	–0.30	1.127	29,29,28	25	0	0	PA Pract. (1–4)	400	39.23	0.90	35.38	539.38	8.80	23,248	6.56	43.02
1	8.20	0.47	1.123	37,37,37	25	0	0	PA Opt. (1–6)	341	56.57	1.29	73.12	855.07	13.50	13,678	8.55	73.12
2	7.60	1.07	1.122	85,85,85	50	0,0,137	0,0,150	PA Pract. (1–6)	400	46.68	1.32	61.37	722.53	9.50	13,859	8.49	72.15
3	3.05	1.54	1.121	116,122,122	100,125,125	224,187,137	200,200,150	PA Opt. (1–8)	356	59.92	1.56	93.29	966.07	13.50	10,718	9.66	93.33
4	4.82	2.20	1.124	116,174,174	100,150,200	224,187,174	200	PA Pract. (1–8)	400	53.38	1.62	86.20	903.00	12.00	10,974	9.55	91.12
5	2.28	2.99	1.125	116,176,225	100,150,200	236,236,236	200	BC Opt. (1–2)	45	281.83	3.79	175.64	1325.25	13.10	5,689	13.25	175.64
6	2.13	3.63	1.122	116,176,225	100,150,200	289,289,289	300	BC Pract. (1–2)	50	259.80	3.81	162.84	1232.00	11.60	5,727	13.22	174.67
7	0.62	4.19	1.119	116,176,225	100,150,200	334,334,334	300	BC Opt. (1–3)	53	276.03	4.07	184.46	1358.23	13.60	5,422	13.58	184.47
8	1.33	4.79	1.116	116,176,225	100,150,200	385,385,385	400,500,500	BC Pract. (1–3)	50	291.40	4.11	196.81	1457.20	15.50	5,483	13.51	182.41
9	0.61	5.82	1.109	116,176,225	100,150,200	449,474,474	400,500,500	BC Opt. (1–4)	73	221.99	3.56	190.14	1378.99	13.40	5,262	13.79	190.14
>9	1.62	7.45	1.098	116,176,225	100,150,200	449,562,547	400,600,600	BC Pract. (1–4)	67	240.30	3.58	206.69	1507.20	15.90	5,317	13.71	188.06

TABLE 10.100
52/104 H17 DAS LS

TC	TC Freq. (%)	EV W/L (%)	Std. Dev.	Play-All Bets Optimal ($)	Play-All Bets Practical ($)	Back-Count 27% 18% 11% Optimal ($)	Back-Count 27% 18% 11% Practical ($)	Bet Spread	Kelly Bank (units)	Avg. Bet ($)	Results W/L (%)	Results W/100 ($)	Risk Measures SD/100 ($)	Risk Measures ROR (%)	Risk Measures N0 (hands)	Rankings DI	Rankings SCORE
<0	37.28	–1.59	1.145	17,17,16	15	0	0	PA Opt. (1–4)	591	28.43	0.54	15.36	391.93	13.50	65,129	3.92	15.36
0	36.01	–0.20	1.134	17,17,16	15	0	0	PA Pract. (1–4)	667	25.28	0.54	13.66	348.56	10.50	65,061	3.92	15.36
1	8.69	0.56	1.128	44,44,44	40	0	0	PA Opt. (1–6)	574	34.14	0.79	26.91	518.81	13.50	37,167	5.19	26.92
2	7.26	1.11	1.126	68,87,87	60,75,75	134,116,102	125,125,100	PA Pract. (1–6)	667	29.60	0.79	23.29	449.06	9.90	37,160	5.19	26.91
3	3.37	1.64	1.127	68,104,129	60,90,120	134,129,130	125	PA Opt. (1–8)	610	36.22	0.97	34.99	591.62	13.50	28,582	5.92	35.00
4	3.83	2.30	1.129	68,104,131	60,90,120	180,180,181	200	PA Pract. (1–8)	667	32.84	0.97	31.73	536.45	11.00	28,592	5.91	34.98
5	1.29	2.99	1.128	68,104,131	60,90,120	235,235,236	200	BC Opt. (1–2)	75	167.78	2.30	69.59	834.21	13.40	14,377	8.34	69.59
6	0.93	3.56	1.128	68,104,131	60,90,120	268,280,281	250,300,300	BC Pract. (1–2)	80	162.00	2.30	67.04	806.63	12.70	14,494	8.31	69.07
7	0.53	4.13	1.125	68,104,131	60,90,120	268,326,327	250,300,300	BC Opt. (1–3)	86	165.84	2.45	73.19	855.46	13.50	13,670	8.56	73.19
8	0.42	4.88	1.120	68,104,131	60,90,120	268,348,390	250,375,400	BC Pract. (1–3)	80	171.63	2.43	75.29	883.25	14.40	13,777	8.52	72.65
9	0.17	5.40	1.116	68,104,131	60,90,120	268,348,410	250,375,400	BC Opt. (1–4)	98	162.79	2.54	74.50	863.15	13.40	13,430	8.63	74.50
>9	0.22	6.61	1.108	68,104,131	60,90,120	268,348,410	250,375,400	BC Pract. (1–4)	100	162.70	2.54	74.38	864.80	13.60	13,511	8.60	73.98

TABLE 10.101
62/104 H17 DAS LS

TC	TC Freq. (%)	EV W/L (%)	Std. Dev.	Play-All Bets Optimal ($)	Play-All Bets Practical ($)	Back-Count 30% 21% 13% Optimal ($)	Back-Count 30% 21% 13% Practical ($)	Bet Spread	Kelly Bank (units)	Avg. Bet ($)	Results W/L (%)	Results W/100 ($)	Risk Measures SD/100 ($)	Risk Measures ROR (%)	Risk Measures N0 (hands)	Rankings DI	Rankings SCORE
<0	38.70	–1.74	1.147	22,23,21	20,25,20	0	0	PA Opt. (1–4)	446	38.25	0.73	27.84	527.86	13.50	35,917	5.28	27.86
0	31.77	–0.19	1.134	22,23,21	20,25,20	0	0	PA Pract. (1–4)	500	35.12	0.73	25.46	483.50	11.30	36,064	5.27	27.74
1	8.69	0.57	1.129	44,44,44	50	0	0	PA Opt. (1–6)	440	44.59	1.02	45.50	674.53	13.50	21,974	6.75	45.50
2	7.38	1.12	1.127	88,88,88	80,100,100	153,133,117	150,125,125	PA Pract. (1–6)	400	48.60	1.01	49.03	728.23	15.70	22,065	6.73	45.34
3	3.78	1.68	1.127	90,132,132	80,125,125	153,133,133	150,125,125	PA Opt. (1–8)	471	46.77	1.23	57.47	757.89	13.50	17,411	7.58	57.44
4	4.17	2.31	1.130	90,136,170	80,150,160	181,181,182	200	PA Pract. (1–8)	500	46.04	1.22	55.95	740.10	12.90	17,491	7.56	57.15
5	1.85	3.02	1.130	90,136,170	80,150,160	236,236,237	200	BC Opt. (1–2)	66	191.37	2.63	104.74	1023.37	13.20	9,556	10.23	104.74
6	1.36	3.66	1.128	90,136,170	80,150,160	288,288,288	300	BC Pract. (1–2)	67	190.80	2.62	104.13	1020.00	13.30	9,601	10.21	104.20
7	0.77	4.34	1.124	90,136,170	80,150,160	305,344,344	300	BC Opt. (1–3)	75	188.85	2.83	111.27	1054.84	13.60	8,988	10.55	111.27
8	0.67	4.95	1.120	90,136,170	80,150,160	305,395,396	300,375,500	BC Pract. (1–3)	80	182.88	2.81	107.16	1019.38	12.60	9,055	10.51	110.49
9	0.35	5.63	1.115	90,136,170	80,150,160	305,398,454	300,375,500	BC Opt. (1–4)	85	186.23	2.93	113.77	1066.61	13.60	8,788	10.67	113.77
>9	0.51	7.11	1.105	90,136,170	80,150,160	305,398,468	300,375,500	BC Pract. (1–4)	80	192.13	2.96	118.32	1116.75	14.90	8,900	10.59	112.25

TABLE 10.102
70/104 H17 DAS LS

TC	TC Freq. (%)	EV W/L (%)	Std. Dev.	Play-All Bets Optimal ($)	Play-All Bets Practical ($)	Back-Count 31% 22% 15% Optimal ($)	Back-Count 31% 22% 15% Practical ($)	Bet Spread	Kelly Bank (units)	Avg. Bet ($)	Results W/L (%)	Results W/100 ($)	Risk Measures SD/100 ($)	Risk Measures ROR (%)	Risk Measures N0 (hands)	Rankings DI	Rankings SCORE
<0	39.71	−1.89	1.148	28,27,25	25	0	0	PA Opt. (1–4)	362	46.58	0.87	40.63	637.73	13.40	24,591	6.38	40.67
0	29.22	−0.18	1.134	28,27,25	25	0	0	PA Pract. (1–4)	400	43.95	0.88	38.57	606.45	12.20	24,716	6.36	40.44
1	8.71	0.60	1.130	47,47,47	50	0	0	PA Opt. (1–6)	366	53.32	1.21	64.44	802.40	13.50	15,524	8.02	64.39
2	7.46	1.19	1.128	93,94,94	100	0,148,132	0,150,125	PA Pract. (1–6)	400	50.55	1.20	60.72	758.05	12.00	15,583	8.01	64.16
3	3.35	1.68	1.127	110,132,132	100,125,125	210,148,132	200,150,125	PA Opt. (1–8)	393	55.63	1.44	80.14	895.04	13.40	12,486	8.95	80.11
4	4.49	2.33	1.130	110,164,183	100,150,200	210,183,183	200	PA Pract. (1–8)	400	56.33	1.45	81.46	911.40	14.00	12,521	8.94	79.88
5	2.28	3.08	1.131	110,164,204	100,150,200	241,241,241	200	BC Opt. (1–2)	48	262.98	3.58	140.21	1184.15	13.30	7,135	11.84	140.21
6	1.71	3.80	1.128	110,164,204	100,150,200	299,299,299	300	BC Pract. (1–2)	50	248.00	3.59	132.76	1123.00	12.10	7,151	11.82	139.75
7	0.68	4.34	1.124	110,164,204	100,150,200	343,343,344	300	BC Opt. (1–3)	67	211.85	3.16	149.63	1223.18	13.60	6,677	12.23	149.63
8	0.96	5.00	1.120	110,164,204	100,150,200	399,399,400	400,450,500	BC Pract. (1–3)	67	213.15	3.16	150.52	1235.10	13.60	6,739	12.19	148.51
9	0.57	5.80	1.114	110,164,204	100,150,200	419,445,468	400,450,500	BC Opt. (1–4)	76	207.89	3.30	153.51	1239.00	13.30	6,518	12.39	153.51
>9	0.86	7.51	1.102	110,164,204	100,150,200	419,445,528	400,450,500	BC Pract. (1–4)	80	206.38	3.34	154.16	1253.50	13.90	6,615	12.30	151.26

TABLE 10.103
78/104 H17 DAS LS

TC	TC Freq. (%)	EV W/L (%)	Std. Dev.	Play-All Bets Optimal ($)	Play-All Bets Practical ($)	Back-Count 32% 24% 16% Optimal ($)	Back-Count 32% 24% 16% Practical ($)	Bet Spread	Kelly Bank (units)	Avg. Bet ($)	Results W/L (%)	Results W/100 ($)	Risk Measures SD/100 ($)	Risk Measures ROR (%)	Risk Measures N0 (hands)	Rankings DI	Rankings SCORE
<0	40.35	−2.05	1.150	34,33,30	25	0	0	PA Opt. (1–4)	298	56.02	1.05	58.68	765.82	13.50	17,043	7.66	58.65
0	27.42	−0.16	1.134	34,33,30	25	0	0	PA Pract. (1–4)	400	45.08	1.05	47.12	621.40	8.70	17,388	7.58	57.51
1	8.20	0.62	1.130	49,49,49	50	0	0	PA Opt. (1–6)	306	63.24	1.43	90.32	950.14	13.50	11,075	9.50	90.28
2	7.60	1.23	1.128	97,97,97	100	0,165,142	0,200,150	PA Pract. (1–6)	400	52.53	1.42	74.31	787.00	9.00	11,219	9.44	89.14
3	3.05	1.68	1.127	132,132,132	100,125,125	232,165,142	200,200,150	PA Opt. (1–8)	329	65.73	1.69	111.19	1054.97	13.40	8,987	10.55	111.30
4	4.81	2.37	1.130	134,186,185	100,150,200	232,185,185	200	PA Pract. (1–8)	400	59.23	1.69	100.21	956.78	11.10	9,118	10.47	109.69
5	2.28	3.15	1.130	134,196,243	100,150,200	246,246,246	200	BC Opt. (1–2)	43	291.07	3.95	188.88	1374.31	13.50	5,290	13.74	188.88
6	2.13	3.80	1.127	134,196,243	100,150,200	299,299,299	300	BC Pract. (1–2)	50	259.80	3.97	169.56	1237.20	10.80	5,321	13.70	187.79
7	0.62	4.34	1.124	134,196,243	100,150,200	343,343,344	300	BC Opt. (1–3)	61	235.81	3.53	200.02	1414.19	13.30	5,003	14.14	200.01
8	1.33	4.96	1.121	134,196,243	100,150,200	395,395,395	400,500,500	BC Pract. (1–3)	50	262.40	3.54	223.03	1589.80	16.90	5,082	14.03	196.81
9	0.61	5.96	1.113	134,196,243	100,150,200	463,482,482	400,500,500	BC Opt. (1–4)	71	230.52	3.71	205.63	1433.98	13.20	4,864	14.34	205.63
>9	1.62	7.60	1.101	134,196,243	100,150,200	463,494,566	400,600,600	BC Pract. (1–4)	67	240.30	3.74	215.78	1513.35	14.90	4,916	14.26	203.30

TABLE 10.104
1–Deck SCORE Summary

		S17	S17 DAS		S17 LS		S17 DAS LS		H17		H17 DAS		H17 LS		H17 DAS LS		H17 D10	
26/52	1–2	**19.09**	**30.37**	11.28	**26.17**	7.08	**39.27**	20.18	**7.69**	–11.40	**15.54**	–3.55	**12.83**	–6.26	**22.66**	3.57	**0.05**	–19.04
	1–3	**39.64**	**54.08**	14.44	**52.22**	12.58	**68.83**	29.19	**23.57**	–16.07	**35.25**	–4.39	**33.99**	–5.65	**47.90**	8.26	**5.82**	–33.82
	1–4	**55.25**	**70.94**	15.69	**71.80**	16.55	**89.83**	34.58	**37.46**	–17.79	**50.76**	–4.49	**51.73**	–3.52	**67.33**	12.08	**14.44**	–40.81
31/52	1–2	**41.41**	**57.34**	15.93	**54.96**	13.55	**73.30**	31.89	**24.16**	–17.25	**36.97**	–4.44	**35.44**	–5.97	**50.76**	9.35	**5.33**	–36.08
		22.32	26.97		28.79		34.03		16.47		21.43		22.61		28.10		5.28	
	1–3	**75.80**	**95.22**	19.42	**98.33**	22.53	**120.61**	44.81	**54.07**	–21.73	**70.98**	–4.82	**73.99**	–1.81	**93.70**	17.90	**23.69**	–52.11
		36.16	41.14		46.11		51.78		30.50		35.73		40.00		45.80		17.87	
	1–4	**101.05**	**121.79**	20.74	**130.19**	29.14	**153.96**	52.91	**77.63**	–23.42	**96.36**	–4.69	**104.07**	3.02	**125.63**	24.58	**41.22**	–59.83
		45.80	50.85		58.39		64.13		40.17		45.60		52.34		58.30		26.78	
35/52	1–2	**49.85**	**67.10**	17.25	**65.35**	15.50	**84.96**	35.11	**29.71**	–20.14	**43.65**	–6.20	**42.55**	–7.30	**59.06**	9.21	**8.08**	–41.77
		30.76	36.73		39.18		45.69		22.02		28.11		29.72		36.40		8.03	
	1–3	**93.26**	**114.50**	21.24	**119.64**	26.38	**143.71**	50.45	**67.53**	–25.73	**86.14**	–7.12	**90.72**	–2.54	**112.23**	18.97	**32.79**	–60.47
		53.62	60.42		67.42		74.88		43.96		50.89		56.73		64.33		26.97	
	1–4	**125.15**	**147.85**	22.70	**159.49**	34.34	**185.20**	60.05	**97.38**	–27.77	**117.96**	–7.19	**128.33**	3.18	**151.87**	26.72	**55.76**	–69.39
		69.90	76.91		87.69		95.37		59.92		67.20		76.60		84.54		41.32	
39/52	1–2	**76.53**	**97.59**	21.06	**99.21**	22.68	**122.93**	46.40	**52.19**	–24.34	**70.14**	–6.39	**71.77**	–4.76	**92.69**	16.16	**21.12**	–55.41
		57.44	67.22		73.04		83.66		44.50		54.60		58.94		70.03		21.07	
	1–3	**136.14**	**161.56**	25.42	**173.88**	37.74	**202.42**	66.28	**106.40**	–29.74	**129.36**	–6.78	**140.31**	4.17	**166.66**	30.52	**60.54**	–75.6
		96.50	107.48		121.66		133.59		82.83		94.11		106.32		118.76		54.72	
	1–4	**178.81**	**205.87**	27.06	**227.63**	48.82	**258.05**	79.24	**147.38**	–31.43	**172.34**	–6.47	**191.80**	12.99	**220.37**	41.56	**94.12**	–84.69
		123.56	134.93		155.83		168.22		109.92		121.58		140.07		153.04		79.68	

Legend: **X** y
z

X = SCORE; y = \$ gain from rules; z = \$ gain from penetration

TABLE 10.105
26/52 S17

				Play-All Bets		Back-Counting					Results		Risk Measures			Rankings	
TC	TC Freq. (%)	EV W/L (%)	Std. Dev.	Optimal ($)	Practical ($)	Optimal ($)	Practical ($)	Bet Spread	Kelly Bank (units)	Avg. Bet ($)	W/L (%)	W/100 ($)	SD/100 ($)	ROR (%)	N0 (hands)	DI	SCORE
<0	33.62	−1.69	1.154	28,32,32	25			PA Opt. (1–2)	355	35.86	0.53	19.09	436.87	13.50	52,398	4.37	19.09
0	39.16	0.10	1.150	28,32,32	25			PA Pract. (1–2)	400	31.80	0.53	16.93	387.43	10.50	52,398	4.37	19.09
1	7.63	0.92	1.152	56,70,70	50			PA Opt. (1–3)	308	48.08	0.82	39.61	629.59	13.50	25,247	6.30	39.64
2	7.50	1.61	1.151	56,97,121	50,75,100			PA Pract. (1–3)	400	36.70	0.82	30.23	480.30	7.20	25,252	6.29	39.62
3								PA Opt. (1–4)	313	53.11	1.04	55.23	743.29	13.50	18,098	7.43	55.25
4	5.28	2.42	1.148	56,97,128	50,75,100			PA Pract. (1–4)	400	41.60	1.05	43.54	586.10	7.90	18,133	7.43	55.18
5	2.44	3.23	1.146	56,97,128	50,75,100			BC Opt. (1–2)									
6	2.16	3.84	1.144	56,97,128	50,75,100			BC Pract. (1–2)									
7								BC Opt. (1–3)									
8	1.16	4.75	1.141	56,97,128	50,75,100			BC Pract. (1–3)									
9	0.25	6.03	1.134	56,97,128	50,75,100			BC Opt. (1–4)									
>9	0.80	6.40	1.130	56,97,128	50,75,100			BC Pract. (1–4)									

TABLE 10.106
31/52 S17

				Play-All Bets		Back-Counting					Results		Risk Measures			Rankings	
TC	TC Freq. (%)	EV W/L (%)	Std. Dev.	Optimal ($)	Practical ($)	Optimal ($)	Practical ($)	Bet Spread	Kelly Bank (units)	Avg. Bet ($)	W/L (%)	W/100 ($)	SD/100 ($)	ROR (%)	N0 (hands)	DI	SCORE
<0	34.19	−1.81	1.155	41,45,44	50			PA Opt. (1–2)	242	53.05	0.78	41.43	643.45	13.50	24,155	6.43	41.41
0	35.35	0.13	1.150	41,45,44	50			PA Pract. (1–2)	200	63.60	0.78	49.54	770.35	18.80	24,171	6.43	41.36
1	6.55	0.92	1.152	70,70,70	75			PA Opt. (1–3)	224	66.70	1.14	75.80	870.62	13.50	13,191	8.71	75.80
2	8.19	1.63	1.151	83,123,123	100,125,125			PA Pract. (1–3)	200	73.50	1.13	83.03	954.40	16.10	13,206	8.70	75.68
3								PA Opt. (1–4)	229	72.53	1.39	101.00	1005.24	13.40	9,898	10.05	101.05
4	6.14	2.50	1.147	83,134,175	100,150,200			PA Pract. (1–4)	200	81.35	1.39	113.18	1126.90	16.70	9,910	10.04	100.87
5	2.09	3.23	1.146	83,134,175	100,150,200			BC Opt. (1–2)									
6	3.18	3.84	1.143	83,134,175	100,150,200			BC Pract. (1–2)									
7								BC Opt. (1–3)									
8	1.98	4.87	1.141	83,134,175	100,150,200			BC Pract. (1–3)									
9	0.22	6.03	1.134	83,134,175	100,150,200			BC Opt. (1–4)									
>9	2.11	7.10	1.126	83,134,175	100,150,200			BC Pract. (1–4)									

TABLE 10.107
35/52 S17

TC	TC Freq. (%)	EV W/L (%)	Std. Dev.	Play-All Bets Optimal ($)	Play-All Bets Practical ($)	Back-Counting Optimal ($)	Back-Counting Practical ($)	Bet Spread	Kelly Bank (units)	Avg. Bet ($)	Results W/L (%)	Results W/100 ($)	Risk Measures SD/100 ($)	Risk Measures ROR (%)	Risk Measures N0 (hands)	Rankings DI	Rankings SCORE
<0	36.14	–2.05	1.158	45,49,48	50			PA Opt. (1–2)	221	58.24	0.86	49.85	706.05	13.50	20,053	7.06	49.85
0	32.64	0.17	1.150	45,49,48	50			PA Pract. (1–2)	200	64.15	0.86	54.94	778.10	16.20	20,051	7.06	49.85
1	5.75	0.92	1.152	70,70,70	75			PA Opt. (1–3)	203	74.01	1.26	93.25	965.72	13.40	10,718	9.66	93.26
2	8.13	1.71	1.151	91,129,129	100,125,125			PA Pract. (1–3)	200	74.85	1.26	94.02	973.80	13.70	10,732	9.65	93.21
3								PA Opt. (1–4)	209	80.54	1.55	125.15	1118.73	13.40	7,992	11.19	125.15
4	6.60	2.63	1.147	91,148,191	100,150,200			PA Pract. (1–4)	200	83.55	1.55	129.62	1159.05	14.40	7,998	11.18	125.06
5	1.84	3.23	1.146	91,148,191	100,150,200			BC Opt. (1–2)									
6	3.29	4.01	1.143	91,148,191	100,150,200			BC Pract. (1–2)									
7								BC Opt. (1–3)									
8	2.47	4.90	1.140	91,148,191	100,150,200			BC Pract. (1–3)									
9	0.19	6.03	1.134	91,148,191	100,150,200			BC Opt. (1–4)									
>9	2.94	7.30	1.126	91,148,191	100,150,200			BC Pract. (1–4)									

TABLE 10.108
39/52 S17

TC	TC Freq. (%)	EV W/L (%)	Std. Dev.	Play-All Bets Optimal ($)	Play-All Bets Practical ($)	Back-Counting Optimal ($)	Back-Counting Practical ($)	Bet Spread	Kelly Bank (units)	Avg. Bet ($)	Results W/L (%)	Results W/100 ($)	Risk Measures SD/100 ($)	Risk Measures ROR (%)	Risk Measures N0 (hands)	Rankings DI	Rankings SCORE
<0	36.33	–2.18	1.160	56,60,58	50			PA Opt. (1–2)	178	72.13	1.06	76.55	874.86	13.50	13,061	8.75	76.53
0	30.96	0.21	1.150	56,60,58	50			PA Pract. (1–2)	200	65.05	1.06	68.91	788.35	10.80	13,092	8.74	76.40
1	5.26	0.92	1.152	70,70,70	75			PA Opt. (1–3)	167	89.69	1.52	136.09	1166.80	13.40	7,347	11.67	136.14
2	7.43	1.71	1.151	112,129,129	100,125,125			PA Pract. (1–3)	200	76.90	1.52	116.49	1000.55	9.70	7,376	11.64	135.56
3								PA Opt. (1–4)	171	96.99	1.84	178.73	1337.19	13.40	5,593	13.37	178.81
4	7.14	2.65	1.146	112,180,202	100,150,200			PA Pract. (1–4)	200	86.90	1.85	160.90	1206.25	10.80	5,620	13.34	177.93
5	1.68	3.23	1.146	112,180,233	100,150,200			BC Opt. (1–2)									
6	3.01	4.01	1.143	112,180,233	100,150,200			BC Pract. (1–2)									
7								BC Opt. (1–3)									
8	3.24	4.68	1.137	112,180,233	100,150,200			BC Pract. (1–3)									
9	0.18	6.03	1.134	112,180,233	100,150,200			BC Opt. (1–4)									
>9	4.76	7.59	1.122	112,180,233	100,150,200			BC Pract. (1–4)									

TABLE 10.109
26/52 S17 DAS

				Play-All Bets		Back-Counting					Results		Risk Measures			Rankings	
TC	TC Freq. (%)	EV W/L (%)	Std. Dev.	Optimal ($)	Practical ($)	Optimal ($)	Practical ($)	Bet Spread	Kelly Bank (units)	Avg. Bet ($)	W/L (%)	W/100 ($)	SD/100 ($)	ROR (%)	N0 (hands)	DI	SCORE
<0	33.62	−1.56	1.163	35,38,36	25,50,25			PA Opt. (1–2)	283	44.99	0.68	30.37	551.07	13.50	32,896	5.51	30.37
0	39.18	0.25	1.156	35,38,36	25,50,25			PA Pract. (1–2)	400	31.80	0.68	21.46	389.50	5.90	32,896	5.51	30.37
1	7.62	1.08	1.158	71,80,80	50,75,50			PA Opt. (1–3)	264	55.88	0.97	54.09	735.43	13.50	18,479	7.35	54.08
2	7.49	1.76	1.157	71,114,132	50,125,100			PA Pract. (1–3)	200	69.60	0.94	65.62	904.50	20.00	19,011	7.26	52.64
3								PA Opt. (1–4)	274	60.15	1.18	71.01	842.22	13.50	14,096	8.42	70.94
4	5.27	2.58	1.153	71,114,146	50,150,100			PA Pract. (1–4)	400	41.60	1.19	49.51	588.83	5.70	14,136	8.41	70.71
5	2.43	3.38	1.151	71,114,146	50,150,100			BC Opt. (1–2)									
6	2.16	3.98	1.148	71,114,146	50,150,100			BC Pract. (1–2)									
7								BC Opt. (1–3)									
8	1.16	4.87	1.145	71,114,146	50,150,100			BC Pract. (1–3)									
9	0.25	6.11	1.137	71,114,146	50,150,100			BC Opt. (1–4)									
>9	0.81	6.51	1.134	71,114,146	50,150,100			BC Pract. (1–4)									

TABLE 10.110
31/52 S17 DAS

				Play-All Bets		Back-Counting					Results		Risk Measures			Rankings	
TC	TC Freq. (%)	EV W/L (%)	Std. Dev.	Optimal ($)	Practical ($)	Optimal ($)	Practical ($)	Bet Spread	Kelly Bank (units)	Avg. Bet ($)	W/L (%)	W/100 ($)	SD/100 ($)	ROR (%)	N0 (hands)	DI	SCORE
<0	34.20	−1.68	1.164	48,50,48	50			PA Opt. (1–2)	206	62.13	0.92	57.33	757.26	13.50	17,447	7.57	57.34
0	35.36	0.28	1.156	48,50,48	50			PA Pract. (1–2)	200	63.60	0.92	58.61	774.35	14.10	17,461	7.57	57.30
1	6.54	1.08	1.158	80,80,80	75			PA Opt. (1–3)	200	74.56	1.28	95.31	975.77	13.40	10,500	9.76	95.22
2	8.19	1.79	1.157	97,134,134	100,125,125			PA Pract. (1–3)	200	73.50	1.27	93.53	958.90	13.00	10,507	9.75	95.14
3								PA Opt. (1–4)	210	79.35	1.54	121.80	1103.61	13.40	8,214	11.04	121.79
4	6.13	2.66	1.152	97,150,191	100,150,200			PA Pract. (1–4)	200	81.35	1.53	124.79	1131.85	14.10	8,225	11.03	121.55
5	2.09	3.38	1.151	97,150,191	100,150,200			BC Opt. (1–2)									
6	3.18	3.99	1.148	97,150,191	100,150,200			BC Pract. (1–2)									
7								BC Opt. (1–3)									
8	1.98	5.00	1.145	97,150,191	100,150,200			BC Pract. (1–3)									
9	0.22	6.11	1.137	97,150,191	100,150,200			BC Opt. (1–4)									
>9	2.11	7.20	1.129	97,150,191	100,150,200			BC Pract. (1–4)									

TABLE 10.111
35/52 S17 DAS

TC	TC Freq. (%)	EV W/L (%)	Std. Dev.	Play-All Bets Optimal ($)	Play-All Bets Practical ($)	Back-Counting Optimal ($)	Back-Counting Practical ($)	Bet Spread	Kelly Bank (units)	Avg. Bet ($)	Results W/L (%)	Results W/100 ($)	Risk Measures SD/100 ($)	Risk Measures ROR (%)	Risk Measures N0 (hands)	Rankings DI	Rankings SCORE
<0	36.15	-1.91	1.167	52,54,52	50			PA Opt. (1–2)	191	67.18	1.00	67.09	819.12	13.50	14,913	8.19	67.10
0	32.66	0.32	1.156	52,54,52	50			PA Pract. (1–2)	200	64.15	1.00	64.06	782.10	12.30	14,910	8.19	67.10
1	5.75	1.08	1.158	80,80,80	75			PA Opt. (1–3)	184	81.77	1.40	114.52	1070.02	13.40	8,731	10.70	114.50
2	8.12	1.87	1.157	105,140,140	100,150,150			PA Pract. (1–3)	200	76.90	1.41	108.46	1015.70	12.10	8,771	10.68	114.04
3								PA Opt. (1–4)	194	87.23	1.70	147.89	1215.96	13.40	6,761	12.16	147.85
4	6.60	2.79	1.152	105,163,207	100,150,200			PA Pract. (1–4)	200	85.55	1.70	145.27	1195.60	13.00	6,775	12.15	147.64
5	1.83	3.38	1.151	105,163,207	100,150,200			BC Opt. (1–2)									
6	3.29	4.16	1.148	105,163,207	100,150,200			BC Pract. (1–2)									
7								BC Opt. (1–3)									
8	2.47	5.03	1.144	105,163,207	100,150,200			BC Pract. (1–3)									
9	0.19	6.11	1.137	105,163,207	100,150,200			BC Opt. (1–4)									
>9	2.93	7.41	1.129	105,163,207	100,150,200			BC Pract. (1–4)									

TABLE 10.112
39/52 S17 DAS

TC	TC Freq. (%)	EV W/L (%)	Std. Dev.	Play-All Bets Optimal ($)	Play-All Bets Practical ($)	Back-Counting Optimal ($)	Back-Counting Practical ($)	Bet Spread	Kelly Bank (units)	Avg. Bet ($)	Results W/L (%)	Results W/100 ($)	Risk Measures SD/100 ($)	Risk Measures ROR (%)	Risk Measures N0 (hands)	Rankings DI	Rankings SCORE
<0	36.34	-2.05	1.169	63,65,62	75,75,50			PA Opt. (1–2)	159	81.07	1.20	97.57	987.90	13.40	10,244	9.88	97.59
0	30.97	0.36	1.156	63,65,62	75,75,50			PA Pract. (1–2)	133	95.55	1.21	115.21	1167.45	18.40	10,270	9.87	97.39
1	5.26	1.08	1.158	80,80,80	75			PA Opt. (1–3)	154	97.31	1.66	161.59	1271.10	13.40	6,190	12.71	161.56
2	7.43	1.87	1.157	126,140,140	150			PA Pract. (1–3)	133	108.83	1.65	178.94	1413.15	16.60	6,236	12.66	160.32
3								PA Opt. (1–4)	160	103.91	1.98	205.91	1434.81	13.40	4,857	14.35	205.87
4	7.14	2.81	1.151	126,195,212	150,200,200			PA Pract. (1–4)	200	88.75	1.99	176.82	1239.15	9.80	4,912	14.27	203.62
5	1.68	3.38	1.151	126,195,250	150,225,200			BC Opt. (1–2)									
6	3.01	4.16	1.148	126,195,250	150,225,200			BC Pract. (1–2)									
7								BC Opt. (1–3)									
8	3.24	4.82	1.141	126,195,250	150,225,200			BC Pract. (1–3)									
9	0.17	6.11	1.137	126,195,250	150,225,200			BC Opt. (1–4)									
>9	4.76	7.70	1.125	126,195,250	150,225,200			BC Pract. (1–4)									

TABLE 10.113
26/52 S17 LS

TC	TC Freq. (%)	EV W/L (%)	Std. Dev.	Play-All Bets Optimal ($)	Play-All Bets Practical ($)	Back-Counting Optimal ($)	Back-Counting Practical ($)	Bet Spread	Kelly Bank (units)	Avg. Bet ($)	Results W/L (%)	Results W/100 ($)	Risk Measures SD/100 ($)	Risk Measures ROR (%)	Risk Measures N0 (hands)	Rankings DI	Rankings SCORE
<0	33.66	−1.76	1.148	33,38,38	25,50,50			PA Opt. (1–2)	299	42.53	0.62	26.17	511.58	13.50	38,192	5.12	26.17
0	39.12	0.13	1.140	33,38,38	25,50,50			PA Pract. (1–2)	400	31.80	0.62	19.57	382.55	6.90	38,192	5.12	26.17
1	7.65	1.00	1.138	67,77,77	50,75,75			PA Opt. (1–3)	262	56.03	0.93	52.21	722.61	13.50	19,155	7.23	52.22
2	7.51	1.76	1.134	67,114,137	50,150,150			PA Pract. (1–3)	200	71.50	0.93	66.49	923.30	21.00	19,277	7.20	51.86
3								PA Opt. (1–4)	266	61.69	1.16	71.87	847.34	13.50	13,925	8.47	71.80
4	5.27	2.66	1.127	67,114,151	50,150,200			PA Pract. (1–4)	200	77.50	1.15	89.11	1057.30	20.20	14,081	8.43	71.04
5	2.43	3.56	1.122	67,114,151	50,150,200			BC Opt. (1–2)									
6	2.16	4.24	1.119	67,114,151	50,150,200			BC Pract. (1–2)									
7								BC Opt. (1–3)									
8	1.16	5.24	1.116	67,114,151	50,150,200			BC Pract. (1–3)									
9	0.25	6.71	1.108	67,114,151	50,150,200			BC Opt. (1–4)									
>9	0.79	7.13	1.104	67,114,151	50,150,200			BC Pract. (1–4)									

TABLE 10.114
31/52 S17 LS

TC	TC Freq. (%)	EV W/L (%)	Std. Dev.	Play-All Bets Optimal ($)	Play-All Bets Practical ($)	Back-Counting Optimal ($)	Back-Counting Practical ($)	Bet Spread	Kelly Bank (units)	Avg. Bet ($)	Results W/L (%)	Results W/100 ($)	Risk Measures SD/100 ($)	Risk Measures ROR (%)	Risk Measures N0 (hands)	Rankings DI	Rankings SCORE
<0	34.20	−1.88	1.150	49,52,51	50			PA Opt. (1–2)	206	62.00	0.89	54.98	741.34	13.50	18,194	7.41	54.96
0	35.34	0.16	1.140	49,52,51	50			PA Pract. (1–2)	200	63.60	0.89	56.32	759.80	14.20	18,213	7.41	54.94
1	6.57	1.00	1.138	77,77,77	75			PA Opt. (1–3)	191	77.36	1.27	98.36	991.61	13.50	10,175	9.92	98.33
2	8.20	1.77	1.134	97,138,138	100,150,150			PA Pract. (1–3)	200	75.55	1.28	96.72	976.85	13.10	10,205	9.90	98.03
3								PA Opt. (1–4)	196	83.73	1.55	130.20	1140.98	13.40	7,684	11.41	130.19
4	6.13	2.74	1.125	97,157,204	100,150,200			PA Pract. (1–4)	200	83.35	1.56	129.87	1139.15	13.40	7,696	11.40	129.98
5	2.09	3.56	1.122	97,157,204	100,150,200			BC Opt. (1–2)									
6	3.17	4.22	1.118	97,157,204	100,150,200			BC Pract. (1–2)									
7								BC Opt. (1–3)									
8	1.98	5.37	1.115	97,157,204	100,150,200			BC Pract. (1–3)									
9	0.22	6.71	1.108	97,157,204	100,150,200			BC Opt. (1–4)									
>9	2.10	7.89	1.099	97,157,204	100,150,200			BC Pract. (1–4)									

TABLE 10.115
35/52 S17 LS

TC	TC Freq. (%)	EV W/L (%)	Std. Dev.	Play-All Bets Optimal ($)	Play-All Bets Practical ($)	Back-Counting Optimal ($)	Back-Counting Practical ($)	Bet Spread	Kelly Bank (units)	Avg. Bet ($)	Results W/L (%)	Results W/100 ($)	Risk Measures SD/100 ($)	Risk Measures ROR (%)	Risk Measures N0 (hands)	Rankings DI	Rankings SCORE
<0	36.15	–2.13	1.152	53,57,55	50			PA Opt. (1–2)	189	67.60	0.97	65.34	808.35	13.40	15,310	8.08	65.35
0	32.62	0.20	1.140	53,57,55	50			PA Pract. (1–2)	200	64.15	0.97	62.03	767.40	12.10	15,295	8.08	65.34
1	5.77	1.00	1.138	77,77,77	75			PA Opt. (1–3)	174	85.35	1.40	119.60	1093.79	13.50	8,357	10.94	119.64
2	8.13	1.84	1.134	106,143,143	100,150,150			PA Pract. (1–3)	200	76.90	1.41	108.52	995.00	11.10	8,410	10.91	118.96
3								PA Opt. (1–4)	181	92.46	1.73	159.51	1262.85	13.30	6,272	12.63	159.49
4	6.60	2.85	1.125	106,172,221	100,150,200			PA Pract. (1–4)	200	85.55	1.72	147.53	1169.90	11.40	6,287	12.61	159.02
5	1.83	3.56	1.122	106,172,221	100,150,200			BC Opt. (1–2)									
6	3.29	4.39	1.118	106,172,221	100,150,200			BC Pract. (1–2)									
7								BC Opt. (1–3)									
8	2.47	5.36	1.114	106,172,221	100,150,200			BC Pract. (1–3)									
9	0.19	6.71	1.108	106,172,221	100,150,200			BC Opt. (1–4)									
>9	2.94	8.07	1.099	106,172,221	100,150,200			BC Pract. (1–4)									

TABLE 10.116
39/52 S17 LS

TC	TC Freq. (%)	EV W/L (%)	Std. Dev.	Play-All Bets Optimal ($)	Play-All Bets Practical ($)	Back-Counting Optimal ($)	Back-Counting Practical ($)	Bet Spread	Kelly Bank (units)	Avg. Bet ($)	Results W/L (%)	Results W/100 ($)	Risk Measures SD/100 ($)	Risk Measures ROR (%)	Risk Measures N0 (hands)	Rankings DI	Rankings SCORE
<0	36.33	–2.27	1.154	65,69,68	75			PA Opt. (1–2)	154	83.29	1.19	99.21	996.02	13.50	10,085	9.96	99.21
0	30.95	0.24	1.140	65,69,68	75			PA Pract. (1–2)	133	95.55	1.19	113.97	1144.80	17.50	10,084	9.96	99.11
1	5.28	1.00	1.138	77,77,77	75			PA Opt. (1–3)	144	103.05	1.69	173.85	1318.66	13.40	5,750	13.19	173.88
2	7.45	1.84	1.134	130,143,143	150			PA Pract. (1–3)	133	110.55	1.69	186.86	1417.28	15.50	5,755	13.18	173.83
3								PA Opt. (1–4)	147	111.64	2.04	227.62	1508.76	13.30	4,393	15.09	227.63
4	7.13	2.85	1.125	130,208,225	150,225,200			PA Pract. (1–4)	133	118.43	2.02	239.45	1592.25	15.00	4,421	15.04	226.15
5	1.68	3.56	1.122	130,208,272	150,225,300			BC Opt. (1–2)									
6	3.01	4.39	1.118	130,208,272	150,225,300			BC Pract. (1–2)									
7								BC Opt. (1–3)									
8	3.24	5.09	1.112	130,208,272	150,225,300			BC Pract. (1–3)									
9	0.17	6.71	1.108	130,208,272	150,225,300			BC Opt. (1–4)									
>9	4.76	8.40	1.095	130,208,272	150,225,300			BC Pract. (1–4)									

TABLE 10.117

26/52 S17 DAS LS

TC	TC Freq. (%)	EV W/L (%)	Std. Dev.	Play-All Bets Optimal ($)	Play-All Bets Practical ($)	Back-Counting Optimal ($)	Back-Counting Practical ($)	Bet Spread	Kelly Bank (units)	Avg. Bet ($)	Results W/L (%)	Results W/100 ($)	Risk Measures SD/100 ($)	Risk Measures ROR (%)	Risk Measures N0 (hands)	Rankings DI	Rankings SCORE
<0	33.66	−1.63	1.157	41,44,42	50			PA Opt. (1–2)	246	51.81	0.76	39.27	626.66	13.40	25,474	6.27	39.27
0	39.14	0.28	1.147	41,44,42	50			PA Pract. (1–2)	200	63.60	0.76	48.21	769.30	19.60	25,474	6.27	39.27
1	7.64	1.15	1.144	81,88,88	100			PA Opt. (1–3)	229	63.99	1.08	68.86	829.63	13.50	14,529	8.30	68.83
2	7.50	1.92	1.140	81,131,147	100,150,150			PA Pract. (1–3)	200	73.40	1.08	78.92	951.25	17.40	14,536	8.30	68.83
3								PA Opt. (1–4)	237	68.91	1.30	89.84	947.77	13.40	11,131	9.48	89.83
4	5.26	2.81	1.132	81,131,169	100,150,200			PA Pract. (1–4)	200	79.40	1.29	102.44	1082.80	17.30	11,170	9.46	89.50
5	2.42	3.72	1.127	81,131,169	100,150,200			BC Opt. (1–2)									
6	2.15	4.39	1.124	81,131,169	100,150,200			BC Pract. (1–2)									
7								BC Opt. (1–3)									
8	1.16	5.38	1.120	81,131,169	100,150,200			BC Pract. (1–3)									
9	0.25	6.84	1.112	81,131,169	100,150,200			BC Opt. (1–4)									
>9	0.80	7.24	1.107	81,131,169	100,150,200			BC Pract. (1–4)									

TABLE 10.118

31/52 S17 DAS LS

TC	TC Freq. (%)	EV W/L (%)	Std. Dev.	Play-All Bets Optimal ($)	Play-All Bets Practical ($)	Back-Counting Optimal ($)	Back-Counting Practical ($)	Bet Spread	Kelly Bank (units)	Avg. Bet ($)	Results W/L (%)	Results W/100 ($)	Risk Measures SD/100 ($)	Risk Measures ROR (%)	Risk Measures N0 (hands)	Rankings DI	Rankings SCORE
<0	34.21	−1.75	1.158	56,58,55	50			PA Opt. (1–2)	179	71.24	1.03	73.33	856.16	13.50	13,641	8.56	73.30
0	35.35	0.30	1.147	56,58,55	50			PA Pract. (1–2)	200	65.20	1.03	67.27	788.10	11.40	13,733	8.54	72.87
1	6.57	1.15	1.144	88,87,88	100			PA Opt. (1–3)	173	85.43	1.41	120.55	1098.20	13.40	8,294	10.98	120.61
2	8.19	1.92	1.139	112,148,148	100,150,150			PA Pract. (1–3)	200	77.15	1.42	109.45	1000.60	11.10	8,358	10.94	119.64
3								PA Opt. (1–4)	182	90.73	1.70	154.05	1240.83	13.50	6,495	12.41	153.96
4	6.12	2.89	1.131	112,174,220	100,150,200			PA Pract. (1–4)	200	85.00	1.69	143.75	1160.75	11.70	6,520	12.38	153.37
5	2.08	3.72	1.127	112,174,220	100,150,200			BC Opt. (1–2)									
6	3.17	4.37	1.123	112,174,220	100,150,200			BC Pract. (1–2)									
7								BC Opt. (1–3)									
8	1.98	5.52	1.119	112,174,220	100,150,200			BC Pract. (1–3)									
9	0.22	6.84	1.112	112,174,220	100,150,200			BC Opt. (1–4)									
>9	2.11	8.00	1.103	112,174,220	100,150,200			BC Pract. (1–4)									

TABLE 10.119
35/52 S17 DAS LS

TC	TC Freq. (%)	EV W/L (%)	Std. Dev.	Play-All Bets Optimal ($)	Play-All Bets Practical ($)	Back-Counting Optimal ($)	Back-Counting Practical ($)	Bet Spread	Kelly Bank (units)	Avg. Bet ($)	Results W/L (%)	Results W/100 ($)	Risk Measures SD/100 ($)	Risk Measures ROR (%)	Risk Measures N0 (hands)	Rankings DI	Rankings SCORE
<0	36.16	-2.00	1.162	60,63,60	50,75,50			PA Opt. (1–2)	167	76.66	1.11	84.97	921.71	13.50	11,765	9.22	84.96
0	32.63	0.35	1.147	60,63,60	50,75,50			PA Pract. (1–2)	200	65.60	1.11	72.76	792.60	9.80	11,870	9.18	84.27
1	5.77	1.15	1.144	88,88,88	100			PA Opt. (1–3)	160	93.17	1.54	143.70	1198.77	13.30	6,956	11.99	143.71
2	8.13	2.00	1.139	120,154,154	100,150,150			PA Pract. (1–3)	133	108.53	1.53	166.19	1389.90	17.80	6,994	11.96	142.97
3								PA Opt. (1–4)	168	99.41	1.86	185.33	1360.89	13.40	5,399	13.61	185.20
4	6.59	3.00	1.131	120,188,235	100,225,200			PA Pract. (1–4)	200	87.00	1.86	161.34	1189.00	10.10	5,430	13.57	184.13
5	1.83	3.72	1.127	120,188,238	100,225,200			BC Opt. (1–2)									
6	3.29	4.55	1.123	120,188,238	100,225,200			BC Pract. (1–2)									
7								BC Opt. (1–3)									
8	2.47	5.51	1.119	120,188,238	100,225,200			BC Pract. (1–3)									
9	0.19	6.84	1.112	120,188,238	100,225,200			BC Opt. (1–4)									
>9	2.94	8.18	1.102	120,188,238	100,225,200			BC Pract. (1–4)									

TABLE 10.120
39/52 S17 DAS LS

TC	TC Freq. (%)	EV W/L (%)	Std. Dev.	Play-All Bets Optimal ($)	Play-All Bets Practical ($)	Back-Counting Optimal ($)	Back-Counting Practical ($)	Bet Spread	Kelly Bank (units)	Avg. Bet ($)	Results W/L (%)	Results W/100 ($)	Risk Measures SD/100 ($)	Risk Measures ROR (%)	Risk Measures N0 (hands)	Rankings DI	Rankings SCORE
<0	36.34	-2.14	1.163	72,74,72	75			PA Opt. (1–2)	139	92.27	1.33	122.95	1108.71	13.50	8,138	11.09	122.93
0	30.96	0.39	1.147	72,74,72	75			PA Pract. (1–2)	133	96.90	1.33	129.01	1163.85	14.80	8,140	11.09	122.88
1	5.28	1.15	1.144	88,88,87	100			PA Opt. (1–3)	134	110.70	1.83	202.43	1422.71	13.40	4,939	14.23	202.42
2	7.44	2.00	1.139	144,154,154	150			PA Pract. (1–3)	133	111.90	1.82	204.03	1434.45	13.60	4,944	14.22	202.30
3								PA Opt. (1–4)	139	118.55	2.18	257.91	1606.39	13.20	3,875	16.06	258.05
4	7.13	3.00	1.130	144,223,235	150,225,200			PA Pract. (1–4)	133	119.78	2.15	257.64	1608.60	13.50	3,899	16.02	256.53
5	1.67	3.72	1.127	144,223,289	150,225,300			BC Opt. (1–2)									
6	3.01	4.55	1.123	144,223,289	150,225,300			BC Pract. (1–2)									
7								BC Opt. (1–3)									
8	3.24	5.25	1.116	144,223,289	150,225,300			BC Pract. (1–3)									
9	0.17	6.84	1.112	144,223,289	150,225,300			BC Opt. (1–4)									
>9	4.76	8.51	1.098	144,223,289	150,225,300			BC Pract. (1–4)									

TABLE 10.121
26/52 H17

TC	TC Freq. (%)	EV W/L (%)	Std. Dev.	Play-All Bets Optimal ($)	Play-All Bets Practical ($)	Back-Counting Optimal ($)	Back-Counting Practical ($)	Bet Spread	Kelly Bank (units)	Avg. Bet ($)	Results W/L (%)	Results W/100 ($)	Risk Measures SD/100 ($)	Risk Measures ROR (%)	Risk Measures N0 (hands)	Rankings DI	Rankings SCORE
<0	33.60	–1.95	1.155	18,25,26	20,25,25			PA Opt. (1–2)	559	22.75	0.34	7.69	277.29	13.50	129,985	2.77	7.69
0	39.24	–0.09	1.151	18,25,26	20,25,25			PA Pract. (1–2)	500	25.44	0.34	8.60	310.06	16.70	129,985	2.77	7.69
1	7.63	0.75	1.153	36,56,56	40,50,50			PA Opt. (1–3)	402	37.01	0.64	23.55	485.53	13.50	42,431	4.86	23.57
2	7.47	1.45	1.152	36,75,104	40,75,100			PA Pract. (1–3)	400	36.68	0.64	23.30	480.28	13.20	42,488	4.85	23.53
3								PA Opt. (1–4)	385	43.45	0.86	37.48	612.03	13.50	26,691	6.12	37.46
4	5.26	2.27	1.148	36,75,104	40,75,100			PA Pract. (1–4)	400	41.55	0.86	35.85	585.90	12.40	26,710	6.12	37.44
5	2.44	3.09	1.147	36,75,104	40,75,100			BC Opt. (1–2)									
6	2.14	3.72	1.145	36,75,104	40,75,100			BC Pract. (1–2)									
7								BC Opt. (1–3)									
8	1.16	4.63	1.143	36,75,104	40,75,100			BC Pract. (1–3)									
9	0.25	5.89	1.135	36,75,104	40,75,100			BC Opt. (1–4)									
>9	0.80	6.31	1.131	36,75,104	40,75,100			BC Pract. (1–4)									

TABLE 10.122
31/52 H17

TC	TC Freq. (%)	EV W/L (%)	Std. Dev.	Play-All Bets Optimal ($)	Play-All Bets Practical ($)	Back-Counting Optimal ($)	Back-Counting Practical ($)	Bet Spread	Kelly Bank (units)	Avg. Bet ($)	Results W/L (%)	Results W/100 ($)	Risk Measures SD/100 ($)	Risk Measures ROR (%)	Risk Measures N0 (hands)	Rankings DI	Rankings SCORE
<0	34.17	–2.06	1.156	31,37,38	25,25,50			PA Opt. (1–2)	319	40.43	0.60	24.16	491.55	13.50	41,337	4.92	24.16
0	35.37	–0.06	1.151	31,37,38	25,25,50			PA Pract. (1–2)	400	32.63	0.60	19.53	397.63	8.40	41,473	4.91	24.13
1	6.54	0.75	1.153	56,56,56	50			PA Opt. (1–3)	270	55.94	0.97	54.08	735.34	13.50	18,505	7.35	54.07
2	8.19	1.47	1.152	63,111,111	50,75,100			PA Pract. (1–3)	400	38.60	0.96	37.14	506.28	5.50	18,572	7.34	53.83
3								PA Opt. (1–4)	262	63.36	1.23	77.65	881.10	13.40	12,876	8.81	77.63
4	6.14	2.36	1.148	63,111,153	50,75,200			PA Pract. (1–4)	200	77.70	1.22	95.01	1088.20	20.00	13,121	8.73	76.22
5	2.09	3.09	1.147	63,111,153	50,75,200			BC Opt. (1–2)									
6	3.18	3.72	1.144	63,111,153	50,75,200			BC Pract. (1–2)									
7								BC Opt. (1–3)									
8	1.99	4.77	1.142	63,111,153	50,75,200			BC Pract. (1–3)									
9	0.22	5.89	1.135	63,111,153	50,75,200			BC Opt. (1–4)									
>9	2.12	7.01	1.128	63,111,153	50,75,200			BC Pract. (1–4)									

TABLE 10.123
35/52 H17

				Play-All Bets		Back-Counting					Results		Risk Measures			Rankings	
TC	TC Freq. (%)	EV W/L (%)	Std. Dev.	Optimal ($)	Practical ($)	Optimal ($)	Practical ($)	Bet Spread	Kelly Bank (units)	Avg. Bet ($)	W/L (%)	W/100 ($)	SD/100 ($)	ROR (%)	N0 (hands)	DI	SCORE
<0	36.15	−2.31	1.159	35,41,42	25,50,50			PA Opt. (1–2)	287	44.90	0.66	29.73	545.03	13.50	33,700	5.45	29.71
0	32.70	−0.02	1.151	35,41,42	25,50,50			PA Pract. (1–2)	400	32.80	0.66	21.75	399.80	6.60	33,788	5.44	29.59
1	5.76	0.75	1.153	56,56,56	50			PA Opt. (1–3)	242	62.64	1.08	67.57	821.80	13.50	14,814	8.22	67.53
2	8.11	1.55	1.151	70,117,117	50,125,125			PA Pract. (1–3)	200	73.35	1.07	78.84	961.50	18.10	14,869	8.20	67.23
3								PA Opt. (1–4)	238	70.81	1.38	97.36	986.79	13.40	10,270	9.87	97.38
4	6.59	2.48	1.147	70,124,168	50,150,200			PA Pract. (1–4)	200	82.00	1.38	113.20	1148.65	17.90	10,296	9.85	97.11
5	1.84	3.09	1.147	70,124,168	50,150,200			BC Opt. (1–2)									
6	3.28	3.89	1.144	70,124,168	50,150,200			BC Pract. (1–2)									
7								BC Opt. (1–3)									
8	2.46	4.77	1.141	70,124,168	50,150,200			BC Pract. (1–3)									
9	0.19	5.89	1.135	70,124,168	50,150,200			BC Opt. (1–4)									
>9	2.92	7.20	1.127	70,124,168	50,150,200			BC Pract. (1–4)									

TABLE 10.124
39/52 H17

				Play-All Bets		Back-Counting					Results		Risk Measures			Rankings	
TC	TC Freq. (%)	EV W/L (%)	Std. Dev.	Optimal ($)	Practical ($)	Optimal ($)	Practical ($)	Bet Spread	Kelly Bank (units)	Avg. Bet ($)	W/L (%)	W/100 ($)	SD/100 ($)	ROR (%)	N0 (hands)	DI	SCORE
<0	36.32	−2.44	1.160	46,53,52	50			PA Opt. (1–2)	216	59.52	0.88	52.18	722.42	13.50	19,146	7.22	52.19
0	30.98	0.02	1.151	46,53,52	50			PA Pract. (1–2)	200	63.70	0.88	55.95	774.95	15.50	19,184	7.22	52.12
1	5.25	0.75	1.153	56,56,56	50			PA Opt. (1–3)	189	79.07	1.35	106.46	1031.49	13.40	9,401	10.31	106.40
2	7.40	1.55	1.151	93,117,117	100,125,125			PA Pract. (1–3)	200	75.60	1.35	102.06	990.45	12.40	9,420	10.30	106.19
3								PA Opt. (1–4)	191	87.53	1.68	147.51	1214.00	13.40	6,784	12.14	147.38
4	7.15	2.50	1.146	93,159,190	100,150,200			PA Pract. (1–4)	200	85.60	1.70	145.31	1198.45	13.10	6,803	12.12	147.01
5	1.68	3.09	1.147	93,159,210	100,150,200			BC Opt. (1–2)									
6	3.00	3.89	1.144	93,159,210	100,150,200			BC Pract. (1–2)									
7								BC Opt. (1–3)									
8	3.26	4.56	1.138	93,159,210	100,150,200			BC Pract. (1–3)									
9	0.18	5.89	1.135	93,159,210	100,150,200			BC Opt. (1–4)									
>9	4.79	7.50	1.124	93,159,210	100,150,200			BC Pract. (1–4)									

TABLE 10.125
26/52 H17 DAS

TC	TC Freq. (%)	EV W/L (%)	Std. Dev.	Play-All Bets Optimal ($)	Play-All Bets Practical ($)	Back-Counting Optimal ($)	Back-Counting Practical ($)	Bet Spread	Kelly Bank (units)	Avg. Bet ($)	Results W/L (%)	Results W/100 ($)	Risk Measures SD/100 ($)	Risk Measures ROR (%)	Risk Measures N0 (hands)	Rankings DI	Rankings SCORE
<0	33.60	−1.81	1.164	25,30,30	25			PA Opt. (1–2)	395	32.15	0.48	15.54	394.22	13.50	64,437	3.94	15.54
0	39.26	0.06	1.157	25,30,30	25			PA Pract. (1–2)	400	31.78	0.48	15.36	389.65	13.20	64,437	3.94	15.54
1	7.63	0.91	1.159	51,67,68	50			PA Opt. (1–3)	329	45.06	0.78	35.21	593.72	13.40	28,367	5.94	35.25
2	7.47	1.60	1.158	51,91,120	50,75,100			PA Pract. (1–3)	400	36.68	0.78	28.64	482.65	8.50	28,380	5.93	35.21
3								PA Opt. (1–4)	333	50.39	1.01	50.84	712.48	13.50	19,711	7.12	50.76
4	5.25	2.43	1.154	51,91,120	50,75,100			PA Pract. (1–4)	400	41.55	1.01	41.92	588.63	8.90	19,712	7.12	50.71
5	2.44	3.24	1.152	51,91,120	50,75,100			BC Opt. (1–2)									
6	2.14	3.87	1.150	51,91,120	50,75,100			BC Pract. (1–2)									
7								BC Opt. (1–3)									
8	1.16	4.77	1.147	51,91,120	50,75,100			BC Pract. (1–3)									
9	0.25	6.07	1.139	51,91,120	50,75,100			BC Opt. (1–4)									
>9	0.81	6.42	1.135	51,91,120	50,75,100			BC Pract. (1–4)									

TABLE 10.126
31/52 H17 DAS

TC	TC Freq. (%)	EV W/L (%)	Std. Dev.	Play-All Bets Optimal ($)	Play-All Bets Practical ($)	Back-Counting Optimal ($)	Back-Counting Practical ($)	Bet Spread	Kelly Bank (units)	Avg. Bet ($)	Results W/L (%)	Results W/100 ($)	Risk Measures SD/100 ($)	Risk Measures ROR (%)	Risk Measures N0 (hands)	Rankings DI	Rankings SCORE
<0	34.18	−1.93	1.165	39,43,42	50			PA Opt. (1–2)	259	49.80	0.74	36.98	607.99	13.50	27,053	6.08	36.97
0	35.38	0.09	1.157	39,43,42	50			PA Pract. (1–2)	200	63.60	0.74	47.07	775.00	20.80	27,132	6.07	36.89
1	6.53	0.91	1.159	68,68,67	75			PA Opt. (1–3)	235	64.04	1.11	70.99	842.52	13.50	14,089	8.43	70.98
2	8.19	1.63	1.157	77,122,122	100,125,125			PA Pract. (1–3)	200	73.50	1.10	80.74	959.80	17.30	14,128	8.41	70.77
3								PA Opt. (1–4)	236	70.40	1.37	96.34	981.63	13.50	10,382	9.82	96.36
4	6.13	2.52	1.153	77,128,169	100,150,200			PA Pract. (1–4)	200	81.35	1.37	111.08	1133.00	17.60	10,400	9.80	96.11
5	2.09	3.24	1.152	77,128,169	100,150,200			BC Opt. (1–2)									
6	3.18	3.87	1.149	77,128,169	100,150,200			BC Pract. (1–2)									
7								BC Opt. (1–3)									
8	1.99	4.91	1.146	77,128,169	100,150,200			BC Pract. (1–3)									
9	0.22	6.07	1.139	77,128,169	100,150,200			BC Opt. (1–4)									
>9	2.12	7.13	1.131	77,128,169	100,150,200			BC Pract. (1–4)									

TABLE 10.127
35/52 H17 DAS

TC	TC Freq. (%)	EV W/L (%)	Std. Dev.	Play-All Bets Optimal ($)	Play-All Bets Practical ($)	Back-Counting Optimal ($)	Back-Counting Practical ($)	Bet Spread	Kelly Bank (units)	Avg. Bet ($)	Results W/L (%)	Results W/100 ($)	Risk Measures SD/100 ($)	Risk Measures ROR (%)	Risk Measures N0 (hands)	Rankings DI	Rankings SCORE
<0	36.15	−2.18	1.168	42,47,46	50			PA Opt. (1–2)	238	54.14	0.81	43.67	660.73	13.50	22,921	6.61	43.65
0	32.71	0.13	1.158	42,47,46	50			PA Pract. (1–2)	200	64.15	0.81	51.68	782.40	18.40	22,902	6.61	43.64
1	5.75	0.91	1.159	68,67,68	75			PA Opt. (1–3)	214	70.61	1.22	86.12	928.14	13.50	11,603	9.28	86.14
2	8.10	1.71	1.157	84,128,128	100,125,125			PA Pract. (1–3)	200	74.80	1.21	90.74	978.35	14.90	11,622	9.27	86.02
3								PA Opt. (1–4)	217	77.74	1.52	117.95	1086.08	13.50	8,477	10.86	117.96
4	6.58	2.63	1.153	84,140,184	100,150,200			PA Pract. (1–4)	200	83.45	1.51	126.33	1163.95	15.40	8,486	10.85	117.81
5	1.84	3.24	1.152	84,140,184	100,150,200			BC Opt. (1–2)									
6	3.28	4.03	1.149	84,140,184	100,150,200			BC Pract. (1–2)									
7								BC Opt. (1–3)									
8	2.46	4.92	1.145	84,140,184	100,150,200			BC Pract. (1–3)									
9	0.19	6.07	1.139	84,140,184	100,150,200			BC Opt. (1–4)									
>9	2.92	7.32	1.130	84,140,184	100,150,200			BC Pract. (1–4)									

TABLE 10.128
39/52 H17 DAS

TC	TC Freq. (%)	EV W/L (%)	Std. Dev.	Play-All Bets Optimal ($)	Play-All Bets Practical ($)	Back-Counting Optimal ($)	Back-Counting Practical ($)	Bet Spread	Kelly Bank (units)	Avg. Bet ($)	Results W/L (%)	Results W/100 ($)	Risk Measures SD/100 ($)	Risk Measures ROR (%)	Risk Measures N0 (hands)	Rankings DI	Rankings SCORE
<0	36.33	−2.31	1.170	53,58,57	50			PA Opt. (1–2)	188	68.65	1.02	70.16	837.49	13.40	14,258	8.37	70.14
0	30.99	0.17	1.157	53,58,57	50			PA Pract. (1–2)	200	65.05	1.02	66.36	792.90	12.10	14,281	8.37	70.04
1	5.25	0.91	1.159	68,68,67	75			PA Opt. (1–3)	172	86.88	1.49	129.38	1137.39	13.50	7,733	11.37	129.36
2	7.40	1.71	1.157	107,128,128	100,125,125			PA Pract. (1–3)	200	76.90	1.49	114.20	1005.90	10.40	7,758	11.35	128.90
3								PA Opt. (1–4)	177	94.57	1.82	172.22	1312.78	13.30	5,802	13.13	172.35
4	7.14	2.66	1.151	107,174,201	100,150,200			PA Pract. (1–4)	200	86.90	1.83	158.89	1212.55	11.40	5,823	13.10	171.71
5	1.68	3.24	1.152	107,174,226	100,150,200			BC Opt. (1–2)									
6	2.99	4.03	1.149	107,174,226	100,150,200			BC Pract. (1–2)									
7								BC Opt. (1–3)									
8	3.26	4.71	1.142	107,174,226	100,150,200			BC Pract. (1–3)									
9	0.17	6.07	1.139	107,174,226	100,150,200			BC Opt. (1–4)									
>9	4.79	7.61	1.126	107,174,226	100,150,200			BC Pract. (1–4)									

TABLE 10.129
26/52 H17 LS

				Play-All Bets		Back-Counting					Results		Risk Measures			Rankings	
TC	TC Freq. (%)	EV W/L (%)	Std. Dev.	Optimal ($)	Practical ($)	Optimal ($)	Practical ($)	Bet Spread	Kelly Bank (units)	Avg. Bet ($)	W/L (%)	W/100 ($)	SD/100 ($)	ROR (%)	N0 (hands)	DI	SCORE
<0	33.66	–1.99	1.148	23,31,31	25			PA Opt. (1–2)	427	29.78	0.43	12.83	358.23	13.50	77,849	3.58	12.83
0	39.20	–0.04	1.140	23,31,31	25			PA Pract. (1–2)	400	31.78	0.43	13.69	382.25	15.30	77,849	3.58	12.83
1	7.65	0.83	1.137	47,64,64	50			PA Opt. (1–3)	326	45.21	0.75	34.00	583.04	13.50	29,443	5.83	33.99
2	7.48	1.60	1.134	47,92,125	50,75,100			PA Pract. (1–3)	400	36.65	0.75	27.55	472.63	8.50	29,430	5.83	33.99
3								PA Opt. (1–4)	321	51.89	1.00	51.75	719.24	13.50	19,326	7.19	51.73
4	5.25	2.52	1.126	47,92,125	50,75,100			PA Pract. (1–4)	400	41.53	1.00	41.41	575.78	8.20	19,342	7.19	51.73
5	2.43	3.43	1.122	47,92,125	50,75,100			BC Opt. (1–2)									
6	2.14	4.10	1.119	47,92,125	50,75,100			BC Pract. (1–2)									
7								BC Opt. (1–3)									
8	1.15	5.13	1.116	47,92,125	50,75,100			BC Pract. (1–3)									
9	0.25	6.60	1.108	47,92,125	50,75,100			BC Opt. (1–4)									
>9	0.79	7.02	1.104	47,92,125	50,75,100			BC Pract. (1–4)									

TABLE 10.130
31/52 H17 LS

				Play-All Bets		Back-Counting					Results		Risk Measures			Rankings	
TC	TC Freq. (%)	EV W/L (%)	Std. Dev.	Optimal ($)	Practical ($)	Optimal ($)	Practical ($)	Bet Spread	Kelly Bank (units)	Avg. Bet ($)	W/L (%)	W/100 ($)	SD/100 ($)	ROR (%)	N0 (hands)	DI	SCORE
<0	34.18	–2.11	1.149	39,45,46	50			PA Opt. (1–2)	257	49.78	0.71	35.45	595.32	13.50	28,218	5.95	35.44
0	35.37	–0.02	1.140	39,45,46	50			PA Pract. (1–2)	200	63.60	0.71	45.20	759.60	20.80	28,242	5.95	35.41
1	6.56	0.83	1.137	64,64,64	75			PA Opt. (1–3)	223	66.83	1.11	73.99	860.19	13.40	13,509	8.60	73.99
2	8.19	1.61	1.133	78,126,126	100,125,125			PA Pract. (1–3)	200	73.50	1.10	80.66	938.85	16.00	13,551	8.59	73.82
3								PA Opt. (1–4)	220	74.73	1.39	104.07	1020.15	13.40	9,612	10.20	104.07
4	6.12	2.60	1.125	78,135,182	100,150,200			PA Pract. (1–4)	200	81.30	1.39	112.82	1106.75	15.70	9,627	10.19	103.92
5	2.09	3.43	1.122	78,135,182	100,150,200			BC Opt. (1–2)									
6	3.17	4.09	1.118	78,135,182	100,150,200			BC Pract. (1–2)									
7								BC Opt. (1–3)									
8	1.98	5.27	1.115	78,135,182	100,150,200			BC Pract. (1–3)									
9	0.22	6.60	1.108	78,135,182	100,150,200			BC Opt. (1–4)									
>9	2.11	7.78	1.100	78,135,182	100,150,200			BC Pract. (1–4)									

TABLE 10.131
35/52 H17 LS

TC	TC Freq. (%)	EV W/L (%)	Std. Dev.	Play-All Bets Optimal ($)	Play-All Bets Practical ($)	Back-Counting Optimal ($)	Back-Counting Practical ($)	Bet Spread	Kelly Bank (units)	Avg. Bet ($)	Results W/L (%)	Results W/100 ($)	Risk Measures SD/100 ($)	Risk Measures ROR (%)	Risk Measures N0 (hands)	Rankings DI	Rankings SCORE
<0	36.17	-2.37	1.152	43,50,50	50			PA Opt. (1–2)	235	54.55	0.78	42.56	652.30	13.50	23,488	6.52	42.55
0	32.67	0.02	1.140	43,50,50	50			PA Pract. (1–2)	200	64.15	0.78	50.02	766.85	18.20	23,522	6.52	42.55
1	5.77	0.83	1.137	64,64,64	75			PA Opt. (1–3)	202	74.09	1.23	90.72	952.48	13.40	11,019	9.52	90.72
2	8.11	1.68	1.133	85,131,131	100,125,125			PA Pract. (1–3)	200	74.80	1.22	91.11	957.25	13.60	11,041	9.52	90.59
3								PA Opt. (1–4)	201	82.83	1.55	128.33	1132.85	13.50	7,794	11.33	128.33
4	6.58	2.70	1.125	85,149,199	100,150,200			PA Pract. (1–4)	200	83.45	1.54	128.79	1137.45	13.50	7,799	11.32	128.20
5	1.84	3.43	1.122	85,149,199	100,150,200			BC Opt. (1–2)									
6	3.28	4.25	1.118	85,149,199	100,150,200			BC Pract. (1–2)									
7								BC Opt. (1–3)									
8	2.46	5.23	1.115	85,149,199	100,150,200			BC Pract. (1–3)									
9	0.19	6.60	1.108	85,149,199	100,150,200			BC Opt. (1–4)									
>9	2.93	7.93	1.099	85,149,199	100,150,200			BC Pract. (1–4)									

TABLE 10.132
39/52 H17 LS

TC	TC Freq. (%)	EV W/L (%)	Std. Dev.	Play-All Bets Optimal ($)	Play-All Bets Practical ($)	Back-Counting Optimal ($)	Back-Counting Practical ($)	Bet Spread	Kelly Bank (units)	Avg. Bet ($)	Results W/L (%)	Results W/100 ($)	Risk Measures SD/100 ($)	Risk Measures ROR (%)	Risk Measures N0 (hands)	Rankings DI	Rankings SCORE
<0	36.32	-2.51	1.154	55,62,62	50			PA Opt. (1–2)	181	70.88	1.01	71.76	847.15	13.50	13,936	8.47	71.77
0	30.97	0.06	1.140	55,62,62	50			PA Pract. (1–2)	200	65.05	1.01	65.70	776.80	11.30	13,979	8.46	71.54
1	5.28	0.83	1.137	64,64,64	75			PA Opt. (1–3)	160	92.42	1.52	140.34	1184.52	13.50	7,124	11.85	140.31
2	7.42	1.68	1.133	110,131,131	100,125,125			PA Pract. (1–3)	200	76.90	1.51	116.10	983.60	9.00	7,177	11.80	139.33
3								PA Opt. (1–4)	161	102.05	1.88	191.90	1384.90	13.40	5,213	13.85	191.80
4	7.13	2.70	1.124	110,187,214	100,150,200			PA Pract. (1–4)	200	86.90	1.88	163.38	1184.10	9.60	5,251	13.80	190.39
5	1.68	3.43	1.122	110,187,248	100,150,200			BC Opt. (1–2)									
6	3.00	4.25	1.118	110,187,248	100,150,200			BC Pract. (1–2)									
7								BC Opt. (1–3)									
8	3.25	4.96	1.112	110,187,248	100,150,200			BC Pract. (1–3)									
9	0.17	6.60	1.108	110,187,248	100,150,200			BC Opt. (1–4)									
>9	4.79	8.26	1.095	110,187,248	100,150,200			BC Pract. (1–4)									

TABLE 10.133
26/52 H17 DAS LS

				Play-All Bets		Back-Counting					Results		Risk Measures			Rankings	
TC	TC Freq. (%)	EV W/L (%)	Std. Dev.	Optimal ($)	Practical ($)	Optimal ($)	Practical ($)	Bet Spread	Kelly Bank (units)	Avg. Bet ($)	W/L (%)	W/100 ($)	SD/100 ($)	ROR (%)	N0 (hands)	DI	SCORE
<0	33.66	−1.86	1.157	31,36,36	25			PA Opt. (1–2)	323	39.35	0.58	22.66	476.03	13.50	44,117	4.76	22.66
0	39.21	0.10	1.147	31,36,36	25			PA Pract. (1–2)	400	31.78	0.58	18.30	384.38	8.40	44,117	4.76	22.66
1	7.65	0.99	1.144	62,75,75	50			PA Opt. (1–3)	276	53.38	0.90	47.90	692.11	13.50	20,877	6.92	47.90
2	7.48	1.76	1.140	62,109,135	50,75,100			PA Pract. (1–3)	400	36.65	0.90	32.87	475.03	5.40	20,879	6.92	47.89
3								PA Opt. (1–4)	279	59.29	1.14	67.35	820.58	13.40	14,851	8.21	67.33
4	5.24	2.66	1.132	62,109,144	50,75,100			PA Pract. (1–4)	400	41.53	1.14	47.45	578.55	5.80	14,867	8.20	67.27
5	2.43	3.59	1.127	62,109,144	50,75,100			BC Opt. (1–2)									
6	2.14	4.24	1.124	62,109,144	50,75,100			BC Pract. (1–2)									
7								BC Opt. (1–3)									
8	1.15	5.27	1.120	62,109,144	50,75,100			BC Pract. (1–3)									
9	0.25	6.78	1.112	62,109,144	50,75,100			BC Opt. (1–4)									
>9	0.80	7.12	1.107	62,109,144	50,75,100			BC Pract. (1–4)									

TABLE 10.134
31/52 H17 DAS LS

				Play-All Bets		Back-Counting					Results		Risk Measures			Rankings	
TC	TC Freq. (%)	EV W/L (%)	Std. Dev.	Optimal ($)	Practical ($)	Optimal ($)	Practical ($)	Bet Spread	Kelly Bank (units)	Avg. Bet ($)	W/L (%)	W/100 ($)	SD/100 ($)	ROR (%)	N0 (hands)	DI	SCORE
<0	34.19	−1.98	1.158	46,51,50	50			PA Opt. (1–2)	216	59.27	0.86	50.78	712.44	13.50	19,707	7.12	50.76
0	35.38	0.13	1.147	46,51,50	50			PA Pract. (1–2)	200	63.55	0.86	54.39	763.65	15.40	19,706	7.12	50.73
1	6.56	0.99	1.144	75,75,75	75			PA Opt. (1–3)	198	75.04	1.25	93.68	968.02	13.40	10,677	9.68	93.70
2	8.19	1.77	1.139	93,136,137	100,150,150			PA Pract. (1–3)	200	75.50	1.26	94.91	981.40	13.80	10,694	9.67	93.52
3								PA Opt. (1–4)	201	81.85	1.54	125.66	1120.84	13.40	7,960	11.21	125.63
4	6.12	2.75	1.130	93,152,199	100,150,200			PA Pract. (1–4)	200	83.35	1.54	128.17	1144.20	14.00	7,972	11.20	125.49
5	2.08	3.59	1.127	93,152,199	100,150,200			BC Opt. (1–2)									
6	3.17	4.23	1.123	93,152,199	100,150,200			BC Pract. (1–2)									
7								BC Opt. (1–3)									
8	1.98	5.41	1.119	93,152,199	100,150,200			BC Pract. (1–3)									
9	0.22	6.78	1.112	93,152,199	100,150,200			BC Opt. (1–4)									
>9	2.11	7.89	1.103	93,152,199	100,150,200			BC Pract. (1–4)									

TABLE 10.135
35/52 H17 DAS LS

TC	TC Freq. (%)	EV W/L (%)	Std. Dev.	Play-All Bets Optimal ($)	Play-All Bets Practical ($)	Back-Counting Optimal ($)	Back-Counting Practical ($)	Bet Spread	Kelly Bank (units)	Avg. Bet ($)	Results W/L (%)	Results W/100 ($)	Risk Measures SD/100 ($)	Risk Measures ROR (%)	Risk Measures N0 (hands)	Rankings DI	Rankings SCORE
<0	36.18	−2.24	1.161	50,55,54	50			PA Opt. (1–2)	201	63.90	0.92	59.03	768.49	13.40	16,924	7.68	59.06
0	32.69	0.18	1.147	50,55,54	50			PA Pract. (1–2)	200	64.15	0.92	59.24	770.90	13.60	16,929	7.68	59.06
1	5.77	0.99	1.144	75,75,75	75			PA Opt. (1–3)	182	82.16	1.37	112.22	1059.40	13.40	8,912	10.59	112.23
2	8.11	1.84	1.139	100,142,142	100,150,150			PA Pract. (1–3)	200	76.80	1.38	105.60	998.65	11.90	8,943	10.57	111.80
3								PA Opt. (1–4)	186	89.85	1.69	151.93	1232.36	13.50	6,586	12.32	151.87
4	6.58	2.85	1.130	100,165,215	100,150,200			PA Pract. (1–4)	200	85.45	1.69	144.49	1173.70	12.10	6,597	12.31	151.55
5	1.83	3.59	1.127	100,165,215	100,150,200			BC Opt. (1–2)									
6	3.28	4.40	1.123	100,165,215	100,150,200			BC Pract. (1–2)									
7								BC Opt. (1–3)									
8	2.46	5.38	1.119	100,165,215	100,150,200			BC Pract. (1–3)									
9	0.19	6.78	1.112	100,165,215	100,150,200			BC Opt. (1–4)									
>9	2.92	8.05	1.102	100,165,215	100,150,200			BC Pract. (1–4)									

TABLE 10.136
39/52 H17 DAS LS

TC	TC Freq. (%)	EV W/L (%)	Std. Dev.	Play-All Bets Optimal ($)	Play-All Bets Practical ($)	Back-Counting Optimal ($)	Back-Counting Practical ($)	Bet Spread	Kelly Bank (units)	Avg. Bet ($)	Results W/L (%)	Results W/100 ($)	Risk Measures SD/100 ($)	Risk Measures ROR (%)	Risk Measures N0 (hands)	Rankings DI	Rankings SCORE
<0	36.34	−2.38	1.163	62,67,66	50,75,75			PA Opt. (1–2)	160	80.11	1.16	92.73	962.75	13.50	10,784	9.63	92.69
0	30.98	0.22	1.146	62,67,66	50,75,75			PA Pract. (1–2)	200	65.00	1.16	75.09	780.85	8.50	10,811	9.62	92.48
1	5.27	0.99	1.144	75,75,75	75			PA Opt. (1–3)	148	100.39	1.66	166.61	1290.92	13.40	6,001	12.91	166.66
2	7.41	1.84	1.139	125,142,142	100,150,150			PA Pract. (1–3)	133	108.75	1.64	178.67	1389.38	15.60	6,048	12.86	165.36
3								PA Opt. (1–4)	151	109.20	2.02	220.50	1484.49	13.30	4,537	14.84	220.37
4	7.12	2.86	1.129	125,202,224	100,200,200			PA Pract. (1–4)	133	118.43	2.00	236.47	1599.30	15.60	4,574	14.79	218.63
5	1.67	3.59	1.127	125,202,265	100,225,300			BC Opt. (1–2)									
6	2.99	4.40	1.123	125,202,265	100,225,300			BC Pract. (1–2)									
7								BC Opt. (1–3)									
8	3.25	5.11	1.116	125,202,265	100,225,300			BC Pract. (1–3)									
9	0.17	6.78	1.112	125,202,265	100,225,300			BC Opt. (1–4)									
>9	4.79	8.38	1.098	125,202,265	100,225,300			BC Pract. (1–4)									

TABLE 10.137
26/52 H17 D10

TC	TC Freq. (%)	EV W/L (%)	Std. Dev.	Play-All Bets Optimal ($)	Play-All Bets Practical ($)	Back-Counting Optimal ($)	Back-Counting Practical ($)	Bet Spread	Kelly Bank (units)	Avg. Bet ($)	Results W/L (%)	Results W/100 ($)	Risk Measures SD/100 ($)	Risk Measures ROR (%)	Risk Measures N0 (hands)	Rankings DI	Rankings SCORE
<0	33.61	−2.10	1.107	1,13,17	1,15,15			PA Opt. (1–2)	6,974	1.82	0.03	0.05	21.36	13.50	21,671,934	0.21	0.05
0	39.24	−0.38	1.103	1,13,17	1,15,15			PA Pract. (1–2)	10,000	1.27	0.03	0.03	14.90	5.70	21,671,934	0.21	0.05
1	7.63	0.39	1.106	3,32,32	2,25,25			PA Opt. (1–3)	791	19.06	0.31	5.82	241.28	13.50	171,991	2.41	5.82
2	7.47	1.04	1.107	3,38,68	2,45,60			PA Pract. (1–3)	667	21.62	0.30	6.52	272.52	17.30	175,241	2.39	5.72
3								PA Opt. (1–4)	591	27.97	0.52	14.45	379.97	13.50	69,242	3.80	14.44
4	5.25	1.82	1.106	3,38,68	2,45,60			PA Pract. (1–4)	667	24.54	0.52	12.71	334.55	10.30	69,336	3.80	14.42
5	2.44	2.61	1.108	3,38,68	2,45,60			BC Opt. (1–2)									
6	2.14	3.21	1.108	3,38,68	2,45,60			BC Pract. (1–2)									
7								BC Opt. (1–3)									
8	1.15	4.10	1.108	3,38,68	2,45,60			BC Pract. (1–3)									
9	0.25	5.38	1.105	3,38,68	2,45,60			BC Opt. (1–4)									
>9	0.80	5.75	1.103	3,38,68	2,45,60			BC Pract. (1–4)									

TABLE 10.138
31/52 H17 D10

TC	TC Freq. (%)	EV W/L (%)	Std. Dev.	Play-All Bets Optimal ($)	Play-All Bets Practical ($)	Back-Counting Optimal ($)	Back-Counting Practical ($)	Bet Spread	Kelly Bank (units)	Avg. Bet ($)	Results W/L (%)	Results W/100 ($)	Risk Measures SD/100 ($)	Risk Measures ROR (%)	Risk Measures N0 (hands)	Rankings DI	Rankings SCORE
<0	34.18	−2.21	1.108	15,26,29	15,25,25			PA Opt. (1–2)	663	19.67	0.27	5.33	230.79	13.50	187,786	2.31	5.33
0	35.38	−0.35	1.103	15,26,29	15,25,25			PA Pract. (1–2)	667	19.56	0.27	5.30	229.46	13.40	187,786	2.31	5.33
1	6.54	0.39	1.106	30,32,32	30,25,25			PA Opt. (1–3)	390	38.29	0.62	23.70	486.74	13.50	42,165	4.87	23.69
2	8.19	1.06	1.107	30,77,86	30,50,50			PA Pract. (1–3)	400	34.90	0.60	20.79	437.43	11.40	44,226	4.75	22.59
3								PA Opt. (1–4)	345	47.51	0.87	41.25	642.07	13.50	24,258	6.42	41.22
4	6.13	1.89	1.106	30,77,116	30,75,100			PA Pract. (1–4)	400	38.83	0.86	33.40	524.85	8.80	24,693	6.36	40.50
5	2.09	2.61	1.108	30,77,116	30,75,100			BC Opt. (1–2)									
6	3.18	3.20	1.107	30,77,116	30,75,100			BC Pract. (1–2)									
7								BC Opt. (1–3)									
8	1.99	4.23	1.108	30,77,116	30,75,100			BC Pract. (1–3)									
9	0.22	5.38	1.105	30,77,116	30,75,100			BC Opt. (1–4)									
>9	2.12	6.44	1.100	30,77,116	30,75,100			BC Pract. (1–4)									

TABLE 10.139
35/52 H17 D10

				Play-All Bets		Back-Counting					Results		Risk Measures			Rankings	
TC	TC Freq. (%)	EV W/L (%)	Std. Dev.	Optimal ($)	Practical ($)	Optimal ($)	Practical ($)	Bet Spread	Kelly Bank (units)	Avg. Bet ($)	W/L (%)	W/100 ($)	SD/100 ($)	ROR (%)	N0 (hands)	DI	SCORE
<0	36.16	–2.44	1.110	19,30,33	20,25,25			PA Opt. (1–2)	532	24.32	0.33	8.09	284.30	13.50	123,773	2.84	8.08
0	32.70	–0.31	1.103	19,30,33	20,25,25			PA Pract. (1–2)	500	25.36	0.33	8.40	296.26	14.70	124,391	2.83	8.03
1	5.76	0.39	1.106	32,32,33	25			PA Opt. (1–3)	337	44.91	0.73	32.80	572.66	13.50	30,505	5.73	32.79
2	8.11	1.12	1.107	38,89,92	40,50,50			PA Pract. (1–3)	400	35.68	0.71	25.28	447.93	8.00	31,407	5.64	31.86
3								PA Opt. (1–4)	301	55.20	1.01	55.78	746.72	13.50	17,933	7.47	55.76
4	6.59	2.00	1.106	38,89,133	40,75,100			PA Pract. (1–4)	400	39.98	1.01	40.21	541.30	6.40	18,131	7.43	55.17
5	1.84	2.61	1.108	38,89,133	40,75,100			BC Opt. (1–2)									
6	3.28	3.35	1.108	38,89,133	40,75,100			BC Pract. (1–2)									
7								BC Opt. (1–3)									
8	2.46	4.22	1.107	38,89,133	40,75,100			BC Pract. (1–3)									
9	0.19	5.38	1.105	38,89,133	40,75,100			BC Opt. (1–4)									
>9	2.92	6.62	1.099	38,89,133	40,75,100			BC Pract. (1–4)									

TABLE 10.140
39/52 H17 D10

				Play-All Bets		Back-Counting					Results		Risk Measures			Rankings	
TC	TC Freq. (%)	EV W/L (%)	Std. Dev.	Optimal ($)	Practical ($)	Optimal ($)	Practical ($)	Bet Spread	Kelly Bank (units)	Avg. Bet ($)	W/L (%)	W/100 ($)	SD/100 ($)	ROR (%)	N0 (hands)	DI	SCORE
<0	36.33	–2.56	1.110	31,41,43	25,50,50			PA Opt. (1–2)	325	39.25	0.54	21.12	459.60	13.50	47,298	4.60	21.12
0	30.98	–0.28	1.103	31,41,43	25,50,50			PA Pract. (1–2)	400	31.85	0.54	17.15	373.08	8.50	47,322	4.60	21.12
1	5.25	0.39	1.106	32,41,43	25,50,50			PA Opt. (1–3)	241	61.75	0.98	60.54	778.07	13.50	16,515	7.78	60.54
2	7.40	1.12	1.107	62,92,92	50,100,100			PA Pract. (1–3)	200	73.75	0.98	72.18	928.25	18.70	16,529	7.78	60.47
3								PA Opt. (1–4)	232	71.97	1.31	94.15	970.14	13.50	10,625	9.70	94.12
4	7.14	2.01	1.105	62,124,165	50,150,200			PA Pract. (1–4)	200	83.75	1.31	110.08	1135.10	18.00	10,629	9.70	94.04
5	1.68	2.61	1.108	62,124,172	50,150,200			BC Opt. (1–2)									
6	2.99	3.35	1.108	62,124,172	50,150,200			BC Pract. (1–2)									
7								BC Opt. (1–3)									
8	3.26	3.99	1.104	62,124,172	50,150,200			BC Pract. (1–3)									
9	0.17	5.38	1.105	62,124,172	50,150,200			BC Opt. (1–4)									
>9	4.79	6.91	1.097	62,124,172	50,150,200			BC Pract. (1–4)									

Chapter 11

Team Play

I have often been asked, "How do you, personally, play the game?" When I am "attacking the shoe," my philosophy and approach are documented, in great detail, in the "Team Handbook" contained in this chapter. It is bookended by an introductory "Gospel" on the concept of a "joint bank," and a concluding study on a novel way to divide the winnings of such a bank among the investors and players. The formula and methodology for this approach had never been presented before their appearance in ***Blackjack Forum.***

From "The Gospel According to Don," *Blackjack Forum*, March 1990:

Q. *On the subject of a joint bank, I always thought I understood the basic formula of dividing a win using the 50% to investors and 50% to the players (25% of which is allocated based on time played and 25% based on winnings) method described by Uston and Snyder. However, an ambiguity arose when I drew the plan out on paper. Maybe you can clear it up for me.*

Let's say that I invest $5,000 in a blackjack bank and that an outside investor also *invests $5,000 in my bank, making a total bank of $10,000. Now, as a worst-case scenario, let's say I lose the entire $10,000. According to the Uston/Snyder formula, the investor would be responsible for 50% of the loss, or $5,000. Since I invested half of the bank, I would be responsible for $2,500 of the investor's loss, right? And since I am the only player, I would also lose the entire $5,000 of my original stake. The outside investor could claim that he is only liable for a $2,500 loss based on the formula and want a $2,500 refund. My true potential liability is $7,500, not $5,000, or is it?*

I realize that if we flip the coin over and win $10,000, then I will benefit by an "extra" $2,500 since my profit will be $7,500 and the outside investor will gain a 50% profit on his $5,000 investment ($2,500), even though the bank was doubled (100% profit).

A. Your question brings up two aspects of team play that need to be clarified. First, your notion of the player-investor responsibility when a loss occurs is erroneous. By definition, it is the investors alone who bear the entire loss; the players cannot be asked for money that they did not initially invest. So, for your example, your maximum risk as a player-investor is the original $5,000 you put into the joint bank. The outside investor loses the other $5,000 and the two of you share the $10,000 loss equally. Simply put, the method for calculating investors' share of a loss is: Each is responsible for that percentage of the loss that equals the percentage of the total investment he made in the bank. Thus, the loss is directly proportional to the investment.

Your description of the split when a win occurs is correct. However, the 50-50 division among players and investors is not "written in stone." It seems fair, but other percentages could, of course, be negotiated (see pp. 298–301). On the other hand, I never agreed with Uston when it came to dividing up the players' shares. Of the 50% they are entitled to, I am strongly opposed to an equal split between hours played and money won.

Your own recent play illustrates perfectly why I feel this way. On one trip to Las Vegas, you played 21 hours and lost $2,200. Shortly thereafter, you went to A.C., played 16 hours, and won $2,900! Each result was excessive in that it greatly exceeded the expected outcome — once in each direction. A violent swing in standard deviation was the culprit on each occasion. Had you been playing on a team, does it really make sense that you be "punished" for the first outcome (with five extra hours of play, to boot!) and "rewarded" for the second? I don't think so.

Rather, I think that each member of the team, after duly qualifying to play, should be treated as a machine that, robotically, puts in hours in an attempt to make money. Never was the dictum "Time is money" more applicable than to the joint-bank, team approach. Players, I feel, should be compensated for their time. Their individual wins and losses are, in the short run, due almost entirely to the whims of standard deviation and should be treated accordingly.

Thus, when I managed a high-stakes team of a dozen players, their 50% split of winnings was apportioned 40% for hours and 10% for contribution to the win. The latter was a concession on my part to those members who felt that, as an incentive, they should be compensated for a win. I have always believed that it is not possible to demonstrate one's blackjack-playing ability by pointing to a period of time during which one won money.

Indeed, if someone had followed me around during 1989 and judged my skills solely on results at the tables, he would have concluded that I was a sorry excuse for a blackjack player! If I then "pleaded" that this was only my second losing year in fifteen, what good would my past record do? I would be penalized (unjustly, in my view) for a short-term aberration that I was powerless to control. *[Note: For more on this very important concept, about which there is much disagreement, see the next chapter for an entirely new and rather novel treatment of the subject. — D.S.]*

On the other hand, I would be very meticulous in my choice of teammates. It is possible, objectively, to evaluate the quality of a prospective team player. But all this must be done before the actual results come rolling in. If there is reader interest in this topic, perhaps I will publish, in a future "Gospel," the "Blackjack Team Handbook" I created and used for my teams of a few years ago. If you would like to see this, please drop me a line. *[Note: It took six years, but eventually, I did, in fact, publish the handbook, which follows. — D.S.]*

"The Blackjack Team Handbook," From *Blackjack Forum,* Spring 1996:

This handbook is intended for the exclusive use of team members. It should not be shown to anyone else. It should be kept in a safe place at all times. Players must return the handbook when the team is disbanded or when the player leaves the team for whatever reason. Please do not reproduce any of this material.

INTRODUCTION

The following principles and guidelines are intended for the use of all team players. Be aware that team play is different from individual play and may require you to change from the style with which you are most familiar or most comfortable. However, to give the team the maximum chance of success, it is vital that all players adhere strictly to the rules and guidelines contained in this handbook, without any deviation whatsoever.

Your ability to play according to the principles outlined herein is a function of discipline and will power more than anything else. There is no room for sloppiness of thought or execution in such an operation. Playing high-stakes, winning blackjack is, in large part, a result of outlook and commitment to excellence. Without these, all the technical knowledge in the world will not guarantee success.

It is understood that, when you are tested on the polygraph, you will be asked if you have been complying with all of the team guidelines as set forth in your handbook. Failure to satisfy the polygraph that you have, indeed, been complying may be grounds for dismissal from the team. (See Section IX.) To facilitate use of the handbook, guidelines have been arranged sequentially, by topic.

I. SELECTION OF TEAM PLAYERS AND QUALIFYING PROCEDURES

1. All team players are expected to play a valid point-count system that counts the ace as a minus card. Players who wish to join the team and who use an ace-neutral count will be expected to demonstrate proficiency in keeping a side count of aces and must use this procedure for all bet-sizing. Preference will be given to players who do not use such counts as, in general, they are inferior to the ace-reckoned counts for multiple-deck play when a large bet spread is employed.

2. As indicated in the "Minutes of the First Team Meeting," playing qualifications and certification criteria are as follows: (a) Players must demonstrate mastery of true-count conversion by computing number of decks in tray, number of remaining decks, and true count for randomly given running counts and deck situations. After true count is established, the player must associate the correct bet with that particular count. (b) Counting ability will, in part, be demonstrated by counting down a single deck (50 cards, two pulled out) in 25 seconds or less with no mistakes, and by counting down four decks (four cards pulled out) in two minutes or less with no mistakes. Faster times are encouraged so that minimum attention is given to the cards at the table. In addition, players will be trained in "scanning" a full table to ascertain the count of the 15 cards in three to four seconds. This technique is particularly useful for back-counting. (c) Playing ability will be demonstrated by accuracy and speed with flash cards. The 63 numbers (including insurance) from –1 to +10 will be utilized. The standard will be five minutes or less, although, once again, players are strongly encouraged to achieve a substantially faster time. For example, an average of two seconds per number (not at all unreasonable) would produce a time of just over two minutes. Strive for it!

3. Players who do not qualify at the certification meeting may request to be certified any time thereafter and will make an appointment with one of the team managers for this purpose.

4. Players who do qualify may nonetheless be asked, at periodic intervals in the future, and at the sole discretion of the team managers, to demonstrate that they have maintained their skills and high level of proficiency. Keep sharp and keep practicing on an ongoing basis.

5. In fairness to original team members, no new players will be accepted for a

period of six (6) months from the date of the initial team play unless they are substitutes for members who drop out. After that time, prospective members must be certified in exactly the same fashion as charter members. (See, also, VIII.6.)

II. SELECTING A CASINO

Before a day's play, one of the team managers will give you an assignment of casinos to play in and an order in which to play them. It is imperative that you do not deviate from this order, as we do not wish two or more team members to be in the same casino at the same time. A lunch and/or dinner break will, of course, be scheduled. We anticipate that six (6) sessions will constitute a day's play.

III. COMPORTMENT INSIDE THE CASINO AND PLAY SELECTION CRITERIA

1. You are expected to utilize your time as expeditiously as possible. Please limit non-playing activities to a reasonable amount of time.

2. Spend all time within the casino area actively playing or back-counting. Time spent walking, but not looking at a table, is wasted time. It is, nonetheless, understood that there is a certain amount of "dead time" associated with a back-counting approach.

3. Try to look like everyone else in the casino as you walk about. For example, walk, don't run, to a table with a good count that you have spotted from across the pit! You won't be alone as you stroll around, so don't feel self-conscious. In addition, avoid back-counting the lower-limit tables. Large bets at a $5 table would be grotesquely out of place.

4. Back-count from afar, not from directly behind the players' seats. Look casual, and divert your attention from the game as frequently as you possibly can. It is foolish to stare at the table if the dealer isn't dealing. It is actually possible to wander a couple of tables away between rounds and be back in time to count the next round. Use this technique.

5. As a general guideline, walk away from a potential game (give up on playing there) if the count is minus and you have seen 1 to 1 1/2 decks. Even if fewer than 1 1/2 decks have been dealt, consider walking if your running count is –8 or lower, as there is little chance that, by the 1 1/2-deck level, the true count will have become sufficiently positive. *[Note: For recent findings in this area, see Chapter 13. — D.S.]*

6. If the true count equals or exceeds +1, consider playing. Your participation may not be automatic, as a true count of exactly +1 at a full table is not yet worth pursuing.

7. Consider standing instead of sitting down as you play the first hand or two. The count may deteriorate, and you may be leaving sooner than you think! It is easier to "get away" from a table if you have not yet made yourself comfortable.

8. Your buy-in should not exceed five to ten times your unit bet size.

9. Set up your departure from a table in advance. You might try looking at your watch habitually as if preoccupied with the time. The implication is you are late for an appointment and may have to leave at any moment. If the count deteriorates during a big win, lose a hand (or two) and then leave, possibly stating out loud, "Well, I guess it's (winning streak) over. I think I'll take a break before I give too much of this back."

10. Squirrel away whatever chips you can, but don't be obvious. As you change tables, if you have a stack of chips, don't put the whole stack on the new table. Play out of your pocket for a while. When leaving, the dealer will ask to color your chips to a higher denomination. Try to keep some, saying, "I'm going to play some more." But if he insists nonetheless (and dealers often do), by all means, give in your chips. This isn't important enough to start an argument over.

11. If you are already known by name at the casino, continue, of course, to use that same name! However, if you have the opportunity to establish a pseudonym, then do so. Remember the name you use as if it were your own. Groping for an answer when someone asks for your name is the ultimate embarrassment! Of course, any comps extended to you (under whatever name) are yours to do with as you see fit.

12. If you spot a teammate in the casino, completely ignore him or her. Make no attempt whatsoever to communicate with the person in any manner while still inside the casino.

13. Spend between 1 and 1 1/4 hours in any one casino.

14. Leave if you have lost your session bankroll. If the situation presents itself, you might lament your loss and let it be known that you have dropped a bundle.

15. Leave if you have won a session bankroll even if this occurs after only fifteen or twenty minutes.

16. Cash in quickly and quietly. When asked, "Do you have any markers?" answer "No" and collect your money. Count it along with the cashier to make sure there is no error.

17. Record your result in writing, so as not to forget the outcome.

18. Stroll to the next casino, rest, clear your head, collect your thoughts, and … start all over again!

IV. MONEY MANAGEMENT AND BETTING PROCEDURES

1. A session bankroll is 30 starting units. You will begin each day's play with two session bankrolls or as large a percentage thereof as possible. The team manager will give you the money at the start of the day.

2. Keep these two bankrolls separate from each other. You don't want to lose both of them at one table!

3. Dip into a second bankroll at the same table after a "wipeout" (session bankroll loss) only if: (a) you need more money to make a correct play (split and/or double), or (b) the count is still high and the shoe is not over. It is thus possible to lose more than one session bankroll at a table (a "super wipeout") given the above conditions.

4. Betting levels are as follows (for our initial play): Below +1, do not play; +1 to +2, 1 unit; +2 to +3, 2 units; +3 to +4, 4 units; +4 to +5, 6 units; +5 and above, two hands of 6 units each.

5. Under no circumstances should a bet be placed from the chips in front of you (but not already in the betting circle) if you have won the previous hand. The next bet is, at most, a parlay of the previous original bet and should never exceed double the prior wager, even after winning a blackjack, double, or split. THERE ARE TO BE NO EXCEPTIONS WHATSOEVER TO THIS RULE!!!

6. Do not increase your bet after losing a hand, even if the count merits a higher wager.

7. Do not decrease your bet after winning a hand, even if the count merits a lower wager.

8. Do not change your bet after a push. (Possible exceptions: After two consecutive pushes at 20, with a deteriorated count, you might pull back half of the wager while stating, "What's the sense; I can't even win with a 20.")

9. Determine your next bet quickly so as not to give the impression of calculating or pondering over what to bet.

10. In general, it is best to leave a table at which you have placed large bets towards the end of the shoe once the shoe is over. If you do stay at the table (what is your reason for being there?), you must bet approximately half of your final bet off the top of the next shoe (two units is fine) so as not to be too obvious. Put this bet in the circle immediately, and leave it there as the dealer is shuffling. Best advice: Leave the table!

V. TIPPING

1. Over-tipping can be a serious drain on your bankroll and can cut deeply into your hourly win rate. If your expected win is 1.25 units per hour, and after a big win, you toss the dealer a toke of about 1/4 your unit size, you have forfeited 20% of your average hourly expectation in one gesture! Do not do this!

2. On the other hand, if you are projecting the image of a high roller (and with large bets, you have no choice!), it is incongruous not to tip at all. As you begin to win, ask for a green chip to be colored and, when you make a big bet, make a $5 wager for the dealer, stating, "Let's see if we can both win this one." Do this no more than twice if you win the bets and, therefore, have effectively tipped the dealer $20.

3. It makes no sense whatsoever to tip while you are losing.

4. Wait to tip until you are reasonably assured of a winning session. It is embarrassing to tip early, then lose all your winnings and more. You have to record a loss for the session and then a tip on top of a loss!

5. Record all tips separately along with gross wins for your final accounting. (See Section VII.5.)

VI. PLAYING PROCEDURE

1. Try to sit in the middle of the table, if possible. In any event, avoid first and third base unless the count is compelling and they are the only seats available. Playing first base puts you "under the gun," and forces you to count all cards to your left before making playing and insurance decisions. Playing third base "brands" you as the table expert and subjects your play to the scrutiny of everyone. You don't need the aggravation!

2. If you have the option, sit where there are two contiguous spots open. If this is not possible, but there is an open spot to the left or right of your neighbor, you can ask him/her to please move over because you'd like to play two hands. (Do this only when it is time to bet two hands, not before. You don't want to be pushy, and you may never get to the two hands, anyway.)

3. Count all the cards on the table before you play your hand. This is important. Basic strategy departures should be based on all the information available to you at the moment the play is made. Anticipate a possible departure. Be ready to make the calculations. If you have the slightest doubt as to the appropriateness of the decision, abstain. There is no sense in guessing to make a departure. When in doubt, don't!

4. Never round a true count computation up for the purpose of deviating from

basic strategy. If you haven't achieved the entire number (without rounding), you don't make the play.

5. The precise Hi-Lo insurance numbers are 3.0 for six decks and 3.1 for eight decks. Thus, in an 8-deck game, if you estimate the true count to be exactly 3, you don't insure. (This is a discipline thing!)

6. If unsure about an insurance play, when the dealer asks for insurance, tell him, "Wait a minute," and fumble with your chips as if you are considering making the bet. All the while, you are gaining extra time and should now know whether or not to insure. If you decide not to, simply state, "No, forget about it."

7. If you're worried about not having insured several times at the same table and then having to take insurance, simply say, "This time, I think you've got it. You're overdue." (Assuming, of course, that when you did not insure, the dealer didn't have a blackjack!)

8. Give a hand signal for every play, every time, no matter what the total. This is mandatory. Signal "stand" for every 20, even in two cards. This way, the dealer will get used to expecting a signal from you. For example, when you have A,7 v. his 10, he will be less likely to pass you by. (Proper procedure for this play is to get your hand out early and to be ready for the "hit" signal before the dealer gets to you.) In addition, be ready to stop the dealer if you intend to split tens or double on A,9.

9. Call attention to a dealer's mistake (when not in your favor) concerning your hand as soon as possible, preferably, the moment it happens. Once your bet is collected and the hands are picked up, it is much more difficult to rectify an error.

10. Again, try to anticipate close plays as the hands are unfolding. Be ready when your turn comes so that your play is natural and casual.

11. Don't give advice to other players. Don't appear to know a lot about the game. Play dumb. Don't explain strange plays that you make. "I had a hunch" will do! Don't make adverse comments about the play of others. Also, if you still spend your time figuring out what you would have done if the player to your right had only done this or what the dealer would have made if only the third baseman had done this or that — it's about time you stopped! Aggravating yourself and making useless calculations of hypothetical situations are counterproductive. Avoid them!

12. There is no such thing as luck, a hunch, a feeling, intuition, or forecasting hot and cold dealers or tables. The latter, in particular, is fiction. You are to play mechanically, according to sound mathematical principles, as if you were a machine. If you do not wish to play this way, DO NOT PLAY AT ALL!

13. Finally, if you are challenged or spotted by a pit boss or "counter-catcher," do not get involved in a dispute. Do not become excited or indignant. You are in a no-win situation and should act accordingly. After cashing your chips, leave the casino promptly and without discussion. After you have left, a description of the personnel and the circumstances involved in the incident should be written down.

Note: A special supplement dealing with additional playing procedures for the Las Vegas game will be provided at a later date. It should be incorporated into your handbook after you receive it.

VII. ACCOUNTING PROCEDURES

1. You will receive your day's bankroll at a designated meeting place prior to the commencement of the team play. (See II.)

2. Keep this money separate from all other money you may have on your person.

3. Carry your wallet (or wad) in an inside jacket breast pocket or a front pants pocket. Do not carry the wallet in a rear pants pocket.

4. As you begin your play, isolate the 30-unit session bankroll so that you will know where it is at all times and if it is gone.

5. Record all results diligently and accurately. Be precise and careful. You are to record time played and the outcome (both gross and net win, after tips).

6. Count your chips before cashing in and know ahead of time how much you will be receiving from the cashier.

7. Midway through a day's play, there will be a brief meeting, the time and location of which will be given to you by the manager at the start of the day. The purpose of the meeting will be to assess the cash flow and balance out the capital among team members. Please be prompt to this get-together, as time is money. It may be possible to combine a joint lunch with the meeting.

8. Convert all odd bills (20s, 10s, etc.) into as many 100s as possible. At the end of the day, only amounts below $100 should not be in $100 bills.

9. All incidental expenses are incurred by the player: Food, lodging, entertainment, tips to the cocktail waitress, and transportation are to be paid as out-of-pocket expenses and will not be borne by the team as a unit. *[Note: This was changed with later teams. — D.S.]*

10. Expenses for polygraph testing, safe deposit box rental, and room rentals for initial and future team meetings will be deducted as overhead from gross team winnings.

11. One or both of the team managers will keep all funds in a safe deposit box at all times when the team is not playing. They will answer, on a polygraph test, any questions relating to the safekeeping of the team's bankroll.

VIII. DISTRIBUTION OF WINNINGS

1. Winnings will be distributed once the initial 480-unit bank is doubled and again each time another 480 units are won. All disbursals will be in cash.

2. Assuming initial equal investments by all players, 90% of the net win will be returned to the player-investors in the same proportion as their contribution to the number of sessions played. See #4, below, for a further discussion of this procedure. (In accordance with III.15, a player will not be penalized for leaving a casino early after a big win. Consider the leisure time as a small bonus for having been so successful!)

3. 10% of the net win will be returned to the winning players in the same proportion as their contribution to the gross win. Although losing players will not participate in this distribution, they will not be penalized for a loss.

4. Assuming an initial investment of 40 units per player-investor, each player-investor will be said to have contributed four (4) shares (of 10 units each) to the joint bank. Each of the two team managers will be credited with four such shares although no monetary contribution to the joint bank will be made by the managers. Thus, for the purposes of calculating the disbursals in #2 above, the eight extra shares of the managers must be included before a division is made.

5. In the event that the team experiences a loss equal to 50% of the initial bankroll, an additional investment of 20 units will be required to permit the team to continue play at the same level. (For those familiar with stock market investments, consider this as a "margin" call.) No additional funds will be called for if the team continues to lose. In such an unlikely event, play will continue on a reduced scale in proportion to the lower bankroll and will be adjusted in either direction according to the outcome of play.

6. If a new player or players join the team, there will be a disbursal of profits (no matter what the size) before the new capital is received. This will permit a start "from scratch," so as to facilitate calculations and share distributions.

IX. POLYGRAPH TESTING

1. Periodic polygraph testing will be performed on all players.

2. Two basic questions will be asked: (1) "Have you reported all results honestly?" and (2) "Have you followed, to the best of your ability, the guidelines set forth in the handbook?"

3. Players who fail a test will be extended the courtesy of a re-test. Players who fail the re-test for question #2 may be excluded from the team. Players who fail the re-test for question #1 will be dismissed summarily from the team. We recognize the fallibility of the polygraph test. A person who fails test #1 twice may actually be innocent of any wrongdoing. However, in order to protect the members of the team, it will be necessary to dismiss any player who does not pass the polygraph test as described above.

X. PLEDGE

I, ________________________, do hereby agree to conform to the guidelines for team play as set forth in the Team Handbook. I realize that failure to comply with these principles may result in my exclusion from the team. I accept the terms and conditions upon which team play will be based.

Date:____________________

Signature:________________________

From "The Gospel According to Don," *Blackjack Forum*, June 1993:

Q. *I have played extensively on blackjack teams and am also currently bankrolling a couple of players with whom I maintain an investor/player relationship. A thorny problem has arisen and I thought you might be able to help me solve it. We have agreed that, since I put up all the money and the players do all the playing, a "fair" number of hours of play to warrant a 50-50 split of any potential profits is 240, or roughly the number of hours for which the expected win equals the size of our 400-unit bank. (Our win rate is 1.68 units per hour.) The players have approached me and requested a "payout" sooner than the 240-hour mark. Specifically, they asked for a division of profits at each successive 40-hour level. My problem is: How do I calculate the percentage that should go to the players for this smaller increment of time? Intuitively, since greater swings and more risk are involved over the shorter period, it is obvious to me that the players should receive less than the original 50%. The question is, how much less?*

A. This is one of the most interesting questions I've received in quite some time. I'm reasonably certain that it has never been discussed before in the literature and that my answer will break new ground. I apologize in advance for the rather complicated math that's involved, but, after researching, with a colleague, the methodology for a solution, there appeared to be no simple route. Before I begin, I would like to thank Bill Margrabe for his guidance and help with the mathematical analysis that follows. Bill's derivation and subsequent solution of the integral needed to resolve this problem were invaluable.

First, I'll describe conceptually what we need to know to find our answer, then I'll provide the math. Not all 40-hour periods of play will produce a win. For those that are losers, the investor bears the entire brunt of the loss. So, we need to know, on average, how often a 40-hour play results in a win. Fortunately, this is very simple.

I have already worked out such questions in my "Ups and Downs of Your Bankroll" article *[now Chapter 2 — D.S.]* and will furnish the formula necessary to do the calculation a little later. What follows is the hard part!

In addition to knowing how frequently our segment of time produces a win, we need to know the magnitude of that win — that is, the average win when, in fact, 40 hours of play results in a win. Understand that this is very different from the average result of 40 hours of play, which adds up all the wins, all the losses, nets them, and divides by the number of 40-hour samples. We know, for example, that at 1.68 units per hour, our average 40-hour period produces a win of 67 units (1.68 x 40). But this is not the number we're interested in finding.

Let's digress for a moment to define all of the terms we'll need for this calculation. Let b = total bankroll (in units); μ = one-hand expectation; σ = one-hand standard deviation; h = number of hands required, on average, to double the bank (for which a 50-50 split is deemed "fair"); n = number of hands after which each split will actually take place; x = probability that n hands of play produces a win; w = average win when n hands produces a win. I've purposely used *per-hand* results, rather than per hour, since the style of play (back-count, play all, etc.) can change the hourly results.

Now, it turns out that, in order to determine the average win, when there is one, we need to find the solution to an integral calculus problem, the details of which I will not burden you with. Suffice it to say, after some complicated integration, we obtain the desired result, w, which is equal to:

$$n\mu + \frac{\sqrt{n}\,\sigma}{\sqrt{2\pi}} \cdot \frac{e^{\frac{-n}{2}\left(\frac{\mu}{\sigma}\right)^2}}{N\left(\sqrt{n} \cdot \frac{\mu}{\sigma}\right)}$$

Don't panic! Some immediate simplification is possible. First, we observe that $n\mu$ represents, simply, the expected win after n hands of play and $\sqrt{n}\,\sigma$ is the standard deviation for the same n hands. $1/\sqrt{2\pi}$ is a constant term, very nearly equal to exactly .4 (actually, .3989). Finally, when n is relatively small and $\frac{\mu}{\sigma}$ is also very small, e (the base of the natural logarithm system) raised to this rather small power is very nearly equal to one, and thus may be ignored.

I promised earlier to provide the formula for calculating the probability of being ahead after n hands of play. It appears in the denominator of our expression for w as $N\left(\sqrt{n}\cdot\frac{\mu}{\sigma}\right)$where N( ) is the area under the normal probability density curve to the left of the number contained in parentheses. What all of this means is that a description in words of the approximate average win produced by the play of n hands would be: "The expectation for n hands of play plus four-tenths of the standard deviation for n hands of play divided by the probability of being ahead after n hands of play."

Finally — and this is the answer you've been waiting for — a little algebra tells us that the correct percentage, p, to pay the players after n hands is (100 x .5bn)/xhw. I'll now provide an illustration of the 6-deck, back-counting, Atlantic City game described in the "Ups and Downs" article in Chapter 2. I've chosen this game because it seems to represent the style of play outlined in your question.

For this approach, b = 400, μ = .063 (win/hr. divided by hands played/hr. = win/hand), σ = 4.70, h = 6,425 (26.77 x 240), n = 1,071 (26.77 x 40), and x = .67 (from Table 2.2, or from the formula in this article). So,

$$w = 67 + \frac{.4\sqrt{1{,}071}\cdot e^{\frac{-1{,}071}{2}\left(\frac{.063}{4.70}\right)^2}}{.67} =$$

$$67 + \frac{.4(32.7)(4.70)(0.9)}{.67} = 67 + 83 = 150.$$

Thus, p = (.5 x 400 x 1,071)/(.67 x 6,425 x 150) = 0.33 = 33%. Note how we can check the answer. We know that after 240 hours, the players should receive half of the doubled (we hope) bank, or 200 units. There are six 40-hour segments in 240 hours, and 67% (2/3) of those six segments, or four, are winners. The average win is 150 units when we win. So the players win 150 x 4 = 600 units. They need to keep 200 to be "even," so 200/600 = 33%. From the investor's viewpoint, he will keep 100 units out of each of the four 150-unit wins, for a total of 400 units. This is necessary because the average loss turns out to be 100 units, and the investor suffers two of those, or 200 units, on his own. When all the smoke clears, after 240 hours, the investor, as the players, nets 200 units, as it should be.

A few final comments. The formula contained in this article permits us to get a "feel" for how much we win, when we win, for any number of hours of play, and also for how much we lose. We all know, instinctively, that although we may win on average, say, two units per hour, our actual hourly results are much more volatile than those two units. For fun, why not try to calculate, for the game described in this article, what the average win is for one hour, when that hour produces a win. I'll provide the answer (upside down) at the end of this chapter so you can check. Hints: 1) Probability of being ahead after one hour is about .53 (for those who don't have access to the normal distribution charts *[Note: Those charts are now provided at the end of Chapter 8 in Tables 8.9 and 8.10. — D.S.]*); and 2) You can ignore the e term.

Peter Griffin may be able to simplify all of the above by dusting off his trusty "Unit Normal Linear Loss Integral" (p. 87, *The Theory of Blackjack*). I showed this article to Prof. Griffin and he both corroborated the accuracy of my approach and suggested an easier solution. Here is his methodology:

1. Determine for the specified number of hours (40, in our case) the expected win and the standard deviation (μ and σ).

2. Standardize the break-even point by finding $z = \frac{0-\mu}{\sigma}$. In our case, we have $z = \frac{0-67}{154} = -.435$.

3. Look up the UNLLI for $|z| = .435$. We'll call the result $U \approx .2170$.

Now, the sum of possible losses times their associated probabilities is given by σ x U = 154 x .2170 = 33.4.

Since we know that the expected win for the 40 hours of play (μ) is simply $\sum$ win x Pr(win) – $\sum$ loss x Pr(loss), it is clear, by substituting and rearranging terms, that $\sum$ win x Pr(win) = μ + (σ x U) = 67 + 33.4 = 100.4.

Finally, the percentage, p, to pay the players must satisfy the equation p x $\sum$ win x Pr(win) = μ/2. (Recall the 50-50 split of profits with investors.) Thus, p x 100.4 = 67/2, whence p = 33.4%.

Notice that it was unnecessary, in order to solve for p, to know separately the average win when we win (150) or the probability of being ahead after 40 hours of play (67%). It suffices that we know the *product* of these two quantities. So, if you use Prof. Griffin's method but are still curious as to the magnitude of the average win, you'll have to do the extra $N\left(\sqrt{n} \cdot \frac{\mu}{\sigma}\right)$ calculation and then simply divide $\sum$ win x Pr(win) by this number.

(Average win = 20 units.)

Postscript: *In 1996, in a three-part series (the second of which was my "Team Handbook" article that you've just read), Arnold Snyder published, in* ***Blackjack Forum,*** *a most revealing study of the high-stakes team approach to blackjack, as it exists today. In my opinion, the series becomes, along with Ken Uston's legendary* ***The Big Player,*** *required reading for anyone contemplating forming his own blackjack team or playing on one.*

Chapter 12

More on Team Play: A Random Walk Down the Strip

The following is an attempt to inject a bit of rationality into the time-honored discussion of compensating team players, at least partially, according to their actual results at the table. It is the offshoot of a spirited debate that took place awhile ago on Richard Reid's *bjmath.com* site, when a highly respected team manager and yours truly exchanged several lively posts on the topic. You may be surprised by some of my conclusions, but I would ask you to be patient, read all the way through, and keep an open mind. Hey, you might actually learn something!

We have a team of five players, Al, Bob, Cal, Don, and Ed, each of whom thinks he is God's gift to the blackjack tables. Each is, in his mind, the greatest player in the world, never makes a mistake, and is honest as the day is long. The team is managed by an experienced, wily veteran, Mark, who has seen it all. He decides that a hefty percentage of the players' compensation will be determined by their individual results at the tables. You've got to give the boys some incentive, no?

The count employed is Hi-Lo, utilizing the I18 and "Fab 4." The game is 6-deck, s17, das, ls (standard Strip fare), and back-counting will be employed at all times. The bank is $100,000, and the decision is to play half-Kelly, making the unit (starting bet) $300, while the top wager will be six units, or $1,800. Playing in such fashion, risk of ruin is kept to just over 2% (see Table 12.1).

TABLE 12.1
The Simulation Settings

BJRM - The One Second Simulator

Menu Help

Sim Source: (•) BJRM ( ) SBA Extras ( ) Purchased Extras [?]

Count System: hilo | Rules: S17DASLS | Load | Bank ($) 99900 | Unit ($) 300.00 | Adjust Unit$

Decks in Play: 6 | Penetration: 4.5 of 6 | Calc | Bank (u) 333 | OptB (u) 173 | <- Cpy OptB

Avg Bet	W/L%	SD/hnd	SD/100	Win/100	DI	Hnds to N0	ROR	Hnds to Dbl	Exp Grwth Rate
1.94	1.94	2.55	13.15	1.00	7.62	17,203	2.11%	33,203	0.002229 %
$ 582.00		$ 765	$ 3,945	$300.00					

Game conditions: I18 (and Fab4). Four players total. We're on 3rd Base. Face Up game.

Sim Raw Data, and Bets

# of Hnds: 1

tc	<= -3	-2	-1	0	1	2	3	4	5	6	7	8	>= 9
bet u	0	0	0	0	1	2	2	3	4	5	6	6	6
bet $	0	0	0	0	300	600	600	900	1200	1500	1800	1800	1800
freq	8.63	7.17	11.91	45.61	11.43	6.91	3.24	2.50	1.15	0.71	0.29	0.24	0.20
ev	-2.808	-1.571	-1.028	-0.300	0.505	1.118	1.704	2.315	3.024	3.857	4.445	5.250	6.473
sd	1.167	1.154	1.148	1.136	1.131	1.125	1.120	1.119	1.122	1.128	1.130	1.128	1.117
o bet	0.0	0.0	0.0	0.0	1.3	2.9	4.5	6.2	8.0	10.1	11.6	13.8	17.3

Quick Bet Options: Opt 1 to MaxBet | Use BJA Chpt 10 Bets | 2 Hnd Equivalents | Print Screen Image

Adjst EV: 0 | **MaxBet (x)**: 6

- DI * DI -

per 100 $58.06 | .95 | per Hr $55.16

Your Factor 100 | Apply | Copy Factor from BJClock

Find Spread Configuration That Maximizes DI

MaxBet(x) By TC/RC of: 9 | **Kelly (%)** 50 | (•) Wong ( ) Play All

BackCount at TC/RC of: 1

Calc Optimum Bets | [x] Use Integer Bets | [ ] 2x Jumps Max

Opt f: = 0.005968 | [ ] Find Best Wong-In | [ ] Play "2 Hands of"

SBA Detected

The guys will play 400 hours a year (100 hours per quarter), for five years: a total of 2,000 hours per player, 2,000 team-hours per year, 10,000 total hours, or exactly one million hands of blackjack back-counted by our peripatetic quintet. The individual hourly e.v. is exactly one unit, or $300 (convenient, no?). The hourly s.d. is 13.15 units, or $3,945. For any one player, the one-year e.v. is $120,000, and the one-year s.d. is $78,900. The five-year e.v. is $600,000, and the five-year s.d. is $176,426. For the team's yearly results, the one-year e.v. is $600,000, and the one-year s.d. is $176,426. Finally, for the team's five-year, 10,000-hour venture, the cumulative e.v. is a cool $3,000,000, while the s.d. is $394,500. So much for the theory; now, for some results (eventually I'll tell you where all the data came from).

In the first year, players Al, Bob, Cal, Don, and Ed win, respectively (in thousands of dollars): 157.8, 169.8, 262.5, 75.9, and –35.1. The team wins 630.9 (Table 12.2). As the guys expected to win 600 grand, what could be more perfect? Of course, the individual results are all over the place, with two players underachieving their $120,000 e.v.s (Ed actually loses), and three overachieving (Cal is the "star"), but, hey, we're talking s.d. here, right? Nonetheless, the members are compensated according to the manager's rather weird concept of "equity," and Cal is deemed to be almost 3.5 times "better" than Don. As for Ed, the rumors are flying, and his future with the team is very shaky indeed. Of course, Al, Bob, Cal, and Don know that all this innuendo is a crock, but they're team players, so they say nothing.

TABLE 12.2

The Team's Year 1 Results

Players	1st Qr	2nd Qr	3rd Qr	4th Qr	Player Total
Al	33,000	25,500	7,800	91,500	157,800
Bob	51,900	64,200	–4,800	58,500	169,800
Cal	22,500	55,500	121,800	62,700	262,500
Don	–13,200	16,800	92,400	–20,100	75,900
Ed	–27,300	–19,200	–61,200	72,600	–35,100
Team Total	66,900	142,800	156,000	265,200	630,900

Year 1 S.D. Events

Players	1st Qr	2nd Qr	3rd Qr	4th Qr	Player Total
Al	0.076	–0.144	–0.563	1.559	0.479
Bob	0.555	0.816	–0.882	0.722	0.631
Cal	–0.190	0.646	2.327	0.829	1.806
Don	–1.095	–0.335	1.582	–1.270	–0.559
Ed	–1.450	–1.247	–2.312	1.080	–1.966

Now, I'm going to let you in on a little secret. Cal and Ed are connivers. They have decided not to respect Mark's edicts, and they play, respectively, for double and half stakes. That's right, Cal is out to show everyone just how great he really is. He's hell-bent on being the best performer, and he wants his results to dazzle. Ed, by contrast, is scared to death of losing big and being thrown off the team. So, he decides, secretly, to play for half-stakes, in the hope of not having any big swings. He just wants to keep his "job." And, after his first-year loss, he's happy he wasn't playing for full stakes. Let's take an even closer look at that first year of play, dissecting the results from several revealing angles.

With $30,000 per quarter as the individual-player e.v., Al earns just what he should for the first half of the year (Table 12.3) before stumbling in the third quarter (Table 12.4). However, his fourth quarter comeback (Table 12.5) more than compensates for his brief misfortune, and he finishes happily above his yearly $120,000 e.v. at +$157,800.

Bob, on the other hand, gets off to a horrific start, with a 16-hour weekend from hell (Table 12.6). He drops a cool $23,100 and can only hope that his bad luck will soon turn around. Not to worry! By the end of the quarter, the red ink has vanished, and Bob is not only a winner, but is comfortably ahead of e.v. (Table 12.7). The second quarter brings more of the same good fortune, and at mid-year, Bob's "report card" shows a stellar double e.v., at +$116,100 (Table 12.8). My main tenet while playing this game: Never celebrate early. The third quarter is a bummer, and Bob actually loses $4,800 (Table 12.9). Ever the steady soul, he perseveres, sucks up his wounded pride, and grinds out another $58,500, to finish at a comfortable +$169,800 for the year — a solid, above-e.v. win (Table 12.10).

And then there is Cal. We already know of his deviousness — he's playing for double stakes. And, coincidentally, his first weekend produces the same $23,100 loss that Bob experienced (Table 12.11). But, hour 47 yields a resounding $12,000 win, and Cal never looks back (Table 12.12). Although his shaky start sets the tone for a mediocre first quarter (Table 12.13), the rest of the year is pure dynamite, especially in the eyes of the manager and Cal's teammates. The doubly duplicitous one reels off second, third, and fourth quarters of +$55,500, +$121,800 (!), and +$62,700, respectively (Tables 12.14–12.16), to end the year at an eye-popping +$262,500. And for this quarter-million-dollar spree, Cal is handsomely rewarded by Mark, who is easily impressed by this sort of positive-e.v. flux. Clearly, Cal made good on his boasts, and, quite literally, when the chips were down, he was The Man.

Not so Don. Nice first and second weekends get him off and running (Tables 12.17 and 12.18), but just when he's feeling his oats (+$44,100 after just 32 hours), a heavy dose of reality sets in and, believe it or not, in the next 54 sessions, Don loses 35 times (Table 12.19). He is now in the red, to the tune of $27,300, and is wondering if he'll ever dig out by quarter's end. The short answer is no. Don finishes at –$13,200 (Table 12.20) and vows to do better in the spring. He does manage to wind up in the black (+$16,800), but after six months of hitting the tables, Don has a mere $3,600 to show for his 200-hour effort (Table 12.21). I said it on p. 6, and I'll say it again: "The days follow one another, but don't resemble one another." How else to explain the torrid third quarter that Don fashions, blazing a trail to $92,400 worth of profits, and ending up at +$96,000, actually slightly ahead of expectation (Table 12.22)? Talk about a comeback! This guy Don may be all right, after all. Or not. For, unfortunately, the fourth quarter is yet another losing one, and Don gives back $20,100, to finish at +$75,900 for the year (Table 12.23) — a negative 0.559-s.d. event (Table 12.2).

But, if you think Don is a bit unhappy, consider good ol' conservative Ed, our half-stakes milquetoast. We already noted that he got pummeled in the first quarter, dropping $27,300 (Table 12.24). As if this isn't bad enough, he can't buy a win and drops $19,200 more in the second quarter (Table 12.25). After a mid-season pep talk from Mark, Ed is determined to do better. I'll let you in on another secret that you've probably figured out by now: If we could simply *will* ourselves to win, this endeavor would be like taking candy from a baby. Unfortunately, although wanting to win is probably a *necessary* condition for ultimate success at this wacky game, it most assuredly is not a *sufficient* condition for winning. And so, not only does Ed lose again in the third quarter, he does so with gusto, getting hammered for a whopping $61,200 (Table 12.26). The red ink is now at $107,700 as we head into the home stretch. Finally, the bad luck turns around, and Ed is able to get off the schneid, pocketing $72,600 for the last quarter (Table 12.27). Still, he is a net loser for the year — the ultimate disgrace. No bonus money for Ed. He figures he'll be lucky to have a job next year.

In that second year, the results are, in order (and again, in thousands of dollars): 149.1, 236.7, 144.0, 97.8, and 122.1. Interesting. The team has managed to win $749,700, and, although above e.v., the numbers are well within statistical norms (Table 12.28). Ed has rebounded nicely, and he's glad he didn't get thrown off the team at the end of last year. Cal has come back into the fold (no, he wasn't really the "better" player in year one), but Mark is still concerned about Don and Ed. Don has underachieved for the second year in a row. Perhaps he's too timid with his betting. Perhaps he just can't get the money out when he should. He will bear watching. He may not be quite the player he's cracked up to be. As for Ed, the manager reasons, "imagine playing 800 hours and winning only 36% of e.v. We're talking about a negative 1.37-s.d. result. These things are supposed to happen just 8.53% of the time, or once every 12 attempts." Although certainly well within the realm of possibility, Mark wonders if Ed isn't pocketing a few purple chips every now and then for his personal "rainy day" fund! Of course, the manager doesn't know Ed is playing for half stakes.

The team's third year is perfectly fine, as well. The results, 123.3, 165.9, 285.3, 70.5, and 26.7, add $671,700 to the till, but the same two players, Don and Ed, "underachieve," as they did in year 1. Cal, on the other hand, is once again the big winner, so maybe there really is something to his swaggering posture after all. He did "save" the team, didn't he? He "deserves" his bigger cut, doesn't he?

TABLE 12.3
Al's Year 1 6-Month Results

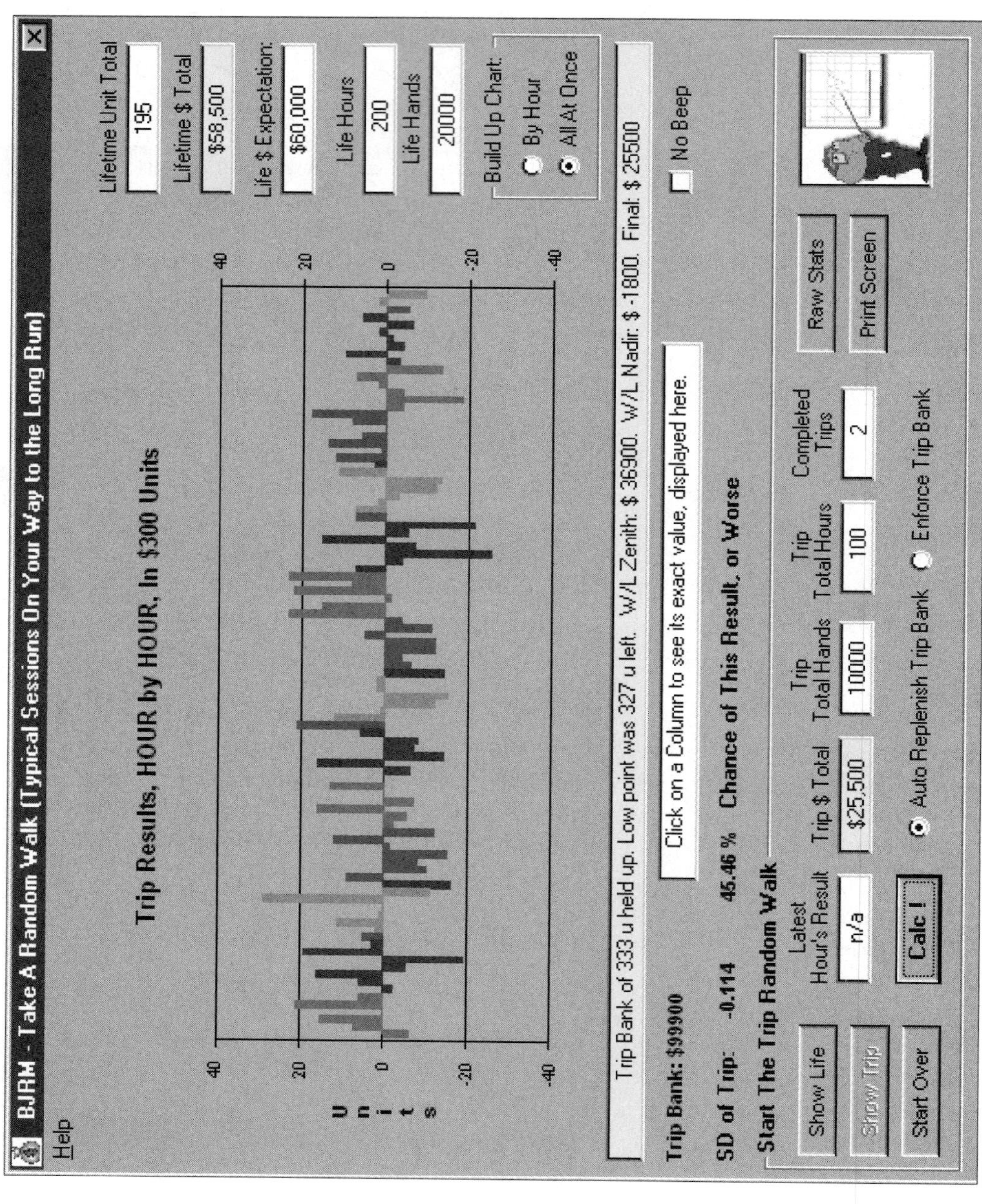

TABLE 12.4
Al's Year 1 Third Quarter Results

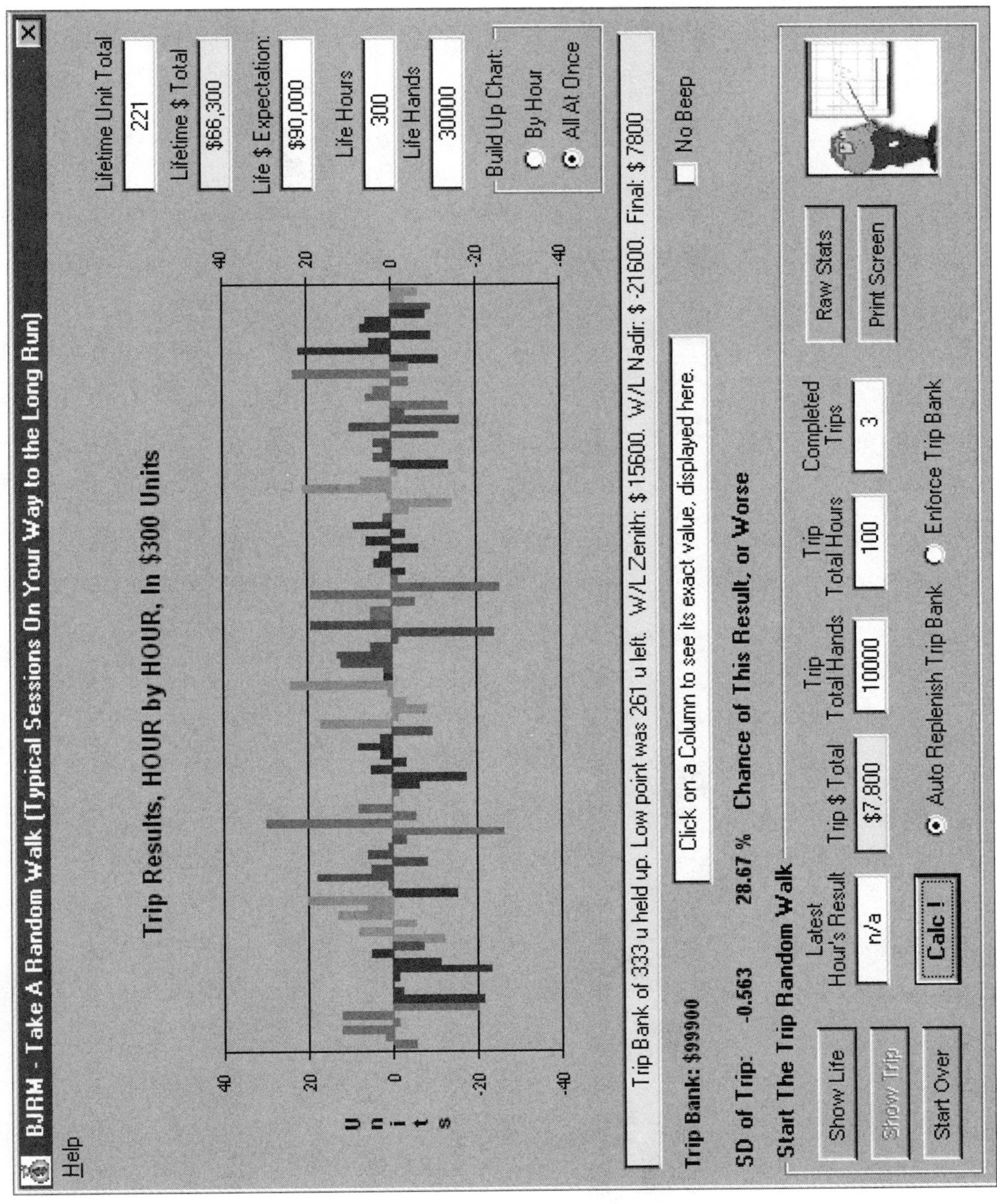

TABLE 12.5
Al's Year 1 Fourth Quarter Results

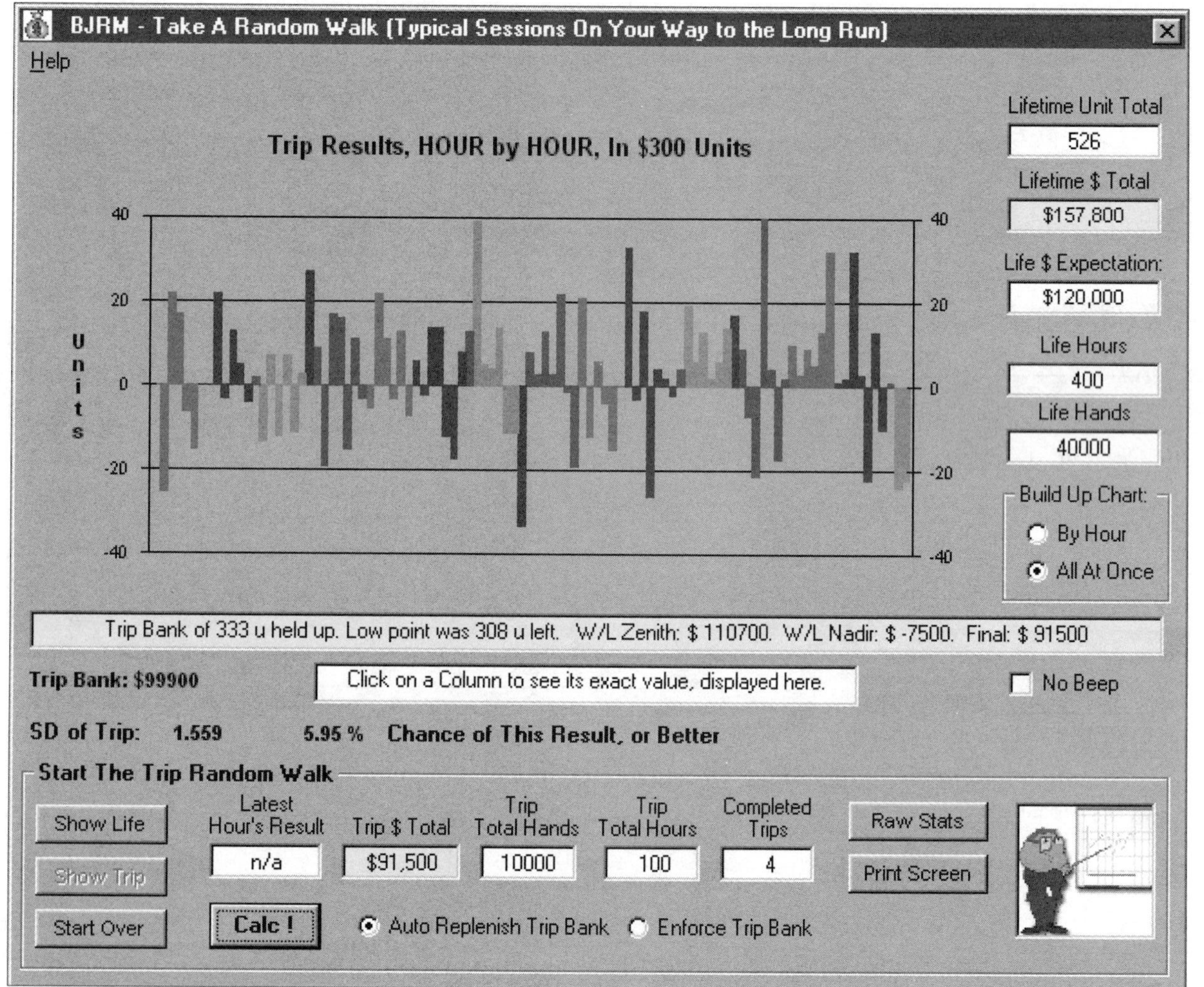

TABLE 12.6
Bob's Year 1 First Weekend Results

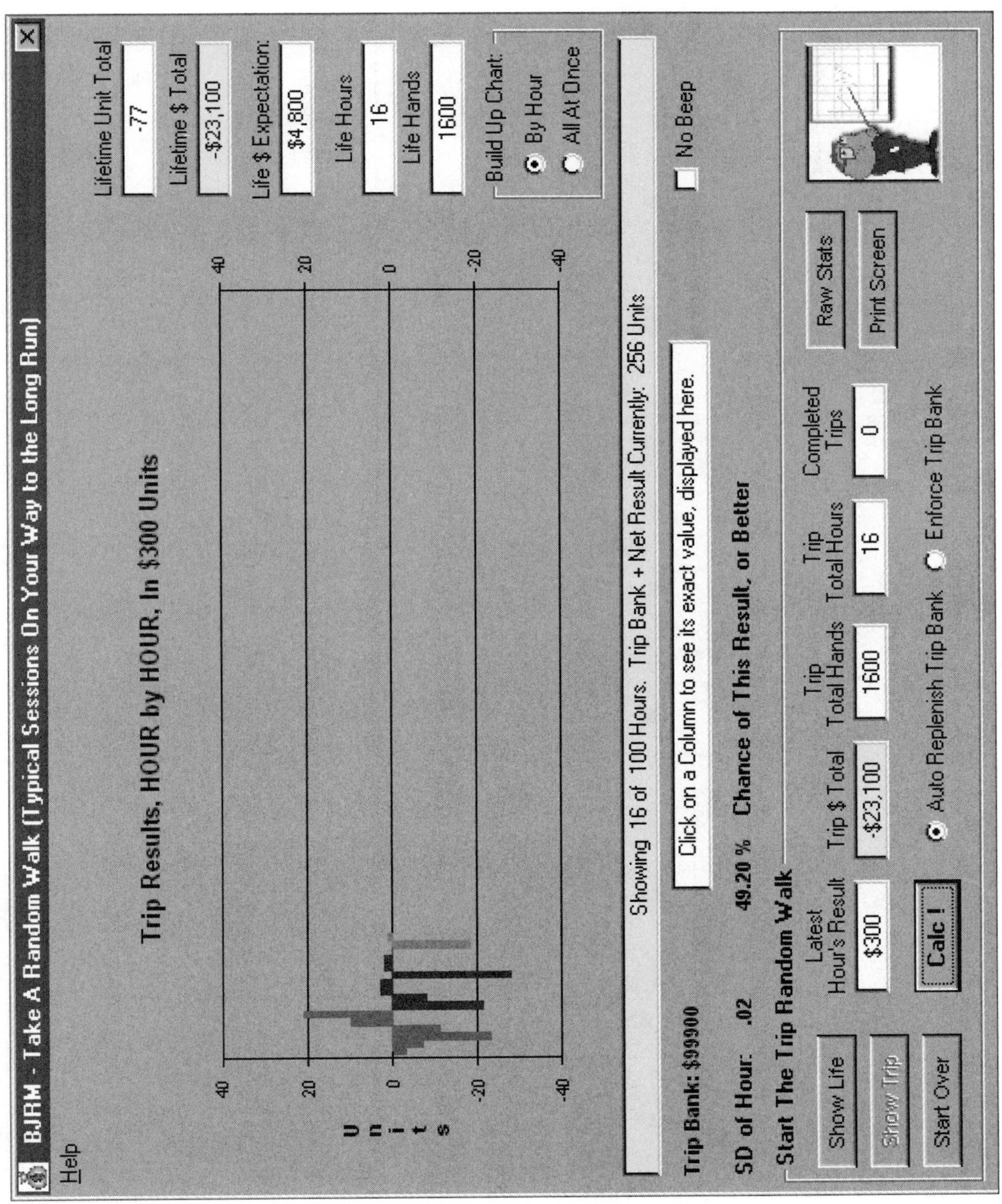

TABLE 12.7
Bob's Year 1 First Quarter Results

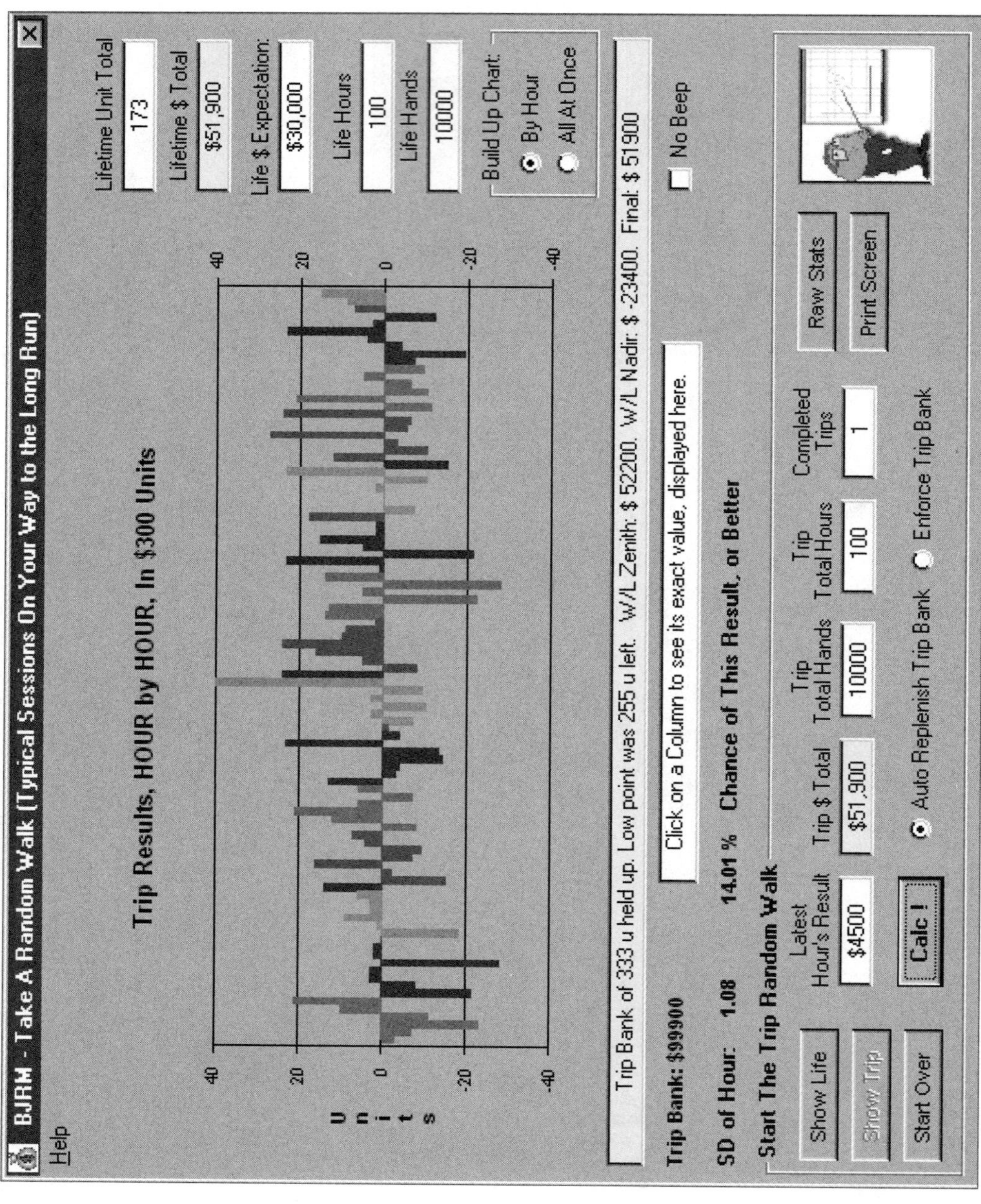

TABLE 12.8
Bob's Year 1 Second Quarter Results

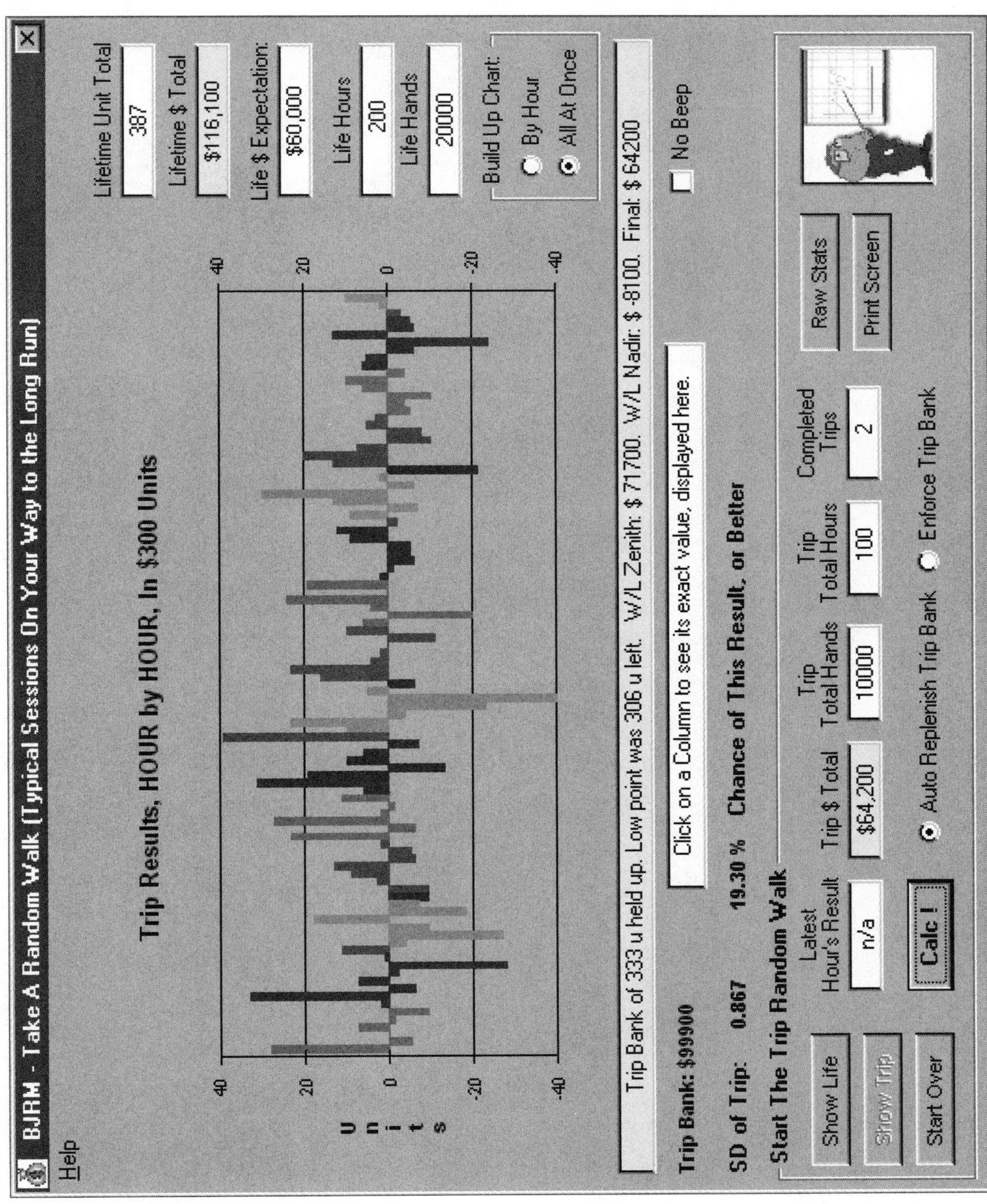

TABLE 12.9
Bob's Year 1 Third Quarter Results

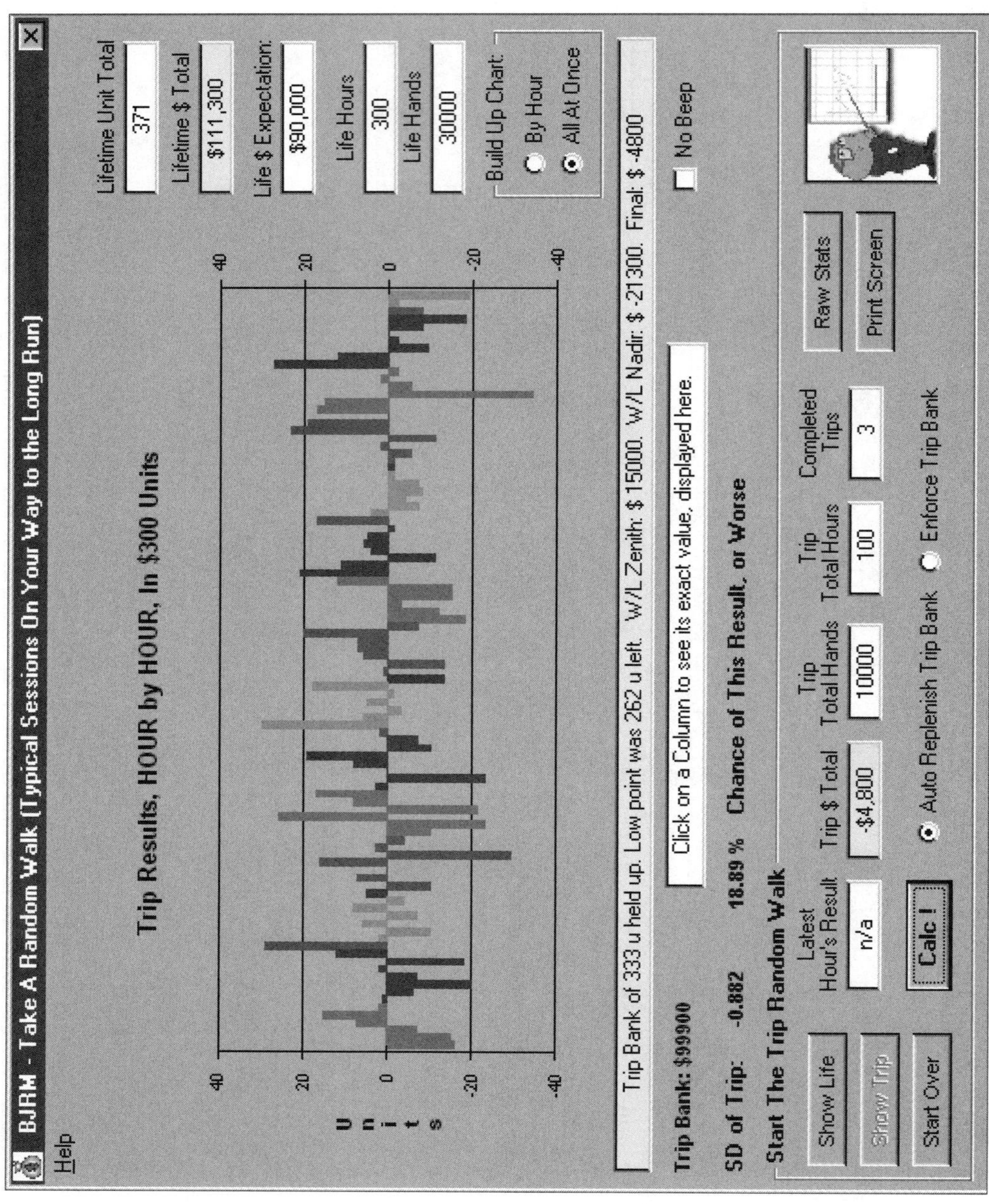

TABLE 12.10
Bob's Year 1 Fourth Quarter Results

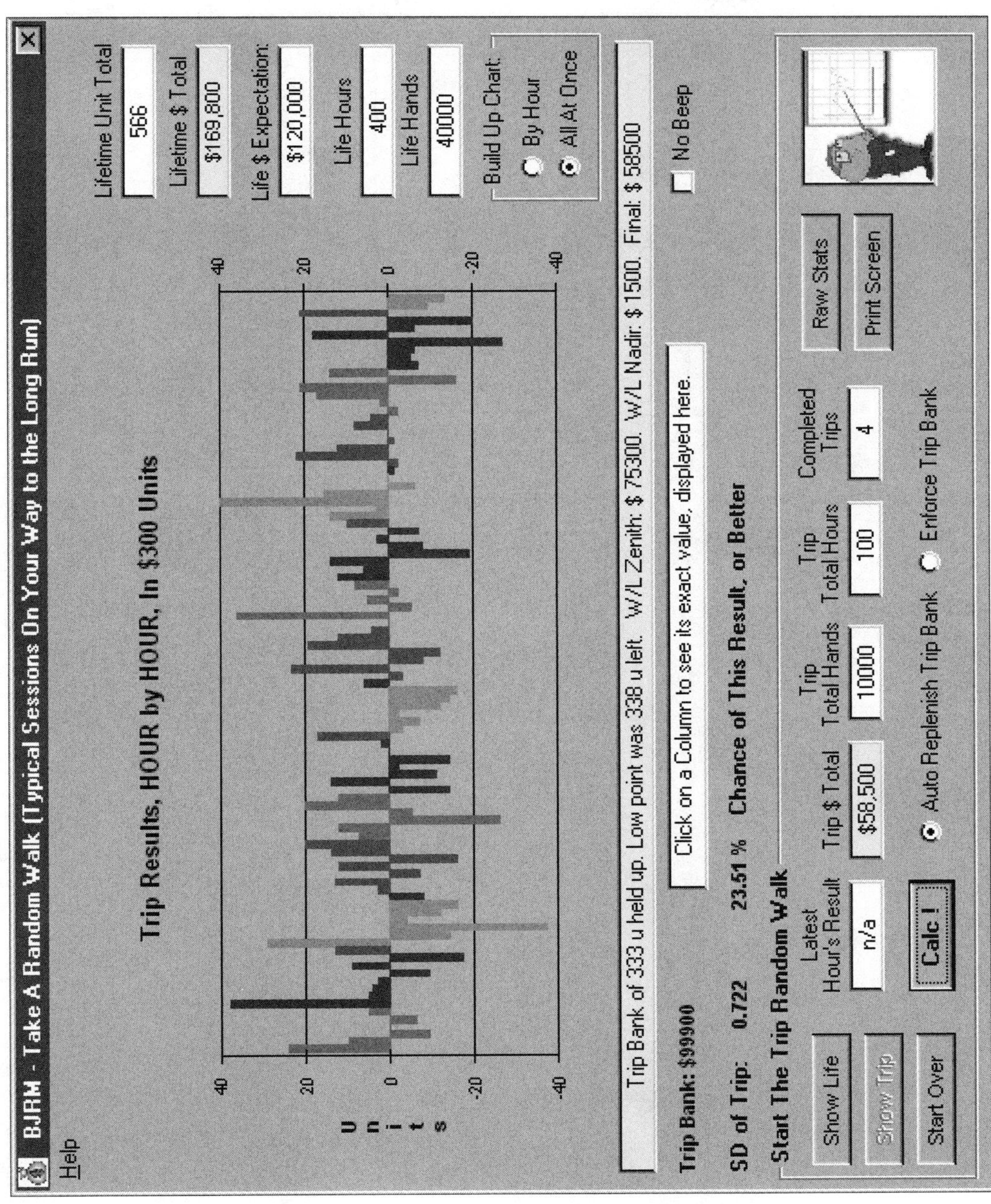

TABLE 12.11
Cal's Year 1 First Weekend Results

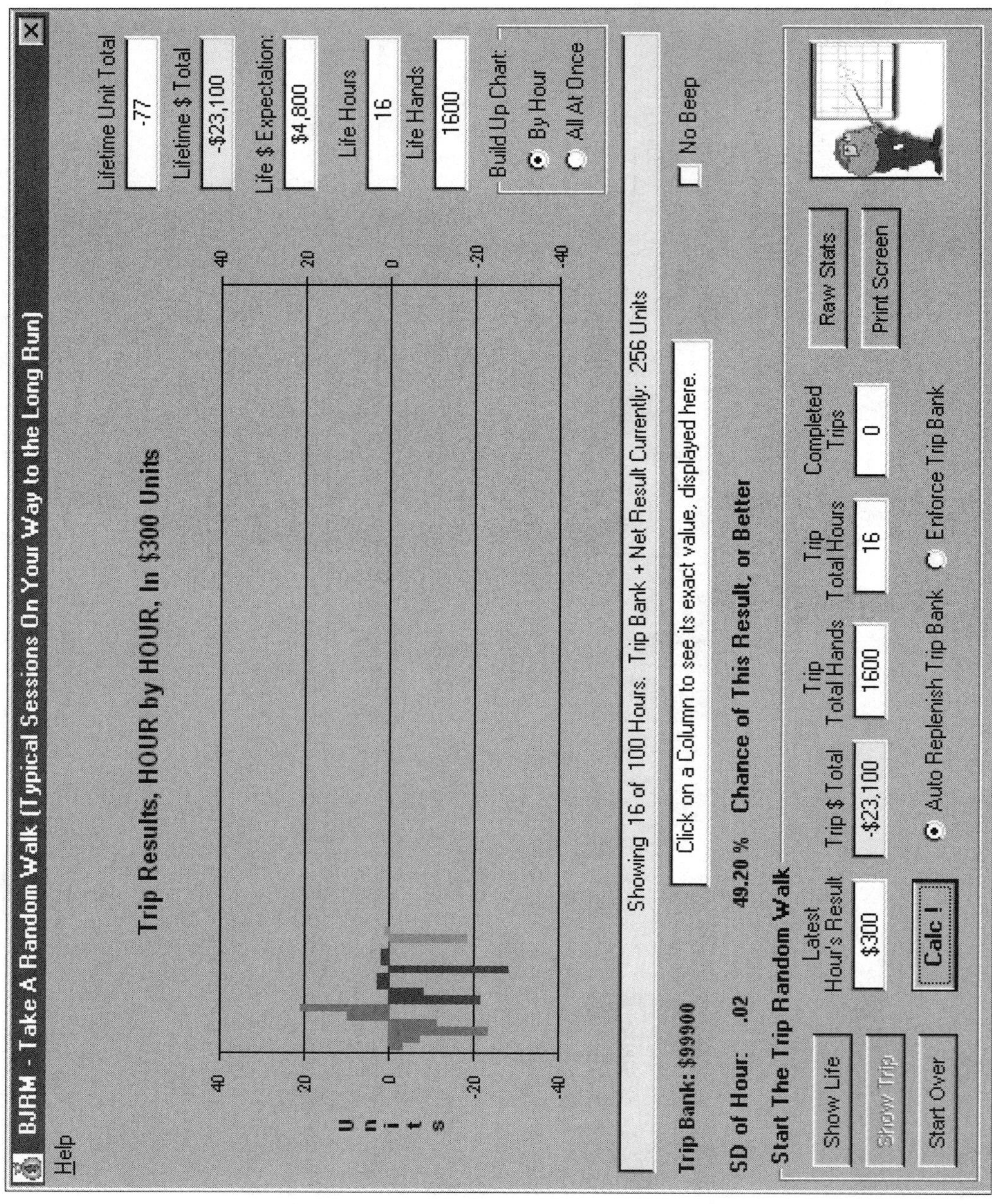

TABLE 12.12
Cal's Year 1 Results after Hour 47

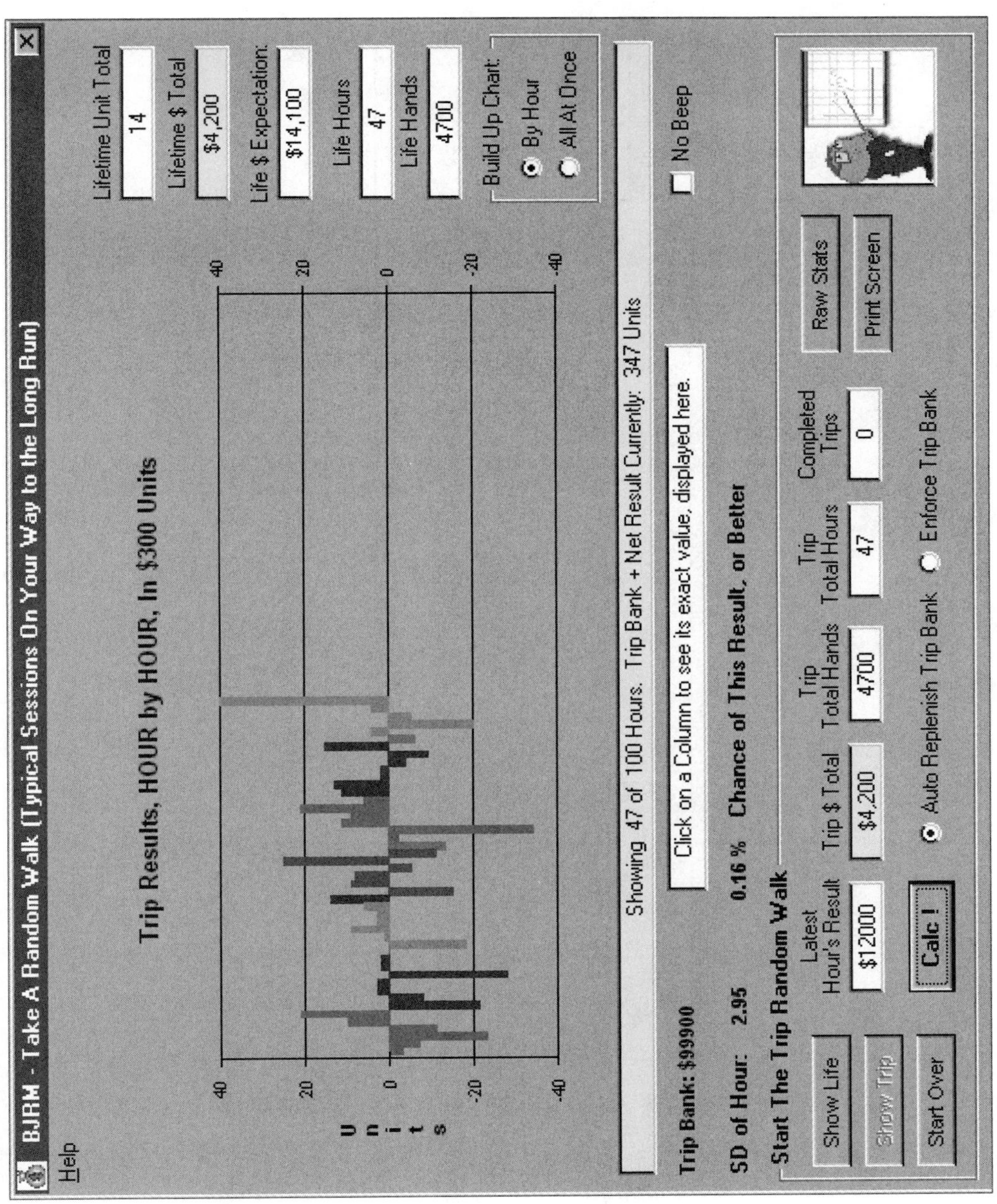

TABLE 12.13
Cal's Year 1 First Quarter Results

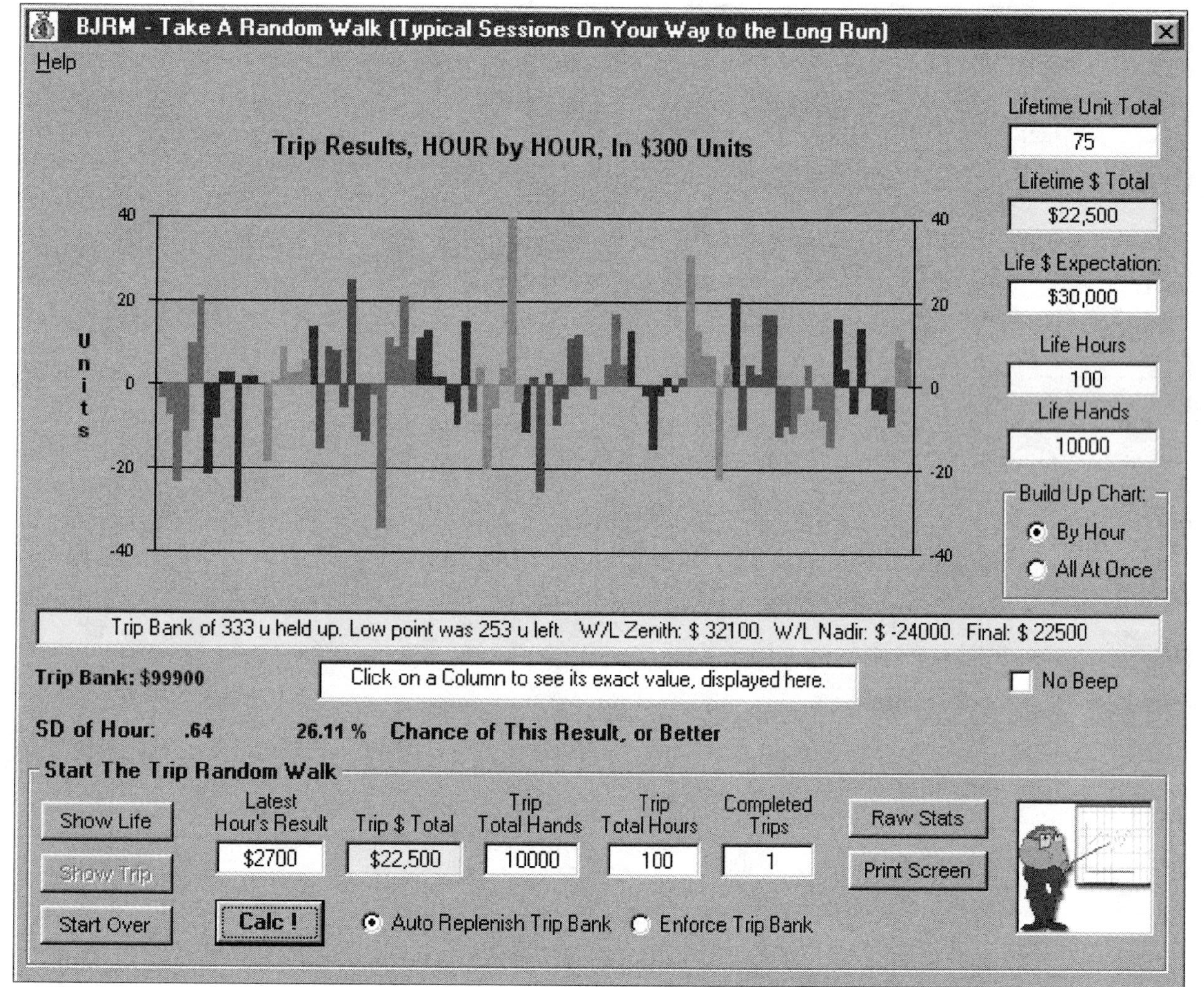

TABLE 12.14
Cal's Year 1 Second Quarter Results

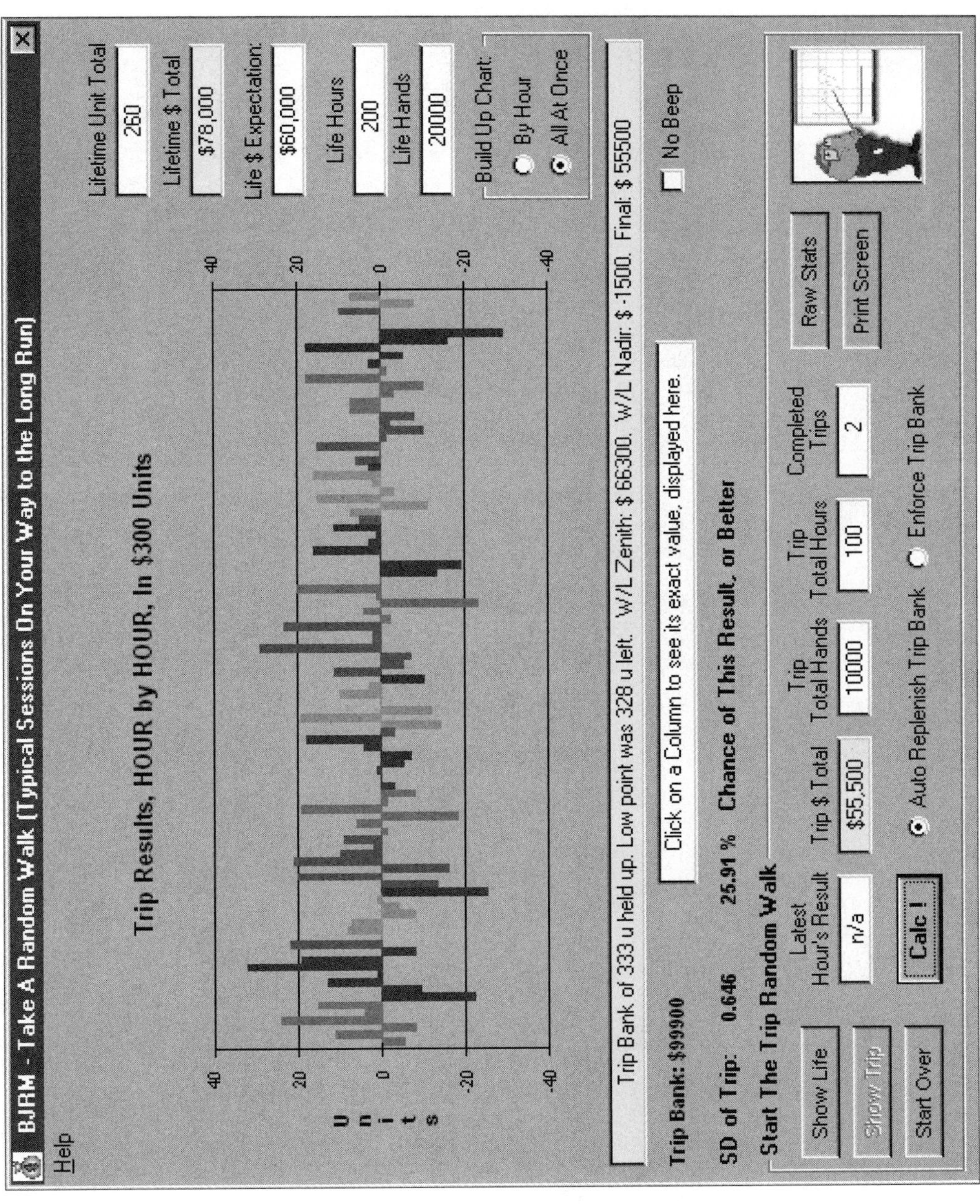

TABLE 12.15
Cal's Year 1 Third Quarter Results

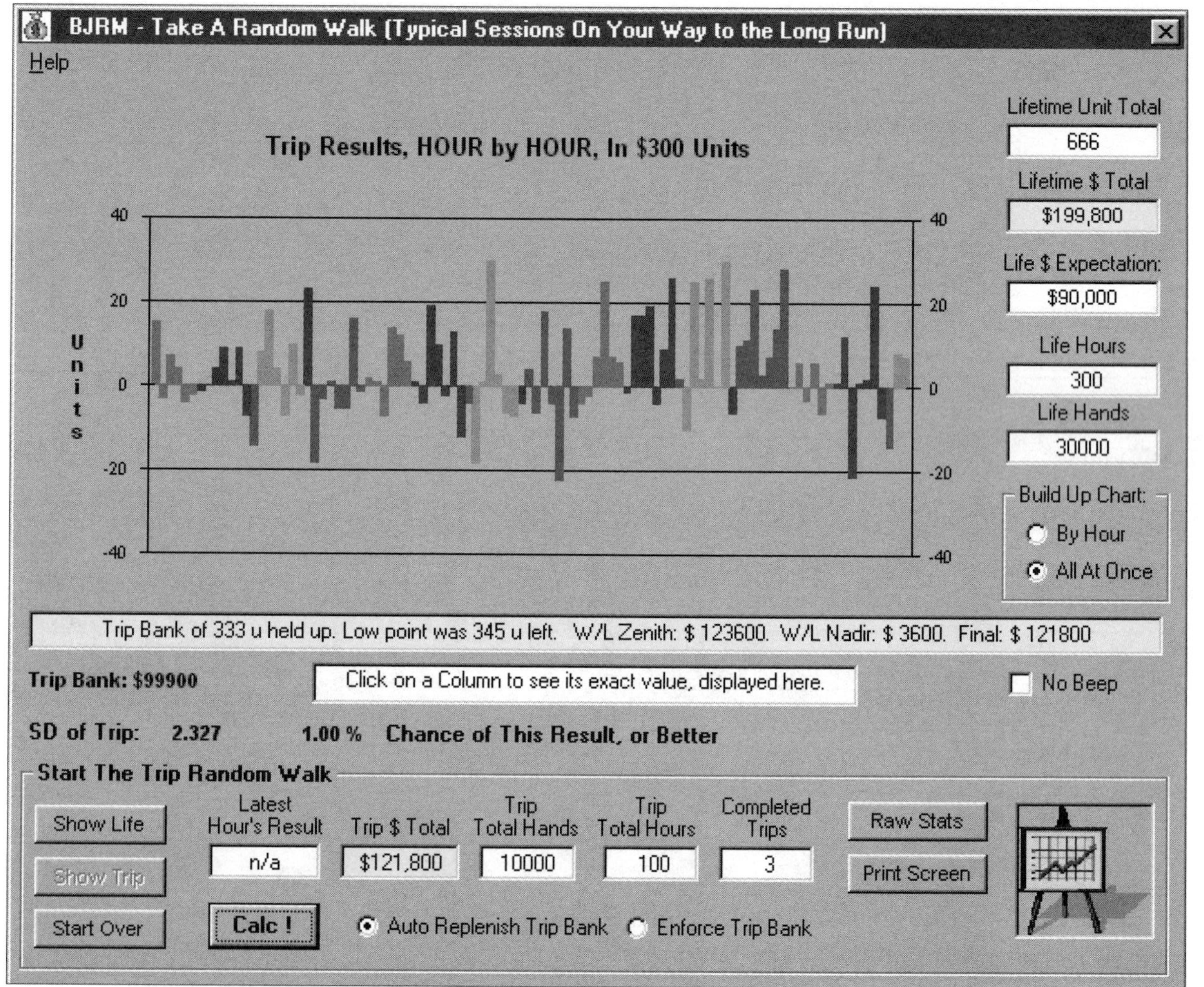

TABLE 12.16
Cal's Year 1 Fourth Quarter Results

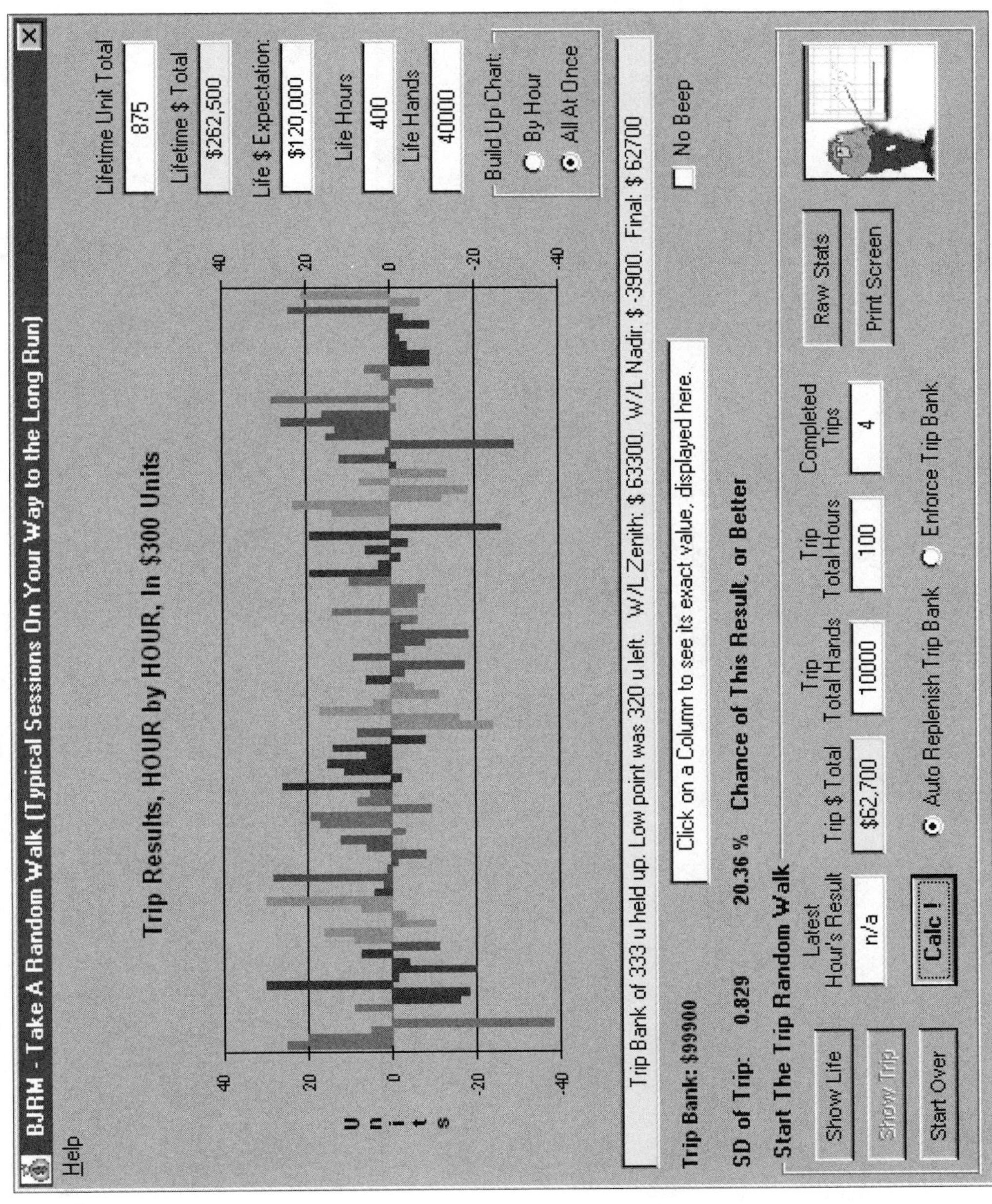

TABLE 12.17
Don's Year 1 First Weekend Results

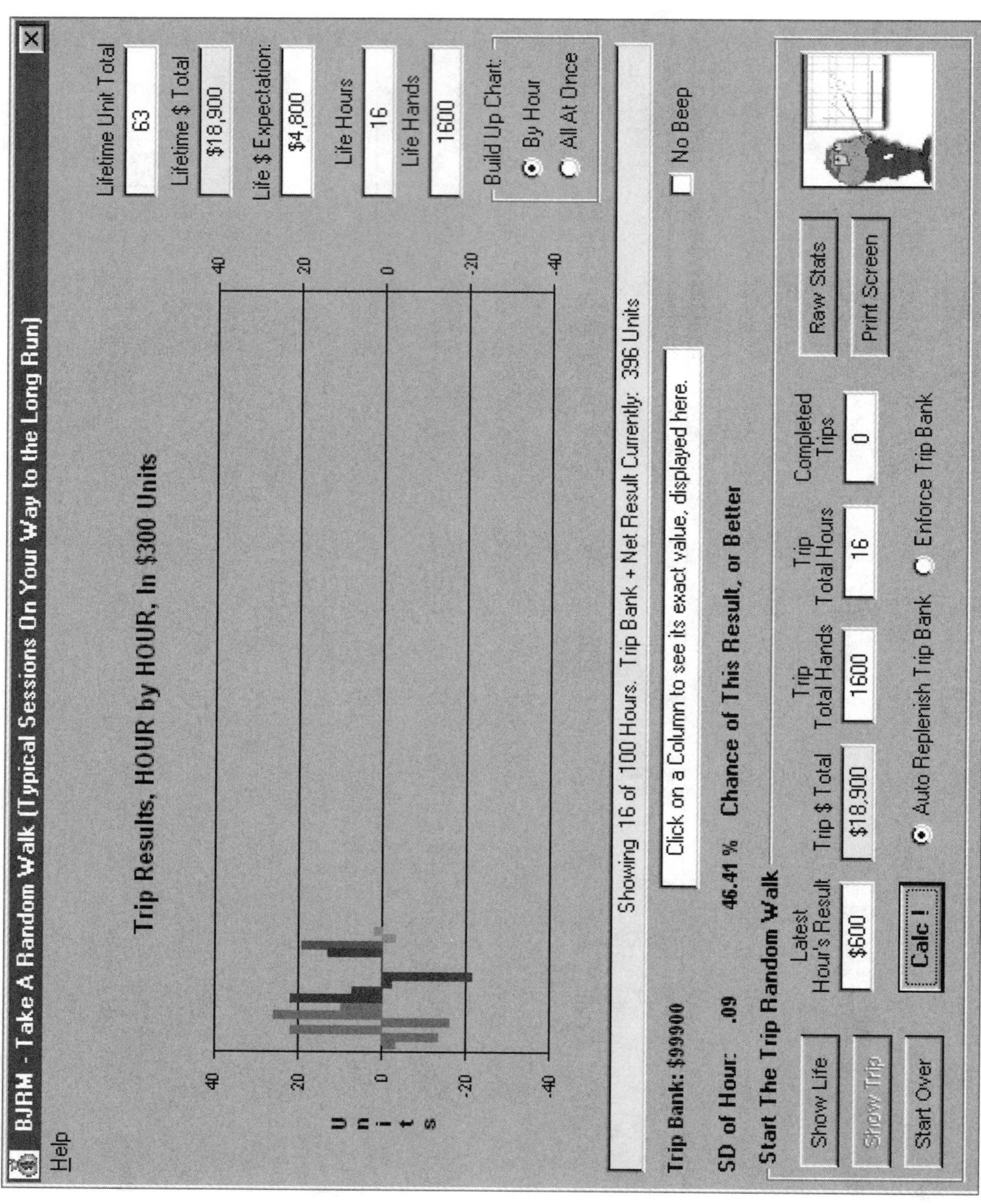

TABLE 12.18
Don's Year 1 Second Weekend Results

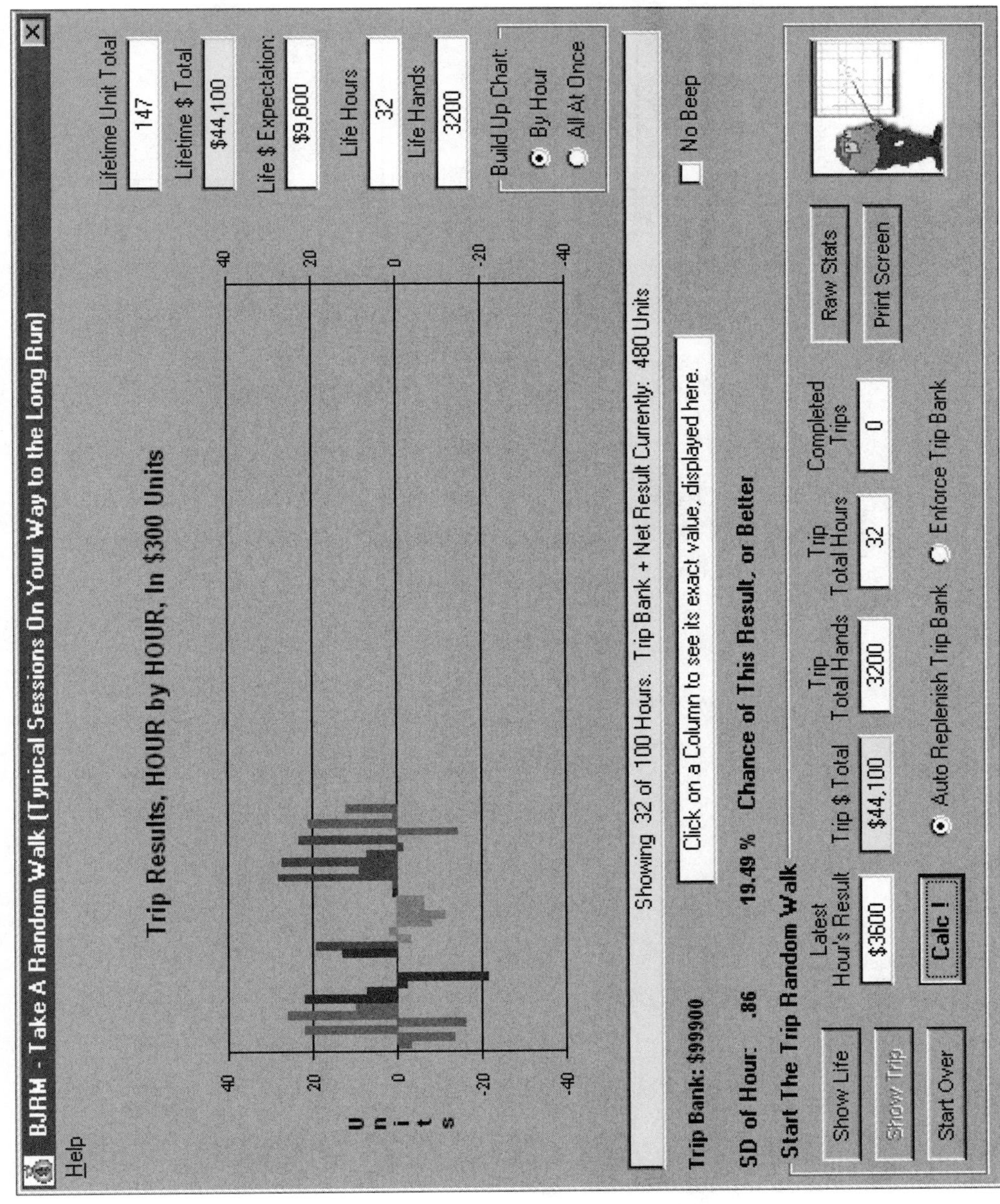

TABLE 12.19
Don's Year 1 Results after 86 Hours

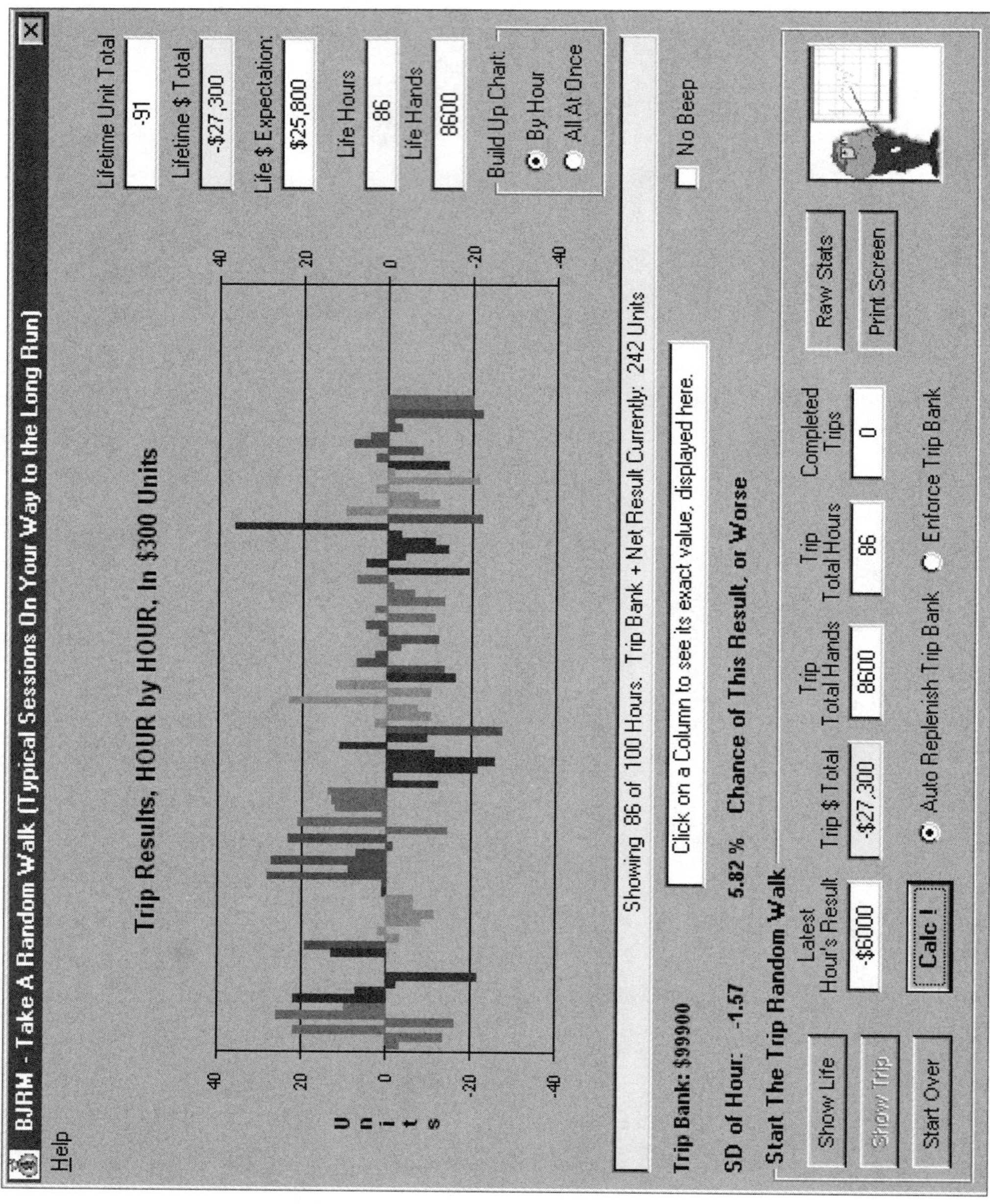

TABLE 12.20
Don's Year 1 First Quarter Results

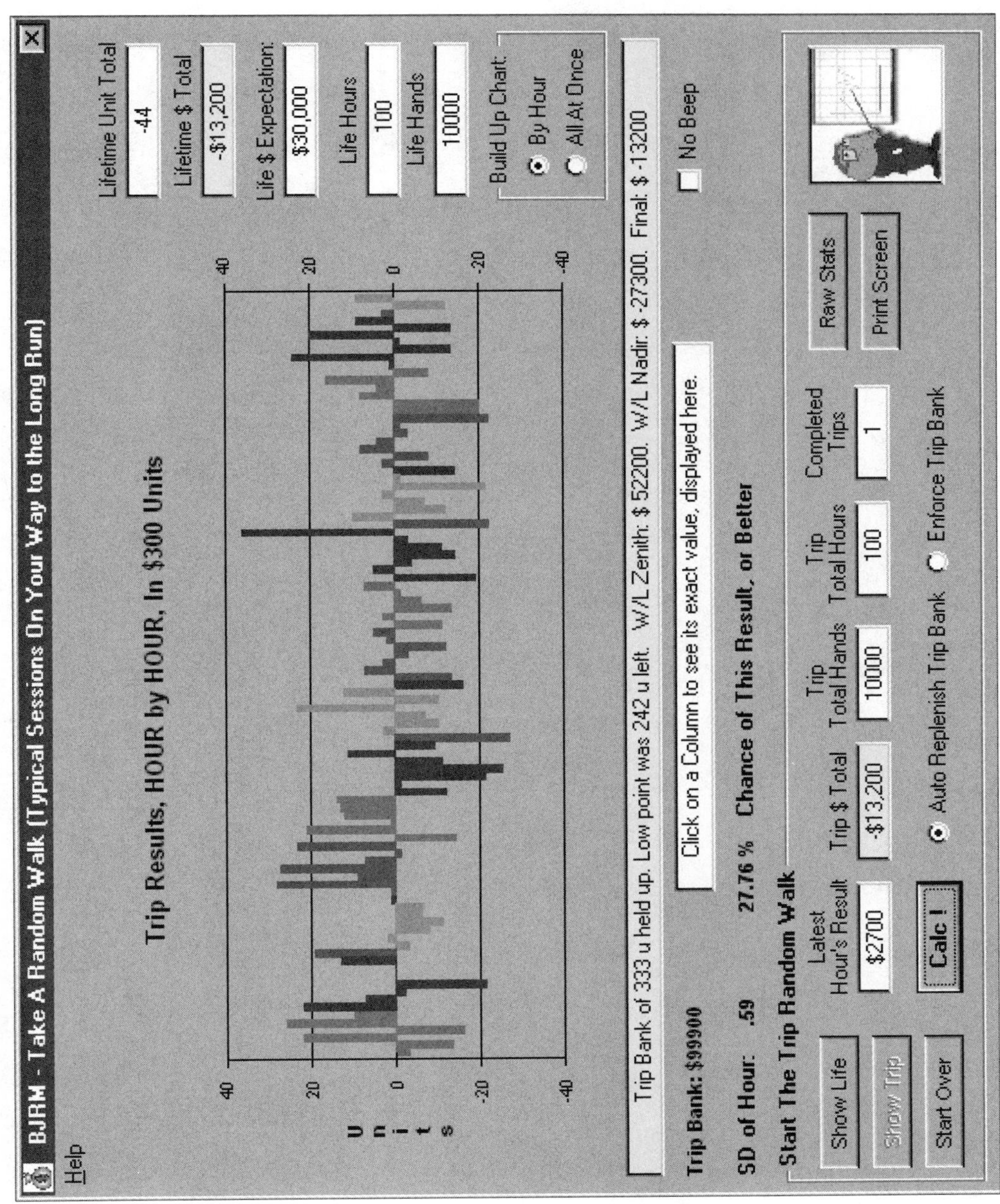

TABLE 12.21
Don's Year 1 Second Quarter Results

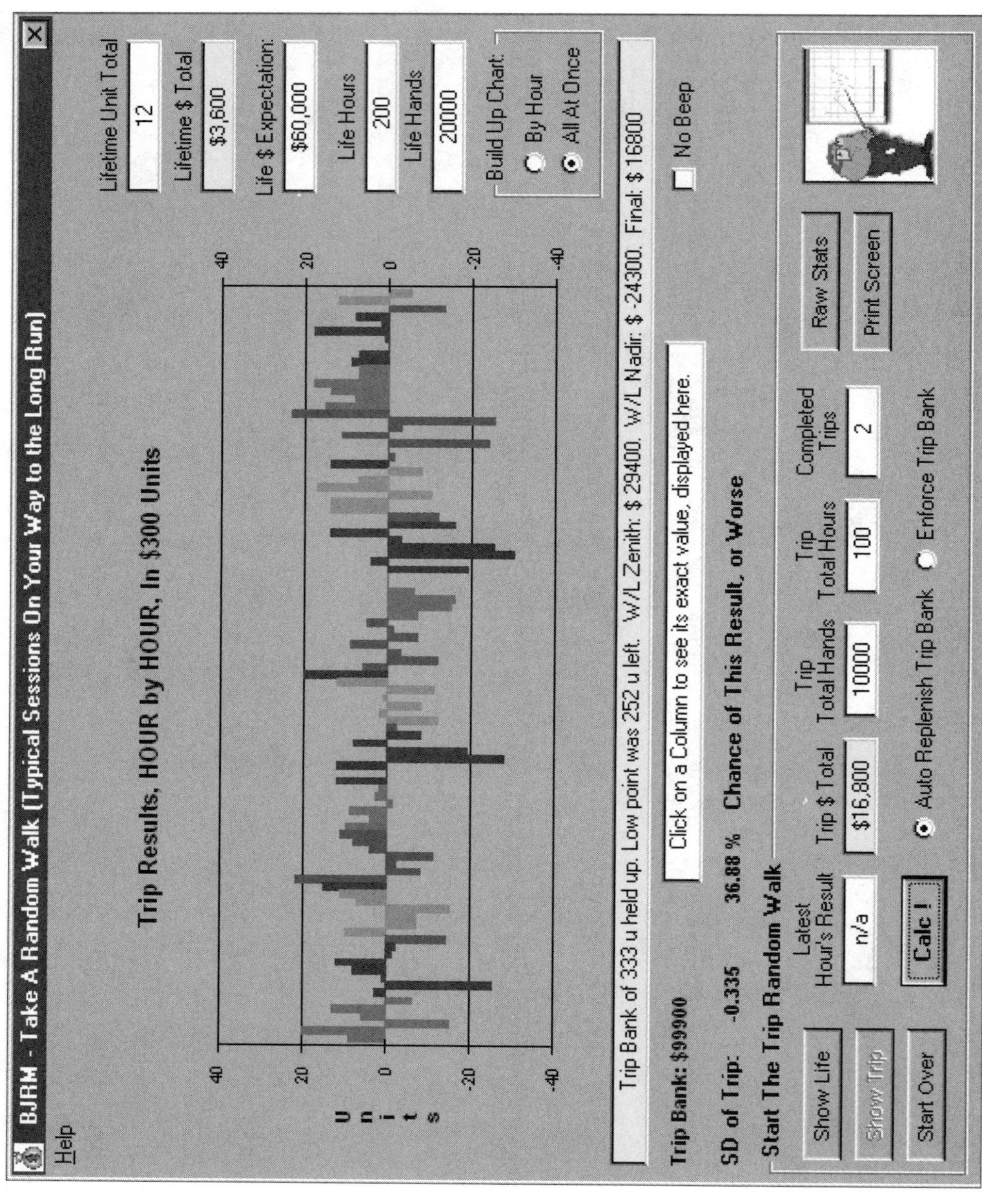

TABLE 12.22
Don's Year 1 Third Quarter Results

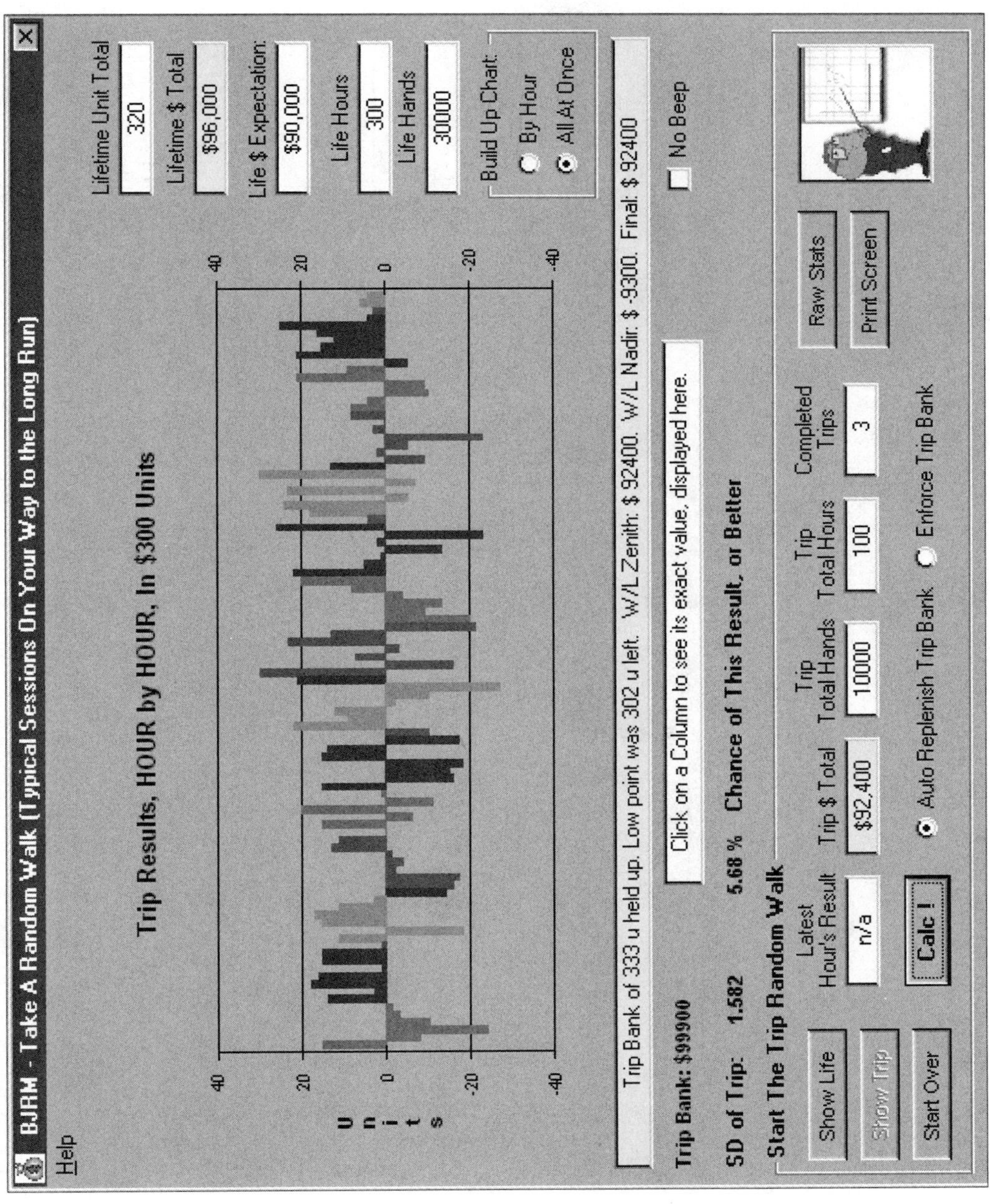

TABLE 12.23
Don's Year 1 Fourth Quarter Results

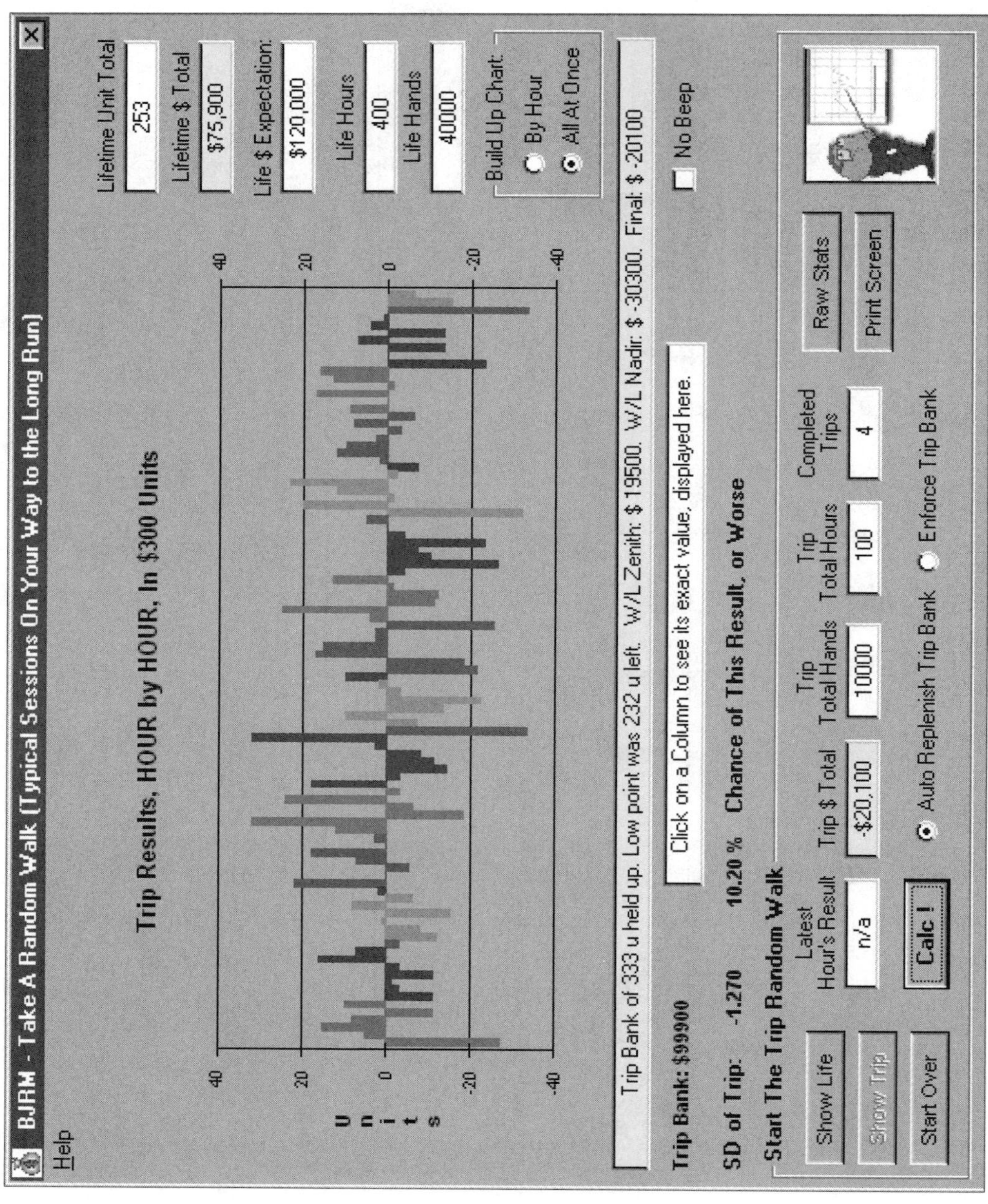

TABLE 12.24
Ed's Year 1 First Quarter Results

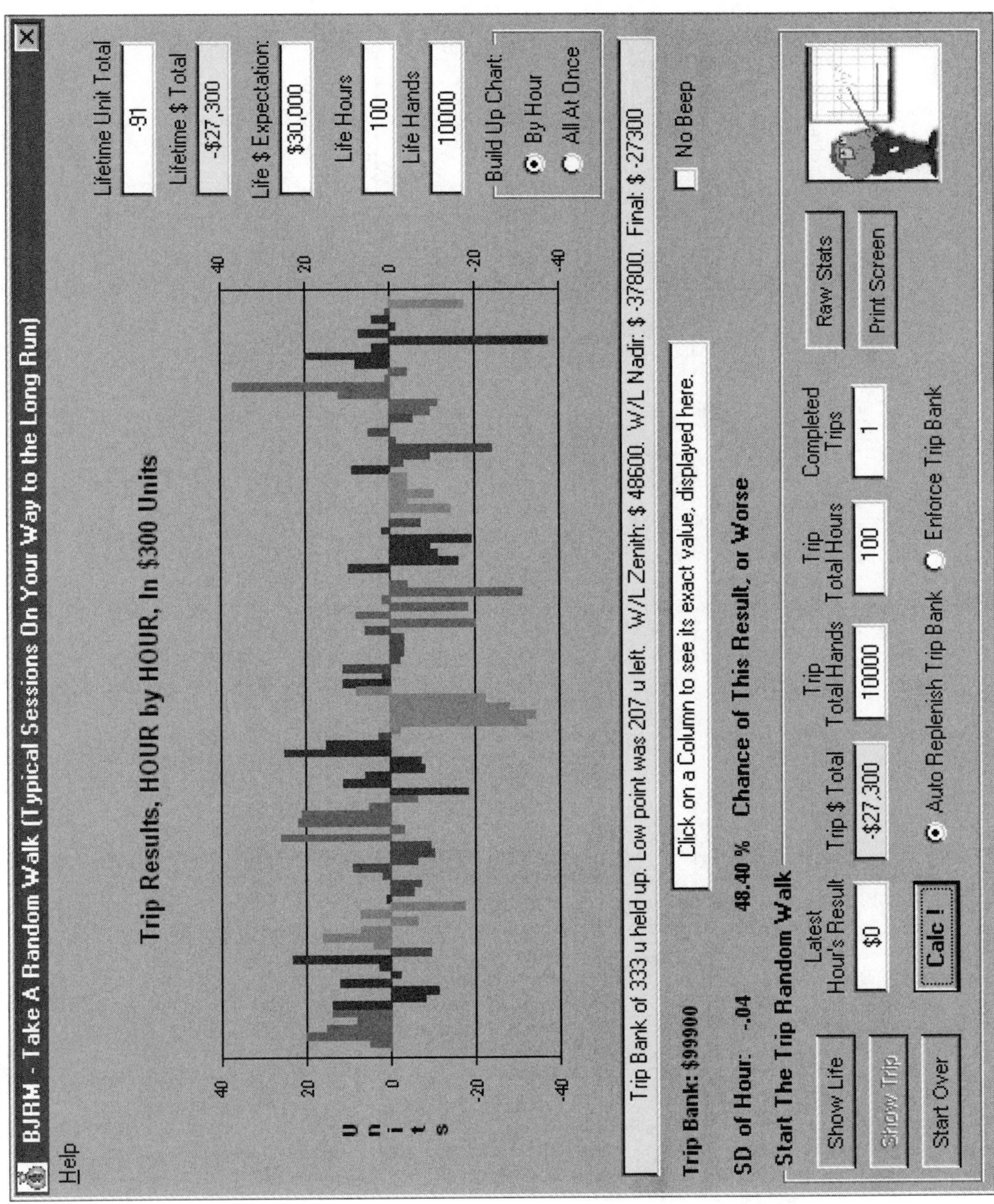

TABLE 12.25
Ed's Year 1 Second Quarter Results

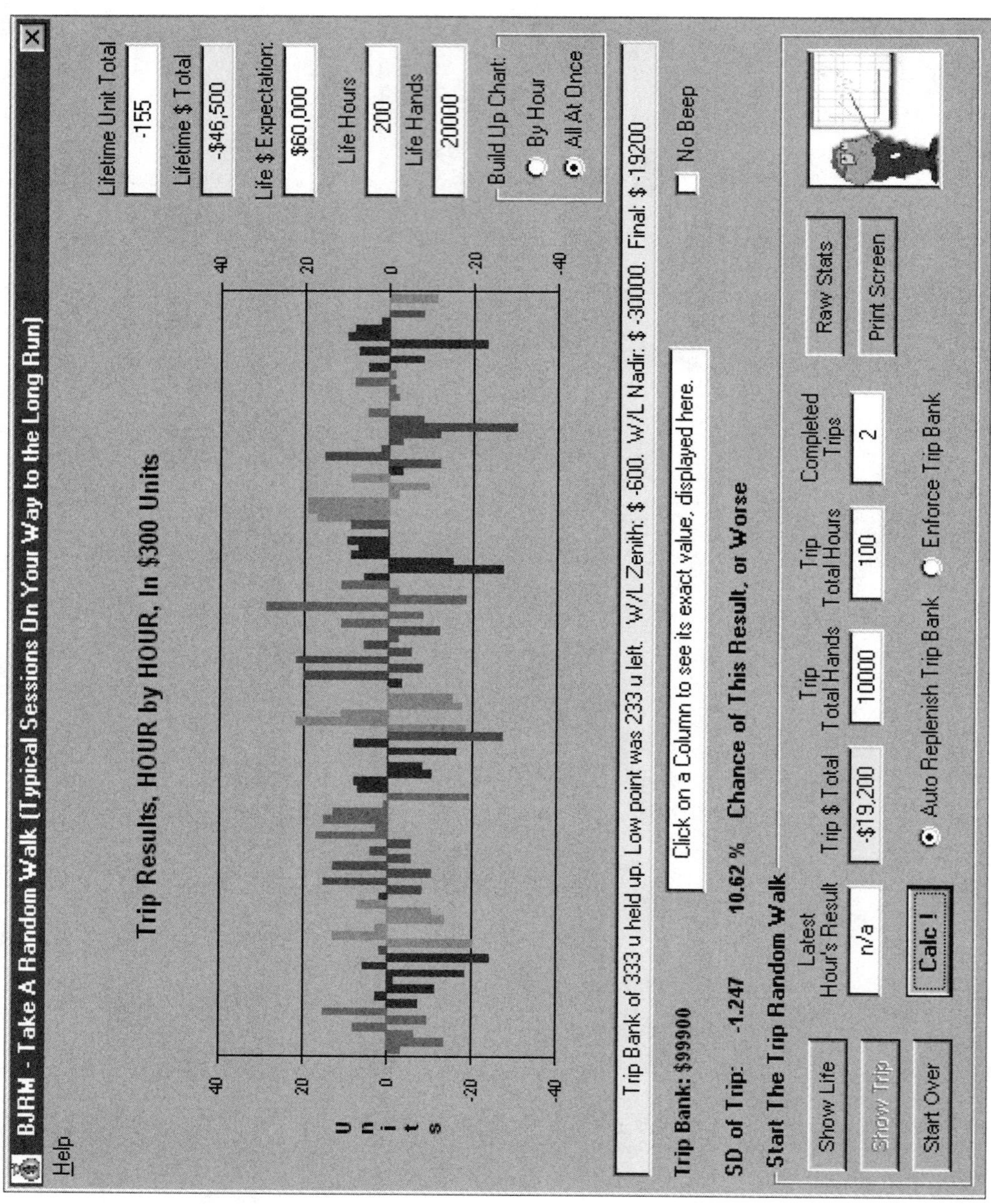

TABLE 12.26
Ed's Year 1 Third Quarter Results

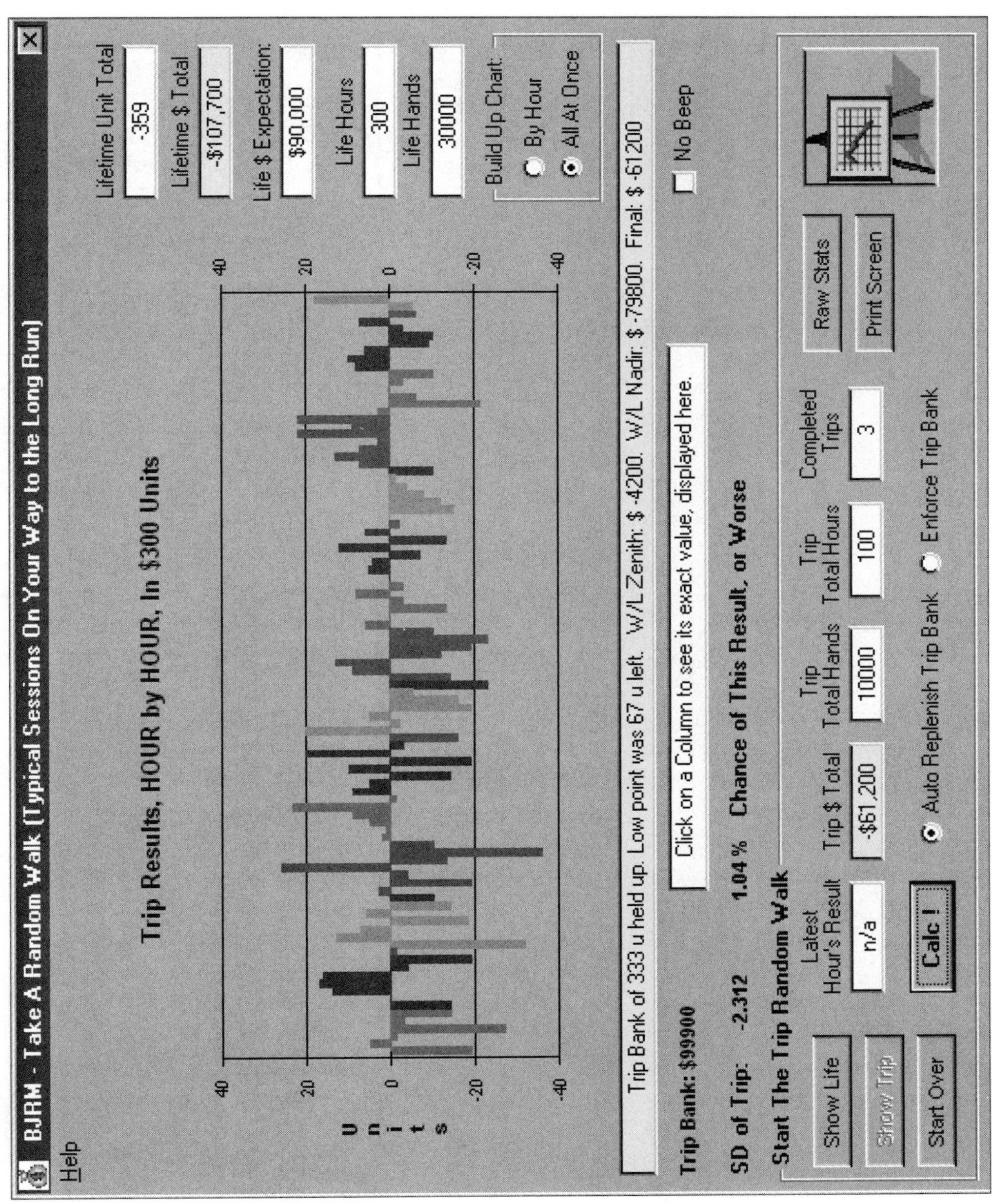

TABLE 12.27
Ed's Year 1 Fourth Quarter Results

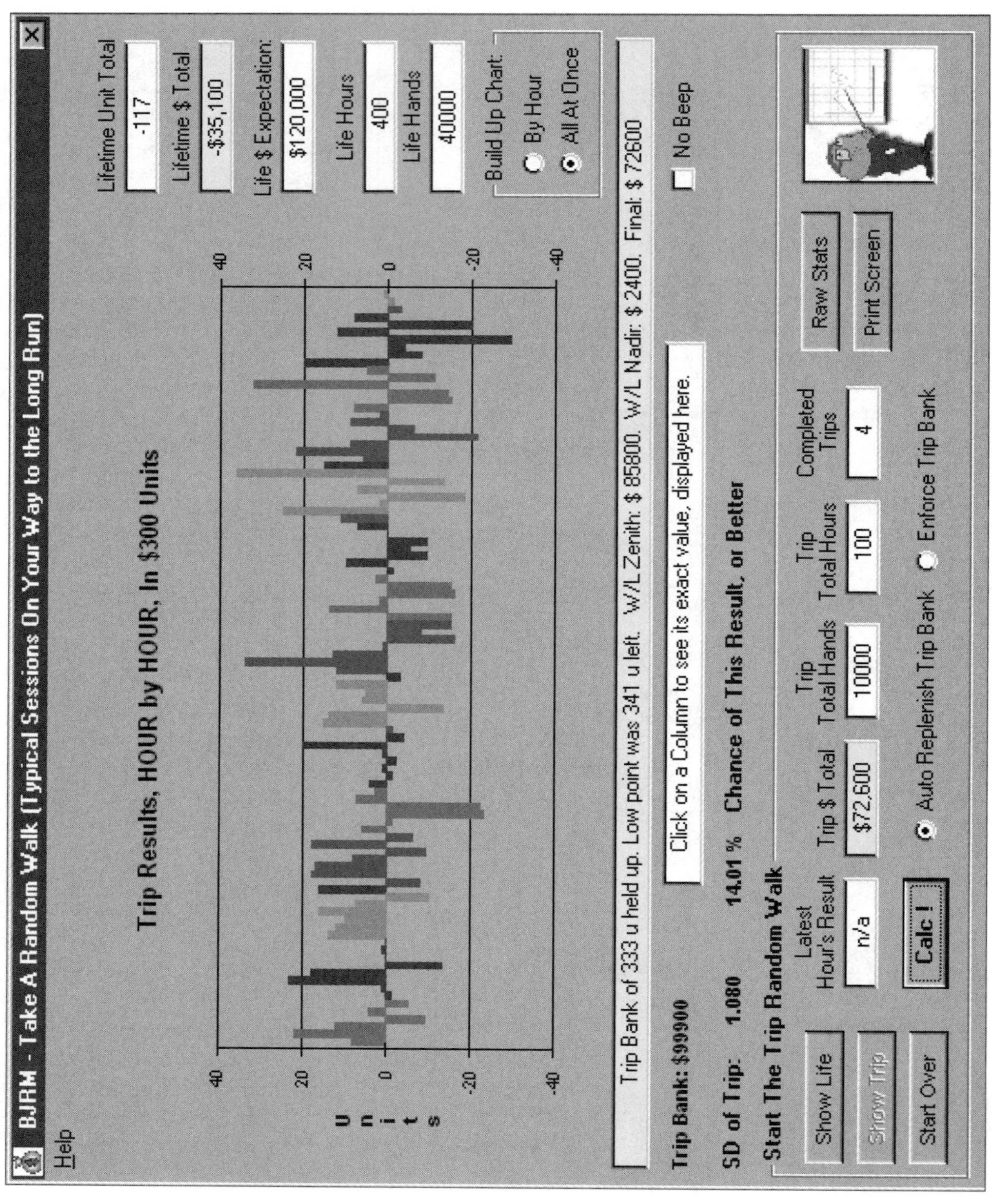

TABLE 12.28
The Team's 5-Year Results

Players	1st Yr	2nd Yr	3rd Yr	4th Yr	5th Yr	Player Total
Al	157,800	149,100	123,300	87,000	170,400	687,600
Bob	169,800	236,700	165,900	152,700	90,600	815,700
Cal	262,500	144,000	285,300	213,900	98,700	1,004,400
Don	75,900	97,800	70,500	105,300	133,800	483,300
Ed	–35,100	122,100	26,700	115,800	108,000	337,500
Team Total	630,900	749,700	671,700	674,700	601,500	3,328,500
Yr S.D. Event	0.175	0.849	0.406	0.423	0.009	0.833
Lance	97,800	141,000	184,800	203,100	196,500	823,200

Years 4 and 5 play out in unremarkable fashion. The team continues to achieve e.v. or better, Cal continues to outperform, while Don and Ed both manage just about e.v., and are not quite as harried as they once were. Clearly, Cal, with his million-dollar overall win, really is the player he claimed to be. This guy is worth every penny that he's "overpaid," no?

Here are the final figures for the individual five players and the team (thousands): Al, 687.6; Bob, 815.7; Cal, 1,004.4; Don, 483.3; and Ed, 337.5. The team wins $3,328,500 (Table 12.28). What's more, the math is validated by the play. According to our model, a player's probability of having a losing quarter is 22.4% (which compares favorably to the 24% that can be inferred from Table 2.2, on p. 21). We should have expected four or five such losses, and, in actuality, we had six, with poor Don and Ed bearing the lion's share (five losing quarters) of the burden. In addition, we shouldn't have been shocked that a player could lose for an entire year (400 hours), as the chance of such an occurrence is 6.4% (again, see p. 21, where 8% can be inferred). With 20 such yearly results, Ed's year 1 loss should not surprise us.

Of course, it would have been nice if we had simply used identical playing tactics for all five of the players so that we could make some further probabilistic statements. Well, guess what, gang? *I lied to you!* That's right. The whole bit about Cal's playing for double stakes and Ed's conservative half-stakes was a complete fabrication. I'm sorry; I've never lied to you before, and I promise I won't do it again. But, the ruse was just too good to pass up. Now you know it: The results from those two players were generated in *exactly the same fashion* as those of their three teammates! In fact, *all* of the results were generated using John Auston's genial "Random Walk" function, from *BJRM 2002. All of the results represent completely random play, with identical parameters for all five teammates!* What happened to them could happen to you. A computer dealt and played these million hands! Now, as a team manager, just how dumb do you feel?

Cal's brilliant five-year results (+2.29 s.d.s; probability, one in a hundred)? Pure luck! Ed's apparently conservative results and relative misfortune, including a losing year? Well, first, they were achieved playing for full stakes, and secondly, they represent a 1.49-s.d. negative swing (6.81% probability), and not "timid" play (unless, of course, the computer blew a fuse!). But, the real eye-opener is that the results of these two players are entirely legit. Call them the revenge of the computer! Variance in its finest hour! But, the results of conniving players? Overly aggressive or overly timid souls? Cheaters? Don't make me laugh. You think you can judge these players by their *results?* Think again.

By the way, one quick parenthetical remark regarding methodology is in order here. A study like this would lose much of its validity if the results shown were culled from a much larger set of simulations, so, of course, John and I agreed to take the first 20 data points, in strict numerical order, and use them for our five players' 20 quarters for the first year. We would repeat the process for the ensuing years and would be honor-bound to accept for the book the precise results, come what may. So, what we saw is what you've now got before your eyes. *There was one such run.* Period.

Conclusions

It is irrational and just plain ridiculous to think that an individual player's results — even taken 400 hours (or 2,000 hours) at a time — could possibly form the basis for a scheme to compensate him for his play. It is equally ridiculous to claim that an honest team's aggregate play can point out patterns of cheating, timidity, aggressiveness, or the like. Now, let's understand one another. The above results could very well have been achieved by cheaters, cowards, or overly exuberant blowhards with big egos. And, they might have inspired all the conclusions of the team manager that we read above. But the results were generated by a computer's Monte Carlo simulation! And, clearly, paying the players differently, according to their "performance," would be the epitome of stupidity. Or, would it? I'll play the devil's advocate for just a moment.

In our online debate, the argument was advanced that players are, in fact, not robots or computers. Rather, they are imperfect human beings who come to the team with their personal strengths and weaknesses. Some truly are aggressive, while others are passive or downright timid. And, some, it was pointed out, may play higher-level counts that, inherently, are worth more than Hi-Lo. Finally, some may have better acts and are more comfortable with the caliber of high-level play that gets the big money. With all of this, I heartily concur. But, I argue, one must take the above into consideration *before* the games begin.

I believe that each player should undergo a complete "workup" before he joins the team. To create this profile, you analyze everything about the count he uses, his predilection for the type of games he plays, his aggressiveness, his extra skills (non-counting), and so on. *Everything.* The player is given an "earnings-per-hour" rating, which can differ from any other. For example, considering just the count system employed, a Hi-Opt II player may rate 10–15% higher than a Hi-Lo player (see the *Postscript* that follows).

Now, they play and put in their hours. And, the players are compensated according to their rating multiplied by hours played. And, if the results fall within acceptable norms, or confidence levels, that's it! You do *not* reward the guy to the right of the curve for winning the money that the guy to the left of the curve lost.

Players accept this methodology before joining the team, or they don't join. Whiners are summarily dismissed from the team. And, to guarantee hours, you insist, up front, that a minimum number of hours will be played before anyone is eligible for a split. This lessens the likelihood that a winning player will quit because a losing player is bringing down the team. In business, you get paid, occasionally, with restricted stock. The stock is worthless to you if it isn't vested, meaning that you have to hold it for a minimum amount of time. Shares may be prorated, as time goes on (the same could be done for team players' hours), but to get the full value, you have to stay the full time. You don't like that arrangement, because you lose money if you quit the firm prematurely? Tough! Don't join that firm. You get the idea.

But, you say, the team is flourishing. Everyone seems to be happy. The $3.3 million is in the bank. So, what's not to like? The model is working. The natives are not restless. The well-oiled machine moves on. Right? Well, it all depends, but I happen to agree with this assessment. Who can argue with success? God bless the free enterprise system! Go do your thing, be happy, and get paid for your productivity. Just don't pretend that your compensation plan has some underlying mathematical principle behind it. Don't suggest that your scheme is "fair." Don't even intimate that a player's performance is an indication of his ability to "get the money." Just *don't* go there. Play, be happy, and win money. But, when it comes time to describe how the players are to be paid, please don't suggest some claptrap, dressed up as mathematics, about how results should be factored into the equation, because the arguments are patently absurd.

And, I've got the statistics to prove it.

Postscript: *Shortly after writing the above concluding paragraphs, it dawned on me that perhaps we should add an actual Hi-Opt II player to our team, just to see if he could "prove" his superiority in a year's worth of play. He couldn't! Lance, a Hi-Opt II aficionado, who played a game reckoned to be worth fully 15% higher e.v. than that of our intrepid quintet (Table 12.29), set out to play 400 hours on his own. The results were altogether ordinary. Fast out of the gate, he won a cool $54,000 in the first two weekends (Table 12.30), well above expectation. It didn't last. Thriving at the end of the first quarter, to the tune of double-e.v. $60,900 (Table 12.31), Lance struggled through a nightmarish second quarter, eking out a mere $3,600 additional profit (Table*

12.32). The third quarter was even worse, as $16,800 was actually lost (Table 12.33). A decent fourth quarter of $50,100 saved the year, but the $97,800 result was nothing to write home about (Table 12.34). Indeed, it was not only below expectation, but would have ranked Lance a humble (pardon the pun) fourth, had he joined the team.

What does this prove? Well, frankly, absolutely nothing at all. Do we want Lance on the team? Without a doubt. Will he add value? You bet. Should we factor in his skills at, roughly, 15% premium to the other players? I think we should. And then, should he be compensated according to how much he wins? Absolutely not! Why? Because, over the course of 400, or even 2,000, hours, it is highly unlikely that the individual results of a counter will tell us anything about his skill level. Your witness. The defense rests.

TABLE 12.29

Hi-Opt II Simulation Results (Lance)

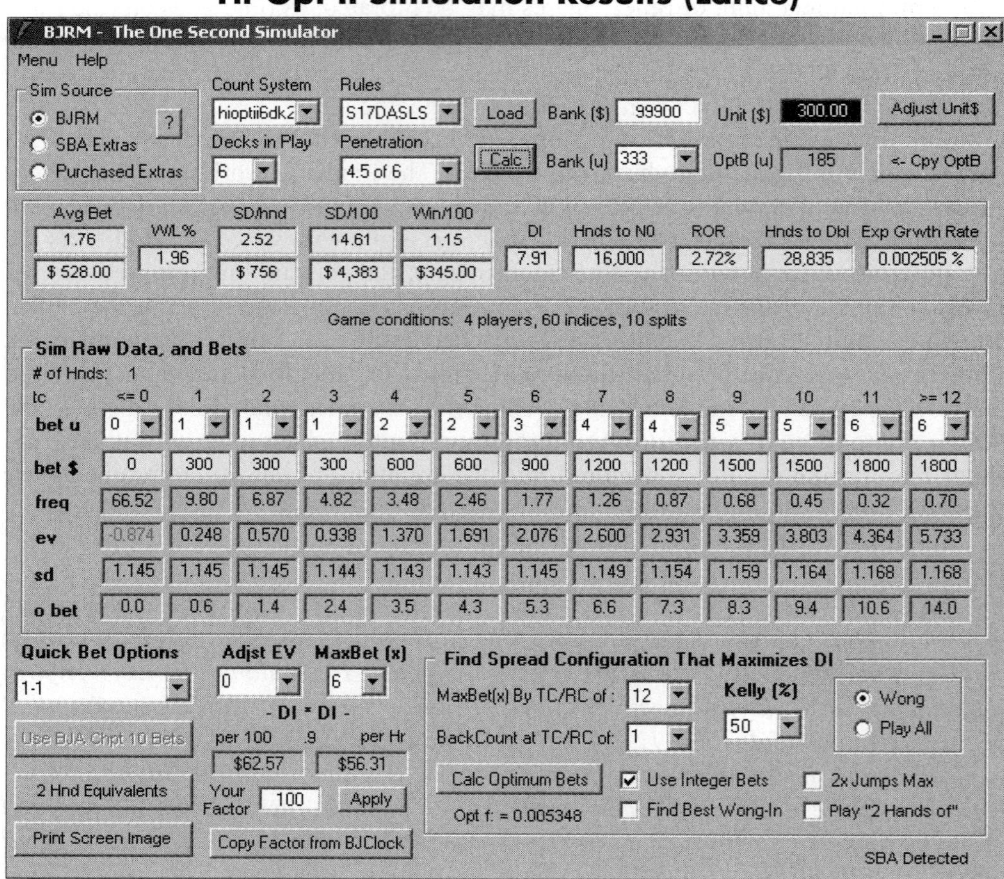

TABLE 12.30
Lance's First Two Weekends Results

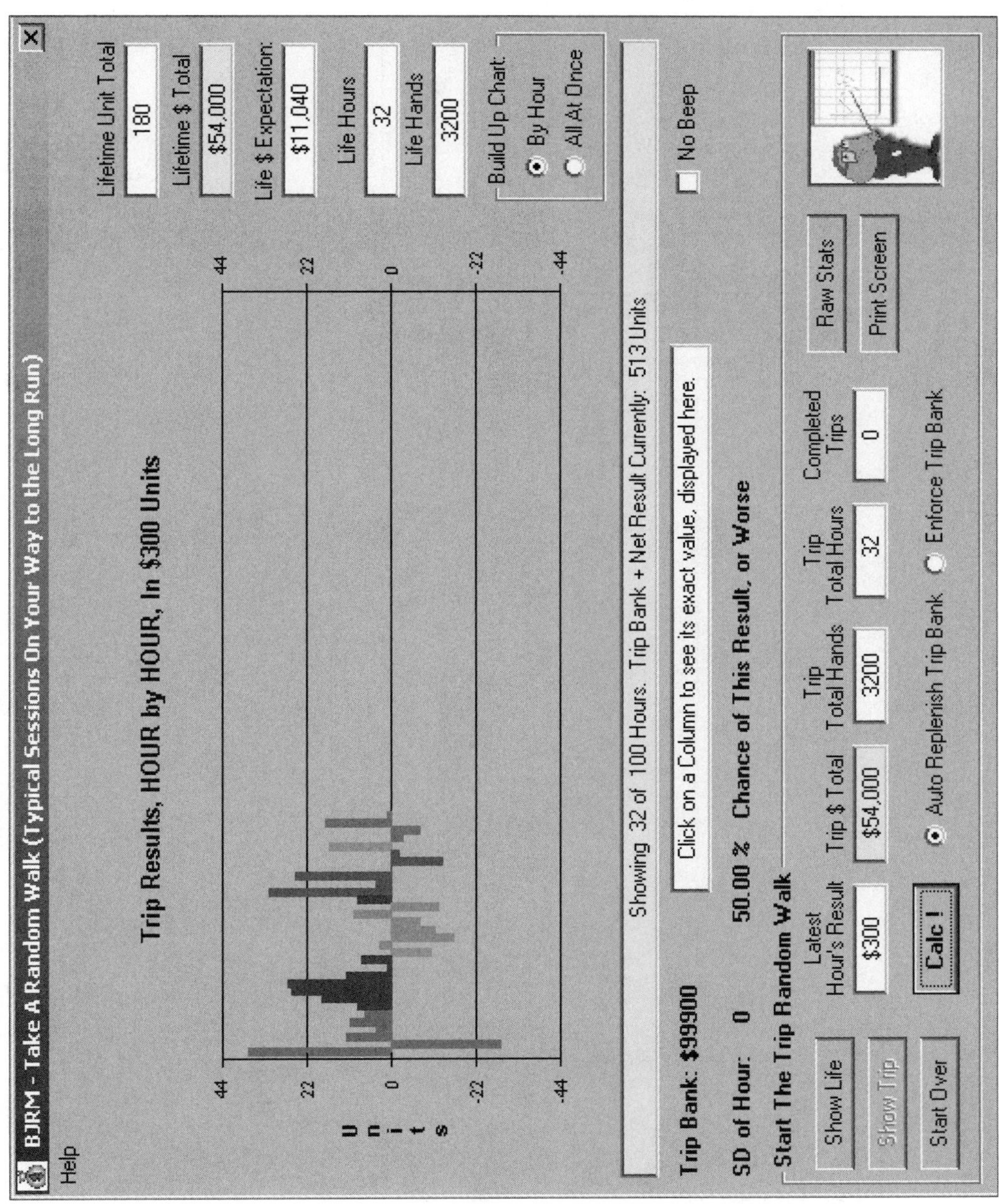

TABLE 12.31
Lance's First Quarter Results

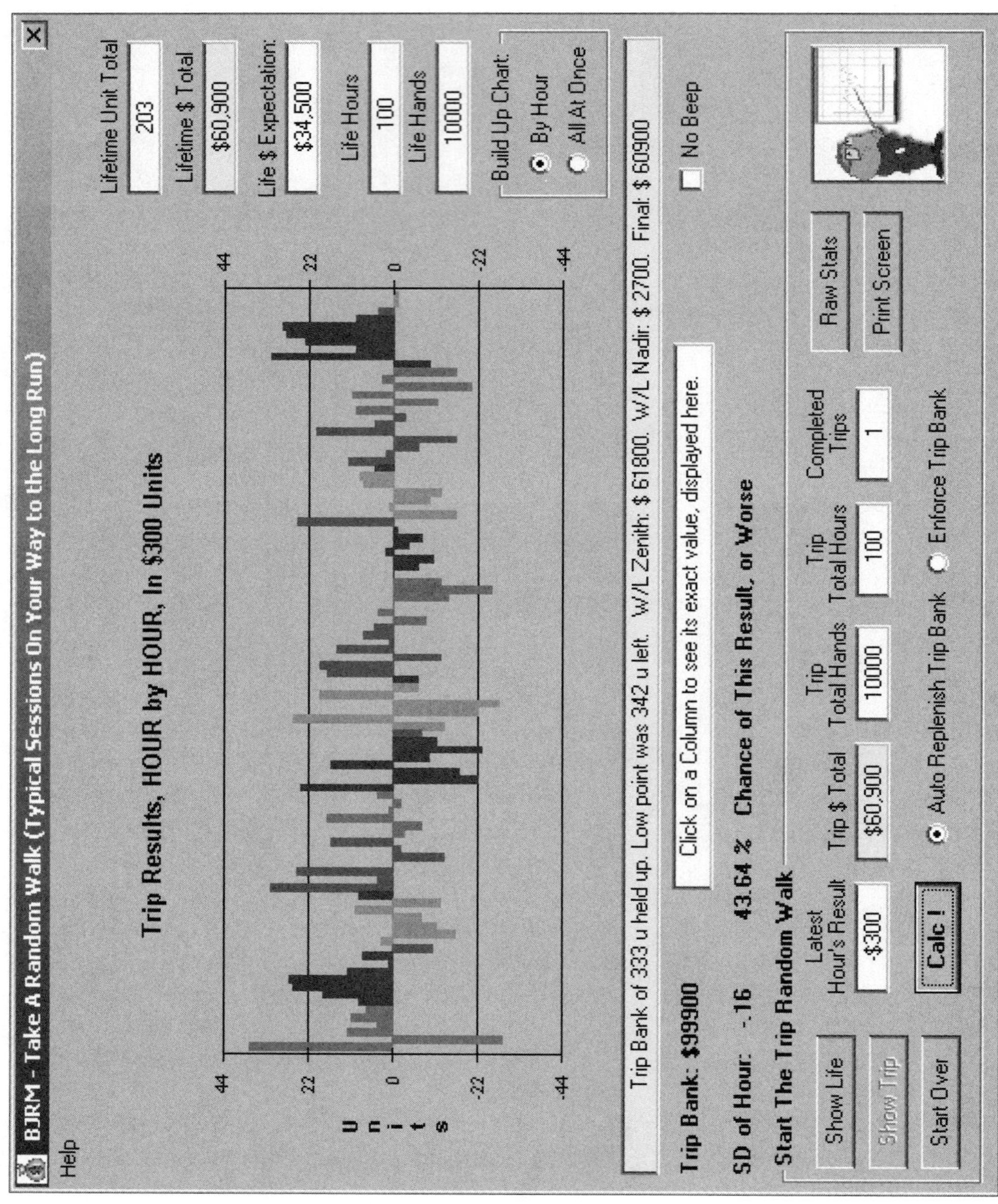

TABLE 12.32
Lance's Second Quarter Results

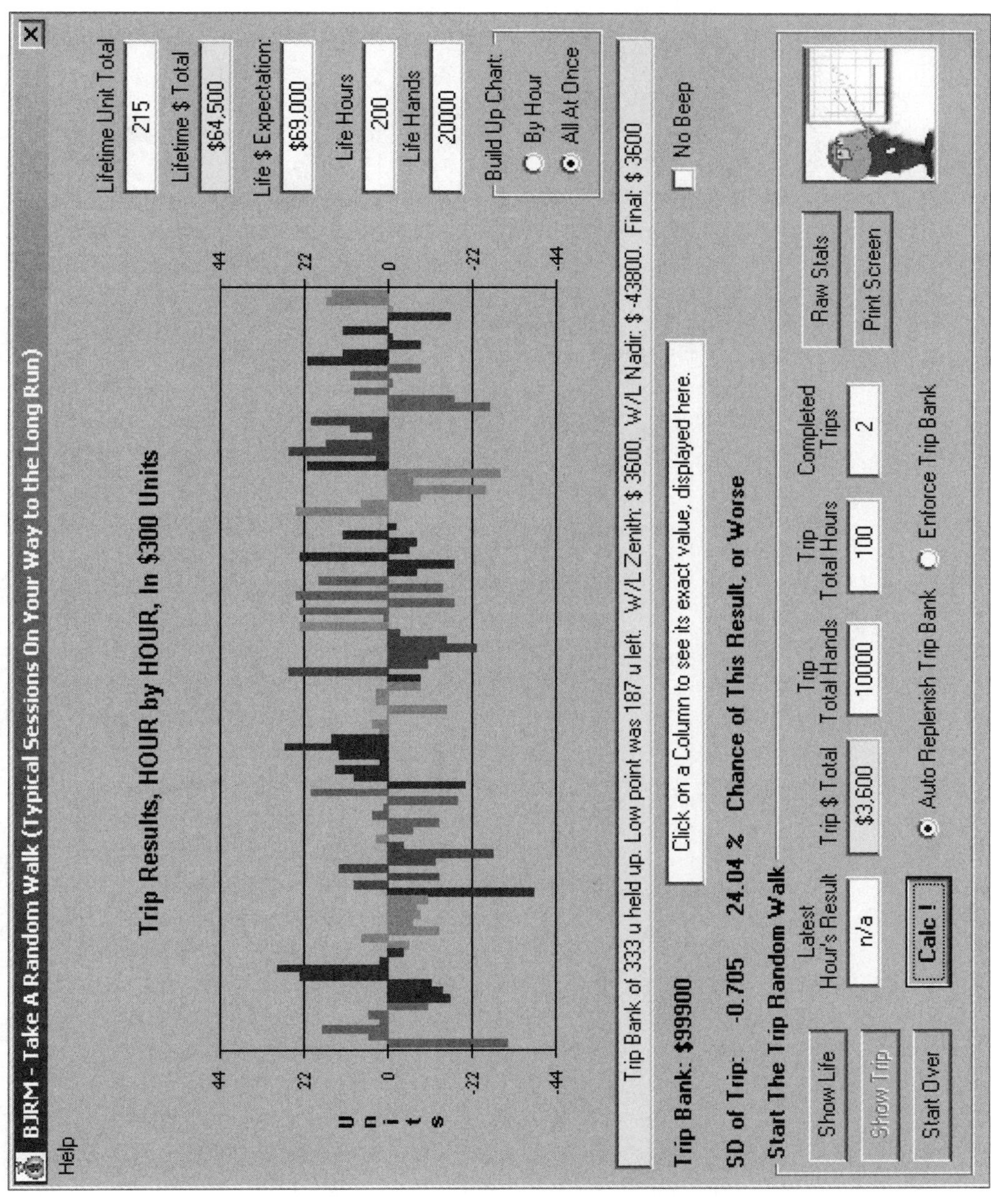

TABLE 12.33
Lance's Third Quarter Results

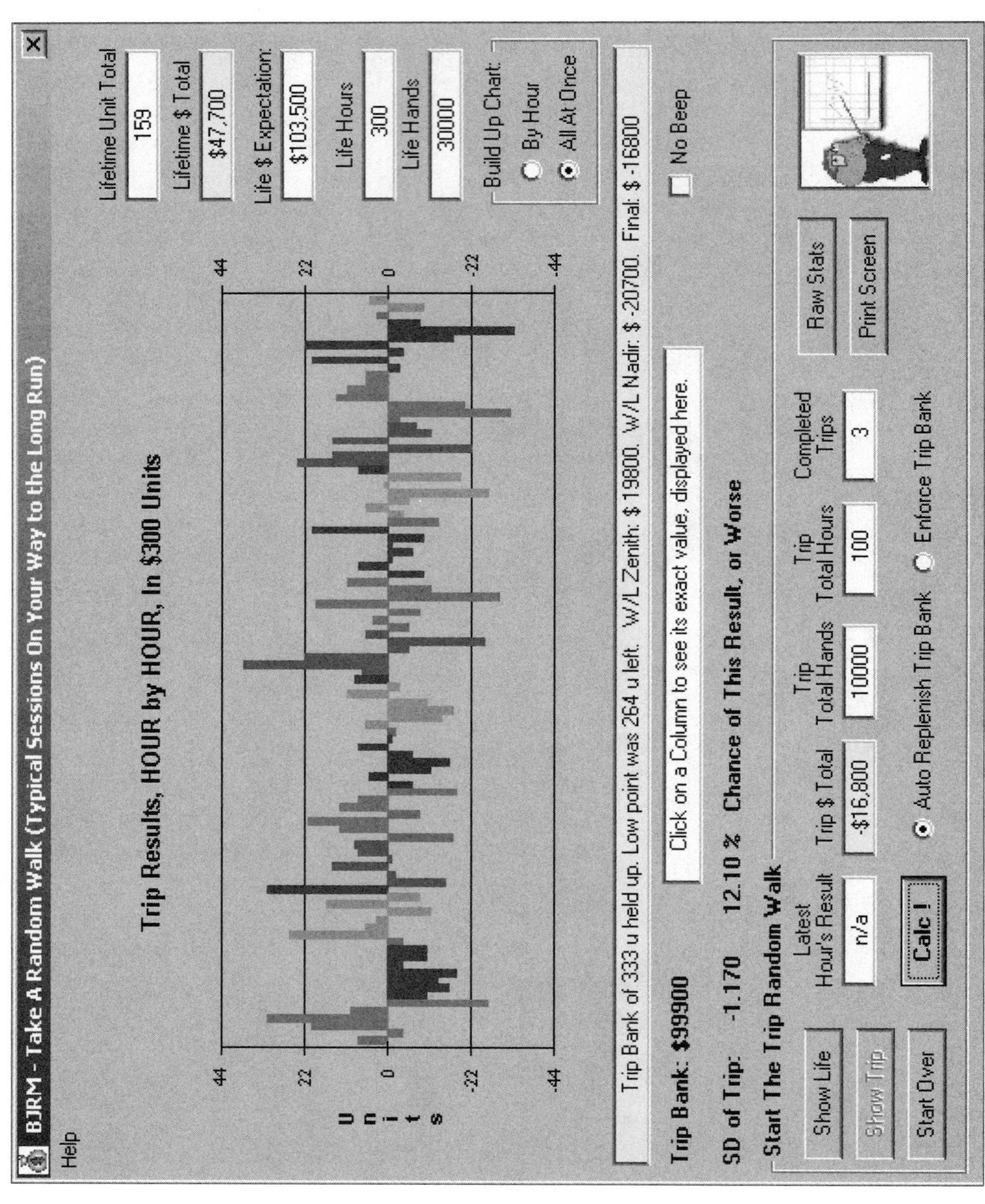

TABLE 12.34
Lance's Fourth Quarter Results

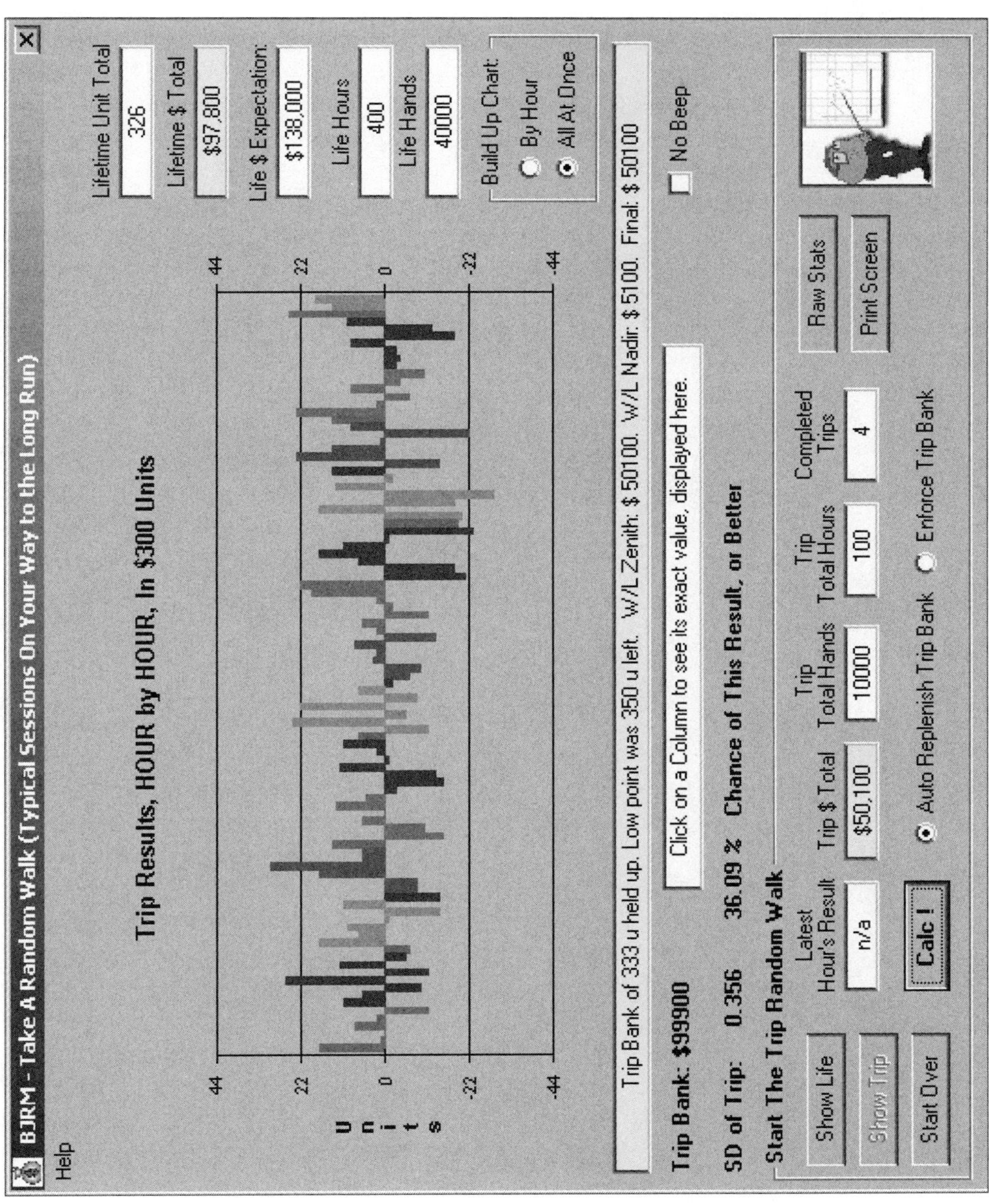

Chapter 13

New Answers to Old Questions

The saying goes, "There's nothing new under the sun," and, truth be told, it isn't very easy these days to find new areas for research concerning blackjack matters. On the other hand, there is certainly no shortage of the "old questions," which, seemingly, get dredged up, ad infinitum, *by posters on the various Internet groups. But here's the novelty of this brand-new, two-part chapter for the revised editions of* ***Blackjack Attack:*** *The questions I'm about to consider are old, but the answers are new and break ground in a number of areas. I've wanted to answer the first question precisely, and mathematically correctly, for as long as I've been playing the game. And, I've wanted to place a value on the usefulness of the second concept for 20 years. So, you can imagine my excitement in presenting the following information, for the first time ever, in a book on blackjack.*

Part I

Section I

You enter a crowded casino, intent on playing the shoe game in the most efficient manner possible, consistent with not being so obvious as to stick out like a sore thumb. You realize that, to accomplish this goal, you're going to have to do two things: 1) back-count the various games, to give yourself the greatest opportunity to be playing with an advantage; and 2) attempt to answer, repeatedly, with each shoe you encounter, our age-old question number one, namely, *"When do I leave?"* Permit me to explain.

A fresh shoe is now shuffled and ready to be dealt, there are a few players seated and about to play, and, according to the principles espoused in Chapter 1 of this book, you are observing the situation, assessing the advisability of jumping in to play at this table. Now, with minor variations on the theme, it is true that your first advantage, depending on the particular rules you're up against, will come at a Hi-Lo true count of about +1. So, it seems logical that, unless the table achieves that threshold, you will not enter the game to play, but will continue to watch the action. As rounds are dealt and time goes by, without any noticeable improvement in the true count, you begin to ask yourself the question that has perplexed more counters than any other in the history of the game: "When should I move on?" Should I wait for a substantially negative count? Since I have no edge at a TC of zero, should I move to another table and start the process all over again if a deck or two have gone by and I'm still in neutral, zero territory? Or, since I've invested some time in this shoe, should I be more tolerant of a slightly negative count, in the hope that, because of the current level to which the pack has been dealt, the true count might revert to positive and afford me the opportunity to play?

Questions, questions, questions! And now, finally, years later, answers! Thanks to the incredibly hard work and dedication of a number of the brightest and most talented blackjack researchers on the scene today, I'm going to tell you, while you're back-counting, when to stay, when to play, and when to walk away. I'm also going to tell you, when seated at the table, when to play, when to sit out, and when to leave. The answers are not always intuitive, and the methodology employed to ascertain them is extremely sophisticated and complex.

Some background: A little more than a year ago, I approached John Auston with the above questions, and the two of us began to wonder if we could simulate our way to the correct answers. After many aborted tries, we agreed that the task was simply too difficult, and we began to seek help with a theoretical, rather than empirical, solution. Enter Kim Lee. A former professor of finance, he's as brilliant as they come and — glory be! — he's also a blackjack player and researcher (how very convenient!). Kim contacted me and said that, independently, he had been working on a similar project and would like to share his approach with me. He sent me a spreadsheet outlining his first attempt at solving the problem.

Now, if you don't already know this about me, I am not a theoretical mathematician. I try very hard to have a solid grasp of the math of the game, and I hold my own in most blackjack circles. But, what Kim was proposing was a bit beyond my math capabilities and light-years past my programming and computer skills (which, for the record, are nonexistent!). Undaunted, Kim painstakingly attempted to explain his methodology, which involved Brownian Bridges, recursive formulas, iterative processes, etc. You get the idea! We tried to find a common ground for further discussion, but, despite valiant attempts, we eventually parted company for a while, each a bit frustrated.

Very shortly thereafter, and completely by chance, another group, led by my dear friend, Chris Cummings, about whom you will soon learn more, began research on the same topic and posted results on Richard Reid's brilliant *bjmath.com* Web site. As Chris had done so much wonderful work for me in previous years, notably on some of the risk of ruin formulas you've read earlier, I immediately contacted him to see if his approach to solving what had now been dubbed the "optimal departure problem" was similar to Kim's. It was. In fact, it was somewhat less rigorous than Lee's algorithm, but to give credit where it is due, when Chris posted his findings to Reid's site, in April of 1999, I believe this was the first-ever published result purporting to give recommendations for true counts at which to abandon watching or playing a shoe, according to penetration and current levels of depletion. Thrilled with the prospect that I might convince Chris to collaborate with me on a full-blown investigation of all the subtleties of this intriguing question, with the work ultimately to find its way into this revised edition, we began to explore the possibilities. Several months went by. And then, the unthinkable happened: Chris died.

It is hard to describe how devastated I was that this friend of 15 years, this young (40), vibrant, most gentle, lovable soul was, suddenly, without any warning, literally here one moment and gone the next. A kinder, more caring individual, you will never hope to meet. Not only had I lost a wonderful friend, but the entire blackjack-playing community had lost one of its most talented and brightest researchers. We shall miss him very much.

But, life goes on, and, intent on bringing to fruition this project that was so dear to Chris, I came full circle, explained the circumstances to Kim Lee, and once again solicited his assistance in making this study a reality. Kim could help with the theory and the computer program, but someone would have to take the reins and funnel all of the data and all of the "raw materials" into intelligible and useful output. And so, I called upon yet another of today's blackjack luminaries, none other than Richard Reid, on whose site Chris had most enjoyed posting, to come to the rescue of this ongoing investigation. Although I am rarely at a loss for words, it is hard even for me to acknowledge adequately the incredible amount of time, effort, and brilliant work

Richard put forth in pursuit of getting everything letter-perfect. His dozens of hours of meticulous programming, exchanging of e-mails with Kim and me, and simply being the perfectionist that he is, are testimony to his dedication to this project. I thank him publicly, as I do Kim (whose baby this is, after all!), for making what you are about to read a reality. And, somewhere, I know Chris is looking down on all of this with a broad smile on his face. More than for anyone, big guy, this one's for you.

Methodology: Quite frankly, I'm not even going to try to explain! The math is beyond the scope of this book, but an explanation of the concepts involved is, I feel, nonetheless, in order. When you're watching a shoe in progress, the decision to continue to watch, or to depart for "greener pastures" (or table-tops!) involves two components: the instantaneous, or "current," value of the very next hand to be dealt, and the potential, or "future," value of the remainder of the shoe, down to the stop-card. In essence, we are attempting to determine if the sum of both of these components can provide a sufficient per-hand "wage," or profit, to induce us to stay where we are, rather than to seek a fresh shoe and begin the process of "stalking" all over gain. Needless to say, to make this assessment in the most intelligent manner, we need to know not only the average per-hand value of a new shoe, but also the aforementioned values of the current shoe under scrutiny. To furnish these numbers, we rely on Kim's brilliant program and Richard's ability to present the data in the tables and charts below.

But, obviously, this is only half of our playing experience — and the less complicated half, at that. For, sooner or later, we take a seat at one of the games and begin to play. Once actively involved in the game, we now expand our options to four possible actions: 1) at true counts that continue to be instantly profitable (read: at least +1), we continue to bet; 2) at counts that equal or fall below the optimal departure point for our current pack level (for, you will learn, these "Wong-out" points are constantly changing), we excuse ourselves and move on to a new shoe. But, what about those counts that, while too low to warrant a bet, remain above the optimal departure levels? Ostensibly, we're talking about a kind of "twilight zone," corresponding, roughly, to trues above ever-so-slightly negative values, yet below the +1 threshold of profitability. When we find ourselves in this zone, two more actions are possible: 3) ideally, we may remain at the table, while sitting out the hands in question (not wagering, but waiting); or 4) more pragmatically, we may make minimum "waiting bets," of the smallest unit size we think we can get away with, while contemplating our next move, namely, to resume betting, or, alternatively, to leave. For, in Lee's elegantly simple description of optimal table-departure strategy, it all boils down to the following: "Leave when you can make more money standing than sitting!"

Now, returning to our back-counter, we have made a tacit assumption that it makes no sense to simply walk up to a table and to begin to play immediately, off the top, without first waiting to establish that some player edge has been attained. And yet, circumstances may dictate this "inferior" style of play. In their never-ending quest for sufficient camouflage to escape detection as skilled players, many counters feel a need to avoid "circling" for long periods of time, as they assess games in progress. Some feel, quite simply, that to do so is to convey the image of, well, a back-counter! In particular, it is often quite difficult, in smaller casinos that don't have a lot of tables, to pull off the "cruising" scene without appearing grotesquely obvious. And so, in deference to such players and the dictates of challenging conditions, we add yet another wrinkle to our repertoire: that of simply walking into the casino and pulling up a chair at the very first game that is about to be dealt. In similar fashion, eschewing "walkabout," such a player, upon leaving a game, will once again head for the nearest available fresh shoe and play right away. (Note that with the current trend, in many casinos, for the "No mid-shoe entry" policy, the "walk-right-up, sit-right-down" style becomes *de rigueur.*) Finally, in extremely crowded casinos, where seats are at a premium, one might have no choice but to enter a game immediately, lest no seat be available when the ideal time to play does come around.

If you've followed all of the above, then it should be clear to you that we've actually created three different categories of player, or styles of play. I'm going to label them: "Mr. Perfect," "Wong-in Wong-out" (WiWo), and the "White Rabbit" — this latter a reference to the *Alice in Wonderland* character who is "Late for a very important date/No time to say 'hello' — 'goodbye!'/I'm late, I'm late, I'm late." Our table-hopping Rabbit (how very felicitous!) is constantly rushing into games ("Hello"), only to leave the moment they get bad, that is, when the optimal departure level is reached ("Goodbye"). The rest of the time, he is making bets according to the count, including minimum, waiting ones, in the twilight zone.

As for the first two players, WiWo stalks the games, waiting for the edge, leaves (while watching) if it's right to do so, and, once seated, does the following: TC $\geq +1$, makes optimal bets; optimal departure point $<$ TC $< +1$, makes waiting bets; and TC $\leq$ optimal departure level, leaves. *[Note: Technically, WiWo would have slightly different optimal departure points — depending on whether he happened to be playing or still watching — because of the higher waiting cost associated with playing in the twilight zone. Because this effect would be negligible in our scenarios, we ignored it to simplify the calculations. — D.S.]* Finally, Mr. Perfect wants the best of all possible worlds. He conducts himself just as WiWo does, with the exception that, once seated, Perfect does not make waiting bets on non-profitable hands in the twilight zone. Instead, he sits out these hands, thus avoiding the drain to his bankroll. Although we are not maintaining that playing in this manner is achievable in a casino setting (too obvious? too bold?), it does, for the record, establish, perhaps for the first time ever in print, the theoretically most efficient way a lone counter can play blackjack in a casino. I have summarized, below, the precise actions taken by each of our participants.

Description of Players' Actions

(ODP = Optimal Departure Point for each specific player)

	Standing	Seated
Mr. Perfect	Enters game if TC ≥ +1 Walks away if TC ≤ ODP	Bets optimally if TC ≥ +1 Sits out if ODP < TC < +1 Leaves table if TC ≤ ODP
WiWo	Enters game if TC ≥ +1 Walks away if TC ≤ ODP	Bets optimally if TC ≥ +1 Waiting bets if ODP < TC < +1 Leaves table if TC ≤ ODP
White Rabbit	Enters game at top of shoe	Bets optimally if TC ≥ +1 Waiting bets if ODP < TC < +1 Leaves table if TC ≤ ODP

Still another problem needs to be considered. In supplying some of the data you are about to read, we assumed "Utopia," in order to establish some baseline comparisons. In Utopia, we play in whatever manner we please, and, when seeking a fresh shoe, one is always instantly available to us; that is, there is no lag in time from the moment we leave a table to the next moment, when we are playing or watching a new game. Obviously, on only the rarest of occasions do we enjoy such attractive conditions. It isn't realistic to assume that, upon leaving one game, we flow seamlessly to the next, without missing a card. Casinos are crowded; it takes time to navigate our way around them. Games are currently at all different stages of the shoe; few are just beginning. Some dealers may be standing waiting for players, the cards spread on the table. Were we to play here (only the Rabbit would do so, anyway), we'd have to wait at least two minutes for a complete shuffle. *[Note: Recently, in an effort to increase efficiency, some casinos have taken to shuffling and loading the cards into the shoe, even while the dealer awaits the first player at the table. This practice, of course, expedites our style of play. — D.S.]*

And so, it is apparent that, the vast majority of the time, when we finally do decide to leave a table, we need to build into our model a realistic lag, expressed as a number of forgone rounds, before we are once again watching or playing. The choice of how many such rounds to specify is somewhat arbitrary and depends greatly on casino size and conditions. We settled upon a penalty of three minutes, which corresponds to approximately six rounds of thirty seconds each, the latter being a fair estimate of the time it takes to deal a round, face-up, to three or four players. New charts needed to then be constructed to reflect these more practical guidelines. Those tables appear after the baseline ones and, in my view, represent a truer picture of what you're likely to encounter in your favorite casinos. (Note that the lag has a decided effect on the optimal departure recommendations. We'll have more to say on this topic in the section that follows.)

Appreciation of the data: For now, we consider shoe games only, both the 6-deck and 8-deck varieties. For the former, s17, das, ls games — with two levels of penetration (4.5/6 and 5/6) and two bet spreads (1–12, and 1–20) — were studied. For the latter, we examined games with similar rules (no surrender) and spreads, and penetrations of 6/8 and 6.5/8. As there was much overlapping of the data, not all the aforementioned charts are presented. Those that we have chosen are more than adequate to convey the essential ideas. In all, eight different scenarios are represented in the tables at the end of Part I.

It shouldn't come as a surprise, as we first peruse the "no lag" charts, that optimal departure points, across the board, for all three players, are slightly negative in the early going. The true counts at which we're advised to leave all cluster in the –0.10 to –0.71 range, over the first half of the 6-deck shoe. The implications are clear: Although we usually do not have an edge in the 0 to +1 interval, we're better off remaining at games that exhibit these counts, for the potential value they possess, than abandoning them for a fresh start. But, the line of demarcation is a rather fine one, and in the case of the shallower penetration, 4.5/6 game, an excess of a single negative card seen, thereby lowering the running count to –1 and the TC to, at most, –1/6 = –0.167 (and lower, later) is sufficient to send us on our way.

And then something very interesting — and perhaps counter-intuitive for many — takes place. You'll remember one of my questions, posed earlier: "Since I've invested some time in this shoe, should I be more tolerant of a slightly negative count, in the hope that, because of the current level to which the pack has been dealt, the true count might revert to positive and afford me the opportunity to play?" For years, players have been answering in the affirmative: "Yes, since the appearance of just a couple of small cards could now turn the true count on a dime, the deeper I am, the more tolerant and patient I must be." In fact, *just the opposite is true!*

For, if you look not only at the charts, but especially at the *graphs* of those charts, you'll find, past the 3- or 3 1/2-deck levels, a fascinating upturn of the optimal departure numbers, allowing only one conclusion: The deeper we get, the *sooner* we walk! I doubt that one seasoned player in ten would have reasoned in that manner prior to reading these lines. And yet, the evidence is incontrovertible. Let's see if we can understand what's taking place.

As the shoe progresses, it becomes clear that, if we don't yet have a +1 or higher true count, we're fast running out of time. Faced with the prospect of continuing to toil in mediocrity (read: not playing), we begin to get more demanding, as if to imply that, even if we *were* to achieve the desired +1 count, it may be a case of too little, too late.

After all, of what use is a small positive count if it comes so deep in the shoe that, by the time we hope to further exploit it, the shuffle-card appears? And that, in a nutshell, is the problem. So, the curve turns upward and — lo and behold! — for the 4.5/6 game, it actually reaches and *surpasses* a TC of +1, between the 4- and 4 1/2-deck levels, as it does between the 4 1/2- and 5-deck levels of the 5/6 games. *Although we have the edge, we leave anyway!* I told you there would be some surprising conclusions along the way. Note that altogether analogous observations can be made for the 8-deck games. (There is a special consideration to be made concerning play of the last hand in a lag situation. We shall discuss this case in Section II.)

I also mentioned, above, that the inclusion of a lag changes the departure values. Just look at Tables 13.3–4, and 13.7–8, and compare them to the no-lag ones. In every instance, when it no longer is possible to simply get up and immediately find another opportunity, we need to be more patient with our current situation. Intuitively, this makes perfectly good sense. So, to take just a single example, note that, in our 5/6, no-lag game (Table 13.2), where, at the 3 1/2-deck level, all three players would get up at trues below –0.84, they are now required, with a six-round lag (Table 13.4), to stay seated until reaching TCs ranging from –1.55 (WiWo) to –1.74 (Perfect).

Next, as we look at the charts horizontally, rather than vertically, another interesting phenomenon presents itself. Note how, in the early segments, Rabbit and WiWo actually both leave slightly *sooner* than Mr. Perfect, whereas, deeper in the shoe, exactly the opposite is true, and Perfect becomes the least patient player of the three. What sense can we make of this? Here, the implication is that, in the latter portion of the shoe, Rabbit and WiWo, despite having lower expectations than Perfect, actually benefit *more* from the increased level of depth (and subsequent increased volatility of the true count) than does Mr. Perfect. As Richard Reid pointed out, "I find this neat to see visually." Indeed, the graphs clearly indicate Rabbit's departure line *above* Perfect's in the early going, only to cross it and wind up *below,* later on.

Preliminary conclusions: I have a confession to make. This study is a work still in progress. I wish it could have been otherwise. It would have been rewarding to be able to report, for all our players, in each of the different game scenarios, the statistics you are accustomed to reading: hands played per hour, Win/Loss %, and the all-important SCOREs. But, as the deadline for publishing the second edition drew near, it became apparent that the programming and simulating needed to complete the job were so complex that we could not finish without compromising the integrity of the data and the accuracy of the findings. I'd rather be 100% correct a little later on than to rush into print and relate potentially incorrect numbers that we might regret down the line.

And so, it is my intention to conclude Part I of this chapter in the upcoming Fall 2000 issue of *Blackjack Forum. [Note: The entire study is now included in this third edition. — D.S.]* For now, we are able to draw at least some preliminary conclusions concerning the win rates and the accuracy of the methods described above.

Compared to the SCOREs of Chapter 9, is there any appreciable difference between the win rates for the Hi-Lo players in this present study and their supposedly less efficient, "traditional" counterparts? Well, for the Chapter 9 play-all participants, we really don't have an apples-to-apples comparison, because it is quite clear that our normal definition of "play all" is just that: We sit down, and we play all the hands of the shoe, regardless of the count. Even our least efficient player, the White Rabbit, is more discriminating in his approach to the game, in that he gets up when it's intelligent to do so; thus, it is obvious that Rabbit's SCOREs will be superior to the Chapter 9, play-all ones. See, for example, the 4.5/6, s17, das, ls, 1–12 value of $32.91, from Table 9.3, and compare that to the $95.33 we found that Rabbit won in this study. In essence, Rabbit most closely resembles what many players refer to as the person who "Wongs out"; that is, he is willing to play right away, he bets all the time, once seated, but he refuses to play below a pre-designated true count. (We now know that, in fact, there is actually an entire series of different optimal-departure counts.)

WiWo has no real counterpart in the SCORE chapter, either, but it is clear that, by Wonging *in* as well as out, he outperforms any type of play we might consider, except the "Perfect" variety. Which, for me, leads to the most fascinating of the comparisons, namely, that of my previous concept of the quintessential back-counter, as described on p. 18, and our current Mr. Perfect. Recall what I said in Chapter 2: "Assume you actually have the right to sit at any given table and count all the rounds for the hour and that, furthermore, you may bet *only* the +1 and higher counts, betting in effect, zero for the others." And earlier: "Granted, by judiciously walking away from poor counts in progress, one ought to be able to do better, in theory. However, in practice, it is not that easy. … Limited availability of higher-stakes tables invalidates the assumption that one can always find *instantly* another dealer ready to shuffle."

And so, it isn't the least bit surprising that Mr. Perfect, *with no lag,* out-SCOREs the player that I have been offering you for many years as the model back-counter. In fact, the comparison isn't even close. For the 4.5/6 game, the SCORE player earns $59.91, while our Mr. Perfect clocks in with a hefty $104.40 — a substantial improvement. Some may be surprised at the closeness of the SCOREs of Mr. Perfect and the White Rabbit (only a $9 differential). For the particular game we are studying, with its very favorable rules (s17, das, ls), the only "penalty" Rabbit incurs, with respect to Perfect, is to make minimum bets in the twilight zone, while Perfect either watches from behind, or sits out those hands, once at the table. In this situation, the negative edge in the twilight zone is virtually nonexistent (or may even be ever-so-slightly positive), and so, having the luxury of not playing those hands does not offer a great deal of extra value to Perfect. He does benefit, however, from a larger unit size than Rabbit's, a by-product of being more selective in the hands he plays.

Our studies bogged down somewhat when we attempted to derive accurate win rates for our three heroes when they encountered the three-minute, six-hand lag we imposed for the various games (see the final versions in Section II, below). But, one thing is clear: No matter what the lag (within reason), our current Mr. Perfect will always be able to outperform the aforementioned SCORE back-counter. Logic dictates that this must be the case. For, the SCORE back-counter is playing, in essence, with an infinite lag. He never leaves a shoe for another table! Rather, he sits out the balance of any shoe whose true count drops below the ODP, and, in so doing, he passes up the opportunity to head for a fresh shoe. As a result, *any* optimal departure schedule must be preferable to no departure plan. It appears true that, the greater the lag, the more likely it is that Perfect's SCOREs will begin to converge to that of the SCORE back-counter. But, they will never be worse.

On the television series *The A-Team,* the leader, Hannibal Smith, played by George Peppard, was fond of saying, when everything worked out smoothly in the end, "I love it when a plan comes together!" In my more than 20 years of conducting blackjack research, I can't think of a project where more dedicated individuals put in more time and effort to make a plan, indeed a dream, come together. And, although we've left a little on our plates, before we can say we're done, much has already been learned from our investigation of this fascinating topic. I hope you enjoy the fruits of our labor.

(One final note: In the latter stages of our work, two more prominent researchers lent their considerable expertise to the project. First, John Auston reconsidered the simulations that he had undertaken more than a year ago and found new approaches to this tantalizing puzzle. Next, James Grosjean came on board, and he, too, provided many helpful suggestions and insights into how we might proceed. I thank them both for their valuable participation and contributions.)

Section II
Blackjack Forum, Fall 2000

The saga continues! Several months have passed since I wrote the above lines. The research has proceeded at a frenetic pace. I know that I owe you some data. And data you shall have — but not before I once again recognize the extraordinary efforts put forth by "Team ODP," namely James Grosjean, John Auston, Richard Reid, and, more recently, Karel Janecek. As the going got tough, these incredibly talented programmers and theoreticians went far beyond the call of duty to get the job done.

Richard reran all of the ODPs, making certain they conformed to the latest information we had obtained. In the process, he even bought a new computer, thus facilitating his task and greatly reducing the turn-around time required to produce the information. Not to be outdone, John, temporarily stymied by the need for a correct algorithm to account properly for the hands actually played by each of our three participants, developed a new piece of software and dubbed it his "BJ Clock." We would use it to verify how many shoes were contemplated, abandoned, and entered by our players, and to determine how many hands per shoe, and, ultimately, per hour, each played. Of course, John also used his vaunted *BJRM* to churn out optimal bet sizes and win rates.

Karel, now back in the U.S. and enrolled in the mathematics doctoral program at the prestigious Carnegie Mellon University, was kind enough to take time out from his busy schedule to "tweak" some of his *SBA* code in an effort to respond to some of our concerns. I thank him, as I do the others, for all of his help with this project.

And then there is James! I often wonder what prompts someone to drop everything he is doing and to spend dozens upon dozens of hours, including a couple of "all-nighters" (as if back in college!), simply to lend a helping hand. As John and I got bogged down in some of the particulars, James decided that he, too, would start from scratch, write tons of code, and build his own simulator to resolve the problems! Why? Because: a) he can (!); b) he's one of the most talented and innovative gaming researchers on the scene today; c) he is intrigued by challenging puzzles and enjoys solving them; d) all of the above! Ultimately, it would be James's program that we would use to produce all of the statistics that follow.

Methodology: The big stumbling block was that, while *SBA* produced theoretical wins for our players, based on 100 hands *actually played,* it didn't provide a "pro-rating" of those hands, needed to furnish the traditional, per-hour SCOREs. That's when John got the idea to create his clock, which would tally the comings and goings of our peripatetic heroes, thus allowing for an accurate accounting of the players' activities.

We decided on the previously mentioned 100 seconds necessary to shuffle a 6-deck shoe and on 140 seconds for the 8-deck variety (see p. 18). That impost was added at the end of every shoe completed by any of our players. On the other hand, while moving from one table to another, either with or without a lag, no shuffle-time penalty was imposed. But, to be as realistic as possible, the delay for a one-time initial buy-in, per casino, was factored in, as well. And, as explained earlier, a lag of 180 seconds was added each time a player abandoned a shoe to find a new one, provided, of course, we were simulating in "lag mode."

As was the case with all prior SCORE determinations, we kept four players at the table at all times. When one of our guys entered a game, someone conveniently got up. Precise bet sizes were used (no rounding), and the true count was estimated to the exact number of cards remaining. I chose this approach simply to generate numbers that were as "perfect" as possible and because we wanted all the parties involved in the research to keep any approximations to a bare minimum. John and James, using different approaches, compared their results, and they were similar enough to inspire the greatest of confidence in me that we had solved, once and for all, perhaps the most complicated blackjack problem I have ever studied.

How to read the tables: Added to the original tables and graphs (our current Tables 13.1–8 which begin on p. 360), which first appeared in the second edition of *BJA,* are several additional sources of data. At the top, you will find the precise unit-bet sizes of each player, by true-count bin, followed by the unit size, in dollars, for each. Below the optimal departure points, we have appended four new lines, presenting each player's hands played per hour, average bet size, win rate, in percent, and the product of these three values, the all-important win per hour, expressed in dollars. Note that our traditional concept of playing or observing 100 hands per hour has been altered for this study. The 100 hands assumes a certain number of shuffles per hour, based on my previously explained methodology for calculating this number. But, we now have a new methodology, and our aggressive style of leaving tables reduces the number of shuffles we are forced to endure. This, in turn, produces anywhere from 13 to 15 extra hands seen, or played, per hour, depending on who is playing and the style of play he is employing.

Two entirely new tables, created specifically for this Section II, are also provided, following the eight we've just discussed. Table 13.9 is devoted to the 6-deck games, while Table 13.10 treats the 8-deck variety. Although I believe most of the headings are self-explanatory, a brief description of the meaning of each statistic will, perhaps, be helpful. Row one (under "Players") reports the total shoes we look upon, during our hour of play. Clearly, we do not enter every one of these shoes, and even when we do sit down, we often leave that table, as well. So, the next row reports, of the total shoes seen, how many we wind up abandoning, whether from the standing or seated position. To break down that statistic into its component parts, row four ("Shoes Never Entered") tells us how many shoes get ditched while we're just watching, while row eight ("Entered Shoes Abandoned") reports the number of shoes we leave after having taken a seat. Obviously, the sum of the two rows can be found in row two.

Next, we consider the tables we *do* get to play, so row six catalogs the number of shoes we actually enter. Once in the game, we look to row ten for our rounds played per shoe. Thus, the product of this row and row six ("Shoes Entered") will give us our final row 13, namely rounds played per hour. To complete the record, row 11 reports the number of rounds actually dealt per hour, whether or not we ever get to see or play them, and row 12 provides the number of rounds actually seen by our players. *[Note: These last two rows discussed are, by definition, the same, when there is no lag, but differ whenever we lose time moving from one table to another. — D.S.]*

Interpreting the data: There are several aspects of the play that need to be discussed, as we try to make sense of all the data. First, it is obvious, upon considering the no-lag "Shoes Seen," "Shoes Abandoned," "Shoes Entered," and "Entered Shoes

Abandoned" rows, in Tables 13.9 and 13.10, that this style of play is almost certainly unattainable in actual casino play. You simply aren't going to be able to jump around that much without getting caught. Consider the 4.5/6 game, with no-lag-Rabbit. To attain his aforementioned $95.33 SCORE, he sees almost 113 hands per hour, plays all of them, but only to the tune of 4.75 rounds per shoe! He abandons fully 91% of the shoes he sees, thereby requiring him to sit at 24 different shoes per hour (!), or a new shoe at the rate of one every 2.5 minutes! Rabbit indeed. *Dead* rabbit, if you ask me!

No, clearly, this is not the way to endear oneself to the resident pit bosses of America! As we move to lag-Rabbit, the numbers are attenuated slightly. We learn that now 80% of the shoes seen are abandoned (still an unacceptably high percentage), while those entered provide, on average, 8.57 rounds per shoe. Thus, our typical hour of table-hopping produces just over 73 hands played, at a total of, roughly, 8.5 different shoes. And, of course, the reduction in SCORE is considerable, falling to 54.7% of the $95.33 no-lag value, or $52.16.

Is there a more normal-looking way to play this game solo, which will, nonetheless, garner a more attractive return? Let's move on to WiWo. You remember him: He scouts the tables, waiting for a +1 TC advantage before sitting down, and, once in the game, bets optimally until in the twilight zone, where he makes minimum waiting bets. If the count descends to the ODP, he leaves. The tables tell us that, without a delay, WiWo will get to see 113 hands per hour (as did Rabbit), but as the more discerning player, WiWo will enter only 25.8% of the 24 shoes he scouts, settling down to just over six shoes per hour, at which an average of almost ten hands per shoe will be played, for an hourly total of 61. The SCORE turns out to be $103.88 — a lofty, although, once again, somewhat unattainable, stipend.

We begin to approach what I consider to be perhaps the most realistic and, therefore, most plausibly attainable scenario for real-world play when we study WiWo with a lag. Our protagonist sees 8.65 shoes per hour (reasonable), or one about every seven minutes. He enters 42.6% of those shoes, or 3.68 per hour, at which he plays 10.4 hands per shoe. The hour's work produces just over 38 hands played (recall that WiWo does *not* sit out the twilight-zone hands), and a SCORE of $59.40, a number about which I'll have more to say in a little while.

Finally, to complete our analysis of the 4.5/6 game, we turn to our presumed "leader," the indomitable Mr. Perfect. As the name implies, Mr. P. is supposedly playing as perfectly as a lone, traditional card counter can. He has similar tactics to WiWo, while stalking the tables, but once in a game, Perfect refuses to play a single hand where he doesn't have the advantage. And, of course, as does WiWo, he walks as soon as the ODP is reached. So, what are Perfect's "specs"? Well, without a delay (I know — Fantasyland!), he sees our now familiar 113 hands per hour and contemplates 23 shoes. Of those watched, he enters 26.1%, or almost precisely six shoes per hour, and manages to play just about 8 hands per shoe, yielding 48.2 hands played per hour. His reward? Our previously mentioned SCORE of $104.38 — the ultimate Hi-Lo compensation for the game conditions under consideration.

But, we all agree that Perfect is unlikely to get away with this style of play in a real-world setting. So, let's see what happens when our six-round lag is imposed. Forced to stick around longer at each table he joins, and delayed by the ongoing three-minute imposts for leaving, Perfect gazes upon only 8.28 shoes per hour and enters 44.3% of them, or 3.67 per hour. On average, in each of those chosen shoes, he plays 7.71 hands, so Perfect manages to play 28.3 hands per hour and wins $61.39. Retain these numbers for just a moment.

I told you before that I was very anxious to see what these last two values would turn out to be for lag-Perfect, as I was convinced that they would be extremely close to the statistics associated with my prototypical back-counter, the one described on p. 18. We reported his SCORE as $59.91, and a brief look at Table 10.51, on p. 240, informs us that, if entering at +1, he plays 27 hands per hour. Indeed, the latest version of *BJRM,* which uses a slightly different software package for calculating SCORE, yields $60.37. What's the point? Well, despite a much less scientific approach to the game than our current Mr. Perfect, the player I've been modeling as a "back-counter" all of these years produces a SCORE within 1.69% of the "official" lag version. I am gratified that the two methodologies have proven to be so compatible. Finally, note that the $59.40 for lag-WiWo, reported above, represents, in my opinion, the most realistic hourly win projection for the modern player. It, too, is less than two percent away from our back-counting standard-bearer's SCORE.

Further observations and fine points: Of course, so far, we have applied our detailed analysis to just one of the four games studied. Many more observations can be made as we peruse the data contained in the tables. I'm not going to rehash the 5/6 numbers, as all the principles outlined above pertain to this more desirable game, as well. Suffice it to say that the lag-Perfect SCORE of $92.80 once again compares admirably to that of the archetypal back-counter, whose hourly win, as reported on p. 157 is $90.21. And, it is WiWo's $88.80 lag SCORE that I believe is attainable in a modern-day casino.

On to the 8-deck games! Inherently worse than the 6-deck variety, our 8-deck representative is further handicapped by the absence of the late surrender option. It is fair to say that in so modeling this particular game, I had our east coast Atlantic City brethren in mind. And now, I'm going to share a secret with you. Before I had even had the opportunity to see the 8-deck data, I had prepared the following introduction for our study of these games:

Why should we now expect increasing disparities among the SCOREs of our players? If we render the twilight zone less inviting, by turning it into truly negative territory, rather than merely the "neutral hangout" it was in our 6-deck game, the gaps between our three protagonists' SCOREs are bound to widen. Clearly, those who hang out in the zone, making waiting bets, are now going to be more severely punished for their "loitering" activities.

Turns out, I was wrong! In fact, there is little difference between the 8-deck disparities and the 6-deck ones. My error came in focusing on the "inflated" value to the counter of surrender, which we had found in Chapter 10. But that value is attributable to the relatively wide spreading of bets that our players engage in and to the few (four) indices we employ to refine our surrendering skills. Virtually none of this activity (except utilizing the 15 v. 10 zero index) is applicable in the twilight zone. And so, with their one-unit waiting bets, the loss of surrender is hardly noticeable to WiWo and Rabbit, as they try to capture for themselves the lion's share of Perfect's higher SCOREs. Furthermore, it is clear that, by judiciously Wonging *in,* as well as out, both Perfect and WiWo keep the time they spend in the "zone" to a minimum. Finally, let us note that, if the aforementioned spreads were smaller, thereby conferring greater importance to the waiting bets, we would surely see a wider gap between Perfect and WiWo.

No, the comparisons among our heroes aren't going to change dramatically, as we examine the results from our 8-deck games. What is clear, of course, is that everyone's win rate is substantially decreased, with respect to the 6-deck offerings. There is, however, an interesting turn of events, as we once again seek to compare lag-Perfect or lag-WiWo to our SCORE back-counter. Our guys are starting to pull away! In the 6.5/8 game, Perfect wins $44.57 and WiWo $42.96. Our SCORE prototype no longer fares as well, earning a mere $37.63 per hour. Why, even lag-Rabbit, with his $35.70, is a fair challenge for the SCORE player. I'm happy to see some extra profits accruing to our modern-day heroes! I'll allow you to peruse the rest of the data on your own, and I trust you will find a few more engaging comparisons.

I mentioned earlier that the play of the last hand of the shoe posed a special problem that I promised to discuss. I have already received, since the publication of the second edition of *BJA,* several questions concerning the advisability of leaving a positive count to play a new shoe off the top. But, the math is compelling, and, when no lag is imposed, the numbers do not lie. There is no change in the recommendations offered in Section I of this article. However, a very definite problem arises when a lag is imposed. I'd like to acknowledge that Bob Fisher was the first to bring this observation to my attention, and, for the record, he was absolutely correct with the objections he raised.

If you look at any of the final ODPs in the 6-deck, 6-round-delay tables — the rather elevated values at the bottom of the charts — you'll notice that they are all approximately +2 true counts. The implication is that, despite the imposition of a three-minute delay to find a fresh shoe, we are nonetheless better off leaving the table and forgoing the play of this current, advantageous hand. Fisher wondered how this could possibly be. He argued that we could play the hand in 30 seconds and, because it was the final hand of the shoe, a shuffle would then produce a new shoe, *right at this table,* in another 100 seconds. So, all told, not only would we get to play the hand (with its edge of greater than 1%) and still be ready for our new shoe, but we'd achieve both in 50 seconds less than it would take us to find a new table (while not having played the hand). And so, we need to consider the final ODPs of each chart as not directly applicable to the final hand, which marches to its own drummer.

Truth be told, the values at each extreme of the ODP charts must be taken with a grain of salt, for what sense does it make to say that, at zero deck level (the beginning of the shoe), the ODP is zero? Does this imply that, as we start to watch a brand-new shoe, we should walk away if the TC is (as it *must* be) zero? Hardly! For, we'd be walking in an eternal circle, without ever playing a single hand! So, it's best you ignore these "barrier" entries and concentrate on the truly important values in between.

Conclusions: It is customary, when I reach this stage of an article, to propose the "Where do we go from here?" question. Before I do that, I'd like to take a moment to summarize the findings from what ought to be characterized as one of the most ambitious and painstaking blackjack research projects ever undertaken (and, trust me, I've been involved in enough of them to recognize the granddaddy of them all when I see it!).

Traditionally, serious card counters have recognized that, to earn a meaningful wage at a shoe game, under ordinary circumstances, it is necessary to back-count. It is fair to say that Stanford Wong started it all by describing, in his earliest version of the classic *Professional Blackjack,* his personal approach to what, logically enough, came to be known as "Wonging." But, as mentioned earlier, Wong's activities almost always involved flat-betting several black chips, and simply entering the games at the first instance of advantage and exiting, ostensibly, at the most convenient moment after perceiving the end to that advantage.

Later, as practitioners sought to refine the original process, variations on the theme surfaced. Some were content to enter a game immediately, preferring to concentrate simply on Wonging *out,* rather than on Wonging *in.* Hence, the need for this new bimodal terminology, not utilized in the early literature. Still others favored the complementary approach, namely, Wonging *into* the games but, once at the table, remaining until at least the completion of that present shoe, regardless of the ensuing counts, mainly for appearance's sake. Of course, many, hoping to win as much as possible, sought the ultimate back-counting experience, contriving to enter the game at the most advantageous moment while also choosing to effect their egress from the table at the most propitious time.

This desire for back-counting "perfection" led, quite naturally, to the posing of the most innocent-sounding of all questions: *When should I leave?* I have to snicker. Often, when someone says to me, "May I ask you a quick question?" I will jokingly reply, "The question may be quick, but the answer might take a little longer!" And so it has been with the above research. I suppose that's why it has taken me, from the first time I wondered about the "correct" answer, about 20 years to track down the elusive response. Not that I could have possibly done it alone. No, to nail down this baby has taken my original curiosity; the vision of Chris Cummings; the theoretical musings of Kim Lee; the practical implementation, by Richard Reid, of Lee's initial efforts; the programming wizardry of John Auston and Karel Janecek; and last, but certainly not least, the extraordinary vision, talent, and perseverance of James Grosjean, without whose genius you most definitely would not be reading this study today. Together, this magnificent "Team ODP," as we called ourselves, has furnished the answers — but, more importantly, in my view, the correct methodology — for attacking both present and future problems concerning optimal table departure.

Which leads, of course, to the aforementioned "Where do we go from here?" Typically, when we embark on research of the kind you have just read, many choices and compromises have to be made. There is, after all, room for just so many pages and so much data in a journal of this type. So, we mandate a count system — invariably, Hi-Lo, because of its mass appeal. Then, we choose games to study. In this instance, it made sense to select the Las Vegas Strip 6-deck game and the standard Atlantic City 8-decker. Then, the details need to be ironed out. So, more choices and compromises are made. In the end, we have just a small (but representative) subset of all the possible scenarios that could apply to the myriad players out there plying their trade; but again, more importantly, we have an algorithm, a methodology, that can now be used to study many more situations.

We should impose different lag-times for the "reconnoitering." Surely, different count systems should be considered. Less advantageous rules should be specified. "Real-world" integral bet sizes can be used, with true-count approximations made to, say, the nearest half-deck. Indeed, less precise ODPs, themselves, might be contemplated, for convenience. Fewer (or more) players could be present at the table. And so it goes. A work in progress? Why, of course. Most of my research marks a beginning, rather than an end, of the answers to the question at hand. In this instance, a rather ambitious beginning, to be sure, but just a start nonetheless. Until we unravel a few more of the mysteries that still lie ahead, good luck, and … good cards!

TABLE 13.1
4.5/6, S17, DAS, LS
(With no delay between shoes)

True Count:	Twil. Zone	1	2	3	4	5	6	7	8	≥9	Unit ($)
Rabbit bet ramp:	1	1.7	3.5	5.7	7.7	10	12	12	12	12	24.15
WiWo bet ramp:	1	1.2	2.6	4.1	5.6	7.3	9	10.5	12	12	33.33
Perfect bet ramp:	0	1	2.1	3.3	4.6	5.9	7.4	8.5	9.6	12	40.82

Current Shoe Depth	Optimal Departure TC Rabbit	WiWo	Perfect
0.0	0.00	0.00	0.00
0.5	–0.10	–0.10	–0.10
1.0	–0.15	–0.15	–0.20
1.5	–0.20	–0.20	–0.25
2.0	–0.25	–0.25	–0.30
2.5	–0.30	–0.30	–0.30
3.0	–0.30	–0.25	–0.25
3.5	–0.15	–0.10	–0.05
4.0	0.30	0.35	0.45
4.5	2.15	2.25	2.35
Hands Played/Hour	112.91	61.04	48.06
Avg. Bet ($)	56.14	87.53	101.18
Win/Loss %	1.50	1.97	2.15
Win/Hour ($)	95.33	105.13	104.40

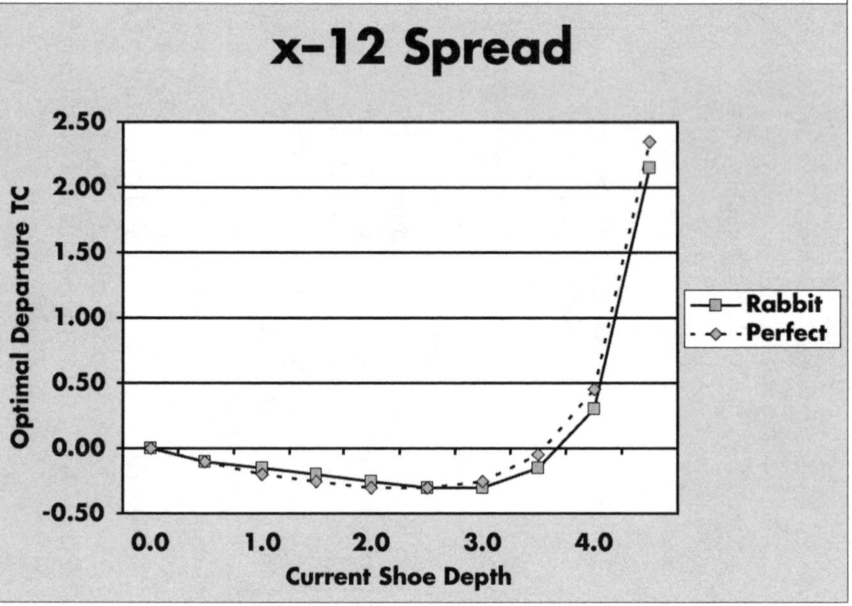

x-12 Spread

Optimal Departure TC
2.50
2.00
1.50
1.00
0.50
0.00
-0.50
0.0
1.0
2.0
3.0
4.0
Current Shoe Depth
Rabbit
Perfect

TABLE 13.2
5/6, S17, DAS, LS
(With no delay between shoes)

True Count:	Twil. Zone	1	2	3	4	5	6	7	8	≥9	Unit ($)
Rabbit bet ramp:	1	1.4	3	4.7	6.6	8.3	10.2	11.7	12	12	29.50
WiWo bet ramp:	1	1.1	2.2	3.4	4.7	6	7.5	8.9	10.4	12	40.16
Perfect bet ramp:	0	1	2	3.1	4.2	5.4	6.7	8	9.3	12	44.64

Current Shoe Depth	Optimal Departure TC Rabbit	WiWo	Perfect
0.0	0.00	0.00	0.00
0.5	–0.13	–0.13	–0.13
1.0	–0.26	–0.26	–0.26
1.5	–0.39	–0.39	–0.39
2.0	–0.52	–0.52	–0.52
2.5	–0.65	–0.65	–0.71
3.0	–0.77	–0.71	–0.77
3.5	–0.84	–0.84	–0.84
4.0	–0.84	–0.77	–0.77
4.5	–0.39	–0.32	–0.26
5.0	2.52	2.71	2.84
Hands Played/Hour	113.16	65.59	49.50
Avg. Bet ($)	65.93	97.69	116.21
Win/Loss %	1.78	2.27	2.54
Win/Hour ($)	132.82	145.53	146.29

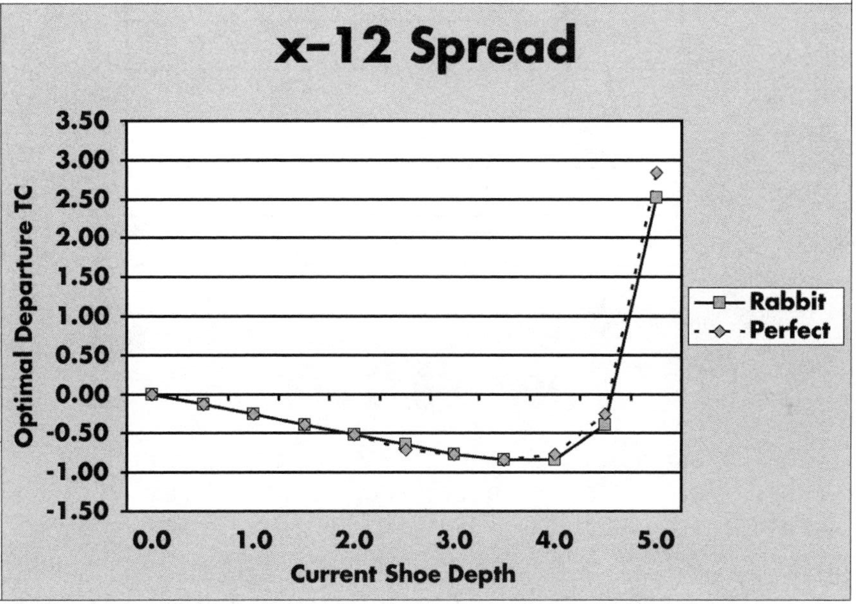

x–12 Spread

TABLE 13.3
4.5/6, S17, DAS, LS
(With delay of 6 rounds between shoes)

True Count:	Twil. Zone	1	2	3	4	5	6	7	8	≥9	Unit ($)
Rabbit bet ramp:	1	1.9	4.1	6.7	8.8	11.7	12	12	12	12	20.88
WiWo bet ramp:	1	1.4	2.9	4.4	6.3	8	10.1	11.3	12	12	30.12
Perfect bet ramp:	0	1.1	2.2	3.5	4.7	6.1	7.5	8.6	10	12	40.16

Current Shoe Depth	Optimal Departure TC Rabbit	WiWo	Perfect
0.0	0.00	0.00	0.00
0.5	–0.85	–0.80	–0.95
1.0	–0.90	–0.85	–1.00
1.5	–0.95	–0.95	–1.10
2.0	–1.00	–0.95	–1.10
2.5	–1.05	–1.00	–1.10
3.0	–1.00	–0.95	–1.05
3.5	–0.85	–0.80	–0.80
4.0	–0.35	–0.30	–0.20
4.5	1.95	1.95	1.95
Hands Played/Hour	73.27	38.44	28.40
Avg. Bet ($)	50.17	82.65	104.23
Win/Loss %	1.42	1.89	2.11
Win/Hour ($)	52.16	59.95	62.36

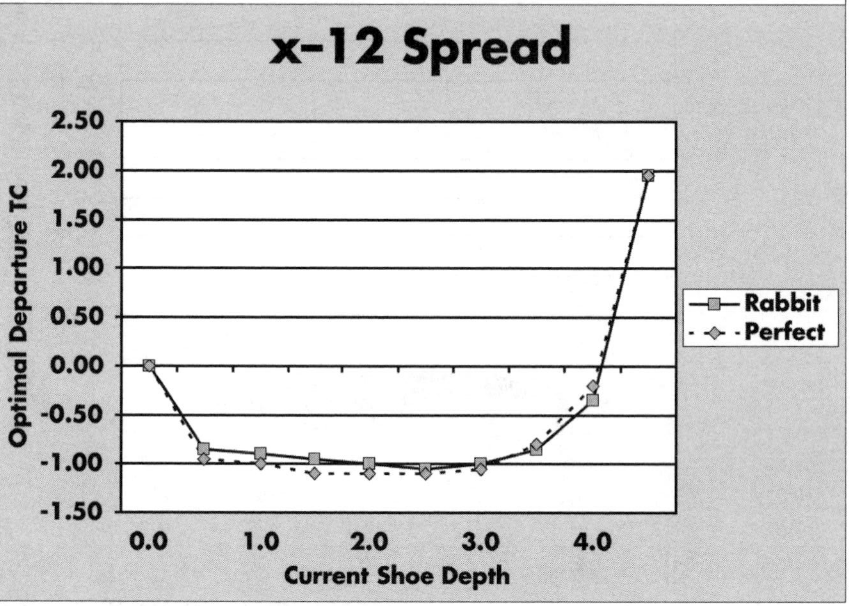

x–12 Spread

Optimal Departure TC: 2.50, 2.00, 1.50, 1.00, 0.50, 0.00, -0.50, -1.00, -1.50

Current Shoe Depth: 0.0, 1.0, 2.0, 3.0, 4.0

Rabbit
Perfect

TABLE 13.4
5/6, S17, DAS, LS
(With delay of 6 rounds between shoes)

True Count:	Twil. Zone	1	2	3	4	5	6	7	8	≥9	Unit ($)
Rabbit bet ramp:	1	1.6	3.6	5.4	7.4	9.2	11.3	12	12	12	26.11
WiWo bet ramp:	1	1.1	2.4	3.7	4.9	6.2	7.8	9.5	10.6	12	38.02
Perfect bet ramp:	0	1	2	3.2	4.3	5.5	6.7	7.9	9.4	12	44.64

Current Shoe Depth	Optimal Departure TC Rabbit	WiWo	Perfect
0.0	0.00	0.00	0.00
0.5	–0.90	–0.90	–1.03
1.0	–1.03	–0.97	–1.16
1.5	–1.16	–1.10	–1.29
2.0	–1.29	–1.23	–1.42
2.5	–1.42	–1.36	–1.55
3.0	–1.55	–1.48	–1.68
3.5	–1.61	–1.55	–1.74
4.0	–1.55	–1.48	–1.61
4.5	–1.16	–1.03	–0.97
5.0	1.94	1.94	2.00
Hands Played/Hour	77.92	42.63	30.49
Avg. Bet ($)	60.16	93.72	118.73
Win/Loss %	1.67	2.23	2.57
Win/Hour ($)	78.41	88.93	93.13

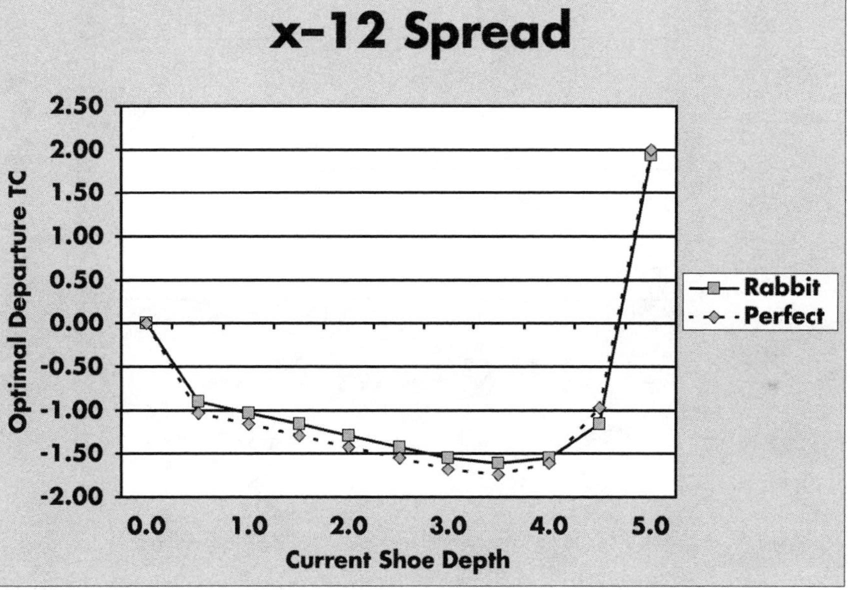

x-12 Spread

Optimal Departure TC
2.50
2.00
1.50
1.00
0.50
0.00
-0.50
-1.00
-1.50
-2.00
0.0 1.0 2.0 3.0 4.0 5.0
Current Shoe Depth
Rabbit
Perfect

TABLE 13.5
6/8, S17, DAS, NS
(With no delay between shoes)

True Count:	Twil. Zone	1	2	3	4	5	6	7	8	≥9	Unit ($)
Rabbit bet ramp:	1	1.9	4.3	6.5	10.1	12	12	12	12	12	15.15
WiWo bet ramp:	1	1.1	2.9	4.5	6.3	8.2	10.1	12	12	12	23.58
Perfect bet ramp:	0	1	2.3	3.6	5.1	6.6	8.1	9.8	11.2	12	29.41

Current Shoe Depth	Optimal Departure TC Rabbit	WiWo	Perfect
0.0	0.00	0.00	0.00
0.5	–0.04	–0.04	–0.04
1.0	–0.04	–0.04	–0.09
1.5	–0.09	–0.09	–0.09
2.0	–0.09	–0.09	–0.13
2.5	–0.09	–0.09	–0.13
3.0	–0.09	–0.09	–0.13
3.5	–0.09	–0.04	–0.09
4.0	–0.04	0.00	–0.04
4.5	0.09	0.13	0.09
5.0	0.30	0.35	0.35
5.5	0.78	0.82	0.82
6.0	2.17	2.25	2.30
Hands Played/Hour	113.38	58.31	45.52
Avg. Bet ($)	35.58	59.22	70.40
Win/Loss %	1.05	1.48	1.63
Win/Hour ($)	42.51	50.98	52.12

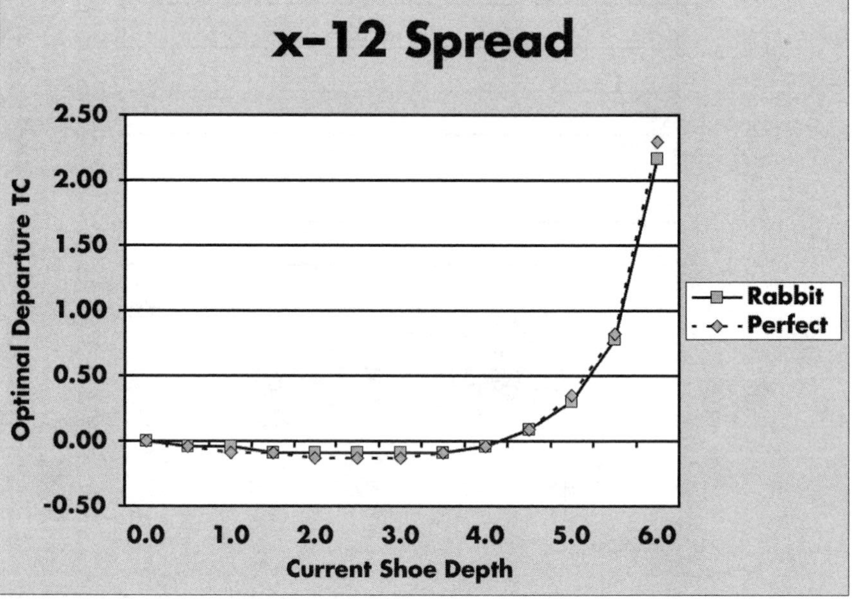

x-12 Spread

Optimal Departure TC

2.50
2.00
1.50
1.00
0.50
0.00
-0.50

0.0 1.0 2.0 3.0 4.0 5.0 6.0

Current Shoe Depth

Rabbit
Perfect

TABLE 13.6

6.5/8, S17, DAS, NS

(With no delay between shoes)

True Count:	Twil. Zone	1	2	3	4	5	6	7	8	≥9	Unit ($)
Rabbit bet ramp:	1	1.5	3.7	6.4	8.6	10.8	12	12	12	12	17.86
WiWo bet ramp:	1	1	2.5	4.1	5.8	7.4	9.1	11.1	12	12	26.74
Perfect bet ramp:	0	1	2	3.2	4.5	5.9	7.3	8.6	10.2	12	33.56

Current Shoe Depth	Optimal Departure TC Rabbit	WiWo	Perfect
0.0	0.00	0.00	0.00
0.5	–0.05	–0.05	–0.10
1.0	–0.10	–0.10	–0.16
1.5	–0.16	–0.16	–0.21
2.0	–0.21	–0.16	–0.26
2.5	–0.21	–0.21	–0.31
3.0	–0.26	–0.26	–0.36
3.5	–0.31	–0.26	–0.36
4.0	–0.31	–0.31	–0.42
4.5	–0.31	–0.31	–0.36
5.0	–0.26	–0.21	–0.31
5.5	–0.10	0.00	–0.05
6.0	0.42	0.52	0.52
6.5	2.50	2.71	2.71
Hands Played/Hour	113.64	61.59	46.69
Avg. Bet ($)	40.82	64.80	79.04
Win/Loss %	1.24	1.68	1.84
Win/Hour ($)	57.55	67.25	68.01

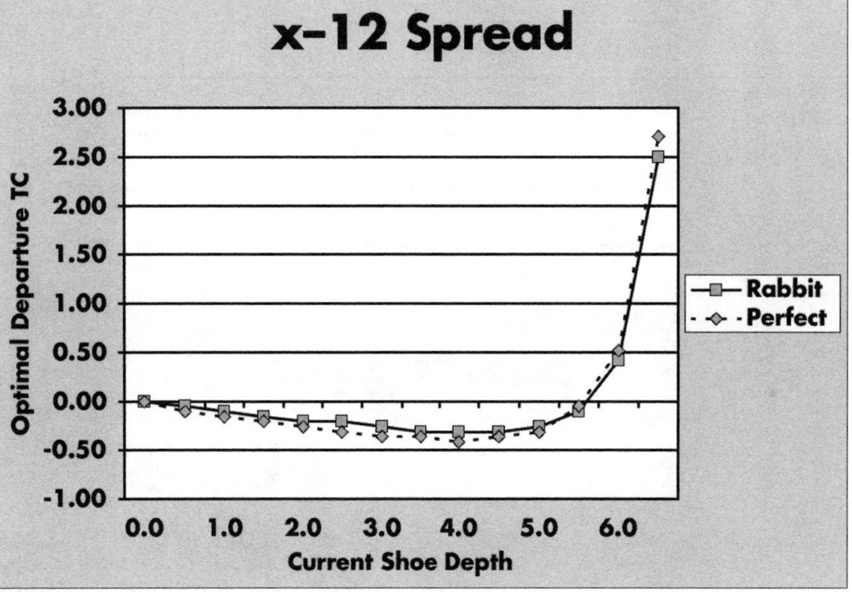

x–12 Spread

Optimal Departure TC

Current Shoe Depth

Rabbit

Perfect

TABLE 13.7
6/8, S17, DAS, NS
(With delay of 6 rounds between shoes)

True Count:	Twil. Zone	1	2	3	4	5	6	7	8	≥9	Unit ($)
Rabbit bet ramp:	1	2.2	5.1	7.6	11.2	12	12	12	12	12	13.48
WiWo bet ramp:	1	1.2	3	5	7.1	9.2	11.1	12	12	12	21.74
Perfect bet ramp:	0	1	2.2	3.5	4.9	6.4	7.7	9.7	11	12	30.49

Current Shoe Depth	Optimal Departure TC Rabbit	WiWo	Perfect
0.0	0.00	0.00	0.00
0.5	–0.65	–0.61	–0.65
1.0	–0.65	–0.61	–0.69
1.5	–0.65	–0.61	–0.74
2.0	–0.65	–0.61	–0.74
2.5	–0.69	–0.61	–0.74
3.0	–0.65	–0.61	–0.74
3.5	–0.65	–0.56	–0.69
4.0	–0.61	–0.52	–0.61
4.5	–0.48	–0.39	–0.48
5.0	–0.26	–0.13	–0.17
5.5	0.26	0.39	0.39
6.0	1.99	1.99	1.99
Hands Played/Hour	75.70	37.82	27.97
Avg. Bet ($)	33.00	56.56	70.94
Win/Loss %	1.02	1.46	1.61
Win/Hour ($)	25.53	31.17	32.00

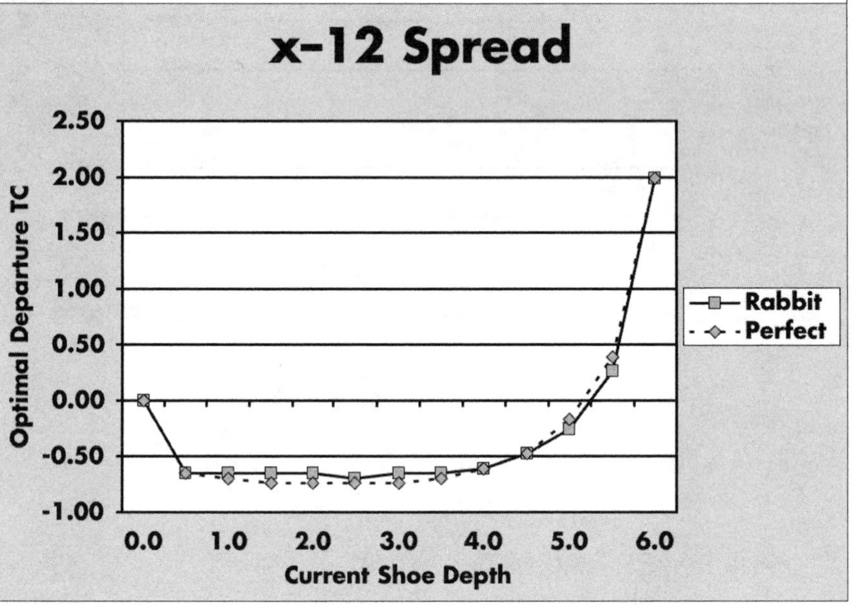

x-12 Spread

TABLE 13.8
6.5/8, S17, DAS, NS
(With delay of 6 rounds between shoes)

True Count:	Twil. Zone	1	2	3	4	5	6	7	8	≥9	Unit ($)
Rabbit bet ramp:	1	1.7	4	6.4	9.3	11.2	12	12	12	12	16.53
WiWo bet ramp:	1	1.1	2.6	4.4	6	7.7	9.8	11.5	12	12	25.51
Perfect bet ramp:	0	1	2.1	3.3	4.7	6.2	7.6	9.1	9.9	12	32.68

Current Shoe Depth	Optimal Departure TC Rabbit	WiWo	Perfect
0.0	0.00	0.00	0.00
0.5	–0.68	–0.62	–0.73
1.0	–0.68	–0.68	–0.78
1.5	–0.73	–0.68	–0.83
2.0	–0.78	–0.73	–0.88
2.5	–0.83	–0.78	–0.94
3.0	–0.83	–0.83	–0.99
3.5	–0.88	–0.83	–1.04
4.0	–0.94	–0.83	–1.04
4.5	–0.88	–0.83	–0.99
5.0	–0.83	–0.78	–0.88
5.5	–0.62	–0.52	–0.62
6.0	–0.10	0.00	0.00
6.5	2.03	2.03	2.03
Hands Played/Hour	78.47	41.13	29.86
Avg. Bet ($)	37.67	63.01	80.03
Win/Loss %	1.20	1.66	1.87
Win/Hour ($)	35.43	42.96	44.57

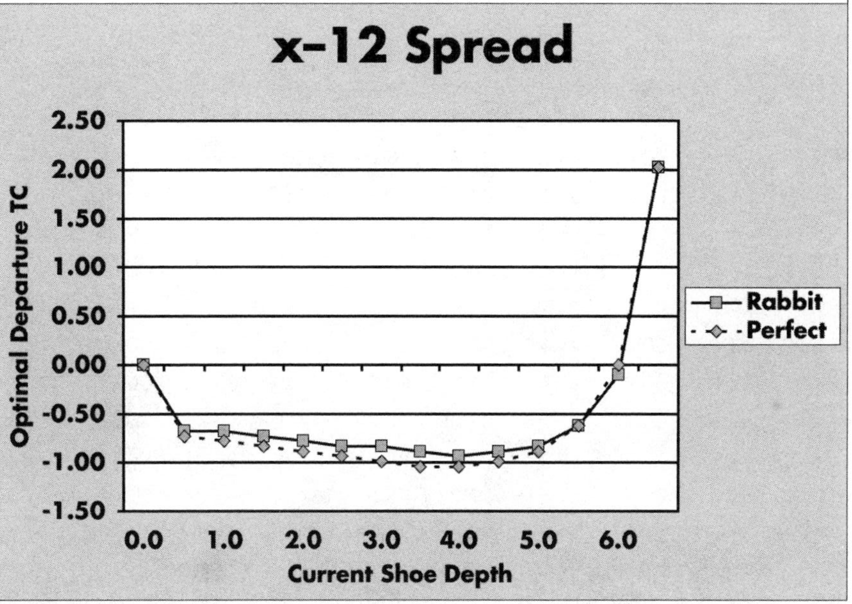

x–12 Spread

Optimal Departure TC
2.50
2.00
1.50
1.00
0.50
0.00
-0.50
-1.00
-1.50
0.0 1.0 2.0 3.0 4.0 5.0 6.0
Current Shoe Depth
Rabbit
Perfect

TABLE 13.9
Hourly 6-Deck Entry and Departure Data

Style of Play	No Delay						6-Round Delay					
Penetration/No. of Decks	**4.5/6 Decks**			**5/6 Decks**			**4.5/6 Decks**			**5/6 Decks**		
Players	**Rabbit**	**WiWo**	**Perfect**	**Rabbit**	**WiWo**	**Perfect**	**Rabbit**	**WiWo**	**Perfect**	**Rabbit**	**WiWo**	**Perfect**
Shoes Seen	23.79	23.86	22.69	20.70	20.74	20.39	8.55	8.63	8.12	7.78	8.06	7.41
Total Shoes Abandoned	21.67	21.77	20.65	18.65	18.71	18.48	6.84	6.96	6.48	6.06	6.43	5.79
(%)	91.07	91.22	91.01	90.08	90.21	90.66	80.06	80.63	79.84	77.93	79.69	78.16
Shoes Never Entered	0.00	17.70	16.68	0.00	15.08	14.82	0.00	4.94	4.42	0.00	4.49	3.83
(%)	0.00	74.19	73.48	0.00	72.72	72.72	0.00	57.26	54.44	0.00	55.72	51.71
Shoes Entered	23.79	6.16	6.02	20.70	5.66	5.56	8.55	3.69	3.70	7.78	3.57	3.58
(%)	100.00	25.81	26.52	100.00	27.28	27.28	100.00	42.74	45.56	100.00	44.28	48.29
Entered Shoes Abandoned	21.67	4.06	3.98	18.65	3.63	3.66	6.84	2.02	2.06	6.06	1.93	1.96
(%)	91.07	65.99	66.09	90.08	64.11	65.75	80.06	54.70	55.75	77.93	54.17	54.81
Rounds Played/Shoe Entered	4.75	9.91	7.99	5.47	11.59	8.90	8.57	10.43	7.68	10.02	11.94	8.52
Rounds Dealt	112.91	113.02	113.20	113.16	113.23	113.65	114.32	114.43	114.54	114.28	114.54	114.60
Rounds Seen	112.91	113.02	113.20	113.16	113.23	113.65	73.27	72.70	75.65	77.92	75.99	79.83
Rounds Played	112.91	61.04	48.06	113.16	65.59	49.50	73.27	38.44	28.40	77.92	42.63	29.86

TABLE 13.10
Hourly 8-Deck Entry and Departure Data

Style of Play	No Delay						6-Round Delay					
Penetration/No. of Decks	6/8 Decks			6.5/8 Decks			6/8 Decks			6.5/8 Decks		
Players	Rabbit	WiWo	Perfect	Rabbit	WiWo	Perfect	Rabbit	WiWo	Perfect	Rabbit	WiWo	Perfect
Shoes Seen	21.13	21.23	20.88	19.56	19.88	18.60	7.64	7.76	7.45	7.17	7.31	6.65
Total Shoes Abandoned	19.71	19.84	19.53	18.19	18.56	17.29	6.48	6.66	6.35	6.04	6.21	5.54
(%)	93.29	93.49	93.56	93.04	93.33	92.95	84.80	85.77	85.23	84.27	85.04	83.20
Shoes Never Entered	0.00	16.62	16.35	0.00	15.53	14.34	0.00	4.89	4.60	0.00	4.50	3.83
(%)	0.00	78.32	78.32	0.00	78.08	77.09	0.00	63.04	61.77	0.00	61.52	57.53
Shoes Entered	21.13	4.60	4.53	19.56	4.36	4.26	7.64	2.87	2.85	7.17	2.81	2.83
(%)	100.00	21.68	21.68	100.00	21.92	22.91	100.00	36.96	38.23	100.00	38.48	42.47
Entered Shoes Abandoned	19.71	3.22	3.18	18.19	3.03	2.95	6.48	1.76	1.75	6.04	1.72	1.71
(%)	93.29	69.96	70.30	93.04	69.56	69.24	84.80	61.49	61.38	84.27	61.12	60.44
Rounds Played/Shoe Entered	5.37	12.67	10.06	5.81	14.13	10.96	9.91	13.19	9.82	10.94	14.62	10.56
Rounds Dealt	113.38	113.55	113.73	113.65	113.81	113.88	114.58	114.85	114.86	114.73	114.90	114.78
Rounds Seen	113.38	113.55	113.73	113.65	113.81	113.88	75.70	74.91	76.75	78.47	77.61	81.56
Rounds Played	113.38	58.31	45.52	113.65	61.59	46.69	75.70	37.82	27.97	78.47	41.13	29.86

Part II

The Hi-Lo true count rises to just above +4, so we plunk down a sizable six-unit bet on the upcoming hand. The dealer flips a ten as his upcard, and, as we gaze at our own holding, a two-card version adding up to the same total of ten, we scan our memory for the appropriate index that would permit us to double. Plus four it is! Grudgingly, we shove out an additional six-unit stack, not at all confident that our one-card draw will be sufficient to beat the dealer, who, unlike us, gets to hit again, should he receive a small card. And yet, we make the play. Why? Because the literature tells us that it is at this precise level of true count that our expectation from doubling surpasses our expectation from simply hitting the hand (perhaps more than once). But, if our original six-unit bet was sized properly for the pre-deal edge we perceived, is it possible that we should now be willing to have twice as much money riding on a play that, in reality, offers only the tiniest bit of extra e.v. over the hitting alternative? Perhaps not. Enter the realm of risk-averse indices, and our second question: *What are risk-averse indices, and should I be using them?*

"Risk" and "reward" — two words, inextricably linked, whether in the world of finance and investment, or … blackjack. Our earlier SCORE chapter emphasized the importance of considering more than solely mathematical expectation when comparing the attractiveness of various blackjack-playing opportunities. Indeed, we learned that evaluating risk and linking the risk component of the game we are contemplating to its potential rewards are vital to our overall success in this endeavor.

But what of the riskiness of several *individual* plays, especially those such as the one mentioned above, where so much is riding on such a relatively razor-thin additional margin? Can the player truly justify the extra money required to double the already large wager, given virtually no appreciable extra advantage? Perhaps there is a better way. Might it not be wiser to wait until a higher count is reached, conferring a greater edge to the double-down, before risking the additional funds? Is it possible that, for some plays, it might never be correct to assume that additional risk, to obtain, ostensibly, small extra profits? Could it be that avoiding certain plays, and thereby reducing our risk, will in fact permit us to be bolder with *all* of our wagers, thereby actually increasing our overall SCORE? Some of these questions are old, but the answers you are about to read are new and fresh, and I am indebted to several individuals (whom I will name along the way) for making this research possible.

Almost 20 years ago to the day that these lines are being written, Joel Friedman presented his seminal work, "Risk Averse Playing Strategies in the Game of Blackjack," to the spring Own Risk and Solvency Assessment (ORSA) Conference, held in Washington, D.C. The description of the paper's content was brief and to the point: "Playing strategies for card-counting blackjack systems are usually computed with the aim of maximizing expected wealth. This paper examines the changes in optimal playing strategy that occur when the more appropriate objective of maximizing the expected value of the log of wealth is used."

But, there was a problem. Truth be told, the paper was read by relatively few people, many of whom were not even blackjack aficionados. And so, this gem and the fascinating information it contained were kept under wraps, as it were, for nearly two decades. Oh, to be sure, I would receive an occasional question or two on the subject and, always, I would refer the player to Joel's work, which, admittedly, was not that easy to find. I was grateful that the author had been kind enough to send me a copy, and I was one of only a few blackjack researchers who, over the years, readily made reference to the Friedman paper.

Fade out, fade in. Again and again, I have told you of the marvelous research going on over the Internet and of the valuable tools that have been developed in the past few years to further our knowledge of the game and to answer the thorniest of questions. And so it is with risk-aversion. We have the answers! For those who require explanations and the mathematical justification for the conclusions we're about to draw, the following is for you. For those who are not only risk-averse, but "math-averse," as well, you are invited to skip what follows and cut directly to the chase, to the tables and findings a bit further down in the chapter.

The passing of Peter Griffin left an enormous void in the blackjack community and a vacancy in the office of "High Priest" of the theoretical mathematics of the game. A professor of mathematics at the University of Detroit Mercy, who posts prolifically to the Web groups under the aptly chosen pseudonym of "MathProf," gets my vote as today's most talented expert on the "theory of blackjack." I am grateful to him for the explanation that follows, which I have liberally and shamelessly adapted from his personal correspondence with me on the subject of risk-averse indices.

Risk-averse (r-a) indices can best be understood using the concept of "Certainty Equivalent" (CE). In CE, we discount expected value by adjusting for risk. For example, suppose we are confronted with a choice of two plays. Play 1 has higher expectation (EV), but also higher variance (risk). Play 2 is the "risk-averse" play. Typically, play 1 will involve doubling, while play 2 will call for hitting, but there can be other applications as well.

We consider the Certainty Equivalents associated with each play:

$$CE_1 = E_1 - (f/2) \bullet V_1$$
$$CE_2 = E_2 - (f/2) \bullet V_2$$

Here, the Es represent "Expected Value," the Vs represent "Variance," and the f is the size of the wager, expressed as a fraction of the Kelly-equivalent (optimal) bankroll. Now, as mentioned above, conventional indices are determined by choosing the play that has the higher E, while r-a indices are computed by selecting the play with the higher CE.

Some readers may have difficulty understanding why we are considering CE and not EV here. However, if they accept SCORE, then this seems more natural. A higher EV raises SCORE (defined as EV squared/variance), but higher variance clearly lowers it. So it should not be surprising that our optimal SCORE will come when we achieve a balance between these two opposing factors. In fact, it can be shown mathematically that our method of selecting plays based on CE will give the optimal SCORE.

If we look at our CE equations, it becomes clear that play 1 (which has both higher E and V) will be better when f is small, but play 2 will yield a higher CE than play 1 when f is large. There is obviously a "critical value" where the two CEs are equal. With a little algebra, we set $CE_1 = CE_2$ and solve:

$$\text{Critical } f = 2\,(E_1 - E_2) / (V_1 - V_2)$$

If your original bet size is bigger than this "critical" wager, you should prefer the risk-averse play to the e.v.-maximizing one. To illustrate, I prepared several examples, which can be found in Tables 13.13–15, at the end of this chapter. *[Note: The plays of 9 v. 7 and 11 v. Ace are furnished as additional information and are for illustrative purposes only. They will not be discussed in the body of the text. — D.S.]*

I am studying the aforementioned 10 v. 10, in a 6-deck, s17, das game. We are comparing the decision of doubling to hitting. (I also assumed that the player was already using the appropriate index numbers of +4 for 15 v. 10 and 0 for 16 v. 10, as per the "Illustrious 18.")

The column labeled "TC" gives the floored true count. *[Note: As we mentioned in Chapter 10, flooring is one of several methodologies for "binning" true counts when generating index numbers. With flooring, one rounds down, to the next-lowest integer, any value. Thus, +1.7 becomes +1 and –1.7 becomes –2. — D.S.]* The column labeled "EV Gain" gives the conventional gain in expectation: $E_1 - E_2$. The next column is the differences in variances ("Diff. Var."). These values are followed by the critical fraction ("Crit. Frct."), computed via the above method.

There is, of course, some sampling error involved in generating the data. To conserve space, we have not listed the standard errors in the tables that follow. However, the errors are small enough such that the critical fractions will be accurate to within 0.1%.

Now suppose the Hi-Lo TC is +4. If we had a negligible bet out, we would double, because the "CE Gain" from so doing is positive. But let us say that we had placed a bet of 1% of our Kelly-equivalent bankroll. This is above the CF (0.6%), so we would be better off not doubling. Doubling here would actually *reduce* the SCORE. From this, we see that if we have a 1% bet out, our (floored) index becomes +5.

However, if we had an even bigger bet, say 1.5% (as was perhaps the case for our six-unit wager, above), then we would not double at +5; rather, we would require a true count of +6. In similar fashion, if we had made a bet of eight units, or 2% of our bank, then our doubling index would now increase to +7. Again, the reason for this is that, in each case, doubling prior to the indicated r-a index gives us a negative CE gain, and thus will reduce our SCORE.

Now, one could, in theory, learn different indices for different bet sizes. But this becomes quite cumbersome and is not very practical. Most players prefer to memorize just one number for all situations.

To find that single index number, we make two simplifying assumptions. First, we assume that the pre-deal count and the count at the moment we play the hand are the same. This is true, on average. Second, we assume that our bet was determined entirely by the pre-deal count (without regard for any camouflage restrictions). Typically, we assume that we have made the optimal bet. With that, we can determine a single risk-averse index number for any given play.

As a preliminary step, I have included, in the "Opt. f" column, the pure Kelly bets as computed by *BJRM.* There are some assumptions behind these wagers; for example, *BJRM* assumes the player uses the "Illustrious 18." A player employing more indices would have slightly higher bets, while one using fewer indices would have lower wagers. Again, there is a little approximating going on here.

Also, *BJRM* does not give separate values for true counts of +9, +10, and +11, preferring to group all the counts above +8 into one bin, for simplicity. To give a complete picture, I have extrapolated these from the other data. Of course, these counts are so rare that this part of the table has little practical significance.

Now let us suppose that the player could bet pure Kelly, raising and lowering his bet without constraint, that is to say, with no predetermined upper limit. Then the bets would be as in the "Spread 1" column. In all cases, they are greater than the corresponding CFs. *So the player would never make this particular double!*

The details of this observation are illustrated in the "Bets" and "CE Gain" columns, under "Spread 1: Pure Opt." Here, using the formulas above, we compute the CE gain as $CE_1 - CE_2$. In all cases, it is negative, so doubling is ill advised.

More realistically, what if the player uses some betting constraints? Then he will not make these very big bets, and hence will have some opportunity to double correctly. I have furnished some examples of this.

In the columns under "Spread 2: Max at +5," I looked at a second spread, where the player placed a maximum bet at 2% of bank (at a TC of +5). If we go back to the CF column, we see that our CF does not exceed 2% until the TC is +7. This tells us that +7 is our (floored) risk-averse index. Thus, the following advice would prevail:

Double at true counts ≥ +7, hit below. This can be verified by computing the CEs (see the "CE Gain" column, for Spread 2). The CE gain is negative for the +5 and +6 bins, but positive thereafter.

If the player had to use a weaker spread, because, for example, he was in a play-all situation or up against a table maximum, the above index would be a bit lower. In the next two columns, I looked at a third spread ("Spread 3: Play All"), whose top bet maxed out at 1.5%. From the CFs, we see that we could double at the slightly lower index of +6. An equivalent way of saying this is that the CE gain is negative at +5 and below, and positive at +6 or above.

Finally, I considered a spread ("Spread 4: 2 Spots") where the player bets on two spots. Here, the optimal bets are about 70–75% of the original Kelly bet, and we see that the proper r-a index is +6. Details of this calculation are found in the last two columns of Table 13.13.

One final comment. Throughout the above, I have conveniently assumed that the player was using Kelly. In fact, this assumption is not really required. If the player were using half-Kelly, then the CFs would be reduced by one-half, but the optimal bets would also be reduced by one-half. So, we end up with exactly the same indices. The same would be true if the player used third-Kelly, quarter-Kelly, or any other fraction of full Kelly.

Thank you, MathProf! The math-shy among you may come back now! How do we apply the above knowledge to the games we are currently playing? What are the ramifications of incorporating risk-averse indices into our repertoire? Is it worth it to learn the new indices? Will the gain justify the effort? Stick around; you're about to find out.

A recent study, performed by yet another talented Internet researcher, who goes by the name of "Cacarulo," uncovered four plays that could, under certain conditions, be ever-so-slightly more valuable than a few of the original "Illustrious 18." In particular, he found that the bottom three negative-index plays, namely, 12 v. 5 and 6, and 13 v. 3, could be replaced, along with our ubiquitous 10 v. 10, by doubling 8 v. 6 (+2); A,8 v. 6 (+1); A,8 v.5 (+1); and 8 v. 5 (+4), to give marginally better results, in shoe games. (Note that, in Chapter 10, we had already suggested using the two hard-8 doubling plays for single-deck.) Rather than quibble over these differences, I think it's best to embrace the newfound "discovery" and to incorporate the new with the old, producing, if you will, a "Catch 22" — the 22 most useful indices to employ, under most blackjack-playing conditions.

I next asked John Auston to fire up his trusty, most recent version of Karel Janecek's *SBA* (v. 5.03), the one that produces risk-averse indices upon demand. (See Table 13.11.) Note that there are seven changes, indicated in gray shading, for risk-averse indices.

TABLE 13.11
The "Catch 22" Indices

Play	EV-Max.	R-A	Play	EV-Max.	R-A
Insurance	3	3	8 v. 5	4	5
12 v. 2	3	3	8 v. 6	2	3
12 v. 3	2	2	9 v. 2	1	1
12 v. 4	0	0	9 v. 7	3	4
12 v. 5	−2	−2	10 v. 10	4	7
12 v. 6	−1	−1	10 v. A	3	4
13 v. 2	−1	−1	11 v. A	1	1
13 v. 3	−2	−2	A,8 v. 5	1	1
15 v. 10	4	4	A,8 v. 6	1	1
16 v. 9	4	4	10,10 v. 5	5	6
16 v. 10	0	0	10,10 v. 6	4	5

TABLE 13.12
Comparative Study of EV-Maximizing and Risk-Averse Approaches

	SCORE	Unit Size	Float Bank	Unit Size	Float ROR	Float Win/100
"Catch 22" – EV-Maximizing						
Play All	$40.58	$11.42	$8760	$10	10.21%	$35.60
Back-Count	$65.61	$37.04	$9450	$35	12.04%	$61.90
"Catch 22" – Risk-Averse						
Play All	$41.60	$11.59	$8630	$10	9.86%	$35.90
Back-Count	$66.91	$37.04	$9450	$35	12.04%	$63.60
"Catch 22" – EV-Maximizing, except Double 10 v. 10 at TC ≥ +7						
Play All	$41.34	$11.56	$8650	$10	9.90%	$35.70
Back-Count	$66.42	$37.59	$9310	$35	11.60%	$61.55

With these new indices established, I proposed the following comparative study. We assumed, as a starting point, a 5/6 game, with s17 and das. The "Catch 22" were used, first in play-all mode, then with back-counting. We also furnished two spreads, 1–16 for the former, and 1–12 for the latter. And, of course, we used both the traditional, e.v.-maximizing, indices and the risk-averse ones. Our findings are summarized in Table 13.12, above.

Remember that the SCORE approach supposes a $10,000 bank in each instance, which, in turn, yields identical risks of ruin (13.5%) for all lines. (Ignore the three "Float" columns for now; we'll get to them in just a moment.) As expected, our risk-averse player improves his SCORE by 2.51%, in one case (play all), and by 1.98% in the other (back-count). We also decided to add a third case, that of using the original indices, except for a lone risk-averse one, namely +7, for doubling 10 v. 10. Here, the idea was to see how much of *all* the gain afforded by r-a indices was supplied by the single most important one. The answer? An astonishing 75% for the play-all scenario, and slightly more than 62% for the back-counter. So, if you're too lazy to learn even the seven r-a changes, you might settle for just the one "biggie" and glean, nonetheless, the lion's share of the extra profits.

What's that? You're not impressed with those extra profits? Well, join the club! John suggested we subtitle the current work, "Much Ado About Nothing"! But, perhaps there is another way of looking at the rather meager improvement risk-averse indices can afford us. Permit me to digress for just a moment. There is an old joke concerning two vaudeville comedians who are onstage performing when, without warning, one of them drops to the ground. Frantic efforts by the one left standing to revive his partner fail. The comedian is dead of a heart attack. All of a sudden, a little old lady shouts down from the balcony: "Give him some chicken soup!" The grief-stricken performer replies: "Madam, this man is dead. What *possible* good could chicken soup do?" Without batting an eye, the lady responds: "Couldn't hurt him!"

What's our point? Simply this: What harm can using the r-a indices do? We unlearn a couple of indices we've been employing in one form and substitute, in virtually all cases, a slightly higher number. And, for this no-brainer, for this slight adjustment to our thinking, we are rewarded with 1.5–2.5% more hourly expectation. Now, MathProf has pointed out, and rightly so, that 2% of 50 weeks of play is exactly one week. So, if you're a regular, give yourself a week's vacation, courtesy of r-a indices!

Finally, I think it's important to point out another practical aspect of our study. In the real world, we don't use units of $11.42, $11.56, and $11.59, or $37.04 and $37.59. In each case, we're virtually certain to round to, say, $10 in the first instance and $35 or $40 in the second. The implications here are that we really aren't going to be able to capitalize very much on the extra e.v. the r-a indices afford us, because we're not going to be able to do the necessary hair-splitting on the unit sizes to implement the strategy in a casino.

Although this is true, it doesn't mean that r-a indices have no use. They do, but we'll have to take our improvement in one or two other forms. Suppose we choose to keep our unit sizes identical, at $10 each, for the play-all approach, or at $35 each for back-counting. Then, for the former, the r-a way will permit us to win about the same amount as the traditional approach ($0.30 more), but with *both* a smaller bankroll (see the "Float Banks," instead of the standard SCORE $10,000 ones) and slightly less risk. For the back-counter, altogether analogous comments apply. Consult Table 13.12 for all the particulars. In either event, there is some tangible compensation for using the r-a indices.

As always, I'd like to leave the door open for further study of this topic. If you use dozens of indices, instead of just 18 or 22, you might broaden your r-a horizon somewhat and learn additional numbers, if you're so inclined. Furthermore, we haven't considered, in this present chapter, the gains attainable from using r-a methodologies in single- or double-deck games, where strategy variation plays a more important role than in shoe games, but where, it is also true, large bet variation is not widely encountered. In any event, there is more work to be done in this area. Perhaps we'll save it for a future article.

In closing, I'd like to make a confession, and then recant, just a little. Twenty years ago, when all the hullabaloo surrounding risk-averse indices first surfaced, I was pretty adamant in my stance against their use, because, at the time, I hadn't considered the variable unit size. I reasoned that we would play the same bet ramp, with identical units, and that we would, quite simply, earn less by using r-a indices. Obviously, I was overlooking the contribution that the risk-reduction factor makes to our optimal bet-sizing schedule. And so, now, I confess: I've been too harsh on risk-averse indices all these years. There, I said it! Having done so, I must admit that, in the final analysis, we probably have the proverbial tempest in a teapot. I won't ever again say anything bad about r-a indices, but let's understand one another — what it really boils down to is the little old lady versus John's suggested subtitle: The indices can't do any harm, but they probably don't do a helluva lot of good, either!

Postscript: *We are saddened to report that, since the last printing of this book, Michael "MathProf" Canjar has passed away. He is sorely missed by the entire blackjack community, who so greatly profited, over the years, from his wisdom and expertise.*

TABLE 13.13

Risk-Averse Analysis for 10 v. 10 (5/6, S17, DAS)

					Spread 1: Pure Opt.		Spread 2: Max at +5		Spread 3: Play All		Spread 4: 2 Spots	
TC	EV Gain	Diff. Var.	Crit. Frct.	Opt. f	Bets	CE Gain	Bets	CE Gain	Bets	CE Gain	Bets	CE Gain
+4	+0.008	2.52	0.6%	1.5%	1.5%	−0.011	1.5%	−0.011	1.5%	−0.011	1.1%	−0.006
+5	+0.015	2.50	1.2%	2.0%	2.0%	−0.010	2.0%	−0.010	1.5%	−0.004	1.4%	−0.003
+6	+0.022	2.48	1.7%	2.4%	2.4%	−0.008	2.0%	−0.003	1.5%	+0.003	1.7%	+0.001
+7	+0.029	2.47	2.3%	2.9%	2.9%	−0.007	2.0%	+0.004	1.5%	+0.010	2.0%	+0.004
+8	+0.035	2.45	2.8%	3.4%	3.4%	−0.007	2.0%	+0.010	1.5%	+0.016	2.4%	+0.006
+9	+0.040	2.43	3.3%	3.9%	3.9%	−0.007	2.0%	+0.016	1.5%	+0.022	2.7%	+0.007
+10	+0.045	2.41	3.7%	4.4%	4.4%	−0.008	2.0%	+0.021	1.5%	+0.027	3.1%	+0.008
+11	+0.049	2.39	4.1%	4.9%	4.9%	−0.001	2.0%	+0.025	1.5%	+0.031	3.4%	+0.008

TABLE 13.14

Risk-Averse Analysis for 9 v. 7 (5/6, S17, DAS)

					Spread 1: Pure Opt.		Spread 2: Max at +5		Spread 3: Play All		Spread 4: 2 Spots	
TC	EV Gain	Diff. Var.	Crit. Frct.	Opt. f	Bets	CE Gain	Bets	CE Gain	Bets	CE Gain	Bets	CE Gain
+2	−0.017	2.82	−1.2%	0.7%	0.7%	−0.026	0.7%	−0.026	0.7%	−0.026	0.5%	−0.024
+3	+0.003	2.81	0.2%	1.1%	1.1%	−0.012	1.1%	−0.012	1.1%	−0.012	0.8%	−0.008
+4	+0.022	2.81	1.6%	1.5%	1.5%	+0.001	1.5%	+0.001	1.5%	+0.001	1.1%	+0.007
+5	+0.041	2.81	2.9%	2.0%	2.0%	+0.013	2.0%	+0.013	1.5%	+0.020	1.4%	+0.021
+6	+0.059	2.81	4.2%	2.4%	2.4%	+0.025	2.0%	+0.031	1.5%	+0.038	1.7%	+0.035
+7	+0.081	2.81	5.7%	2.9%	2.9%	+0.040	2.0%	+0.052	1.5%	+0.059	2.0%	+0.052
+8	+0.091	2.81	6.5%	3.4%	3.4%	+0.044	2.0%	+0.063	1.5%	+0.070	2.4%	+0.058

Table 13.15

Risk-Averse Analysis for 11 v. Ace (5/6, S17, DAS)

					Spread 1: Pure Opt.		Spread 2: Max at +5		Spread 3: Play All		Spread 4: 2 Spots	
TC	EV Gain	Diff. Var.	Crit. Frct.	Opt. f	Bets	CE Gain	Bets	CE Gain	Bets	CE Gain	Bets	CE Gain
+0	−0.012	2.79	−0.9%	0.0%	0.0%	−0.012	0.0%	−0.012	0.1%	−0.014	0.0%	−0.012
+1	+0.013	2.78	0.9%	0.4%	0.4%	+0.007	0.4%	+0.007	0.4%	+0.007	0.3%	+0.009
+2	+0.038	2.78	2.8%	0.7%	0.7%	+0.029	0.7%	+0.029	0.7%	+0.029	0.5%	+0.032
+3	+0.062	2.78	4.5%	1.1%	1.1%	+0.047	1.1%	+0.047	1.1%	+0.047	0.8%	+0.051
+4	+0.078	2.78	5.6%	1.5%	1.5%	+0.057	1.5%	+0.057	1.5%	+0.057	1.1%	+0.064

Chapter 14

Some Final "Words of Wisdom" from Cyberspace

As mentioned in Chapter 10, I first became acquainted with John Auston through e-mail correspondence. When he learned that I would be writing a book, he communicated the following to me:

Don,

You mentioned that part of your book will be a collection of past writings for *Blackjack Forum.* Over the last couple of years, you've sent me *lots* of advice and opinions on a wide range of bj topics. It occurred to me that some of them might be worthy of inclusion in the book. Since I've saved all your e-mail responses, it would be fairly easy for me to send them back to you. For example, here is some stuff from late 1995 and early 1996. …

I never dreamed that I had written so much through e-mail. Now, I realize that what you're about to read is only a small fraction of the "e-missives" I've produced in the past two or three years. I hope you'll enjoy these writings from cyberspace!

On the Zen count and "semi-neutralizing" the ace

Snyder's concept, in keeping with the spirit of Zen, was to "seek the middle ground." In a level-2 count, if the ace were to be counted, one would assume it would count as –2 (e.g., Revere Point Count). If the same count were to be ace-neutralized (ace not counted), by definition, the ace would count as 0 (e.g., Hi-Opt II). As we know, if one plays mostly multi-deck games, with wide bet variation, betting efficiency is far more important than playing efficiency, and the ace ought to be reckoned, or counted. On the other hand, if one plays mostly single-deck, with minimal bet variation, playing strategy indices become much more important and ace-neutral counts tend to outperform ace-reckoned ones under these circumstances.

Snyder's "brainstorm," as it were, was to strike a balance between the two approaches (hence "Zen") and to "semi-neutralize" the ace, that is, count it not as –2 or as 0, but rather as –1 (while counting the tens as –2 and the small cards as +2). The count is a very fine one and, in essence, attempts to be "all things to all people."

On choosing a counting system

I have used one count exclusively — the Revere Point Count (level 2) — for 20 years. Slightly more complicated would be Wong's Halves (level 3). All of the indices are in his book, but if you were to use Halves, I would double all the values (no fractions that way) and convert to true count by dividing by half-decks (same as Revere).

I am not a fan of unbalanced counts. As you stray from the pivot point, you lose accuracy for bet estimation. There's no free lunch. Unbalanced counts are "easier" because you don't have to convert to true count, but betting according to running count can never be as accurate as betting with true count. Learning to eyeball the discard tray for the number of decks remaining is much easier than you think. You practice it — with actual decks — like anything else. You can get more accurate than you would ever dream, simply by spending a little time practicing. As far as dividing one relatively low whole number by another low whole number — I've never had too much trouble doing that, either. *[Note: Since writing these lines, I've softened my opinion of unbalanced counts after seeing the impressive results obtained by Olaf Vancura and Ken Fuchs, with their K-O Count. — D.S.]*

Best possible advice I can give you, once you have a good, solid count behind you: Be patient! Blackjack can really try your soul at times. You have to have an enormous amount of self-restraint to put up with all that goes on — most of all, the inevitable losing streaks. Just keep the faith and be adequately capitalized.

On reacting to "bad beats"

After 20 years of playing, I could fill a book (unfortunately) all by myself! Actually, I find that it's better, after a bad beat, to show the same kind of emotion that a regular player would show. Playing like a machine is fine for the count and decisions, but normal people *do* react to perverse outcomes, and to show no emotions sets you apart from what is considered normal.

On the "Illustrious 18" and replacing splitting 10s

I don't remember how many indices I calculated in order to generate the final list, but as a pure guess, I'd add 15 v. 9 as the next one. *[Note: It was a bad guess! See the "Catch 22" in Chapter 13. — D.S.]* By the way, it was my original intent to call the list the "Sweet 16" and to exclude the ten-splitting for the exact reason you mentioned (heat). Peter Griffin talked me into including them because they *are* valuable, and he thought I shouldn't "censor" the list. He said, "Give the readers the information, and let them do what they want with it." Unfortunately, now with 18 numbers, my catchy "Sweet 16" was spoiled. If I had it to do over again, I would call them the "Elite 18," taking a cue from the "Elite (round of) 8" in the NCAA basketball tournament.

On your preference for playing 1-deck over 6-deck

Forget 1-deck! The cheating "problem" is *way* overdone. That's not the main concern. My remark pertains to higher-stakes players. It is virtually impossible to play high-stakes, single-deck blackjack and make the kind of money that can be made at multi-deck games. If you play in Las Vegas, the penetration is so terrible that win rate suffers dramatically. If you play in Northern Nevada, and are a small-stakes player, single-deck is possible, but the rules stink. I don't think single-deck is the most profitable way to play blackjack. Period. And I know of very few people who disagree with me. Having said all of that, I suppose that with a great many "ifs," single-deck could be an alternative, but I'm convinced that you can and should do better.

On the mechanics of counting the 2-deck game

Two-deck is different; it has a different feel. If you've never played it before, it takes some getting used to. You should try it; but don't play if you're not comfortable or are having trouble with accuracy. Count dealer's upcard and your hand, then players' hit cards and the hole cards of anyone who busts, has a natural, doubles, or splits. Then count your own hit cards, then dealer's hole card, hit cards, and finally, remaining players' hole cards. Be careful when dealer has a natural and everyone tosses his cards at him in disgust! You have to be on your toes then! Be careful not to double-count hit cards when players' hole cards are spread and revealed. You want to count only the two cards that are now *closest* to the dealer (the player's original two cards). The other cards you have counted already.

On what constitutes "good" penetration

As Wong and I have pointed out on several occasions, the "goodness" of a game is inversely related to the number of decks (or, alternatively, cards) that remain undealt. Thus, rather than give penetration percentages, I convinced Wong to switch over to actual number of decks left undealt.

Having said that, 2-deck blackjack is "poor" if, say, only 50–55% of the cards are dealt. It becomes "good" (or at least acceptable) in the 65–75% range. Anything above 75% is "excellent." I realize I've just gone and given you penetration levels, but you can go and do the conversions to cards or decks.

For 6-deck, the "standard" is to cut off 1 1/2 decks (75% penetration). But this is really not a very good game. It is "acceptable." Cutting 2 decks or more is unacceptable, although exactly 2 might be "borderline." Dealing 5 decks out of 6 is "very good," and anything more than this is "excellent" but, alas, also virtually nonexistent!

On "benchmarks" for average number of hands per unit of time

I've always used 100 hands per hour as a perfectly reasonable benchmark, not only because it is attainable, but also because of the obvious convenience of the number 100. It represents the number of hands one can play per hour with a dealer of decent speed and two or three other players at the table. Playing alone will produce 180–200 hands/hr., and at a full table, one rarely does better than 60–70 hands/hr.

300 hands/day seems a bit low. Of course, this depends on whether or not you play every day of your life (God forbid!) or just occasionally. But, on the average, I would think that a well-motivated player would want to play four to five hours per day (I used to do five or six), and therefore 400–500 hands/day.

The yearly number is wildly variable. Some "pros" play easily 500 hours or more. This would yield 50,000 hands annually. I, personally, never played that much, because blackjack has never been a job for me; I've been gainfully employed for the past 27 years and play blackjack avocationally. In my best years, I may have played around 200 hours. So, call this whatever you like; there just can't really be a "norm." For the occasional or recreational player, I would guess that 50–100 hours would be quite a lot.

On calculating the gain from seat position at 6-deck

In Griffin's book are charts of exact gain from "perfect" play at various levels of the deck. He has this for both single- and multiple-deck. In order to obtain the overall, average gain from variation, you sum all the gains at all the levels, and divide by the number of levels.

In a face-up game, the last player sees the hit cards of the first six players before he plays. As the average hand uses 2.7 cards, in essence, he sees .7 cards extra from six players, or roughly 4.2 cards. So, when calculating the cumulative gain from strategy variation for seat 7 versus seat 1, all of our "levels" in Griffin's charts start four cards "later" (deeper). So, to compare, all the middle values remain the same, and, in essence, we drop the first value for player 7 and substitute the last (more advantageous) one in its place.

Piece of cake! No sim required. There will be a large difference between the first and last value, but it will become quite small when divided by *all* of the levels. The averaging process will dilute the discrepancy to a point where, I believe, it will be negligible and uninteresting.

More on the advisability of splitting 10s

Although I included the gains available from splitting a pair of tens against the dealer's five or six, when I wrote "Attacking the Shoe," I don't recommend use of these departures. I believe that making these plays, along with jumping one's bet, is one of the biggest giveaways that you are a counter, and thus should be avoided despite the decent amount of gain available from the maneuver.

I used to split tens earlier in my playing career, but it made me very uncomfortable and I stopped. Last year, at double-deck, just for the hell of it, towards the end of a trip, I did it again. I won both hands, but drew the line at re-splitting the first 20 that I got. The count was so high that I probably would have gotten to four hands with ease. I also seemingly got away with it (at least the pit boss's back was turned, as I remember, despite my large bets), but it goes against my grain.

What is ten-splitting worth? Well, in the shoe game (4-deck) that the article was based on, and for my 1–12 spread, the two plays contributed 0.034% to the total gain. I think that you probably get to make this play about once every seven hours or so (based on 100 hands per hour). The actual hands (10,10 v. 5 or 6) come up about 1,455 times per 100,000 hands; but the count needs to be +5 in one instance and +4 in the other. To the black-chip player, the two plays might be worth about \$6–\$7 per hour. Even for the quarter player (\$1.75 at most), it just doesn't make sense to me to take the risk.

Others will say, "If I play a lot of blackjack, this is real money to me, in the long run." Maybe. But longevity is real money to you, also. And if, because of ten-splitting, you get thrown out of casino after casino, there is an implicit price tag attached to that, also. To each his own. For the great majority of the time, I'll pass.

On the tendency of the true count to remain the same, while the running count varies

Here would be my attempt to make the layman understand. The running count tends to get lower if it starts out positive, and higher if it starts out negative. This must be true, since, for balanced counts, the running count must restore to zero when no cards remain.

But the true count shows no such tendency to "revert" to any number, zero or otherwise. Rather, it tends, on average, to stay right where it was before the new round was dealt. This is true because the true count is calculated by taking the running count and dividing it by the number of decks remaining to be dealt. So, for example, if the running count is, say, +9 and three decks remain, the true count is +9/3 = +3. If one more deck is dealt, we would expect our +9 running count to go down to +6. But, as this is happening, the number of decks remaining has also gone down — in this case, to two. So, +6/2 = +3 — the same true count we had before.

In similar fashion, the same reasoning holds, even as you go from the first-base player to the third-base player. On average, we can expect the third-base player to be playing against the identical true count that existed when his first-base friend played 30 seconds earlier.

On resisting the temptation to make some close-to-zero Basic Strategy departures based on running count alone

Suppose the correct index for a play is precisely 0.2. Your index is zero. Your running count is +1. You decide to ignore the true-count conversion and use running count. At the start of six decks, the true count is +1/6 = +0.1667. You're making a mistake by standing here. On the other hand, at the start of 2-deck, you've got +0.5, so you stand, and you're right. Now, suppose the correct number for a departure is –0.2. Again, the system seller gives zero for the index. You decide –1 running count will work just as well. At the top of six decks, true count is –1/6 = –0.1667. You stand, and this time, you're right. For 2-deck, you stand, and because the real true is –0.5, you're wrong. So, it all depends.

I've seen index numbers to two or three decimal places. I think 12 v. 4, 16 v. 10, and A,2 v. 5 are among the closest to zero.

On very high counts

What people are continually misunderstanding is that whether we win or *lose* huge amounts of money, it's *always* because of what happens at the high counts. Nobody loses 80 units, one at a time, during an hour session. He loses 80 units because the count got high, and for the seven or eight hands he played at those levels, he lost all or most of them. How else could it be?

On "special rules" for "special situations"

Q. *Are there such things as "special rules" that apply only VERY LATE in "shoes" and, say, only with very high counts? Things such as not taking an extra card very late in a high-count shoe when the hand you are hitting has very little chance given dealer upcard on this round, in essence, "saving" the cards for the next round? Hope this isn't a stupid question!*

A. On the contrary, this is an important question. Whereas it seems clear that the taking of a single card doesn't alter penetration enough for it to matter, certainly, if that card forces a shuffle and deprives you of a hand in a great situation, it *does* matter. It's not easy to tell if the stop-card is about to come out in a shoe game until you actually see it, and of course, by then, it's too late!

One clear rule of thumb works in the opposite direction. In negative counts, when you're trying to "eat" (waste) cards, if you're right on the number, trying to decide if it's worth hitting or not, you hit, to use up the cards. But purposely playing a hand incorrectly (by standing), so as to preserve another round in a favorable situation, can be quantified as follows:

1. What is the cost of your "mistake"?
2. What is your advantage on the next hand?
3. What is the probability that, by taking a card, you will lose the opportunity to exploit the advantage (translation: the dealer shuffles)?
4. Multiply the edge in (2) by the number 1 minus the probability of (3) (i.e., by the probability that you will get another round).
5. Compare the cost of (1) to the expected gain of (4). If (1) is greater, don't make the wrong play; if less, do make the wrong play.

What's that? You say you can't do all of this in your head in the two seconds you have to make the decision? Well, in all likelihood, nobody else can either!

From a rec.gambling.blackjack poster

Q. *Griffin talks about the optimum tens density on p. 147 of the elephant edition of **The Theory of Blackjack.** The player's advantage ... "reaches its zenith (almost 13%) when 73% of the cards are tens." Thus, more tens than 73% lowers the player advantage. The optimum dealer break ratio, however, happens at 41% tens in the deck (break ratio of .295, compared to a normal .286 and a .185 for no tens at all).*

A. If you read this, you would assume that, with 41% tens remaining, the count was high (usual density is 31%). So, maybe now you figure that higher counts imply that the dealer breaks more frequently. In fact, he doesn't. He breaks less frequently. The numbers above aren't useful because:

1) They talk about *specific* points and deck compositions, as opposed to what happens around those points; and

2) For systems that count the ace as a negative card, your depiction of what the count would be in the above situation is very different from what it would be in ace-neutral counts.

For all intents and purposes, there is no practical positive count that is normally attainable in regular play where the edge starts to decrease.

Postscript: *While many blackjack forums have come and gone over the years, two that have stood the test of time are Norm Wattenberger's* ***blackjacktheforum.com*** *and Stanford Wong's* ***bj21.com****. On those sites, you can find me, along with many other authorities, dispensing wisdom on a variety of blackjack topics, on a daily basis.*

Once upon a time, I envisioned founding a "National Center for Blackjack Research," which would serve as a kind of clearinghouse for all blackjack-related studies. I would "run" the center, but whenever I couldn't answer a question personally, I would "reach out," over my "network" of resources and expert friends, to obtain the desired response. Well, someone (Wong) beat me to the punch, but no matter. The dream is a reality, and, today, more so than at any other time in history, blackjack information is readily available and flows freely on the "information superhighway." I urge you to participate. See you in cyberspace!

Appendix A

Complete Basic Strategy EVs for the 1-, 2-, 4-, 6-, and 8-Deck Games

What Is Basic Strategy?

From time to time, I've engaged in conversations about a topic that people actually get very passionate over. So, I thought I'd bring it to our pages here. What should be the definition of "basic strategy"? Now, before you react: "What a silly question; everyone knows what basic strategy is," permit me to point out that, technically, that simply isn't true.

As a start, we observe that one "classic" definition of basic strategy (BS) is the decision, off the top of the pack, that maximizes the expectation (e.v.) of the hand we are playing, given knowledge of only our holding and the dealer's upcard. This, of course, incorporates the "zero-memory" aspect of BS into the definition. We aren't permitted to specify which individual cards comprise our total, nor may we take note of the cards in any other player's hand. I'll return to this idea a little later.

Now, the above seems simple enough, until we realize that this *total-dependent* basic strategy (T-D BS) is different from *composition-dependent* basic strategy (C-D BS). Already, a new wrinkle! We all know that we're supposed to stand with a *total* of 12 v. 4; however, that rule assumes that we're blind to the *nature* of that 12. Is it 7,5; 8,4; 9,3; or T,2 (not to mention 6,6, which is an entirely different kettle of fish!)? Or, might we have a multi-card holding, such as 6,2,4 or 2,6,2,2? So, as the name implies, T-D BS does not take into account the actual cards that comprise the hand, but rather just the *total* of those cards.

If, instead, we treat those cards as separate entities, to be differentiated, one from the other, we learn, for example, that not all two-card 12s are created equal. Indeed, in a single-deck, s17 game, we note that a holding of T,2 v. 4 should be hit, the e.v. being –0.193955, compared to –0.211839 for standing. And, of course, there are several such composition-dependent plays that differ from their total-dependent counterparts (see the e.v. tables that follow, as well as the basic strategy charts in Appendix B).

But, where do we draw the line? If we're "permitted" to distinguish both cards in our initial holding, for the purposes of applying "basic" strategy, are we also permitted to observe a third, hit card? A fourth? Where does the reckoning end? Purists might say, "never," but this seems quite absurd to me. The dictionary defines "basic" as: "fundamental; serving as a starting point; an essential element or basis." Is it possible that learning dozens of different 3-, 4-, 5-, or 6-or-more-card combinations versus various dealer upcards could ever remotely qualify as "basic" in any reasonable person's mind? I think not.

So, what is realistic? How "basic" does C-D BS have to be to remain *basic* strategy? That's the first question, to which I'll propose an answer in just a moment. The second question is even thornier.

What do we do about pairs? I think we need to accept that, for pairs, we shouldn't consider holdings such as 6,6; 8,8; or 3,3 as a "composition" of the hand. We need to specify, right at the outset, that these simply are not 12, or 16, or 6. So, pairs get a kind of exemption; they're "excused" from being "composition-dependent," in the strict sense of the term. They get their own section of the BS charts we all cut our teeth on (more on these later), and they are learned separately, as are the soft totals, such as A,7. Indeed, we should treat soft totals in the same fashion that we treat hard hands. That is, they can be T-D as well as C-D. As an example, in the two-deck, s17 game, the generic, possibly multi-card, "soft 18" v. A requires us to *stand,* but if we note that our soft 18 is, in fact, A,7, then that holding versus the dealer's ace should be hit. (As a quick aside, I have always been strongly opposed to calling hands such as A,7 "soft eighteen," as the holding has nothing whatever to do with "18," and should never be considered as such, lest we make foolish mistakes in the play of the hand. But, I digress.)

So now, we move on to the following: If pairs get an "exemption" from being considered as composition-dependent, it's because we just want to consider, for any holding, if the cards should be split or not split. If the answer is don't split, then we revert back to the total (12, 16, or 6) and play the hand according to T-D BS, right? Well, almost. What about 7,7 v. T in single-deck (1D)? Once we decide that we're not splitting this pair, what gives us the right to no longer consider this a "regular" 14 (which we would hit), but to stand, instead? Well, once we're not splitting the pair, we should continue with the *original* strategy. Had it been T-D, we would have hit that total of 14, but had it been C-D, we would have stood. If we don't split, we consider the hand as any other hand (hard or soft), and we follow either a T-D strategy or a C-D strategy (the latter being the choice for the methodology to be presented here).

Finally, and here's where things get slightly out of control, how should we treat the hit cards we see *after* we split pairs, or the split cards themselves, if we resplit? Suppose I know that it's correct T-D BS to stand on multi-card holdings of 16 v. T. Suppose I split a pair of 8s against the dealer's T and hit the first 8 with a 4 and then a 5. On the second 8, I draw another 8, but I am not permitted to resplit. Does BS say to hit or stand on my 8,8 16? Well, T-D BS says to hit a two-card total of 16 v. T, *off the top of the pack.* So does C-D BS if I put on blinders and look squarely at my second hand (8,8), without regard to the cards to my right (remember, I can't resplit). But, those other cards are *my* cards, in *my* hand, so why can't I look at them? If I do, then I stand on my 8,8 v. T, because I also see another 8, a 4, and a 5, and I hit solely when the *only* cards lying on the table are my *original two* cards and the dealer's upcard. Clearly, in this instance, more than those three cards are lying on the table, so we stand. We'll return to this fine point in a little while.

So, how many cards in split hands am I allowed to reckon before "basic" turns hopelessly complicated? Again, the purists say, All of them. So, if I'm allowed splits to four hands, and, after playing the first three hands, there are, say, 13 cards lying on the table, and I now am playing my fourth hand, am I supposed to include knowledge of all of those previous cards in the play of the current hand while still invoking "basic" strategy? Gimme a break! Permit me to claim how utterly ridiculous that concept is to me.

The purists rebel: Where do we draw the line? May you have knowledge, at the very least, of the card you originally split away from the beginning hand? Perhaps. It was, after all, one of the first two cards you received. What about one hit card to that new hand? Hmm. Problem. If I allow one, why not two? If two, why not three? Do you see where we're going? Why not 13?! Why not, indeed? And, if you permit the 13, then I ask, when I'm at the table with one other player, and he plays his hand, and I see the cards in it, how ridiculous is it to claim that I absolutely may not use knowledge of *those* cards in the play of my own hand and still invoke "basic" strategy, but, alternatively, if I happen to split pairs three times and have 15 or 16 cards lying around, it's fine and dandy for me to reckon them all in the play of my hands, because, er, they *belong* to me?!

We return to the original question: Just what would you want us researchers and authors to call "basic strategy"? Enter Cacarulo. You've met him before, in Chapter 13, p. 374. As a combinatorial analyst, there simply is no one sharper or more reliable on the scene today. As one of the "Masters," in "Don's Domain" on *www.AdvantagePlayer.com,* here's what I wrote about the inimitable Cacarulo: "'Cacarulo' is the pseudonym of one of the foremost computer analysts and authorities on combinatorial analysis in the industry today. He is regarded as one of our leading experts on the subject of index generation and determination of the precise values of rules variation for all varieties of blackjack games. His tables of expected values for blackjack hands, featured on Richard Reid's *www.bjmath.com* Web site, are perhaps the most accurate and extensive study ever done on the subject. It was Cacarulo who provided much of the insight that led to the expansion of the "Illustrious 18" indices into the "Catch 22," and it's safe to say that when he fires up his vaunted combinatorial analyzer, players take notice!"

Well, it gives me great pleasure to present, in the following three appendixes, the most accurate expectations (e.v.s), basic strategy charts, and values of basic strategy rules variations ever to be found in a blackjack book — all courtesy of Cacarulo.

Let's permit the Master to explain his methodology to us in his own words. Along the way, we'll see how both he and I decided to answer the questions originally posed a while ago:

"'CD-EV' on the Table A1 Summary page is the expected value generated by using the optimal two-card decision that appears in the relevant table (the composition-dependent part) plus a total-dependent (T-D) strategy for three or more cards (including pair splits). Using two fine points, this T-D EV (which is invoked once we draw to our original two-card holding) has been improved: For all numbers of decks, it stands on all multi-card 16s v. T, since this is the best strategy when the hand is comprised of three or more cards. It also stands on, rather than hits, 12 v. 3 in single-deck, when that 12 is an *original hand,* comprising three or more cards. The extra value gleaned from these plays has been included, since although we may have a multi-card holding, we are still reckoning only the *total* of the hand and not the individual cards that comprise it.

[Thus, the answer to our first question is that we consider basic strategy as the optimal play considering the dealer's upcard and the individual *first two cards in our hand, combined with the optimal play for the* total *of our holding, once the hand comprises three or more cards. — D. S.]*

"On the other hand, we have 'TD-EV,' which is the e.v. generated by using only a T-D BS. We never reckon the individual cards that comprise the holding. This T-D BS is the best strategy overall where we don't distinguish whether the total is comprised of two, three, or more cards. We can get the same e.v. if we run a simulation with the use of a commercial simulator and a fixed number of rounds. The difference is that a simulation has its limitations (standard error, time to generate sufficiently accurate values), whereas the answer provided by combinatorial analysis is exact.

"The expected value takes into account every card removed. Now, when you split 8,8 v. T, there are three cards removed (8,8, and T). Suppose that in your first hand you receive a 3. What would you normally do? You would simply think that you have 11 v. T (independently of the other 8) and that T-D strategy (for 'generic' 11) dictates doubling down. So you double and get a T. Now you play your second hand and receive a 7. T-D strategy says hit 15 v. T, so you hit, independently of the other cards. Correct? Well, that's the way the algorithm works. You consider only the cards in your hand. The program calculates the expectation considering all possible cards that may appear on the table, but will consider a T-D strategy for each hand played.

"Remember that these tables reflect the way [virtually all] basic strategists play [in a real casino]. You can compare the e.v.s against the ones in *bjmath.com,* which are for optimal play." *[We did. The values are, as one can imagine, virtually identical. — D.S.]*

So, there you have it. In presenting the tables that follow, we come full circle in answering our original questions. Now, you don't have to agree with me, but, for my money (and for Cacarulo's, as well), we consider a *practical, workable* definition of "basic strategy" to be: that play of the original two-card holding that maximizes the e.v. of the hand, combined with the subsequent play of the three-or-more-card holding that maximizes the e.v. for the *total of the hand,* without regard to the individual cards that comprise the total. Finally, for pair splitting, we treat each *original* pair the same as any original holding; that is, we reckon the two cards in the paired hand and play according to the optimal two-card C-D strategy (including our 7,7 v. T example), but then revert to T-D strategy, once any split hands are created. In addition (and this is important!), since a hand created by a pair-split is, by definition, no longer an *original* hand, any total represented by that holding is considered the equivalent of a *multi-card* total, since more cards than the original three (player's two cards and dealer's upcard) are now lying on the table. (Note, however, that, in single-deck, if a two-card or multi-card total of 12 is the result of a pair-split v. dealer's 3 upcard, we hit.)

A few words about the charts you're about to peruse. Most of the data are self-explanatory. However, we should point out that, for each holding, the play that maximizes e.v. is furnished under the "BS" column, to the far right, so that the correct play will be evident without the need for the reader to compare the individual expectations. In addition, to determine the frequency of a total, rather than of a specific hand, we have summed the individual rows that contribute to that particular total and provided the overall frequency on the same line as the last entry in the group. Thus, for example, frequencies for T,2; 9,3; 8,4; and 7,5 v. say, dealer's A (1D), are summed on the last line (7,5) to give the overall frequency of a generic holding of 12 (in this case, 0.00462954). To this total, the reader might want to add the frequency (0.00024379) of the non-splitting 6,6 = 12, to be completely accurate. The same logic applies with the total of 20 and the A,9 line. To the latter 0.00048758 frequency, we must add the 0.00517130 value from the T,T line, giving a grand total of 0.00565888 for the frequency of 20 v. A. Finally, for the pairs entries, we note two complete sets of values, the first for ndas and the second for das. In addition, there is no "cumulative frequency" column necessary ("frequency" suffices), since each holding is unique. We should also mention that all frequencies against dealer tens or aces are for *playable* hands, after the dealer has ascertained that he does not have a natural.

At this point, you may be wondering about surrender. To the right, under the "BS" heading, you will note, for plays involving surrender, designations where the strategy such as hit ("H") or stand ("S") is preceded by the letter "E" or "R" (for "early surrender only" or "early or late surrender," respectively). Thus, for example, when holding 3,2 v. A, multi-deck, we might read: "EH" (early surrender, if permitted, otherwise hit). And, for 15 v. T, we read "RH," which means that we late surrender, if permitted, otherwise hit. (It is to be understood that if we would late surrender, we would certainly also early surrender, but, of course, the reverse is not always true.) As an alternative to finding, from the "BS" column, hands that should be surrendered, we might consult the tables' e.v. entries directly. The following methodology can be used to analyze late surrender (LS). We all know that the e.v. of surrender is –0.5 (we forfeit one-half of our wager), so, if a game happens to offer LS, we can easily determine when it's better to take that option than to play out the hand. A single example will suffice. Consider T,6 v. A in a 1D, h17 game. This play has an e.v. of –0.528965, which is worse than (less than) –0.5. In this case, correct BS would be to surrender.

The precise e.v.s can also be used to analyze early surrender (ES), but for that we'll need to do some extra calculations. First we need to know the e.v. *before* the dealer has checked for a natural. This can be obtained as follows: EV (before) = P(BJ) x (–1) + [1 – P(BJ)] x EV (after), where P(BJ) = the probability of the dealer's obtaining a natural, and EV (after) is simply the best of the expectations that we can read directly from the tables.

Again, a single example will suffice. Suppose we'd like to know whether it is correct BS to early surrender T,4 v. T in 6-deck. Of the 309 cards remaining after we remove our two-card holding and the dealer's upcard, 24 are aces, which will give the dealer a natural. So, P(BJ) = 24/309. It follows that 1 – (24/309) = 285/309. Finally, from the tables, the best play available, before considering the early surrender option, is to hit, with e.v. of –0.463098. So, our e.v. (before) is: (24/309) x (–1) + (285/309) x (–0.463098) = –0.504799. As this value is more negative than (worse than) –0.5, we opt for early surrender.

Now sit back, relax, and enjoy the most accurate e.v. tables that you have ever seen in print.

TABLE A1

Summary of Basic Strategy EVs for the 1-, 2-, 4-, 6-, and 8-Deck Games

Base Rules: DOA, SPA1, SPL3, NS

1 Deck	S17	NDAS	CD-EV	=	0.032548%	
			TD-EV	=	0.002255%	(0.003058%)*
		DAS	CD-EV	=	0.173486%	
			TD-EV	=	0.143263%	(0.144066%)*
	H17	NDAS	CD-EV	=	–0.162287%	
			TD-EV	=	–0.188476%	(–0.187674%)*
		DAS	CD-EV	=	–0.018666%	
			TD-EV	=	–0.044780%	(–0.043977%)*
2 Decks	S17	NDAS	CD-EV	=	–0.324338%	
			TD-EV	=	–0.334864%	
		DAS	CD-EV	=	–0.182027%	
			TD-EV	=	–0.192515%	
	H17	NDAS	CD-EV	=	–0.528687%	
			TD-EV	=	–0.536688%	
		DAS	CD-EV	=	–0.383687%	
			TD-EV	=	–0.391650%	
4 Decks	S17	NDAS	CD-EV	=	–0.491129%	
			TD-EV	=	–0.494956%	
		DAS	CD-EV	=	–0.349202%	
			TD-EV	=	–0.353011%	
	H17	NDAS	CD-EV	=	–0.703545%	
			TD-EV	=	–0.706490%	
		DAS	CD-EV	=	–0.558806%	
			TD-EV	=	–0.561733%	
6 Decks	S17	NDAS	CD-EV	=	–0.545766%	
			TD-EV	=	–0.547629%	
		DAS	CD-EV	=	–0.404071%	
			TD-EV	=	–0.405922%	
	H17	NDAS	CD-EV	=	–0.761033%	
			TD-EV	=	–0.762620%	
		DAS	CD-EV	=	–0.616573%	
			TD-EV	=	–0.618149%	
8 Decks	S17	NDAS	CD-EV	=	–0.573088%	
			TD-EV	=	–0.574005%	
		DAS	CD-EV	=	–0.431570%	
			TD-EV	=	–0.432478%	
	H17	NDAS	CD-EV	=	–0.789800%	
			TD-EV	=	–0.790716%	
		DAS	CD-EV	=	–0.645542%	
			TD-EV	=	–0.646450%	

* EV when standing on 7,7 v. the dealer's T is incorporated into the overall strategy

TABLE A2
1D S17: Dealer's Upcard A

Hand	Stand	Hit	Double	Freq.	Cum. Freq.	BS
3,2	−0.659375	−0.291565	−1.318751	0.00065011	0.00065011	EH
4,2	−0.658712	−0.334901	−1.295655	0.00065011	0.00065011	EH
5,2	−0.657529	−0.331115	−1.125237	0.00065011		EH
4,3	−0.657967	−0.345048	−1.126423	0.00065011	0.00130021	EH
6,2	−0.639160	−0.225772	−0.799095	0.00065011		H
5,3	−0.656761	−0.217133	−0.793964	0.00065011	0.00130021	H
7,2	−0.645559	−0.082785	−0.402057	0.00065011		H
6,3	−0.638344	−0.078495	−0.401329	0.00065011		H
5,4	−0.654533	−0.070215	−0.395771	0.00065011	0.00195032	H
8,2	−0.651889	0.086448	0.025083	0.00065011		H
7,3	−0.644759	0.086843	0.040179	0.00065011		H
6,4	−0.636118	0.081506	0.048615	0.00065011	0.00195032	H
9,2	−0.657160	0.169644	0.188935	0.00065011		D
8,3	−0.650336	0.170920	0.205672	0.00065011		D
7,4	−0.641782	0.171306	0.222041	0.00065011		D
6,5	−0.634013	0.172946	0.240259	0.00065011	0.00260042	D
T,2	−0.672788	−0.348923	−0.810930	0.00267923		EH
9,3	−0.656577	−0.374483	−0.858217	0.00065011		EH
8,4	−0.648266	−0.374915	−0.846660	0.00065011		EH
7,5	−0.640619	−0.376769	−0.837456	0.00065011	0.00462954	EH
T,3	−0.672168	−0.392503	−0.859049	0.00267923		EH
9,4	−0.654534	−0.383075	−0.838019	0.00065011		EH
8,5	−0.647201	−0.433030	−0.930925	0.00065011		EH
7,6	−0.623767	−0.443347	−0.937178	0.00065011	0.00462954	EH
T,4	−0.670237	−0.444961	−0.931613	0.00267923		EH
9,5	−0.653430	−0.441215	−0.920605	0.00065011		EH
8,6	−0.630436	−0.453389	−0.941763	0.00065011	0.00397944	EH
T,5	−0.669157	−0.498741	−1.012900	0.00267923		EH
9,6	−0.636809	−0.495853	−1.008370	0.00065011		EH
8,7	−0.637158	−0.455656	−0.928015	0.00065011	0.00397944	EH
T,6	−0.652909	−0.508750	−1.017500	0.00267923		RH
9,7	−0.643247	−0.495493	−0.990987	0.00065011	0.00332933	EH
T,7	−0.467041	−0.555802	−1.111604	0.00267923		ES
9,8	−0.451875	−0.544608	−1.089217	0.00065011	0.00332933	ES
T,8	−0.082020	−0.633177	−1.266355	0.00267923	0.00267923	S
T,9	0.307676	−0.742645	−1.485290	0.00267923	0.00267923	S
A,2	−0.661732	−0.067808	−0.593281	0.00048758	0.00048758	H
A,3	−0.661047	−0.100568	−0.595803	0.00048758	0.00048758	H
A,4	−0.658818	−0.153931	−0.619882	0.00048758	0.00048758	H
A,5	−0.659151	−0.206061	−0.656070	0.00048758	0.00048758	H
A,6	−0.482815	−0.199924	−0.527213	0.00048758	0.00048758	H
A,7	−0.101015	−0.108386	−0.357778	0.00048758	0.00048758	S
A,8	0.289743	−0.017242	−0.192057	0.00048758	0.00048758	S
A,9	0.680745	0.067505	−0.040657	0.00048758	0.00048758	S
A,T	1.500000	0.152526	0.145194	0.00200942	0.00200942	S

Hand	Stand	Hit	Double	Spl1	Spl2	Spl3	Freq.	BS
				NDAS				
A,A	−0.663142	−0.030684	−0.598755	0.223931	0.261074	0.261074	0.00012189	P
2,2	−0.660049	−0.258901	−1.320098	−0.427834	−0.441545	−0.442274	0.00024379	EH
3,3	−0.660147	−0.333963	−1.298180	−0.471347	−0.480954	−0.481366	0.00024379	EH
4,4	−0.655759	−0.208972	−0.796496	−0.547373	−0.575658	−0.577198	0.00024379	H
5,5	−0.653287	0.090561	0.055149	−0.648455	−0.711842	−0.715393	0.00024379	H
6,6	−0.617010	−0.386204	−0.834602	−0.637988	−0.657692	−0.658691	0.00024379	EH
7,7	−0.630502	−0.494721	−1.029343	−0.606933	−0.611805	−0.611818	0.00024379	EH
8,8	−0.643532	−0.494905	−0.989811	−0.340317	−0.325769	−0.324889	0.00024379	EP
9,9	−0.055174	−0.624843	−1.249686	−0.072145	−0.071777	−0.071659	0.00024379	S
T,T	0.650097	−0.883183	−1.766366	0.098543	−0.186974	−0.354749	0.00517130	S
				DAS				
A,A	−0.663142	−0.030684	−0.598755	0.223931	0.261074	0.261074	0.00012189	P
2,2	−0.660049	−0.258901	−1.320098	−0.421365	−0.434266	−0.434936	0.00024379	EH
3,3	−0.660147	−0.333963	−1.298180	−0.462243	−0.470825	−0.471167	0.00024379	EH
4,4	−0.655759	−0.208972	−0.796496	−0.535358	−0.562365	−0.563821	0.00024379	H
5,5	−0.653287	0.090561	0.055149	−0.633159	−0.694962	−0.698412	0.00024379	H
6,6	−0.617010	−0.386204	−0.834602	−0.621394	−0.639234	−0.640104	0.00024379	EH
7,7	−0.630502	−0.494721	−1.029343	−0.595329	−0.598957	−0.598886	0.00024379	EH
8,8	−0.643532	−0.494905	−0.989811	−0.333718	−0.318552	−0.317635	0.00024379	EP
9,9	−0.055174	−0.624843	−1.249686	−0.070544	−0.070186	−0.070078	0.00024379	S
T,T	0.650097	−0.883183	−1.766366	0.098543	−0.186974	−0.354749	0.00517130	S

TABLE A3
1D S17: Dealer's Upcard 2

Hand	Stand	Hit	Double	Freq.	Cum. Freq.	BS
3,2	–0.293289	–0.131387	–0.586579	0.00072398	0.00072398	H
4,2	–0.290967	–0.150708	–0.560134	0.00072398	0.00072398	H
5,2	–0.274153	–0.099369	–0.401783	0.00072398		H
4,3	–0.292521	–0.122957	–0.429083	0.00096531	0.00168929	H
6,2	–0.277108	–0.012997	–0.162783	0.00072398		H
5,3	–0.275744	–0.016523	–0.167800	0.00096531	0.00168929	H
7,2	–0.280229	0.083488	0.117402	0.00072398		D
6,3	–0.278474	0.092149	0.111874	0.00096531		D
5,4	–0.273342	0.093338	0.113509	0.00096531	0.00265460	D
8,2	–0.282092	0.213045	0.424480	0.00072398		D
7,3	–0.280805	0.214421	0.425980	0.00096531		D
6,4	–0.275236	0.217494	0.432187	0.00096531	0.00265460	D
9,2	–0.286593	0.263585	0.524828	0.00072398		D
8,3	–0.283577	0.268218	0.533572	0.00096531		D
7,4	–0.278423	0.272341	0.541765	0.00096531		D
6,5	–0.258788	0.284031	0.567187	0.00096531	0.00361991	D
T,2	–0.310993	–0.243412	–0.486825	0.00289593		H
9,3	–0.288030	–0.266255	–0.532510	0.00096531		H
8,4	–0.281306	–0.259826	–0.519652	0.00096531		H
7,5	–0.262070	–0.252632	–0.505264	0.00096531	0.00579186	H
T,3	–0.312390	–0.304215	–0.608429	0.00386124		H
9,4	–0.285726	–0.293008	–0.586015	0.00096531		S
8,5	–0.264895	–0.331281	–0.662563	0.00096531		S
7,6	–0.265046	–0.331966	–0.663932	0.00096531	0.00675716	S
T,4	–0.310107	–0.368778	–0.737556	0.00386124		S
9,5	–0.269314	–0.358926	–0.717852	0.00096531		S
8,6	–0.267800	–0.362615	–0.725231	0.00096531	0.00579186	S
T,5	–0.294783	–0.436219	–0.872437	0.00386124		S
9,6	–0.273299	–0.430257	–0.860513	0.00096531		S
8,7	–0.272058	–0.390377	–0.780755	0.00096531	0.00579186	S
T,6	–0.297664	–0.465396	–0.930791	0.00386124		S
9,7	–0.276548	–0.455974	–0.911949	0.00096531	0.00482655	S
T,7	–0.158128	–0.538453	–1.076907	0.00386124		S
9,8	–0.136521	–0.530171	–1.060342	0.00096531	0.00482655	S
T,8	0.118877	–0.632537	–1.265075	0.00386124	0.00386124	S
T,9	0.384834	–0.749661	–1.499321	0.00386124	0.00386124	S
A,2	–0.282713	0.039247	–0.042020	0.00072398	0.00072398	H
A,3	–0.284211	0.016914	–0.046786	0.00096531	0.00096531	H
A,4	–0.283157	–0.011697	–0.069980	0.00096531	0.00096531	H
A,5	–0.266382	–0.031724	–0.081847	0.00096531	0.00096531	H
A,6	–0.131767	0.007098	0.013321	0.00096531	0.00096531	D
A,7	0.135802	0.065248	0.127578	0.00096531	0.00096531	S
A,8	0.401625	0.120029	0.237194	0.00096531	0.00096531	S
A,9	0.655985	0.190762	0.379768	0.00096531	0.00096531	S
A,T	1.500000	0.247091	0.486979	0.00386124	0.00386124	S

Hand	Stand	Hit	Double	Spl1	Spl2	Spl3	Freq.	BS
				NDAS				
A,A	–0.274327	0.094777	–0.019360	0.565704	0.635469	0.639247	0.00036199	P
2,2	–0.290803	–0.113174	–0.581605	–0.128397	–0.128950	–0.128950	0.00018100	H
3,3	–0.294866	–0.152948	–0.567470	–0.202293	–0.205682	–0.205826	0.00036199	H
4,4	–0.290133	–0.012616	–0.184577	–0.235939	–0.254725	–0.255762	0.00036199	H
5,5	–0.255915	0.223862	0.446442	–0.231714	–0.268438	–0.270359	0.00036199	D
6,6	–0.261714	–0.252671	–0.505342	–0.218894	–0.212877	–0.212365	0.00036199	P
7,7	–0.268309	–0.406388	–0.812776	–0.162582	–0.152512	–0.151896	0.00036199	P
8,8	–0.274814	–0.454093	–0.908186	0.016727	0.042268	0.043727	0.00036199	P
9,9	0.137057	–0.627497	–1.254994	0.170071	0.172790	0.172933	0.00036199	P
T,T	0.627226	–0.846661	–1.693322	0.310083	0.138349	0.033475	0.00723982	S
				DAS				
A,A	–0.274327	0.094777	–0.019360	0.565704	0.635469	0.639247	0.00036199	P
2,2	–0.290803	–0.113174	–0.581605	–0.039310	–0.035872	–0.035872	0.00018100	P
3,3	–0.294866	–0.152948	–0.567470	–0.117460	–0.113558	–0.113294	0.00036199	P
4,4	–0.290133	–0.012616	–0.184577	–0.149757	–0.161158	–0.161785	0.00036199	H
5,5	–0.255915	0.223862	0.446442	–0.156769	–0.206996	–0.209721	0.00036199	D
6,6	–0.261714	–0.252671	–0.505342	–0.121722	–0.106685	–0.105630	0.00036199	P
7,7	–0.268309	–0.406388	–0.812776	–0.072861	–0.054788	–0.053707	0.00036199	P
8,8	–0.274814	–0.454093	–0.908186	0.089088	0.120859	0.122667	0.00036199	P
9,9	0.137057	–0.627497	–1.254994	0.202475	0.207896	0.208186	0.00036199	P
T,T	0.627226	–0.846661	–1.693322	0.310083	0.138349	0.033475	0.00723982	S

TABLE A4
1D S17: Dealer's Upcard 3

Hand	Stand	Hit	Double	Freq.	Cum. Freq.	BS
3,2	–0.248452	–0.098608	–0.496903	0.00072398	0.00072398	H
4,2	–0.232274	–0.106155	–0.445516	0.00096531	0.00096531	H
5,2	–0.228858	–0.061505	–0.310676	0.00096531		H
4,3	–0.229953	–0.081528	–0.323683	0.00072398	0.00168929	H
6,2	–0.231142	0.023779	–0.078252	0.00096531		H
5,3	–0.226529	0.019448	–0.083156	0.00072398	0.00168929	H
7,2	–0.233898	0.130645	0.194602	0.00096531		D
6,3	–0.227965	0.138911	0.189258	0.00072398		D
5,4	–0.209737	0.141082	0.195605	0.00096531	0.00265460	D
8,2	–0.238252	0.241239	0.482478	0.00096531		D
7,3	–0.231628	0.246071	0.492141	0.00072398		D
6,4	–0.212068	0.250439	0.500878	0.00096531	0.00265460	D
9,2	–0.264051	0.269022	0.538044	0.00096531		D
8,3	–0.235978	0.295198	0.590396	0.00072398		D
7,4	–0.215710	0.305578	0.611156	0.00096531		D
6,5	–0.208559	0.314733	0.629467	0.00096531	0.00361991	D
T,2	–0.267867	–0.219298	–0.438596	0.00386124		H
9,3	–0.261815	–0.255712	–0.511424	0.00072398		H
8,4	–0.220209	–0.229755	–0.459509	0.00096531		S
7,5	–0.212344	–0.221955	–0.443909	0.00096531	0.00651584	S
T,3	–0.265648	–0.283224	–0.566448	0.00289593		S
9,4	–0.245979	–0.280361	–0.560722	0.00096531		S
8,5	–0.216700	–0.305572	–0.611145	0.00096531		S
7,6	–0.214771	–0.304029	–0.608059	0.00096531	0.00579186	S
T,4	–0.250870	–0.355683	–0.711366	0.00386124		S
9,5	–0.243558	–0.356045	–0.712089	0.00096531		S
8,6	–0.220210	–0.346969	–0.693938	0.00096531	0.00579186	S
T,5	–0.247378	–0.429758	–0.859516	0.00386124		S
9,6	–0.245987	–0.432276	–0.864553	0.00096531		S
8,7	–0.223854	–0.381276	–0.762552	0.00096531	0.00579186	S
T,6	–0.249876	–0.461402	–0.922803	0.00386124		S
9,7	–0.249700	–0.459006	–0.918012	0.00096531	0.00482655	S
T,7	–0.118943	–0.536375	–1.072751	0.00386124		S
9,8	–0.120664	–0.535486	–1.070972	0.00096531	0.00482655	S
T,8	0.144414	–0.633653	–1.267307	0.00386124	0.00386124	S
T,9	0.383557	–0.712572	–1.425143	0.00386124	0.00386124	S
A,2	–0.241149	0.070342	0.028414	0.00096531	0.00096531	H
A,3	–0.240149	0.044157	0.010855	0.00072398	0.00072398	H
A,4	–0.223953	0.023369	0.002627	0.00096531	0.00096531	H
A,5	–0.220511	–0.001897	–0.019208	0.00096531	0.00096531	H
A,6	–0.093231	0.036948	0.073896	0.00096531	0.00096531	D
A,7	0.166816	0.094467	0.188934	0.00096531	0.00096531	D
A,8	0.419872	0.173024	0.346048	0.00096531	0.00096531	S
A,9	0.644126	0.196117	0.392234	0.00096531	0.00096531	S
A,T	1.500000	0.268175	0.536351	0.00386124	0.00386124	S

Hand	Stand	Hit	Double	Spl1	Spl2	Spl3	Freq.	BS
				NDAS				
A,A	–0.232311	0.107290	0.054882	0.612856	0.683041	0.686842	0.00036199	P
2,2	–0.250722	–0.082240	–0.501443	–0.070143	–0.069094	–0.069033	0.00036199	P
3,3	–0.246186	–0.118294	–0.472052	–0.127583	–0.127644	–0.127644	0.00018100	H
4,4	–0.213730	0.028836	–0.082465	–0.127138	–0.139049	–0.139643	0.00036199	H
5,5	–0.206172	0.254782	0.509563	–0.150489	–0.182690	–0.184349	0.00036199	D
6,6	–0.211033	–0.222134	–0.444267	–0.135171	–0.124384	–0.123590	0.00036199	P
7,7	–0.219400	–0.388265	–0.776530	–0.083622	–0.070617	–0.069814	0.00036199	P
8,8	–0.228354	–0.449931	–0.899863	0.076808	0.103265	0.104756	0.00036199	P
9,9	0.122553	–0.638281	–1.276562	0.170370	0.172662	0.172687	0.00036199	P
T,T	0.636134	–0.846301	–1.692602	0.364450	0.215812	0.124253	0.00723982	S
				DAS				
A,A	–0.232311	0.107290	0.054882	0.612856	0.683041	0.686842	0.00036199	P
2,2	–0.250722	–0.082240	–0.501443	0.029069	0.038768	0.039321	0.00036199	P
3,3	–0.246186	–0.118294	–0.472052	–0.027521	–0.023281	–0.023281	0.00018100	P
4,4	–0.213730	0.028836	–0.082465	–0.019940	–0.022439	–0.022497	0.00036199	H
5,5	–0.206172	0.254782	0.509563	–0.060338	–0.107712	–0.110278	0.00036199	D
6,6	–0.211033	–0.222134	–0.444267	–0.016185	0.005641	0.007101	0.00036199	P
7,7	–0.219400	–0.388265	–0.776530	0.024857	0.048036	0.049458	0.00036199	P
8,8	–0.228354	–0.449931	–0.899863	0.156890	0.190341	0.192232	0.00036199	P
9,9	0.122553	–0.638281	–1.276562	0.211552	0.216994	0.217173	0.00036199	P
T,T	0.636134	–0.846301	–1.692602	0.364450	0.215812	0.124253	0.00723982	S

TABLE A5
1D S17: Dealer's Upcard 4

Hand	Stand	Hit	Double	Freq.	Cum. Freq.	BS
3,2	−0.175850	−0.041018	−0.351700	0.00096531	0.00096531	H
4,2	−0.172106	−0.055247	−0.323649	0.00072398	0.00072398	H
5,2	−0.169574	−0.010320	−0.194641	0.00096531		H
4,3	−0.155934	−0.016569	−0.177708	0.00072398	0.00168929	H
6,2	−0.171327	0.081062	0.023309	0.00096531		H
5,3	−0.152839	0.086611	0.041092	0.00096531	0.00193062	H
7,2	−0.176727	0.167601	0.271815	0.00096531		D
6,3	−0.155475	0.186021	0.289638	0.00096531		D
5,4	−0.148879	0.189608	0.297004	0.00072398	0.00265460	D
8,2	−0.202468	0.252464	0.504929	0.00096531		D
7,3	−0.160912	0.285179	0.570357	0.00096531		D
6,4	−0.151621	0.292761	0.585523	0.00072398	0.00265460	D
9,2	−0.207032	0.300049	0.600097	0.00096531		D
8,3	−0.186681	0.312585	0.625171	0.00096531		D
7,4	−0.157223	0.342543	0.685086	0.00072398		D
6,5	−0.149063	0.351969	0.703939	0.00096531	0.00361991	D
T,2	−0.211839	−0.193955	−0.387911	0.00386124		H
9,3	−0.191249	−0.225500	−0.451000	0.00096531		S
8,4	−0.182857	−0.214040	−0.428081	0.00072398		S
7,5	−0.154588	−0.191775	−0.383549	0.00096531	0.00651584	S
T,3	−0.197109	−0.257952	−0.515904	0.00386124		S
9,4	−0.188473	−0.254283	−0.508566	0.00072398		S
8,5	−0.181302	−0.291820	−0.583640	0.00096531		S
7,6	−0.158372	−0.277040	−0.554079	0.00096531	0.00651584	S
T,4	−0.193331	−0.336459	−0.672917	0.00289593		S
9,5	−0.185886	−0.335518	−0.671036	0.00096531		S
8,6	−0.184078	−0.337542	−0.675083	0.00096531	0.00482655	S
T,5	−0.190687	−0.417990	−0.835980	0.00386124		S
9,6	−0.188622	−0.419615	−0.839231	0.00096531		S
8,7	−0.189590	−0.380044	−0.760089	0.00096531	0.00579186	S
T,6	−0.193442	−0.454566	−0.909133	0.00386124		S
9,7	−0.194143	−0.452965	−0.905930	0.00096531	0.00482655	S
T,7	−0.064395	−0.535108	−1.070216	0.00386124		S
9,8	−0.084414	−0.541140	−1.082279	0.00096531	0.00482655	S
T,8	0.164240	−0.597317	−1.194633	0.00386124	0.00386124	S
T,9	0.404114	−0.711696	−1.423391	0.00386124	0.00386124	S
A,2	−0.186079	0.110213	0.115097	0.00096531	0.00096531	D
A,3	−0.169927	0.090755	0.109144	0.00096531	0.00096531	D
A,4	−0.166175	0.061445	0.084883	0.00072398	0.00072398	D
A,5	−0.163592	0.037976	0.062592	0.00096531	0.00096531	D
A,6	−0.036667	0.077274	0.154548	0.00096531	0.00096531	D
A,7	0.203974	0.156342	0.312684	0.00096531	0.00096531	D
A,8	0.415490	0.186537	0.373075	0.00096531	0.00096531	S
A,9	0.653883	0.229578	0.459157	0.00096531	0.00096531	S
A,T	1.500000	0.297535	0.595071	0.00386124	0.00386124	S

Hand	Stand	Hit	Double	Spl1	Spl2	Spl3	Freq.	BS
				NDAS				
A,A	−0.178249	0.142136	0.136650	0.668582	0.738687	0.742469	0.00036199	P
2,2	−0.191999	−0.034842	−0.383998	−0.007191	−0.004979	−0.004868	0.00036199	P
3,3	−0.159691	−0.047439	−0.302167	0.009340	0.015381	0.015779	0.00036199	P
4,4	−0.151694	0.097855	0.044128	−0.013396	−0.017428	−0.017428	0.00018100	H
5,5	−0.146291	0.294911	0.589822	−0.038239	−0.064079	−0.065369	0.00036199	D
6,6	−0.151883	−0.190113	−0.380227	−0.027556	−0.014890	−0.014074	0.00036199	P
7,7	−0.163937	−0.368803	−0.737605	0.016269	0.033094	0.034107	0.00036199	P
8,8	−0.215266	−0.461146	−0.922291	0.105596	0.131836	0.133230	0.00036199	P
9,9	0.166978	−0.597001	−1.194002	0.252688	0.258493	0.258737	0.00036199	P
T,T	0.644848	−0.846273	−1.692545	0.417460	0.288940	0.207681	0.00723982	S
				DAS				
A,A	−0.178249	0.142136	0.136650	0.668582	0.738687	0.742469	0.00036199	P
2,2	−0.191999	−0.034842	−0.383998	0.107127	0.119271	0.119944	0.00036199	P
3,3	−0.159691	−0.047439	−0.302167	0.135242	0.152336	0.153364	0.00036199	P
4,4	−0.151694	0.097855	0.044128	0.122532	0.124641	0.124641	0.00018100	P
5,5	−0.146291	0.294911	0.589822	0.070369	0.027435	0.025124	0.00036199	D
6,6	−0.151883	−0.190113	−0.380227	0.111606	0.137161	0.138753	0.00036199	P
7,7	−0.163937	−0.368803	−0.737605	0.155108	0.184719	0.186498	0.00036199	P
8,8	−0.215266	−0.461146	−0.922291	0.195731	0.229200	0.230970	0.00036199	P
9,9	0.166978	−0.597001	−1.194002	0.299235	0.308672	0.309100	0.00036199	P
T,T	0.644848	−0.846273	−1.692545	0.417460	0.288940	0.207681	0.00723982	S

TABLE A6
1D S17: Dealer's Upcard 5

Hand	Stand	Hit	Double	Freq.	Cum. Freq.	BS
3,2	–0.104257	0.021505	–0.208514	0.00096531	0.00096531	H
4,2	–0.101371	0.008890	–0.185536	0.00096531	0.00096531	H
5,2	–0.098373	0.056768	–0.070318	0.00072398		H
4,3	–0.097993	0.048986	–0.061355	0.00096531	0.00168929	H
6,2	–0.103819	0.130630	0.130583	0.00096531		H
5,3	–0.095495	0.141067	0.154983	0.00072398	0.00168929	D
7,2	–0.130567	0.195515	0.332121	0.00096531		D
6,3	–0.100904	0.234973	0.391603	0.00096531		D
5,4	–0.092477	0.246436	0.414995	0.00072398	0.00265460	D
8,2	–0.135542	0.295987	0.591974	0.00096531		D
7,3	–0.127734	0.307505	0.615011	0.00096531		D
6,4	–0.097979	0.338825	0.677649	0.00096531	0.00289593	D
9,2	–0.138632	0.339824	0.679649	0.00096531		D
8,3	–0.132654	0.349536	0.699071	0.00096531		D
7,4	–0.124742	0.361679	0.723359	0.00096531		D
6,5	–0.095545	0.393653	0.787307	0.00072398	0.00361991	D
T,2	–0.144042	–0.163570	–0.327141	0.00386124		S
9,3	–0.136834	–0.196612	–0.393224	0.00096531		S
8,4	–0.130749	–0.186621	–0.373241	0.00096531		S
7,5	–0.123356	–0.177736	–0.355472	0.00072398	0.00651584	S
T,3	–0.141141	–0.235303	–0.470606	0.00386124		S
9,4	–0.133859	–0.232059	–0.464118	0.00096531		S
8,5	–0.128264	–0.270015	–0.540029	0.00072398		S
7,6	–0.128821	–0.269028	–0.538057	0.00096531	0.00651584	S
T,4	–0.138177	–0.314638	–0.629277	0.00386124		S
9,5	–0.131352	–0.313835	–0.627670	0.00072398		S
8,6	–0.133747	–0.317192	–0.634384	0.00096531	0.00555053	S
T,5	–0.135683	–0.399584	–0.799167	0.00289593		S
9,6	–0.136855	–0.402035	–0.804070	0.00096531		S
8,7	–0.160467	–0.373118	–0.746237	0.00096531	0.00482655	S
T,6	–0.141194	–0.444223	–0.888446	0.00386124		S
9,7	–0.163593	–0.452057	–0.904114	0.00096531	0.00482655	S
T,7	–0.043148	–0.492434	–0.984867	0.00386124		S
9,8	–0.044394	–0.492582	–0.985164	0.00096531	0.00482655	S
T,8	0.202290	–0.590755	–1.181511	0.00386124	0.00386124	S
T,9	0.447849	–0.708740	–1.417480	0.00386124	0.00386124	S
A,2	–0.118846	0.158731	0.212295	0.00096531	0.00096531	D
A,3	–0.116020	0.136564	0.203578	0.00096531	0.00096531	D
A,4	–0.113097	0.107857	0.174985	0.00096531	0.00096531	D
A,5	–0.110736	0.082113	0.148256	0.00072398	0.00072398	D
A,6	0.004662	0.140017	0.280033	0.00096531	0.00096531	D
A,7	0.222295	0.174547	0.349094	0.00096531	0.00096531	D
A,8	0.460792	0.226529	0.453059	0.00096531	0.00096531	S
A,9	0.682074	0.268101	0.536203	0.00096531	0.00096531	S
A,T	1.500000	0.331496	0.662991	0.00386124	0.00386124	S

Hand	Stand	Hit	Double	Spl1	Spl2	Spl3	Freq.	BS
				NDAS				
A,A	–0.130086	0.182014	0.215727	0.732160	0.803601	0.807466	0.00036199	P
2,2	–0.107014	0.035944	–0.214028	0.127678	0.136001	0.136489	0.00036199	P
3,3	–0.101455	0.008318	–0.183982	0.116658	0.127251	0.127911	0.00036199	P
4,4	–0.094897	0.153927	0.162314	0.094717	0.090943	0.090804	0.00036199	D
5,5	–0.090048	0.347346	0.694692	0.068117	0.057221	0.057221	0.00018100	D
6,6	–0.102166	–0.162360	–0.324720	0.067714	0.084118	0.085134	0.00036199	P
7,7	–0.155509	–0.370327	–0.740653	0.038792	0.054908	0.055777	0.00036199	P
8,8	–0.165443	–0.452874	–0.905749	0.188016	0.216845	0.218373	0.00036199	P
9,9	0.202893	–0.590310	–1.180620	0.339220	0.349245	0.349719	0.00036199	P
T,T	0.673675	–0.845597	–1.691194	0.496722	0.393479	0.326618	0.00723982	S
				DAS				
A,A	–0.130086	0.182014	0.215727	0.732160	0.803601	0.807466	0.00036199	P
2,2	–0.107014	0.035944	–0.214028	0.276378	0.298090	0.299360	0.00036199	P
3,3	–0.101455	0.008318	–0.183982	0.276757	0.301742	0.303241	0.00036199	P
4,4	–0.094897	0.153927	0.162314	0.256441	0.264210	0.264619	0.00036199	P
5,5	–0.090048	0.347346	0.694692	0.205292	0.184731	0.184731	0.00018100	D
6,6	–0.102166	–0.162360	–0.324720	0.241531	0.274150	0.276148	0.00036199	P
7,7	–0.155509	–0.370327	–0.740653	0.193212	0.222213	0.223782	0.00036199	P
8,8	–0.165443	–0.452874	–0.905749	0.291005	0.328089	0.330046	0.00036199	P
9,9	0.202893	–0.590310	–1.180620	0.392304	0.406503	0.407189	0.00036199	P
T,T	0.673675	–0.845597	–1.691194	0.496722	0.393479	0.326618	0.00723982	S

TABLE A7
1D S17: Dealer's Upcard 6

Hand	Stand	Hit	Double	Freq.	Cum. Freq.	BS
3,2	–0.121915	0.019233	–0.243829	0.00096531	0.00096531	H
4,2	–0.119123	0.014113	–0.215036	0.00096531	0.00096531	H
5,2	–0.117900	0.069596	–0.061038	0.00096531		H
4,3	–0.116458	0.059192	–0.056865	0.00096531	0.00193062	H
6,2	–0.144181	0.132139	0.124194	0.00072398		H
5,3	–0.115244	0.163696	0.189929	0.00096531	0.00168929	D
7,2	–0.148738	0.209711	0.357361	0.00096531		D
6,3	–0.141538	0.231326	0.379215	0.00072398		D
5,4	–0.112881	0.263316	0.443310	0.00096531	0.00265460	D
8,2	–0.151323	0.307533	0.615067	0.00096531		D
7,3	–0.146112	0.318050	0.636099	0.00096531		D
6,4	–0.139190	0.331648	0.663295	0.00072398	0.00265460	D
9,2	–0.156394	0.345816	0.691632	0.00096531		D
8,3	–0.149705	0.356117	0.712234	0.00096531		D
7,4	–0.144697	0.365741	0.731482	0.00096531		D
6,5	–0.138999	0.380696	0.761392	0.00072398	0.00361991	D
T,2	–0.160379	–0.159436	–0.318872	0.00386124		H
9,3	–0.153745	–0.193794	–0.387588	0.00096531		S
8,4	–0.147293	–0.184088	–0.368175	0.00096531		S
7,5	–0.143473	–0.178846	–0.357692	0.00096531	0.00675716	S
T,3	–0.157773	–0.228909	–0.457819	0.00386124		S
9,4	–0.151345	–0.228134	–0.456268	0.00096531		S
8,5	–0.146040	–0.267603	–0.535207	0.00096531		S
7,6	–0.169692	–0.280666	–0.561333	0.00072398	0.00651584	S
T,4	–0.155349	–0.308093	–0.616186	0.00386124		S
9,5	–0.150100	–0.308744	–0.617488	0.00096531		S
8,6	–0.172288	–0.323254	–0.646508	0.00072398	0.00555053	S
T,5	–0.154176	–0.387327	–0.774654	0.00386124		S
9,6	–0.176503	–0.400876	–0.801753	0.00072398		S
8,7	–0.176976	–0.362116	–0.724232	0.00096531	0.00555053	S
T,6	–0.179023	–0.396278	–0.792555	0.00289593		S
9,7	–0.179634	–0.397210	–0.794420	0.00096531	0.00386124	S
T,7	–0.011287	–0.483343	–0.966686	0.00386124		S
9,8	–0.011411	–0.483777	–0.967554	0.00096531	0.00482655	S
T,8	0.268101	–0.586115	–1.172230	0.00386124	0.00386124	S
T,9	0.484093	–0.706989	–1.413978	0.00386124	0.00386124	S
A,2	–0.114167	0.168495	0.230211	0.00096531	0.00096531	D
A,3	–0.111568	0.147164	0.221860	0.00096531	0.00096531	D
A,4	–0.109194	0.120257	0.200721	0.00096531	0.00096531	D
A,5	–0.107596	0.115918	0.216665	0.00096531	0.00096531	D
A,6	0.010435	0.133243	0.266486	0.00072398	0.00072398	D
A,7	0.262174	0.192430	0.384860	0.00096531	0.00096531	D
A,8	0.482354	0.241313	0.482626	0.00096531	0.00096531	D
A,9	0.694187	0.279876	0.559752	0.00096531	0.00096531	S
A,T	1.500000	0.341117	0.682234	0.00386124	0.00386124	S

Hand	Stand	Hit	Double	Spl1	Spl2	Spl3	Freq.	BS
				NDAS				
A,A	–0.103505	0.199607	0.247914	0.758276	0.828269	0.831965	0.00036199	P
2,2	–0.124502	0.032055	–0.249003	0.125925	0.134595	0.135111	0.00036199	P
3,3	–0.118893	0.013896	–0.214511	0.111575	0.121528	0.122171	0.00036199	P
4,4	–0.114034	0.175290	0.193184	0.083438	0.076810	0.076506	0.00036199	D
5,5	–0.111669	0.361823	0.723645	0.055851	0.031548	0.030293	0.00036199	D
6,6	–0.165187	–0.193568	–0.387137	–0.010651	–0.003899	–0.003899	0.00018100	P
7,7	–0.174225	–0.366941	–0.733882	0.053024	0.071754	0.072757	0.00036199	P
8,8	–0.178171	–0.396692	–0.793384	0.233716	0.267683	0.269513	0.00036199	P
9,9	0.265195	–0.586713	–1.173426	0.359377	0.365690	0.365950	0.00036199	P
T,T	0.697403	–0.845027	–1.690053	0.520374	0.417420	0.350904	0.00723982	S
				DAS				
A,A	–0.103505	0.199607	0.247914	0.758276	0.828269	0.831965	0.00036199	P
2,2	–0.124502	0.032055	–0.249003	0.281275	0.304028	0.305372	0.00036199	P
3,3	–0.118893	0.013896	–0.214511	0.274115	0.298993	0.300527	0.00036199	P
4,4	–0.114034	0.175290	0.193184	0.244333	0.248557	0.248774	0.00036199	P
5,5	–0.111669	0.361823	0.723645	0.200282	0.156001	0.153548	0.00036199	D
6,6	–0.165187	–0.193568	–0.387137	0.154148	0.167915	0.167915	0.00018100	P
7,7	–0.174225	–0.366941	–0.733882	0.209924	0.241645	0.243348	0.00036199	P
8,8	–0.178171	–0.396692	–0.793384	0.340370	0.383004	0.385294	0.00036199	P
9,9	0.265195	–0.586713	–1.173426	0.413268	0.423806	0.424280	0.00036199	P
T,T	0.697403	–0.845027	–1.690053	0.520374	0.417420	0.350904	0.00723982	S

TABLE A8

1D S17: Dealer's Upcard 7

Hand	Stand	Hit	Double	Freq.	Cum. Freq.	BS
3,2	–0.468910	–0.119255	–0.937820	0.00096531	0.00096531	H
4,2	–0.466638	–0.163911	–0.870342	0.00096531	0.00096531	H
5,2	–0.477078	–0.067371	–0.567247	0.00096531		H
4,3	–0.464879	–0.070634	–0.542577	0.00096531	0.00193062	H
6,2	–0.481633	0.091839	–0.148120	0.00096531		H
5,3	–0.475347	0.092588	–0.132299	0.00096531	0.00193062	H
7,2	–0.485803	0.183612	0.160008	0.00072398		H
6,3	–0.479852	0.197716	0.175697	0.00096531		H
5,4	–0.473057	0.201306	0.190459	0.00096531	0.00265460	H
8,2	–0.487638	0.267639	0.416638	0.00096531		D
7,3	–0.485024	0.277201	0.444697	0.00072398		D
6,4	–0.478620	0.285586	0.475371	0.00096531	0.00265460	D
9,2	–0.493342	0.288889	0.455401	0.00096531		D
8,3	–0.485835	0.291794	0.472587	0.00096531		D
7,4	–0.482795	0.293769	0.487454	0.00072398		D
6,5	–0.488938	0.297369	0.500525	0.00096531	0.00361991	D
T,2	–0.472817	–0.212020	–0.495828	0.00386124		H
9,3	–0.491583	–0.247104	–0.578757	0.00096531		H
8,4	–0.483611	–0.245545	–0.563593	0.00096531		H
7,5	–0.493091	–0.258201	–0.585151	0.00072398	0.00651584	H
T,3	–0.471067	–0.270391	–0.582990	0.00386124		H
9,4	–0.489331	–0.274092	–0.600140	0.00096531		H
8,5	–0.493922	–0.327475	–0.706532	0.00096531		H
7,6	–0.497620	–0.330723	–0.714109	0.00072398	0.00651584	H
T,4	–0.468887	–0.342199	–0.706808	0.00386124		H
9,5	–0.499719	–0.348015	–0.727641	0.00096531		H
8,6	–0.498604	–0.348533	–0.734594	0.00096531	0.00579186	H
T,5	–0.478497	–0.364503	–0.736265	0.00386124		H
9,6	–0.502845	–0.363208	–0.744596	0.00096531		H
8,7	–0.501385	–0.324114	–0.666031	0.00072398	0.00555053	H
T,6	–0.483324	–0.376194	–0.752387	0.00386124		H
9,7	–0.507327	–0.374945	–0.749890	0.00072398	0.00458522	H
T,7	–0.121287	–0.451963	–0.903925	0.00289593		S
9,8	–0.122900	–0.448819	–0.897639	0.00096531	0.00386124	S
T,8	0.388746	–0.567259	–1.134517	0.00386124	0.00386124	S
T,9	0.610120	–0.698546	–1.397092	0.00386124	0.00386124	S
A,2	–0.461877	0.107343	–0.157238	0.00096531	0.00096531	H
A,3	–0.460210	0.060426	–0.174539	0.00096531	0.00096531	H
A,4	–0.457617	0.033732	–0.140937	0.00096531	0.00096531	H
A,5	–0.467986	–0.023784	–0.189067	0.00096531	0.00096531	H
A,6	–0.089639	0.059646	0.014177	0.00096531	0.00096531	H
A,7	0.411952	0.174667	0.240224	0.00072398	0.00072398	S
A,8	0.614504	0.221936	0.325331	0.00096531	0.00096531	S
A,9	0.773194	0.242711	0.351279	0.00096531	0.00096531	S
A,T	1.500000	0.285776	0.467593	0.00386124	0.00386124	S

Hand	Stand	Hit	Double	Spl1	Spl2	Spl3	Freq.	BS
				NDAS				
A,A	–0.452479	0.158433	–0.136975	0.540712	0.623186	0.627650	0.00036199	P
2,2	–0.471027	–0.091753	–0.942054	–0.054464	–0.051372	–0.051212	0.00036199	P
3,3	–0.467213	–0.164471	–0.871217	–0.115328	–0.109162	–0.108699	0.00036199	P
4,4	–0.462716	0.110632	–0.108353	–0.222838	–0.252296	–0.253997	0.00036199	H
5,5	–0.484469	0.278602	0.466340	–0.298829	–0.349824	–0.352765	0.00036199	D
6,6	–0.493437	–0.264854	–0.598514	–0.269899	–0.268091	–0.267863	0.00036199	H
7,7	–0.501956	–0.389227	–0.823012	–0.122529	–0.110318	–0.110318	0.00018100	P
8,8	–0.502514	–0.373561	–0.747122	0.202329	0.249110	0.251584	0.00036199	P
9,9	0.401060	–0.566049	–1.132097	0.340574	0.335270	0.334969	0.00036199	S
T,T	0.764677	–0.843026	–1.686052	0.482229	0.336773	0.251783	0.00723982	S
				DAS				
A,A	–0.452479	0.158433	–0.136975	0.540712	0.623186	0.627650	0.00036199	P
2,2	–0.471027	–0.091753	–0.942054	0.004752	0.013371	0.013864	0.00036199	P
3,3	–0.467213	–0.164471	–0.871217	–0.057250	–0.045608	–0.044810	0.00036199	P
4,4	–0.462716	0.110632	–0.108353	–0.158732	–0.182108	–0.183437	0.00036199	H
5,5	–0.484469	0.278602	0.466340	–0.244667	–0.308565	–0.312308	0.00036199	D
6,6	–0.493437	–0.264854	–0.598514	–0.201370	–0.193535	–0.192962	0.00036199	P
7,7	–0.501956	–0.389227	–0.823012	–0.065747	–0.051285	–0.051285	0.00018100	P
8,8	–0.502514	–0.373561	–0.747122	0.250712	0.301026	0.303664	0.00036199	P
9,9	0.401060	–0.566049	–1.132097	0.362756	0.358851	0.358600	0.00036199	S
T,T	0.764677	–0.843026	–1.686052	0.482229	0.336773	0.251783	0.00723982	S

TABLE A9
1D S17: Dealer's Upcard 8

Hand	Stand	Hit	Double	Freq.	Cum. Freq.	BS
3,2	–0.512722	–0.180835	–1.025445	0.00096531	0.00096531	H
4,2	–0.522155	–0.234151	–1.023132	0.00096531	0.00096531	H
5,2	–0.522459	–0.217784	–0.853460	0.00096531		H
4,3	–0.519886	–0.228382	–0.848583	0.00096531	0.00193062	H
6,2	–0.524927	–0.055928	–0.437135	0.00096531		H
5,3	–0.520171	–0.056988	–0.428185	0.00096531	0.00193062	H
7,2	–0.528340	0.107594	0.007362	0.00096531		H
6,3	–0.523699	0.117522	0.014199	0.00096531		H
5,4	–0.530618	0.108081	0.000739	0.00096531	0.00289593	H
8,2	–0.534064	0.207288	0.294474	0.00072398		D
7,3	–0.526077	0.217139	0.326864	0.00096531		D
6,4	–0.533077	0.207506	0.317153	0.00096531	0.00265460	D
9,2	–0.513558	0.215256	0.327731	0.00096531		D
8,3	–0.531845	0.220266	0.329975	0.00072398		D
7,4	–0.535471	0.221712	0.340213	0.00096531		D
6,5	–0.533297	0.229714	0.365672	0.00096531	0.00361991	D
T,2	–0.517357	–0.274473	–0.625855	0.00386124		H
9,3	–0.511343	–0.316143	–0.703700	0.00096531		H
8,4	–0.541217	–0.319226	–0.719327	0.00072398		H
7,5	–0.535706	–0.320983	–0.711203	0.00096531	0.00651584	H
T,3	–0.515179	–0.328235	–0.705640	0.00386124		H
9,4	–0.520792	–0.338911	–0.724110	0.00096531		H
8,5	–0.541529	–0.386852	–0.829134	0.00072398		H
7,6	–0.538333	–0.394373	–0.842908	0.00096531	0.00651584	H
T,4	–0.523850	–0.357370	–0.739262	0.00386124		H
9,5	–0.520326	–0.370139	–0.761928	0.00096531		H
8,6	–0.542601	–0.369092	–0.774545	0.00072398	0.00555053	H
T,5	–0.524234	–0.417965	–0.844148	0.00386124		H
9,6	–0.523099	–0.420733	–0.855589	0.00096531		H
8,7	–0.545308	–0.379554	–0.777379	0.00072398	0.00555053	H
T,6	–0.527007	–0.424823	–0.849645	0.00386124		H
9,7	–0.525806	–0.427790	–0.855581	0.00096531	0.00482655	H
T,7	–0.394240	–0.473803	–0.947605	0.00386124		S
9,8	–0.414899	–0.475296	–0.950593	0.00072398	0.00458522	S
T,8	0.095530	–0.565137	–1.130274	0.00289593	0.00289593	S
T,9	0.576828	–0.697331	–1.394661	0.00386124	0.00386124	S
A,2	–0.507685	0.039113	–0.312387	0.00096531	0.00096531	H
A,3	–0.505092	0.035024	–0.254214	0.00096531	0.00096531	H
A,4	–0.514496	–0.035484	–0.314134	0.00096531	0.00096531	H
A,5	–0.514869	–0.084303	–0.332550	0.00096531	0.00096531	H
A,6	–0.385254	–0.064896	–0.229744	0.00096531	0.00096531	H
A,7	0.120931	0.047500	–0.015314	0.00096531	0.00096531	S
A,8	0.607840	0.157747	0.190197	0.00072398	0.00072398	S
A,9	0.784814	0.171549	0.229784	0.00096531	0.00096531	S
A,T	1.500000	0.221207	0.332681	0.00386124	0.00386124	S

NDAS

Hand	Stand	Hit	Double	Spl1	Spl2	Spl3	Freq.	BS
A,A	–0.499700	0.093045	–0.295648	0.406468	0.482240	0.486371	0.00036199	P
2,2	–0.514917	–0.141047	–1.029833	–0.212598	–0.218269	–0.218566	0.00036199	H
3,3	–0.510419	–0.231120	–0.999660	–0.265388	–0.265953	–0.265843	0.00036199	H
4,4	–0.529261	–0.054803	–0.447062	–0.342754	–0.367918	–0.369353	0.00036199	H
5,5	–0.530853	0.207500	0.322885	–0.448230	–0.505818	–0.509123	0.00036199	D
6,6	–0.535771	–0.321707	–0.711354	–0.412582	–0.418398	–0.418613	0.00036199	H
7,7	–0.539340	–0.407893	–0.857865	–0.423471	–0.422466	–0.422274	0.00036199	H
8,8	–0.551276	–0.426315	–0.852630	–0.100087	–0.086971	–0.086971	0.00018100	P
9,9	0.064518	–0.566265	–1.132530	0.179367	0.189670	0.190276	0.00036199	P
T,T	0.783251	–0.842727	–1.685455	0.353853	0.136617	0.011765	0.00723982	S

DAS

Hand	Stand	Hit	Double	Spl1	Spl2	Spl3	Freq.	BS
A,A	–0.499700	0.093045	–0.295648	0.406468	0.482240	0.486371	0.00036199	P
2,2	–0.514917	–0.141047	–1.029833	–0.177104	–0.179293	–0.179370	0.00036199	H
3,3	–0.510419	–0.231120	–0.999660	–0.226050	–0.222676	–0.222314	0.00036199	P
4,4	–0.529261	–0.054803	–0.447062	–0.299949	–0.321123	–0.322317	0.00036199	H
5,5	–0.530853	0.207500	0.322885	–0.412012	–0.477572	–0.481386	0.00036199	D
6,6	–0.535771	–0.321707	–0.711354	–0.367710	–0.369416	–0.369387	0.00036199	H
7,7	–0.539340	–0.407893	–0.857865	–0.385238	–0.380967	–0.380591	0.00036199	P
8,8	–0.551276	–0.426315	–0.852630	–0.073509	–0.059525	–0.059525	0.00018100	P
9,9	0.064518	–0.566265	–1.132530	0.195409	0.206836	0.207491	0.00036199	P
T,T	0.783251	–0.842727	–1.685455	0.353853	0.136617	0.011765	0.00723982	S

TABLE A10
1D S17: Dealer's Upcard 9

Hand	Stand	Hit	Double	Freq.	Cum. Freq.	BS
3,2	–0.532954	–0.262470	–1.065907	0.00096531	0.00096531	H
4,2	–0.532529	–0.304594	–1.045309	0.00096531	0.00096531	H
5,2	–0.531595	–0.284938	–0.926277	0.00096531		H
4,3	–0.542010	–0.304698	–0.946164	0.00096531	0.00193062	H
6,2	–0.536503	–0.208348	–0.691586	0.00096531		H
5,3	–0.542147	–0.218486	–0.702933	0.00096531	0.00193062	H
7,2	–0.540546	–0.052384	–0.274062	0.00096531		H
6,3	–0.545956	–0.051082	–0.289689	0.00096531		H
5,4	–0.541653	–0.051981	–0.278961	0.00096531	0.00289593	H
8,2	–0.520050	0.120543	0.174428	0.00096531		D
7,3	–0.550044	0.117569	0.153705	0.00096531		D
6,4	–0.545472	0.117402	0.164633	0.00096531	0.00289593	D
9,2	–0.523864	0.142527	0.213840	0.00072398		D
8,3	–0.529562	0.139171	0.215208	0.00096531		D
7,4	–0.549545	0.149504	0.224729	0.00096531		D
6,5	–0.544508	0.151713	0.239911	0.00096531	0.00361991	D
T,2	–0.527482	–0.344351	–0.745531	0.00386124		H
9,3	–0.533417	–0.392178	–0.840364	0.00072398		H
8,4	–0.529140	–0.394885	–0.833734	0.00096531		H
7,5	–0.548658	–0.389559	–0.826051	0.00096531	0.00651584	H
T,3	–0.536257	–0.358832	–0.752138	0.00386124		H
9,4	–0.532217	–0.365253	–0.758677	0.00072398		H
8,5	–0.527475	–0.413421	–0.847602	0.00096531		H
7,6	–0.551096	–0.418484	–0.869390	0.00096531	0.00651584	H
T,4	–0.535908	–0.413520	–0.839394	0.00386124		H
9,5	–0.531402	–0.425895	–0.859833	0.00072398		H
8,6	–0.531614	–0.437187	–0.887881	0.00096531	0.00555053	H
T,5	–0.535093	–0.475256	–0.951924	0.00386124		H
9,6	–0.535541	–0.479661	–0.966804	0.00072398		H
8,7	–0.535909	–0.443182	–0.893558	0.00096531	0.00555053	H
T,6	–0.539232	–0.479306	–0.958613	0.00386124		H
9,7	–0.539836	–0.482037	–0.964075	0.00072398	0.00458522	H
T,7	–0.416111	–0.526459	–1.052918	0.00386124		S
9,8	–0.411646	–0.531882	–1.063764	0.00072398	0.00458522	S
T,8	–0.196136	–0.593227	–1.186454	0.00386124	0.00386124	S
T,9	0.264279	–0.697795	–1.395590	0.00289593	0.00289593	S
A,2	–0.517198	–0.014062	–0.372682	0.00096531	0.00096531	H
A,3	–0.526764	–0.059995	–0.393525	0.00096531	0.00096531	H
A,4	–0.526348	–0.113691	–0.421784	0.00096531	0.00096531	H
A,5	–0.525482	–0.166825	–0.452038	0.00096531	0.00096531	H
A,6	–0.407070	–0.134736	–0.345244	0.00096531	0.00096531	H
A,7	–0.178832	–0.086958	–0.254497	0.00096531	0.00096531	H
A,8	0.287946	0.004693	–0.060207	0.00096531	0.00096531	S
A,9	0.765635	0.096418	0.110516	0.00072398	0.00072398	S
A,T	1.500000	0.148864	0.216490	0.00386124	0.00386124	S

Hand	Stand	Hit	Double	Spl1	Spl2	Spl3	Freq.	BS
				NDAS				
A,A	–0.510996	–0.002731	–0.420608	0.289770	0.357091	0.360781	0.00036199	P
2,2	–0.523455	–0.222489	–1.046910	–0.383380	–0.396163	–0.396825	0.00036199	H
3,3	–0.542453	–0.309972	–1.065288	–0.418153	–0.425736	–0.426061	0.00036199	H
4,4	–0.542638	–0.206479	–0.701087	–0.493182	–0.517465	–0.518807	0.00036199	H
5,5	–0.540668	0.118869	0.174553	–0.607597	–0.670990	–0.674607	0.00036199	D
6,6	–0.548502	–0.386242	–0.817364	–0.569709	–0.583792	–0.584492	0.00036199	H
7,7	–0.555391	–0.474654	–0.978238	–0.563840	–0.568740	–0.568857	0.00036199	H
8,8	–0.516426	–0.487124	–0.974248	–0.429934	–0.427106	–0.427063	0.00036199	P
9,9	–0.196372	–0.594732	–1.189465	–0.112424	–0.108836	–0.108836	0.00018100	P
T,T	0.743970	–0.842055	–1.684111	0.183754	–0.096498	–0.255830	0.00723982	S
				DAS				
A,A	–0.510996	–0.002731	–0.420608	0.289770	0.357091	0.360781	0.00036199	P
2,2	–0.523455	–0.222489	–1.046910	–0.359454	–0.369711	–0.370205	0.00036199	H
3,3	–0.542453	–0.309972	–1.065288	–0.396141	–0.401593	–0.401785	0.00036199	H
4,4	–0.542638	–0.206479	–0.701087	–0.469024	–0.490985	–0.492185	0.00036199	H
5,5	–0.540668	0.118869	0.174553	–0.584760	–0.651999	–0.655885	0.00036199	D
6,6	–0.548502	–0.386242	–0.817364	–0.544299	–0.556016	–0.556574	0.00036199	H
7,7	–0.555391	–0.474654	–0.978238	–0.545047	–0.548318	–0.548341	0.00036199	H
8,8	–0.516426	–0.487124	–0.974248	–0.406325	–0.401319	–0.401144	0.00036199	P
9,9	–0.196372	–0.594732	–1.189465	–0.102933	–0.099043	–0.099043	0.00018100	P
T,T	0.743970	–0.842055	–1.684111	0.183754	–0.096498	–0.255830	0.00723982	S

TABLE A11
1D S17: Dealer's Upcard T

Hand	Stand	Hit	Double	Freq.	Cum. Freq.	BS
3,2	-0.546375	-0.307857	-1.092750	0.00354603	0.00354603	H
4,2	-0.546578	-0.345111	-1.073122	0.00354603	0.00354603	H
5,2	-0.546774	-0.322021	-0.955173	0.00354603		H
4,3	-0.547042	-0.334779	-0.955483	0.00354603	0.00709207	H
6,2	-0.551019	-0.252153	-0.745151	0.00354603		H
5,3	-0.546016	-0.250864	-0.738023	0.00354603	0.00709207	H
7,2	-0.528701	-0.154112	-0.435521	0.00354603		H
6,3	-0.550310	-0.141660	-0.444546	0.00354603		H
5,4	-0.546165	-0.138056	-0.438369	0.00354603	0.01063810	H
8,2	-0.532853	0.029291	0.005950	0.00354603		H
7,3	-0.528008	0.029609	0.014484	0.00354603		H
6,4	-0.550442	0.033407	0.011219	0.00354603	0.01063810	H
9,2	-0.536793	0.103847	0.142776	0.00354603		D
8,3	-0.532205	0.109187	0.162257	0.00354603		D
7,4	-0.528224	0.109571	0.170991	0.00354603		D
6,5	-0.549500	0.113266	0.170737	0.00354603	0.01418414	D
T,2	-0.539571	-0.346638	-0.739114	0.01329763		H
9,3	-0.535298	-0.388543	-0.821117	0.00354603		H
8,4	-0.531574	-0.385856	-0.805908	0.00354603		H
7,5	-0.526434	-0.384545	-0.790615	0.00354603	0.02393573	H
T,3	-0.539002	-0.393091	-0.811084	0.01329763		H
9,4	-0.535593	-0.398702	-0.818901	0.00354603		H
8,5	-0.530710	-0.446362	-0.909614	0.00354603		H
7,6	-0.530953	-0.451149	-0.908270	0.00354603	0.02393573	H
T,4	-0.539296	-0.445788	-0.899256	0.01329763		H
9,5	-0.534729	-0.453066	-0.911135	0.00354603		H
8,6	-0.535229	-0.466555	-0.936341	0.00354603	0.02038969	EH
T,5	-0.538433	-0.501091	-1.002183	0.01329763		RH
9,6	-0.539248	-0.511965	-1.023930	0.00354603		RH
8,7	-0.514015	-0.474794	-0.949587	0.00354603	0.02038969	EH
T,6	-0.542952	-0.506929	-1.013858	0.01329763		RH
9,7	-0.518034	-0.512009	-1.024017	0.00354603	0.01684366	RH
T,7	-0.412341	-0.556704	-1.113407	0.01329763		S
9,8	-0.390691	-0.558165	-1.116330	0.00354603	0.01684366	S
T,8	-0.155192	-0.622794	-1.245589	0.01329763	0.01329763	S
T,9	0.102517	-0.710037	-1.420073	0.01329763	0.01329763	S
A,2	-0.538705	-0.088917	-0.478997	0.00362483	0.00362483	H
A,3	-0.538043	-0.123590	-0.479655	0.00362483	0.00362483	H
A,4	-0.538314	-0.170424	-0.496264	0.00362483	0.00362483	H
A,5	-0.537311	-0.223362	-0.536694	0.00362483	0.00362483	H
A,6	-0.417783	-0.190991	-0.432804	0.00362483	0.00362483	H
A,7	-0.186187	-0.138734	-0.321341	0.00362483	0.00362483	H
A,8	0.064313	-0.085611	-0.224373	0.00362483	0.00362483	S
A,9	0.554551	0.006555	-0.052039	0.00362483	0.00362483	S
A,T	1.500000	0.104681	0.139083	0.01359313	0.01359313	S

Hand	Stand	Hit	Double	Spl1	Spl2	Spl3	Freq.	BS
				NDAS				
A,A	-0.530674	-0.047033	-0.468319	0.194251	0.256082	0.259524	0.00138886	P
2,2	-0.547137	-0.275621	-1.094274	-0.459601	-0.474748	-0.475563	0.00132976	H
3,3	-0.545672	-0.343655	-1.071267	-0.504229	-0.516567	-0.517177	0.00132976	H
4,4	-0.547191	-0.241025	-0.738996	-0.570489	-0.598025	-0.599527	0.00132976	H
5,5	-0.545139	0.037781	0.018005	-0.671747	-0.732934	-0.736384	0.00132976	H
6,6	-0.552167	-0.389097	-0.803852	-0.687613	-0.711933	-0.713216	0.00132976	H
7,7	-0.509739	-0.514818	-1.034724	-0.629402	-0.636119	-0.636319	0.00132976	RS
8,8	-0.518291	-0.511755	-1.023510	-0.462839	-0.458694	-0.458498	0.00132976	EP
9,9	-0.133285	-0.624465	-1.248929	-0.267816	-0.279668	-0.280351	0.00132976	S
T,T	0.583154	-0.836969	-1.673938	0.051976	-0.198286	-0.332369	0.02327085	S
				DAS				
A,A	-0.530674	-0.047033	-0.468319	0.194251	0.256082	0.259524	0.00138886	P
2,2	-0.547137	-0.275621	-1.094274	-0.451895	-0.466291	-0.467059	0.00132976	H
3,3	-0.545672	-0.343655	-1.071267	-0.494165	-0.505554	-0.506106	0.00132976	H
4,4	-0.547191	-0.241025	-0.738996	-0.559246	-0.585765	-0.587208	0.00132976	H
5,5	-0.545139	0.037781	0.018005	-0.660597	-0.720723	-0.724108	0.00132976	H
6,6	-0.552167	-0.389097	-0.803852	-0.676669	-0.699937	-0.701154	0.00132976	H
7,7	-0.509739	-0.514818	-1.034724	-0.616575	-0.621976	-0.622090	0.00132976	RS
8,8	-0.518291	-0.511755	-1.023510	-0.452851	-0.447756	-0.447501	0.00132976	P
9,9	-0.133285	-0.624465	-1.248929	-0.262645	-0.274168	-0.274839	0.00132976	S
T,T	0.583154	-0.836969	-1.673938	0.051976	-0.198286	-0.332369	0.02327085	S

TABLE A12
1D H17: Dealer's Upcard A

Hand	Stand	Hit	Double	Freq.	Cum. Freq.	BS
3,2	−0.584988	−0.323390	−1.169976	0.00065011	0.00065011	EH
4,2	−0.583756	−0.363665	−1.157797	0.00065011	0.00065011	EH
5,2	−0.585491	−0.365739	−1.073457	0.00065011		EH
4,3	−0.582464	−0.376697	−1.067304	0.00065011	0.00130021	EH
6,2	−0.579622	−0.278269	−0.817393	0.00065011		EH
5,3	−0.583566	−0.279644	−0.813846	0.00065011	0.00130021	EH
7,2	−0.573681	−0.141859	−0.433412	0.00065011		H
6,3	−0.577736	−0.125146	−0.422809	0.00065011		H
5,4	−0.581270	−0.124710	−0.419143	0.00065011	0.00195032	H
8,2	−0.580706	0.038685	0.002385	0.00065011		H
7,3	−0.571831	0.038922	0.017585	0.00065011		H
6,4	−0.575556	0.044447	0.030867	0.00065011	0.00195032	H
9,2	−0.586701	0.130469	0.181921	0.00065011		D
8,3	−0.578030	0.133295	0.202358	0.00065011		D
7,4	−0.568782	0.136548	0.224107	0.00065011		D
6,5	−0.575374	0.140999	0.234296	0.00065011	0.00260042	D
T,2	−0.600796	−0.380064	−0.817531	0.00267923		EH
9,3	−0.585004	−0.405024	−0.864053	0.00065011		EH
8,4	−0.575884	−0.403544	−0.848853	0.00065011		EH
7,5	−0.570148	−0.406425	−0.838128	0.00065011	0.00462954	EH
T,3	−0.599060	−0.420999	−0.876377	0.00267923		EH
9,4	−0.582881	−0.410744	−0.850753	0.00065011		EH
8,5	−0.577345	−0.460491	−0.943102	0.00065011		EH
7,6	−0.565103	−0.466478	−0.947653	0.00065011	0.00462954	EH
T,4	−0.597063	−0.468678	−0.951894	0.00267923		EH
9,5	−0.584302	−0.467809	−0.947125	0.00065011		EH
8,6	−0.572297	−0.476332	−0.965356	0.00065011	0.00397944	EH
T,5	−0.598257	−0.523957	−1.051636	0.00267923		RH
9,6	−0.579070	−0.517921	−1.043006	0.00065011		RH
8,7	−0.567389	−0.483180	−0.971421	0.00065011	0.00397944	EH
T,6	−0.593889	−0.528965	−1.057930	0.00267923		RH
9,7	−0.574249	−0.520784	−1.041569	0.00065011	0.00332933	RH
T,7	−0.503886	−0.576778	−1.153557	0.00267923		RS
9,8	−0.490839	−0.565750	−1.131499	0.00065011	0.00332933	ES
T,8	−0.211146	−0.645242	−1.290484	0.00267923	0.00267923	S
T,9	0.212689	−0.747687	−1.495373	0.00267923	0.00267923	S
A,2	−0.589084	−0.108133	−0.552040	0.00048758	0.00048758	H
A,3	−0.587263	−0.139310	−0.548792	0.00048758	0.00048758	H
A,4	−0.585081	−0.188524	−0.569570	0.00048758	0.00048758	H
A,5	−0.587177	−0.242524	−0.607984	0.00048758	0.00048758	H
A,6	−0.507702	−0.229258	−0.529713	0.00048758	0.00048758	H
A,7	−0.223909	−0.173387	−0.414678	0.00048758	0.00048758	H
A,8	0.197058	−0.076035	−0.232282	0.00048758	0.00048758	S
A,9	0.625139	0.017967	−0.062818	0.00048758	0.00048758	S
A,T	1.500000	0.112270	0.137215	0.00200942	0.00200942	S

Hand	Stand	Hit	Double	Spl1	Spl2	Spl3	Freq.	BS
				NDAS				
A,A	−0.591119	−0.071368	−0.558494	0.215092	0.248830	0.248830	0.00012189	P
2,2	−0.586848	−0.293920	−1.173695	−0.502039	−0.518792	−0.519673	0.00024379	EH
3,3	−0.584067	−0.363077	−1.158259	−0.541679	−0.554607	−0.555195	0.00024379	EH
4,4	−0.580201	−0.267254	−0.808571	−0.604566	−0.632370	−0.633864	0.00024379	EH
5,5	−0.582567	0.042732	0.026604	−0.715792	−0.780594	−0.784210	0.00024379	H
6,6	−0.570140	−0.404037	−0.832166	−0.683118	−0.704507	−0.705562	0.00024379	EH
7,7	−0.560383	−0.523378	−1.059340	−0.686353	−0.695539	−0.695793	0.00024379	RH
8,8	−0.574417	−0.520115	−1.040230	−0.482754	−0.478089	−0.477752	0.00024379	EP
9,9	−0.186130	−0.637010	−1.274020	−0.192828	−0.191417	−0.191231	0.00024379	S
T,T	0.593596	−0.884336	−1.768673	−0.004047	−0.313679	−0.495791	0.00517130	S
				DAS				
A,A	−0.591119	−0.071368	−0.558494	0.215092	0.248830	0.248830	0.00012189	P
2,2	−0.586848	−0.293920	−1.173695	−0.490595	−0.506152	−0.506955	0.00024379	EH
3,3	−0.584067	−0.363077	−1.158259	−0.527061	−0.538514	−0.539008	0.00024379	EH
4,4	−0.580201	−0.267254	−0.808571	−0.586402	−0.612408	−0.613789	0.00024379	EH
5,5	−0.582567	0.042732	0.026604	−0.696684	−0.759624	−0.763126	0.00024379	H
6,6	−0.570140	−0.404037	−0.832166	−0.663140	−0.682461	−0.683379	0.00024379	EH
7,7	−0.560383	−0.523378	−1.059340	−0.668344	−0.675727	−0.675865	0.00024379	RH
8,8	−0.574417	−0.520115	−1.040230	−0.470321	−0.464537	−0.464135	0.00024379	EP
9,9	−0.186130	−0.637010	−1.274020	−0.185971	−0.184139	−0.183940	0.00024379	P
T,T	0.593596	−0.884336	−1.768673	−0.004047	−0.313679	−0.495791	0.00517130	S

TABLE A13
1D H17: Dealer's Upcard 2

Hand	Stand	Hit	Double	Freq.	Cum. Freq.	BS
3,2	–0.285836	–0.127700	–0.571673	0.00072398	0.00072398	H
4,2	–0.285022	–0.147360	–0.549146	0.00072398	0.00072398	H
5,2	–0.266787	–0.098615	–0.395929	0.00072398		H
4,3	–0.286480	–0.122094	–0.423946	0.00096531	0.00168929	H
6,2	–0.269843	–0.015204	–0.164091	0.00072398		H
5,3	–0.268275	–0.018657	–0.168844	0.00096531	0.00168929	H
7,2	–0.273051	0.081403	0.115180	0.00072398		D
6,3	–0.271108	0.090258	0.110074	0.00096531		D
5,4	–0.267337	0.091789	0.112286	0.00096531	0.00265460	D
8,2	–0.274989	0.211945	0.422928	0.00072398		D
7,3	–0.273536	0.213165	0.424426	0.00096531		D
6,4	–0.269321	0.216507	0.430971	0.00096531	0.00265460	D
9,2	–0.279580	0.262883	0.524387	0.00072398		D
8,3	–0.276377	0.267697	0.533501	0.00096531		D
7,4	–0.272580	0.272126	0.542094	0.00096531		D
6,5	–0.251544	0.283327	0.566721	0.00096531	0.00361991	D
T,2	–0.303613	–0.243855	–0.487710	0.00289593		H
9,3	–0.280919	–0.266647	–0.533294	0.00096531		H
8,4	–0.275517	–0.259998	–0.519996	0.00096531		H
7,5	–0.254915	–0.252773	–0.505546	0.00096531	0.00579186	H
T,3	–0.304909	–0.305033	–0.610067	0.00386124		S
9,4	–0.280008	–0.293478	–0.586956	0.00096531		S
8,5	–0.257809	–0.331827	–0.663653	0.00096531		S
7,6	–0.257996	–0.332521	–0.665041	0.00096531	0.00675716	S
T,4	–0.304094	–0.369450	–0.738901	0.00386124		S
9,5	–0.262319	–0.360026	–0.720051	0.00096531		S
8,6	–0.260819	–0.363780	–0.727560	0.00096531	0.00579186	S
T,5	–0.287395	–0.437794	–0.875589	0.00386124		S
9,6	–0.266381	–0.431908	–0.863816	0.00096531		S
8,7	–0.265141	–0.392055	–0.784111	0.00096531	0.00579186	S
T,6	–0.290383	–0.467345	–0.934691	0.00386124		S
9,7	–0.269723	–0.457901	–0.915802	0.00096531	0.00482655	S
T,7	–0.161066	–0.540185	–1.080371	0.00386124		S
9,8	–0.139571	–0.531861	–1.063722	0.00096531	0.00482655	S
T,8	0.107288	–0.633540	–1.267079	0.00386124	0.00386124	S
T,9	0.376238	–0.750089	–1.500178	0.00386124	0.00386124	S
A,2	–0.277228	0.039915	–0.038722	0.00072398	0.00072398	H
A,3	–0.278614	0.018179	–0.042787	0.00096531	0.00096531	H
A,4	–0.278727	–0.010303	–0.066388	0.00096531	0.00096531	H
A,5	–0.260892	–0.029729	–0.077071	0.00096531	0.00096531	H
A,6	–0.133508	0.007066	0.014198	0.00096531	0.00096531	D
A,7	0.127707	0.063306	0.124454	0.00096531	0.00096531	S
A,8	0.395473	0.118492	0.235092	0.00096531	0.00096531	S
A,9	0.652277	0.189816	0.378598	0.00096531	0.00096531	S
A,T	1.500000	0.246619	0.486540	0.00386124	0.00386124	S

Hand	Stand	Hit	Double	Spl1	Spl2	Spl3	Freq.	BS
				NDAS				
A,A	–0.270691	0.093896	–0.017397	0.565372	0.634965	0.638743	0.00036199	P
2,2	–0.283340	–0.110110	–0.566680	–0.127987	–0.128657	–0.128657	0.00018100	H
3,3	–0.287310	–0.148694	–0.553531	–0.196704	–0.199978	–0.200116	0.00036199	H
4,4	–0.285624	–0.013517	–0.184721	–0.231736	–0.250241	–0.251272	0.00036199	H
5,5	–0.248567	0.222272	0.444231	–0.224785	–0.260765	–0.262643	0.00036199	D
6,6	–0.254573	–0.252577	–0.505153	–0.211449	–0.204838	–0.204295	0.00036199	P
7,7	–0.261350	–0.407605	–0.815210	–0.162567	–0.153108	–0.152527	0.00036199	P
8,8	–0.267966	–0.456002	–0.912005	0.010992	0.035445	0.036843	0.00036199	P
9,9	0.125626	–0.628481	–1.256962	0.166085	0.169446	0.169624	0.00036199	P
T,T	0.621940	–0.846767	–1.693533	0.307283	0.136859	0.032768	0.00723982	S
				DAS				
A,A	–0.270691	0.093896	–0.017397	0.565372	0.634965	0.638743	0.00036199	P
2,2	–0.283340	–0.110110	–0.566680	–0.039010	–0.035695	–0.035695	0.00018100	P
3,3	–0.287310	–0.148694	–0.553531	–0.111785	–0.107758	–0.107487	0.00036199	P
4,4	–0.285624	–0.013517	–0.184721	–0.145395	–0.156502	–0.157123	0.00036199	H
5,5	–0.248567	0.222272	0.444231	–0.149838	–0.199258	–0.201937	0.00036199	D
6,6	–0.254573	–0.252577	–0.505153	–0.114184	–0.098553	–0.097466	0.00036199	P
7,7	–0.261350	–0.407605	–0.815210	–0.072848	–0.055389	–0.054343	0.00036199	P
8,8	–0.267966	–0.456002	–0.912005	0.083397	0.114086	0.115834	0.00036199	P
9,9	0.125626	–0.628481	–1.256962	0.198514	0.204579	0.204905	0.00036199	P
T,T	0.621940	–0.846767	–1.693533	0.307283	0.136859	0.032768	0.00723982	S

TABLE A14
1D H17: Dealer's Upcard 3

Hand	Stand	Hit	Double	Freq.	Cum. Freq.	BS
3,2	–0.244538	–0.096197	–0.489076	0.00072398	0.00072398	H
4,2	–0.226627	–0.102372	–0.435087	0.00096531	0.00096531	H
5,2	–0.223426	–0.060526	–0.306360	0.00096531		H
4,3	–0.225831	–0.080526	–0.320149	0.00072398	0.00168929	H
6,2	–0.225779	0.022662	–0.079188	0.00096531		H
5,3	–0.222563	0.018710	–0.083667	0.00072398	0.00168929	H
7,2	–0.228612	0.129548	0.193097	0.00096531		D
6,3	–0.224049	0.138148	0.188409	0.00072398		D
5,4	–0.204084	0.140087	0.194600	0.00096531	0.00265460	D
8,2	–0.233015	0.240666	0.481332	0.00096531		D
7,3	–0.227773	0.245660	0.491319	0.00072398		D
6,4	–0.206493	0.249879	0.499758	0.00096531	0.00265460	D
9,2	–0.258880	0.268804	0.537608	0.00096531		D
8,3	–0.232154	0.295130	0.590260	0.00072398		D
7,4	–0.210215	0.305685	0.611371	0.00096531		D
6,5	–0.203214	0.314508	0.629017	0.00096531	0.00361991	D
T,2	–0.262425	–0.219625	–0.439251	0.00386124		H
9,3	–0.258039	–0.255924	–0.511848	0.00072398		H
8,4	–0.214758	–0.229913	–0.459825	0.00096531		S
7,5	–0.207078	–0.222064	–0.444129	0.00096531	0.00651584	S
T,3	–0.261675	–0.283766	–0.567531	0.00289593		S
9,4	–0.240596	–0.280926	–0.561853	0.00096531		S
8,5	–0.211478	–0.306103	–0.612207	0.00096531		S
7,6	–0.209576	–0.304575	–0.609149	0.00096531	0.00579186	S
T,4	–0.245208	–0.356272	–0.712545	0.00386124		S
9,5	–0.238402	–0.356830	–0.713660	0.00096531		S
8,6	–0.215060	–0.347797	–0.695594	0.00096531	0.00579186	S
T,5	–0.241933	–0.430760	–0.861519	0.00386124		S
9,6	–0.240883	–0.433334	–0.866667	0.00096531		S
8,7	–0.218761	–0.382357	–0.764715	0.00096531	0.00579186	S
T,6	–0.244504	–0.462853	–0.925707	0.00386124		S
9,7	–0.244676	–0.460439	–0.920878	0.00096531	0.00482655	S
T,7	–0.121108	–0.537656	–1.075312	0.00386124		S
9,8	–0.122920	–0.536738	–1.073475	0.00096531	0.00482655	S
T,8	0.135861	–0.634420	–1.268839	0.00386124	0.00386124	S
T,9	0.377418	–0.712929	–1.425859	0.00386124	0.00386124	S
A,2	–0.237101	0.071245	0.030644	0.00096531	0.00096531	H
A,3	–0.237271	0.045042	0.012926	0.00072398	0.00072398	H
A,4	–0.219780	0.025231	0.006237	0.00096531	0.00096531	H
A,5	–0.216499	–0.000090	–0.015728	0.00096531	0.00096531	H
A,6	–0.094505	0.037264	0.074528	0.00096531	0.00096531	D
A,7	0.160897	0.093377	0.186753	0.00096531	0.00096531	D
A,8	0.415527	0.172193	0.344387	0.00096531	0.00096531	S
A,9	0.641414	0.195695	0.391390	0.00096531	0.00096531	S
A,T	1.500000	0.267978	0.535956	0.00386124	0.00386124	S

Hand	Stand	Hit	Double	Spl1	Spl2	Spl3	Freq.	BS
				NDAS				
A,A	–0.229679	0.107702	0.056303	0.612564	0.682623	0.686424	0.00036199	P
2,2	–0.245297	–0.079329	–0.490594	–0.069856	–0.068972	–0.068920	0.00036199	P
3,3	–0.243798	–0.116683	–0.467619	–0.125857	–0.125913	–0.125913	0.00018100	H
4,4	–0.207855	0.027952	–0.082649	–0.121599	–0.132912	–0.133471	0.00036199	H
5,5	–0.200756	0.253973	0.507946	–0.145279	–0.176913	–0.178540	0.00036199	D
6,6	–0.205760	–0.222063	–0.444125	–0.129683	–0.118458	–0.117641	0.00036199	P
7,7	–0.214283	–0.389122	–0.778245	–0.083600	–0.071043	–0.070266	0.00036199	P
8,8	–0.223307	–0.451354	–0.902708	0.072570	0.098224	0.099670	0.00036199	P
9,9	0.114107	–0.639033	–1.278067	0.167755	0.170549	0.170602	0.00036199	P
T,T	0.632230	–0.846392	–1.692784	0.362148	0.214355	0.123301	0.00723982	S
				DAS				
A,A	–0.229679	0.107702	0.056303	0.612564	0.682623	0.686424	0.00036199	P
2,2	–0.245297	–0.079329	–0.490594	0.029281	0.038804	0.039347	0.00036199	P
3,3	–0.243798	–0.116683	–0.467619	–0.025771	–0.021525	–0.021525	0.00018100	P
4,4	–0.207855	0.027952	–0.082649	–0.014174	–0.016045	–0.016066	0.00036199	H
5,5	–0.200756	0.253973	0.507946	–0.055119	–0.101889	–0.104421	0.00036199	D
6,6	–0.205760	–0.222063	–0.444125	–0.010624	0.011641	0.013124	0.00036199	P
7,7	–0.214283	–0.389122	–0.778245	0.025487	0.048264	0.049661	0.00036199	P
8,8	–0.223307	–0.451354	–0.902708	0.152655	0.185306	0.187152	0.00036199	P
9,9	0.114107	–0.639033	–1.278067	0.208952	0.214897	0.215105	0.00036199	P
T,T	0.632230	–0.846392	–1.692784	0.362148	0.214355	0.123301	0.00723982	S

TABLE A15
1D H17: Dealer's Upcard 4

Hand	Stand	Hit	Double	Freq.	Cum. Freq.	BS
3,2	–0.170623	–0.037830	–0.341245	0.00096531	0.00096531	H
4,2	–0.166851	–0.051734	–0.313959	0.00072398	0.00072398	H
5,2	–0.164521	–0.009423	–0.190654	0.00096531		H
4,3	–0.148930	–0.014907	–0.171802	0.00072398	0.00168929	H
6,2	–0.166343	0.079904	0.022391	0.00096531		H
5,3	–0.146102	0.085175	0.040128	0.00096531	0.00193062	H
7,2	–0.171803	0.166619	0.270291	0.00096531		D
6,3	–0.148832	0.184771	0.288025	0.00096531		D
5,4	–0.142106	0.188471	0.295635	0.00072398	0.00265460	D
8,2	–0.197595	0.251944	0.503889	0.00096531		D
7,3	–0.154356	0.284502	0.569003	0.00096531		D
6,4	–0.144950	0.292106	0.584213	0.00072398	0.00265460	D
9,2	–0.202220	0.299913	0.599827	0.00096531		D
8,3	–0.180185	0.312578	0.625157	0.00096531		D
7,4	–0.150632	0.342768	0.685536	0.00072398		D
6,5	–0.142668	0.351794	0.703588	0.00096531	0.00361991	D
T,2	–0.206776	–0.194247	–0.388494	0.00386124		H
9,3	–0.184834	–0.225837	–0.451674	0.00096531		S
8,4	–0.176327	–0.214210	–0.428419	0.00072398		S
7,5	–0.148272	–0.191879	–0.383759	0.00096531	0.00651584	S
T,3	–0.190360	–0.258650	–0.517300	0.00386124		S
9,4	–0.182023	–0.254754	–0.509507	0.00072398		S
8,5	–0.175045	–0.292257	–0.584514	0.00096531		S
7,6	–0.152149	–0.277486	–0.554972	0.00096531	0.00651584	S
T,4	–0.186549	–0.337146	–0.674293	0.00289593		S
9,5	–0.179709	–0.336438	–0.672875	0.00096531		S
8,6	–0.177915	–0.338520	–0.677040	0.00096531	0.00482655	S
T,5	–0.184164	–0.419369	–0.838737	0.00386124		S
9,6	–0.182514	–0.421061	–0.842122	0.00096531		S
8,7	–0.183482	–0.381513	–0.763027	0.00096531	0.00579186	S
T,6	–0.187014	–0.456305	–0.912610	0.00386124		S
9,7	–0.188117	–0.454683	–0.909367	0.00096531	0.00482655	S
T,7	–0.067021	–0.536662	–1.073324	0.00386124		S
9,8	–0.087137	–0.542657	–1.085314	0.00096531	0.00482655	S
T,8	0.153963	–0.598217	–1.196433	0.00386124	0.00386124	S
T,9	0.396530	–0.712077	–1.424155	0.00386124	0.00386124	S
A,2	–0.182251	0.111106	0.117425	0.00096531	0.00096531	D
A,3	–0.164807	0.092459	0.112854	0.00096531	0.00096531	D
A,4	–0.161030	0.063560	0.089066	0.00072398	0.00072398	D
A,5	–0.158644	0.040197	0.066862	0.00096531	0.00096531	D
A,6	–0.038255	0.077644	0.155288	0.00096531	0.00096531	D
A,7	0.196651	0.154903	0.309807	0.00096531	0.00096531	D
A,8	0.409950	0.185581	0.371163	0.00096531	0.00096531	S
A,9	0.650575	0.229065	0.458130	0.00096531	0.00096531	S
A,T	1.500000	0.297356	0.594712	0.00386124	0.00386124	S

Hand	Stand	Hit	Double	Spl1	Spl2	Spl3	Freq.	BS
				NDAS				
A,A	–0.174886	0.142669	0.138475	0.668294	0.738235	0.742017	0.00036199	P
2,2	–0.188494	–0.032968	–0.376988	–0.004054	–0.001782	–0.001671	0.00036199	P
3,3	–0.152722	–0.042778	–0.289322	0.016358	0.022602	0.023011	0.00036199	P
4,4	–0.144654	0.096628	0.043836	–0.005046	–0.008641	–0.008641	0.00018100	H
5,5	–0.139802	0.293960	0.587920	–0.030646	–0.055746	–0.056995	0.00036199	D
6,6	–0.145580	–0.190010	–0.380020	–0.020176	–0.007440	–0.006621	0.00036199	P
7,7	–0.157794	–0.369828	–0.739655	0.017472	0.033856	0.034844	0.00036199	P
8,8	–0.209220	–0.462849	–0.925698	0.101430	0.126787	0.128131	0.00036199	P
9,9	0.156842	–0.597885	–1.195770	0.250300	0.256768	0.257049	0.00036199	P
T,T	0.640239	–0.846366	–1.692732	0.415927	0.289026	0.208732	0.00723982	S
				DAS				
A,A	–0.174886	0.142669	0.138475	0.668294	0.738235	0.742017	0.00036199	P
2,2	–0.188494	–0.032968	–0.376988	0.110270	0.122473	0.123145	0.00036199	P
3,3	–0.152722	–0.042778	–0.289322	0.142527	0.159856	0.160898	0.00036199	P
4,4	–0.144654	0.096628	0.043836	0.131310	0.133883	0.133883	0.00018100	P
5,5	–0.139802	0.293960	0.587920	0.078149	0.036061	0.033798	0.00036199	D
6,6	–0.145580	–0.190010	–0.380020	0.118810	0.144415	0.146009	0.00036199	P
7,7	–0.157794	–0.369828	–0.739655	0.156801	0.186006	0.187760	0.00036199	P
8,8	–0.209220	–0.462849	–0.925698	0.191501	0.224083	0.225803	0.00036199	P
9,9	0.156842	–0.597885	–1.195770	0.296817	0.306915	0.307379	0.00036199	P
T,T	0.640239	–0.846366	–1.692732	0.415927	0.289026	0.208732	0.00723982	S

TABLE A16
1D H17: Dealer's Upcard 5

Hand	Stand	Hit	Double	Freq.	Cum. Freq.	BS
3,2	–0.101809	0.022930	–0.203618	0.00096531	0.00096531	H
4,2	–0.098911	0.010470	–0.180887	0.00096531	0.00096531	H
5,2	–0.096017	0.057090	–0.068370	0.00072398		H
4,3	–0.095520	0.049485	–0.059167	0.00096531	0.00168929	H
6,2	–0.101494	0.129947	0.130021	0.00096531		D
5,3	–0.093126	0.140423	0.154514	0.00072398	0.00168929	D
7,2	–0.128271	0.194894	0.331200	0.00096531		D
6,3	–0.098567	0.234370	0.390834	0.00096531		D
5,4	–0.090096	0.245872	0.414311	0.00072398	0.00265460	D
8,2	–0.133267	0.295645	0.591290	0.00096531		D
7,3	–0.125429	0.307167	0.614334	0.00096531		D
6,4	–0.095632	0.338493	0.676986	0.00096531	0.00289593	D
9,2	–0.136389	0.339627	0.679254	0.00096531		D
8,3	–0.130367	0.349405	0.698810	0.00096531		D
7,4	–0.122425	0.361633	0.723267	0.00096531		D
6,5	–0.093306	0.393455	0.786911	0.00072398	0.00361991	D
T,2	–0.141673	–0.163741	–0.327482	0.00386124		S
9,3	–0.134579	–0.196768	–0.393536	0.00096531		S
8,4	–0.128450	–0.186713	–0.373426	0.00096531		S
7,5	–0.121146	–0.177804	–0.355609	0.00072398	0.00651584	S
T,3	–0.138759	–0.235585	–0.471170	0.00386124		S
9,4	–0.131592	–0.232262	–0.464525	0.00096531		S
8,5	–0.126072	–0.270202	–0.540404	0.00072398		S
7,6	–0.126642	–0.269218	–0.538436	0.00096531	0.00651584	S
T,4	–0.135782	–0.314985	–0.629970	0.00386124		S
9,5	–0.129192	–0.314270	–0.628539	0.00072398		S
8,6	–0.131586	–0.317650	–0.635301	0.00096531	0.00555053	S
T,5	–0.133391	–0.400250	–0.800500	0.00289593		S
9,6	–0.134717	–0.402728	–0.805456	0.00096531		S
8,7	–0.158326	–0.373822	–0.747644	0.00096531	0.00482655	S
T,6	–0.138934	–0.445008	–0.890016	0.00386124		S
9,7	–0.161486	–0.452834	–0.905668	0.00096531	0.00482655	S
T,7	–0.044237	–0.493048	–0.986096	0.00386124		S
9,8	–0.045521	–0.493184	–0.986369	0.00096531	0.00482655	S
T,8	0.198262	–0.591096	–1.182192	0.00386124	0.00386124	S
T,9	0.444808	–0.708872	–1.417744	0.00386124	0.00386124	S
A,2	–0.117634	0.158994	0.212994	0.00096531	0.00096531	D
A,3	–0.114802	0.136909	0.204352	0.00096531	0.00096531	D
A,4	–0.111873	0.108295	0.175804	0.00096531	0.00096531	D
A,5	–0.109565	0.082543	0.149096	0.00072398	0.00072398	D
A,6	0.004205	0.140020	0.280040	0.00096531	0.00096531	D
A,7	0.220355	0.174120	0.348240	0.00096531	0.00096531	D
A,8	0.459289	0.226219	0.452437	0.00096531	0.00096531	S
A,9	0.681170	0.267929	0.535859	0.00096531	0.00096531	S
A,T	1.500000	0.331386	0.662773	0.00386124	0.00386124	S

Hand	Stand	Hit	Double	Spl1	Spl2	Spl3	Freq.	BS
				NDAS				
A,A	–0.129684	0.182058	0.215933	0.732081	0.803521	0.807386	0.00036199	P
2,2	–0.104579	0.037203	–0.209157	0.129807	0.138207	0.138701	0.00036199	P
3,3	–0.098994	0.009903	–0.179332	0.119086	0.129752	0.130415	0.00036199	P
4,4	–0.092410	0.153362	0.162098	0.097580	0.094114	0.093994	0.00036199	D
5,5	–0.087779	0.346903	0.693805	0.070616	0.059848	0.059848	0.00018100	D
6,6	–0.099958	–0.162349	–0.324698	0.070145	0.086559	0.087575	0.00036199	P
7,7	–0.153359	–0.370802	–0.741605	0.039012	0.054956	0.055815	0.00036199	P
8,8	–0.163322	–0.453646	–0.907293	0.186257	0.214750	0.216259	0.00036199	P
9,9	0.198919	–0.590647	–1.181293	0.338061	0.348327	0.348814	0.00036199	P
T,T	0.671820	–0.845624	–1.691247	0.495996	0.393348	0.326844	0.00723982	S
				DAS				
A,A	–0.129684	0.182058	0.215933	0.732081	0.803521	0.807386	0.00036199	P
2,2	–0.104579	0.037203	–0.209157	0.278443	0.300226	0.301500	0.00036199	P
3,3	–0.098994	0.009903	–0.179332	0.279207	0.304270	0.305773	0.00036199	P
4,4	–0.092410	0.153362	0.162098	0.259384	0.267434	0.267860	0.00036199	P
5,5	–0.087779	0.346903	0.693805	0.207802	0.187390	0.187390	0.00018100	D
6,6	–0.099958	–0.162349	–0.324698	0.243835	0.276450	0.278447	0.00036199	P
7,7	–0.153359	–0.370802	–0.741605	0.193472	0.222301	0.223859	0.00036199	P
8,8	–0.163322	–0.453646	–0.907293	0.289175	0.325919	0.327857	0.00036199	P
9,9	0.198919	–0.590647	–1.181293	0.391108	0.405544	0.406243	0.00036199	P
T,T	0.671820	–0.845624	–1.691247	0.495996	0.393348	0.326844	0.00723982	S

TABLE A17
1D H17: Dealer's Upcard 6

Hand	Stand	Hit	Double	Freq.	Cum. Freq.	BS
3,2	–0.084250	0.040538	–0.168499	0.00096531	0.00096531	H
4,2	–0.081310	0.036856	–0.145841	0.00096531	0.00096531	H
5,2	–0.081684	0.072187	–0.036149	0.00096531		H
4,3	–0.078446	0.064515	–0.028223	0.00096531	0.00193062	H
6,2	–0.108450	0.119677	0.111237	0.00072398		H
5,3	–0.078830	0.151810	0.178469	0.00096531	0.00168929	D
7,2	–0.113461	0.199017	0.340632	0.00096531		D
6,3	–0.105616	0.220973	0.364869	0.00072398		D
5,4	–0.076320	0.253610	0.430362	0.00096531	0.00265460	D
8,2	–0.116385	0.301488	0.602977	0.00096531		D
7,3	–0.110684	0.312070	0.624139	0.00096531		D
6,4	–0.103164	0.325778	0.651556	0.00072398	0.00265460	D
9,2	–0.121773	0.344152	0.688303	0.00096531		D
8,3	–0.114577	0.355332	0.710664	0.00096531		D
7,4	–0.109126	0.366308	0.732615	0.00096531		D
6,5	–0.104621	0.378909	0.757817	0.00072398	0.00361991	D
T,2	–0.123948	–0.160845	–0.321691	0.00386124		S
9,3	–0.118933	–0.194949	–0.389897	0.00096531		S
8,4	–0.112021	–0.184372	–0.368744	0.00096531		S
7,5	–0.109542	–0.178756	–0.357511	0.00096531	0.00675716	S
T,3	–0.121151	–0.233285	–0.466569	0.00386124		S
9,4	–0.116390	–0.231222	–0.462445	0.00096531		S
8,5	–0.112403	–0.270568	–0.541136	0.00096531		S
7,6	–0.136248	–0.283701	–0.567402	0.00072398	0.00651584	S
T,4	–0.118584	–0.313423	–0.626847	0.00386124		S
9,5	–0.116774	–0.315459	–0.630918	0.00096531		S
8,6	–0.139141	–0.330418	–0.660837	0.00072398	0.00555053	S
T,5	–0.118969	–0.397574	–0.795149	0.00386124		S
9,6	–0.143531	–0.411515	–0.823030	0.00072398		S
8,7	–0.144146	–0.372986	–0.745971	0.00096531	0.00555053	S
T,6	–0.144315	–0.409144	–0.818288	0.00289593		S
9,7	–0.147125	–0.409917	–0.819833	0.00096531	0.00386124	S
T,7	–0.030494	–0.494208	–0.988415	0.00386124		S
9,8	–0.031055	–0.494406	–0.988813	0.00096531	0.00482655	S
T,8	0.203123	–0.592371	–1.184742	0.00386124	0.00386124	S
T,9	0.436459	–0.709602	–1.419204	0.00386124	0.00386124	S
A,2	–0.086202	0.173799	0.245963	0.00096531	0.00096531	D
A,3	–0.083459	0.154385	0.239239	0.00096531	0.00096531	D
A,4	–0.080980	0.129642	0.219050	0.00096531	0.00096531	D
A,5	–0.080578	0.124604	0.233116	0.00096531	0.00096531	D
A,6	–0.001906	0.131582	0.263165	0.00072398	0.00072398	D
A,7	0.215166	0.181045	0.362090	0.00096531	0.00096531	D
A,8	0.446965	0.233260	0.466520	0.00096531	0.00096531	D
A,9	0.673021	0.275419	0.550838	0.00096531	0.00096531	S
A,T	1.500000	0.339582	0.679165	0.00386124	0.00386124	S

NDAS

Hand	Stand	Hit	Double	Spl1	Spl2	Spl3	Freq.	BS
A,A	–0.084955	0.201887	0.258415	0.755927	0.825003	0.828699	0.00036199	P
2,2	–0.087026	0.050745	–0.174053	0.157711	0.167538	0.168121	0.00036199	P
3,3	–0.081029	0.036723	–0.145223	0.147530	0.158603	0.159308	0.00036199	P
4,4	–0.075873	0.164731	0.185687	0.125980	0.124034	0.124000	0.00036199	D
5,5	–0.076807	0.354212	0.708425	0.092770	0.072218	0.071169	0.00036199	D
6,6	–0.131294	–0.192225	–0.384450	0.023565	0.030264	0.030264	0.00018100	P
7,7	–0.141234	–0.374367	–0.748733	0.051649	0.067398	0.068233	0.00036199	P
8,8	–0.145647	–0.409377	–0.818754	0.202707	0.231295	0.232828	0.00036199	P
9,9	0.201291	–0.592827	–1.185654	0.340087	0.350152	0.350618	0.00036199	P
T,T	0.668303	–0.845644	–1.691287	0.507561	0.413036	0.351471	0.00723982	S

DAS

Hand	Stand	Hit	Double	Spl1	Spl2	Spl3	Freq.	BS
A,A	–0.084955	0.201887	0.258415	0.755927	0.825003	0.828699	0.00036199	P
2,2	–0.087026	0.050745	–0.174053	0.312213	0.336031	0.337435	0.00036199	P
3,3	–0.081029	0.036723	–0.145223	0.310796	0.336907	0.338511	0.00036199	P
4,4	–0.075873	0.164731	0.185687	0.288536	0.297290	0.297766	0.00036199	P
5,5	–0.076807	0.354212	0.708425	0.236913	0.197055	0.194848	0.00036199	D
6,6	–0.131294	–0.192225	–0.384450	0.185416	0.198974	0.198974	0.00018100	P
7,7	–0.141234	–0.374367	–0.748733	0.210090	0.238906	0.240441	0.00036199	P
8,8	–0.145647	–0.409377	–0.818754	0.311458	0.348893	0.350896	0.00036199	P
9,9	0.201291	–0.592827	–1.185654	0.393660	0.407920	0.408599	0.00036199	P
T,T	0.668303	–0.845644	–1.691287	0.507561	0.413036	0.351471	0.00723982	S

TABLE A18
2D S17: Dealer's Upcard A

Hand	Stand	Hit	Double	Freq.	Cum. Freq.	BS
3,2	–0.663299	–0.285578	–1.326599	0.00064026	0.00064026	EH
4,2	–0.662671	–0.319289	–1.299930	0.00064026	0.00064026	EH
5,2	–0.661890	–0.320599	–1.127296	0.00064026		EH
4,3	–0.662066	–0.326938	–1.127698	0.00064026	0.00128052	EH
6,2	–0.653821	–0.211121	–0.805273	0.00064026		H
5,3	–0.661280	–0.206673	–0.801968	0.00064026	0.00128052	H
7,2	–0.656789	–0.073622	–0.418039	0.00064026		H
6,3	–0.653202	–0.072416	–0.417815	0.00064026		H
5,4	–0.660323	–0.067698	–0.413790	0.00064026	0.00192078	H
8,2	–0.659758	0.083968	0.004847	0.00064026		H
7,3	–0.656175	0.084047	0.011980	0.00064026		H
6,4	–0.652241	0.081243	0.016240	0.00064026	0.00192078	H
9,2	–0.662504	0.155791	0.147401	0.00064026		H
8,3	–0.658982	0.156258	0.155381	0.00064026		H
7,4	–0.655051	0.156471	0.163389	0.00064026		D
6,5	–0.651264	0.157240	0.172433	0.00064026	0.00256104	D
T,2	–0.670122	–0.349577	–0.820297	0.00259816		EH
9,3	–0.661939	–0.362141	–0.843287	0.00064026		EH
8,4	–0.658063	–0.362534	–0.837471	0.00064026		EH
7,5	–0.654283	–0.363070	–0.832919	0.00064026	0.00451894	EH
T,3	–0.669563	–0.394894	–0.870100	0.00259816		EH
9,4	–0.661026	–0.389964	–0.859604	0.00064026		EH
8,5	–0.657308	–0.414176	–0.904482	0.00064026		EH
7,6	–0.646546	–0.419926	–0.908861	0.00064026	0.00451894	EH
T,4	–0.668670	–0.442320	–0.931519	0.00259816		EH
9,5	–0.660269	–0.440271	–0.925624	0.00064026		EH
8,6	–0.649597	–0.446625	–0.936751	0.00064026	0.00387868	EH
T,5	–0.667926	–0.489007	–0.997133	0.00259816		EH
9,6	–0.652589	–0.487665	–0.995234	0.00064026		EH
8,7	–0.652642	–0.468091	–0.956136	0.00064026	0.00387868	EH
T,6	–0.660328	–0.513042	–1.026085	0.00259816		RH
9,7	–0.655571	–0.506503	–1.013005	0.00064026	0.00323842	RH
T,7	–0.472736	–0.556549	–1.113099	0.00259816		ES
9,8	–0.465325	–0.550940	–1.101880	0.00064026	0.00323842	ES
T,8	–0.091302	–0.629699	–1.259397	0.00259816	0.00259816	S
T,9	0.292376	–0.733384	–1.466767	0.00259816	0.00259816	S
A,2	–0.664614	–0.062819	–0.609422	0.00056023	0.00056023	H
A,3	–0.664036	–0.097891	–0.610525	0.00056023	0.00056023	H
A,4	–0.663103	–0.142028	–0.622335	0.00056023	0.00056023	H
A,5	–0.662640	–0.185372	–0.639449	0.00056023	0.00056023	H
A,6	–0.480381	–0.190002	–0.533008	0.00056023	0.00056023	H
A,7	–0.100502	–0.100359	–0.360929	0.00056023	0.00056023	H
A,8	0.283692	–0.011208	–0.190714	0.00056023	0.00056023	S
A,9	0.667985	0.074708	–0.027343	0.00056023	0.00056023	S
A,T	1.500000	0.147647	0.126302	0.00227339	0.00227339	S

NDAS

Hand	Stand	Hit	Double	Spl1	Spl2	Spl3	Freq.	BS
A,A	–0.665322	–0.025783	–0.611808	0.164548	0.246111	0.254238	0.00021009	P
2,2	–0.663878	–0.256368	–1.327756	–0.416884	–0.435449	–0.437713	0.00028011	H
3,3	–0.663014	–0.319065	–1.300533	–0.464880	–0.480922	–0.482768	0.00028011	EH
4,4	–0.661114	–0.202472	–0.802584	–0.525748	–0.563828	–0.568555	0.00028011	H
5,5	–0.659528	0.085850	0.020818	–0.599731	–0.681418	–0.691677	0.00028011	H
6,6	–0.643494	–0.368407	–0.833531	–0.624056	–0.653285	–0.656803	0.00028011	EH
7,7	–0.649594	–0.466588	–0.979387	–0.616023	–0.630713	–0.632173	0.00028011	EH
8,8	–0.655627	–0.506225	–1.012450	–0.367828	–0.350398	–0.348097	0.00028011	EP
9,9	–0.078374	–0.625411	–1.250822	–0.102473	–0.104102	–0.104156	0.00028011	S
T,T	0.652797	–0.867286	–1.734572	0.131546	–0.139316	–0.301586	0.00510585	S

DAS

Hand	Stand	Hit	Double	Spl1	Spl2	Spl3	Freq.	BS
A,A	–0.665322	–0.025783	–0.611808	0.164548	0.246111	0.254238	0.00021009	P
2,2	–0.663878	–0.256368	–1.327756	–0.416717	–0.435090	–0.437309	0.00028011	H
3,3	–0.663014	–0.319065	–1.300533	–0.463421	–0.479112	–0.480893	0.00028011	EH
4,4	–0.661114	–0.202472	–0.802584	–0.522852	–0.560393	–0.565029	0.00028011	H
5,5	–0.659528	0.085850	0.020818	–0.595172	–0.676086	–0.686222	0.00028011	H
6,6	–0.643494	–0.368407	–0.833531	–0.619241	–0.647621	–0.650998	0.00028011	EH
7,7	–0.649594	–0.466588	–0.979387	–0.613519	–0.627751	–0.629134	0.00028011	EH
8,8	–0.655627	–0.506225	–1.012450	–0.367614	–0.350120	–0.347806	0.00028011	EP
9,9	–0.078374	–0.625411	–1.250822	–0.104552	–0.106516	–0.106622	0.00028011	S
T,T	0.652797	–0.867286	–1.734572	0.131546	–0.139316	–0.301586	0.00510585	S

TABLE A19

2D S17: Dealer's Upcard 2

Hand	Stand	Hit	Double	Freq.	Cum. Freq.	BS
3,2	−0.292351	−0.129507	−0.584701	0.00082004	0.00082004	H
4,2	−0.291341	−0.145047	−0.561064	0.00082004	0.00082004	H
5,2	−0.283790	−0.104574	−0.419082	0.00082004		H
4,3	−0.291503	−0.114691	−0.430417	0.00093719	0.00175724	H
6,2	−0.285228	−0.017231	−0.184049	0.00082004		H
5,3	−0.283952	−0.018463	−0.185449	0.00093719	0.00175724	H
7,2	−0.286745	0.079254	0.088700	0.00082004		D
6,3	−0.285365	0.083495	0.086625	0.00093719		D
5,4	−0.282928	0.084468	0.087997	0.00093719	0.00269443	D
8,2	−0.287955	0.196786	0.389971	0.00082004		D
7,3	−0.286695	0.198063	0.391767	0.00093719		D
6,4	−0.284151	0.199749	0.395121	0.00093719	0.00269443	D
9,2	−0.290042	0.250350	0.496301	0.00082004		D
8,3	−0.288107	0.252694	0.500818	0.00093719		D
7,4	−0.285679	0.255053	0.505502	0.00093719		D
6,5	−0.276683	0.260707	0.517673	0.00093719	0.00363162	D
T,2	−0.301915	−0.248636	−0.497272	0.00328018		H
9,3	−0.290193	−0.259673	−0.519346	0.00093719		H
8,4	−0.287108	−0.256480	−0.512959	0.00093719		H
7,5	−0.278224	−0.253047	−0.506094	0.00093719	0.00609176	H
T,3	−0.302064	−0.305893	−0.611786	0.00374877		S
9,4	−0.289194	−0.300447	−0.600893	0.00093719		S
8,5	−0.279649	−0.319275	−0.638550	0.00093719		S
7,6	−0.279679	−0.319640	−0.639279	0.00093719	0.00656035	S
T,4	−0.301062	−0.365213	−0.730427	0.00374877		S
9,5	−0.281728	−0.360624	−0.721248	0.00093719		S
8,6	−0.281082	−0.362539	−0.725079	0.00093719	0.00562316	S
T,5	−0.293842	−0.426093	−0.852186	0.00374877		S
9,6	−0.283417	−0.423290	−0.846580	0.00093719		S
8,7	−0.282861	−0.404069	−0.808138	0.00093719	0.00562316	S
T,6	−0.295283	−0.468273	−0.936546	0.00374877		S
9,7	−0.284957	−0.463808	−0.927616	0.00093719	0.00468597	S
T,7	−0.155550	−0.537254	−1.074508	0.00374877		S
9,8	−0.145084	−0.533426	−1.066852	0.00093719	0.00468597	S
T,8	0.120145	−0.627392	−1.254785	0.00374877	0.00374877	S
T,9	0.385790	−0.739082	−1.478164	0.00374877	0.00374877	S
A,2	−0.287971	0.043116	−0.056759	0.00082004	0.00082004	H
A,3	−0.288130	0.019708	−0.058773	0.00093719	0.00093719	H
A,4	−0.287392	−0.005643	−0.069976	0.00093719	0.00093719	H
A,5	−0.279842	−0.026423	−0.077004	0.00093719	0.00093719	H
A,6	−0.142686	0.003148	0.002556	0.00093719	0.00093719	H
A,7	0.128632	0.064072	0.123539	0.00093719	0.00093719	S
A,8	0.393503	0.121926	0.239282	0.00093719	0.00093719	S
A,9	0.647449	0.186157	0.368465	0.00093719	0.00093719	S
A,T	1.500000	0.242549	0.478266	0.00374877	0.00374877	S

Hand	Stand	Hit	Double	Spl1	Spl2	Spl3	Freq.	BS
				NDAS				
A,A	−0.283862	0.088038	−0.045768	0.517050	0.611110	0.622750	0.00041002	P
2,2	−0.291988	−0.114288	−0.583976	−0.140722	−0.143216	−0.143454	0.00030752	H
3,3	−0.292519	−0.145707	−0.563234	−0.200175	−0.206066	−0.206733	0.00041002	H
4,4	−0.290478	−0.016406	−0.192747	−0.230667	−0.255849	−0.258971	0.00041002	H
5,5	−0.275251	0.202855	0.402068	−0.244071	−0.295835	−0.302148	0.00041002	D
6,6	−0.278126	−0.253133	−0.506266	−0.252332	−0.250199	−0.249678	0.00041002	P
7,7	−0.281210	−0.383726	−0.767452	−0.191524	−0.180102	−0.178579	0.00041002	P
8,8	−0.284276	−0.463093	−0.926185	−0.015171	0.017261	0.021377	0.00041002	P
9,9	0.129292	−0.624946	−1.249892	0.159542	0.163110	0.163551	0.00041002	P
T,T	0.633650	−0.851075	−1.702151	0.338544	0.181498	0.085421	0.00726325	S
				DAS				
A,A	−0.283862	0.088038	−0.045768	0.517050	0.611110	0.622750	0.00041002	P
2,2	−0.291988	−0.114288	−0.583976	−0.066852	−0.061832	−0.061296	0.00030752	P
3,3	−0.292519	−0.145707	−0.563234	−0.127839	−0.125061	−0.124634	0.00041002	P
4,4	−0.290478	−0.016406	−0.192747	−0.157469	−0.173866	−0.175880	0.00041002	H
5,5	−0.275251	0.202855	0.402068	−0.175406	−0.243261	−0.251662	0.00041002	D
6,6	−0.278126	−0.253133	−0.506266	−0.176274	−0.164788	−0.163057	0.00041002	P
7,7	−0.281210	−0.383726	−0.767452	−0.116492	−0.095900	−0.093197	0.00041002	P
8,8	−0.284276	−0.463093	−0.926185	0.052391	0.092896	0.098029	0.00041002	P
9,9	0.129292	−0.624946	−1.249892	0.193813	0.201408	0.202348	0.00041002	P
T,T	0.633650	−0.851075	−1.702151	0.338544	0.181498	0.085421	0.00726325	S

TABLE A20
2D S17: Dealer's Upcard 3

Hand	Stand	Hit	Double	Freq.	Cum. Freq.	BS
3,2	−0.249151	−0.095976	−0.498301	0.00082004	0.00082004	H
4,2	−0.241727	−0.106269	−0.463485	0.00093719	0.00093719	H
5,2	−0.240269	−0.068930	−0.334666	0.00093719		H
4,3	−0.240720	−0.078107	−0.340391	0.00082004	0.00175724	H
6,2	−0.241600	0.015863	−0.107412	0.00093719		H
5,3	−0.239263	0.014311	−0.108794	0.00082004	0.00175724	H
7,2	−0.242993	0.115558	0.156613	0.00093719		D
6,3	−0.240403	0.119563	0.154514	0.00082004		D
5,4	−0.231714	0.120857	0.158382	0.00093719	0.00269443	D
8,2	−0.245060	0.223137	0.444811	0.00093719		D
7,3	−0.241998	0.225624	0.449597	0.00082004		D
6,4	−0.233052	0.227652	0.454387	0.00093719	0.00269443	D
9,2	−0.257258	0.265096	0.528455	0.00093719		D
8,3	−0.244071	0.277219	0.553175	0.00082004		D
7,4	−0.234649	0.281844	0.563123	0.00093719		D
6,5	−0.231581	0.286269	0.572015	0.00093719	0.00363162	D
T,2	−0.259456	−0.226608	−0.453215	0.00374877		H
9,3	−0.256279	−0.244013	−0.488026	0.00082004		H
8,4	−0.236742	−0.231696	−0.463391	0.00093719		H
7,5	−0.233196	−0.228124	−0.456247	0.00093719	0.00644320	H
T,3	−0.258473	−0.287062	−0.574125	0.00328018		S
9,4	−0.248934	−0.285580	−0.571160	0.00093719		S
8,5	−0.235266	−0.298249	−0.596498	0.00093719		S
7,6	−0.234541	−0.297699	−0.595398	0.00093719	0.00609176	S
T,4	−0.251371	−0.351911	−0.703821	0.00374877		S
9,5	−0.247705	−0.351964	−0.703929	0.00093719		S
8,6	−0.236867	−0.348022	−0.696044	0.00093719	0.00562316	S
T,5	−0.249898	−0.417521	−0.835042	0.00374877		S
9,6	−0.249063	−0.418669	−0.837338	0.00093719		S
8,7	−0.238466	−0.394344	−0.788688	0.00093719	0.00562316	S
T,6	−0.251265	−0.462601	−0.925202	0.00374877		S
9,7	−0.250671	−0.461374	−0.922747	0.00093719	0.00468597	S
T,7	−0.118078	−0.534054	−1.068109	0.00374877		S
9,8	−0.118292	−0.533478	−1.066955	0.00093719	0.00468597	S
T,8	0.146843	−0.626688	−1.253375	0.00374877	0.00374877	S
T,9	0.394575	−0.720499	−1.440999	0.00374877	0.00374877	S
A,2	−0.246019	0.072623	0.011561	0.00093719	0.00093719	H
A,3	−0.245298	0.047784	0.003422	0.00082004	0.00082004	H
A,4	−0.237872	0.025949	−0.001697	0.00093719	0.00093719	H
A,5	−0.236410	0.003169	−0.012975	0.00093719	0.00093719	H
A,6	−0.105529	0.032292	0.064061	0.00093719	0.00093719	D
A,7	0.157066	0.091750	0.182934	0.00093719	0.00093719	D
A,8	0.411521	0.160480	0.319854	0.00093719	0.00093719	S
A,9	0.647448	0.201679	0.401620	0.00093719	0.00093719	S
A,T	1.500000	0.265419	0.526803	0.00374877	0.00374877	S

Hand	Stand	Hit	Double	Spl1	Spl2	Spl3	Freq.	BS
				NDAS				
A,A	−0.242022	0.111397	0.023580	0.564482	0.659158	0.670869	0.00041002	P
2,2	−0.250141	−0.081419	−0.500282	−0.084518	−0.084817	−0.084847	0.00041002	H
3,3	−0.248154	−0.111120	−0.475815	−0.131309	−0.132909	−0.133031	0.00030752	H
4,4	−0.233290	0.018454	−0.108279	−0.145577	−0.164133	−0.166343	0.00041002	H
5,5	−0.230232	0.229862	0.458808	−0.170726	−0.216891	−0.222491	0.00041002	D
6,6	−0.232945	−0.228296	−0.456592	−0.177479	−0.169309	−0.168018	0.00041002	P
7,7	−0.236374	−0.368221	−0.736441	−0.119858	−0.105079	−0.103113	0.00041002	P
8,8	−0.240562	−0.457223	−0.914446	0.046012	0.080433	0.084786	0.00041002	P
9,9	0.136895	−0.628673	−1.257346	0.188392	0.193382	0.193865	0.00041002	P
T,T	0.643327	−0.850776	−1.701552	0.388890	0.252693	0.168950	0.00726325	S
				DAS				
A,A	−0.242022	0.111397	0.023580	0.564482	0.659158	0.670869	0.00041002	P
2,2	−0.250141	−0.081419	−0.500282	0.001401	0.011448	0.012730	0.00041002	P
3,3	−0.248154	−0.111120	−0.475815	−0.044814	−0.037776	−0.037023	0.00030752	P
4,4	−0.233290	0.018454	−0.108279	−0.055725	−0.063380	−0.064198	0.00041002	H
5,5	−0.230232	0.229862	0.458808	−0.087943	−0.152140	−0.160083	0.00041002	D
6,6	−0.232945	−0.228296	−0.456592	−0.079992	−0.059664	−0.056778	0.00041002	P
7,7	−0.236374	−0.368221	−0.736441	−0.027066	−0.000665	0.002831	0.00041002	P
8,8	−0.240562	−0.457223	−0.914446	0.121468	0.164959	0.170461	0.00041002	P
9,9	0.136895	−0.628673	−1.257346	0.228919	0.238506	0.239537	0.00041002	P
T,T	0.643327	−0.850776	−1.701552	0.388890	0.252693	0.168950	0.00726325	S

TABLE A21
2D S17: Dealer's Upcard 4

Hand	Stand	Hit	Double	Freq.	Cum. Freq.	BS
3,2	–0.193707	–0.051289	–0.387414	0.00093719	0.00093719	H
4,2	–0.192171	–0.064105	–0.364064	0.00082004	0.00082004	H
5,2	–0.190986	–0.026981	–0.239811	0.00093719		H
4,3	–0.184757	–0.030159	–0.232010	0.00082004	0.00175724	H
6,2	–0.192144	0.059196	–0.023330	0.00093719		H
5,3	–0.183453	0.061992	–0.014615	0.00093719	0.00187439	H
7,2	–0.194453	0.147757	0.225479	0.00093719		D
6,3	–0.184807	0.156631	0.234050	0.00093719		D
5,4	–0.181887	0.158124	0.237088	0.00082004	0.00269443	D
8,2	–0.206627	0.241485	0.482971	0.00093719		D
7,3	–0.187123	0.257118	0.514237	0.00093719		D
6,4	–0.183255	0.260417	0.520834	0.00082004	0.00269443	D
9,2	–0.208998	0.291313	0.582625	0.00093719		D
8,3	–0.199305	0.297574	0.595148	0.00093719		D
7,4	–0.185594	0.311606	0.623212	0.00082004		D
6,5	–0.182061	0.316054	0.632109	0.00093719	0.00363162	D
T,2	–0.211303	–0.203909	–0.407818	0.00374877		H
9,3	–0.201669	–0.219254	–0.438508	0.00093719		S
8,4	–0.197754	–0.213895	–0.427789	0.00082004		S
7,5	–0.184385	–0.203351	–0.406702	0.00093719	0.00644320	S
T,3	–0.204217	–0.266109	–0.532218	0.00374877		S
9,4	–0.200360	–0.264586	–0.529172	0.00082004		S
8,5	–0.196790	–0.282888	–0.565776	0.00093719		S
7,6	–0.185996	–0.275956	–0.551912	0.00093719	0.00644320	S
T,4	–0.202671	–0.335691	–0.671383	0.00328018		S
9,5	–0.199156	–0.335434	–0.670869	0.00093719		S
8,6	–0.198167	–0.336384	–0.672769	0.00093719	0.00515456	S
T,5	–0.201463	–0.406617	–0.813233	0.00374877		S
9,6	–0.200532	–0.407487	–0.814974	0.00093719		S
8,7	–0.200502	–0.388150	–0.776299	0.00093719	0.00562316	S
T,6	–0.202842	–0.455539	–0.911078	0.00374877		S
9,7	–0.202868	–0.454804	–0.909607	0.00093719	0.00468597	S
T,7	–0.072689	–0.530970	–1.061940	0.00374877		S
9,8	–0.082337	–0.533662	–1.067324	0.00093719	0.00468597	S
T,8	0.170305	–0.607589	–1.215177	0.00374877	0.00374877	S
T,9	0.413684	–0.719545	–1.439091	0.00374877	0.00374877	S
A,2	–0.198301	0.106243	0.087099	0.00093719	0.00093719	H
A,3	–0.190888	0.085205	0.083857	0.00093719	0.00093719	H
A,4	–0.189352	0.059893	0.070813	0.00082004	0.00082004	D
A,5	–0.188156	0.038439	0.059559	0.00093719	0.00093719	D
A,6	–0.059450	0.067796	0.135592	0.00093719	0.00093719	D
A,7	0.189355	0.136662	0.273325	0.00093719	0.00093719	D
A,8	0.419907	0.181043	0.362085	0.00093719	0.00093719	S
A,9	0.657397	0.229848	0.459697	0.00093719	0.00093719	S
A,T	1.500000	0.290094	0.580187	0.00374877	0.00374877	S

Hand	Stand	Hit	Double	Spl1	Spl2	Spl3	Freq.	BS
				NDAS				
A,A	–0.195062	0.134028	0.096965	0.616546	0.711730	0.723492	0.00041002	P
2,2	–0.201117	–0.041765	–0.402234	–0.025241	–0.023270	–0.023026	0.00041002	P
3,3	–0.186291	–0.060388	–0.353751	–0.032921	–0.028782	–0.028157	0.00041002	P
4,4	–0.183112	0.067218	–0.013696	–0.057870	–0.069499	–0.070590	0.00030752	H
5,5	–0.180685	0.261478	0.522956	–0.083080	–0.122420	–0.127145	0.00041002	D
6,6	–0.183429	–0.202617	–0.405233	–0.089939	–0.077422	–0.075681	0.00041002	P
7,7	–0.188334	–0.351727	–0.703455	–0.036668	–0.017616	–0.015105	0.00041002	P
8,8	–0.212673	–0.458364	–0.916728	0.092243	0.127846	0.132223	0.00041002	P
9,9	0.170851	–0.607518	–1.215036	0.254659	0.263602	0.264599	0.00041002	P
T,T	0.653132	–0.850614	–1.701228	0.439801	0.323712	0.251329	0.00726325	S
				DAS				
A,A	–0.195062	0.134028	0.096965	0.616546	0.711730	0.723492	0.00041002	P
2,2	–0.201117	–0.041765	–0.402234	0.073848	0.087680	0.089420	0.00041002	P
3,3	–0.186291	–0.060388	–0.353751	0.070084	0.086596	0.088786	0.00041002	P
4,4	–0.183112	0.067218	–0.013696	0.051753	0.051335	0.051405	0.00030752	H
5,5	–0.180685	0.261478	0.522956	0.017276	–0.041940	–0.049247	0.00041002	D
6,6	–0.183429	–0.202617	–0.405233	0.026702	0.053740	0.057381	0.00041002	P
7,7	–0.188334	–0.351727	–0.703455	0.079975	0.113501	0.117901	0.00041002	P
8,8	–0.212673	–0.458364	–0.916728	0.176686	0.222049	0.227614	0.00041002	P
9,9	0.170851	–0.607518	–1.215036	0.299621	0.313686	0.315297	0.00041002	P
T,T	0.653132	–0.850614	–1.701228	0.439801	0.323712	0.251329	0.00726325	S

TABLE A22

2D S17: Dealer's Upcard 5

Hand	Stand	Hit	Double	Freq.	Cum. Freq.	BS
3,2	–0.137274	–0.002331	–0.274549	0.00093719	0.00093719	H
4,2	–0.136003	–0.014063	–0.253915	0.00093719	0.00093719	H
5,2	–0.134750	0.023249	–0.140025	0.00082004		H
4,3	–0.134557	0.019478	–0.135851	0.00093719	0.00175724	H
6,2	–0.137053	0.099520	0.064213	0.00093719		H
5,3	–0.133406	0.104298	0.075361	0.00082004	0.00175724	H
7,2	–0.149467	0.176412	0.286537	0.00093719		D
6,3	–0.135708	0.194917	0.314106	0.00093719		D
5,4	–0.132110	0.200126	0.324678	0.00082004	0.00269443	D
8,2	–0.151938	0.275645	0.551291	0.00093719		D
7,3	–0.148136	0.281189	0.562379	0.00093719		D
6,4	–0.134426	0.295709	0.591418	0.00093719	0.00281158	D
9,2	–0.153851	0.322873	0.645746	0.00093719		D
8,3	–0.150594	0.327851	0.655703	0.00093719		D
7,4	–0.146840	0.333599	0.667199	0.00093719		D
6,5	–0.133275	0.348473	0.696946	0.00082004	0.00363162	D
T,2	–0.156154	–0.178955	–0.357911	0.00374877		S
9,3	–0.152755	–0.195065	–0.390131	0.00093719		S
8,4	–0.149544	–0.190121	–0.380241	0.00093719		S
7,5	–0.145932	–0.185810	–0.371620	0.00082004	0.00644320	S
T,3	–0.154810	–0.246726	–0.493451	0.00374877		S
9,4	–0.151461	–0.245298	–0.490597	0.00093719		S
8,5	–0.148389	–0.263657	–0.527313	0.00082004		S
7,6	–0.148249	–0.263054	–0.526108	0.00093719	0.00644320	S
T,4	–0.153519	–0.318209	–0.636418	0.00374877		S
9,5	–0.150305	–0.317947	–0.635895	0.00082004		S
8,6	–0.150711	–0.319520	–0.639040	0.00093719	0.00550601	S
T,5	–0.152366	–0.392463	–0.784925	0.00328018		S
9,6	–0.152631	–0.393815	–0.787630	0.00093719		S
8,7	–0.163128	–0.379390	–0.758780	0.00093719	0.00515456	S
T,6	–0.154693	–0.447218	–0.894436	0.00374877		S
9,7	–0.165053	–0.450718	–0.901435	0.00093719	0.00468597	S
T,7	–0.044055	–0.508232	–1.016464	0.00374877		S
9,8	–0.044564	–0.508274	–1.016549	0.00093719	0.00468597	S
T,8	0.201020	–0.603364	–1.206727	0.00374877	0.00374877	S
T,9	0.443429	–0.717644	–1.435288	0.00374877	0.00374877	S
A,2	–0.143515	0.145574	0.168229	0.00093719	0.00093719	D
A,3	–0.142185	0.123718	0.163932	0.00093719	0.00093719	D
A,4	–0.140905	0.099430	0.149697	0.00093719	0.00093719	D
A,5	–0.139777	0.077246	0.135928	0.00082004	0.00082004	D
A,6	–0.021250	0.114814	0.229627	0.00093719	0.00093719	D
A,7	0.210518	0.160980	0.321959	0.00093719	0.00093719	D
A,8	0.449866	0.214682	0.429365	0.00093719	0.00093719	S
A,9	0.676012	0.261907	0.523814	0.00093719	0.00093719	S
A,T	1.500000	0.319067	0.638134	0.00374877	0.00374877	S

Hand	Stand	Hit	Double	Spl1	Spl2	Spl3	Freq.	BS
				NDAS				
A,A	–0.148301	0.169241	0.170637	0.672798	0.769064	0.780959	0.00041002	P
2,2	–0.138596	0.010588	–0.277191	0.075500	0.083620	0.084686	0.00041002	P
3,3	–0.135949	–0.014189	–0.253041	0.055614	0.064866	0.066141	0.00041002	P
4,4	–0.133254	0.110439	0.078651	0.032085	0.023622	0.022666	0.00041002	H
5,5	–0.130965	0.299561	0.599122	0.006778	–0.021159	–0.023857	0.00030752	D
6,6	–0.135845	–0.178613	–0.357226	–0.003579	0.013473	0.015771	0.00041002	P
7,7	–0.160661	–0.345179	–0.690358	0.012595	0.032981	0.035507	0.00041002	P
8,8	–0.165601	–0.450917	–0.901834	0.165339	0.203907	0.208640	0.00041002	P
9,9	0.201095	–0.603264	–1.206527	0.326639	0.340459	0.342057	0.00041002	P
T,T	0.671980	–0.850167	–1.700333	0.504865	0.412438	0.354036	0.00726325	S
				DAS				
A,A	–0.148301	0.169241	0.170637	0.672798	0.769064	0.780959	0.00041002	P
2,2	–0.138596	0.010588	–0.277191	0.196387	0.219222	0.222172	0.00041002	P
3,3	–0.135949	–0.014189	–0.253041	0.182481	0.207157	0.210405	0.00041002	P
4,4	–0.133254	0.110439	0.078651	0.162789	0.170291	0.171390	0.00041002	P
5,5	–0.130965	0.299561	0.599122	0.127026	0.080444	0.075782	0.00030752	D
6,6	–0.135845	–0.178613	–0.357226	0.136739	0.171151	0.175709	0.00041002	P
7,7	–0.160661	–0.345179	–0.690358	0.146164	0.182289	0.186768	0.00041002	P
8,8	–0.165601	–0.450917	–0.901834	0.259895	0.309394	0.315457	0.00041002	P
9,9	0.201095	–0.603264	–1.206527	0.376614	0.396144	0.398428	0.00041002	P
T,T	0.671980	–0.850167	–1.700333	0.504865	0.412438	0.354036	0.00726325	S

TABLE A23

2D S17: Dealer's Upcard 6

Hand	Stand	Hit	Double	Freq.	Cum. Freq.	BS
3,2	–0.138195	0.008933	–0.276390	0.00093719	0.00093719	H
4,2	–0.137013	0.000278	–0.249561	0.00093719	0.00093719	H
5,2	–0.136188	0.048935	–0.100542	0.00093719		H
4,3	–0.135831	0.043849	–0.099087	0.00093719	0.00187439	H
6,2	–0.148395	0.123995	0.106321	0.00082004		H
5,3	–0.135006	0.138577	0.136682	0.00093719	0.00175724	H
7,2	–0.150676	0.203411	0.337949	0.00093719		D
6,3	–0.147218	0.213436	0.347835	0.00082004		D
5,4	–0.133918	0.228390	0.377777	0.00093719	0.00269443	D
8,2	–0.152382	0.297625	0.595250	0.00093719		D
7,3	–0.149489	0.302910	0.605820	0.00093719		D
6,4	–0.146119	0.309327	0.618653	0.00082004	0.00269443	D
9,2	–0.154535	0.339941	0.679882	0.00093719		D
8,3	–0.151438	0.344733	0.689467	0.00093719		D
7,4	–0.148626	0.349631	0.699262	0.00093719		D
6,5	–0.145530	0.356848	0.713696	0.00082004	0.00363162	D
T,2	–0.156818	–0.165123	–0.330246	0.00374877		S
9,3	–0.153351	–0.181639	–0.363278	0.00093719		S
8,4	–0.150339	–0.177017	–0.354035	0.00093719		S
7,5	–0.147796	–0.174154	–0.348309	0.00093719	0.00656035	S
T,3	–0.155641	–0.232503	–0.465006	0.00374877		S
9,4	–0.152256	–0.231913	–0.463826	0.00093719		S
8,5	–0.149507	–0.251039	–0.502079	0.00093719		S
7,6	–0.160000	–0.256970	–0.513939	0.00082004	0.00644320	S
T,4	–0.154544	–0.304424	–0.608848	0.00374877		S
9,5	–0.151425	–0.304513	–0.609025	0.00093719		S
8,6	–0.161716	–0.311433	–0.622866	0.00082004	0.00550601	S
T,5	–0.153729	–0.376364	–0.752728	0.00374877		S
9,6	–0.163669	–0.382634	–0.765268	0.00082004		S
8,7	–0.164030	–0.363854	–0.727707	0.00093719	0.00550601	S
T,6	–0.165609	–0.414113	–0.828226	0.00328018		S
9,7	–0.165618	–0.414394	–0.828788	0.00093719	0.00421737	S
T,7	0.001024	–0.496273	–0.992546	0.00374877		S
9,8	0.000912	–0.496420	–0.992840	0.00093719	0.00468597	S
T,8	0.276027	–0.597068	–1.194136	0.00374877	0.00374877	S
T,9	0.490271	–0.714945	–1.429890	0.00374877	0.00374877	S
A,2	–0.134355	0.164810	0.204564	0.00093719	0.00093719	D
A,3	–0.133179	0.142659	0.200079	0.00093719	0.00093719	D
A,4	–0.132096	0.118918	0.189631	0.00093719	0.00093719	D
A,5	–0.131183	0.107088	0.197579	0.00093719	0.00093719	D
A,6	0.012003	0.131284	0.262569	0.00082004	0.00082004	D
A,7	0.273910	0.192289	0.384579	0.00093719	0.00093719	D
A,8	0.489571	0.240709	0.481418	0.00093719	0.00093719	S
A,9	0.699584	0.284227	0.568454	0.00093719	0.00093719	S
A,T	1.500000	0.337395	0.674791	0.00374877	0.00374877	S

Hand	Stand	Hit	Double	Spl1	Spl2	Spl3	Freq.	BS
				NDAS				
A,A	–0.129268	0.192311	0.213109	0.712562	0.810449	0.822437	0.00041002	P
2,2	–0.139357	0.021616	–0.278715	0.101677	0.111681	0.112993	0.00041002	P
3,3	–0.136925	0.000220	–0.249373	0.079192	0.089641	0.091082	0.00041002	P
4,4	–0.134736	0.143985	0.137949	0.052192	0.042095	0.040930	0.00041002	H
5,5	–0.133084	0.323497	0.646993	0.026841	–0.007201	–0.011315	0.00041002	D
6,6	–0.157726	–0.181043	–0.362087	–0.016361	–0.002132	–0.000682	0.00030752	P
7,7	–0.162279	–0.332449	–0.664898	0.058321	0.084201	0.087398	0.00041002	P
8,8	–0.165417	–0.414294	–0.828588	0.231878	0.278317	0.284039	0.00041002	P
9,9	0.275070	–0.597222	–1.194444	0.377062	0.388019	0.389250	0.00041002	P
T,T	0.700605	–0.849453	–1.698906	0.548677	0.464088	0.410353	0.00726325	S
				DAS				
A,A	–0.129268	0.192311	0.213109	0.712562	0.810449	0.822437	0.00041002	P
2,2	–0.139357	0.021616	–0.278715	0.236220	0.262642	0.266061	0.00041002	P
3,3	–0.136925	0.000220	–0.249373	0.215341	0.242392	0.245963	0.00041002	P
4,4	–0.134736	0.143985	0.137949	0.191940	0.198966	0.200010	0.00041002	P
5,5	–0.133084	0.323497	0.646993	0.159550	0.101753	0.094526	0.00041002	D
6,6	–0.157726	–0.181043	–0.362087	0.133284	0.162366	0.165311	0.00030752	P
7,7	–0.162279	–0.332449	–0.664898	0.201342	0.244033	0.249311	0.00041002	P
8,8	–0.165417	–0.414294	–0.828588	0.332625	0.390733	0.397880	0.00041002	P
9,9	0.275070	–0.597222	–1.194444	0.429729	0.446709	0.448665	0.00041002	P
T,T	0.700605	–0.849453	–1.698906	0.548677	0.464088	0.410353	0.00726325	S

TABLE A24
2D S17: Dealer's Upcard 7

Hand	Stand	Hit	Double	Freq.	Cum. Freq.	BS
3,2	−0.472119	−0.119475	−0.944238	0.00093719	0.00093719	H
4,2	−0.471060	−0.157582	−0.882431	0.00093719	0.00093719	H
5,2	−0.475744	−0.068246	−0.577813	0.00093719		H
4,3	−0.470122	−0.069734	−0.566481	0.00093719	0.00187439	H
6,2	−0.478031	0.087274	−0.167517	0.00093719		H
5,3	−0.474810	0.087228	−0.160229	0.00093719	0.00187439	H
7,2	−0.480113	0.178602	0.132272	0.00082004		H
6,3	−0.477079	0.184615	0.139619	0.00093719		H
5,4	−0.473738	0.185696	0.146485	0.00093719	0.00269443	H
8,2	−0.481516	0.262056	0.404348	0.00093719		D
7,3	−0.479403	0.267143	0.419128	0.00082004		D
6,4	−0.476256	0.271053	0.433498	0.00093719	0.00269443	D
9,2	−0.484204	0.290775	0.459573	0.00093719		D
8,3	−0.480568	0.292116	0.467800	0.00093719		D
7,4	−0.478344	0.293804	0.476435	0.00082004		D
6,5	−0.480916	0.295496	0.482822	0.00093719	0.00363162	D
T,2	−0.474068	−0.212296	−0.501185	0.00374877		H
9,3	−0.483263	−0.229396	−0.541524	0.00093719		H
8,4	−0.479510	−0.228519	−0.534108	0.00093719		H
7,5	−0.483004	−0.233995	−0.542892	0.00082004	0.00644320	H
T,3	−0.473130	−0.269685	−0.585144	0.00374877		H
9,4	−0.482203	−0.271343	−0.593311	0.00093719		H
8,5	−0.484173	−0.297051	−0.644387	0.00093719		H
7,6	−0.485288	−0.298211	−0.647041	0.00082004	0.00644320	H
T,4	−0.472085	−0.331372	−0.686850	0.00374877		H
9,5	−0.486883	−0.334046	−0.696404	0.00093719		H
8,6	−0.486492	−0.334295	−0.699743	0.00093719	0.00562316	H
T,5	−0.476583	−0.367207	−0.742763	0.00374877		H
9,6	−0.488837	−0.366618	−0.746713	0.00093719		H
8,7	−0.488255	−0.347657	−0.708659	0.00082004	0.00550601	H
T,6	−0.478934	−0.395986	−0.791972	0.00374877		H
9,7	−0.490997	−0.395393	−0.790786	0.00082004	0.00456882	H
T,7	−0.113423	−0.467984	−0.935968	0.00328018		S
9,8	−0.115314	−0.466834	−0.933668	0.00093719	0.00421737	S
T,8	0.394125	−0.579526	−1.159052	0.00374877	0.00374877	S
T,9	0.613132	−0.707224	−1.414448	0.00374877	0.00374877	S
A,2	−0.468732	0.115420	−0.170646	0.00093719	0.00093719	H
A,3	−0.467811	0.070361	−0.179066	0.00093719	0.00093719	H
A,4	−0.466670	0.035241	−0.163232	0.00093719	0.00093719	H
A,5	−0.471345	−0.014048	−0.185731	0.00093719	0.00093719	H
A,6	−0.097990	0.056572	0.000677	0.00093719	0.00093719	H
A,7	0.406211	0.173677	0.230736	0.00082004	0.00082004	S
A,8	0.615245	0.221501	0.322230	0.00093719	0.00093719	S
A,9	0.773306	0.250001	0.372549	0.00093719	0.00093719	S
A,T	1.500000	0.289207	0.465413	0.00374877	0.00374877	S

NDAS

Hand	Stand	Hit	Double	Spl1	Spl2	Spl3	Freq.	BS
A,A	−0.464261	0.162012	−0.161064	0.501347	0.614739	0.628769	0.00041002	P
2,2	−0.473143	−0.089962	−0.946285	−0.053836	−0.049517	−0.048977	0.00041002	P
3,3	−0.471189	−0.157778	−0.882620	−0.115520	−0.109353	−0.108433	0.00041002	P
4,4	−0.469075	0.095815	−0.149153	−0.197625	−0.233088	−0.237606	0.00041002	H
5,5	−0.478645	0.267524	0.429428	−0.266946	−0.331564	−0.339795	0.00041002	D
6,6	−0.483194	−0.237102	−0.549126	−0.289078	−0.293854	−0.294276	0.00041002	H
7,7	−0.487416	−0.353743	−0.741839	−0.130546	−0.107502	−0.105094	0.00030752	P
8,8	−0.489491	−0.395130	−0.790259	0.184075	0.251430	0.259682	0.00041002	P
9,9	0.399873	−0.578992	−1.157985	0.342372	0.335424	0.334541	0.00041002	S
T,T	0.769342	−0.847556	−1.695112	0.498351	0.357732	0.273615	0.00726325	S

DAS

Hand	Stand	Hit	Double	Spl1	Spl2	Spl3	Freq.	BS
A,A	−0.464261	0.162012	−0.161064	0.501347	0.614739	0.628769	0.00041002	P
2,2	−0.473143	−0.089962	−0.946285	−0.000928	0.009997	0.011404	0.00041002	P
3,3	−0.471189	−0.157778	−0.882620	−0.062622	−0.049801	−0.048002	0.00041002	P
4,4	−0.469075	0.095815	−0.149153	−0.141876	−0.170274	−0.173852	0.00041002	H
5,5	−0.478645	0.267524	0.429428	−0.214819	−0.293479	−0.303594	0.00041002	D
6,6	−0.483194	−0.237102	−0.549126	−0.231642	−0.229464	−0.229000	0.00041002	P
7,7	−0.487416	−0.353743	−0.741839	−0.078299	−0.050218	−0.047320	0.00030752	P
8,8	−0.489491	−0.395130	−0.790259	0.231816	0.304451	0.313315	0.00041002	P
9,9	0.399873	−0.578992	−1.157985	0.366775	0.362407	0.361808	0.00041002	S
T,T	0.769342	−0.847556	−1.695112	0.498351	0.357732	0.273615	0.00726325	S

TABLE A25
2D S17: Dealer's Upcard 8

Hand	Stand	Hit	Double	Freq.	Cum. Freq.	BS
3,2	–0.511663	–0.184992	–1.023327	0.00093719	0.00093719	H
4,2	–0.516122	–0.225061	–1.011794	0.00093719	0.00093719	H
5,2	–0.516059	–0.213954	–0.849797	0.00093719		H
4,3	–0.515063	–0.218958	–0.847889	0.00093719	0.00187439	H
6,2	–0.517750	–0.057871	–0.445343	0.00093719		H
5,3	–0.514988	–0.058521	–0.440119	0.00093719	0.00187439	H
7,2	–0.519528	0.102664	–0.011180	0.00093719		H
6,3	–0.516927	0.107275	–0.006958	0.00093719		H
5,4	–0.519673	0.103089	–0.012421	0.00093719	0.00281158	H
8,2	–0.522226	0.201999	0.289844	0.00082004		D
7,3	–0.518464	0.206785	0.305370	0.00093719		D
6,4	–0.521368	0.202873	0.301684	0.00093719	0.00269443	D
9,2	–0.512099	0.222775	0.339415	0.00093719		D
8,3	–0.521169	0.225495	0.340640	0.00082004		D
7,4	–0.522910	0.226186	0.345736	0.00093719		D
6,5	–0.521285	0.230293	0.358694	0.00093719	0.00363162	D
T,2	–0.513895	–0.272918	–0.620728	0.00374877		H
9,3	–0.511043	–0.293204	–0.658421	0.00093719		H
8,4	–0.525614	–0.294556	–0.665672	0.00082004		H
7,5	–0.522830	–0.295155	–0.661379	0.00093719	0.00644320	H
T,3	–0.512852	–0.326012	–0.698186	0.00374877		H
9,4	–0.515505	–0.330989	–0.707004	0.00093719		H
8,5	–0.525551	–0.354027	–0.757300	0.00082004		H
7,6	–0.524566	–0.357661	–0.764108	0.00093719	0.00644320	H
T,4	–0.517131	–0.364851	–0.753157	0.00374877		H
9,5	–0.515260	–0.370900	–0.763867	0.00093719		H
8,6	–0.526923	–0.370560	–0.770247	0.00082004	0.00550601	H
T,5	–0.517084	–0.417242	–0.842575	0.00374877		H
9,6	–0.517028	–0.418707	–0.848320	0.00093719		H
8,7	–0.528533	–0.398844	–0.810584	0.00082004	0.00550601	H
T,6	–0.518852	–0.442171	–0.884343	0.00374877		H
9,7	–0.518639	–0.443663	–0.887326	0.00093719	0.00468597	H
T,7	–0.388541	–0.490539	–0.981079	0.00374877		S
9,8	–0.398522	–0.491245	–0.982490	0.00082004	0.00456882	S
T,8	0.100441	–0.578564	–1.157128	0.00328018	0.00328018	S
T,9	0.585476	–0.705728	–1.411456	0.00374877	0.00374877	S
A,2	–0.509038	0.047092	–0.313413	0.00093719	0.00093719	H
A,3	–0.507898	0.023364	–0.285539	0.00093719	0.00093719	H
A,4	–0.512353	–0.031171	–0.314136	0.00093719	0.00093719	H
A,5	–0.512296	–0.074952	–0.322914	0.00093719	0.00093719	H
A,6	–0.383757	–0.069230	–0.243292	0.00093719	0.00093719	H
A,7	0.112712	0.043676	–0.023938	0.00093719	0.00093719	S
A,8	0.600309	0.154490	0.191933	0.00082004	0.00082004	S
A,9	0.788331	0.185043	0.258780	0.00093719	0.00093719	S
A,T	1.500000	0.225584	0.341862	0.00374877	0.00374877	S

Hand	Stand	Hit	Double	Spl1	Spl2	Spl3	Freq.	BS
				NDAS				
A,A	–0.505189	0.093723	–0.305396	0.378075	0.482130	0.495045	0.00041002	P
2,2	–0.512714	–0.151141	–1.025427	–0.208836	–0.215337	–0.216110	0.00041002	H
3,3	–0.510601	–0.223474	–1.000756	–0.264676	–0.268170	–0.268428	0.00041002	H
4,4	–0.519505	–0.057427	–0.449287	–0.329897	–0.362635	–0.366779	0.00041002	H
5,5	–0.519589	0.202720	0.305454	–0.409927	–0.483761	–0.493139	0.00041002	D
6,6	–0.522988	–0.295476	–0.661531	–0.425245	–0.439453	–0.441081	0.00041002	H
7,7	–0.525781	–0.389533	–0.810925	–0.423697	–0.426337	–0.426487	0.00041002	H
8,8	–0.531286	–0.442962	–0.885925	–0.109412	–0.077207	–0.074056	0.00030752	P
9,9	0.085625	–0.579074	–1.158148	0.187257	0.199613	0.201198	0.00041002	P
T,T	0.787591	–0.847237	–1.694474	0.374847	0.162608	0.036700	0.00726325	S
				DAS				
A,A	–0.505189	0.093723	–0.305396	0.378075	0.482130	0.495045	0.00041002	P
2,2	–0.512714	–0.151141	–1.025427	–0.174935	–0.177091	–0.177280	0.00041002	H
3,3	–0.510601	–0.223474	–1.000756	–0.229164	–0.228042	–0.227672	0.00041002	H
4,4	–0.519505	–0.057427	–0.449287	–0.292590	–0.320651	–0.324180	0.00041002	H
5,5	–0.519589	0.202720	0.305454	–0.374665	–0.457364	–0.467956	0.00041002	D
6,6	–0.522988	–0.295476	–0.661531	–0.387115	–0.396614	–0.397630	0.00041002	H
7,7	–0.525781	–0.389533	–0.810925	–0.388898	–0.387431	–0.387069	0.00041002	P
8,8	–0.531286	–0.442962	–0.885925	–0.079889	–0.045047	–0.041661	0.00030752	P
9,9	0.085625	–0.579074	–1.158148	0.204642	0.218905	0.220708	0.00041002	P
T,T	0.787591	–0.847237	–1.694474	0.374847	0.162608	0.036700	0.00726325	S

TABLE A26

2D S17: Dealer's Upcard 9

Hand	Stand	Hit	Double	Freq.	Cum. Freq.	BS
3,2	–0.538281	–0.264663	–1.076562	0.00093719	0.00093719	H
4,2	–0.538044	–0.298422	–1.057028	0.00093719	0.00093719	H
5,2	–0.537681	–0.285435	–0.942305	0.00093719		H
4,3	–0.542516	–0.294789	–0.951743	0.00093719	0.00187439	H
6,2	–0.539807	–0.209270	–0.705207	0.00093719		H
5,3	–0.542395	–0.213994	–0.710402	0.00093719	0.00187439	H
7,2	–0.542119	–0.052203	–0.288469	0.00093719		H
6,3	–0.544275	–0.051695	–0.294881	0.00093719		H
5,4	–0.542137	–0.052165	–0.290132	0.00093719	0.00281158	H
8,2	–0.532000	0.117940	0.158168	0.00093719		D
7,3	–0.546594	0.117026	0.148805	0.00093719		D
6,4	–0.544020	0.117088	0.154562	0.00093719	0.00281158	D
9,2	–0.533804	0.150757	0.220808	0.00082004		D
8,3	–0.536478	0.149199	0.221597	0.00093719		D
7,4	–0.546339	0.154012	0.226106	0.00093719		D
6,5	–0.543651	0.155447	0.234280	0.00093719	0.00363162	D
T,2	–0.535500	–0.342118	–0.741526	0.00374877		H
9,3	–0.538295	–0.365198	–0.787128	0.00082004		H
8,4	–0.536240	–0.366689	–0.784010	0.00093719		H
7,5	–0.545988	–0.363849	–0.780357	0.00093719	0.00644320	H
T,3	–0.539809	–0.373504	–0.780838	0.00374877		H
9,4	–0.537875	–0.376489	–0.783989	0.00082004		H
8,5	–0.535707	–0.399877	–0.827419	0.00093719		H
7,6	–0.547545	–0.402296	–0.837776	0.00093719	0.00644320	H
T,4	–0.539587	–0.422586	–0.859574	0.00374877		H
9,5	–0.537540	–0.428554	–0.869564	0.00082004		H
8,6	–0.537660	–0.433777	–0.882562	0.00093719	0.00550601	H
T,5	–0.539252	–0.473409	–0.950271	0.00374877		H
9,6	–0.539493	–0.475410	–0.957130	0.00082004		H
8,7	–0.540028	–0.457744	–0.921685	0.00093719	0.00550601	H
T,6	–0.541205	–0.494730	–0.989459	0.00374877		H
9,7	–0.541862	–0.496081	–0.992162	0.00082004	0.00456882	H
T,7	–0.419976	–0.540537	–1.081073	0.00374877		S
9,8	–0.417488	–0.542975	–1.085950	0.00082004	0.00456882	S
T,8	–0.189406	–0.605222	–1.210445	0.00374877	0.00374877	S
T,9	0.276287	–0.706964	–1.413927	0.00328018	0.00328018	S
A,2	–0.530702	–0.026497	–0.415935	0.00093719	0.00093719	H
A,3	–0.535194	–0.067780	–0.425809	0.00093719	0.00093719	H
A,4	–0.534956	–0.112901	–0.439725	0.00093719	0.00093719	H
A,5	–0.534609	–0.157257	–0.454436	0.00093719	0.00093719	H
A,6	–0.415301	–0.142237	–0.373972	0.00093719	0.00093719	H
A,7	–0.181370	–0.093889	–0.273400	0.00093719	0.00093719	H
A,8	0.287702	0.006449	–0.067212	0.00093719	0.00093719	S
A,9	0.761849	0.106504	0.127087	0.00082004	0.00082004	S
A,T	1.500000	0.153536	0.222117	0.00374877	0.00374877	S

Hand	Stand	Hit	Double	Spl1	Spl2	Spl3	Freq.	BS
				NDAS				
A,A	–0.527616	–0.001207	–0.439365	0.258121	0.350950	0.362504	0.00041002	P
2,2	–0.533796	–0.232011	–1.067592	–0.381706	–0.398847	–0.400915	0.00041002	H
3,3	–0.542765	–0.300815	–1.066499	–0.425474	–0.439271	–0.440869	0.00041002	H
4,4	–0.542509	–0.208846	–0.709968	–0.486845	–0.519745	–0.523848	0.00041002	H
5,5	–0.541765	0.117198	0.159119	–0.568129	–0.650433	–0.660851	0.00041002	D
6,6	–0.545572	–0.362092	–0.775819	–0.578573	–0.603268	–0.606234	0.00041002	H
7,7	–0.549913	–0.452235	–0.926791	–0.568627	–0.580770	–0.582078	0.00041002	H
8,8	–0.530143	–0.498373	–0.996747	–0.423635	–0.416058	–0.415272	0.00041002	P
9,9	–0.189527	–0.605915	–1.211831	–0.108256	–0.100208	–0.099399	0.00030752	P
T,T	0.751376	–0.846569	–1.693139	0.208339	–0.069368	–0.233274	0.00726325	S
				DAS				
A,A	–0.527616	–0.001207	–0.439365	0.258121	0.350950	0.362504	0.00041002	P
2,2	–0.533796	–0.232011	–1.067592	–0.362434	–0.376946	–0.378643	0.00041002	H
3,3	–0.542765	–0.300815	–1.066499	–0.407176	–0.418623	–0.419905	0.00041002	H
4,4	–0.542509	–0.208846	–0.709968	–0.467393	–0.497782	–0.501547	0.00041002	H
5,5	–0.541765	0.117198	0.159119	–0.548381	–0.634035	–0.644961	0.00041002	D
6,6	–0.545572	–0.362092	–0.775819	–0.558551	–0.580721	–0.583353	0.00041002	H
7,7	–0.549913	–0.452235	–0.926791	–0.552038	–0.562218	–0.563279	0.00041002	H
8,8	–0.530143	–0.498373	–0.996747	–0.404886	–0.395003	–0.393917	0.00041002	P
9,9	–0.189527	–0.605915	–1.211831	–0.098187	–0.089253	–0.088367	0.00030752	P
T,T	0.751376	–0.846569	–1.693139	0.208339	–0.069368	–0.233274	0.00726325	S

TABLE A27
2D S17: Dealer's Upcard T

Hand	Stand	Hit	Double	Freq.	Cum. Freq.	BS
3,2	–0.543554	–0.311077	–1.087108	0.00345184	0.00345184	H
4,2	–0.543433	–0.341426	–1.067579	0.00345184	0.00345184	H
5,2	–0.543298	–0.320408	–0.952695	0.00345184		H
4,3	–0.543391	–0.326571	–0.952805	0.00345184	0.00690368	H
6,2	–0.545775	–0.250988	–0.746489	0.00345184		H
5,3	–0.542982	–0.250003	–0.742177	0.00345184	0.00690368	H
7,2	–0.534785	–0.153370	–0.451553	0.00345184		H
6,3	–0.545466	–0.147853	–0.456584	0.00345184		H
5,4	–0.542851	–0.145657	–0.452508	0.00345184	0.01035552	H
8,2	–0.536744	0.027216	–0.001603	0.00345184		H
7,3	–0.534480	0.027351	0.002467	0.00345184		H
6,4	–0.545336	0.029075	0.000806	0.00345184	0.01035552	H
9,2	–0.538586	0.112044	0.161948	0.00345184		D
8,3	–0.536454	0.114553	0.171219	0.00345184		D
7,4	–0.534368	0.114704	0.175585	0.00345184		D
6,5	–0.544946	0.116699	0.175697	0.00345184	0.01380736	D
T,2	–0.540092	–0.364461	–0.769083	0.01337588		H
9,3	–0.538098	–0.384825	–0.808804	0.00345184		H
8,4	–0.536144	–0.383430	–0.801325	0.00345184		H
7,5	–0.533780	–0.382742	–0.794091	0.00345184	0.02373141	H
T,3	–0.539818	–0.409859	–0.840385	0.01337588		H
9,4	–0.538003	–0.412387	–0.843825	0.00345184		H
8,5	–0.535771	–0.435487	–0.887908	0.00345184		H
7,6	–0.536318	–0.437877	–0.887738	0.00345184	0.02373141	H
T,4	–0.539724	–0.456381	–0.919267	0.01337588		H
9,5	–0.537631	–0.459797	–0.924890	0.00345184		EH
8,6	–0.538309	–0.466506	–0.937535	0.00345184	0.02027957	EH
T,5	–0.539351	–0.502775	–1.005550	0.01337588		RH
9,6	–0.540169	–0.508185	–1.016371	0.00345184		RH
8,7	–0.527574	–0.490065	–0.980131	0.00345184	0.02027957	EH
T,6	–0.541889	–0.523992	–1.047984	0.01337588		RH
9,7	–0.529433	–0.526286	–1.052572	0.00345184	0.01682773	RH
T,7	–0.415965	–0.571003	–1.142007	0.01337588		S
9,8	–0.405483	–0.571672	–1.143344	0.00345184	0.01682773	S
T,8	–0.166918	–0.635636	–1.271272	0.01337588	0.01337588	S
T,9	0.082381	–0.720065	–1.440130	0.01337588	0.01337588	S
A,2	–0.539648	–0.097094	–0.497210	0.00348896	0.00348896	H
A,3	–0.539358	–0.132039	–0.497628	0.00348896	0.00348896	H
A,4	–0.539255	–0.172109	–0.505362	0.00348896	0.00348896	H
A,5	–0.538853	–0.215154	–0.525040	0.00348896	0.00348896	H
A,6	–0.419050	–0.194160	–0.446523	0.00348896	0.00348896	H
A,7	–0.182147	–0.141132	–0.334322	0.00348896	0.00348896	H
A,8	0.063724	–0.086693	–0.229870	0.00348896	0.00348896	S
A,9	0.554572	0.016271	–0.029496	0.00348896	0.00348896	S
A,T	1.500000	0.112205	0.159892	0.01351971	0.01351971	S

Hand	Stand	Hit	Double	Spl1	Spl2	Spl3	Freq.	BS
				NDAS				
A,A	–0.535536	–0.058754	–0.491614	0.186732	0.272592	0.283347	0.00154266	P
2,2	–0.543856	–0.282860	–1.087711	–0.472010	–0.494078	–0.496792	0.00151018	H
3,3	–0.543248	–0.340745	–1.067197	–0.518490	–0.538730	–0.541154	0.00151018	H
4,4	–0.543260	–0.245136	–0.742386	–0.574123	–0.612819	–0.617616	0.00151018	H
5,5	–0.542443	0.031176	0.004965	–0.647359	–0.728348	–0.738538	0.00151018	H
6,6	–0.547053	–0.385028	–0.800867	–0.682697	–0.717464	–0.721737	0.00151018	H
7,7	–0.525583	–0.489910	–0.985179	–0.635109	–0.650410	–0.652081	0.00151018	EH
8,8	–0.529565	–0.526181	–1.052362	–0.480781	–0.475311	–0.474668	0.00151018	EP
9,9	–0.156304	–0.636400	–1.272801	–0.286581	–0.302256	–0.304245	0.00151018	S
T,T	0.568553	–0.843225	–1.686450	0.051241	–0.206038	–0.353944	0.02507978	S
				DAS				
A,A	–0.535536	–0.058754	–0.491614	0.186732	0.272592	0.283347	0.00154266	P
2,2	–0.543856	–0.282860	–1.087711	–0.463460	–0.484454	–0.487026	0.00151018	H
3,3	–0.543248	–0.340745	–1.067197	–0.508855	–0.527890	–0.530157	0.00151018	H
4,4	–0.543260	–0.245136	–0.742386	–0.563819	–0.601238	–0.605869	0.00151018	H
5,5	–0.542443	0.031176	0.004965	–0.637116	–0.716805	–0.726822	0.00151018	H
6,6	–0.547053	–0.385028	–0.800867	–0.672682	–0.706194	–0.710302	0.00151018	H
7,7	–0.525583	–0.489910	–0.985179	–0.624252	–0.638136	–0.639614	0.00151018	EH
8,8	–0.529565	–0.526181	–1.052362	–0.471245	–0.464591	–0.463793	0.00151018	EP
9,9	–0.156304	–0.636400	–1.272801	–0.279227	–0.294104	–0.296002	0.00151018	S
T,T	0.568553	–0.843225	–1.686450	0.051241	–0.206038	–0.353944	0.02507978	S

TABLE A28
2D H17: Dealer's Upcard A

Hand	Stand	Hit	Double	Freq.	Cum. Freq.	BS
3,2	−0.592126	−0.320313	−1.184253	0.00064026	0.00064026	EH
4,2	−0.591222	−0.351868	−1.171187	0.00064026	0.00064026	EH
5,2	−0.591782	−0.357274	−1.080591	0.00064026		EH
4,3	−0.590338	−0.362213	−1.077577	0.00064026	0.00128052	EH
6,2	−0.589526	−0.270784	−0.826132	0.00064026		EH
5,3	−0.590764	−0.271028	−0.823487	0.00064026	0.00128052	EH
7,2	−0.586678	−0.132352	−0.446152	0.00064026		H
6,3	−0.588515	−0.125084	−0.441418	0.00064026		H
5,4	−0.589636	−0.124219	−0.438228	0.00064026	0.00192078	H
8,2	−0.590011	0.035845	−0.017464	0.00064026		H
7,3	−0.585677	0.035903	−0.010256	0.00064026		H
6,4	−0.587412	0.038312	−0.003909	0.00064026	0.00192078	H
9,2	−0.593114	0.116041	0.141065	0.00064026		D
8,3	−0.588832	0.117332	0.150867	0.00064026		D
7,4	−0.584383	0.118882	0.161500	0.00064026		D
6,5	−0.587564	0.121079	0.166650	0.00064026	0.00256104	D
T,2	−0.599990	−0.381947	−0.827562	0.00259816		EH
9,3	−0.592147	−0.394213	−0.850257	0.00064026		EH
8,4	−0.587742	−0.393719	−0.842649	0.00064026		EH
7,5	−0.584874	−0.394705	−0.837264	0.00064026	0.00451894	EH
T,3	−0.599028	−0.424713	−0.887981	0.00259816		EH
9,4	−0.591063	−0.419451	−0.875330	0.00064026		EH
8,5	−0.588245	−0.443541	−0.919889	0.00064026		EH
7,6	−0.582801	−0.447078	−0.923146	0.00064026	0.00451894	EH
T,4	−0.597967	−0.468715	−0.956232	0.00259816		EH
9,5	−0.591563	−0.468095	−0.953423	0.00064026		EH
8,6	−0.586172	−0.472648	−0.963082	0.00064026	0.00387868	EH
T,5	−0.598424	−0.515081	−1.036243	0.00259816		RH
9,6	−0.589440	−0.512262	−1.032531	0.00064026		RH
8,7	−0.583562	−0.495281	−0.997462	0.00064026	0.00387868	EH
T,6	−0.596497	−0.535711	−1.071423	0.00259816		RH
9,7	−0.586859	−0.531624	−1.063248	0.00064026	0.00323842	RH
T,7	−0.509914	−0.578076	−1.156152	0.00259816		RS
9,8	−0.503604	−0.572583	−1.145167	0.00064026	0.00323842	RS
T,8	−0.218577	−0.641955	−1.283911	0.00259816	0.00259816	S
T,9	0.200106	−0.738688	−1.477375	0.00259816	0.00259816	S
A,2	−0.594150	−0.103429	−0.571816	0.00056023	0.00056023	H
A,3	−0.593169	−0.137018	−0.570172	0.00056023	0.00056023	H
A,4	−0.592080	−0.178451	−0.580261	0.00056023	0.00056023	H
A,5	−0.592718	−0.221872	−0.598489	0.00056023	0.00056023	H
A,6	−0.511565	−0.224617	−0.541393	0.00056023	0.00056023	H
A,7	−0.224703	−0.165428	−0.416886	0.00056023	0.00056023	H
A,8	0.192556	−0.068979	−0.230106	0.00056023	0.00056023	S
A,9	0.613241	0.025607	−0.049596	0.00056023	0.00056023	S
A,T	1.500000	0.107390	0.119512	0.00227339	0.00227339	S

Hand	Stand	Hit	Double	Spl1	Spl2	Spl3	Freq.	BS
				NDAS				
A,A	−0.595164	−0.067327	−0.574721	0.157378	0.231323	0.238688	0.00021009	P
2,2	−0.593113	−0.293373	−1.186226	−0.495649	−0.518975	−0.521811	0.00028011	EH
3,3	−0.591323	−0.351766	−1.171354	−0.540143	−0.561084	−0.563525	0.00028011	EH
4,4	−0.589214	−0.264933	−0.820424	−0.593662	−0.632098	−0.636834	0.00028011	H
5,5	−0.590103	0.037716	−0.004346	−0.671274	−0.755576	−0.766142	0.00028011	H
6,6	−0.585447	−0.393801	−0.835230	−0.683285	−0.716145	−0.720068	0.00028011	EH
7,7	−0.580226	−0.495371	−1.008716	−0.694186	−0.714718	−0.716907	0.00028011	EH
8,8	−0.586903	−0.531321	−1.062642	−0.505294	−0.501152	−0.500507	0.00028011	RP
9,9	−0.206613	−0.637736	−1.275473	−0.221462	−0.221886	−0.221775	0.00028011	S
T,T	0.597669	−0.868561	−1.737122	0.031759	−0.262413	−0.438709	0.00510585	S
				DAS				
A,A	−0.595164	−0.067327	−0.574721	0.157378	0.231323	0.238688	0.00021009	P
2,2	−0.593113	−0.293373	−1.186226	−0.490313	−0.512857	−0.515576	0.00028011	EH
3,3	−0.591323	−0.351766	−1.171354	−0.533289	−0.553253	−0.555552	0.00028011	EH
4,4	−0.589214	−0.264933	−0.820424	−0.585057	−0.622279	−0.626841	0.00028011	H
5,5	−0.590103	0.037716	−0.004346	−0.662116	−0.745133	−0.755515	0.00028011	H
6,6	−0.585447	−0.393801	−0.835230	−0.674050	−0.705591	−0.709321	0.00028011	EH
7,7	−0.580226	−0.495371	−1.008716	−0.685826	−0.705192	−0.707214	0.00028011	EH
8,8	−0.586903	−0.531321	−1.062642	−0.499517	−0.494647	−0.493906	0.00028011	EP
9,9	−0.206613	−0.637736	−1.275473	−0.218242	−0.218379	−0.218245	0.00028011	S
T,T	0.597669	−0.868561	−1.737122	0.031759	−0.262413	−0.438709	0.00510585	S

TABLE A29
2D H17: Dealer's Upcard 2

Hand	Stand	Hit	Double	Freq.	Cum. Freq.	BS
3,2	–0.285594	–0.126538	–0.571189	0.00082004	0.00082004	H
4,2	–0.285221	–0.141992	–0.550009	0.00082004	0.00082004	H
5,2	–0.277073	–0.104443	–0.414360	0.00082004		H
4,3	–0.285306	–0.114432	–0.425872	0.00093719	0.00175724	H
6,2	–0.278555	–0.019699	–0.185795	0.00082004		H
5,3	–0.277158	–0.020915	–0.187089	0.00093719	0.00175724	H
7,2	–0.280112	0.077111	0.086424	0.00082004		D
6,3	–0.278615	0.081427	0.084527	0.00093719		D
5,4	–0.276760	0.082575	0.086204	0.00093719	0.00269443	D
8,2	–0.281358	0.195461	0.388165	0.00082004		D
7,3	–0.279987	0.196663	0.389953	0.00093719		D
6,4	–0.278025	0.198485	0.393480	0.00093719	0.00269443	D
9,2	–0.283482	0.249610	0.495811	0.00082004		D
8,3	–0.281434	0.252036	0.500498	0.00093719		D
7,4	–0.279590	0.254568	0.505428	0.00093719		D
6,5	–0.269980	0.259967	0.517177	0.00093719	0.00363162	D
T,2	–0.295189	–0.249031	–0.498063	0.00328018		H
9,3	–0.283557	–0.260052	–0.520103	0.00093719		H
8,4	–0.281050	–0.256746	–0.513492	0.00093719		H
7,5	–0.271561	–0.253303	–0.506607	0.00093719	0.00609176	H
T,3	–0.295261	–0.306724	–0.613447	0.00374877		S
9,4	–0.283170	–0.301099	–0.602198	0.00093719		S
8,5	–0.273021	–0.319981	–0.639962	0.00093719		S
7,6	–0.273062	–0.320347	–0.640694	0.00093719	0.00656035	S
T,4	–0.294887	–0.366201	–0.732403	0.00374877		S
9,5	–0.275138	–0.361847	–0.723694	0.00093719		S
8,6	–0.274499	–0.363793	–0.727586	0.00093719	0.00562316	S
T,5	–0.287080	–0.427789	–0.855578	0.00374877		S
9,6	–0.276866	–0.425021	–0.850043	0.00093719		S
8,7	–0.276313	–0.405814	–0.811628	0.00093719	0.00562316	S
T,6	–0.288567	–0.470392	–0.940783	0.00374877		S
9,7	–0.278447	–0.465915	–0.931830	0.00093719	0.00468597	S
T,7	–0.158751	–0.539132	–1.078264	0.00374877		S
9,8	–0.148342	–0.535286	–1.070573	0.00093719	0.00468597	S
T,8	0.108622	–0.628466	–1.256931	0.00374877	0.00374877	S
T,9	0.377408	–0.739552	–1.479103	0.00374877	0.00374877	S
A,2	–0.282107	0.043477	–0.053532	0.00082004	0.00082004	H
A,3	–0.282189	0.020601	–0.055156	0.00093719	0.00093719	H
A,4	–0.282017	–0.004415	–0.066429	0.00093719	0.00093719	H
A,5	–0.273946	–0.024758	–0.072938	0.00093719	0.00093719	H
A,6	–0.145231	0.002585	0.002413	0.00093719	0.00093719	H
A,7	0.118882	0.061594	0.119481	0.00093719	0.00093719	D
A,8	0.386326	0.120025	0.236476	0.00093719	0.00093719	S
A,9	0.643124	0.184924	0.366865	0.00093719	0.00093719	S
A,T	1.500000	0.241925	0.477783	0.00374877	0.00374877	S

Hand	Stand	Hit	Double	Spl1	Spl2	Spl3	Freq.	BS
				NDAS				
A,A	–0.278818	0.086622	–0.043149	0.516612	0.610124	0.621708	0.00041002	P
2,2	–0.285287	–0.111793	–0.570575	–0.140668	–0.143403	–0.143666	0.00030752	H
3,3	–0.285685	–0.142295	–0.550893	–0.195720	–0.201486	–0.202137	0.00041002	H
4,4	–0.284922	–0.018168	–0.193777	–0.226212	–0.250735	–0.253786	0.00041002	H
5,5	–0.268503	0.201293	0.399936	–0.238348	–0.289236	–0.295438	0.00041002	D
6,6	–0.271468	–0.253276	–0.506553	–0.245992	–0.243108	–0.242495	0.00041002	P
7,7	–0.274633	–0.385000	–0.769999	–0.191924	–0.181343	–0.179926	0.00041002	P
8,8	–0.277763	–0.465195	–0.930391	–0.020752	0.010231	0.014165	0.00041002	P
9,9	0.117843	–0.626010	–1.252021	0.155328	0.159762	0.160312	0.00041002	P
T,T	0.628556	–0.851193	–1.702385	0.335599	0.179679	0.084283	0.00726325	S
				DAS				
A,A	–0.278818	0.086622	–0.043149	0.516612	0.610124	0.621708	0.00041002	P
2,2	–0.285287	–0.111793	–0.570575	–0.066876	–0.062108	–0.061597	0.00030752	P
3,3	–0.285685	–0.142295	–0.550893	–0.123382	–0.120476	–0.120033	0.00041002	P
4,4	–0.284922	–0.018168	–0.193777	–0.152947	–0.168674	–0.170614	0.00041002	H
5,5	–0.268503	0.201293	0.399936	–0.169719	–0.236629	–0.244909	0.00041002	D
6,6	–0.271468	–0.253276	–0.506553	–0.169986	–0.157759	–0.155937	0.00041002	P
7,7	–0.274633	–0.385000	–0.769999	–0.115262	–0.095137	–0.092471	0.00041002	P
8,8	–0.277763	–0.465195	–0.930391	0.046807	0.085865	0.090816	0.00041002	P
9,9	0.117843	–0.626010	–1.252021	0.189630	0.198094	0.199143	0.00041002	P
T,T	0.628556	–0.851193	–1.702385	0.335599	0.179679	0.084283	0.00726325	S

TABLE A30
2D H17: Dealer's Upcard 3

Hand	Stand	Hit	Double	Freq.	Cum. Freq.	BS
3,2	–0.244219	–0.093791	–0.488438	0.00082004	0.00082004	H
4,2	–0.236052	–0.103439	–0.453236	0.00093719	0.00093719	H
5,2	–0.234705	–0.068825	–0.330757	0.00093719		H
4,3	–0.235686	–0.077896	–0.336694	0.00082004	0.00175724	H
6,2	–0.236071	0.013816	–0.108844	0.00093719		H
5,3	–0.234327	0.012528	–0.109967	0.00082004	0.00175724	H
7,2	–0.237500	0.113862	0.154799	0.00093719		D
6,3	–0.235498	0.118129	0.153057	0.00082004		D
5,4	–0.226046	0.119194	0.156810	0.00093719	0.00269443	D
8,2	–0.239595	0.221984	0.443322	0.00093719		D
7,3	–0.237126	0.224558	0.448288	0.00082004		D
6,4	–0.227421	0.226436	0.452894	0.00093719	0.00269443	D
9,2	–0.251824	0.264460	0.528002	0.00093719		D
8,3	–0.239223	0.276720	0.552896	0.00082004		D
7,4	–0.229054	0.281377	0.563012	0.00093719		D
6,5	–0.226063	0.285639	0.571561	0.00093719	0.00363162	D
T,2	–0.253885	–0.226933	–0.453867	0.00374877		H
9,3	–0.251458	–0.244286	–0.488572	0.00082004		H
8,4	–0.231174	–0.231937	–0.463873	0.00093719		S
7,5	–0.227715	–0.228334	–0.456667	0.00093719	0.00644320	S
T,3	–0.253531	–0.287726	–0.575451	0.00328018		S
9,4	–0.243398	–0.286242	–0.572484	0.00093719		S
8,5	–0.229812	–0.298894	–0.597788	0.00093719		S
7,6	–0.229096	–0.298347	–0.596694	0.00093719	0.00609176	S
T,4	–0.245696	–0.352790	–0.705581	0.00374877		S
9,5	–0.242282	–0.352948	–0.705896	0.00093719		S
8,6	–0.231449	–0.349030	–0.698060	0.00093719	0.00562316	S
T,5	–0.244333	–0.418831	–0.837661	0.00374877		S
9,6	–0.243671	–0.420008	–0.840016	0.00093719		S
8,7	–0.233079	–0.395695	–0.791390	0.00093719	0.00562316	S
T,6	–0.245737	–0.464355	–0.928710	0.00374877		S
9,7	–0.245316	–0.463117	–0.926235	0.00093719	0.00468597	S
T,7	–0.120718	–0.535606	–1.071212	0.00374877		S
9,8	–0.120980	–0.535015	–1.070031	0.00093719	0.00468597	S
T,8	0.137349	–0.627587	–1.255175	0.00374877	0.00374877	S
T,9	0.387781	–0.720908	–1.441815	0.00374877	0.00374877	S
A,2	–0.241160	0.072904	0.014117	0.00093719	0.00093719	H
A,3	–0.241006	0.048392	0.006064	0.00082004	0.00082004	H
A,4	–0.232931	0.027168	0.001707	0.00093719	0.00093719	H
A,5	–0.231565	0.004529	–0.009651	0.00093719	0.00093719	H
A,6	–0.107626	0.031822	0.063928	0.00093719	0.00093719	D
A,7	0.149046	0.089694	0.179646	0.00093719	0.00093719	D
A,8	0.405716	0.158974	0.317473	0.00093719	0.00093719	S
A,9	0.643900	0.200617	0.400317	0.00093719	0.00093719	S
A,T	1.500000	0.264888	0.526367	0.00374877	0.00374877	S

Hand	Stand	Hit	Double	Spl1	Spl2	Spl3	Freq.	BS
				NDAS				
A,A	–0.237884	0.110236	0.025730	0.564088	0.658312	0.669978	0.00041002	P
2,2	–0.244572	–0.079331	–0.489143	–0.084473	–0.085014	–0.085074	0.00041002	H
3,3	–0.243863	–0.108982	–0.468059	–0.128552	–0.130100	–0.130217	0.00030752	H
4,4	–0.227510	0.016612	–0.109341	–0.140907	–0.158672	–0.160781	0.00041002	H
5,5	–0.224679	0.228521	0.457065	–0.165971	–0.211403	–0.216910	0.00041002	D
6,6	–0.227464	–0.228411	–0.456822	–0.172267	–0.163479	–0.162114	0.00041002	P
7,7	–0.230965	–0.369242	–0.738485	–0.120191	–0.106104	–0.104226	0.00041002	P
8,8	–0.235203	–0.458963	–0.917927	0.041409	0.074638	0.078841	0.00041002	P
9,9	0.127459	–0.629565	–1.259130	0.185080	0.190802	0.191377	0.00041002	P
T,T	0.639140	–0.850879	–1.701759	0.386352	0.251021	0.167799	0.00726325	S
				DAS				
A,A	–0.237884	0.110236	0.025730	0.564088	0.658312	0.669978	0.00041002	P
2,2	–0.244572	–0.079331	–0.489143	0.001384	0.011180	0.012431	0.00041002	P
3,3	–0.243863	–0.108982	–0.468059	–0.042052	–0.034960	–0.034203	0.00030752	P
4,4	–0.227510	0.016612	–0.109341	–0.050977	–0.057826	–0.058540	0.00041002	H
5,5	–0.224679	0.228521	0.457065	–0.083213	–0.146627	–0.154470	0.00041002	D
6,6	–0.227464	–0.228411	–0.456822	–0.074774	–0.053832	–0.050872	0.00041002	P
7,7	–0.230965	–0.369242	–0.738485	–0.026671	–0.000879	0.002539	0.00041002	P
8,8	–0.235203	–0.458963	–0.917927	0.116851	0.159150	0.164501	0.00041002	P
9,9	0.127459	–0.629565	–1.259130	0.225632	0.235954	0.237077	0.00041002	P
T,T	0.639140	–0.850879	–1.701759	0.386352	0.251021	0.167799	0.00726325	S

TABLE A31

2D H17: Dealer's Upcard 4

Hand	Stand	Hit	Double	Freq.	Cum. Freq.	BS
3,2	–0.188445	–0.048353	–0.376890	0.00093719	0.00093719	H
4,2	–0.186897	–0.060899	–0.354544	0.00082004	0.00082004	H
5,2	–0.185815	–0.026426	–0.236190	0.00093719		H
4,3	–0.178737	–0.029323	–0.227615	0.00082004	0.00175724	H
6,2	–0.187007	0.057719	–0.024684	0.00093719		H
5,3	–0.177550	0.060352	–0.016050	0.00093719	0.00187439	H
7,2	–0.189346	0.146561	0.223733	0.00093719		D
6,3	–0.178943	0.155291	0.232237	0.00093719		D
5,4	–0.175970	0.156833	0.235379	0.00082004	0.00269443	D
8,2	–0.201547	0.240801	0.481603	0.00093719		D
7,3	–0.181295	0.256346	0.512691	0.00093719		D
6,4	–0.177379	0.259647	0.519293	0.00082004	0.00269443	D
9,2	–0.203947	0.291135	0.582271	0.00093719		D
8,3	–0.193507	0.297449	0.594899	0.00093719		D
7,4	–0.179753	0.311587	0.623175	0.00082004		D
6,5	–0.176304	0.315857	0.631713	0.00093719	0.00363162	D
T,2	–0.206125	–0.204205	–0.408411	0.00374877		H
9,3	–0.195904	–0.219573	–0.439147	0.00093719		S
8,4	–0.191942	–0.214139	–0.428278	0.00082004		S
7,5	–0.178662	–0.203561	–0.407122	0.00093719	0.00644320	S
T,3	–0.198306	–0.266807	–0.533613	0.00374877		S
9,4	–0.194580	–0.265184	–0.530368	0.00082004		S
8,5	–0.191097	–0.283470	–0.566941	0.00093719		S
7,6	–0.180312	–0.276540	–0.553080	0.00093719	0.00644320	S
T,4	–0.196747	–0.336606	–0.673211	0.00328018		S
9,5	–0.193495	–0.336456	–0.672911	0.00093719		S
8,6	–0.192512	–0.337433	–0.674865	0.00093719	0.00515456	S
T,5	–0.195654	–0.408069	–0.816137	0.00374877		S
9,6	–0.194905	–0.408969	–0.817939	0.00093719		S
8,7	–0.194877	–0.389644	–0.779288	0.00093719	0.00562316	S
T,6	–0.197073	–0.457372	–0.914743	0.00374877		S
9,7	–0.197277	–0.456626	–0.913252	0.00093719	0.00468597	S
T,7	–0.075457	–0.532597	–1.065194	0.00374877		S
9,8	–0.085153	–0.535274	–1.070547	0.00093719	0.00468597	S
T,8	0.160381	–0.608518	–1.217035	0.00374877	0.00374877	S
T,9	0.406488	–0.719949	–1.439898	0.00374877	0.00374877	S
A,2	–0.193757	0.107026	0.089626	0.00093719	0.00093719	H
A,3	–0.185697	0.086531	0.087059	0.00093719	0.00093719	D
A,4	–0.184149	0.061577	0.074250	0.00082004	0.00082004	D
A,5	–0.183054	0.040315	0.063048	0.00093719	0.00093719	D
A,6	–0.061667	0.067717	0.135433	0.00093719	0.00093719	D
A,7	0.180897	0.134886	0.269771	0.00093719	0.00093719	D
A,8	0.413699	0.179824	0.359647	0.00093719	0.00093719	S
A,9	0.653683	0.229168	0.458337	0.00093719	0.00093719	S
A,T	1.500000	0.289897	0.579795	0.00374877	0.00374877	S

Hand	Stand	Hit	Double	Spl1	Spl2	Spl3	Freq.	BS
				NDAS				
A,A	–0.190661	0.134467	0.099262	0.616184	0.710888	0.722601	0.00041002	P
2,2	–0.196595	–0.039502	–0.393190	–0.021523	–0.019401	–0.019141	0.00041002	P
3,3	–0.180285	–0.056736	–0.342912	–0.027348	–0.022978	–0.022325	0.00041002	P
4,4	–0.177078	0.065674	–0.014834	–0.051486	–0.062304	–0.063311	0.00030752	H
5,5	–0.174888	0.260577	0.521155	–0.076798	–0.115279	–0.119895	0.00041002	D
6,6	–0.177710	–0.202730	–0.405459	–0.083479	–0.070887	–0.069139	0.00041002	P
7,7	–0.182685	–0.352793	–0.705586	–0.035898	–0.017436	–0.015000	0.00041002	P
8,8	–0.207078	–0.460183	–0.920366	0.088392	0.122864	0.127099	0.00041002	P
9,9	0.160990	–0.608440	–1.216879	0.252051	0.261859	0.262965	0.00041002	P
T,T	0.648790	–0.850714	–1.701427	0.438172	0.323503	0.251977	0.00726325	S
				DAS				
A,A	–0.190661	0.134467	0.099262	0.616184	0.710888	0.722601	0.00041002	P
2,2	–0.196595	–0.039502	–0.393190	0.077360	0.091320	0.093073	0.00041002	P
3,3	–0.180285	–0.056736	–0.342912	0.076767	0.093784	0.096053	0.00041002	P
4,4	–0.177078	0.065674	–0.014834	0.058304	0.058721	0.058879	0.00030752	H
5,5	–0.174888	0.260577	0.521155	0.023605	–0.034633	–0.041816	0.00041002	D
6,6	–0.177710	–0.202730	–0.405459	0.032901	0.059980	0.063624	0.00041002	P
7,7	–0.182685	–0.352793	–0.705586	0.081283	0.114277	0.118608	0.00041002	P
8,8	–0.207078	–0.460183	–0.920366	0.172716	0.216934	0.222355	0.00041002	P
9,9	0.160990	–0.608440	–1.216879	0.296981	0.311907	0.313626	0.00041002	P
T,T	0.648790	–0.850714	–1.701427	0.438172	0.323503	0.251977	0.00726325	S

TABLE A32

2D H17: Dealer's Upcard 5

Hand	Stand	Hit	Double	Freq.	Cum. Freq.	BS
3,2	–0.134777	–0.000976	–0.269554	0.00093719	0.00093719	H
4,2	–0.133500	–0.012577	–0.249341	0.00093719	0.00093719	H
5,2	–0.132298	0.023462	–0.138268	0.00082004		H
4,3	–0.132048	0.019779	–0.133975	0.00093719	0.00175724	H
6,2	–0.134617	0.098745	0.063497	0.00093719		H
5,3	–0.130948	0.103542	0.074691	0.00082004	0.00175724	H
7,2	–0.147046	0.175762	0.285601	0.00093719		D
6,3	–0.133267	0.194276	0.313244	0.00093719		D
5,4	–0.129646	0.199504	0.323860	0.00082004	0.00269443	D
8,2	–0.149529	0.275268	0.550537	0.00093719		D
7,3	–0.145709	0.280815	0.561629	0.00093719		D
6,4	–0.131979	0.295335	0.590670	0.00093719	0.00281158	D
9,2	–0.151457	0.322721	0.645443	0.00093719		D
8,3	–0.148179	0.327733	0.655466	0.00093719		D
7,4	–0.144408	0.333526	0.667053	0.00093719		D
6,5	–0.130880	0.348322	0.696645	0.00082004	0.00363162	D
T,2	–0.153697	–0.179113	–0.358225	0.00374877		S
9,3	–0.150354	–0.195216	–0.390432	0.00093719		S
8,4	–0.147124	–0.190239	–0.380478	0.00093719		S
7,5	–0.143552	–0.185913	–0.371827	0.00082004	0.00644320	S
T,3	–0.152346	–0.247037	–0.494073	0.00374877		S
9,4	–0.149055	–0.245568	–0.491136	0.00093719		S
8,5	–0.146020	–0.263919	–0.527838	0.00082004		S
7,6	–0.145885	–0.263317	–0.526633	0.00093719	0.00644320	S
T,4	–0.151050	–0.318649	–0.637299	0.00374877		S
9,5	–0.147951	–0.318434	–0.636868	0.00082004		S
8,6	–0.148358	–0.320019	–0.640037	0.00093719	0.00550601	S
T,5	–0.149947	–0.393164	–0.786329	0.00328018		S
9,6	–0.150290	–0.394530	–0.789061	0.00093719		S
8,7	–0.160788	–0.380111	–0.760222	0.00093719	0.00515456	S
T,6	–0.152291	–0.448074	–0.896148	0.00374877		S
9,7	–0.162727	–0.451569	–0.903138	0.00093719	0.00468597	S
T,7	–0.045297	–0.508948	–1.017896	0.00374877		S
9,8	–0.045826	–0.508984	–1.017969	0.00093719	0.00468597	S
T,8	0.196670	–0.603766	–1.207532	0.00374877	0.00374877	S
T,9	0.440237	–0.717813	–1.435626	0.00374877	0.00374877	S
A,2	–0.141651	0.145875	0.169235	0.00093719	0.00093719	D
A,3	–0.140316	0.124142	0.164993	0.00093719	0.00093719	D
A,4	–0.139032	0.099979	0.150790	0.00093719	0.00093719	D
A,5	–0.137943	0.077844	0.137044	0.00082004	0.00082004	D
A,6	–0.022113	0.114717	0.229434	0.00093719	0.00093719	D
A,7	0.207314	0.160272	0.320545	0.00093719	0.00093719	D
A,8	0.447486	0.214179	0.428357	0.00093719	0.00093719	S
A,9	0.674583	0.261623	0.523246	0.00093719	0.00093719	S
A,T	1.500000	0.318946	0.637891	0.00374877	0.00374877	S

Hand	Stand	Hit	Double	Spl1	Spl2	Spl3	Freq.	BS
				NDAS				
A,A	–0.146973	0.169339	0.171311	0.672616	0.768759	0.780644	0.00041002	P
2,2	–0.136105	0.011809	–0.272209	0.077520	0.085737	0.086815	0.00041002	P
3,3	–0.133446	–0.012701	–0.248466	0.057902	0.067251	0.068538	0.00041002	P
4,4	–0.130739	0.109724	0.078108	0.034694	0.026636	0.025733	0.00041002	H
5,5	–0.128554	0.299131	0.598262	0.009308	–0.018332	–0.021001	0.00030752	D
6,6	–0.133466	–0.178674	–0.357349	–0.000969	0.016105	0.018405	0.00041002	P
7,7	–0.158311	–0.345685	–0.691370	0.012813	0.032941	0.035433	0.00041002	P
8,8	–0.163273	–0.451767	–0.903535	0.163585	0.201664	0.206336	0.00041002	P
9,9	0.196771	–0.603663	–1.207326	0.325388	0.339574	0.341218	0.00041002	P
T,T	0.670048	–0.850206	–1.700412	0.504084	0.412260	0.354222	0.00726325	S
				DAS				
A,A	–0.146973	0.169339	0.171311	0.672616	0.768759	0.780644	0.00041002	P
2,2	–0.136105	0.011809	–0.272209	0.198382	0.221310	0.224272	0.00041002	P
3,3	–0.133446	–0.012701	–0.248466	0.184774	0.209550	0.212810	0.00041002	P
4,4	–0.130739	0.109724	0.078108	0.165416	0.173331	0.174482	0.00041002	P
5,5	–0.128554	0.299131	0.598262	0.129534	0.083290	0.078663	0.00030752	D
6,6	–0.133466	–0.178674	–0.357349	0.139203	0.173617	0.178175	0.00041002	P
7,7	–0.158311	–0.345685	–0.691370	0.146536	0.182420	0.186867	0.00041002	P
8,8	–0.163273	–0.451767	–0.903535	0.258067	0.307067	0.313068	0.00041002	P
9,9	0.196771	–0.603663	–1.207326	0.375337	0.395230	0.397559	0.00041002	P
T,T	0.670048	–0.850206	–1.700412	0.504084	0.412260	0.354222	0.00726325	S

TABLE A33
2D H17: Dealer's Upcard 6

Hand	Stand	Hit	Double	Freq.	Cum. Freq.	BS
3,2	–0.103191	0.027607	–0.206381	0.00093719	0.00093719	H
4,2	–0.101935	0.020378	–0.186556	0.00093719	0.00093719	H
5,2	–0.101826	0.050816	–0.078271	0.00093719		H
4,3	–0.100667	0.046958	–0.075130	0.00093719	0.00187439	H
6,2	–0.114257	0.112260	0.094362	0.00082004		H
5,3	–0.100559	0.127095	0.125379	0.00093719	0.00175724	H
7,2	–0.116747	0.193795	0.323687	0.00093719		D
6,3	–0.112996	0.203963	0.334636	0.00082004		D
5,4	–0.099397	0.219201	0.365198	0.00093719	0.00269443	D
8,2	–0.118628	0.291989	0.583978	0.00093719		D
7,3	–0.115485	0.297307	0.594615	0.00093719		D
6,4	–0.111834	0.303736	0.607471	0.00082004	0.00269443	D
9,2	–0.120945	0.338435	0.676869	0.00093719		D
8,3	–0.117601	0.343664	0.687328	0.00093719		D
7,4	–0.114550	0.349204	0.698407	0.00093719		D
6,5	–0.111971	0.355332	0.710664	0.00082004	0.00363162	D
T,2	–0.122380	–0.166777	–0.333555	0.00374877		S
9,3	–0.119678	–0.183190	–0.366380	0.00093719		S
8,4	–0.116430	–0.178139	–0.356278	0.00093719		S
7,5	–0.114445	–0.175071	–0.350143	0.00093719	0.00656035	S
T,3	–0.121119	–0.236895	–0.473790	0.00374877		S
9,4	–0.118510	–0.235709	–0.471419	0.00093719		S
8,5	–0.116321	–0.254757	–0.509513	0.00093719		S
7,6	–0.126873	–0.260701	–0.521402	0.00082004	0.00644320	S
T,4	–0.119950	–0.310606	–0.621212	0.00374877		S
9,5	–0.118402	–0.311350	–0.622700	0.00093719		S
8,6	–0.128756	–0.318456	–0.636912	0.00082004	0.00550601	S
T,5	–0.119841	–0.386197	–0.772395	0.00374877		S
9,6	–0.130841	–0.392652	–0.785304	0.00082004		S
8,7	–0.131248	–0.373963	–0.747927	0.00093719	0.00550601	S
T,6	–0.131948	–0.426495	–0.852990	0.00328018		S
9,7	–0.133001	–0.426706	–0.853412	0.00093719	0.00421737	S
T,7	–0.017509	–0.506975	–1.013950	0.00374877		S
9,8	–0.017880	–0.507026	–1.014052	0.00093719	0.00468597	S
T,8	0.213657	–0.603162	–1.206325	0.00374877	0.00374877	S
T,9	0.445120	–0.717582	–1.435163	0.00374877	0.00374877	S
A,2	–0.103873	0.169299	0.220744	0.00093719	0.00093719	D
A,3	–0.102624	0.149176	0.217147	0.00093719	0.00093719	D
A,4	–0.101478	0.127506	0.207197	0.00093719	0.00093719	D
A,5	–0.101190	0.116167	0.214204	0.00093719	0.00093719	D
A,6	–0.003090	0.128756	0.257512	0.00082004	0.00082004	D
A,7	0.220300	0.179925	0.359850	0.00093719	0.00093719	D
A,8	0.450275	0.232007	0.464014	0.00093719	0.00093719	D
A,9	0.676013	0.279298	0.558595	0.00093719	0.00093719	S
A,T	1.500000	0.335943	0.671886	0.00374877	0.00374877	S

Hand	Stand	Hit	Double	Spl1	Spl2	Spl3	Freq.	BS
				NDAS				
A,A	–0.103203	0.194397	0.227011	0.709899	0.804904	0.816599	0.00041002	P
2,2	–0.104437	0.038384	–0.208874	0.129480	0.140820	0.142302	0.00041002	P
3,3	–0.101836	0.020343	–0.186344	0.110545	0.122335	0.123943	0.00041002	P
4,4	–0.099499	0.133111	0.128449	0.088049	0.083602	0.083157	0.00041002	H
5,5	–0.099300	0.317115	0.634230	0.061612	0.032438	0.028931	0.00041002	D
6,6	–0.124391	–0.181364	–0.362729	0.018754	0.033091	0.034546	0.00030752	P
7,7	–0.129361	–0.339575	–0.679151	0.059162	0.081203	0.083916	0.00041002	P
8,8	–0.132803	–0.426598	–0.853196	0.205534	0.244983	0.249832	0.00041002	P
9,9	0.213138	–0.603261	–1.206521	0.358710	0.374813	0.376685	0.00041002	P
T,T	0.673275	–0.850096	–1.700192	0.537013	0.460559	0.411694	0.00726325	S
				DAS				
A,A	–0.103203	0.194397	0.227011	0.709899	0.804904	0.816599	0.00041002	P
2,2	–0.104437	0.038384	–0.208874	0.263909	0.291643	0.295228	0.00041002	P
3,3	–0.101836	0.020343	–0.186344	0.247048	0.275509	0.279262	0.00041002	P
4,4	–0.099499	0.133111	0.128449	0.228293	0.241081	0.242865	0.00041002	P
5,5	–0.099300	0.317115	0.634230	0.193915	0.141728	0.135205	0.00041002	D
6,6	–0.124391	–0.181364	–0.362729	0.166059	0.194998	0.197919	0.00030752	P
7,7	–0.129361	–0.339575	–0.679151	0.204840	0.243975	0.248799	0.00041002	P
8,8	–0.132803	–0.426598	–0.853196	0.307512	0.358792	0.365088	0.00041002	P
9,9	0.213138	–0.603261	–1.206521	0.411114	0.433207	0.435799	0.00041002	P
T,T	0.673275	–0.850096	–1.700192	0.537013	0.460559	0.411694	0.00726325	S

TABLE A34
4D S17: Dealer's Upcard A

Hand	Stand	Hit	Double	Freq.	Cum. Freq.	BS
3,2	–0.665146	–0.282310	–1.330292	0.00063526	0.00063526	EH
4,2	–0.664774	–0.311679	–1.302311	0.00063526	0.00063526	EH
5,2	–0.664349	–0.315555	–1.128767	0.00063526		EH
4,3	–0.664426	–0.318705	–1.128919	0.00063526	0.00127053	EH
6,2	–0.660560	–0.204007	–0.808090	0.00063526		H
5,3	–0.664000	–0.201757	–0.806288	0.00063526	0.00127053	H
7,2	–0.661992	–0.069512	–0.425610	0.00063526		H
6,3	–0.660208	–0.069118	–0.425523	0.00063526		H
5,4	–0.663552	–0.066634	–0.423258	0.00063526	0.00190579	H
8,2	–0.663434	0.082708	–0.004763	0.00063526		H
7,3	–0.661642	0.082724	–0.001287	0.00063526		H
6,4	–0.659759	0.081295	0.000844	0.00063526	0.00190579	H
9,2	–0.664827	0.149272	0.127853	0.00063526		H
8,3	–0.663046	0.149469	0.131759	0.00063526		H
7,4	–0.661155	0.149578	0.135720	0.00063526		H
6,5	–0.659290	0.149949	0.140210	0.00063526	0.00254106	H
T,2	–0.668593	–0.350025	–0.824863	0.00255908		EH
9,3	–0.664489	–0.356252	–0.836196	0.00063526		EH
8,4	–0.662608	–0.356489	–0.833283	0.00063526		EH
7,5	–0.660734	–0.356669	–0.831025	0.00063526	0.00446487	EH
T,3	–0.668257	–0.395954	–0.875406	0.00255908		EH
9,4	–0.664052	–0.393442	–0.870173	0.00063526		EH
8,5	–0.662189	–0.405365	–0.892240	0.00063526		EH
7,6	–0.657023	–0.408371	–0.894703	0.00063526	0.00446487	EH
T,4	–0.667825	–0.441125	–0.931624	0.00255908		EH
9,5	–0.663634	–0.440065	–0.928599	0.00063526		EH
8,6	–0.658485	–0.443296	–0.934274	0.00063526	0.00382961	EH
T,5	–0.667411	–0.484420	–0.989897	0.00255908		EH
9,6	–0.659937	–0.483772	–0.989034	0.00063526		EH
8,7	–0.659935	–0.474112	–0.969744	0.00063526	0.00382961	EH
T,6	–0.663732	–0.515118	–1.030236	0.00255908		RH
9,7	–0.661373	–0.511872	–1.023744	0.00063526	0.00319434	RH
T,7	–0.475431	–0.556924	–1.113849	0.00255908		ES
9,8	–0.471769	–0.554120	–1.108241	0.00063526	0.00319434	ES
T,8	–0.095796	–0.628072	–1.256144	0.00255908	0.00255908	S
T,9	0.284937	–0.729009	–1.458019	0.00255908	0.00255908	S
A,2	–0.665842	–0.060532	–0.617034	0.00059556	0.00059556	H
A,3	–0.665502	–0.096578	–0.617555	0.00059556	0.00059556	H
A,4	–0.665062	–0.136428	–0.623385	0.00059556	0.00059556	H
A,5	–0.664711	–0.175627	–0.631749	0.00059556	0.00059556	H
A,6	–0.479196	–0.184673	–0.535285	0.00059556	0.00059556	H
A,7	–0.100326	–0.096573	–0.362009	0.00059556	0.00059556	H
A,8	0.280663	–0.008408	–0.189670	0.00059556	0.00059556	S
A,9	0.661689	0.078130	–0.020689	0.00059556	0.00059556	S
A,T	1.500000	0.145297	0.117483	0.00239914	0.00239914	S

NDAS

Hand	Stand	Hit	Double	Spl1	Spl2	Spl3	Freq.	BS
A,A	–0.666194	–0.023116	–0.618147	0.136338	0.235767	0.249959	0.00026056	P
2,2	–0.665487	–0.254950	–1.330973	–0.411961	–0.432699	–0.435870	0.00029778	H
3,3	–0.664871	–0.311664	–1.302485	–0.460980	–0.480309	–0.483202	0.00029778	EH
4,4	–0.663979	–0.199637	–0.806452	–0.515825	–0.557978	–0.564485	0.00029778	H
5,5	–0.663125	0.083604	0.003395	–0.577829	–0.666494	–0.680255	0.00029778	H
6,6	–0.655570	–0.359475	–0.831783	–0.616330	–0.650089	–0.655225	0.00029778	EH
7,7	–0.658474	–0.453109	–0.955310	–0.618513	–0.638985	–0.641902	0.00029778	EH
8,8	–0.661381	–0.511736	–1.023472	–0.381095	–0.363043	–0.360161	0.00029778	EP
9,9	–0.089453	–0.625904	–1.251808	–0.117093	–0.120107	–0.120469	0.00029778	S
T,T	0.654136	–0.859623	–1.719247	0.147418	–0.116219	–0.275395	0.00507366	S

DAS

Hand	Stand	Hit	Double	Spl1	Spl2	Spl3	Freq.	BS
A,A	–0.666194	–0.023116	–0.618147	0.136338	0.235767	0.249959	0.00026056	P
2,2	–0.665487	–0.254950	–1.330973	–0.411961	–0.432699	–0.435870	0.00029778	H
3,3	–0.664871	–0.311664	–1.302485	–0.460980	–0.480309	–0.483202	0.00029778	EH
4,4	–0.663979	–0.199637	–0.806452	–0.515825	–0.557978	–0.564485	0.00029778	H
5,5	–0.663125	0.083604	0.003395	–0.577829	–0.666494	–0.680255	0.00029778	H
6,6	–0.655570	–0.359475	–0.831783	–0.616330	–0.650089	–0.655225	0.00029778	EH
7,7	–0.658474	–0.453109	–0.955310	–0.618513	–0.638985	–0.641902	0.00029778	EH
8,8	–0.661381	–0.511736	–1.023472	–0.381095	–0.363043	–0.360161	0.00029778	EP
9,9	–0.089453	–0.625904	–1.251808	–0.117093	–0.120107	–0.120469	0.00029778	S
T,T	0.654136	–0.859623	–1.719247	0.147418	–0.116219	–0.275395	0.00507366	S

TABLE A35
4D S17: Dealer's Upcard 2

Hand	Stand	Hit	Double	Freq.	Cum. Freq.	BS
3,2	–0.292406	–0.128787	–0.584812	0.00086589	0.00086589	H
4,2	–0.291936	–0.142738	–0.562316	0.00086589	0.00086589	H
5,2	–0.288347	–0.106927	–0.427470	0.00086589		H
4,3	–0.291893	–0.111628	–0.432649	0.00092361	0.00178950	H
6,2	–0.289067	–0.019475	–0.194365	0.00086589		H
5,3	–0.288304	–0.019972	–0.194821	0.00092361	0.00178950	H
7,2	–0.289814	0.076910	0.074766	0.00086589		H
6,3	–0.289021	0.079013	0.073894	0.00092361		H
5,4	–0.287831	0.079581	0.074693	0.00092361	0.00271311	H
8,2	–0.290474	0.189422	0.374064	0.00086589		D
7,3	–0.289723	0.190187	0.375187	0.00092361		D
6,4	–0.288502	0.191062	0.376916	0.00092361	0.00271311	D
9,2	–0.291487	0.244208	0.483144	0.00086589		D
8,3	–0.290430	0.245388	0.485436	0.00092361		D
7,4	–0.289251	0.246627	0.487895	0.00092361		D
6,5	–0.284934	0.249411	0.493862	0.00092361	0.00363672	D
T,2	–0.297344	–0.251066	–0.502132	0.00346354		H
9,3	–0.291444	–0.256493	–0.512986	0.00092361		H
8,4	–0.289961	–0.254906	–0.509811	0.00092361		H
7,5	–0.285685	–0.253225	–0.506450	0.00092361	0.00623437	H
T,3	–0.297301	–0.306815	–0.613631	0.00369444		S
9,4	–0.290977	–0.304134	–0.608268	0.00092361		S
8,5	–0.286395	–0.313467	–0.626935	0.00092361		S
7,6	–0.286410	–0.313654	–0.627309	0.00092361	0.00646528	S
T,4	–0.296832	–0.363641	–0.727281	0.00369444		S
9,5	–0.287408	–0.361422	–0.722844	0.00092361		S
8,6	–0.287114	–0.362395	–0.724791	0.00092361	0.00554167	S
T,5	–0.293322	–0.421269	–0.842538	0.00369444		S
9,6	–0.288190	–0.419909	–0.839819	0.00092361		S
8,7	–0.287924	–0.410466	–0.820932	0.00092361	0.00554167	S
T,6	–0.294045	–0.469652	–0.939305	0.00369444		S
9,7	–0.288941	–0.467478	–0.934956	0.00092361	0.00461805	S
T,7	–0.154260	–0.536693	–1.073386	0.00369444		S
9,8	–0.149103	–0.534848	–1.069696	0.00092361	0.00461805	S
T,8	0.120910	–0.624892	–1.249785	0.00369444	0.00369444	S
T,9	0.386095	–0.734011	–1.468022	0.00369444	0.00369444	S
A,2	–0.290409	0.044916	–0.064142	0.00086589	0.00086589	H
A,3	–0.290367	0.021065	–0.065064	0.00092361	0.00092361	H
A,4	–0.289959	–0.002824	–0.070592	0.00092361	0.00092361	H
A,5	–0.286371	–0.023729	–0.074339	0.00092361	0.00092361	H
A,6	–0.147903	0.001295	–0.002371	0.00092361	0.00092361	H
A,7	0.125145	0.063484	0.121608	0.00092361	0.00092361	S
A,8	0.389794	0.122929	0.240518	0.00092361	0.00092361	S
A,9	0.643593	0.184222	0.363501	0.00092361	0.00092361	S
A,T	1.500000	0.240410	0.474326	0.00369444	0.00369444	S

Hand	Stand	Hit	Double	Spl1	Spl2	Spl3	Freq.	BS
				NDAS				
A,A	–0.288391	0.084868	–0.058748	0.493574	0.597683	0.613730	0.00043294	P
2,2	–0.292402	–0.114641	–0.584803	–0.146374	–0.150238	–0.150780	0.00037882	H
3,3	–0.292366	–0.142977	–0.563092	–0.200272	–0.207611	–0.208698	0.00043294	H
4,4	–0.291419	–0.018929	–0.198235	–0.229762	–0.257807	–0.262128	0.00043294	H
5,5	–0.284214	0.192598	0.380351	–0.250231	–0.308790	–0.317754	0.00043294	D
6,6	–0.285656	–0.253283	–0.506565	–0.267409	–0.268191	–0.268142	0.00043294	H
7,7	–0.287158	–0.372824	–0.745647	–0.205212	–0.193807	–0.191974	0.00043294	P
8,8	–0.288633	–0.467167	–0.934334	–0.029770	0.005137	0.010585	0.00043294	P
9,9	0.125487	–0.623687	–1.247374	0.154221	0.158063	0.158656	0.00043294	P
T,T	0.636830	–0.853184	–1.706368	0.352011	0.201996	0.110425	0.00727344	S
				DAS				
A,A	–0.288391	0.084868	–0.058748	0.493574	0.597683	0.613730	0.00043294	P
2,2	–0.292402	–0.114641	–0.584803	–0.079497	–0.074998	–0.074335	0.00037882	P
3,3	–0.292366	–0.142977	–0.563092	–0.133194	–0.131528	–0.131216	0.00043294	P
4,4	–0.291419	–0.018929	–0.198235	–0.162294	–0.181274	–0.184185	0.00043294	H
5,5	–0.284214	0.192598	0.380351	–0.185115	–0.260475	–0.272102	0.00043294	D
6,6	–0.285656	–0.253283	–0.506565	–0.198941	–0.190457	–0.188957	0.00043294	P
7,7	–0.287158	–0.372824	–0.745647	–0.137673	–0.117165	–0.113911	0.00043294	P
8,8	–0.288633	–0.467167	–0.934334	0.035437	0.079100	0.085910	0.00043294	P
9,9	0.125487	–0.623687	–1.247374	0.189268	0.197776	0.199089	0.00043294	P
T,T	0.636830	–0.853184	–1.706368	0.352011	0.201996	0.110425	0.00727344	S

TABLE A36
4D S17: Dealer's Upcard 3

Hand	Stand	Hit	Double	Freq.	Cum. Freq.	BS
3,2	−0.250442	−0.095461	−0.500883	0.00086589	0.00086589	H
4,2	−0.246883	−0.106586	−0.473380	0.00092361	0.00092361	H
5,2	−0.246208	−0.072673	−0.347120	0.00092361		H
4,3	−0.246415	−0.077047	−0.349816	0.00086589	0.00178950	H
6,2	−0.246910	0.012000	−0.121865	0.00092361		H
5,3	−0.245741	0.011365	−0.122335	0.00086589	0.00178950	H
7,2	−0.247610	0.108301	0.138466	0.00092361		D
6,3	−0.246397	0.110287	0.137534	0.00086589		D
5,4	−0.242154	0.111051	0.139607	0.00092361	0.00271311	D
8,2	−0.248619	0.214463	0.426813	0.00092361		D
7,3	−0.247145	0.215722	0.429189	0.00086589		D
6,4	−0.242857	0.216800	0.431672	0.00092361	0.00271311	D
9,2	−0.254554	0.262771	0.523244	0.00092361		D
8,3	−0.248156	0.268664	0.535256	0.00086589		D
7,4	−0.243606	0.270943	0.540132	0.00092361		D
6,5	−0.242179	0.273134	0.544523	0.00092361	0.00363672	D
T,2	−0.255715	−0.230175	−0.460350	0.00369444		H
9,3	−0.254093	−0.238694	−0.477387	0.00086589		H
8,4	−0.244620	−0.232693	−0.465386	0.00092361		H
7,5	−0.242931	−0.230978	−0.461956	0.00092361	0.00640755	H
T,3	−0.255253	−0.289105	−0.578209	0.00346354		S
9,4	−0.250554	−0.288354	−0.576708	0.00092361		S
8,5	−0.243941	−0.294694	−0.589387	0.00092361		S
7,6	−0.243634	−0.294470	−0.588941	0.00092361	0.00623437	S
T,4	−0.251772	−0.350253	−0.700507	0.00369444		S
9,5	−0.249933	−0.350253	−0.700505	0.00092361		S
8,6	−0.244707	−0.348415	−0.696830	0.00092361	0.00554167	S
T,5	−0.251093	−0.411772	−0.823544	0.00369444		S
9,6	−0.250641	−0.412320	−0.824640	0.00092361		S
8,7	−0.245456	−0.400432	−0.800863	0.00092361	0.00554167	S
T,6	−0.251802	−0.463189	−0.926377	0.00369444		S
9,7	−0.251392	−0.462570	−0.925140	0.00092361	0.00461805	S
T,7	−0.117647	−0.532869	−1.065738	0.00369444		S
9,8	−0.117610	−0.532554	−1.065107	0.00092361	0.00461805	S
T,8	0.147677	−0.623311	−1.246622	0.00369444	0.00369444	S
T,9	0.399611	−0.724314	−1.448628	0.00369444	0.00369444	S
A,2	−0.248982	0.073436	0.002390	0.00092361	0.00092361	H
A,3	−0.248581	0.049381	−0.001551	0.00086589	0.00086589	H
A,4	−0.245022	0.027614	−0.004340	0.00092361	0.00092361	H
A,5	−0.244346	0.006157	−0.010051	0.00092361	0.00092361	H
A,6	−0.111445	0.030612	0.059478	0.00092361	0.00092361	D
A,7	0.152566	0.090978	0.180202	0.00092361	0.00092361	D
A,8	0.407804	0.154763	0.307584	0.00092361	0.00092361	S
A,9	0.648917	0.203979	0.405660	0.00092361	0.00092361	S
A,T	1.500000	0.262854	0.522236	0.00369444	0.00369444	S

Hand	Stand	Hit	Double	Spl1	Spl2	Spl3	Freq.	BS
				NDAS				
A,A	−0.247084	0.107339	0.008117	0.540940	0.645827	0.661990	0.00043294	P
2,2	−0.250905	−0.081843	−0.501811	−0.091934	−0.093236	−0.093431	0.00043294	H
3,3	−0.249976	−0.108859	−0.479328	−0.134228	−0.137130	−0.137510	0.00037882	H
4,4	−0.242856	0.013335	−0.122047	−0.155292	−0.177338	−0.180674	0.00043294	H
5,5	−0.241474	0.217915	0.433895	−0.180702	−0.233295	−0.241328	0.00043294	D
6,6	−0.242888	−0.231077	−0.462154	−0.196612	−0.190849	−0.189779	0.00043294	P
7,7	−0.244441	−0.358404	−0.716809	−0.136846	−0.121931	−0.119541	0.00043294	P
8,8	−0.246472	−0.460584	−0.921167	0.030893	0.068248	0.074070	0.00043294	P
9,9	0.142932	−0.624231	−1.248461	0.195905	0.202391	0.203303	0.00043294	P
T,T	0.646830	−0.852909	−1.705818	0.400670	0.270612	0.191003	0.00727344	S
				DAS				
A,A	−0.247084	0.107339	0.008117	0.540940	0.645827	0.661990	0.00043294	P
2,2	−0.250905	−0.081843	−0.501811	−0.012175	−0.002720	−0.001235	0.00043294	P
3,3	−0.249976	−0.108859	−0.479328	−0.054160	−0.047055	−0.045995	0.00037882	P
4,4	−0.242856	0.013335	−0.122047	−0.073597	−0.084574	−0.086175	0.00043294	H
5,5	−0.241474	0.217915	0.433895	−0.102198	−0.173547	−0.184541	0.00043294	D
6,6	−0.242888	−0.231077	−0.462154	−0.109131	−0.091344	−0.088362	0.00043294	P
7,7	−0.244441	−0.358404	−0.716809	−0.051397	−0.024718	−0.020455	0.00043294	P
8,8	−0.246472	−0.460584	−0.921167	0.104052	0.151257	0.158615	0.00043294	P
9,9	0.142932	−0.624231	−1.248461	0.236000	0.247739	0.249447	0.00043294	P
T,T	0.646830	−0.852909	−1.705818	0.400670	0.270612	0.191003	0.00727344	S

TABLE A37

4D S17: Dealer's Upcard 4

Hand	Stand	Hit	Double	Freq.	Cum. Freq.	BS
3,2	–0.202456	–0.056398	–0.404912	0.00092361	0.00092361	H
4,2	–0.201763	–0.068519	–0.383358	0.00086589	0.00086589	H
5,2	–0.201184	–0.035076	–0.261375	0.00092361		H
4,3	–0.198209	–0.036671	–0.257640	0.00086589	0.00178950	H
6,2	–0.201826	0.048818	–0.045273	0.00092361		H
5,3	–0.197602	0.050221	–0.040958	0.00092361	0.00184722	H
7,2	–0.202892	0.138243	0.203383	0.00092361		D
6,3	–0.198291	0.142594	0.207579	0.00092361		D
5,4	–0.196904	0.143280	0.208968	0.00086589	0.00271311	D
8,2	–0.208819	0.235977	0.471954	0.00092361		D
7,3	–0.199359	0.243622	0.487244	0.00092361		D
6,4	–0.197595	0.245162	0.490325	0.00086589	0.00271311	D
9,2	–0.210021	0.287113	0.574227	0.00092361		D
8,3	–0.205287	0.290239	0.580479	0.00092361		D
7,4	–0.198667	0.297036	0.594072	0.00086589		D
6,5	–0.197013	0.299204	0.598409	0.00092361	0.00363672	D
T,2	–0.211151	–0.208764	–0.417528	0.00369444		H
9,3	–0.206487	–0.216336	–0.432672	0.00092361		S
8,4	–0.204591	–0.213737	–0.427474	0.00086589		S
7,5	–0.198081	–0.208602	–0.417205	0.00092361	0.00640755	S
T,3	–0.207676	–0.270178	–0.540356	0.00369444		S
9,4	–0.205849	–0.269488	–0.538976	0.00086589		S
8,5	–0.204064	–0.278521	–0.557043	0.00092361		S
7,6	–0.198831	–0.275166	–0.550331	0.00092361	0.00640755	S
T,4	–0.206980	–0.335306	–0.670612	0.00346354		S
9,5	–0.205264	–0.335223	–0.670446	0.00092361		S
8,6	–0.204757	–0.335682	–0.671364	0.00092361	0.00531076	S
T,5	–0.206395	–0.401063	–0.802127	0.00369444		S
9,6	–0.205958	–0.401510	–0.803019	0.00092361		S
8,7	–0.205830	–0.391951	–0.783901	0.00092361	0.00554167	S
T,6	–0.207089	–0.455938	–0.911877	0.00369444		S
9,7	–0.207031	–0.455585	–0.911171	0.00092361	0.00461805	S
T,7	–0.076682	–0.528969	–1.057938	0.00369444		S
9,8	–0.081417	–0.530243	–1.060485	0.00092361	0.00461805	S
T,8	0.173142	–0.612575	–1.225149	0.00369444	0.00369444	S
T,9	0.418446	–0.723295	–1.446590	0.00369444	0.00369444	S
A,2	–0.204628	0.104331	0.072829	0.00092361	0.00092361	H
A,3	–0.201074	0.082588	0.071144	0.00092361	0.00092361	H
A,4	–0.200381	0.059483	0.064430	0.00086589	0.00086589	D
A,5	–0.199799	0.039093	0.058790	0.00092361	0.00092361	D
A,6	–0.070209	0.063444	0.126889	0.00092361	0.00092361	D
A,7	0.182473	0.127404	0.254809	0.00092361	0.00092361	D
A,8	0.421677	0.178307	0.356613	0.00092361	0.00092361	S
A,9	0.659211	0.230121	0.460242	0.00092361	0.00092361	S
A,T	1.500000	0.286513	0.573026	0.00369444	0.00369444	S

Hand	Stand	Hit	Double	Spl1	Spl2	Spl3	Freq.	BS
				NDAS				
A,A	–0.203147	0.130233	0.077564	0.591106	0.696781	0.713059	0.00043294	P
2,2	–0.206009	–0.045498	–0.412018	–0.034634	–0.033149	–0.032915	0.00043294	P
3,3	–0.198901	–0.066701	–0.378308	–0.053051	–0.050726	–0.050290	0.00043294	P
4,4	–0.197491	0.052735	–0.040638	–0.078718	–0.094602	–0.096811	0.00037882	H
5,5	–0.196320	0.245694	0.491389	–0.103551	–0.149446	–0.156427	0.00043294	D
6,6	–0.197701	–0.208262	–0.416523	–0.118655	–0.107291	–0.105411	0.00043294	P
7,7	–0.199903	–0.343285	–0.686570	–0.061789	–0.042716	–0.039672	0.00043294	P
8,8	–0.211757	–0.457247	–0.914494	0.085046	0.124464	0.130524	0.00043294	P
9,9	0.173230	–0.612558	–1.225116	0.256162	0.266701	0.268249	0.00043294	P
T,T	0.657136	–0.852691	–1.705382	0.450513	0.340435	0.272563	0.00727344	S
				DAS				
A,A	–0.203147	0.130233	0.077564	0.591106	0.696781	0.713059	0.00043294	P
2,2	–0.206009	–0.045498	–0.412018	0.058345	0.072332	0.074511	0.00043294	P
3,3	–0.198901	–0.066701	–0.378308	0.041828	0.056938	0.059368	0.00043294	P
4,4	–0.197491	0.052735	–0.040638	0.019183	0.015680	0.015273	0.00037882	H
5,5	–0.196320	0.245694	0.491389	–0.008141	–0.074538	–0.084756	0.00043294	D
6,6	–0.197701	–0.208262	–0.416523	–0.012447	0.013495	0.017689	0.00043294	P
7,7	–0.199903	–0.343285	–0.686570	0.044565	0.078210	0.083564	0.00043294	P
8,8	–0.211757	–0.457247	–0.914494	0.166734	0.216946	0.224656	0.00043294	P
9,9	0.173230	–0.612558	–1.225116	0.300395	0.316735	0.319164	0.00043294	P
T,T	0.657136	–0.852691	–1.705382	0.450513	0.340435	0.272563	0.00727344	S

TABLE A38
4D S17: Dealer's Upcard 5

Hand	Stand	Hit	Double	Freq.	Cum. Freq.	BS
3,2	−0.152592	−0.013403	−0.305183	0.00092361	0.00092361	H
4,2	−0.151991	−0.024723	−0.285513	0.00092361	0.00092361	H
5,2	−0.151416	0.007645	−0.172172	0.00086589		H
4,3	−0.151321	0.005793	−0.170158	0.00092361	0.00178950	H
6,2	−0.152481	0.084884	0.033185	0.00092361		H
5,3	−0.150768	0.087177	0.038529	0.00086589	0.00178950	H
7,2	−0.158468	0.167137	0.264547	0.00092361		D
6,3	−0.151834	0.176105	0.277834	0.00092361		D
5,4	−0.150162	0.178597	0.282882	0.00086589	0.00271311	D
8,2	−0.159696	0.265838	0.531677	0.00092361		D
7,3	−0.157824	0.268558	0.537116	0.00092361		D
6,4	−0.151230	0.275562	0.551125	0.00092361	0.00277083	D
9,2	−0.160734	0.314945	0.629891	0.00092361		D
8,3	−0.159049	0.317460	0.634919	0.00092361		D
7,4	−0.157217	0.320261	0.640521	0.00092361		D
6,5	−0.150677	0.327443	0.654886	0.00086589	0.00363672	D
T,2	−0.161797	−0.186239	−0.372477	0.00369444		S
9,3	−0.160146	−0.194195	−0.388391	0.00092361		S
8,4	−0.158501	−0.191732	−0.383465	0.00092361		S
7,5	−0.156722	−0.189616	−0.379232	0.00086589	0.00640755	S
T,3	−0.161150	−0.252125	−0.504250	0.00369444		S
9,4	−0.159540	−0.251454	−0.502907	0.00092361		S
8,5	−0.157947	−0.260486	−0.520972	0.00086589		S
7,6	−0.157791	−0.260173	−0.520345	0.00092361	0.00640755	S
T,4	−0.160544	−0.319845	−0.639690	0.00369444		S
9,5	−0.158986	−0.319746	−0.639492	0.00086589		S
8,6	−0.159018	−0.320513	−0.641026	0.00092361	0.00548394	S
T,5	−0.159992	−0.388942	−0.777883	0.00346354		S
9,6	−0.160058	−0.389648	−0.779296	0.00092361		S
8,7	−0.165007	−0.382453	−0.764906	0.00092361	0.00531076	S
T,6	−0.161064	−0.448446	−0.896891	0.00369444		S
9,7	−0.166048	−0.450104	−0.900207	0.00092361	0.00461805	S
T,7	−0.044502	−0.515732	−1.031464	0.00369444		S
9,8	−0.044729	−0.515746	−1.031492	0.00092361	0.00461805	S
T,8	0.200313	−0.609397	−1.218794	0.00369444	0.00369444	S
T,9	0.441414	−0.721883	−1.443767	0.00369444	0.00369444	S
A,2	−0.155477	0.139355	0.146873	0.00092361	0.00092361	D
A,3	−0.154833	0.117682	0.144734	0.00092361	0.00092361	D
A,4	−0.154230	0.095584	0.137644	0.00092361	0.00092361	D
A,5	−0.153683	0.075227	0.130673	0.00086589	0.00086589	D
A,6	−0.033358	0.102810	0.205619	0.00092361	0.00092361	D
A,7	0.204943	0.154268	0.308536	0.00092361	0.00092361	D
A,8	0.444620	0.208811	0.417621	0.00092361	0.00092361	S
A,9	0.673137	0.259017	0.518035	0.00092361	0.00092361	S
A,T	1.500000	0.313124	0.626248	0.00369444	0.00369444	S

Hand	Stand	Hit	Double	Spl1	Spl2	Spl3	Freq.	BS
				NDAS				
A,A	−0.157688	0.162849	0.148228	0.643588	0.750170	0.766583	0.00043294	P
2,2	−0.153234	−0.001167	−0.306468	0.051006	0.058234	0.059391	0.00043294	P
3,3	−0.151949	−0.024753	−0.285067	0.027312	0.034821	0.036067	0.00043294	P
4,4	−0.150714	0.090177	0.040095	0.003027	−0.008168	−0.009832	0.00043294	H
5,5	−0.149611	0.277401	0.554803	−0.021325	−0.057829	−0.062967	0.00037882	D
6,6	−0.151806	−0.186140	−0.372279	−0.037316	−0.021237	−0.018631	0.00043294	P
7,7	−0.163779	−0.333129	−0.666257	−0.000882	0.020842	0.024195	0.00043294	P
8,8	−0.166236	−0.450158	−0.900316	0.153583	0.196014	0.202530	0.00043294	P
9,9	0.200299	−0.609373	−1.218745	0.321126	0.336689	0.339008	0.00043294	P
T,T	0.671161	−0.852354	−1.704708	0.508750	0.421512	0.367341	0.00727344	S
				DAS				
A,A	−0.157688	0.162849	0.148228	0.643588	0.750170	0.766583	0.00043294	P
2,2	−0.153234	−0.001167	−0.306468	0.160480	0.182567	0.186057	0.00043294	P
3,3	−0.151949	−0.024753	−0.285067	0.141358	0.164337	0.168010	0.00043294	P
4,4	−0.150714	0.090177	0.040095	0.120615	0.125401	0.126251	0.00043294	P
5,5	−0.149611	0.277401	0.554803	0.093041	0.035761	0.027574	0.00037882	D
6,6	−0.151806	−0.186140	−0.372279	0.089412	0.122830	0.128181	0.00043294	P
7,7	−0.163779	−0.333129	−0.666257	0.122943	0.161185	0.167089	0.00043294	P
8,8	−0.166236	−0.450158	−0.900316	0.244145	0.298542	0.306888	0.00043294	P
9,9	0.200299	−0.609373	−1.218745	0.369709	0.391652	0.394941	0.00043294	P
T,T	0.671161	−0.852354	−1.704708	0.508750	0.421512	0.367341	0.00727344	S

TABLE A39

4D S17: Dealer's Upcard 6

Hand	Stand	Hit	Double	Freq.	Cum. Freq.	BS
3,2	−0.146040	0.003852	−0.292079	0.00092361	0.00092361	H
4,2	−0.145497	−0.006429	−0.266007	0.00092361	0.00092361	H
5,2	−0.145039	0.038950	−0.119648	0.00092361		H
4,3	−0.144936	0.036440	−0.119054	0.00092361	0.00184722	H
6,2	−0.150928	0.119576	0.096823	0.00086589		H
5,3	−0.144478	0.126595	0.111435	0.00092361	0.00178950	H
7,2	−0.152067	0.199835	0.327660	0.00092361		D
6,3	−0.150368	0.204673	0.332377	0.00086589		D
5,4	−0.143958	0.211908	0.346859	0.00092361	0.00271311	D
8,2	−0.153013	0.292696	0.585391	0.00092361		D
7,3	−0.151503	0.295345	0.590691	0.00092361		D
6,4	−0.149845	0.298463	0.596925	0.00086589	0.00271311	D
9,2	−0.154002	0.336855	0.673709	0.00092361		D
8,3	−0.152509	0.339171	0.678341	0.00092361		D
7,4	−0.151038	0.341636	0.683273	0.00092361		D
6,5	−0.149443	0.345182	0.690365	0.00086589	0.00363672	D
T,2	−0.155208	−0.167860	−0.335719	0.00369444		S
9,3	−0.153441	−0.175959	−0.351918	0.00092361		S
8,4	−0.151987	−0.173705	−0.347410	0.00092361		S
7,5	−0.150578	−0.172221	−0.344443	0.00092361	0.00646528	S
T,3	−0.154647	−0.234119	−0.468239	0.00369444		S
9,4	−0.152919	−0.233779	−0.467558	0.00092361		S
8,5	−0.151526	−0.243201	−0.486402	0.00092361		S
7,6	−0.156468	−0.246032	−0.492065	0.00086589	0.00640755	S
T,4	−0.154126	−0.302576	−0.605152	0.00369444		S
9,5	−0.152459	−0.302569	−0.605138	0.00092361		S
8,6	−0.157417	−0.305955	−0.611909	0.00086589	0.00548394	S
T,5	−0.153670	−0.371043	−0.742085	0.00369444		S
9,6	−0.158358	−0.374064	−0.748128	0.00086589		S
8,7	−0.158564	−0.364819	−0.729638	0.00092361	0.00548394	S
T,6	−0.159481	−0.422642	−0.845285	0.00346354		S
9,7	−0.159416	−0.422744	−0.845488	0.00092361	0.00438715	S
T,7	0.006566	−0.502570	−1.005140	0.00369444		S
9,8	0.006500	−0.502628	−1.005257	0.00092361	0.00461805	S
T,8	0.279797	−0.602339	−1.204678	0.00369444	0.00369444	S
T,9	0.493179	−0.718793	−1.437585	0.00369444	0.00369444	S
A,2	−0.144131	0.163183	0.192050	0.00092361	0.00092361	D
A,3	−0.143571	0.140795	0.189743	0.00092361	0.00092361	D
A,4	−0.143053	0.118505	0.184548	0.00092361	0.00092361	D
A,5	−0.142573	0.102886	0.188512	0.00092361	0.00092361	D
A,6	0.012075	0.129810	0.259619	0.00086589	0.00086589	D
A,7	0.278929	0.191679	0.383357	0.00092361	0.00092361	D
A,8	0.492868	0.240289	0.480578	0.00092361	0.00092361	S
A,9	0.701886	0.286099	0.572197	0.00092361	0.00092361	S
A,T	1.500000	0.335539	0.671078	0.00369444	0.00369444	S

Hand	Stand	Hit	Double	Spl1	Spl2	Spl3	Freq.	BS
				NDAS				
A,A	−0.141645	0.189020	0.196249	0.689889	0.800078	0.816981	0.00043294	P
2,2	−0.146596	0.016371	−0.293191	0.089672	0.099768	0.101375	0.00043294	P
3,3	−0.145457	−0.006444	−0.265922	0.063830	0.073822	0.075459	0.00043294	P
4,4	−0.144416	0.129206	0.111986	0.037078	0.025197	0.023424	0.00043294	H
5,5	−0.143496	0.305341	0.610682	0.012235	−0.026335	−0.032211	0.00043294	D
6,6	−0.155330	−0.175559	−0.351118	−0.020740	−0.003851	−0.001411	0.00037882	P
7,7	−0.157606	−0.316272	−0.632543	0.058938	0.087771	0.092215	0.00043294	P
8,8	−0.159434	−0.422727	−0.845454	0.230912	0.282746	0.290715	0.00043294	P
9,9	0.279430	−0.602385	−1.204771	0.384862	0.398332	0.400322	0.00043294	P
T,T	0.702265	−0.851570	−1.703141	0.562301	0.486664	0.439456	0.00727344	S
				DAS				
A,A	−0.141645	0.189020	0.196249	0.689889	0.800078	0.816981	0.00043294	P
2,2	−0.146596	0.016371	−0.293191	0.215280	0.242435	0.246723	0.00043294	P
3,3	−0.145457	−0.006444	−0.265922	0.192041	0.219434	0.223805	0.00043294	P
4,4	−0.144416	0.129206	0.111986	0.168642	0.174654	0.175698	0.00043294	P
5,5	−0.143496	0.305341	0.610682	0.142301	0.079701	0.070013	0.00043294	D
6,6	−0.155330	−0.175559	−0.351118	0.120945	0.155479	0.160452	0.00037882	P
7,7	−0.157606	−0.316272	−0.632543	0.195072	0.242047	0.249289	0.00043294	P
8,8	−0.159434	−0.422727	−0.845454	0.329041	0.393847	0.403803	0.00043294	P
9,9	0.279430	−0.602385	−1.204771	0.436874	0.457181	0.460211	0.00043294	P
T,T	0.702265	−0.851570	−1.703141	0.562301	0.486664	0.439456	0.00727344	S

TABLE A40
4D S17: Dealer's Upcard 7

Hand	Stand	Hit	Double	Freq.	Cum. Freq.	BS
3,2	–0.473744	–0.119489	–0.947488	0.00092361	0.00092361	H
4,2	–0.473230	–0.154678	–0.888294	0.00092361	0.00092361	H
5,2	–0.475452	–0.068561	–0.583473	0.00092361		H
4,3	–0.472747	–0.069274	–0.578029	0.00092361	0.00184722	H
6,2	–0.476595	0.084790	–0.177548	0.00092361		H
5,3	–0.474970	0.084676	–0.174035	0.00092361	0.00184722	H
7,2	–0.477635	0.175421	0.118269	0.00086589		H
6,3	–0.476108	0.178192	0.121837	0.00092361		H
5,4	–0.474451	0.178579	0.125159	0.00092361	0.00271311	H
8,2	–0.478446	0.259428	0.398326	0.00092361		D
7,3	–0.477207	0.262042	0.405886	0.00086589		D
6,4	–0.475650	0.263931	0.412852	0.00092361	0.00271311	D
9,2	–0.479752	0.291519	0.461327	0.00092361		D
8,3	–0.477960	0.292163	0.465353	0.00092361		D
7,4	–0.476692	0.293167	0.469940	0.00086589		D
6,5	–0.477867	0.293989	0.473101	0.00092361	0.00363672	D
T,2	–0.474716	–0.212542	–0.503932	0.00369444		H
9,3	–0.479268	–0.220984	–0.523830	0.00092361		H
8,4	–0.477445	–0.220527	–0.520166	0.00092361		H
7,5	–0.478909	–0.223059	–0.524096	0.00086589	0.00640755	H
T,3	–0.474233	–0.269367	–0.586268	0.00369444		H
9,4	–0.478752	–0.270153	–0.590256	0.00092361		H
8,5	–0.479663	–0.282775	–0.615300	0.00092361		H
7,6	–0.480052	–0.283237	–0.616348	0.00086589	0.00640755	H
T,4	–0.473721	–0.326238	–0.677342	0.00369444		H
9,5	–0.480975	–0.327523	–0.681924	0.00092361		H
8,6	–0.480814	–0.327645	–0.683560	0.00092361	0.00554167	H
T,5	–0.475899	–0.368503	–0.745856	0.00369444		H
9,6	–0.482037	–0.368223	–0.747781	0.00092361		H
8,7	–0.481778	–0.358883	–0.729046	0.00086589	0.00548394	H
T,6	–0.477058	–0.405505	–0.811011	0.00369444		H
9,7	–0.483097	–0.405216	–0.810432	0.00086589	0.00456033	H
T,7	–0.109970	–0.475803	–0.951607	0.00346354		S
9,8	–0.111160	–0.475319	–0.950637	0.00092361	0.00438715	S
T,8	0.396838	–0.585413	–1.170826	0.00369444	0.00369444	S
T,9	0.614575	–0.711392	–1.422783	0.00369444	0.00369444	S
A,2	–0.472081	0.119029	–0.177283	0.00092361	0.00092361	H
A,3	–0.471601	0.075026	–0.181438	0.00092361	0.00092361	H
A,4	–0.471066	0.036104	–0.173744	0.00092361	0.00092361	H
A,5	–0.473288	–0.009400	–0.184635	0.00092361	0.00092361	H
A,6	–0.102349	0.055160	–0.006438	0.00092361	0.00092361	H
A,7	0.402983	0.172399	0.225480	0.00086589	0.00086589	S
A,8	0.615612	0.221109	0.320950	0.00092361	0.00092361	S
A,9	0.773289	0.253501	0.382647	0.00092361	0.00092361	S
A,T	1.500000	0.290734	0.464189	0.00369444	0.00369444	S

NDAS

Hand	Stand	Hit	Double	Spl1	Spl2	Spl3	Freq.	BS
A,A	–0.469898	0.163757	–0.172621	0.481997	0.608724	0.628256	0.00043294	P
2,2	–0.474248	–0.089107	–0.948496	–0.054045	–0.049320	–0.048584	0.00043294	P
3,3	–0.473262	–0.154760	–0.888339	–0.115280	–0.109402	–0.108399	0.00043294	P
4,4	–0.472234	0.088868	–0.168701	–0.186617	–0.223805	–0.229617	0.00043294	H
5,5	–0.476727	0.262157	0.410917	–0.252477	–0.321970	–0.332832	0.00043294	D
6,6	–0.479008	–0.224556	–0.527096	–0.296957	–0.305903	–0.307175	0.00043294	H
7,7	–0.481104	–0.337152	–0.704102	–0.134199	–0.108389	–0.104611	0.00037882	P
8,8	–0.482548	–0.405174	–0.810348	0.174398	0.251201	0.262983	0.00043294	P
9,9	0.399617	–0.585162	–1.170324	0.343101	0.335466	0.334273	0.00043294	S
T,T	0.771374	–0.849732	–1.699465	0.506167	0.368243	0.285005	0.00727344	S

DAS

Hand	Stand	Hit	Double	Spl1	Spl2	Spl3	Freq.	BS
A,A	–0.469898	0.163757	–0.172621	0.481997	0.608724	0.628256	0.00043294	P
2,2	–0.474248	–0.089107	–0.948496	–0.004099	0.007490	0.009316	0.00043294	P
3,3	–0.473262	–0.154760	–0.888339	–0.065203	–0.052414	–0.050308	0.00043294	P
4,4	–0.472234	0.088868	–0.168701	–0.135156	–0.165203	–0.169870	0.00043294	H
5,5	–0.476727	0.262157	0.410917	–0.202502	–0.285765	–0.298845	0.00043294	D
6,6	–0.479008	–0.224556	–0.527096	–0.244776	–0.246663	–0.246829	0.00043294	H
7,7	–0.481104	–0.337152	–0.704102	–0.084465	–0.052544	–0.047901	0.00037882	P
8,8	–0.482548	–0.405174	–0.810348	0.221822	0.304758	0.317459	0.00043294	P
9,9	0.399617	–0.585162	–1.170324	0.368477	0.364066	0.363346	0.00043294	S
T,T	0.771374	–0.849732	–1.699465	0.506167	0.368243	0.285005	0.00727344	S

TABLE A41
4D S17: Dealer's Upcard 8

Hand	Stand	Hit	Double	Freq.	Cum. Freq.	BS
3,2	–0.511100	–0.186664	–1.022199	0.00092361	0.00092361	H
4,2	–0.513268	–0.221006	–1.006429	0.00092361	0.00092361	H
5,2	–0.513188	–0.212223	–0.848329	0.00092361		H
4,3	–0.512754	–0.214655	–0.847489	0.00092361	0.00184722	H
6,2	–0.514137	–0.058875	–0.448845	0.00092361		H
5,3	–0.512670	–0.059228	–0.446067	0.00092361	0.00184722	H
7,2	–0.515041	0.100447	–0.019183	0.00092361		H
6,3	–0.513679	0.102668	–0.016898	0.00092361		H
5,4	–0.514892	0.100702	–0.019341	0.00092361	0.00277083	H
8,2	–0.516352	0.199835	0.288079	0.00086589		D
7,3	–0.514524	0.202193	0.295685	0.00092361		D
6,4	–0.515843	0.200444	0.294105	0.00092361	0.00271311	D
9,2	–0.511321	0.226418	0.345105	0.00092361		D
8,3	–0.515837	0.227823	0.345738	0.00086589		D
7,4	–0.516689	0.228163	0.348281	0.00092361		D
6,5	–0.515758	0.230237	0.354804	0.00092361	0.00363672	D
T,2	–0.512196	–0.272222	–0.618185	0.00369444		H
9,3	–0.510806	–0.282234	–0.636741	0.00092361		H
8,4	–0.518002	–0.282865	–0.640233	0.00086589		H
7,5	–0.516606	–0.283101	–0.638037	0.00092361	0.00640755	H
T,3	–0.511684	–0.324828	–0.694547	0.00369444		H
9,4	–0.512975	–0.327234	–0.698857	0.00092361		H
8,5	–0.517922	–0.338532	–0.723484	0.00086589		H
7,6	–0.517566	–0.340317	–0.726862	0.00092361	0.00640755	H
T,4	–0.513809	–0.368434	–0.759805	0.00369444		H
9,5	–0.512851	–0.371382	–0.765021	0.00092361		H
8,6	–0.518794	–0.371252	–0.768217	0.00086589	0.00548394	H
T,5	–0.513733	–0.416983	–0.842012	0.00369444		H
9,6	–0.513819	–0.417733	–0.844886	0.00092361		H
8,7	–0.519657	–0.407971	–0.826337	0.00086589	0.00548394	H
T,6	–0.514701	–0.450434	–0.900867	0.00369444		H
9,7	–0.514682	–0.451179	–0.902358	0.00092361	0.00461805	H
T,7	–0.385343	–0.498411	–0.996822	0.00369444		S
9,8	–0.390252	–0.498754	–0.997508	0.00086589	0.00456033	S
T,8	0.103131	–0.584919	–1.169839	0.00346354	0.00346354	S
T,9	0.589698	–0.709749	–1.419499	0.00369444	0.00369444	S
A,2	–0.509765	0.050688	–0.313928	0.00092361	0.00092361	H
A,3	–0.509230	0.018138	–0.300276	0.00092361	0.00092361	H
A,4	–0.511398	–0.029088	–0.314247	0.00092361	0.00092361	H
A,5	–0.511319	–0.070735	–0.318548	0.00092361	0.00092361	H
A,6	–0.382888	–0.071144	–0.249396	0.00092361	0.00092361	H
A,7	0.109163	0.041697	–0.027229	0.00092361	0.00092361	S
A,8	0.596957	0.153259	0.193419	0.00086589	0.00086589	S
A,9	0.790078	0.191569	0.272846	0.00092361	0.00092361	S
A,T	1.500000	0.227782	0.346321	0.00369444	0.00369444	S

Hand	Stand	Hit	Double	Spl1	Spl2	Spl3	Freq.	BS
				NDAS				
A,A	–0.507875	0.094344	–0.310003	0.364258	0.480545	0.498495	0.00043294	P
2,2	–0.511613	–0.155451	–1.023226	–0.207458	–0.214211	–0.215226	0.00043294	H
3,3	–0.510585	–0.220202	–1.001065	–0.263490	–0.268552	–0.269223	0.00043294	H
4,4	–0.514918	–0.058680	–0.450579	–0.324091	–0.359810	–0.365374	0.00043294	H
5,5	–0.514807	0.200331	0.296188	–0.392536	–0.472459	–0.484931	0.00043294	D
6,6	–0.516710	–0.283253	–0.638127	–0.430243	–0.449273	–0.452124	0.00043294	H
7,7	–0.518333	–0.380634	–0.788316	–0.422741	–0.427630	–0.428272	0.00043294	H
8,8	–0.520981	–0.450837	–0.901675	–0.114475	–0.073149	–0.067300	0.00037882	P
9,9	0.095884	–0.585162	–1.170324	0.191822	0.204804	0.206838	0.00043294	P
T,T	0.789718	–0.849395	–1.698790	0.385378	0.176062	0.050266	0.00727344	S
				DAS				
A,A	–0.507875	0.094344	–0.310003	0.364258	0.480545	0.498495	0.00043294	P
2,2	–0.511613	–0.155451	–1.023226	–0.174390	–0.176538	–0.176812	0.00043294	H
3,3	–0.510585	–0.220202	–1.001065	–0.229687	–0.230004	–0.229905	0.00043294	H
4,4	–0.514918	–0.058680	–0.450579	–0.289381	–0.320314	–0.325116	0.00043294	H
5,5	–0.514807	0.200331	0.296188	–0.358542	–0.447216	–0.461118	0.00043294	D
6,6	–0.516710	–0.283253	–0.638127	–0.395166	–0.409405	–0.411497	0.00043294	H
7,7	–0.518333	–0.380634	–0.788316	–0.389326	–0.389764	–0.389719	0.00043294	H
8,8	–0.520981	–0.450837	–0.901675	–0.083575	–0.038581	–0.032231	0.00037882	P
9,9	0.095884	–0.585162	–1.170324	0.209819	0.225123	0.227503	0.00043294	P
T,T	0.789718	–0.849395	–1.698790	0.385378	0.176062	0.050266	0.00727344	S

TABLE A42
4D S17: Dealer's Upcard 9

Hand	Stand	Hit	Double	Freq.	Cum. Freq.	BS
3,2	–0.540768	–0.265667	–1.081536	0.00092361	0.00092361	H
4,2	–0.540645	–0.295486	–1.062540	0.00092361	0.00092361	H
5,2	–0.540486	–0.285462	–0.949837	0.00092361		H
4,3	–0.542817	–0.290016	–0.954433	0.00092361	0.00184722	H
6,2	–0.541476	–0.209728	–0.711897	0.00092361		H
5,3	–0.542716	–0.212011	–0.714383	0.00092361	0.00184722	H
7,2	–0.542697	–0.052173	–0.294949	0.00092361		H
6,3	–0.543647	–0.051952	–0.297836	0.00092361		H
5,4	–0.542587	–0.052189	–0.295603	0.00092361	0.00277083	H
8,2	–0.537670	0.117102	0.150972	0.00092361		D
7,3	–0.544869	0.116772	0.146517	0.00092361		D
6,4	–0.543520	0.116835	0.149459	0.00092361	0.00277083	D
9,2	–0.538549	0.154592	0.224297	0.00086589		D
8,3	–0.539843	0.153842	0.224717	0.00092361		D
7,4	–0.544742	0.156165	0.226910	0.00092361		D
6,5	–0.543360	0.156958	0.231125	0.00092361	0.00363672	D
T,2	–0.539371	–0.341050	–0.739516	0.00369444		H
9,3	–0.540726	–0.352393	–0.761887	0.00086589		H
8,4	–0.539720	–0.353168	–0.760375	0.00092361		H
7,5	–0.544586	–0.351710	–0.758593	0.00092361	0.00640755	H
T,3	–0.541503	–0.380451	–0.794521	0.00369444		H
9,4	–0.540559	–0.381891	–0.796071	0.00086589		H
8,5	–0.539520	–0.393419	–0.817534	0.00092361		H
7,6	–0.545439	–0.394601	–0.822584	0.00092361	0.00640755	H
T,4	–0.541384	–0.426843	–0.869023	0.00369444		H
9,5	–0.540406	–0.429775	–0.873961	0.00086589		H
8,6	–0.540469	–0.432288	–0.880220	0.00092361	0.00548394	H
T,5	–0.541232	–0.472492	–0.949345	0.00369444		H
9,6	–0.541355	–0.473445	–0.952638	0.00086589		H
8,7	–0.541703	–0.464749	–0.935196	0.00092361	0.00548394	H
T,6	–0.542181	–0.502126	–1.004252	0.00369444		RH
9,7	–0.542589	–0.502797	–1.005594	0.00086589	0.00456033	RH
T,7	–0.421644	–0.547224	–1.094448	0.00369444		S
9,8	–0.420344	–0.548382	–1.096764	0.00086589	0.00456033	S
T,8	–0.186228	–0.610958	–1.221915	0.00369444	0.00369444	S
T,9	0.282025	–0.711335	–1.422671	0.00346354	0.00346354	S
A,2	–0.537051	–0.032240	–0.436487	0.00092361	0.00092361	H
A,3	–0.539227	–0.071521	–0.441293	0.00092361	0.00092361	H
A,4	–0.539104	–0.112536	–0.448202	0.00092361	0.00092361	H
A,5	–0.538949	–0.152841	–0.455455	0.00092361	0.00092361	H
A,6	–0.419274	–0.146009	–0.387684	0.00092361	0.00092361	H
A,7	–0.182351	–0.097328	–0.282050	0.00092361	0.00092361	H
A,8	0.287634	0.007206	–0.070222	0.00092361	0.00092361	S
A,9	0.760068	0.111528	0.135641	0.00086589	0.00086589	S
A,T	1.500000	0.155893	0.224949	0.00369444	0.00369444	S

Hand	Stand	Hit	Double	Spl1	Spl2	Spl3	Freq.	BS
				NDAS				
A,A	–0.535511	–0.000541	–0.448071	0.242792	0.346719	0.362784	0.00043294	P
2,2	–0.538592	–0.236442	–1.077184	–0.380892	–0.399868	–0.402754	0.00043294	H
3,3	–0.542943	–0.296613	–1.067142	–0.428104	–0.445167	–0.447728	0.00043294	H
4,4	–0.542748	–0.209639	–0.714277	–0.484006	–0.520658	–0.526326	0.00043294	H
5,5	–0.542426	0.116751	0.151645	–0.550169	–0.639911	–0.653892	0.00043294	D
6,6	–0.544301	–0.350806	–0.756276	–0.582171	–0.612583	–0.617211	0.00043294	H
7,7	–0.546673	–0.441447	–0.902100	–0.569976	–0.586283	–0.588671	0.00043294	H
8,8	–0.536732	–0.503887	–1.007774	–0.421654	–0.411347	–0.409857	0.00043294	P
9,9	–0.186289	–0.611291	–1.222581	–0.106273	–0.096328	–0.094904	0.00037882	P
T,T	0.754917	–0.848731	–1.697462	0.220695	–0.055104	–0.220441	0.00727344	S
				DAS				
A,A	–0.535511	–0.000541	–0.448071	0.242792	0.346719	0.362784	0.00043294	P
2,2	–0.538592	–0.236442	–1.077184	–0.363814	–0.380315	–0.382788	0.00043294	H
3,3	–0.542943	–0.296613	–1.067142	–0.411513	–0.426257	–0.428443	0.00043294	H
4,4	–0.542748	–0.209639	–0.714277	–0.466823	–0.501058	–0.506335	0.00043294	H
5,5	–0.542426	0.116751	0.151645	–0.532606	–0.625064	–0.639526	0.00043294	D
6,6	–0.544301	–0.350806	–0.756276	–0.564718	–0.592710	–0.596950	0.00043294	H
7,7	–0.546673	–0.441447	–0.902100	–0.554258	–0.568471	–0.570536	0.00043294	H
8,8	–0.536732	–0.503887	–1.007774	–0.404912	–0.392327	–0.390478	0.00043294	P
9,9	–0.186289	–0.611291	–1.222581	–0.095895	–0.084727	–0.083137	0.00037882	P
T,T	0.754917	–0.848731	–1.697462	0.220695	–0.055104	–0.220441	0.00727344	S

TABLE A43
4D S17: Dealer's Upcard T

Hand	Stand	Hit	Double	Freq.	Cum. Freq.	BS
3,2	−0.542021	−0.312369	−1.084042	0.00340610	0.00340610	H
4,2	−0.541913	−0.339672	−1.064902	0.00340610	0.00340610	H
5,2	−0.541797	−0.319734	−0.951709	0.00340610		H
4,3	−0.541832	−0.322763	−0.951751	0.00340610	0.00681219	H
6,2	−0.543112	−0.250258	−0.746678	0.00340610		H
5,3	−0.541651	−0.249688	−0.744354	0.00340610	0.00681219	H
7,2	−0.537660	−0.153151	−0.459236	0.00340610		H
6,3	−0.542967	−0.150553	−0.461865	0.00340610		H
5,4	−0.541541	−0.149368	−0.459610	0.00340610	0.01021829	H
8,2	−0.538613	0.026218	−0.005191	0.00340610		H
7,3	−0.537516	0.026282	−0.003201	0.00340610		H
6,4	−0.542858	0.027102	−0.004039	0.00340610	0.01021829	H
9,2	−0.539505	0.115827	0.170985	0.00340610		D
8,3	−0.538473	0.117046	0.175515	0.00340610		D
7,4	−0.537411	0.117111	0.177694	0.00340610		D
6,5	−0.542681	0.118138	0.177799	0.00340610	0.01362439	D
T,2	−0.540283	−0.372921	−0.783223	0.01341151		H
9,3	−0.539317	−0.382961	−0.802777	0.00340610		H
8,4	−0.538321	−0.382253	−0.799070	0.00340610		H
7,5	−0.537186	−0.381902	−0.795549	0.00340610	0.02362980	H
T,3	−0.540147	−0.417738	−0.854216	0.01341151		H
9,4	−0.539216	−0.418939	−0.855830	0.00340610		H
8,5	−0.538148	−0.430314	−0.877566	0.00340610		H
7,6	−0.538516	−0.431509	−0.877595	0.00340610	0.02362980	H
T,4	−0.540046	−0.461443	−0.928883	0.01341151		EH
9,5	−0.539043	−0.463099	−0.931619	0.00340610		EH
8,6	−0.539478	−0.466443	−0.937945	0.00340610	0.02022370	EH
T,5	−0.539873	−0.503624	−1.007248	0.01341151		RH
9,6	−0.540373	−0.506320	−1.012641	0.00340610		RH
8,7	−0.534086	−0.497373	−0.994746	0.00340610	0.02022370	EH
T,6	−0.541203	−0.532053	−1.064107	0.01341151		RH
9,7	−0.534982	−0.533145	−1.066291	0.00340610	0.01681760	RH
T,7	−0.417830	−0.577834	−1.155668	0.01341151		S
9,8	−0.412670	−0.578154	−1.156309	0.00340610	0.01681760	S
T,8	−0.172653	−0.641750	−1.283500	0.01341151	0.01341151	S
T,9	0.072642	−0.724834	−1.449668	0.01341151	0.01341151	S
A,2	−0.540060	−0.101045	−0.505785	0.00342412	0.00342412	H
A,3	−0.539920	−0.135892	−0.506008	0.00342412	0.00342412	H
A,4	−0.539817	−0.172939	−0.509750	0.00342412	0.00342412	H
A,5	−0.539637	−0.211321	−0.519456	0.00342412	0.00342412	H
A,6	−0.419454	−0.195593	−0.452642	0.00342412	0.00342412	H
A,7	−0.180201	−0.142464	−0.340658	0.00342412	0.00342412	H
A,8	0.063422	−0.087383	−0.232658	0.00342412	0.00342412	S
A,9	0.554561	0.020846	−0.018876	0.00342412	0.00342412	S
A,T	1.500000	0.115850	0.169912	0.01348247	0.01348247	S

Hand	Stand	Hit	Double	Spl1	Spl2	Spl3	Freq.	BS
				NDAS				
A,A	−0.537981	−0.064455	−0.502927	0.183155	0.279724	0.294695	0.00161350	P
2,2	−0.542162	−0.286163	−1.084324	−0.478121	−0.503583	−0.507494	0.00159661	H
3,3	−0.541877	−0.339339	−1.064825	−0.524863	−0.549220	−0.552922	0.00159661	H
4,4	−0.541723	−0.247273	−0.744401	−0.576247	−0.620065	−0.626822	0.00159661	H
5,5	−0.541359	0.028133	−0.001792	−0.636716	−0.725900	−0.739751	0.00159661	H
6,6	−0.543907	−0.383053	−0.798973	−0.679519	−0.718912	−0.724971	0.00159661	H
7,7	−0.533125	−0.477970	−0.961397	−0.636983	−0.657218	−0.660192	0.00159661	EH
8,8	−0.535048	−0.533098	−1.066195	−0.489824	−0.483932	−0.483045	0.00159661	EP
9,9	−0.167428	−0.642116	−1.284232	−0.296241	−0.313576	−0.316278	0.00159661	S
T,T	0.561473	−0.846180	−1.692361	0.050867	−0.209209	−0.363187	0.02598479	S
				DAS				
A,A	−0.537981	−0.064455	−0.502927	0.183155	0.279724	0.294695	0.00161350	P
2,2	−0.542162	−0.286163	−1.084324	−0.469197	−0.493433	−0.497148	0.00159661	H
3,3	−0.541877	−0.339339	−1.064825	−0.515416	−0.538476	−0.541972	0.00159661	H
4,4	−0.541723	−0.247273	−0.744401	−0.566451	−0.608924	−0.615468	0.00159661	H
5,5	−0.541359	0.028133	−0.001792	−0.626955	−0.714784	−0.728417	0.00159661	H
6,6	−0.543907	−0.383053	−0.798973	−0.669896	−0.707969	−0.713818	0.00159661	H
7,7	−0.533125	−0.477970	−0.961397	−0.626961	−0.645789	−0.648536	0.00159661	EH
8,8	−0.535048	−0.533098	−1.066195	−0.480440	−0.473267	−0.472177	0.00159661	EP
9,9	−0.167428	−0.642116	−1.284232	−0.287896	−0.304157	−0.306699	0.00159661	S
T,T	0.561473	−0.846180	−1.692361	0.050867	−0.209209	−0.363187	0.02598479	S

TABLE A44
4D H17: Dealer's Upcard A

Hand	Stand	Hit	Double	Freq.	Cum. Freq.	BS
3,2	-0.595457	-0.318446	-1.190914	0.00063526	0.00063526	EH
4,2	-0.594949	-0.346188	-1.177880	0.00063526	0.00063526	EH
5,2	-0.595170	-0.353184	-1.084618	0.00063526		EH
4,3	-0.594461	-0.355533	-1.083119	0.00063526	0.00127053	EH
6,2	-0.594186	-0.267109	-0.830183	0.00063526		H
5,3	-0.594650	-0.267129	-0.828680	0.00063526	0.00127053	H
7,2	-0.592793	-0.128143	-0.452358	0.00063526		H
6,3	-0.593668	-0.124752	-0.450110	0.00063526		H
5,4	-0.594091	-0.124178	-0.448239	0.00063526	0.00190579	H
8,2	-0.594420	0.034369	-0.027004	0.00063526		H
7,3	-0.592278	0.034387	-0.023485	0.00063526		H
6,4	-0.593115	0.035513	-0.020386	0.00063526	0.00190579	H
9,2	-0.595990	0.109244	0.121847	0.00063526		D
8,3	-0.593863	0.109865	0.126656	0.00063526		D
7,4	-0.591680	0.110619	0.131911	0.00063526		D
6,5	-0.593241	0.111710	0.134492	0.00063526	0.00254106	D
T,2	-0.599391	-0.383001	-0.832454	0.00255908		EH
9,3	-0.595483	-0.389079	-0.843659	0.00063526		EH
8,4	-0.593314	-0.388890	-0.839859	0.00063526		EH
7,5	-0.591886	-0.389279	-0.837167	0.00063526	0.00446487	EH
T,3	-0.598886	-0.426448	-0.893557	0.00255908		EH
9,4	-0.594935	-0.423791	-0.887279	0.00063526		EH
8,5	-0.593521	-0.435648	-0.909178	0.00063526		EH
7,6	-0.590945	-0.437537	-0.911017	0.00063526	0.00446487	EH
T,4	-0.598343	-0.468814	-0.958468	0.00255908		EH
9,5	-0.595142	-0.468465	-0.956977	0.00063526		EH
8,6	-0.592581	-0.470803	-0.961917	0.00063526	0.00382961	EH
T,5	-0.598541	-0.510904	-1.029171	0.00255908		RH
9,6	-0.594189	-0.509542	-1.027457	0.00063526		RH
8,7	-0.591245	-0.501146	-1.010099	0.00063526	0.00382961	RH
T,6	-0.597634	-0.538953	-1.077906	0.00255908		RH
9,7	-0.592862	-0.536913	-1.073826	0.00063526	0.00319434	RH
T,7	-0.512809	-0.578724	-1.157449	0.00255908		RS
9,8	-0.509705	-0.575986	-1.151972	0.00063526	0.00319434	RS
T,8	-0.222188	-0.640412	-1.280824	0.00255908	0.00255908	S
T,9	0.193983	-0.734426	-1.468852	0.00255908	0.00255908	S
A,2	-0.596474	-0.101172	-0.581196	0.00059556	0.00059556	H
A,3	-0.595965	-0.135655	-0.580368	0.00059556	0.00059556	H
A,4	-0.595416	-0.173615	-0.585335	0.00059556	0.00059556	H
A,5	-0.595655	-0.212274	-0.594241	0.00059556	0.00059556	H
A,6	-0.513575	-0.222214	-0.546802	0.00059556	0.00059556	H
A,7	-0.225191	-0.161674	-0.417624	0.00059556	0.00059556	H
A,8	0.190271	-0.065708	-0.228736	0.00059556	0.00059556	S
A,9	0.607410	0.029283	-0.042944	0.00059556	0.00059556	S
A,T	1.500000	0.105025	0.111257	0.00239914	0.00239914	S

Hand	Stand	Hit	Double	Spl1	Spl2	Spl3	Freq.	BS
				NDAS				
A,A	-0.596977	-0.065388	-0.582574	0.129936	0.219987	0.232838	0.00026056	P
2,2	-0.595967	-0.292870	-1.191935	-0.492823	-0.519192	-0.523218	0.00029778	EH
3,3	-0.594987	-0.346177	-1.177947	-0.538664	-0.563660	-0.567416	0.00029778	EH
4,4	-0.593902	-0.264117	-0.827063	-0.588865	-0.631996	-0.638628	0.00029778	H
5,5	-0.594290	0.035263	-0.020254	-0.651352	-0.743294	-0.757548	0.00029778	H
6,6	-0.592290	-0.388892	-0.836346	-0.682528	-0.721007	-0.726844	0.00029778	EH
7,7	-0.589617	-0.481972	-0.984407	-0.696701	-0.723760	-0.727696	0.00029778	EH
8,8	-0.592875	-0.536768	-1.073536	-0.516230	-0.512886	-0.512278	0.00029778	RP
9,9	-0.216334	-0.638282	-1.276564	-0.235175	-0.236953	-0.237116	0.00029778	S
T,T	0.599679	-0.860952	-1.721905	0.048969	-0.237602	-0.410651	0.00507366	S
				DAS				
A,A	-0.596977	-0.065388	-0.582574	0.129936	0.219987	0.232838	0.00026056	P
2,2	-0.595967	-0.292870	-1.191935	-0.490204	-0.516140	-0.520087	0.00029778	EH
3,3	-0.594987	-0.346177	-1.177947	-0.535302	-0.559756	-0.563414	0.00029778	EH
4,4	-0.593902	-0.264117	-0.827063	-0.584636	-0.627089	-0.633600	0.00029778	H
5,5	-0.594290	0.035263	-0.020254	-0.646833	-0.738053	-0.752177	0.00029778	H
6,6	-0.592290	-0.388892	-0.836346	-0.678040	-0.715802	-0.721509	0.00029778	EH
7,7	-0.589617	-0.481972	-0.984407	-0.692627	-0.719047	-0.722870	0.00029778	EH
8,8	-0.592875	-0.536768	-1.073536	-0.513401	-0.509655	-0.508981	0.00029778	RP
9,9	-0.216334	-0.638282	-1.276564	-0.233573	-0.235186	-0.235330	0.00029778	S
T,T	0.599679	-0.860952	-1.721905	0.048969	-0.237602	-0.410651	0.00507366	S

TABLE A45

4D H17: Dealer's Upcard 2

Hand	Stand	Hit	Double	Freq.	Cum. Freq.	BS
3,2	−0.285926	−0.126132	−0.571853	0.00086589	0.00086589	H
4,2	−0.285749	−0.139859	−0.551276	0.00086589	0.00086589	H
5,2	−0.281886	−0.107068	−0.423230	0.00086589		H
4,3	−0.285663	−0.111691	−0.428461	0.00092361	0.00178950	H
6,2	−0.282627	−0.022065	−0.196309	0.00086589		H
5,3	−0.281800	−0.022559	−0.196717	0.00092361	0.00178950	H
7,2	−0.283393	0.074734	0.072461	0.00086589		H
6,3	−0.282538	0.076868	0.071671	0.00092361		H
5,4	−0.281617	0.077524	0.072626	0.00092361	0.00271311	H
8,2	−0.284070	0.187995	0.372141	0.00086589		D
7,3	−0.283259	0.188722	0.373259	0.00092361		D
6,4	−0.282309	0.189665	0.375075	0.00092361	0.00271311	D
9,2	−0.285101	0.243453	0.482637	0.00086589		D
8,3	−0.283983	0.244672	0.485011	0.00092361		D
7,4	−0.283077	0.246002	0.487602	0.00092361		D
6,5	−0.278471	0.248657	0.493353	0.00092361	0.00363672	D
T,2	−0.290878	−0.251443	−0.502887	0.00346354		H
9,3	−0.285014	−0.256863	−0.513727	0.00092361		H
8,4	−0.283803	−0.255220	−0.510439	0.00092361		H
7,5	−0.279241	−0.253535	−0.507070	0.00092361	0.00623437	H
T,3	−0.290792	−0.307652	−0.615303	0.00369444		S
9,4	−0.284835	−0.304881	−0.609761	0.00092361		S
8,5	−0.279969	−0.314244	−0.628489	0.00092361		S
7,6	−0.279988	−0.314431	−0.628862	0.00092361	0.00646528	S
T,4	−0.290614	−0.364794	−0.729587	0.00369444		S
9,5	−0.280999	−0.362697	−0.725395	0.00092361		S
8,6	−0.280710	−0.363686	−0.727371	0.00092361	0.00554167	S
T,5	−0.286832	−0.423019	−0.846038	0.00369444		S
9,6	−0.281801	−0.421677	−0.843354	0.00092361		S
8,7	−0.281537	−0.412240	−0.824480	0.00092361	0.00554167	S
T,6	−0.287576	−0.471853	−0.943706	0.00369444		S
9,7	−0.282572	−0.469673	−0.939345	0.00092361	0.00461805	S
T,7	−0.157582	−0.538641	−1.077282	0.00369444		S
9,8	−0.152453	−0.536788	−1.073577	0.00092361	0.00461805	S
T,8	0.109414	−0.625997	−1.251995	0.00369444	0.00369444	S
T,9	0.377810	−0.734499	−1.468997	0.00369444	0.00369444	S
A,2	−0.284357	0.045095	−0.060966	0.00086589	0.00086589	H
A,3	−0.284272	0.021735	−0.061684	0.00092361	0.00092361	H
A,4	−0.284141	−0.001763	−0.067212	0.00092361	0.00092361	H
A,5	−0.280296	−0.022271	−0.070719	0.00092361	0.00092361	H
A,6	−0.150884	0.000433	−0.003095	0.00092361	0.00092361	H
A,7	0.114542	0.060729	0.117057	0.00092361	0.00092361	D
A,8	0.382107	0.120840	0.237347	0.00092361	0.00092361	S
A,9	0.638969	0.182844	0.361686	0.00092361	0.00092361	S
A,T	1.500000	0.239713	0.473824	0.00369444	0.00369444	S

Hand	Stand	Hit	Double	Spl1	Spl2	Spl3	Freq.	BS
				NDAS				
A,A	−0.282726	0.083156	−0.055862	0.493092	0.596434	0.612371	0.00043294	P
2,2	−0.285960	−0.112376	−0.571921	−0.146497	−0.150657	−0.151241	0.00037882	H
3,3	−0.285842	−0.139940	−0.551452	−0.196331	−0.203548	−0.204616	0.00043294	H
4,4	−0.285483	−0.021159	−0.199796	−0.225416	−0.252619	−0.256815	0.00043294	H
5,5	−0.277730	0.191055	0.378266	−0.245031	−0.302681	−0.311504	0.00043294	D
6,6	−0.279215	−0.253537	−0.507075	−0.261568	−0.261543	−0.261371	0.00043294	H
7,7	−0.280755	−0.374122	−0.748245	−0.205810	−0.195348	−0.193662	0.00043294	P
8,8	−0.282263	−0.469361	−0.938721	−0.035271	−0.001959	0.003241	0.00043294	P
9,9	0.114027	−0.624788	−1.249576	0.149899	0.154699	0.155442	0.00043294	P
T,T	0.631826	−0.853306	−1.706611	0.349010	0.200042	0.109105	0.00727344	S
				DAS				
A,A	−0.282726	0.083156	−0.055862	0.493092	0.596434	0.612371	0.00043294	P
2,2	−0.285960	−0.112376	−0.571921	−0.079666	−0.075468	−0.074849	0.00037882	P
3,3	−0.285842	−0.139940	−0.551452	−0.129274	−0.127488	−0.127157	0.00043294	P
4,4	−0.285483	−0.021159	−0.199796	−0.157949	−0.176085	−0.178871	0.00043294	H
5,5	−0.277730	0.191055	0.378266	−0.179958	−0.254340	−0.265815	0.00043294	D
6,6	−0.279215	−0.253537	−0.507075	−0.193137	−0.183852	−0.182230	0.00043294	P
7,7	−0.280755	−0.374122	−0.748245	−0.137135	−0.117319	−0.114159	0.00043294	P
8,8	−0.282263	−0.469361	−0.938721	0.029915	0.071980	0.078541	0.00043294	P
9,9	0.114027	−0.624788	−1.249576	0.184980	0.194451	0.195914	0.00043294	P
T,T	0.631826	−0.853306	−1.706611	0.349010	0.200042	0.109105	0.00727344	S

TABLE A46
4D H17: Dealer's Upcard 3

Hand	Stand	Hit	Double	Freq.	Cum. Freq.	BS
3,2	−0.245133	−0.093277	−0.490266	0.00086589	0.00086589	H
4,2	−0.241230	−0.103957	−0.463292	0.00092361	0.00092361	H
5,2	−0.240610	−0.072797	−0.343447	0.00092361		H
4,3	−0.241056	−0.077103	−0.346213	0.00086589	0.00178950	H
6,2	−0.241330	0.009754	−0.123541	0.00092361		H
5,3	−0.240434	0.009252	−0.123873	0.00086589	0.00178950	H
7,2	−0.242047	0.106454	0.136505	0.00092361		D
6,3	−0.241108	0.108573	0.135757	0.00086589		D
5,4	−0.236505	0.109220	0.137766	0.00092361	0.00271311	D
8,2	−0.243071	0.213200	0.425153	0.00092361		D
7,3	−0.241872	0.214502	0.427622	0.00086589		D
6,4	−0.237227	0.215503	0.430006	0.00092361	0.00271311	D
9,2	−0.249021	0.262106	0.522783	0.00092361		D
8,3	−0.242897	0.268070	0.534888	0.00086589		D
7,4	−0.237993	0.270365	0.539846	0.00092361		D
6,5	−0.236606	0.272474	0.544064	0.00092361	0.00363672	D
T,2	−0.250114	−0.230500	−0.461000	0.00369444		H
9,3	−0.248848	−0.238994	−0.477988	0.00086589		H
8,4	−0.239022	−0.232977	−0.465953	0.00092361		H
7,5	−0.237375	−0.231244	−0.462489	0.00092361	0.00640755	H
T,3	−0.249943	−0.289818	−0.579635	0.00346354		S
9,4	−0.244970	−0.289064	−0.578127	0.00092361		S
8,5	−0.238399	−0.295395	−0.590791	0.00092361		S
7,6	−0.238096	−0.295172	−0.590345	0.00092361	0.00623437	S
T,4	−0.246120	−0.351286	−0.702572	0.00369444		S
9,5	−0.244406	−0.351338	−0.702677	0.00092361		S
8,6	−0.239183	−0.349513	−0.699027	0.00092361	0.00554167	S
T,5	−0.245496	−0.413237	−0.826474	0.00369444		S
9,6	−0.245131	−0.413800	−0.827601	0.00092361		S
8,7	−0.239949	−0.401918	−0.803836	0.00092361	0.00554167	S
T,6	−0.246223	−0.465092	−0.930185	0.00369444		S
9,7	−0.245899	−0.464469	−0.928938	0.00092361	0.00461805	S
T,7	−0.120516	−0.534554	−1.069107	0.00369444		S
9,8	−0.120503	−0.534231	−1.068463	0.00092361	0.00461805	S
T,8	0.137755	−0.624273	−1.248545	0.00369444	0.00369444	S
T,9	0.392518	−0.724746	−1.449491	0.00369444	0.00369444	S
A,2	−0.243738	0.073581	0.005080	0.00092361	0.00092361	H
A,3	−0.243614	0.049907	0.001226	0.00086589	0.00086589	H
A,4	−0.239733	0.028628	−0.001187	0.00092361	0.00092361	H
A,5	−0.239109	0.007407	−0.006943	0.00092361	0.00092361	H
A,6	−0.114018	0.029863	0.058843	0.00092361	0.00092361	D
A,7	0.143418	0.088592	0.176300	0.00092361	0.00092361	D
A,8	0.401226	0.152991	0.304809	0.00092361	0.00092361	S
A,9	0.644935	0.202765	0.404101	0.00092361	0.00092361	S
A,T	1.500000	0.262244	0.521784	0.00369444	0.00369444	S

Hand	Stand	Hit	Double	Spl1	Spl2	Spl3	Freq.	BS
NDAS								
A,A	−0.242201	0.105864	0.010606	0.540506	0.644730	0.660798	0.00043294	P
2,2	−0.245303	−0.079865	−0.490606	−0.092040	−0.093620	−0.093857	0.00043294	H
3,3	−0.244961	−0.106527	−0.470379	−0.131223	−0.134045	−0.134415	0.00037882	H
4,4	−0.237152	0.011189	−0.123541	−0.151097	−0.172284	−0.175486	0.00043294	H
5,5	−0.235882	0.216556	0.432104	−0.176196	−0.227999	−0.235909	0.00043294	D
6,6	−0.237332	−0.231295	−0.462589	−0.191578	−0.185121	−0.183945	0.00043294	P
7,7	−0.238920	−0.359510	−0.719019	−0.137366	−0.123264	−0.121001	0.00043294	P
8,8	−0.240979	−0.462481	−0.924963	0.026143	0.062122	0.067731	0.00043294	P
9,9	0.133040	−0.625189	−1.250378	0.192253	0.199577	0.200620	0.00043294	P
T,T	0.642519	−0.853018	−1.706035	0.398026	0.268840	0.189758	0.00727344	S
DAS								
A,A	−0.242201	0.105864	0.010606	0.540506	0.644730	0.660798	0.00043294	P
2,2	−0.245303	−0.079865	−0.490606	−0.012337	−0.003168	−0.001728	0.00043294	P
3,3	−0.244961	−0.106527	−0.470379	−0.051175	−0.043992	−0.042921	0.00037882	P
4,4	−0.237152	0.011189	−0.123541	−0.069393	−0.079506	−0.080971	0.00043294	H
5,5	−0.235882	0.216556	0.432104	−0.097731	−0.168236	−0.179098	0.00043294	D
6,6	−0.237332	−0.231295	−0.462589	−0.104132	−0.085656	−0.082570	0.00043294	P
7,7	−0.238920	−0.359510	−0.719019	−0.051133	−0.025164	−0.021013	0.00043294	P
8,8	−0.240979	−0.462481	−0.924963	0.099279	0.145105	0.152248	0.00043294	P
9,9	0.133040	−0.625189	−1.250378	0.232379	0.244960	0.246799	0.00043294	P
T,T	0.642519	−0.853018	−1.706035	0.398026	0.268840	0.189758	0.00727344	S

TABLE A47

4D H17: Dealer's Upcard 4

Hand	Stand	Hit	Double	Freq.	Cum. Freq.	BS
3,2	−0.197210	−0.053618	−0.394420	0.00092361	0.00092361	H
4,2	−0.196511	−0.065491	−0.373989	0.00086589	0.00086589	H
5,2	−0.195983	−0.034700	−0.257969	0.00092361		H
4,3	−0.192612	−0.036176	−0.253886	0.00086589	0.00178950	H
6,2	−0.196642	0.047189	−0.046842	0.00092361		H
5,3	−0.192060	0.048507	−0.042578	0.00092361	0.00184722	H
7,2	−0.197724	0.136946	0.201533	0.00092361		D
6,3	−0.192766	0.141224	0.205692	0.00092361		D
5,4	−0.191356	0.141932	0.207128	0.00086589	0.00271311	D
8,2	−0.203664	0.235210	0.470420	0.00092361		D
7,3	−0.193851	0.242809	0.485617	0.00092361		D
6,4	−0.192064	0.244349	0.488698	0.00086589	0.00271311	D
9,2	−0.204880	0.286916	0.573832	0.00092361		D
8,3	−0.199794	0.290066	0.580131	0.00092361		D
7,4	−0.193153	0.296913	0.593827	0.00086589		D
6,5	−0.191538	0.298997	0.597993	0.00092361	0.00363672	D
T,2	−0.205947	−0.209063	−0.418127	0.00369444		S
9,3	−0.201009	−0.216647	−0.433293	0.00092361		S
8,4	−0.199092	−0.214012	−0.428024	0.00086589		S
7,5	−0.192622	−0.208860	−0.417720	0.00092361	0.00640755	S
T,3	−0.202129	−0.270878	−0.541756	0.00369444		S
9,4	−0.200365	−0.270141	−0.540282	0.00086589		S
8,5	−0.198620	−0.279167	−0.558334	0.00092361		S
7,6	−0.193390	−0.275811	−0.551622	0.00092361	0.00640755	S
T,4	−0.201428	−0.336319	−0.672639	0.00346354		S
9,5	−0.199835	−0.336288	−0.672576	0.00092361		S
8,6	−0.199330	−0.336760	−0.673520	0.00092361	0.00531076	S
T,5	−0.200897	−0.402544	−0.805088	0.00369444		S
9,6	−0.200545	−0.403005	−0.806010	0.00092361		S
8,7	−0.200418	−0.393452	−0.786904	0.00092361	0.00554167	S
T,6	−0.201609	−0.457810	−0.915620	0.00369444		S
9,7	−0.201635	−0.457452	−0.914905	0.00092361	0.00461805	S
T,7	−0.079505	−0.530627	−1.061254	0.00369444		S
9,8	−0.084264	−0.531893	−1.063787	0.00092361	0.00461805	S
T,8	0.163389	−0.613514	−1.227029	0.00369444	0.00369444	S
T,9	0.411428	−0.723708	−1.447417	0.00369444	0.00369444	S
A,2	−0.199743	0.105013	0.075410	0.00092361	0.00092361	H
A,3	−0.195867	0.083700	0.074056	0.00092361	0.00092361	H
A,4	−0.195168	0.060928	0.067459	0.00086589	0.00086589	D
A,5	−0.194637	0.040776	0.061847	0.00092361	0.00092361	D
A,6	−0.072750	0.063127	0.126253	0.00092361	0.00092361	D
A,7	0.173450	0.125459	0.250918	0.00092361	0.00092361	D
A,8	0.415146	0.176956	0.353913	0.00092361	0.00092361	S
A,9	0.655298	0.229357	0.458714	0.00092361	0.00092361	S
A,T	1.500000	0.286307	0.572613	0.00369444	0.00369444	S

Hand	Stand	Hit	Double	Spl1	Spl2	Spl3	Freq.	BS
				NDAS				
A,A	−0.198317	0.130565	0.080030	0.590708	0.695729	0.711913	0.00043294	P
2,2	−0.201106	−0.043137	−0.402213	−0.030790	−0.029117	−0.028856	0.00043294	P
3,3	−0.193311	−0.063477	−0.368335	−0.048110	−0.045553	−0.045081	0.00043294	P
4,4	−0.191887	0.051068	−0.042119	−0.073164	−0.088134	−0.090211	0.00037882	H
5,5	−0.190827	0.244820	0.489639	−0.097827	−0.142836	−0.149679	0.00043294	D
6,6	−0.192244	−0.208472	−0.416945	−0.112601	−0.101164	−0.099274	0.00043294	P
7,7	−0.194479	−0.344370	−0.688740	−0.061214	−0.042796	−0.039854	0.00043294	P
8,8	−0.206360	−0.459113	−0.918225	0.081349	0.119545	0.125415	0.00043294	P
9,9	0.163507	−0.613494	−1.226987	0.253461	0.264942	0.266636	0.00043294	P
T,T	0.652914	−0.852794	−1.705588	0.448847	0.340102	0.273039	0.00727344	S
				DAS				
A,A	−0.198317	0.130565	0.080030	0.590708	0.695729	0.711913	0.00043294	P
2,2	−0.201106	−0.043137	−0.402213	0.061959	0.076105	0.078308	0.00043294	P
3,3	−0.193311	−0.063477	−0.368335	0.046541	0.061854	0.064316	0.00043294	P
4,4	−0.191887	0.051068	−0.042119	0.024798	0.022220	0.021949	0.00037882	H
5,5	−0.190827	0.244820	0.489639	−0.002434	−0.067827	−0.077888	0.00043294	D
6,6	−0.192244	−0.208472	−0.416945	−0.006692	0.019280	0.023478	0.00043294	P
7,7	−0.194479	−0.344370	−0.688740	0.045697	0.078759	0.084021	0.00043294	P
8,8	−0.206360	−0.459113	−0.918225	0.162894	0.211865	0.219382	0.00043294	P
9,9	0.163507	−0.613494	−1.226987	0.297661	0.314939	0.317513	0.00043294	P
T,T	0.652914	−0.852794	−1.705588	0.448847	0.340102	0.273039	0.00727344	S

TABLE A48
4D H17: Dealer's Upcard 5

Hand	Stand	Hit	Double	Freq.	Cum. Freq.	BS
3,2	–0.150084	–0.012095	–0.300168	0.00092361	0.00092361	H
4,2	–0.149480	–0.023294	–0.281007	0.00092361	0.00092361	H
5,2	–0.148930	0.007799	–0.170525	0.00086589		H
4,3	–0.148807	0.005990	–0.168453	0.00092361	0.00178950	H
6,2	–0.150004	0.084068	0.032397	0.00092361		H
5,3	–0.148280	0.086370	0.037764	0.00086589	0.00178950	H
7,2	–0.155998	0.166475	0.263608	0.00092361		D
6,3	–0.149354	0.175448	0.276933	0.00092361		D
5,4	–0.147671	0.177950	0.282001	0.00086589	0.00271311	D
8,2	–0.157233	0.265445	0.530890	0.00092361		D
7,3	–0.155351	0.268166	0.536332	0.00092361		D
6,4	–0.148748	0.275170	0.550340	0.00092361	0.00277083	D
9,2	–0.158278	0.314817	0.629635	0.00092361		D
8,3	–0.156583	0.317348	0.634696	0.00092361		D
7,4	–0.154742	0.320172	0.640344	0.00092361		D
6,5	–0.148220	0.327316	0.654632	0.00086589	0.00363672	D
T,2	–0.159310	–0.186390	–0.372780	0.00369444		S
9,3	–0.157687	–0.194343	–0.388686	0.00092361		S
8,4	–0.156032	–0.191864	–0.383728	0.00092361		S
7,5	–0.154272	–0.189740	–0.379479	0.00086589	0.00640755	S
T,3	–0.158660	–0.252450	–0.504900	0.00369444		S
9,4	–0.157078	–0.251758	–0.503515	0.00092361		S
8,5	–0.155504	–0.260786	–0.521573	0.00086589		S
7,6	–0.155349	–0.260473	–0.520946	0.00092361	0.00640755	S
T,4	–0.158051	–0.320331	–0.640663	0.00369444		S
9,5	–0.156550	–0.320256	–0.640512	0.00086589		S
8,6	–0.156582	–0.321029	–0.642057	0.00092361	0.00548394	S
T,5	–0.157523	–0.389656	–0.779313	0.00346354		S
9,6	–0.157629	–0.390370	–0.780739	0.00092361		S
8,7	–0.162578	–0.383177	–0.766354	0.00092361	0.00531076	S
T,6	–0.158603	–0.449333	–0.898667	0.00369444		S
9,7	–0.163626	–0.450989	–0.901978	0.00092361	0.00461805	S
T,7	–0.045816	–0.516496	–1.032993	0.00369444		S
9,8	–0.046053	–0.516508	–1.033015	0.00092361	0.00461805	S
T,8	0.195823	–0.609827	–1.219655	0.00369444	0.00369444	S
T,9	0.438165	–0.722070	–1.444140	0.00369444	0.00369444	S
A,2	–0.153288	0.139648	0.148011	0.00092361	0.00092361	D
A,3	–0.152642	0.118117	0.145904	0.00092361	0.00092361	D
A,4	–0.152036	0.096157	0.138832	0.00092361	0.00092361	D
A,5	–0.151511	0.075888	0.131876	0.00086589	0.00086589	D
A,6	–0.034467	0.102635	0.205270	0.00092361	0.00092361	D
A,7	0.201050	0.153409	0.306817	0.00092361	0.00092361	D
A,8	0.441785	0.208204	0.416408	0.00092361	0.00092361	S
A,9	0.671436	0.258672	0.517344	0.00092361	0.00092361	S
A,T	1.500000	0.313008	0.626016	0.00369444	0.00369444	S

Hand	Stand	Hit	Double	Spl1	Spl2	Spl3	Freq.	BS
				NDAS				
A,A	–0.155794	0.162955	0.149183	0.643381	0.749718	0.766099	0.00043294	P
2,2	–0.150729	0.000025	–0.301459	0.052954	0.060285	0.061458	0.00043294	P
3,3	–0.149439	–0.023324	–0.280561	0.029511	0.037124	0.038386	0.00043294	P
4,4	–0.148198	0.089391	0.039393	0.005493	–0.005261	–0.006855	0.00043294	H
5,5	–0.147145	0.276981	0.553963	–0.018798	–0.054934	–0.060019	0.00037882	D
6,6	–0.149356	–0.186242	–0.372484	–0.034637	–0.018531	–0.015921	0.00043294	P
7,7	–0.161344	–0.333648	–0.667296	–0.000676	0.020747	0.024053	0.00043294	P
8,8	–0.163814	–0.451043	–0.902086	0.151848	0.193720	0.200149	0.00043294	P
9,9	0.195823	–0.609801	–1.219603	0.319830	0.335819	0.338205	0.00043294	P
T,T	0.669203	–0.852399	–1.704799	0.507949	0.421315	0.367510	0.00727344	S
				DAS				
A,A	–0.155794	0.162955	0.149183	0.643381	0.749718	0.766099	0.00043294	P
2,2	–0.150729	0.000025	–0.301459	0.162432	0.184622	0.188128	0.00043294	P
3,3	–0.149439	–0.023324	–0.280561	0.143564	0.166649	0.170339	0.00043294	P
4,4	–0.148198	0.089391	0.039393	0.123084	0.128314	0.129234	0.00043294	P
5,5	–0.147145	0.276981	0.553963	0.095539	0.038677	0.030551	0.00037882	D
6,6	–0.149356	–0.186242	–0.372484	0.091937	0.125362	0.130713	0.00043294	P
7,7	–0.161344	–0.333648	–0.667296	0.123363	0.161333	0.167194	0.00043294	P
8,8	–0.163814	–0.451043	–0.902086	0.242333	0.296161	0.304419	0.00043294	P
9,9	0.195823	–0.609801	–1.219603	0.368392	0.390758	0.394113	0.00043294	P
T,T	0.669203	–0.852399	–1.704799	0.507949	0.421315	0.367510	0.00727344	S

TABLE A49
4D H17: Dealer's Upcard 6

Hand	Stand	Hit	Double	Freq.	Cum. Freq.	BS
3,2	–0.112255	0.021325	–0.224509	0.00092361	0.00092361	H
4,2	–0.111676	0.012471	–0.205841	0.00092361	0.00092361	H
5,2	–0.111557	0.040496	–0.098602	0.00092361		H
4,3	–0.111075	0.038569	–0.097207	0.00092361	0.00184722	H
6,2	–0.117554	0.108168	0.085294	0.00086589		H
5,3	–0.110956	0.115305	0.100214	0.00092361	0.00178950	H
7,2	–0.118793	0.190690	0.314474	0.00092361		D
6,3	–0.116955	0.195592	0.319694	0.00086589		D
5,4	–0.110400	0.202961	0.334470	0.00092361	0.00271311	D
8,2	–0.119828	0.287230	0.574460	0.00092361		D
7,3	–0.118192	0.289896	0.579793	0.00092361		D
6,4	–0.116397	0.293011	0.586022	0.00086589	0.00271311	D
9,2	–0.120900	0.335424	0.670847	0.00092361		D
8,3	–0.119285	0.337956	0.675913	0.00092361		D
7,4	–0.117691	0.340736	0.681471	0.00092361		D
6,5	–0.116337	0.343758	0.687516	0.00086589	0.00363672	D
T,2	–0.121695	–0.169633	–0.339266	0.00369444		S
9,3	–0.120299	–0.177686	–0.355373	0.00092361		S
8,4	–0.118726	–0.175220	–0.350439	0.00092361		S
7,5	–0.117572	–0.173630	–0.347260	0.00092361	0.00646528	S
T,3	–0.121095	–0.238517	–0.477033	0.00369444		S
9,4	–0.119741	–0.237889	–0.475778	0.00092361		S
8,5	–0.118607	–0.247268	–0.494537	0.00092361		S
7,6	–0.123570	–0.250102	–0.500204	0.00086589	0.00640755	S
T,4	–0.120537	–0.309137	–0.618274	0.00369444		S
9,5	–0.119622	–0.309449	–0.618897	0.00092361		S
8,6	–0.124606	–0.312919	–0.625837	0.00086589	0.00548394	S
T,5	–0.120418	–0.380669	–0.761338	0.00369444		S
9,6	–0.125622	–0.383780	–0.767560	0.00086589		S
8,7	–0.125846	–0.374576	–0.749152	0.00092361	0.00548394	S
T,6	–0.126337	–0.434795	–0.869590	0.00346354		S
9,7	–0.126781	–0.434864	–0.869727	0.00092361	0.00438715	S
T,7	–0.011673	–0.513194	–1.026389	0.00369444		S
9,8	–0.011876	–0.513210	–1.026419	0.00092361	0.00461805	S
T,8	0.218639	–0.608353	–1.216705	0.00369444	0.00369444	S
T,9	0.449198	–0.721432	–1.442864	0.00369444	0.00369444	S
A,2	–0.112532	0.167175	0.208324	0.00092361	0.00092361	D
A,3	–0.111935	0.146848	0.206477	0.00092361	0.00092361	D
A,4	–0.111383	0.126567	0.201535	0.00092361	0.00092361	D
A,5	–0.111221	0.112080	0.205032	0.00092361	0.00092361	D
A,6	–0.004442	0.126797	0.253595	0.00086589	0.00086589	D
A,7	0.222101	0.178870	0.357740	0.00092361	0.00092361	D
A,8	0.451752	0.231294	0.462588	0.00092361	0.00092361	D
A,9	0.677213	0.280966	0.561933	0.00092361	0.00092361	S
A,T	1.500000	0.334133	0.668266	0.00369444	0.00369444	S

Hand	Stand	Hit	Double	Spl1	Spl2	Spl3	Freq.	BS
				NDAS				
A,A	–0.112187	0.190740	0.211466	0.687177	0.793341	0.809664	0.00043294	P
2,2	–0.112850	0.032260	–0.225700	0.115663	0.127125	0.128947	0.00043294	P
3,3	–0.111632	0.012463	–0.205750	0.093099	0.104481	0.106335	0.00043294	P
4,4	–0.110518	0.118208	0.101627	0.069958	0.064002	0.063156	0.00043294	H
5,5	–0.110282	0.299512	0.599024	0.045951	0.012664	0.007606	0.00043294	D
6,6	–0.122332	–0.176676	–0.353352	0.014641	0.031786	0.034258	0.00037882	P
7,7	–0.124808	–0.323281	–0.646562	0.060649	0.085306	0.089101	0.00043294	P
8,8	–0.126803	–0.434843	–0.869687	0.206699	0.250928	0.257720	0.00043294	P
9,9	0.218471	–0.608375	–1.216749	0.366973	0.386193	0.389071	0.00043294	P
T,T	0.675762	–0.852223	–1.704446	0.551167	0.483524	0.441140	0.00727344	S
				DAS				
A,A	–0.112187	0.190740	0.211466	0.687177	0.793341	0.809664	0.00043294	P
2,2	–0.112850	0.032260	–0.225700	0.241437	0.269974	0.274478	0.00043294	P
3,3	–0.111632	0.012463	–0.205750	0.221534	0.250365	0.254962	0.00043294	P
4,4	–0.110518	0.118208	0.101627	0.201656	0.213642	0.215623	0.00043294	P
5,5	–0.110282	0.299512	0.599024	0.175569	0.118990	0.110236	0.00043294	D
6,6	–0.122332	–0.176676	–0.353352	0.154116	0.188621	0.193581	0.00037882	P
7,7	–0.124808	–0.323281	–0.646562	0.199928	0.243132	0.249785	0.00043294	P
8,8	–0.126803	–0.434843	–0.869687	0.305520	0.362821	0.371615	0.00043294	P
9,9	0.218471	–0.608375	–1.216749	0.418750	0.444773	0.448687	0.00043294	P
T,T	0.675762	–0.852223	–1.704446	0.551167	0.483524	0.441140	0.00727344	S

TABLE A50
6D S17: Dealer's Upcard A

Hand	Stand	Hit	Double	Freq.	Cum. Freq.	BS
3,2	–0.665751	–0.281089	–1.331502	0.00063359	0.00063359	EH
4,2	–0.665491	–0.309156	–1.303138	0.00063359	0.00063359	EH
5,2	–0.665202	–0.313715	–1.129307	0.00063359		EH
4,3	–0.665250	–0.315788	–1.129398	0.00063359	0.00126718	EH
6,2	–0.662727	–0.201666	–0.808992	0.00063359		H
5,3	–0.664960	–0.200161	–0.807761	0.00063359	0.00126718	H
7,2	–0.663671	–0.068205	–0.428073	0.00063359		H
6,3	–0.662485	–0.067987	–0.428020	0.00063359		H
5,4	–0.664668	–0.066304	–0.426459	0.00063359	0.00190077	H
8,2	–0.664623	0.082288	–0.007893	0.00063359		H
7,3	–0.663429	0.082293	–0.005594	0.00063359		H
6,4	–0.662192	0.081336	–0.004174	0.00063359	0.00190077	H
9,2	–0.665556	0.147155	0.121507	0.00063359		H
8,3	–0.664365	0.147279	0.124094	0.00063359		H
7,4	–0.663120	0.147352	0.126725	0.00063359		H
6,5	–0.661883	0.147596	0.129710	0.00063359	0.00253436	H
T,2	–0.668057	–0.350190	–0.826367	0.00254626		EH
9,3	–0.665319	–0.354329	–0.833886	0.00063359		EH
8,4	–0.664077	–0.354495	–0.831943	0.00063359		EH
7,5	–0.662832	–0.354597	–0.830442	0.00063359	0.00444703	EH
T,3	–0.667822	–0.396288	–0.877145	0.00254626		EH
9,4	–0.665032	–0.394604	–0.873661	0.00063359		EH
8,5	–0.663790	–0.402513	–0.888292	0.00063359		EH
7,6	–0.660391	–0.404544	–0.889991	0.00063359	0.00444703	EH
T,4	–0.667536	–0.440744	–0.931679	0.00254626		EH
9,5	–0.664746	–0.440030	–0.929648	0.00063359		EH
8,6	–0.661352	–0.442195	–0.933453	0.00063359	0.00381344	EH
T,5	–0.667252	–0.482930	–0.987574	0.00254626		EH
9,6	–0.662311	–0.482503	–0.987018	0.00063359		EH
8,7	–0.662304	–0.476090	–0.974215	0.00063359	0.00381344	EH
T,6	–0.664825	–0.515800	–1.031600	0.00254626		RH
9,7	–0.663256	–0.513641	–1.027283	0.00063359	0.00317985	RH
T,7	–0.476308	–0.557049	–1.114099	0.00254626		ES
9,8	–0.473877	–0.555181	–1.110361	0.00063359	0.00317985	ES
T,8	–0.097274	–0.627545	–1.255091	0.00254626	0.00254626	S
T,9	0.282488	–0.727587	–1.455174	0.00254626	0.00254626	S
A,2	–0.666224	–0.059450	–0.619513	0.00060719	0.00060719	H
A,3	–0.665987	–0.095695	–0.619855	0.00060719	0.00060719	H
A,4	–0.665698	–0.134277	–0.623723	0.00060719	0.00060719	H
A,5	–0.665441	–0.172254	–0.629261	0.00060719	0.00060719	H
A,6	–0.478806	–0.182973	–0.535962	0.00060719	0.00060719	H
A,7	–0.100278	–0.095345	–0.362306	0.00060719	0.00060719	H
A,8	0.279653	–0.007505	–0.189276	0.00060719	0.00060719	S
A,9	0.659607	0.079247	–0.018472	0.00060719	0.00060719	S
A,T	1.500000	0.144526	0.114632	0.00244016	0.00244016	S

Hand	Stand	Hit	Double	Spl1	Spl2	Spl3	Freq.	BS
				NDAS				
A,A	–0.666458	–0.022229	–0.620239	0.127144	0.231959	0.248335	0.00027830	P
2,2	–0.665989	–0.254345	–1.331978	–0.410196	–0.431613	–0.435100	0.00030360	H
3,3	–0.665541	–0.309159	–1.303229	–0.459339	–0.479718	–0.482993	0.00030360	EH
4,4	–0.664958	–0.198744	–0.807840	–0.512547	–0.555941	–0.563050	0.00030360	H
5,5	–0.664377	0.082875	–0.002421	–0.570784	–0.661500	–0.676416	0.00030360	H
6,6	–0.659438	–0.356496	–0.831043	–0.613693	–0.648930	–0.654649	0.00030360	EH
7,7	–0.661343	–0.448700	–0.947415	–0.619153	–0.641648	–0.645159	0.00030360	EH
8,8	–0.663258	–0.513551	–1.027102	–0.385449	–0.367296	–0.364252	0.00030360	EP
9,9	–0.093071	–0.626095	–1.252191	–0.121888	–0.125411	–0.125914	0.00030360	S
T,T	0.654582	–0.857109	–1.714218	0.152621	–0.108624	–0.266728	0.00506302	S
				DAS				
A,A	–0.666458	–0.022229	–0.620239	0.127144	0.231959	0.248335	0.00027830	P
2,2	–0.665989	–0.254345	–1.331978	–0.410196	–0.431613	–0.435100	0.00030360	H
3,3	–0.665541	–0.309159	–1.303229	–0.459339	–0.479718	–0.482993	0.00030360	EH
4,4	–0.664958	–0.198744	–0.807840	–0.512547	–0.555941	–0.563050	0.00030360	H
5,5	–0.664377	0.082875	–0.002421	–0.570784	–0.661500	–0.676416	0.00030360	H
6,6	–0.659438	–0.356496	–0.831043	–0.613693	–0.648930	–0.654649	0.00030360	EH
7,7	–0.661343	–0.448700	–0.947415	–0.619153	–0.641648	–0.645159	0.00030360	EH
8,8	–0.663258	–0.513551	–1.027102	–0.385449	–0.367296	–0.364252	0.00030360	EP
9,9	–0.093071	–0.626095	–1.252191	–0.121888	–0.125411	–0.125914	0.00030360	S
T,T	0.654582	–0.857109	–1.714218	0.152621	–0.108624	–0.266728	0.00506302	S

TABLE A51
6D S17: Dealer's Upcard 2

Hand	Stand	Hit	Double	Freq.	Cum. Freq.	BS
3,2	–0.292497	–0.128580	–0.584995	0.00088085	0.00088085	H
4,2	–0.292191	–0.142043	–0.562844	0.00088085	0.00088085	H
5,2	–0.289838	–0.107688	–0.430242	0.00088085		H
4,3	–0.292138	–0.110747	–0.433593	0.00091915	0.00180000	H
6,2	–0.290319	–0.020241	–0.197761	0.00088085		H
5,3	–0.289785	–0.020547	–0.198013	0.00091915	0.00180000	H
7,2	–0.290814	0.076102	0.070185	0.00088085		H
6,3	–0.290265	0.077499	0.069639	0.00091915		H
5,4	–0.289477	0.077895	0.070195	0.00091915	0.00271916	H
8,2	–0.291265	0.187068	0.368939	0.00088085		D
7,3	–0.290740	0.187604	0.369735	0.00091915		D
6,4	–0.289937	0.188194	0.370898	0.00091915	0.00271916	D
9,2	–0.291935	0.242225	0.478906	0.00088085		D
8,3	–0.291211	0.243014	0.480441	0.00091915		D
7,4	–0.290433	0.243852	0.482105	0.00091915		D
6,5	–0.287593	0.245699	0.486057	0.00091915	0.00363831	D
T,2	–0.295821	–0.251852	–0.503704	0.00352341		H
9,3	–0.291881	–0.255450	–0.510900	0.00091915		H
8,4	–0.290906	–0.254394	–0.508788	0.00091915		H
7,5	–0.288090	–0.253281	–0.506562	0.00091915	0.00628087	H
T,3	–0.295769	–0.307135	–0.614270	0.00367661		S
9,4	–0.291577	–0.305357	–0.610714	0.00091915		S
8,5	–0.288562	–0.311561	–0.623122	0.00091915		S
7,6	–0.288573	–0.311686	–0.623373	0.00091915	0.00643406	S
T,4	–0.295463	–0.363144	–0.726289	0.00367661		S
9,5	–0.289232	–0.361682	–0.723364	0.00091915		S
8,6	–0.289043	–0.362334	–0.724668	0.00091915	0.00551491	S
T,5	–0.293144	–0.419694	–0.839388	0.00367661		S
9,6	–0.289740	–0.418797	–0.837594	0.00091915		S
8,7	–0.289566	–0.412537	–0.825075	0.00091915	0.00551491	S
T,6	–0.293627	–0.470104	–0.940208	0.00367661		S
9,7	–0.290238	–0.468667	–0.937333	0.00091915	0.00459576	S
T,7	–0.153831	–0.536510	–1.073020	0.00367661		S
9,8	–0.150409	–0.535295	–1.070590	0.00091915	0.00459576	S
T,8	0.121180	–0.624069	–1.248139	0.00367661	0.00367661	S
T,9	0.386175	–0.732352	–1.464703	0.00367661	0.00367661	S
A,2	–0.291206	0.045498	–0.066612	0.00088085	0.00088085	H
A,3	–0.291153	0.021511	–0.067209	0.00091915	0.00091915	H
A,4	–0.290874	–0.001911	–0.070879	0.00091915	0.00091915	H
A,5	–0.288520	–0.022828	–0.073426	0.00091915	0.00091915	H
A,6	–0.149609	0.000693	–0.003955	0.00091915	0.00091915	H
A,7	0.124001	0.063289	0.120980	0.00091915	0.00091915	S
A,8	0.388607	0.123270	0.240954	0.00091915	0.00091915	S
A,9	0.642364	0.183625	0.361938	0.00091915	0.00091915	S
A,T	1.500000	0.239715	0.473071	0.00367661	0.00367661	S

Hand	Stand	Hit	Double	Spl1	Spl2	Spl3	Freq.	BS
				NDAS				
A,A	–0.289870	0.083842	–0.063039	0.485870	0.593051	0.610596	0.00044043	P
2,2	–0.292530	–0.114739	–0.585060	–0.148198	–0.152568	–0.153244	0.00040372	H
3,3	–0.292445	–0.142183	–0.563298	–0.200458	–0.208309	–0.209562	0.00044043	H
4,4	–0.291830	–0.019850	–0.200240	–0.229682	–0.258649	–0.263388	0.00044043	H
5,5	–0.287112	0.189215	0.373180	–0.252292	–0.313019	–0.322915	0.00044043	D
6,6	–0.288075	–0.253323	–0.506645	–0.272215	–0.274077	–0.274260	0.00044043	H
7,7	–0.289069	–0.369250	–0.738501	–0.209657	–0.198344	–0.196438	0.00044043	P
8,8	–0.290038	–0.468469	–0.936938	–0.034461	0.001152	0.007030	0.00044043	P
9,9	0.124232	–0.623270	–1.246540	0.152445	0.156355	0.156997	0.00044043	P
T,T	0.637885	–0.853873	–1.707745	0.356392	0.208676	0.118614	0.00727662	S
				DAS				
A,A	–0.289870	0.083842	–0.063039	0.485870	0.593051	0.610596	0.00044043	P
2,2	–0.292530	–0.114739	–0.585060	–0.082685	–0.078349	–0.077656	0.00040372	P
3,3	–0.292445	–0.142183	–0.563298	–0.134802	–0.133530	–0.133281	0.00044043	P
4,4	–0.291830	–0.019850	–0.200240	–0.163762	–0.183565	–0.186794	0.00044043	H
5,5	–0.287112	0.189215	0.373180	–0.187876	–0.265441	–0.278146	0.00044043	D
6,6	–0.288075	–0.253323	–0.506645	–0.205645	–0.198208	–0.196852	0.00044043	P
7,7	–0.289069	–0.369250	–0.738501	–0.143700	–0.123199	–0.119776	0.00044043	P
8,8	–0.290038	–0.468469	–0.936938	0.029968	0.074533	0.081885	0.00044043	P
9,9	0.124232	–0.623270	–1.246540	0.187731	0.196516	0.197956	0.00044043	P
T,T	0.637885	–0.853873	–1.707745	0.356392	0.208676	0.118614	0.00727662	S

TABLE A52
6D S17: Dealer's Upcard 3

Hand	Stand	Hit	Double	Freq.	Cum. Freq.	BS
3,2	−0.250991	−0.095372	−0.501981	0.00088085	0.00088085	H
4,2	−0.248651	−0.106780	−0.476783	0.00091915	0.00091915	H
5,2	−0.248212	−0.073959	−0.351320	0.00091915		H
4,3	−0.248347	−0.076830	−0.353083	0.00088085	0.00180000	H
6,2	−0.248688	0.010682	−0.126661	0.00091915		H
5,3	−0.247908	0.010288	−0.126928	0.00088085	0.00180000	H
7,2	−0.249155	0.105932	0.132529	0.00091915		D
6,3	−0.248363	0.107252	0.131932	0.00088085		D
5,4	−0.245557	0.107786	0.133343	0.00091915	0.00271916	D
8,2	−0.249823	0.211639	0.420927	0.00091915		D
7,3	−0.248852	0.212481	0.422507	0.00088085		D
6,4	−0.246032	0.213213	0.424181	0.00091915	0.00271916	D
9,2	−0.253743	0.261968	0.521453	0.00091915		D
8,3	−0.249520	0.265860	0.529385	0.00088085		D
7,4	−0.246521	0.267373	0.532615	0.00091915		D
6,5	−0.245592	0.268829	0.535531	0.00091915	0.00363831	D
T,2	−0.254531	−0.231352	−0.462705	0.00367661		H
9,3	−0.253442	−0.236992	−0.473984	0.00088085		H
8,4	−0.247191	−0.233026	−0.466053	0.00091915		H
7,5	−0.246082	−0.231898	−0.463795	0.00091915	0.00639576	H
T,3	−0.254229	−0.289800	−0.579600	0.00352341		S
9,4	−0.251111	−0.289298	−0.578596	0.00091915		S
8,5	−0.246750	−0.293525	−0.587049	0.00091915		S
7,6	−0.246558	−0.293387	−0.586773	0.00091915	0.00628087	S
T,4	−0.251923	−0.349731	−0.699462	0.00367661		S
9,5	−0.250696	−0.349725	−0.699450	0.00091915		S
8,6	−0.247253	−0.348528	−0.697056	0.00091915	0.00551491	S
T,5	−0.251483	−0.409906	−0.819813	0.00367661		S
9,6	−0.251174	−0.410266	−0.820533	0.00091915		S
8,7	−0.247742	−0.402400	−0.804800	0.00091915	0.00551491	S
T,6	−0.251961	−0.463383	−0.926765	0.00367661		S
9,7	−0.251664	−0.462970	−0.925939	0.00091915	0.00459576	S
T,7	−0.117503	−0.532472	−1.064943	0.00367661		S
9,8	−0.117448	−0.532256	−1.064512	0.00091915	0.00459576	S
T,8	0.147907	−0.622201	−1.244402	0.00367661	0.00367661	S
T,9	0.401225	−0.725564	−1.451128	0.00367661	0.00367661	S
A,2	−0.250039	0.073677	−0.000768	0.00091915	0.00091915	H
A,3	−0.249764	0.049873	−0.003369	0.00088085	0.00088085	H
A,4	−0.247424	0.028141	−0.005278	0.00091915	0.00091915	H
A,5	−0.246985	0.007133	−0.009099	0.00091915	0.00091915	H
A,6	−0.113384	0.030062	0.057996	0.00091915	0.00091915	D
A,7	0.151119	0.090730	0.179330	0.00091915	0.00091915	D
A,8	0.406627	0.152923	0.303609	0.00091915	0.00091915	S
A,9	0.649381	0.204702	0.406920	0.00091915	0.00091915	S
A,T	1.500000	0.262007	0.520742	0.00367661	0.00367661	S

Hand	Stand	Hit	Double	Spl1	Spl2	Spl3	Freq.	BS
				NDAS				
A,A	−0.248796	0.106038	0.002989	0.533182	0.641204	0.658883	0.00044043	P
2,2	−0.251293	−0.082064	−0.502586	−0.094440	−0.096120	−0.096390	0.00044043	H
3,3	−0.250687	−0.108263	−0.480703	−0.135339	−0.138744	−0.139249	0.00040372	H
4,4	−0.246007	0.011580	−0.126731	−0.158582	−0.181812	−0.185570	0.00044043	H
5,5	−0.245116	0.213958	0.425665	−0.184014	−0.238673	−0.247567	0.00044043	D
6,6	−0.246070	−0.231966	−0.463931	−0.202725	−0.197888	−0.196967	0.00044043	P
7,7	−0.247072	−0.355164	−0.710327	−0.142358	−0.127488	−0.124988	0.00044043	P
8,8	−0.248413	−0.461664	−0.923328	0.025904	0.064099	0.070399	0.00044043	P
9,9	0.144792	−0.622799	−1.245598	0.198208	0.205203	0.206285	0.00044043	P
T,T	0.647984	−0.853605	−1.707210	0.404535	0.276510	0.198306	0.00727662	S
				DAS				
A,A	−0.248796	0.106038	0.002989	0.533182	0.641204	0.658883	0.00044043	P
2,2	−0.251293	−0.082064	−0.502586	−0.016670	−0.007510	−0.005991	0.00044043	P
3,3	−0.250687	−0.108263	−0.480703	−0.057359	−0.050403	−0.049280	0.00040372	P
4,4	−0.246007	0.011580	−0.126731	−0.079529	−0.091705	−0.093631	0.00044043	H
5,5	−0.245116	0.213958	0.425665	−0.107013	−0.180589	−0.192630	0.00044043	D
6,6	−0.246070	−0.231966	−0.463931	−0.118482	−0.101750	−0.098839	0.00044043	P
7,7	−0.247072	−0.355164	−0.710327	−0.059290	−0.032679	−0.028215	0.00044043	P
8,8	−0.248413	−0.461664	−0.923328	0.098301	0.146573	0.154534	0.00044043	P
9,9	0.144792	−0.622799	−1.245598	0.238147	0.250603	0.252570	0.00044043	P
T,T	0.647984	−0.853605	−1.707210	0.404535	0.276510	0.198306	0.00727662	S

TABLE A53
6D S17: Dealer's Upcard 4

Hand	Stand	Hit	Double	Freq.	Cum. Freq.	BS
3,2	–0.205342	–0.058096	–0.410683	0.00091915	0.00091915	H
4,2	–0.204896	–0.069987	–0.389655	0.00088085	0.00088085	H
5,2	–0.204511	–0.037741	–0.268419	0.00091915		H
4,3	–0.202559	–0.038805	–0.265965	0.00088085	0.00180000	H
6,2	–0.204953	0.045436	–0.052398	0.00091915		H
5,3	–0.202163	0.046373	–0.049530	0.00091915	0.00183830	H
7,2	–0.205645	0.135128	0.196167	0.00091915		D
6,3	–0.202625	0.138010	0.198944	0.00091915		D
5,4	–0.201715	0.138455	0.199843	0.00088085	0.00271916	D
8,2	–0.209562	0.234141	0.468281	0.00091915		D
7,3	–0.203317	0.239200	0.478400	0.00091915		D
6,4	–0.202177	0.240204	0.480408	0.00088085	0.00271916	D
9,2	–0.210367	0.285737	0.571475	0.00091915		D
8,3	–0.207235	0.287820	0.575640	0.00091915		D
7,4	–0.202872	0.292304	0.584609	0.00088085		D
6,5	–0.201792	0.293738	0.587476	0.00091915	0.00363831	D
T,2	–0.211115	–0.210364	–0.420729	0.00367661		H
9,3	–0.208039	–0.215390	–0.430780	0.00091915		S
8,4	–0.206788	–0.213675	–0.427349	0.00088085		S
7,5	–0.202485	–0.210281	–0.420562	0.00091915	0.00639576	S
T,3	–0.208813	–0.271530	–0.543059	0.00367661		S
9,4	–0.207617	–0.271085	–0.542170	0.00088085		S
8,5	–0.206426	–0.277081	–0.554162	0.00091915		S
7,6	–0.202973	–0.274868	–0.549736	0.00091915	0.00639576	S
T,4	–0.208366	–0.335176	–0.670351	0.00352341		S
9,5	–0.207230	–0.335130	–0.670260	0.00091915		S
8,6	–0.206890	–0.335432	–0.670865	0.00091915	0.00536172	S
T,5	–0.207979	–0.399232	–0.798464	0.00367661		S
9,6	–0.207694	–0.399532	–0.799063	0.00091915		S
8,7	–0.207585	–0.393183	–0.786366	0.00091915	0.00551491	S
T,6	–0.208443	–0.456060	–0.912120	0.00367661		S
9,7	–0.208389	–0.455828	–0.911656	0.00091915	0.00459576	S
T,7	–0.077990	–0.528312	–1.056623	0.00367661		S
9,8	–0.081128	–0.529145	–1.058291	0.00091915	0.00459576	S
T,8	0.174060	–0.614215	–1.228429	0.00367661	0.00367661	S
T,9	0.420027	–0.724521	–1.449042	0.00367661	0.00367661	S
A,2	–0.206762	0.103706	0.068041	0.00091915	0.00091915	H
A,3	–0.204426	0.081740	0.066904	0.00091915	0.00091915	H
A,4	–0.203980	0.059395	0.062389	0.00088085	0.00088085	D
A,5	–0.203594	0.039363	0.058627	0.00091915	0.00091915	D
A,6	–0.073706	0.062047	0.124093	0.00091915	0.00091915	D
A,7	0.180238	0.124398	0.248797	0.00091915	0.00091915	D
A,8	0.422206	0.177396	0.354792	0.00091915	0.00091915	S
A,9	0.659822	0.230229	0.460459	0.00091915	0.00091915	S
A,T	1.500000	0.285339	0.570678	0.00367661	0.00367661	S

Hand	Stand	Hit	Double	Spl1	Spl2	Spl3	Freq.	BS
				NDAS				
A,A	–0.205803	0.129004	0.071156	0.582709	0.691608	0.709426	0.00044043	P
2,2	–0.207678	–0.046774	–0.415355	–0.037806	–0.036536	–0.036323	0.00044043	P
3,3	–0.203005	–0.068783	–0.386312	–0.059614	–0.057998	–0.057677	0.00044043	P
4,4	–0.202102	0.048027	–0.049346	–0.085480	–0.102830	–0.105496	0.00040372	H
5,5	–0.201329	0.240560	0.481119	–0.110131	–0.158174	–0.165972	0.00044043	D
6,6	–0.202253	–0.210060	–0.420120	–0.127881	–0.117032	–0.115163	0.00044043	P
7,7	–0.203668	–0.340486	–0.680972	–0.069969	–0.051026	–0.047850	0.00044043	P
8,8	–0.211502	–0.456910	–0.913820	0.082582	0.123151	0.129782	0.00044043	P
9,9	0.174080	–0.614207	–1.228414	0.256731	0.267804	0.269561	0.00044043	P
T,T	0.658450	–0.853371	–1.706741	0.454020	0.345916	0.279554	0.00727662	S
				DAS				
A,A	–0.205803	0.129004	0.071156	0.582709	0.691608	0.709426	0.00044043	P
2,2	–0.207678	–0.046774	–0.415355	0.053199	0.067127	0.069425	0.00044043	P
3,3	–0.203005	–0.068783	–0.386312	0.032644	0.047113	0.049555	0.00044043	P
4,4	–0.202102	0.048027	–0.049346	0.008704	0.003967	0.003298	0.00040372	H
5,5	–0.201329	0.240560	0.481119	–0.016478	–0.085151	–0.096379	0.00044043	D
6,6	–0.202253	–0.210060	–0.420120	–0.025038	0.000317	0.004609	0.00044043	P
7,7	–0.203668	–0.340486	–0.680972	0.033066	0.066524	0.072122	0.00044043	P
8,8	–0.211502	–0.456910	–0.913820	0.163367	0.215042	0.223481	0.00044043	P
9,9	0.174080	–0.614207	–1.228414	0.300729	0.317822	0.320554	0.00044043	P
T,T	0.658450	–0.853371	–1.706741	0.454020	0.345916	0.279554	0.00727662	S

TABLE A54
6D S17: Dealer's Upcard 5

Hand	Stand	Hit	Double	Freq.	Cum. Freq.	BS
3,2	–0.157535	–0.016982	–0.315071	0.00091915	0.00091915	H
4,2	–0.157142	–0.028170	–0.295698	0.00091915	0.00091915	H
5,2	–0.156769	0.002599	–0.182527	0.00088085		H
4,3	–0.156706	0.001372	–0.181201	0.00091915	0.00180000	H
6,2	–0.157462	0.080131	0.023136	0.00091915		H
5,3	–0.156343	0.081640	0.026650	0.00088085	0.00180000	H
7,2	–0.161406	0.164083	0.257330	0.00091915		D
6,3	–0.157036	0.170000	0.266081	0.00091915		D
5,4	–0.155947	0.171639	0.269397	0.00088085	0.00271916	D
8,2	–0.162223	0.262620	0.525241	0.00091915		D
7,3	–0.160981	0.264422	0.528844	0.00091915		D
6,4	–0.156641	0.269038	0.538075	0.00091915	0.00275745	D
9,2	–0.162932	0.312378	0.624756	0.00091915		D
8,3	–0.161797	0.314059	0.628118	0.00091915		D
7,4	–0.160585	0.315911	0.631821	0.00091915		D
6,5	–0.156277	0.320645	0.641289	0.00088085	0.00363831	D
T,2	–0.163622	–0.188610	–0.377220	0.00367661		S
9,3	–0.162532	–0.193893	–0.387786	0.00091915		S
8,4	–0.161427	–0.192253	–0.384505	0.00091915		S
7,5	–0.160247	–0.190851	–0.381702	0.00088085	0.00639576	S
T,3	–0.163196	–0.253882	–0.507764	0.00367661		S
9,4	–0.162136	–0.253443	–0.506886	0.00091915		S
8,5	–0.161063	–0.259433	–0.518866	0.00088085		S
7,6	–0.160941	–0.259222	–0.518445	0.00091915	0.00639576	S
T,4	–0.162801	–0.320371	–0.640742	0.00367661		S
9,5	–0.161773	–0.320312	–0.640623	0.00088085		S
8,6	–0.161758	–0.320819	–0.641638	0.00091915	0.00547661	S
T,5	–0.162437	–0.387776	–0.775552	0.00352341		S
9,6	–0.162468	–0.388253	–0.776507	0.00091915		S
8,7	–0.165703	–0.383461	–0.766922	0.00091915	0.00536172	S
T,6	–0.163133	–0.448820	–0.897639	0.00367661		S
9,7	–0.166414	–0.449906	–0.899811	0.00091915	0.00459576	S
T,7	–0.044650	–0.518176	–1.036353	0.00367661		S
9,8	–0.044795	–0.518184	–1.036369	0.00091915	0.00459576	S
T,8	0.200067	–0.611370	–1.222740	0.00367661	0.00367661	S
T,9	0.440768	–0.723267	–1.446534	0.00367661	0.00367661	S
A,2	–0.159409	0.137333	0.139852	0.00091915	0.00091915	D
A,3	–0.158985	0.115726	0.138428	0.00091915	0.00091915	D
A,4	–0.158591	0.094352	0.133708	0.00091915	0.00091915	D
A,5	–0.158230	0.074608	0.129043	0.00088085	0.00088085	D
A,6	–0.037276	0.098894	0.197788	0.00091915	0.00091915	D
A,7	0.203128	0.152043	0.304086	0.00091915	0.00091915	D
A,8	0.442903	0.206863	0.413727	0.00091915	0.00091915	S
A,9	0.672201	0.258083	0.516167	0.00091915	0.00091915	S
A,T	1.500000	0.311181	0.622362	0.00367661	0.00367661	S

Hand	Stand	Hit	Double	Spl1	Spl2	Spl3	Freq.	BS
				NDAS				
A,A	–0.160846	0.160724	0.140787	0.633923	0.743663	0.761616	0.00044043	P
2,2	–0.157959	–0.004963	–0.315919	0.043055	0.049886	0.051035	0.00044043	P
3,3	–0.157112	–0.028184	–0.295400	0.018178	0.024979	0.026157	0.00044043	P
4,4	–0.156310	0.083624	0.027678	–0.006369	–0.018523	–0.020467	0.00044043	H
5,5	–0.155585	0.270245	0.540491	–0.030358	–0.069705	–0.075797	0.00040372	D
6,6	–0.156998	–0.188560	–0.377121	–0.048288	–0.032695	–0.030050	0.00044043	P
7,7	–0.164886	–0.329183	–0.658367	–0.005417	0.016644	0.020259	0.00044043	P
8,8	–0.166521	–0.449933	–0.899865	0.149614	0.193194	0.200311	0.00044043	P
9,9	0.200047	–0.611359	–1.222718	0.319391	0.335518	0.338099	0.00044043	P
T,T	0.670892	–0.853070	–1.706139	0.510019	0.424478	0.371714	0.00727662	S
				DAS				
A,A	–0.160846	0.160724	0.140787	0.633923	0.743663	0.761616	0.00044043	P
2,2	–0.157959	–0.004963	–0.315919	0.148858	0.170500	0.174105	0.00044043	P
3,3	–0.157112	–0.028184	–0.295400	0.128115	0.150301	0.154027	0.00044043	P
4,4	–0.156310	0.083624	0.027678	0.106971	0.110697	0.111389	0.00044043	P
5,5	–0.155585	0.270245	0.540491	0.081906	0.021298	0.011823	0.00040372	D
6,6	–0.156998	–0.188560	–0.377121	0.074064	0.106878	0.112394	0.00044043	P
7,7	–0.164886	–0.329183	–0.658367	0.115255	0.154012	0.160364	0.00044043	P
8,8	–0.166521	–0.449933	–0.899865	0.238876	0.294728	0.303844	0.00044043	P
9,9	0.200047	–0.611359	–1.222718	0.367531	0.390249	0.393899	0.00044043	P
T,T	0.670892	–0.853070	–1.706139	0.510019	0.424478	0.371714	0.00727662	S

TABLE A55
6D S17: Dealer's Upcard 6

Hand	Stand	Hit	Double	Freq.	Cum. Freq.	BS
3,2	−0.148613	0.002168	−0.297226	0.00091915	0.00091915	H
4,2	−0.148262	−0.008635	−0.271375	0.00091915	0.00091915	H
5,2	−0.147947	0.035671	−0.125924	0.00091915		H
4,3	−0.147894	0.034005	−0.125555	0.00091915	0.00183830	H
6,2	−0.151827	0.118058	0.093586	0.00088085		H
5,3	−0.147579	0.122679	0.103206	0.00091915	0.00180000	H
7,2	−0.152586	0.198589	0.324158	0.00091915		D
6,3	−0.151460	0.201776	0.327255	0.00088085		D
5,4	−0.147238	0.206548	0.336805	0.00091915	0.00271916	D
8,2	−0.153236	0.291058	0.582117	0.00091915		D
7,3	−0.152216	0.292827	0.585653	0.00091915		D
6,4	−0.151116	0.294885	0.589770	0.00088085	0.00271916	D
9,2	−0.153877	0.335808	0.671616	0.00091915		D
8,3	−0.152893	0.337335	0.674669	0.00091915		D
7,4	−0.151899	0.338982	0.677964	0.00091915		D
6,5	−0.150826	0.341332	0.682665	0.00088085	0.00363831	D
T,2	−0.154694	−0.168756	−0.337513	0.00367661		S
9,3	−0.153509	−0.174121	−0.348242	0.00091915		S
8,4	−0.152551	−0.172631	−0.345262	0.00091915		S
7,5	−0.151583	−0.171631	−0.343262	0.00091915	0.00643406	S
T,3	−0.154326	−0.234633	−0.469267	0.00367661		S
9,4	−0.153167	−0.234397	−0.468793	0.00091915		S
8,5	−0.152235	−0.240648	−0.481295	0.00091915		S
7,6	−0.155464	−0.242507	−0.485014	0.00088085	0.00639576	S
T,4	−0.153984	−0.301960	−0.603919	0.00367661		S
9,5	−0.152851	−0.301944	−0.603888	0.00091915		S
8,6	−0.156116	−0.304185	−0.608371	0.00088085	0.00547661	S
T,5	−0.153670	−0.369292	−0.738584	0.00367661		S
9,6	−0.156735	−0.371282	−0.742565	0.00088085		S
8,7	−0.156878	−0.365151	−0.730302	0.00091915	0.00547661	S
T,6	−0.157516	−0.425430	−0.850861	0.00352341		S
9,7	−0.157459	−0.425490	−0.850980	0.00091915	0.00444257	S
T,7	0.008330	−0.504644	−1.009287	0.00367661		S
9,8	0.008284	−0.504679	−1.009359	0.00091915	0.00459576	S
T,8	0.281026	−0.604067	−1.208133	0.00367661	0.00367661	S
T,9	0.494123	−0.720056	−1.440112	0.00367661	0.00367661	S
A,2	−0.147343	0.162671	0.187925	0.00091915	0.00091915	D
A,3	−0.146976	0.140226	0.186374	0.00091915	0.00091915	D
A,4	−0.146635	0.118402	0.182917	0.00091915	0.00091915	D
A,5	−0.146311	0.101516	0.185557	0.00091915	0.00091915	D
A,6	0.012006	0.129254	0.258507	0.00088085	0.00088085	D
A,7	0.280488	0.191403	0.382807	0.00091915	0.00091915	D
A,8	0.493924	0.240133	0.480266	0.00091915	0.00091915	S
A,9	0.702601	0.286683	0.573366	0.00091915	0.00091915	S
A,T	1.500000	0.334922	0.669843	0.00367661	0.00367661	S

Hand	Stand	Hit	Double	Spl1	Spl2	Spl3	Freq.	BS
				NDAS				
A,A	−0.145699	0.187974	0.190709	0.682366	0.796424	0.815037	0.00044043	P
2,2	−0.148979	0.014623	−0.297957	0.085692	0.095743	0.097426	0.00044043	P
3,3	−0.148235	−0.008642	−0.271320	0.058817	0.068571	0.070238	0.00044043	P
4,4	−0.147552	0.124399	0.103556	0.032107	0.019621	0.017620	0.00044043	H
5,5	−0.146921	0.299427	0.598853	0.007364	−0.032675	−0.039179	0.00044043	D
6,6	−0.154706	−0.173833	−0.347667	−0.022403	−0.004776	−0.001999	0.00040372	P
7,7	−0.156222	−0.311022	−0.622045	0.058875	0.088600	0.093466	0.00044043	P
8,8	−0.157495	−0.425486	−0.850971	0.230584	0.284107	0.292853	0.00044043	P
9,9	0.280805	−0.604091	−1.208183	0.387321	0.401648	0.403931	0.00044043	P
T,T	0.702826	−0.852263	−1.704526	0.566767	0.494080	0.449060	0.00727662	S
				DAS				
A,A	−0.145699	0.187974	0.190709	0.682366	0.796424	0.815037	0.00044043	P
2,2	−0.148979	0.014623	−0.297957	0.208397	0.235631	0.240164	0.00044043	P
3,3	−0.148235	−0.008642	−0.271320	0.184428	0.211764	0.216345	0.00044043	P
4,4	−0.147552	0.124399	0.103556	0.161015	0.166595	0.167593	0.00044043	P
5,5	−0.146921	0.299427	0.598853	0.136381	0.072332	0.061822	0.00044043	D
6,6	−0.154706	−0.173833	−0.347667	0.116676	0.152736	0.158403	0.00040372	P
7,7	−0.156222	−0.311022	−0.622045	0.192728	0.240961	0.248859	0.00044043	P
8,8	−0.157495	−0.425486	−0.850971	0.327864	0.394763	0.405689	0.00044043	P
9,9	0.280805	−0.604091	−1.208183	0.439110	0.460533	0.463967	0.00044043	P
T,T	0.702826	−0.852263	−1.704526	0.566767	0.494080	0.449060	0.00727662	S

TABLE A56
6D S17: Dealer's Upcard 7

Hand	Stand	Hit	Double	Freq.	Cum. Freq.	BS
3,2	−0.474287	−0.119481	−0.948574	0.00091915	0.00091915	H
4,2	−0.473947	−0.153746	−0.890224	0.00091915	0.00091915	H
5,2	−0.475404	−0.068651	−0.585406	0.00091915		H
4,3	−0.473622	−0.069120	−0.581824	0.00091915	0.00183830	H
6,2	−0.476165	0.083939	−0.180927	0.00091915		H
5,3	−0.475079	0.083844	−0.178613	0.00091915	0.00183830	H
7,2	−0.476858	0.174276	0.113595	0.00088085		H
6,3	−0.475838	0.176073	0.115952	0.00091915		H
5,4	−0.474737	0.176299	0.118144	0.00091915	0.00271916	H
8,2	−0.477422	0.258576	0.396342	0.00091915		D
7,3	−0.476558	0.260334	0.401418	0.00088085		D
6,4	−0.475523	0.261579	0.406015	0.00091915	0.00271916	D
9,2	−0.478284	0.291741	0.461868	0.00091915		D
8,3	−0.477096	0.292164	0.464533	0.00091915		D
7,4	−0.476217	0.292868	0.467648	0.00088085		D
6,5	−0.476977	0.293411	0.469749	0.00091915	0.00363831	D
T,2	−0.474935	−0.212638	−0.504855	0.00367661		H
9,3	−0.477959	−0.218241	−0.518061	0.00091915		H
8,4	−0.476756	−0.217934	−0.515629	0.00091915		H
7,5	−0.477672	−0.219577	−0.518149	0.00088085	0.00639576	H
T,3	−0.474609	−0.269267	−0.586650	0.00367661		H
9,4	−0.477619	−0.269781	−0.589288	0.00091915		H
8,5	−0.478210	−0.278145	−0.605877	0.00091915		H
7,6	−0.478433	−0.278428	−0.606515	0.00088085	0.00639576	H
T,4	−0.474271	−0.324567	−0.674240	0.00367661		H
9,5	−0.479075	−0.325412	−0.677253	0.00091915		H
8,6	−0.478975	−0.325493	−0.678337	0.00091915	0.00551491	H
T,5	−0.475708	−0.368927	−0.746864	0.00367661		H
9,6	−0.479801	−0.368743	−0.748137	0.00091915		H
8,7	−0.479635	−0.362547	−0.735710	0.00088085	0.00547661	H
T,6	−0.476476	−0.408624	−0.817247	0.00367661		H
9,7	−0.480503	−0.408432	−0.816865	0.00088085	0.00455746	H
T,7	−0.108885	−0.478380	−0.956760	0.00352341		S
9,8	−0.109730	−0.478075	−0.956150	0.00091915	0.00444257	S
T,8	0.397743	−0.587340	−1.174681	0.00367661	0.00367661	S
T,9	0.615047	−0.712756	−1.425513	0.00367661	0.00367661	S
A,2	−0.473185	0.120174	−0.179483	0.00091915	0.00091915	H
A,3	−0.472861	0.076540	−0.182241	0.00091915	0.00091915	H
A,4	−0.472512	0.036406	−0.177160	0.00091915	0.00091915	H
A,5	−0.473968	−0.007882	−0.184344	0.00091915	0.00091915	H
A,6	−0.103826	0.054706	−0.008857	0.00091915	0.00091915	H
A,7	0.401861	0.171872	0.223664	0.00088085	0.00088085	S
A,8	0.615734	0.220956	0.320560	0.00091915	0.00091915	S
A,9	0.773273	0.254647	0.385938	0.00091915	0.00091915	S
A,T	1.500000	0.291217	0.463763	0.00367661	0.00367661	S

Hand	Stand	Hit	Double	Spl1	Spl2	Spl3	Freq.	BS
				NDAS				
A,A	−0.471742	0.164333	−0.176403	0.475600	0.606486	0.627910	0.00044043	P
2,2	−0.474621	−0.088828	−0.949243	−0.054175	−0.049345	−0.048548	0.00044043	P
3,3	−0.473963	−0.153797	−0.890244	−0.115161	−0.109406	−0.108396	0.00044043	P
4,4	−0.473283	0.086616	−0.175100	−0.183163	−0.220778	−0.226991	0.00044043	H
5,5	−0.476217	0.260394	0.404746	−0.247854	−0.318736	−0.330447	0.00044043	D
6,6	−0.477738	−0.220562	−0.520123	−0.299362	−0.309783	−0.311412	0.00044043	H
7,7	−0.479132	−0.331784	−0.691921	−0.135364	−0.108956	−0.104759	0.00040372	P
8,8	−0.480181	−0.408423	−0.816845	0.171102	0.250927	0.263954	0.00044043	P
9,9	0.399576	−0.587176	−1.174353	0.343323	0.335474	0.334177	0.00044043	S
T,T	0.772011	−0.850445	−1.700890	0.508736	0.371745	0.288857	0.00727662	S
				DAS				
A,A	−0.471742	0.164333	−0.176403	0.475600	0.606486	0.627910	0.00044043	P
2,2	−0.474621	−0.088828	−0.949243	−0.005187	0.006557	0.008509	0.00044043	P
3,3	−0.473963	−0.153797	−0.890244	−0.066057	−0.053349	−0.051174	0.00044043	P
4,4	−0.473283	0.086616	−0.175100	−0.133145	−0.163650	−0.168667	0.00044043	H
5,5	−0.476217	0.260394	0.404746	−0.198756	−0.283218	−0.297220	0.00044043	D
6,6	−0.477738	−0.220562	−0.520123	−0.248890	−0.252261	−0.252725	0.00044043	H
7,7	−0.479132	−0.331784	−0.691921	−0.086493	−0.053665	−0.048469	0.00040372	P
8,8	−0.480181	−0.408423	−0.816845	0.218422	0.304662	0.318719	0.00044043	P
9,9	0.399576	−0.587176	−1.174353	0.369004	0.364599	0.363848	0.00044043	S
T,T	0.772011	−0.850445	−1.700890	0.508736	0.371745	0.288857	0.00727662	S

TABLE A57

6D S17: Dealer's Upcard 8

Hand	Stand	Hit	Double	Freq.	Cum. Freq.	BS
3,2	–0.510907	–0.187166	–1.021815	0.00091915	0.00091915	H
4,2	–0.512340	–0.219720	–1.004683	0.00091915	0.00091915	H
5,2	–0.512277	–0.211671	–0.847888	0.00091915		H
4,3	–0.512000	–0.213278	–0.847352	0.00091915	0.00183830	H
6,2	–0.512931	–0.059214	–0.449931	0.00091915		H
5,3	–0.511935	–0.059455	–0.448043	0.00091915	0.00183830	H
7,2	–0.513537	0.099741	–0.021682	0.00091915		H
6,3	–0.512615	0.101203	–0.020122	0.00091915		H
5,4	–0.513391	0.099920	–0.021689	0.00091915	0.00275745	H
8,2	–0.514402	0.199178	0.287564	0.00088085		D
7,3	–0.513196	0.200742	0.292600	0.00091915		D
6,4	–0.514046	0.199620	0.291603	0.00091915	0.00271916	D
9,2	–0.511055	0.227615	0.346979	0.00091915		D
8,3	–0.514062	0.228561	0.347405	0.00088085		D
7,4	–0.514626	0.228786	0.349100	0.00091915		D
6,5	–0.513980	0.230173	0.353456	0.00091915	0.00363831	D
T,2	–0.511634	–0.272002	–0.617342	0.00367661		H
9,3	–0.510715	–0.278647	–0.629649	0.00091915		H
8,4	–0.515493	–0.279058	–0.631948	0.00088085		H
7,5	–0.514561	–0.279202	–0.630474	0.00091915	0.00639576	H
T,3	–0.511295	–0.324425	–0.693346	0.00367661		H
9,4	–0.512148	–0.326010	–0.696199	0.00091915		H
8,5	–0.515429	–0.333494	–0.712504	0.00088085		H
7,6	–0.515221	–0.334678	–0.714750	0.00091915	0.00639576	H
T,4	–0.512708	–0.369606	–0.761980	0.00367661		H
9,5	–0.512065	–0.371556	–0.765428	0.00091915		H
8,6	–0.516050	–0.371477	–0.767559	0.00088085	0.00547661	H
T,5	–0.512647	–0.416910	–0.841853	0.00367661		H
9,6	–0.512728	–0.417413	–0.843770	0.00091915		H
8,7	–0.516638	–0.410942	–0.831472	0.00088085	0.00547661	H
T,6	–0.513309	–0.453130	–0.906261	0.00367661		H
9,7	–0.513316	–0.453627	–0.907253	0.00091915	0.00459576	H
T,7	–0.384232	–0.500967	–1.001934	0.00367661		S
9,8	–0.387487	–0.501194	–1.002387	0.00088085	0.00455746	S
T,8	0.104057	–0.586988	–1.173977	0.00352341	0.00352341	S
T,9	0.591091	–0.711065	–1.422130	0.00367661	0.00367661	S
A,2	–0.510013	0.051835	–0.314099	0.00091915	0.00091915	H
A,3	–0.509664	0.016478	–0.305060	0.00091915	0.00091915	H
A,4	–0.511097	–0.028404	–0.314303	0.00091915	0.00091915	H
A,5	–0.511033	–0.069392	–0.317151	0.00091915	0.00091915	H
A,6	–0.382583	–0.071750	–0.251340	0.00091915	0.00091915	H
A,7	0.108057	0.041028	–0.028189	0.00091915	0.00091915	S
A,8	0.595896	0.152904	0.193997	0.00088085	0.00088085	S
A,9	0.790658	0.193713	0.277473	0.00091915	0.00091915	S
A,T	1.500000	0.228515	0.347788	0.00367661	0.00367661	S

Hand	Stand	Hit	Double	Spl1	Spl2	Spl3	Freq.	BS
				NDAS				
A,A	–0.508761	0.094585	–0.311501	0.359708	0.479818	0.499499	0.00044043	P
2,2	–0.511247	–0.156790	–1.022494	–0.207056	–0.213879	–0.214977	0.00044043	H
3,3	–0.510567	–0.219182	–1.001140	–0.262997	–0.268579	–0.269415	0.00044043	H
4,4	–0.513431	–0.059090	–0.451036	–0.322242	–0.358857	–0.364893	0.00044043	H
5,5	–0.513325	0.199537	0.293033	–0.386975	–0.468679	–0.482162	0.00044043	D
6,6	–0.514636	–0.279301	–0.630537	–0.431736	–0.452434	–0.455763	0.00044043	H
7,7	–0.515766	–0.377709	–0.780906	–0.422290	–0.427967	–0.428817	0.00044043	H
8,8	–0.517509	–0.453401	–0.906802	–0.116221	–0.071930	–0.065048	0.00040372	P
9,9	0.099261	–0.587148	–1.174296	0.193427	0.206575	0.208751	0.00044043	P
T,T	0.790420	–0.850101	–1.700201	0.388889	0.180602	0.054928	0.00727662	S
				DAS				
A,A	–0.508761	0.094585	–0.311501	0.359708	0.479818	0.499499	0.00044043	P
2,2	–0.511247	–0.156790	–1.022494	–0.174270	–0.176424	–0.176735	0.00044043	H
3,3	–0.510567	–0.219182	–1.001140	–0.229737	–0.230555	–0.230584	0.00044043	H
4,4	–0.513431	–0.059090	–0.451036	–0.288375	–0.320198	–0.325431	0.00044043	H
5,5	–0.513325	0.199537	0.293033	–0.353518	–0.443869	–0.458826	0.00044043	D
6,6	–0.514636	–0.279301	–0.630537	–0.397634	–0.413538	–0.416069	0.00044043	H
7,7	–0.515766	–0.377709	–0.780906	–0.389294	–0.390412	–0.390516	0.00044043	H
8,8	–0.517509	–0.453401	–0.906802	–0.084875	–0.036558	–0.029064	0.00040372	P
9,9	0.099261	–0.587148	–1.174296	0.211619	0.227230	0.229801	0.00044043	P
T,T	0.790420	–0.850101	–1.700201	0.388889	0.180602	0.054928	0.00727662	S

TABLE A58
6D S17: Dealer's Upcard 9

Hand	Stand	Hit	Double	Freq.	Cum. Freq.	BS
3,2	−0.541573	−0.265989	−1.083146	0.00091915	0.00091915	H
4,2	−0.541491	−0.294528	−1.064329	0.00091915	0.00091915	H
5,2	−0.541389	−0.285443	−0.952282	0.00091915		H
4,3	−0.542924	−0.288453	−0.955318	0.00091915	0.00183830	H
6,2	−0.542034	−0.209881	−0.714108	0.00091915		H
5,3	−0.542848	−0.211386	−0.715741	0.00091915	0.00183830	H
7,2	−0.542861	−0.052171	−0.297011	0.00091915		H
6,3	−0.543467	−0.052030	−0.298867	0.00091915		H
5,4	−0.542763	−0.052189	−0.297409	0.00091915	0.00275745	H
8,2	−0.539517	0.116883	0.148698	0.00091915		D
7,3	−0.544295	0.116689	0.145777	0.00091915		D
6,4	−0.543383	0.116739	0.147751	0.00091915	0.00275745	D
9,2	−0.540098	0.155832	0.225460	0.00088085		D
8,3	−0.540952	0.155339	0.225746	0.00091915		D
7,4	−0.544211	0.156869	0.227194	0.00091915		D
6,5	−0.543281	0.157414	0.230031	0.00091915	0.00363831	D
T,2	−0.540641	−0.340701	−0.738846	0.00367661		H
9,3	−0.541534	−0.348220	−0.753666	0.00088085		H
8,4	−0.540869	−0.348743	−0.752669	0.00091915		H
7,5	−0.544111	−0.347763	−0.751490	0.00091915	0.00639576	H
T,3	−0.542057	−0.382712	−0.798989	0.00367661		H
9,4	−0.541432	−0.383661	−0.800017	0.00088085		H
8,5	−0.540750	−0.391310	−0.814270	0.00091915		H
7,6	−0.544695	−0.392092	−0.817609	0.00091915	0.00639576	H
T,4	−0.541976	−0.428224	−0.872084	0.00367661		H
9,5	−0.541334	−0.430167	−0.875363	0.00088085		H
8,6	−0.541376	−0.431821	−0.879484	0.00091915	0.00547661	H
T,5	−0.541878	−0.472187	−0.949023	0.00367661		H
9,6	−0.541961	−0.472812	−0.951189	0.00088085		H
8,7	−0.542209	−0.467044	−0.939621	0.00091915	0.00547661	H
T,6	−0.542505	−0.504547	−1.009094	0.00367661		RH
9,7	−0.542794	−0.504993	−1.009985	0.00088085	0.00455746	RH
T,7	−0.422164	−0.549404	−1.098808	0.00367661		S
9,8	−0.421286	−0.550164	−1.100327	0.00088085	0.00455746	S
T,8	−0.185194	−0.612833	−1.225665	0.00367661	0.00367661	S
T,9	0.283901	−0.712763	−1.425526	0.00352341	0.00352341	S
A,2	−0.539111	−0.034089	−0.443187	0.00091915	0.00091915	H
A,3	−0.540547	−0.072746	−0.446362	0.00091915	0.00091915	H
A,4	−0.540464	−0.112418	−0.450958	0.00091915	0.00091915	H
A,5	−0.540365	−0.151418	−0.455770	0.00091915	0.00091915	H
A,6	−0.420578	−0.147268	−0.392162	0.00091915	0.00091915	H
A,7	−0.182640	−0.098469	−0.284825	0.00091915	0.00091915	H
A,8	0.287618	0.007442	−0.071161	0.00091915	0.00091915	S
A,9	0.759490	0.113198	0.138523	0.00088085	0.00088085	S
A,T	1.500000	0.156681	0.225894	0.00367661	0.00367661	S

Hand	Stand	Hit	Double	Spl1	Spl2	Spl3	Freq.	BS
				NDAS				
A,A	−0.538085	−0.000333	−0.450881	0.237754	0.345158	0.362773	0.00044043	P
2,2	−0.540137	−0.237873	−1.080275	−0.380624	−0.400170	−0.403340	0.00044043	H
3,3	−0.543008	−0.295264	−1.067368	−0.428848	−0.447005	−0.449928	0.00044043	H
4,4	−0.542865	−0.209848	−0.715694	−0.483098	−0.520933	−0.527140	0.00044043	H
5,5	−0.542661	0.116653	0.149188	−0.544414	−0.636392	−0.651553	0.00044043	D
6,6	−0.543905	−0.347155	−0.749935	−0.583260	−0.615615	−0.620868	0.00044043	H
7,7	−0.545528	−0.437912	−0.894019	−0.570290	−0.588036	−0.590852	0.00044043	H
8,8	−0.538890	−0.505707	−1.011415	−0.421152	−0.409904	−0.408134	0.00044043	P
9,9	−0.185235	−0.613052	−1.226103	−0.105627	−0.095092	−0.093442	0.00040372	P
T,T	0.756075	−0.849438	−1.698876	0.224817	−0.050264	−0.215964	0.00727662	S
				DAS				
A,A	−0.538085	−0.000333	−0.450881	0.237754	0.345158	0.362773	0.00044043	P
2,2	−0.540137	−0.237873	−1.080275	−0.364258	−0.381403	−0.384159	0.00044043	H
3,3	−0.543008	−0.295264	−1.067368	−0.412806	−0.428671	−0.431206	0.00044043	H
4,4	−0.542865	−0.209848	−0.715694	−0.466658	−0.502133	−0.507939	0.00044043	H
5,5	−0.542661	0.116653	0.149188	−0.527671	−0.622109	−0.637718	0.00044043	D
6,6	−0.543905	−0.347155	−0.749935	−0.566643	−0.596637	−0.601493	0.00044043	H
7,7	−0.545528	−0.437912	−0.894019	−0.554834	−0.570445	−0.572911	0.00044043	H
8,8	−0.538890	−0.505707	−1.011415	−0.405024	−0.391516	−0.389371	0.00044043	P
9,9	−0.185235	−0.613052	−1.226103	−0.095143	−0.083268	−0.081416	0.00040372	P
T,T	0.756075	−0.849438	−1.698876	0.224817	−0.050264	−0.215964	0.00727662	S

TABLE A59
6D S17: Dealer's Upcard T

Hand	Stand	Hit	Double	Freq.	Cum. Freq.	BS
3,2	–0.541496	–0.312759	–1.082993	0.00339104	0.00339104	H
4,2	–0.541415	–0.339101	–1.064024	0.00339104	0.00339104	H
5,2	–0.541327	–0.319527	–0.951413	0.00339104		H
4,3	–0.541348	–0.321535	–0.951438	0.00339104	0.00678209	H
6,2	–0.542220	–0.249996	–0.746677	0.00339104		H
5,3	–0.541232	–0.249600	–0.745093	0.00339104	0.00678209	H
7,2	–0.538595	–0.153098	–0.461749	0.00339104		H
6,3	–0.542126	–0.151400	–0.463526	0.00339104		H
5,4	–0.541150	–0.150592	–0.461977	0.00339104	0.01017313	H
8,2	–0.539225	0.025892	–0.006360	0.00339104		H
7,3	–0.538501	0.025933	–0.005043	0.00339104		H
6,4	–0.542043	0.026470	–0.005603	0.00339104	0.01017313	H
9,2	–0.539813	0.117045	0.173923	0.00339104		D
8,3	–0.539133	0.117850	0.176919	0.00339104		D
7,4	–0.538420	0.117891	0.178371	0.00339104		D
6,5	–0.541929	0.118582	0.178452	0.00339104	0.01356418	D
T,2	–0.540337	–0.375678	–0.787819	0.01342288		H
9,3	–0.539699	–0.382341	–0.800788	0.00339104		H
8,4	–0.539031	–0.381867	–0.798324	0.00339104		H
7,5	–0.538285	–0.381631	–0.795997	0.00339104	0.02359601	H
T,3	–0.540246	–0.420295	–0.858713	0.01342288		H
9,4	–0.539621	–0.421082	–0.859767	0.00339104		H
8,5	–0.538919	–0.428628	–0.874190	0.00339104		H
7,6	–0.539184	–0.429424	–0.874234	0.00339104	0.02359601	H
T,4	–0.540168	–0.463098	–0.932032	0.01342288		EH
9,5	–0.539508	–0.464190	–0.933839	0.00339104		EH
8,6	–0.539818	–0.466417	–0.938057	0.00339104	0.02020497	EH
T,5	–0.540055	–0.503907	–1.007815	0.01342288		RH
9,6	–0.540408	–0.505703	–1.011406	0.00339104		RH
8,7	–0.536220	–0.499763	–0.999525	0.00339104	0.02020497	EH
T,6	–0.540954	–0.534676	–1.069351	0.01342288		RH
9,7	–0.536809	–0.535392	–1.070784	0.00339104	0.01681393	RH
T,7	–0.418457	–0.580066	–1.160132	0.01342288		S
9,8	–0.415036	–0.580276	–1.160553	0.00339104	0.01681393	S
T,8	–0.174546	–0.643745	–1.287489	0.01342288	0.01342288	S
T,9	0.069444	–0.726389	–1.452778	0.01342288	0.01342288	S
A,2	–0.540188	–0.102342	–0.508569	0.00340294	0.00340294	H
A,3	–0.540096	–0.137125	–0.508720	0.00340294	0.00340294	H
A,4	–0.540016	–0.173213	–0.511188	0.00340294	0.00340294	H
A,5	–0.539901	–0.210080	–0.517630	0.00340294	0.00340294	H
A,6	–0.419557	–0.196050	–0.454581	0.00340294	0.00340294	H
A,7	–0.179563	–0.142925	–0.342747	0.00340294	0.00340294	H
A,8	0.063321	–0.087632	–0.233592	0.00340294	0.00340294	S
A,9	0.554555	0.022332	–0.015427	0.00340294	0.00340294	S
A,T	1.500000	0.117048	0.173198	0.01346998	0.01346998	S

Hand	Stand	Hit	Double	Spl1	Spl2	Spl3	Freq.	BS
				NDAS				
A,A	–0.538797	–0.066334	–0.506650	0.181988	0.281956	0.298382	0.00163628	P
2,2	–0.541589	–0.287220	–1.083178	–0.480140	–0.506727	–0.511068	0.00162488	H
3,3	–0.541402	–0.338880	–1.063997	–0.526890	–0.552632	–0.556808	0.00162488	H
4,4	–0.541266	–0.247994	–0.745112	–0.576994	–0.622460	–0.629904	0.00162488	H
5,5	–0.541033	0.027154	–0.004071	–0.633371	–0.725080	–0.740165	0.00162488	H
6,6	–0.542783	–0.382400	–0.798288	–0.678371	–0.719216	–0.725894	0.00162488	H
7,7	–0.535586	–0.474064	–0.953611	–0.637476	–0.659419	–0.662907	0.00162488	EH
8,8	–0.536853	–0.535361	–1.070722	–0.492845	–0.486850	–0.485883	0.00162488	EP
9,9	–0.171080	–0.643985	–1.287970	–0.299493	–0.317358	–0.320303	0.00162488	S
T,T	0.559145	–0.847142	–1.694283	0.050742	–0.210173	–0.366060	0.02628648	S
				DAS				
A,A	–0.538797	–0.066334	–0.506650	0.181988	0.281956	0.298382	0.00163628	P
2,2	–0.541589	–0.287220	–1.083178	–0.471097	–0.496409	–0.500537	0.00162488	H
3,3	–0.541402	–0.338880	–1.063997	–0.517502	–0.541921	–0.545876	0.00162488	H
4,4	–0.541266	–0.247994	–0.745112	–0.567371	–0.611479	–0.618696	0.00162488	H
5,5	–0.541033	0.027154	–0.004071	–0.623772	–0.714116	–0.728972	0.00162488	H
6,6	–0.542783	–0.382400	–0.798288	–0.668869	–0.708376	–0.714830	0.00162488	H
7,7	–0.535586	–0.474064	–0.953611	–0.627713	–0.648258	–0.651510	0.00162488	EH
8,8	–0.536853	–0.535361	–1.070722	–0.483502	–0.476196	–0.475011	0.00162488	EP
9,9	–0.171080	–0.643985	–1.287970	–0.290833	–0.307526	–0.310283	0.00162488	S
T,T	0.559145	–0.847142	–1.694283	0.050742	–0.210173	–0.366060	0.02628648	S

TABLE A60
6D H17: Dealer's Upcard A

Hand	Stand	Hit	Double	Freq.	Cum. Freq.	BS
3,2	–0.596541	–0.317783	–1.193081	0.00063359	0.00063359	EH
4,2	–0.596191	–0.344326	–1.180111	0.00063359	0.00063359	EH
5,2	–0.596327	–0.351841	–1.086013	0.00063359		EH
4,3	–0.595856	–0.353382	–1.085015	0.00063359	0.00126718	EH
6,2	–0.595700	–0.265893	–0.831492	0.00063359		H
5,3	–0.595978	–0.265885	–0.830452	0.00063359	0.00126718	H
7,2	–0.594779	–0.126812	–0.454400	0.00063359		H
6,3	–0.595352	–0.124603	–0.452926	0.00063359		H
5,4	–0.595607	–0.124191	–0.451622	0.00063359	0.00190077	H
8,2	–0.595855	0.033872	–0.030126	0.00063359		H
7,3	–0.594433	0.033882	–0.027798	0.00063359		H
6,4	–0.594984	0.034616	–0.025748	0.00063359	0.00190077	H
9,2	–0.596906	0.107036	0.115609	0.00063359		D
8,3	–0.595490	0.107445	0.118796	0.00063359		D
7,4	–0.594044	0.107943	0.122285	0.00063359		D
6,5	–0.595078	0.108668	0.124007	0.00063359	0.00253436	D
T,2	–0.599164	–0.383367	–0.834065	0.00254626		EH
9,3	–0.596562	–0.387407	–0.841503	0.00063359		EH
8,4	–0.595123	–0.387293	–0.838971	0.00063359		EH
7,5	–0.594173	–0.387530	–0.837178	0.00063359	0.00444703	EH
T,3	–0.598823	–0.427010	–0.895385	0.00254626		EH
9,4	–0.596196	–0.425233	–0.891211	0.00063359		EH
8,5	–0.595253	–0.433097	–0.905727	0.00063359		EH
7,6	–0.593566	–0.434381	–0.906997	0.00063359	0.00444703	EH
T,4	–0.598458	–0.468857	–0.959222	0.00254626		EH
9,5	–0.596325	–0.468617	–0.958210	0.00063359		EH
8,6	–0.594646	–0.470188	–0.961526	0.00063359	0.00381344	EH
T,5	–0.598584	–0.509549	–1.026900	0.00254626		RH
9,6	–0.595712	–0.508651	–1.025788	0.00063359		RH
8,7	–0.593749	–0.503074	–1.014255	0.00063359	0.00381344	RH
T,6	–0.597991	–0.540015	–1.080030	0.00254626		RH
9,7	–0.594820	–0.538656	–1.077313	0.00063359	0.00317985	RH
T,7	–0.513758	–0.578940	–1.157881	0.00254626		RS
9,8	–0.511699	–0.577117	–1.154234	0.00063359	0.00317985	RS
T,8	–0.223377	–0.639911	–1.279823	0.00254626	0.00254626	S
T,9	0.191966	–0.733039	–1.466077	0.00254626	0.00254626	S
A,2	–0.597220	–0.100433	–0.584256	0.00060719	0.00060719	H
A,3	–0.596877	–0.135170	–0.583703	0.00060719	0.00060719	H
A,4	–0.596510	–0.172032	–0.586998	0.00060719	0.00060719	H
A,5	–0.596653	–0.209173	–0.592891	0.00060719	0.00060719	H
A,6	–0.514255	–0.221401	–0.548548	0.00060719	0.00060719	H
A,7	–0.225366	–0.160455	–0.417824	0.00060719	0.00060719	H
A,8	0.189506	–0.064654	–0.228245	0.00060719	0.00060719	S
A,9	0.605486	0.030489	–0.040723	0.00060719	0.00060719	S
A,T	1.500000	0.104246	0.108590	0.00244016	0.00244016	S

Hand	Stand	Hit	Double	Spl1	Spl2	Spl3	Freq.	BS
				NDAS				
A,A	–0.597555	–0.064753	–0.585159	0.120987	0.215882	0.230706	0.00027830	P
2,2	–0.596885	–0.292671	–1.193770	–0.491920	–0.519273	–0.523722	0.00030360	EH
3,3	–0.596214	–0.344327	–1.180152	–0.538080	–0.564421	–0.568664	0.00030360	EH
4,4	–0.595485	–0.263885	–0.829357	–0.587354	–0.631971	–0.639263	0.00030360	H
5,5	–0.595732	0.034458	–0.025588	–0.645022	–0.739255	–0.754738	0.00030360	H
6,6	–0.594466	–0.387285	–0.836670	–0.682167	–0.722491	–0.729024	0.00030360	EH
7,7	–0.592673	–0.477592	–0.976448	–0.697349	–0.726669	–0.731299	0.00030360	EH
8,8	–0.594827	–0.538561	–1.077122	–0.519828	–0.516819	–0.516259	0.00030360	RP
9,9	–0.219502	–0.638487	–1.276974	–0.239661	–0.241945	–0.242237	0.00030360	S
T,T	0.600344	–0.858455	–1.716911	0.054608	–0.229448	–0.401375	0.00506302	S
				DAS				
A,A	–0.597555	–0.064753	–0.585159	0.120987	0.215882	0.230706	0.00027830	P
2,2	–0.596885	–0.292671	–1.193770	–0.490161	–0.517214	–0.521605	0.00030360	EH
3,3	–0.596214	–0.344327	–1.180152	–0.535828	–0.561795	–0.565965	0.00030360	EH
4,4	–0.595485	–0.263885	–0.829357	–0.584527	–0.628674	–0.635876	0.00030360	H
5,5	–0.595732	0.034458	–0.025588	–0.641999	–0.735731	–0.751119	0.00030360	H
6,6	–0.594466	–0.387285	–0.836670	–0.679180	–0.719010	–0.725449	0.00030360	EH
7,7	–0.592673	–0.477592	–0.976448	–0.694633	–0.723512	–0.728059	0.00030360	EH
8,8	–0.594827	–0.538561	–1.077122	–0.517931	–0.514643	–0.514035	0.00030360	RP
9,9	–0.219502	–0.638487	–1.276974	–0.238571	–0.240737	–0.241015	0.00030360	S
T,T	0.600344	–0.858455	–1.716911	0.054608	–0.229448	–0.401375	0.00506302	S

TABLE A61

6D H17: Dealer's Upcard 2

Hand	Stand	Hit	Double	Freq.	Cum. Freq.	BS
3,2	–0.286101	–0.126025	–0.572202	0.00088085	0.00088085	H
4,2	–0.285985	–0.139226	–0.551817	0.00088085	0.00088085	H
5,2	–0.283454	–0.107915	–0.426153	0.00088085		H
4,3	–0.285902	–0.110920	–0.429532	0.00091915	0.00180000	H
6,2	–0.283949	–0.022871	–0.199767	0.00088085		H
5,3	–0.283371	–0.023176	–0.199989	0.00091915	0.00180000	H
7,2	–0.284457	0.073913	0.067870	0.00088085		H
6,3	–0.283865	0.075331	0.067378	0.00091915		H
5,4	–0.283253	0.075786	0.068039	0.00091915	0.00271916	H
8,2	–0.284919	0.185608	0.366979	0.00088085		D
7,3	–0.284353	0.186119	0.367771	0.00091915		D
6,4	–0.283726	0.186754	0.368992	0.00091915	0.00271916	D
9,2	–0.285600	0.241465	0.478395	0.00088085		D
8,3	–0.284836	0.242280	0.479984	0.00091915		D
7,4	–0.284235	0.243179	0.481737	0.00091915		D
6,5	–0.281206	0.244940	0.485545	0.00091915	0.00363831	D
T,2	–0.289435	–0.252224	–0.504447	0.00352341		H
9,3	–0.285517	–0.255818	–0.511635	0.00091915		H
8,4	–0.284718	–0.254724	–0.509448	0.00091915		H
7,5	–0.281716	–0.253609	–0.507218	0.00091915	0.00628087	H
T,3	–0.289352	–0.307973	–0.615946	0.00367661		S
9,4	–0.285400	–0.306135	–0.612270	0.00091915		S
8,5	–0.282200	–0.312360	–0.624720	0.00091915		S
7,6	–0.282213	–0.312485	–0.624970	0.00091915	0.00643406	S
T,4	–0.289236	–0.364353	–0.728707	0.00367661		S
9,5	–0.282881	–0.362973	–0.725946	0.00091915		S
8,6	–0.282695	–0.363635	–0.727270	0.00091915	0.00551491	S
T,5	–0.286740	–0.421462	–0.842924	0.00367661		S
9,6	–0.283402	–0.420576	–0.841152	0.00091915		S
8,7	–0.283230	–0.414321	–0.828642	0.00091915	0.00551491	S
T,6	–0.287237	–0.472332	–0.944663	0.00367661		S
9,7	–0.283912	–0.470891	–0.941781	0.00091915	0.00459576	S
T,7	–0.157192	–0.538482	–1.076963	0.00367661		S
9,8	–0.153788	–0.537261	–1.074522	0.00091915	0.00459576	S
T,8	0.109693	–0.625184	–1.250369	0.00367661	0.00367661	S
T,9	0.377921	–0.732845	–1.465690	0.00367661	0.00367661	S
A,2	–0.285091	0.045611	–0.063455	0.00088085	0.00088085	H
A,3	–0.285009	0.022101	–0.063913	0.00091915	0.00091915	H
A,4	–0.284913	–0.000918	–0.067577	0.00091915	0.00091915	H
A,5	–0.282389	–0.021446	–0.069968	0.00091915	0.00091915	H
A,6	–0.152739	–0.000274	–0.004882	0.00091915	0.00091915	H
A,7	0.113110	0.060441	0.116262	0.00091915	0.00091915	D
A,8	0.380751	0.121117	0.237658	0.00091915	0.00091915	S
A,9	0.637642	0.182199	0.360050	0.00091915	0.00091915	S
A,T	1.500000	0.238994	0.472563	0.00367661	0.00367661	S

Hand	Stand	Hit	Double	Spl1	Spl2	Spl3	Freq.	BS
				NDAS				
A,A	–0.284008	0.082026	–0.060073	0.485375	0.591713	0.609125	0.00044043	P
2,2	–0.286162	–0.112545	–0.572323	–0.148381	–0.153065	–0.153789	0.00040372	H
3,3	–0.286019	–0.139266	–0.551880	–0.196682	–0.204413	–0.205646	0.00044043	H
4,4	–0.285785	–0.022238	–0.201987	–0.225400	–0.253468	–0.258062	0.00044043	H
5,5	–0.280711	0.187678	0.371112	–0.247256	–0.307068	–0.316813	0.00044043	D
6,6	–0.281702	–0.253614	–0.507227	–0.266534	–0.267574	–0.267623	0.00044043	H
7,7	–0.282721	–0.370557	–0.741114	–0.210319	–0.199982	–0.198236	0.00044043	P
8,8	–0.283713	–0.470692	–0.941385	–0.039934	–0.005961	–0.000353	0.00044043	P
9,9	0.112768	–0.624382	–1.248764	0.148087	0.152986	0.153790	0.00044043	P
T,T	0.632910	–0.853996	–1.707992	0.353375	0.206680	0.117238	0.00727662	S
				DAS				
A,A	–0.284008	0.082026	–0.060073	0.485375	0.591713	0.609125	0.00044043	P
2,2	–0.286162	–0.112545	–0.572323	–0.082911	–0.078895	–0.078252	0.00040372	P
3,3	–0.286019	–0.139266	–0.551880	–0.131054	–0.129665	–0.129397	0.00044043	P
4,4	–0.285785	–0.022238	–0.201987	–0.159494	–0.178398	–0.181483	0.00044043	H
5,5	–0.280711	0.187678	0.371112	–0.182882	–0.259463	–0.272005	0.00044043	D
6,6	–0.281702	–0.253614	–0.507227	–0.200001	–0.191748	–0.190260	0.00044043	P
7,7	–0.282721	–0.370557	–0.741114	–0.143395	–0.123670	–0.120364	0.00044043	P
8,8	–0.283713	–0.470692	–0.941385	0.024467	0.067389	0.074470	0.00044043	P
9,9	0.112768	–0.624382	–1.248764	0.183409	0.193187	0.194790	0.00044043	P
T,T	0.632910	–0.853996	–1.707992	0.353375	0.206680	0.117238	0.00727662	S

TABLE A62
6D H17: Dealer's Upcard 3

Hand	Stand	Hit	Double	Freq.	Cum. Freq.	BS
3,2	–0.245573	–0.093202	–0.491145	0.00088085	0.00088085	H
4,2	–0.243009	–0.104221	–0.466758	0.00091915	0.00091915	H
5,2	–0.242607	–0.074160	–0.347730	0.00091915		H
4,3	–0.242896	–0.076982	–0.349532	0.00088085	0.00180000	H
6,2	–0.243094	0.008372	–0.128418	0.00091915		H
5,3	–0.242493	0.008067	–0.128592	0.00088085	0.00180000	H
7,2	–0.243573	0.104036	0.130522	0.00091915		D
6,3	–0.242959	0.105445	0.130048	0.00088085		D
5,4	–0.239918	0.105900	0.131414	0.00091915	0.00271916	D
8,2	–0.244251	0.210340	0.419211	0.00091915		D
7,3	–0.243459	0.211210	0.420854	0.00088085		D
6,4	–0.240406	0.211891	0.422460	0.00091915	0.00271916	D
9,2	–0.248181	0.261295	0.520989	0.00091915		D
8,3	–0.244137	0.265234	0.528985	0.00088085		D
7,4	–0.240906	0.266757	0.532269	0.00091915		D
6,5	–0.240003	0.268158	0.535069	0.00091915	0.00363831	D
T,2	–0.248923	–0.231678	–0.463355	0.00367661		H
9,3	–0.248068	–0.237301	–0.474602	0.00088085		H
8,4	–0.241586	–0.233324	–0.466647	0.00091915		H
7,5	–0.240505	–0.232183	–0.464366	0.00091915	0.00639576	H
T,3	–0.248811	–0.290528	–0.581057	0.00352341		S
9,4	–0.245516	–0.290023	–0.580047	0.00091915		S
8,5	–0.241183	–0.294245	–0.588490	0.00091915		S
7,6	–0.240992	–0.294107	–0.588214	0.00091915	0.00628087	S
T,4	–0.246282	–0.350816	–0.701632	0.00367661		S
9,5	–0.245138	–0.350845	–0.701690	0.00091915		S
8,6	–0.241697	–0.349657	–0.699314	0.00091915	0.00551491	S
T,5	–0.245879	–0.411424	–0.822847	0.00367661		S
9,6	–0.245627	–0.411794	–0.823587	0.00091915		S
8,7	–0.242197	–0.403931	–0.807863	0.00091915	0.00551491	S
T,6	–0.246369	–0.465336	–0.930673	0.00367661		S
9,7	–0.246129	–0.464920	–0.929840	0.00091915	0.00459576	S
T,7	–0.120447	–0.534200	–1.068400	0.00367661		S
9,8	–0.120409	–0.533980	–1.067959	0.00091915	0.00459576	S
T,8	0.137849	–0.623183	–1.246366	0.00367661	0.00367661	S
T,9	0.394037	–0.726003	–1.452006	0.00367661	0.00367661	S
A,2	–0.244670	0.073770	0.001962	0.00091915	0.00091915	H
A,3	–0.244578	0.050358	–0.000572	0.00088085	0.00088085	H
A,4	–0.242024	0.029075	–0.002231	0.00091915	0.00091915	H
A,5	–0.241620	0.008338	–0.006083	0.00091915	0.00091915	H
A,6	–0.116126	0.029212	0.057176	0.00091915	0.00091915	D
A,7	0.141584	0.088230	0.175215	0.00091915	0.00091915	D
A,8	0.399787	0.151059	0.300697	0.00091915	0.00091915	S
A,9	0.645253	0.203437	0.405273	0.00091915	0.00091915	S
A,T	1.500000	0.261370	0.520285	0.00367661	0.00367661	S

NDAS

Hand	Stand	Hit	Double	Spl1	Spl2	Spl3	Freq.	BS
A,A	–0.243667	0.104450	0.005586	0.532737	0.640019	0.657582	0.00044043	P
2,2	–0.245684	–0.080126	–0.491369	–0.094600	–0.096570	–0.096888	0.00044043	H
3,3	–0.245460	–0.105892	–0.471414	–0.132284	–0.135601	–0.136093	0.00040372	H
4,4	–0.240332	0.009335	–0.128367	–0.154550	–0.176903	–0.180515	0.00044043	H
5,5	–0.239515	0.212595	0.423861	–0.179593	–0.233447	–0.242208	0.00044043	D
6,6	–0.240493	–0.232219	–0.464438	–0.197756	–0.192201	–0.191162	0.00044043	P
7,7	–0.241518	–0.356297	–0.712593	–0.142940	–0.128924	–0.126566	0.00044043	P
8,8	–0.242878	–0.463614	–0.927228	0.021111	0.057870	0.063934	0.00044043	P
9,9	0.134754	–0.623778	–1.247557	0.194445	0.202312	0.203538	0.00044043	P
T,T	0.643634	–0.853716	–1.707431	0.401858	0.274706	0.197031	0.00727662	S

DAS

Hand	Stand	Hit	Double	Spl1	Spl2	Spl3	Freq.	BS
A,A	–0.243667	0.104450	0.005586	0.532737	0.640019	0.657582	0.00044043	P
2,2	–0.245684	–0.080126	–0.491369	–0.016884	–0.008023	–0.006553	0.00044043	P
3,3	–0.245460	–0.105892	–0.471414	–0.054335	–0.047294	–0.046158	0.00040372	P
4,4	–0.240332	0.009335	–0.128367	–0.075508	–0.086806	–0.088587	0.00044043	H
5,5	–0.239515	0.212595	0.423861	–0.102636	–0.175351	–0.187249	0.00044043	D
6,6	–0.240493	–0.232219	–0.464438	–0.113561	–0.096118	–0.093092	0.00044043	P
7,7	–0.241518	–0.356297	–0.712593	–0.059071	–0.033205	–0.028864	0.00044043	P
8,8	–0.242878	–0.463614	–0.927228	0.093480	0.140312	0.148037	0.00044043	P
9,9	0.134754	–0.623778	–1.247557	0.234416	0.247748	0.249859	0.00044043	P
T,T	0.643634	–0.853716	–1.707431	0.401858	0.274706	0.197031	0.00727662	S

TABLE A63
6D H17: Dealer's Upcard 4

Hand	Stand	Hit	Double	Freq.	Cum. Freq.	BS
3,2	–0.200104	–0.055370	–0.400209	0.00091915	0.00091915	H
4,2	–0.199655	–0.067021	–0.380345	0.00088085	0.00088085	H
5,2	–0.199305	–0.037425	–0.265089	0.00091915		H
4,3	–0.197094	–0.038414	–0.262411	0.00088085	0.00180000	H
6,2	–0.199757	0.043758	–0.054037	0.00091915		H
5,3	–0.196734	0.044637	–0.051206	0.00091915	0.00183830	H
7,2	–0.200460	0.133799	0.194283	0.00091915		D
6,3	–0.197207	0.136632	0.197034	0.00091915		D
5,4	–0.196281	0.137091	0.197964	0.00088085	0.00271916	D
8,2	–0.204386	0.233346	0.466693	0.00091915		D
7,3	–0.197911	0.238374	0.476749	0.00091915		D
6,4	–0.196756	0.239378	0.478756	0.00088085	0.00271916	D
9,2	–0.205200	0.285533	0.571067	0.00091915		D
8,3	–0.201838	0.287631	0.575263	0.00091915		D
7,4	–0.197461	0.292149	0.584298	0.00088085		D
6,5	–0.196406	0.293527	0.587055	0.00091915	0.00363831	D
T,2	–0.205906	–0.210664	–0.421329	0.00367661		S
9,3	–0.202651	–0.215698	–0.431396	0.00091915		S
8,4	–0.201386	–0.213959	–0.427918	0.00088085		S
7,5	–0.197110	–0.210554	–0.421108	0.00091915	0.00639576	S
T,3	–0.203381	–0.272231	–0.544461	0.00367661		S
9,4	–0.202225	–0.271755	–0.543511	0.00088085		S
8,5	–0.201061	–0.277746	–0.555493	0.00091915		S
7,6	–0.197610	–0.275533	–0.551066	0.00091915	0.00639576	S
T,4	–0.202930	–0.336220	–0.672440	0.00352341		S
9,5	–0.201874	–0.336208	–0.672417	0.00091915		S
8,6	–0.201536	–0.336519	–0.673038	0.00091915	0.00536172	S
T,5	–0.202579	–0.400721	–0.801443	0.00367661		S
9,6	–0.202349	–0.401031	–0.802061	0.00091915		S
8,7	–0.202241	–0.394686	–0.789372	0.00091915	0.00551491	S
T,6	–0.203055	–0.457944	–0.915887	0.00367661		S
9,7	–0.203055	–0.457708	–0.915416	0.00091915	0.00459576	S
T,7	–0.080830	–0.529979	–1.059958	0.00367661		S
9,8	–0.083983	–0.530808	–1.061617	0.00091915	0.00459576	S
T,8	0.164363	–0.615157	–1.230315	0.00367661	0.00367661	S
T,9	0.413067	–0.724937	–1.449874	0.00367661	0.00367661	S
A,2	–0.201766	0.104348	0.070634	0.00091915	0.00091915	H
A,3	–0.199215	0.082777	0.069716	0.00091915	0.00091915	H
A,4	–0.198766	0.060757	0.065278	0.00088085	0.00088085	D
A,5	–0.198414	0.040979	0.061535	0.00091915	0.00091915	D
A,6	–0.076356	0.061648	0.123295	0.00091915	0.00091915	D
A,7	0.171028	0.122397	0.244794	0.00091915	0.00091915	D
A,8	0.415570	0.176002	0.352004	0.00091915	0.00091915	S
A,9	0.655844	0.229437	0.458874	0.00091915	0.00091915	S
A,T	1.500000	0.285129	0.570259	0.00367661	0.00367661	S

Hand	Stand	Hit	Double	Spl1	Spl2	Spl3	Freq.	BS
				NDAS				
A,A	–0.200841	0.129292	0.073672	0.582300	0.690484	0.708190	0.00044043	P
2,2	–0.202664	–0.044392	–0.405328	–0.033941	–0.032471	–0.032227	0.00044043	P
3,3	–0.197544	–0.065692	–0.376611	–0.054871	–0.053025	–0.052665	0.00044043	P
4,4	–0.196633	0.046322	–0.050931	–0.080182	–0.096594	–0.099113	0.00040372	H
5,5	–0.195931	0.239693	0.479387	–0.104581	–0.151732	–0.159383	0.00044043	D
6,6	–0.196879	–0.210302	–0.420604	–0.121956	–0.111034	–0.109154	0.00044043	P
7,7	–0.198315	–0.341577	–0.683155	–0.069457	–0.051189	–0.048124	0.00044043	P
8,8	–0.206168	–0.458790	–0.917580	0.078936	0.118256	0.124681	0.00044043	P
9,9	0.164403	–0.615147	–1.230295	0.254001	0.266039	0.267954	0.00044043	P
T,T	0.654267	–0.853474	–1.706948	0.452343	0.345545	0.279976	0.00727662	S
				DAS				
A,A	–0.200841	0.129292	0.073672	0.582300	0.690484	0.708190	0.00044043	P
2,2	–0.202664	–0.044392	–0.405328	0.056828	0.070923	0.073248	0.00044043	P
3,3	–0.197544	–0.065692	–0.376611	0.037152	0.051819	0.054293	0.00044043	P
4,4	–0.196633	0.046322	–0.050931	0.014030	0.010237	0.009718	0.00040372	H
5,5	–0.195931	0.239693	0.479387	–0.010965	–0.078628	–0.089690	0.00044043	D
6,6	–0.196879	–0.210302	–0.420604	–0.019425	0.005958	0.010253	0.00044043	P
7,7	–0.198315	–0.341577	–0.683155	0.034141	0.067000	0.072499	0.00044043	P
8,8	–0.206168	–0.458790	–0.917580	0.159570	0.209976	0.218206	0.00044043	P
9,9	0.164403	–0.615147	–1.230295	0.297966	0.316018	0.318909	0.00044043	P
T,T	0.654267	–0.853474	–1.706948	0.452343	0.345545	0.279976	0.00727662	S

TABLE A64
6D H17: Dealer's Upcard 5

Hand	Stand	Hit	Double	Freq.	Cum. Freq.	BS
3,2	–0.155026	–0.015690	–0.310053	0.00091915	0.00091915	H
4,2	–0.154631	–0.026761	–0.291219	0.00091915	0.00091915	H
5,2	–0.154275	0.002733	–0.180919	0.00088085		H
4,3	–0.154193	0.001535	–0.179553	0.00091915	0.00180000	H
6,2	–0.154973	0.079302	0.022325	0.00091915		H
5,3	–0.153847	0.080817	0.025854	0.00088085	0.00180000	H
7,2	–0.158922	0.163419	0.256391	0.00091915		D
6,3	–0.154545	0.169339	0.265167	0.00091915		D
5,4	–0.153449	0.170984	0.268497	0.00088085	0.00271916	D
8,2	–0.159743	0.262222	0.524444	0.00091915		D
7,3	–0.158495	0.264024	0.528049	0.00091915		D
6,4	–0.154149	0.268640	0.537279	0.00091915	0.00275745	D
9,2	–0.160457	0.312258	0.624515	0.00091915		D
8,3	–0.159316	0.313950	0.627900	0.00091915		D
7,4	–0.158097	0.315817	0.631634	0.00091915		D
6,5	–0.153802	0.320525	0.641050	0.00088085	0.00363831	D
T,2	–0.161126	–0.188759	–0.377518	0.00367661		S
9,3	–0.160055	–0.194040	–0.388080	0.00091915		S
8,4	–0.158944	–0.192389	–0.384778	0.00091915		S
7,5	–0.157776	–0.190982	–0.381963	0.00088085	0.00639576	S
T,3	–0.160698	–0.254211	–0.508423	0.00367661		S
9,4	–0.159657	–0.253758	–0.507517	0.00091915		S
8,5	–0.158596	–0.259746	–0.519492	0.00088085		S
7,6	–0.158476	–0.259536	–0.519071	0.00091915	0.00639576	S
T,4	–0.160301	–0.320872	–0.641745	0.00367661		S
9,5	–0.159310	–0.320829	–0.641658	0.00088085		S
8,6	–0.159297	–0.321340	–0.642681	0.00091915	0.00547661	S
T,5	–0.159954	–0.388494	–0.776988	0.00352341		S
9,6	–0.160011	–0.388976	–0.777952	0.00091915		S
8,7	–0.163247	–0.384186	–0.768372	0.00091915	0.00536172	S
T,6	–0.160655	–0.449718	–0.899436	0.00367661		S
9,7	–0.163962	–0.450802	–0.901604	0.00091915	0.00459576	S
T,7	–0.045986	–0.518957	–1.037913	0.00367661		S
9,8	–0.046138	–0.518963	–1.037926	0.00091915	0.00459576	S
T,8	0.195534	–0.611809	–1.223618	0.00367661	0.00367661	S
T,9	0.437502	–0.723459	–1.446918	0.00367661	0.00367661	S
A,2	–0.157113	0.137618	0.141030	0.00091915	0.00091915	D
A,3	–0.156687	0.116160	0.139629	0.00091915	0.00091915	D
A,4	–0.156291	0.094928	0.134921	0.00091915	0.00091915	D
A,5	–0.155945	0.075289	0.130267	0.00088085	0.00088085	D
A,6	–0.038473	0.098689	0.197379	0.00091915	0.00091915	D
A,7	0.198998	0.151132	0.302263	0.00091915	0.00091915	D
A,8	0.439914	0.206222	0.412443	0.00091915	0.00091915	S
A,9	0.670408	0.257717	0.515435	0.00091915	0.00091915	S
A,T	1.500000	0.311068	0.622136	0.00367661	0.00367661	S

Hand	Stand	Hit	Double	Spl1	Spl2	Spl3	Freq.	BS
				NDAS				
A,A	–0.158751	0.160827	0.141840	0.633712	0.743159	0.761068	0.00044043	P
2,2	–0.155452	–0.003781	–0.310904	0.044978	0.051912	0.053079	0.00044043	P
3,3	–0.154601	–0.026775	–0.290921	0.020346	0.027253	0.028447	0.00044043	P
4,4	–0.153795	0.082815	0.026924	–0.003952	–0.015655	–0.017524	0.00044043	H
5,5	–0.153103	0.269829	0.539658	–0.027834	–0.066791	–0.072822	0.00040372	D
6,6	–0.154527	–0.188677	–0.377354	–0.045589	–0.029967	–0.027318	0.00044043	P
7,7	–0.162425	–0.329707	–0.659414	–0.005217	0.016529	0.020093	0.00044043	P
8,8	–0.164069	–0.450829	–0.901658	0.147887	0.190886	0.197908	0.00044043	P
9,9	0.195523	–0.611797	–1.223594	0.318079	0.334653	0.337308	0.00044043	P
T,T	0.668927	–0.853117	–1.706233	0.509212	0.424276	0.371878	0.00727662	S
				DAS				
A,A	–0.158751	0.160827	0.141840	0.633712	0.743159	0.761068	0.00044043	P
2,2	–0.155452	–0.003781	–0.310904	0.150794	0.172541	0.176164	0.00044043	P
3,3	–0.154601	–0.026775	–0.290921	0.130290	0.152583	0.156328	0.00044043	P
4,4	–0.153795	0.082815	0.026924	0.109385	0.113563	0.114330	0.00044043	P
5,5	–0.153103	0.269829	0.539658	0.084399	0.024233	0.014827	0.00040372	D
6,6	–0.154527	–0.188677	–0.377354	0.076606	0.109428	0.114944	0.00044043	P
7,7	–0.162425	–0.329707	–0.659414	0.115691	0.154165	0.160470	0.00044043	P
8,8	–0.164069	–0.450829	–0.901658	0.237071	0.292331	0.301350	0.00044043	P
9,9	0.195523	–0.611797	–1.223594	0.366200	0.389361	0.393085	0.00044043	P
T,T	0.668927	–0.853117	–1.706233	0.509212	0.424276	0.371878	0.00727662	S

TABLE A65

6D H17: Dealer's Upcard 6

Hand	Stand	Hit	Double	Freq.	Cum. Freq.	BS
3,2	–0.115219	0.019256	–0.230439	0.00091915	0.00091915	H
4,2	–0.114844	0.009881	–0.212121	0.00091915	0.00091915	H
5,2	–0.114751	0.037108	–0.105273	0.00091915		H
4,3	–0.114450	0.035824	–0.104380	0.00091915	0.00183830	H
6,2	–0.118703	0.106754	0.082192	0.00088085		H
5,3	–0.114357	0.111453	0.092013	0.00091915	0.00180000	H
7,2	–0.119527	0.189590	0.311310	0.00091915		D
6,3	–0.118309	0.192820	0.314736	0.00088085		D
5,4	–0.113992	0.197680	0.324478	0.00091915	0.00271916	D
8,2	–0.120237	0.285645	0.571290	0.00091915		D
7,3	–0.119133	0.287425	0.574849	0.00091915		D
6,4	–0.117943	0.289479	0.578959	0.00088085	0.00271916	D
9,2	–0.120933	0.334401	0.668803	0.00091915		D
8,3	–0.119868	0.336072	0.672145	0.00091915		D
7,4	–0.118792	0.337927	0.675854	0.00091915		D
6,5	–0.117876	0.339933	0.679865	0.00088085	0.00363831	D
T,2	–0.121479	–0.170569	–0.341138	0.00367661		S
9,3	–0.120539	–0.175904	–0.351809	0.00091915		S
8,4	–0.119502	–0.174273	–0.348546	0.00091915		S
7,5	–0.118699	–0.173201	–0.346403	0.00091915	0.00643406	S
T,3	–0.121086	–0.239032	–0.478064	0.00367661		S
9,4	–0.120173	–0.238607	–0.477213	0.00091915		S
8,5	–0.119409	–0.244828	–0.489656	0.00091915		S
7,6	–0.122651	–0.246688	–0.493376	0.00088085	0.00639576	S
T,4	–0.120720	–0.308640	–0.617280	0.00367661		S
9,5	–0.120080	–0.308835	–0.617671	0.00091915		S
8,6	–0.123361	–0.311131	–0.622262	0.00088085	0.00547661	S
T,5	–0.120627	–0.378850	–0.757700	0.00367661		S
9,6	–0.124033	–0.380899	–0.761798	0.00088085		S
8,7	–0.124186	–0.374794	–0.749588	0.00091915	0.00547661	S
T,6	–0.124544	–0.437508	–0.875017	0.00352341		S
9,7	–0.124822	–0.437546	–0.875093	0.00091915	0.00444257	S
T,7	–0.009816	–0.515243	–1.030485	0.00367661		S
9,8	–0.009955	–0.515251	–1.030501	0.00091915	0.00459576	S
T,8	0.220261	–0.610054	–1.220107	0.00367661	0.00367661	S
T,9	0.450523	–0.722695	–1.445390	0.00367661	0.00367661	S
A,2	–0.115391	0.166485	0.204216	0.00091915	0.00091915	D
A,3	–0.114998	0.146111	0.202975	0.00091915	0.00091915	D
A,4	–0.114635	0.126272	0.199687	0.00091915	0.00091915	D
A,5	–0.114523	0.110738	0.202017	0.00091915	0.00091915	D
A,6	–0.004992	0.126073	0.252146	0.00088085	0.00088085	D
A,7	0.222598	0.178453	0.356906	0.00091915	0.00091915	D
A,8	0.452220	0.231045	0.462089	0.00091915	0.00091915	D
A,9	0.677574	0.281487	0.562975	0.00091915	0.00091915	S
A,T	1.500000	0.333531	0.667063	0.00367661	0.00367661	S

Hand	Stand	Hit	Double	Spl1	Spl2	Spl3	Freq.	BS
				NDAS				
A,A	–0.115159	0.189536	0.206312	0.679649	0.789286	0.807204	0.00044043	P
2,2	–0.115611	0.030230	–0.231222	0.111104	0.122524	0.124433	0.00044043	P
3,3	–0.114816	0.009878	–0.212063	0.087420	0.098575	0.100473	0.00044043	P
4,4	–0.114085	0.113365	0.092929	0.064043	0.057551	0.056540	0.00044043	H
5,5	–0.113900	0.293773	0.587547	0.040732	0.006097	0.000480	0.00044043	D
6,6	–0.121827	–0.175211	–0.350422	0.013044	0.030983	0.033805	0.00040372	P
7,7	–0.123475	–0.317997	–0.635994	0.060845	0.086293	0.090455	0.00044043	P
8,8	–0.124862	–0.437540	–0.875080	0.207053	0.252792	0.260260	0.00044043	P
9,9	0.220168	–0.610063	–1.220125	0.369586	0.389851	0.393107	0.00044043	P
T,T	0.676590	–0.852918	–1.705835	0.555805	0.491064	0.450852	0.00727662	S
				DAS				
A,A	–0.115159	0.189536	0.206312	0.679649	0.789286	0.807204	0.00044043	P
2,2	–0.115611	0.030230	–0.231222	0.234061	0.262695	0.267458	0.00044043	P
3,3	–0.114816	0.009878	–0.212063	0.213213	0.241985	0.246806	0.00044043	P
4,4	–0.114085	0.113365	0.092929	0.192970	0.204567	0.206562	0.00044043	P
5,5	–0.113900	0.293773	0.587547	0.169290	0.111376	0.101875	0.00044043	D
6,6	–0.121827	–0.175211	–0.350422	0.149949	0.186027	0.191691	0.00040372	P
7,7	–0.123475	–0.317997	–0.635994	0.197998	0.242401	0.249666	0.00044043	P
8,8	–0.124862	–0.437540	–0.875080	0.304855	0.364046	0.373707	0.00044043	P
9,9	0.220168	–0.610063	–1.220125	0.421148	0.448476	0.452878	0.00044043	P
T,T	0.676590	–0.852918	–1.705835	0.555805	0.491064	0.450852	0.00727662	S

TABLE A66
8D S17: Dealer's Upcard A

Hand	Stand	Hit	Double	Freq.	Cum. Freq.	BS
3,2	–0.666052	–0.280469	–1.332104	0.00063275	0.00063275	EH
4,2	–0.665853	–0.307899	–1.303557	0.00063275	0.00063275	EH
5,2	–0.665634	–0.312800	–1.129586	0.00063275		EH
4,3	–0.665669	–0.314345	–1.129650	0.00063275	0.00126550	EH
6,2	–0.663797	–0.200501	–0.809437	0.00063275		H
5,3	–0.665450	–0.199370	–0.808503	0.00063275	0.00126550	H
7,2	–0.664500	–0.067563	–0.429293	0.00063275		H
6,3	–0.663612	–0.067415	–0.429256	0.00063275		H
5,4	–0.665233	–0.066144	–0.428066	0.00063275	0.00189825	H
8,2	–0.665211	0.082078	–0.009444	0.00063275		H
7,3	–0.664316	0.082080	–0.007727	0.00063275		H
6,4	–0.663394	0.081360	–0.006662	0.00063275	0.00189825	H
9,2	–0.665913	0.146106	0.118365	0.00063275		H
8,3	–0.665018	0.146196	0.120299	0.00063275		H
7,4	–0.664089	0.146252	0.122268	0.00063275		H
6,5	–0.663164	0.146433	0.124504	0.00063275	0.00253100	H
T,2	–0.667785	–0.350275	–0.827115	0.00253989		EH
9,3	–0.665731	–0.353375	–0.832741	0.00063275		EH
8,4	–0.664803	–0.353502	–0.831283	0.00063275		EH
7,5	–0.663871	–0.353572	–0.830160	0.00063275	0.00443814	EH
T,3	–0.667604	–0.396452	–0.878009	0.00253989		EH
9,4	–0.665517	–0.395185	–0.875398	0.00063275		EH
8,5	–0.664585	–0.401102	–0.886341	0.00063275		EH
7,6	–0.662053	–0.402636	–0.887637	0.00063275	0.00443814	EH
T,4	–0.667391	–0.440557	–0.931711	0.00253989		EH
9,5	–0.665299	–0.440019	–0.930182	0.00063275		EH
8,6	–0.662769	–0.441646	–0.933044	0.00063275	0.00380539	EH
T,5	–0.667174	–0.482192	–0.986429	0.00253989		EH
9,6	–0.663484	–0.481874	–0.986019	0.00063275		EH
8,7	–0.663477	–0.477074	–0.976438	0.00063275	0.00380539	EH
T,6	–0.665364	–0.516139	–1.032278	0.00253989		RH
9,7	–0.664189	–0.514522	–1.029044	0.00063275	0.00317264	RH
T,7	–0.476743	–0.557112	–1.114224	0.00253989		ES
9,8	–0.474924	–0.555711	–1.111421	0.00063275	0.00317264	ES
T,8	–0.098009	–0.627285	–1.254570	0.00253989	0.00253989	S
T,9	0.281269	–0.726883	–1.453765	0.00253989	0.00253989	S
A,2	–0.666410	–0.058912	–0.620742	0.00061298	0.00061298	H
A,3	–0.666228	–0.095246	–0.620996	0.00061298	0.00061298	H
A,4	–0.666014	–0.133208	–0.623890	0.00061298	0.00061298	H
A,5	–0.665812	–0.170584	–0.628030	0.00061298	0.00061298	H
A,6	–0.478612	–0.182123	–0.536285	0.00061298	0.00061298	H
A,7	–0.100256	–0.094736	–0.362443	0.00061298	0.00061298	H
A,8	0.279149	–0.007059	–0.189071	0.00061298	0.00061298	S
A,9	0.658569	0.079802	–0.017364	0.00061298	0.00061298	S
A,T	1.500000	0.144143	0.113223	0.00246051	0.00246051	S

Hand	Stand	Hit	Double	Spl1	Spl2	Spl3	Freq.	BS
				NDAS				
A,A	–0.666586	–0.021789	–0.621280	0.122585	0.229993	0.247481	0.00028733	P
2,2	–0.666234	–0.254035	–1.332469	–0.409319	–0.431069	–0.434716	0.00030649	H
3,3	–0.665885	–0.307906	–1.303616	–0.458496	–0.479396	–0.482868	0.00030649	EH
4,4	–0.665452	–0.198307	–0.808550	–0.510929	–0.554924	–0.562336	0.00030649	H
5,5	–0.665013	0.082515	–0.005328	–0.567319	–0.659016	–0.674506	0.00030649	H
6,6	–0.661344	–0.355007	–0.830645	–0.612359	–0.648327	–0.654345	0.00030649	EH
7,7	–0.662761	–0.446511	–0.943493	–0.619437	–0.642957	–0.646784	0.00030649	EH
8,8	–0.664189	–0.514455	–1.028909	–0.387614	–0.369429	–0.366310	0.00030649	EP
9,9	–0.094866	–0.626196	–1.252392	–0.124271	–0.128058	–0.128638	0.00030649	S
T,T	0.654804	–0.855859	–1.711719	0.155206	–0.104846	–0.262407	0.00505771	S
				DAS				
A,A	–0.666586	–0.021789	–0.621280	0.122585	0.229993	0.247481	0.00028733	P
2,2	–0.666234	–0.254035	–1.332469	–0.409319	–0.431069	–0.434716	0.00030649	H
3,3	–0.665885	–0.307906	–1.303616	–0.458496	–0.479396	–0.482868	0.00030649	EH
4,4	–0.665452	–0.198307	–0.808550	–0.510929	–0.554924	–0.562336	0.00030649	H
5,5	–0.665013	0.082515	–0.005328	–0.567319	–0.659016	–0.674506	0.00030649	H
6,6	–0.661344	–0.355007	–0.830645	–0.612359	–0.648327	–0.654345	0.00030649	EH
7,7	–0.662761	–0.446511	–0.943493	–0.619437	–0.642957	–0.646784	0.00030649	EH
8,8	–0.664189	–0.514455	–1.028909	–0.387614	–0.369429	–0.366310	0.00030649	EP
9,9	–0.094866	–0.626196	–1.252392	–0.124271	–0.128058	–0.128638	0.00030649	S
T,T	0.654804	–0.855859	–1.711719	0.155206	–0.104846	–0.262407	0.00505771	S

TABLE A67
8D S17: Dealer's Upcard 2

Hand	Stand	Hit	Double	Freq.	Cum. Freq.	BS
3,2	–0.292556	–0.128482	–0.585112	0.00088828	0.00088828	H
4,2	–0.292329	–0.141708	–0.563128	0.00088828	0.00088828	H
5,2	–0.290579	–0.108065	–0.431624	0.00088828		H
4,3	–0.292280	–0.110332	–0.434101	0.00091693	0.00180521	H
6,2	–0.290940	–0.020627	–0.199451	0.00088828		H
5,3	–0.290530	–0.020848	–0.199621	0.00091693	0.00180521	H
7,2	–0.291310	0.075692	0.067906	0.00088828		H
6,3	–0.290891	0.076739	0.067510	0.00091693		H
5,4	–0.290302	0.077042	0.067936	0.00091693	0.00272215	H
8,2	–0.291652	0.185909	0.366409	0.00088828		D
7,3	–0.291250	0.186320	0.367022	0.00091693		D
6,4	–0.290652	0.186765	0.367899	0.00091693	0.00272215	D
9,2	–0.292152	0.241245	0.476814	0.00088828		D
8,3	–0.291603	0.241837	0.477968	0.00091693		D
7,4	–0.291023	0.242470	0.479224	0.00091693		D
6,5	–0.288906	0.243852	0.482179	0.00091693	0.00363908	D
T,2	–0.295061	–0.252241	–0.504481	0.00355312		H
9,3	–0.292104	–0.254932	–0.509864	0.00091693		H
8,4	–0.291376	–0.254141	–0.508282	0.00091693		H
7,5	–0.289277	–0.253309	–0.506618	0.00091693	0.00630392	H
T,3	–0.295013	–0.307297	–0.614594	0.00366774		S
9,4	–0.291877	–0.305967	–0.611934	0.00091693		S
8,5	–0.289632	–0.310613	–0.621226	0.00091693		S
7,6	–0.289640	–0.310707	–0.621415	0.00091693	0.00641854	S
T,4	–0.294786	–0.362902	–0.725803	0.00366774		S
9,5	–0.290132	–0.361811	–0.723621	0.00091693		S
8,6	–0.289993	–0.362301	–0.724602	0.00091693	0.00550161	S
T,5	–0.293055	–0.418913	–0.837826	0.00366774		S
9,6	–0.290508	–0.418244	–0.836487	0.00091693		S
8,7	–0.290379	–0.413562	–0.827124	0.00091693	0.00550161	S
T,6	–0.293417	–0.470328	–0.940656	0.00366774		S
9,7	–0.290880	–0.469255	–0.938510	0.00091693	0.00458467	S
T,7	–0.153617	–0.536420	–1.072839	0.00366774		S
9,8	–0.151056	–0.535514	–1.071027	0.00091693	0.00458467	S
T,8	0.121318	–0.623660	–1.247320	0.00366774	0.00366774	S
T,9	0.386211	–0.731527	–1.463055	0.00366774	0.00366774	S
A,2	–0.291602	0.045786	–0.067849	0.00088828	0.00088828	H
A,3	–0.291553	0.021732	–0.068290	0.00091693	0.00091693	H
A,4	–0.291341	–0.001459	–0.071038	0.00091693	0.00091693	H
A,5	–0.289591	–0.022378	–0.072965	0.00091693	0.00091693	H
A,6	–0.150456	0.000394	–0.004736	0.00091693	0.00091693	H
A,7	0.123432	0.063193	0.120669	0.00091693	0.00091693	S
A,8	0.388023	0.123441	0.241175	0.00091693	0.00091693	S
A,9	0.641759	0.183336	0.361173	0.00091693	0.00091693	S
A,T	1.500000	0.239370	0.472453	0.00366774	0.00366774	S

Hand	Stand	Hit	Double	Spl1	Spl2	Spl3	Freq.	BS
				NDAS				
A,A	–0.290603	0.083335	–0.065179	0.482040	0.590708	0.609004	0.00044414	P
2,2	–0.292593	–0.114785	–0.585187	–0.149100	–0.153732	–0.154481	0.00041638	H
3,3	–0.292508	–0.141807	–0.563445	–0.200578	–0.208690	–0.210031	0.00044414	H
4,4	–0.292052	–0.020323	–0.201274	–0.229681	–0.259104	–0.264055	0.00044414	H
5,5	–0.288545	0.187530	0.369607	–0.253324	–0.315119	–0.325488	0.00044414	D
6,6	–0.289268	–0.253341	–0.506682	–0.274579	–0.276999	–0.277312	0.00044414	H
7,7	–0.290011	–0.367475	–0.734950	–0.211857	–0.200606	–0.198669	0.00044414	P
8,8	–0.290732	–0.469110	–0.938220	–0.036775	–0.000830	0.005260	0.00044414	P
9,9	0.123607	–0.623062	–1.246123	0.151556	0.155497	0.156162	0.00044414	P
T,T	0.638411	–0.854215	–1.708429	0.358564	0.211988	0.122682	0.00727817	S
				DAS				
A,A	–0.290603	0.083335	–0.065179	0.482040	0.590708	0.609004	0.00044414	P
2,2	–0.292593	–0.114785	–0.585187	–0.084257	–0.080023	–0.079321	0.00041638	P
3,3	–0.292508	–0.141807	–0.563445	–0.135624	–0.134562	–0.134352	0.00044414	P
4,4	–0.292052	–0.020323	–0.201274	–0.164528	–0.184744	–0.188138	0.00044414	H
5,5	–0.288545	0.187530	0.369607	–0.189275	–0.267912	–0.281158	0.00044414	D
6,6	–0.289268	–0.253341	–0.506682	–0.208944	–0.202057	–0.200790	0.00044414	P
7,7	–0.290011	–0.367475	–0.734950	–0.146679	–0.126206	–0.122706	0.00044414	P
8,8	–0.290732	–0.469110	–0.938220	0.027266	0.072256	0.079875	0.00044414	P
9,9	0.123607	–0.623062	–1.246123	0.186958	0.195878	0.197381	0.00044414	P
T,T	0.638411	–0.854215	–1.708429	0.358564	0.211988	0.122682	0.00727817	S

TABLE A68
8D S17: Dealer's Upcard 3

Hand	Stand	Hit	Double	Freq.	Cum. Freq.	BS
3,2	−0.251286	−0.095342	−0.502572	0.00088828	0.00088828	H
4,2	−0.249544	−0.106893	−0.478503	0.00091693	0.00091693	H
5,2	−0.249218	−0.074609	−0.353428	0.00091693		H
4,3	−0.249318	−0.076745	−0.354737	0.00088828	0.00180521	H
6,2	−0.249577	0.010018	−0.129054	0.00091693		H
5,3	−0.248993	0.009733	−0.129238	0.00088828	0.00180521	H
7,2	−0.249928	0.104757	0.129581	0.00091693		D
6,3	−0.249341	0.105745	0.129142	0.00088828		D
5,4	−0.247244	0.106154	0.130210	0.00091693	0.00272215	D
8,2	−0.250427	0.210240	0.418005	0.00091693		D
7,3	−0.249703	0.210872	0.419189	0.00088828		D
6,4	−0.247603	0.211426	0.420451	0.00091693	0.00272215	D
9,2	−0.253354	0.261562	0.520547	0.00091693		D
8,3	−0.250203	0.264468	0.526469	0.00088828		D
7,4	−0.247966	0.265599	0.528884	0.00091693		D
6,5	−0.247277	0.266690	0.531066	0.00091693	0.00363908	D
T,2	−0.253950	−0.231939	−0.463878	0.00366774		H
9,3	−0.253131	−0.236154	−0.472309	0.00088828		H
8,4	−0.248466	−0.233193	−0.466385	0.00091693		H
7,5	−0.247641	−0.232352	−0.464703	0.00091693	0.00638989	H
T,3	−0.253726	−0.290150	−0.580301	0.00355312		S
9,4	−0.251393	−0.289773	−0.579546	0.00091693		S
8,5	−0.248140	−0.292943	−0.585886	0.00091693		S
7,6	−0.248000	−0.292844	−0.585687	0.00091693	0.00630392	S
T,4	−0.252002	−0.349475	−0.698951	0.00366774		S
9,5	−0.251081	−0.349469	−0.698937	0.00091693		S
8,6	−0.248514	−0.348582	−0.697163	0.00091693	0.00550161	S
T,5	−0.251676	−0.408983	−0.817966	0.00366774		S
9,6	−0.251442	−0.409251	−0.818502	0.00091693		S
8,7	−0.248877	−0.403373	−0.806746	0.00091693	0.00550161	S
T,6	−0.252037	−0.463479	−0.926958	0.00366774		S
9,7	−0.251805	−0.463169	−0.926338	0.00091693	0.00458467	S
T,7	−0.117432	−0.532273	−1.064545	0.00366774		S
9,8	−0.117379	−0.532109	−1.064218	0.00091693	0.00458467	S
T,8	0.148013	−0.621649	−1.243298	0.00366774	0.00366774	S
T,9	0.402021	−0.726185	−1.452370	0.00366774	0.00366774	S
A,2	−0.250580	0.073793	−0.002365	0.00091693	0.00091693	H
A,3	−0.250371	0.050113	−0.004307	0.00088828	0.00088828	H
A,4	−0.248628	0.028400	−0.005756	0.00091693	0.00091693	H
A,5	−0.248303	0.007618	−0.008628	0.00091693	0.00091693	H
A,6	−0.114348	0.029788	0.057263	0.00091693	0.00091693	D
A,7	0.150405	0.090608	0.178901	0.00091693	0.00091693	D
A,8	0.406050	0.152016	0.301642	0.00091693	0.00091693	S
A,9	0.649609	0.205056	0.407535	0.00091693	0.00091693	S
A,T	1.500000	0.261585	0.520001	0.00366774	0.00366774	S

Hand	Stand	Hit	Double	Spl1	Spl2	Spl3	Freq.	BS
				NDAS				
A,A	−0.249656	0.105396	0.000430	0.529319	0.638860	0.657301	0.00044414	P
2,2	−0.251510	−0.082188	−0.503021	−0.095700	−0.097576	−0.097888	0.00044414	H
3,3	−0.251061	−0.107994	−0.481425	−0.135919	−0.139587	−0.140163	0.00041638	H
4,4	−0.247575	0.010694	−0.129088	−0.160236	−0.184061	−0.188037	0.00044414	H
5,5	−0.246918	0.211985	0.421564	−0.185667	−0.241346	−0.250680	0.00044414	D
6,6	−0.247638	−0.232403	−0.464807	−0.205735	−0.201381	−0.200548	0.00044414	P
7,7	−0.248377	−0.353549	−0.707098	−0.145086	−0.130254	−0.127706	0.00044414	P
8,8	−0.249378	−0.462197	−0.924394	0.023420	0.062011	0.068545	0.00044414	P
9,9	0.145696	−0.622092	−1.244184	0.199323	0.206575	0.207747	0.00044414	P
T,T	0.648558	−0.853951	−1.707901	0.406456	0.279446	0.201948	0.00727817	S
				DAS				
A,A	−0.249656	0.105396	0.000430	0.529319	0.638860	0.657301	0.00044414	P
2,2	−0.251510	−0.082188	−0.503021	−0.018912	−0.009917	−0.008387	0.00044414	P
3,3	−0.251061	−0.107994	−0.481425	−0.058973	−0.052121	−0.050976	0.00041638	P
4,4	−0.247575	0.010694	−0.129088	−0.082489	−0.095280	−0.097381	0.00044414	H
5,5	−0.246918	0.211985	0.421564	−0.109432	−0.184095	−0.196663	0.00044414	D
6,6	−0.247638	−0.232403	−0.464807	−0.123093	−0.106924	−0.104068	0.00044414	P
7,7	−0.248377	−0.353549	−0.707098	−0.063197	−0.036649	−0.032095	0.00044414	P
8,8	−0.249378	−0.462197	−0.924394	0.095437	0.144212	0.152470	0.00044414	P
9,9	0.145696	−0.622092	−1.244184	0.239183	0.251996	0.254097	0.00044414	P
T,T	0.648558	−0.853951	−1.707901	0.406456	0.279446	0.201948	0.00727817	S

TABLE A69
8D S17: Dealer's Upcard 4

Hand	Stand	Hit	Double	Freq.	Cum. Freq.	BS
3,2	–0.206778	–0.058943	–0.413556	0.00091693	0.00091693	H
4,2	–0.206450	–0.070721	–0.392779	0.00088828	0.00088828	H
5,2	–0.206162	–0.039067	–0.271915	0.00091693		H
4,3	–0.204709	–0.039866	–0.270088	0.00088828	0.00180521	H
6,2	–0.206498	0.043759	–0.055925	0.00091693		H
5,3	–0.204415	0.044462	–0.053778	0.00091693	0.00183387	H
7,2	–0.207010	0.133581	0.192586	0.00091693		D
6,3	–0.204762	0.135735	0.194661	0.00091693		D
5,4	–0.204086	0.136064	0.195326	0.00088828	0.00272215	D
8,2	–0.209935	0.233223	0.466445	0.00091693		D
7,3	–0.205275	0.237003	0.474007	0.00091693		D
6,4	–0.204433	0.237748	0.475496	0.00088828	0.00272215	D
9,2	–0.210540	0.285054	0.570108	0.00091693		D
8,3	–0.208200	0.286615	0.573231	0.00091693		D
7,4	–0.204947	0.289961	0.579922	0.00088828		D
6,5	–0.204145	0.291032	0.582064	0.00091693	0.00363908	D
T,2	–0.211100	–0.211161	–0.422322	0.00366774		S
9,3	–0.208805	–0.214922	–0.429844	0.00091693		S
8,4	–0.207871	–0.213642	–0.427283	0.00088828		S
7,5	–0.204658	–0.211107	–0.422215	0.00091693	0.00638989	S
T,3	–0.209379	–0.272204	–0.544409	0.00366774		S
9,4	–0.208489	–0.271877	–0.543753	0.00088828		S
8,5	–0.207596	–0.276364	–0.552728	0.00091693		S
7,6	–0.205020	–0.274713	–0.549426	0.00091693	0.00638989	S
T,4	–0.209049	–0.335110	–0.670220	0.00355312		S
9,5	–0.208201	–0.335079	–0.670159	0.00091693		S
8,6	–0.207945	–0.335305	–0.670609	0.00091693	0.00538699	S
T,5	–0.208761	–0.398320	–0.796640	0.00366774		S
9,6	–0.208549	–0.398546	–0.797091	0.00091693		S
8,7	–0.208458	–0.393793	–0.787586	0.00091693	0.00550161	S
T,6	–0.209109	–0.456118	–0.912237	0.00366774		S
9,7	–0.209063	–0.455946	–0.911891	0.00091693	0.00458467	S
T,7	–0.078640	–0.527985	–1.055970	0.00366774		S
9,8	–0.080986	–0.528605	–1.057209	0.00091693	0.00458467	S
T,8	0.174513	–0.615031	–1.230061	0.00366774	0.00366774	S
T,9	0.420816	–0.725129	–1.450259	0.00366774	0.00366774	S
A,2	–0.207834	0.103396	0.065642	0.00091693	0.00091693	H
A,3	–0.206093	0.081321	0.064784	0.00091693	0.00091693	H
A,4	–0.205765	0.059359	0.061384	0.00088828	0.00088828	D
A,5	–0.205477	0.039507	0.058561	0.00091693	0.00091693	D
A,6	–0.075439	0.061357	0.122714	0.00091693	0.00091693	D
A,7	0.179132	0.122910	0.245820	0.00091693	0.00091693	D
A,8	0.422460	0.176941	0.353882	0.00091693	0.00091693	S
A,9	0.660128	0.230287	0.460573	0.00091693	0.00091693	S
A,T	1.500000	0.284756	0.569512	0.00366774	0.00366774	S

Hand	Stand	Hit	Double	Spl1	Spl2	Spl3	Freq.	BS
NDAS								
A,A	–0.207125	0.128395	0.067963	0.578527	0.688987	0.707580	0.00044414	P
2,2	–0.208518	–0.047417	–0.417036	–0.039399	–0.038245	–0.038047	0.00044414	P
3,3	–0.205038	–0.069820	–0.390280	–0.062868	–0.061625	–0.061371	0.00044414	P
4,4	–0.204374	0.045695	–0.053651	–0.088826	–0.106917	–0.109828	0.00041638	H
5,5	–0.203797	0.238015	0.476030	–0.113379	–0.162489	–0.170706	0.00044414	D
6,6	–0.204492	–0.210944	–0.421888	–0.132431	–0.121861	–0.120009	0.00044414	P
7,7	–0.205534	–0.339089	–0.678178	–0.074023	–0.055170	–0.051936	0.00044414	P
8,8	–0.211384	–0.456748	–0.913496	0.081339	0.122462	0.129378	0.00044414	P
9,9	0.174514	–0.615026	–1.230053	0.257027	0.268368	0.270234	0.00044414	P
T,T	0.659104	–0.853708	–1.707416	0.455762	0.348639	0.283032	0.00727817	S
DAS								
A,A	–0.207125	0.128395	0.067963	0.578527	0.688987	0.707580	0.00044414	P
2,2	–0.208518	–0.047417	–0.417036	0.050631	0.064510	0.066861	0.00044414	P
3,3	–0.205038	–0.069820	–0.390280	0.028096	0.042214	0.044648	0.00044414	P
4,4	–0.204374	0.045695	–0.053651	0.003532	–0.001857	–0.002677	0.00041638	H
5,5	–0.203797	0.238015	0.476030	–0.020624	–0.090413	–0.102154	0.00044414	D
6,6	–0.204492	–0.210944	–0.421888	–0.031250	–0.006226	–0.001903	0.00044414	P
7,7	–0.205534	–0.339089	–0.678178	0.027373	0.060697	0.066402	0.00044414	P
8,8	–0.211384	–0.456748	–0.913496	0.161675	0.214054	0.222858	0.00044414	P
9,9	0.174514	–0.615026	–1.230053	0.300910	0.318377	0.321267	0.00044414	P
T,T	0.659104	–0.853708	–1.707416	0.455762	0.348639	0.283032	0.00727817	S

TABLE A70

8D S17: Dealer's Upcard 5

Hand	Stand	Hit	Double	Freq.	Cum. Freq.	BS
3,2	–0.159978	–0.018750	–0.319957	0.00091693	0.00091693	H
4,2	–0.159686	–0.029875	–0.300729	0.00091693	0.00091693	H
5,2	–0.159410	0.000104	–0.187640	0.00088828		H
4,3	–0.159363	–0.000813	–0.186651	0.00091693	0.00180521	H
6,2	–0.159923	0.077777	0.018164	0.00091693		H
5,3	–0.159093	0.078901	0.020782	0.00088828	0.00180521	H
7,2	–0.162864	0.162564	0.253741	0.00091693		D
6,3	–0.159606	0.166979	0.260266	0.00091693		D
5,4	–0.158799	0.168199	0.262734	0.00088828	0.00272215	D
8,2	–0.163476	0.261021	0.522041	0.00091693		D
7,3	–0.162547	0.262368	0.524736	0.00091693		D
6,4	–0.159312	0.265809	0.531619	0.00091693	0.00275080	D
9,2	–0.164014	0.311107	0.622215	0.00091693		D
8,3	–0.163158	0.312370	0.624740	0.00091693		D
7,4	–0.162253	0.313753	0.627507	0.00091693		D
6,5	–0.159042	0.317284	0.634567	0.00088828	0.00363908	D
T,2	–0.164524	–0.189785	–0.379571	0.00366774		S
9,3	–0.163711	–0.193740	–0.387479	0.00091693		S
8,4	–0.162879	–0.192510	–0.385020	0.00091693		S
7,5	–0.161996	–0.191462	–0.382924	0.00088828	0.00638989	S
T,3	–0.164207	–0.254752	–0.509505	0.00366774		S
9,4	–0.163417	–0.254427	–0.508853	0.00091693		S
8,5	–0.162608	–0.258907	–0.517814	0.00088828		S
7,6	–0.162511	–0.258749	–0.517498	0.00091693	0.00638989	S
T,4	–0.163913	–0.320630	–0.641261	0.00366774		S
9,5	–0.163146	–0.320588	–0.641177	0.00088828		S
8,6	–0.163122	–0.320968	–0.641935	0.00091693	0.00547295	S
T,5	–0.163643	–0.387195	–0.774389	0.00355312		S
9,6	–0.163661	–0.387555	–0.775110	0.00091693		S
8,7	–0.166064	–0.383963	–0.767926	0.00091693	0.00538699	S
T,6	–0.164158	–0.449001	–0.898001	0.00366774		S
9,7	–0.166603	–0.449808	–0.899616	0.00091693	0.00458467	S
T,7	–0.044723	–0.519389	–1.038777	0.00366774		S
9,8	–0.044829	–0.519394	–1.038788	0.00091693	0.00458467	S
T,8	0.199942	–0.612349	–1.224699	0.00366774	0.00366774	S
T,9	0.440450	–0.723953	–1.447907	0.00366774	0.00366774	S
A,2	–0.161365	0.136331	0.136360	0.00091693	0.00091693	D
A,3	–0.161049	0.114758	0.135292	0.00091693	0.00091693	D
A,4	–0.160756	0.093745	0.131755	0.00091693	0.00091693	D
A,5	–0.160487	0.074309	0.128250	0.00088828	0.00088828	D
A,6	–0.039214	0.096952	0.193904	0.00091693	0.00091693	D
A,7	0.202229	0.150933	0.301866	0.00091693	0.00091693	D
A,8	0.442049	0.205892	0.411784	0.00091693	0.00091693	S
A,9	0.671737	0.257622	0.515244	0.00091693	0.00091693	S
A,T	1.500000	0.310216	0.620433	0.00366774	0.00366774	S

Hand	Stand	Hit	Double	Spl1	Spl2	Spl3	Freq.	BS
				NDAS				
A,A	–0.162429	0.159662	0.137072	0.629103	0.740373	0.759099	0.00044414	P
2,2	–0.160295	–0.006838	–0.320589	0.039119	0.045732	0.046870	0.00044414	P
3,3	–0.159662	–0.029883	–0.300505	0.013666	0.020091	0.021223	0.00044414	P
4,4	–0.159069	0.080384	0.021547	–0.011016	–0.023658	–0.025751	0.00044414	H
5,5	–0.158528	0.266709	0.533417	–0.034815	–0.075580	–0.082174	0.00041638	D
6,6	–0.159570	–0.189754	–0.379509	–0.053724	–0.038402	–0.035750	0.00044414	P
7,7	–0.165451	–0.327224	–0.654448	–0.007692	0.014518	0.018261	0.00044414	P
8,8	–0.166676	–0.449825	–0.899649	0.147621	0.191751	0.199168	0.00044414	P
9,9	0.199923	–0.612343	–1.224686	0.318541	0.334948	0.337664	0.00044414	P
T,T	0.670758	–0.853425	–1.706850	0.510648	0.425951	0.373889	0.00727817	S
				DAS				
A,A	–0.162429	0.159662	0.137072	0.629103	0.740373	0.759099	0.00044414	P
2,2	–0.160295	–0.006838	–0.320589	0.143110	0.164495	0.168145	0.00044414	P
3,3	–0.159662	–0.029883	–0.300505	0.121578	0.143327	0.147064	0.00044414	P
4,4	–0.159069	0.080384	0.021547	0.100222	0.103392	0.103989	0.00044414	P
5,5	–0.158528	0.266709	0.533417	0.076374	0.014142	0.004004	0.00041638	D
6,6	–0.159570	–0.189754	–0.379509	0.066468	0.098933	0.104511	0.00044414	P
7,7	–0.165451	–0.327224	–0.654448	0.111422	0.150402	0.156972	0.00044414	P
8,8	–0.166676	–0.449825	–0.899649	0.236239	0.292786	0.302286	0.00044414	P
9,9	0.199923	–0.612343	–1.224686	0.366464	0.389564	0.393400	0.00044414	P
T,T	0.670758	–0.853425	–1.706850	0.510648	0.425951	0.373889	0.00727817	S

TABLE A71
8D S17: Dealer's Upcard 6

Hand	Stand	Hit	Double	Freq.	Cum. Freq.	BS
3,2	−0.149892	0.001328	−0.299784	0.00091693	0.00091693	H
4,2	−0.149633	−0.009733	−0.274038	0.00091693	0.00091693	H
5,2	−0.149393	0.034041	−0.129044	0.00091693		H
4,3	−0.149359	0.032794	−0.128778	0.00091693	0.00183387	H
6,2	−0.152286	0.117291	0.091955	0.00088828		H
5,3	−0.149119	0.120735	0.099125	0.00091693	0.00180521	H
7,2	−0.152855	0.197955	0.322394	0.00091693		D
6,3	−0.152012	0.200333	0.324700	0.00088828		D
5,4	−0.148865	0.203892	0.331823	0.00091693	0.00272215	D
8,2	−0.153350	0.290241	0.580482	0.00091693		D
7,3	−0.152580	0.291568	0.583136	0.00091693		D
6,4	−0.151757	0.293104	0.586209	0.00088828	0.00272215	D
9,2	−0.153824	0.335281	0.670563	0.00091693		D
8,3	−0.153090	0.336420	0.672840	0.00091693		D
7,4	−0.152340	0.337657	0.675314	0.00091693		D
6,5	−0.151531	0.339415	0.678829	0.00088828	0.00363908	D
T,2	−0.154442	−0.169202	−0.338403	0.00366774		S
9,3	−0.153549	−0.173212	−0.346425	0.00091693		S
8,4	−0.152835	−0.172100	−0.344199	0.00091693		S
7,5	−0.152099	−0.171346	−0.342691	0.00091693	0.00641854	S
T,3	−0.154168	−0.234886	−0.469772	0.00366774		S
9,4	−0.153295	−0.234705	−0.469410	0.00091693		S
8,5	−0.152595	−0.239382	−0.478764	0.00091693		S
7,6	−0.154993	−0.240766	−0.481532	0.00088828	0.00638989	S
T,4	−0.153913	−0.301651	−0.603302	0.00366774		S
9,5	−0.153055	−0.301636	−0.603271	0.00091693		S
8,6	−0.155489	−0.303311	−0.606622	0.00088828	0.00547295	S
T,5	−0.153674	−0.368421	−0.736843	0.00366774		S
9,6	−0.155951	−0.369905	−0.739810	0.00088828		S
8,7	−0.156059	−0.365318	−0.730637	0.00091693	0.00547295	S
T,6	−0.156548	−0.426814	−0.853629	0.00355312		S
9,7	−0.156500	−0.426856	−0.853712	0.00091693	0.00447005	S
T,7	0.009197	−0.505676	−1.011351	0.00366774		S
9,8	0.009162	−0.505701	−1.011403	0.00091693	0.00458467	S
T,8	0.281635	−0.604925	−1.209850	0.00366774	0.00366774	S
T,9	0.494591	−0.720684	−1.441368	0.00366774	0.00366774	S
A,2	−0.148941	0.162421	0.185872	0.00091693	0.00091693	D
A,3	−0.148667	0.139951	0.184704	0.00091693	0.00091693	D
A,4	−0.148414	0.118357	0.182114	0.00091693	0.00091693	D
A,5	−0.148169	0.100837	0.184093	0.00091693	0.00091693	D
A,6	0.011955	0.128964	0.257928	0.00088828	0.00088828	D
A,7	0.281247	0.191253	0.382506	0.00091693	0.00091693	D
A,8	0.494445	0.240052	0.480105	0.00091693	0.00091693	S
A,9	0.702950	0.286968	0.573935	0.00091693	0.00091693	S
A,T	1.500000	0.334613	0.669226	0.00366774	0.00366774	S

Hand	Stand	Hit	Double	Spl1	Spl2	Spl3	Freq.	BS
				NDAS				
A,A	−0.147712	0.187459	0.187954	0.678612	0.794562	0.814040	0.00044414	P
2,2	−0.150165	0.013750	−0.300329	0.083706	0.093722	0.095437	0.00044414	P
3,3	−0.149613	−0.009737	−0.273997	0.056330	0.065950	0.067625	0.00044414	P
4,4	−0.149105	0.122018	0.099381	0.029634	0.016843	0.014723	0.00044414	H
5,5	−0.148624	0.296494	0.592989	0.004929	−0.035838	−0.042662	0.00044414	D
6,6	−0.154424	−0.172989	−0.345978	−0.023270	−0.005301	−0.002357	0.00041638	P
7,7	−0.155561	−0.308423	−0.616847	0.058795	0.088949	0.094027	0.00044414	P
8,8	−0.156536	−0.426855	−0.853710	0.230419	0.284767	0.293905	0.00044414	P
9,9	0.281479	−0.604941	−1.209883	0.388525	0.403284	0.405720	0.00044414	P
T,T	0.703108	−0.852607	−1.705213	0.568987	0.497767	0.453843	0.00727817	S
				DAS				
A,A	−0.147712	0.187459	0.187954	0.678612	0.794562	0.814040	0.00044414	P
2,2	−0.150165	0.013750	−0.300329	0.204974	0.232219	0.236864	0.00044414	P
3,3	−0.149613	−0.009737	−0.273997	0.180650	0.207926	0.212602	0.00044414	P
4,4	−0.149105	0.122018	0.099381	0.157226	0.162574	0.163540	0.00044414	P
5,5	−0.148624	0.296494	0.592989	0.133391	0.068642	0.057722	0.00044414	D
6,6	−0.154424	−0.172989	−0.345978	0.114514	0.151285	0.157299	0.00041638	P
7,7	−0.155561	−0.308423	−0.616847	0.191512	0.240342	0.248566	0.00044414	P
8,8	−0.156536	−0.426855	−0.853710	0.327278	0.395199	0.406615	0.00044414	P
9,9	0.281479	−0.604941	−1.209883	0.440202	0.462183	0.465827	0.00044414	P
T,T	0.703108	−0.852607	−1.705213	0.568987	0.497767	0.453843	0.00727817	S

TABLE A72
8D S17: Dealer's Upcard 7

Hand	Stand	Hit	Double	Freq.	Cum. Freq.	BS
3,2	−0.474559	−0.119475	−0.949118	0.00091693	0.00091693	H
4,2	−0.474305	−0.153286	−0.891184	0.00091693	0.00091693	H
5,2	−0.475388	−0.068693	−0.586381	0.00091693		H
4,3	−0.474060	−0.069042	−0.583711	0.00091693	0.00183387	H
6,2	−0.475959	0.083509	−0.182622	0.00091693		H
5,3	−0.475143	0.083432	−0.180897	0.00091693	0.00183387	H
7,2	−0.476479	0.173688	0.111258	0.00088828		H
6,3	−0.475713	0.175017	0.113018	0.00091693		H
5,4	−0.474889	0.175175	0.114653	0.00091693	0.00272215	H
8,2	−0.476910	0.258155	0.395355	0.00091693		D
7,3	−0.476248	0.259479	0.399175	0.00088828		D
6,4	−0.475473	0.260407	0.402606	0.00091693	0.00272215	D
9,2	−0.477554	0.291847	0.462131	0.00091693		D
8,3	−0.476665	0.292162	0.464122	0.00091693		D
7,4	−0.475994	0.292702	0.466480	0.00088828		D
6,5	−0.476555	0.293108	0.468053	0.00091693	0.00363908	D
T,2	−0.475044	−0.212688	−0.505318	0.00366774		H
9,3	−0.477309	−0.216882	−0.515200	0.00091693		H
8,4	−0.476411	−0.216650	−0.513380	0.00091693		H
7,5	−0.477075	−0.217865	−0.515234	0.00088828	0.00638989	H
T,3	−0.474799	−0.269217	−0.586842	0.00366774		H
9,4	−0.477055	−0.269599	−0.588813	0.00091693		H
8,5	−0.477493	−0.275854	−0.601215	0.00091693		H
7,6	−0.477646	−0.276056	−0.601670	0.00088828	0.00638989	H
T,4	−0.474546	−0.323738	−0.672701	0.00366774		H
9,5	−0.478138	−0.324368	−0.674946	0.00091693		H
8,6	−0.478066	−0.324429	−0.675756	0.00091693	0.00550161	H
T,5	−0.475618	−0.369137	−0.747364	0.00366774		H
9,6	−0.478689	−0.369001	−0.748315	0.00091693		H
8,7	−0.478567	−0.364365	−0.739018	0.00088828	0.00547295	H
T,6	−0.476193	−0.410173	−0.820345	0.00366774		H
9,7	−0.479214	−0.410030	−0.820059	0.00088828	0.00455602	H
T,7	−0.108354	−0.479662	−0.959324	0.00355312		S
9,8	−0.109007	−0.479440	−0.958881	0.00091693	0.00447005	S
T,8	0.398196	−0.588298	−1.176595	0.00366774	0.00366774	S
T,9	0.615281	−0.713434	−1.426868	0.00366774	0.00366774	S
A,2	−0.473735	0.120737	−0.180581	0.00091693	0.00091693	H
A,3	−0.473491	0.077289	−0.182645	0.00091693	0.00091693	H
A,4	−0.473232	0.036559	−0.178852	0.00091693	0.00091693	H
A,5	−0.474314	−0.007128	−0.184212	0.00091693	0.00091693	H
A,6	−0.104568	0.054483	−0.010074	0.00091693	0.00091693	H
A,7	0.401292	0.171590	0.222746	0.00088828	0.00088828	S
A,8	0.615795	0.220876	0.320372	0.00091693	0.00091693	S
A,9	0.773263	0.255216	0.387570	0.00091693	0.00091693	S
A,T	1.500000	0.291454	0.463548	0.00366774	0.00366774	S

Hand	Stand	Hit	Double	Spl1	Spl2	Spl3	Freq.	BS
				NDAS				
A,A	−0.472657	0.164619	−0.178282	0.472412	0.605326	0.627704	0.00044414	P
2,2	−0.474809	−0.088690	−0.949618	−0.054251	−0.049374	−0.048547	0.00044414	P
3,3	−0.474314	−0.153324	−0.891195	−0.115096	−0.109406	−0.108395	0.00044414	P
4,4	−0.473807	0.085502	−0.178278	−0.181474	−0.219278	−0.225686	0.00044414	H
5,5	−0.475985	0.259518	0.401662	−0.245579	−0.317115	−0.329243	0.00044414	D
6,6	−0.477126	−0.218600	−0.516704	−0.300525	−0.311696	−0.313518	0.00044414	H
7,7	−0.478169	−0.329129	−0.685903	−0.135936	−0.109286	−0.104889	0.00041638	P
8,8	−0.478988	−0.410029	−0.820058	0.169442	0.250755	0.264414	0.00044414	P
9,9	0.399563	−0.588176	−1.176352	0.343430	0.335476	0.334128	0.00044414	S
T,T	0.772322	−0.850799	−1.701598	0.510013	0.373495	0.290793	0.00727817	S
				DAS				
A,A	−0.472657	0.164619	−0.178282	0.472412	0.605326	0.627704	0.00044414	P
2,2	−0.474809	−0.088690	−0.949618	−0.005736	0.006074	0.008086	0.00044414	P
3,3	−0.474314	−0.153324	−0.891195	−0.066483	−0.053828	−0.051624	0.00044414	P
4,4	−0.473807	0.085502	−0.178278	−0.132180	−0.162899	−0.168089	0.00044414	H
5,5	−0.475985	0.259518	0.401662	−0.196948	−0.281953	−0.296402	0.00044414	D
6,6	−0.477126	−0.218600	−0.516704	−0.250900	−0.255034	−0.255662	0.00044414	H
7,7	−0.478169	−0.329129	−0.685903	−0.087501	−0.054284	−0.048822	0.00041638	P
8,8	−0.478988	−0.410029	−0.820058	0.216711	0.304577	0.319324	0.00044414	P
9,9	0.399563	−0.588176	−1.176352	0.369261	0.364861	0.364097	0.00044414	S
T,T	0.772322	−0.850799	−1.701598	0.510013	0.373495	0.290793	0.00727817	S

TABLE A73
8D S17: Dealer's Upcard 8

Hand	Stand	Hit	Double	Freq.	Cum. Freq.	BS
3,2	–0.510811	–0.187408	–1.021621	0.00091693	0.00091693	H
4,2	–0.511880	–0.219089	–1.003819	0.00091693	0.00091693	H
5,2	–0.511829	–0.211400	–0.847677	0.00091693		H
4,3	–0.511626	–0.212600	–0.847284	0.00091693	0.00183387	H
6,2	–0.512328	–0.059384	–0.450460	0.00091693		H
5,3	–0.511574	–0.059567	–0.449031	0.00091693	0.00183387	H
7,2	–0.512783	0.099394	–0.022901	0.00091693		H
6,3	–0.512087	0.100484	–0.021718	0.00091693		H
5,4	–0.512657	0.099532	–0.022871	0.00091693	0.00275080	H
8,2	–0.513429	0.198861	0.287319	0.00088828		D
7,3	–0.512528	0.200030	0.291084	0.00091693		D
6,4	–0.513156	0.199206	0.290356	0.00091693	0.00272215	D
9,2	–0.510922	0.228210	0.347912	0.00091693		D
8,3	–0.513175	0.228923	0.348233	0.00088828		D
7,4	–0.513597	0.229092	0.349503	0.00091693		D
6,5	–0.513103	0.230133	0.352774	0.00091693	0.00363908	D
T,2	–0.511354	–0.271893	–0.616921	0.00366774		H
9,3	–0.510668	–0.276866	–0.626128	0.00091693		H
8,4	–0.514244	–0.277171	–0.627842	0.00088828		H
7,5	–0.513545	–0.277274	–0.626732	0.00091693	0.00638989	H
T,3	–0.511101	–0.324221	–0.692749	0.00366774		H
9,4	–0.511737	–0.325404	–0.694880	0.00091693		H
8,5	–0.514192	–0.330999	–0.707067	0.00088828		H
7,6	–0.514046	–0.331884	–0.708749	0.00091693	0.00638989	H
T,4	–0.512159	–0.370188	–0.763060	0.00366774		H
9,5	–0.511675	–0.371645	–0.765635	0.00091693		H
8,6	–0.514672	–0.371589	–0.767233	0.00088828	0.00547295	H
T,5	–0.512109	–0.416875	–0.841779	0.00366774		H
9,6	–0.512179	–0.417255	–0.843216	0.00091693		H
8,7	–0.515118	–0.412415	–0.834019	0.00088828	0.00547295	H
T,6	–0.512612	–0.454468	–0.908936	0.00366774		H
9,7	–0.512624	–0.454840	–0.909681	0.00091693	0.00458467	H
T,7	–0.383670	–0.502233	–1.004466	0.00366774		S
9,8	–0.386104	–0.502402	–1.004804	0.00088828	0.00455602	S
T,8	0.104526	–0.588014	–1.176028	0.00355312	0.00355312	S
T,9	0.591784	–0.711718	–1.423436	0.00366774	0.00366774	S
A,2	–0.510138	0.052400	–0.314185	0.00091693	0.00091693	H
A,3	–0.509879	0.015663	–0.307428	0.00091693	0.00091693	H
A,4	–0.510949	–0.028065	–0.314334	0.00091693	0.00091693	H
A,5	–0.510897	–0.068732	–0.316464	0.00091693	0.00091693	H
A,6	–0.382427	–0.072047	–0.252297	0.00091693	0.00091693	H
A,7	0.107517	0.040692	–0.028645	0.00091693	0.00091693	S
A,8	0.595375	0.152736	0.194301	0.00088828	0.00088828	S
A,9	0.790948	0.194779	0.279775	0.00091693	0.00091693	S
A,T	1.500000	0.228882	0.348517	0.00366774	0.00366774	S

NDAS

Hand	Stand	Hit	Double	Spl1	Spl2	Spl3	Freq.	BS
A,A	–0.509202	0.094712	–0.312243	0.357443	0.479420	0.499972	0.00044414	P
2,2	–0.511064	–0.157443	–1.022129	–0.206864	–0.213721	–0.214860	0.00044414	H
3,3	–0.510557	–0.218685	–1.001173	–0.262734	–0.268574	–0.269497	0.00044414	H
4,4	–0.512695	–0.059294	–0.451269	–0.321334	–0.358380	–0.364650	0.00044414	H
5,5	–0.512604	0.199141	0.291445	–0.384237	–0.466788	–0.480771	0.00044414	D
6,6	–0.513603	–0.277348	–0.626780	–0.432452	–0.453993	–0.457574	0.00044414	H
7,7	–0.514468	–0.376254	–0.777224	–0.422042	–0.428117	–0.429079	0.00044414	H
8,8	–0.515767	–0.454672	–0.909343	–0.117104	–0.071347	–0.063926	0.00041638	P
9,9	0.100941	–0.588133	–1.176265	0.194244	0.207469	0.209713	0.00044414	P
T,T	0.790770	–0.850451	–1.700902	0.390644	0.182881	0.057284	0.00727817	S

DAS

Hand	Stand	Hit	Double	Spl1	Spl2	Spl3	Freq.	BS
A,A	–0.509202	0.094712	–0.312243	0.357443	0.479420	0.499972	0.00044414	P
2,2	–0.511064	–0.157443	–1.022129	–0.174220	–0.176380	–0.176711	0.00044414	H
3,3	–0.510557	–0.218685	–1.001173	–0.229740	–0.230812	–0.230911	0.00044414	H
4,4	–0.512695	–0.059294	–0.451269	–0.287884	–0.320140	–0.325589	0.00044414	H
5,5	–0.512604	0.199141	0.291445	–0.351069	–0.442205	–0.457679	0.00044414	D
6,6	–0.513603	–0.277348	–0.626780	–0.398830	–0.415579	–0.418343	0.00044414	H
7,7	–0.514468	–0.376254	–0.777224	–0.389248	–0.390713	–0.390898	0.00044414	H
8,8	–0.515767	–0.454672	–0.909343	–0.085539	–0.035572	–0.027481	0.00041638	P
9,9	0.100941	–0.588133	–1.176265	0.212532	0.228291	0.230956	0.00044414	P
T,T	0.790770	–0.850451	–1.700902	0.390644	0.182881	0.057284	0.00727817	S

TABLE A74
8D S17: Dealer's Upcard 9

Hand	Stand	Hit	Double	Freq.	Cum. Freq.	BS
3,2	−0.541971	−0.266148	−1.083943	0.00091693	0.00091693	H
4,2	−0.541909	−0.294053	−1.065215	0.00091693	0.00091693	H
5,2	−0.541835	−0.285428	−0.953491	0.00091693		H
4,3	−0.542979	−0.287676	−0.955758	0.00091693	0.00183387	H
6,2	−0.542313	−0.209957	−0.715210	0.00091693		H
5,3	−0.542919	−0.211080	−0.716426	0.00091693	0.00183387	H
7,2	−0.542938	−0.052171	−0.298024	0.00091693		H
6,3	−0.543383	−0.052069	−0.299392	0.00091693		H
5,4	−0.542856	−0.052188	−0.298309	0.00091693	0.00275080	H
8,2	−0.540433	0.116784	0.147584	0.00091693		D
7,3	−0.544008	0.116649	0.145410	0.00091693		D
6,4	−0.543320	0.116689	0.146896	0.00091693	0.00275080	D
9,2	−0.540867	0.156445	0.226041	0.00088828		D
8,3	−0.541503	0.156077	0.226257	0.00091693		D
7,4	−0.543945	0.157219	0.227339	0.00091693		D
6,5	−0.543245	0.157633	0.229476	0.00091693	0.00363908	D
T,2	−0.541272	−0.340528	−0.738511	0.00366774		H
9,3	−0.541939	−0.346151	−0.749591	0.00088828		H
8,4	−0.541441	−0.346546	−0.748847	0.00091693		H
7,5	−0.543872	−0.345808	−0.747967	0.00091693	0.00638989	H
T,3	−0.542332	−0.383833	−0.801206	0.00366774		H
9,4	−0.541866	−0.384540	−0.801975	0.00088828		H
8,5	−0.541357	−0.390263	−0.812644	0.00091693		H
7,6	−0.544316	−0.390848	−0.815138	0.00091693	0.00638989	H
T,4	−0.542271	−0.428907	−0.873597	0.00366774		H
9,5	−0.541793	−0.430361	−0.876052	0.00088828		H
8,6	−0.541825	−0.431593	−0.879124	0.00091693	0.00547295	H
T,5	−0.542198	−0.472034	−0.948859	0.00366774		H
9,6	−0.542261	−0.472500	−0.950473	0.00088828		H
8,7	−0.542453	−0.468185	−0.941820	0.00091693	0.00547295	H
T,6	−0.542666	−0.505749	−1.011498	0.00366774		RH
9,7	−0.542889	−0.506083	−1.012165	0.00088828	0.00455602	RH
T,7	−0.422418	−0.550486	−1.100971	0.00366774		S
9,8	−0.421755	−0.551050	−1.102101	0.00088828	0.00455602	S
T,8	−0.184682	−0.613763	−1.227526	0.00366774	0.00366774	S
T,9	0.284831	−0.713471	−1.426942	0.00355312	0.00355312	S
A,2	−0.540131	−0.035002	−0.446509	0.00091693	0.00091693	H
A,3	−0.541202	−0.073354	−0.448880	0.00091693	0.00091693	H
A,4	−0.541140	−0.112360	−0.452323	0.00091693	0.00091693	H
A,5	−0.541066	−0.150716	−0.455924	0.00091693	0.00091693	H
A,6	−0.421226	−0.147898	−0.394384	0.00091693	0.00091693	H
A,7	−0.182777	−0.099039	−0.286193	0.00091693	0.00091693	H
A,8	0.287611	0.007557	−0.071618	0.00091693	0.00091693	S
A,9	0.759204	0.114032	0.139970	0.00088828	0.00088828	S
A,T	1.500000	0.157075	0.226367	0.00366774	0.00366774	S

Hand	Stand	Hit	Double	Spl1	Spl2	Spl3	Freq.	BS
				NDAS				
A,A	−0.539362	−0.000231	−0.452269	0.235248	0.344351	0.362747	0.00044414	P
2,2	−0.540900	−0.238579	−1.081800	−0.380491	−0.400313	−0.403628	0.00044414	H
3,3	−0.543042	−0.294599	−1.067483	−0.429197	−0.447901	−0.451012	0.00044414	H
4,4	−0.542930	−0.209943	−0.716400	−0.482650	−0.521065	−0.527544	0.00044414	H
5,5	−0.542781	0.116614	0.147967	−0.541578	−0.634631	−0.650379	0.00044414	D
6,6	−0.543711	−0.345349	−0.746796	−0.583784	−0.617117	−0.622695	0.00044414	H
7,7	−0.544944	−0.436155	−0.890006	−0.570423	−0.588896	−0.591938	0.00044414	H
8,8	−0.539962	−0.506614	−1.013229	−0.420930	−0.409207	−0.407288	0.00044414	P
9,9	−0.184712	−0.613927	−1.227853	−0.105307	−0.094484	−0.092719	0.00041638	P
T,T	0.756649	−0.849789	−1.699578	0.226878	−0.047829	−0.213690	0.00727817	S
				DAS				
A,A	−0.539362	−0.000231	−0.452269	0.235248	0.344351	0.362747	0.00044414	P
2,2	−0.540900	−0.238579	−1.081800	−0.364477	−0.381941	−0.384842	0.00044414	H
3,3	−0.543042	−0.294599	−1.067483	−0.413425	−0.429854	−0.432572	0.00044414	H
4,4	−0.542930	−0.209943	−0.716400	−0.466580	−0.502667	−0.508741	0.00044414	H
5,5	−0.542781	0.116614	0.147967	−0.525261	−0.620640	−0.636815	0.00044414	D
6,6	−0.543711	−0.345349	−0.746796	−0.567582	−0.598586	−0.603764	0.00044414	H
7,7	−0.544944	−0.436155	−0.890006	−0.555093	−0.571411	−0.574089	0.00044414	H
8,8	−0.539962	−0.506614	−1.013229	−0.405099	−0.391128	−0.388826	0.00044414	P
9,9	−0.184712	−0.613927	−1.227853	−0.094770	−0.082547	−0.080560	0.00041638	P
T,T	0.756649	−0.849789	−1.699578	0.226878	−0.047829	−0.213690	0.00727817	S

TABLE A75
8D S17: Dealer's Upcard T

Hand	Stand	Hit	Double	Freq.	Cum. Freq.	BS
3,2	–0.541232	–0.312946	–1.082464	0.00338355	0.00338355	H
4,2	–0.541167	–0.338817	–1.063588	0.00338355	0.00338355	H
5,2	–0.541097	–0.319426	–0.951270	0.00338355		H
4,3	–0.541113	–0.320928	–0.951288	0.00338355	0.00676711	H
6,2	–0.541773	–0.249862	–0.746666	0.00338355		H
5,3	–0.541027	–0.249559	–0.745464	0.00338355	0.00676711	H
7,2	–0.539058	–0.153076	–0.462997	0.00338355		H
6,3	–0.541703	–0.151815	–0.464339	0.00338355		H
5,4	–0.540962	–0.151202	–0.463160	0.00338355	0.01015066	H
8,2	–0.539528	0.025729	–0.006939	0.00338355		H
7,3	–0.538988	0.025760	–0.005956	0.00338355		H
6,4	–0.541638	0.026160	–0.006376	0.00338355	0.01015066	H
9,2	–0.539967	0.117646	0.175378	0.00338355		D
8,3	–0.539459	0.118247	0.177617	0.00338355		D
7,4	–0.538924	0.118277	0.178705	0.00338355		D
6,5	–0.541554	0.118798	0.178769	0.00338355	0.01353421	D
T,2	–0.540362	–0.377045	–0.790095	0.01342848		H
9,3	–0.539886	–0.382030	–0.799797	0.00338355		H
8,4	–0.539384	–0.381674	–0.797952	0.00338355		H
7,5	–0.538828	–0.381497	–0.796214	0.00338355	0.02357914	H
T,3	–0.540294	–0.421561	–0.860941	0.01342848		H
9,4	–0.539823	–0.422146	–0.861723	0.00338355		H
8,5	–0.539300	–0.427791	–0.872516	0.00338355		H
7,6	–0.539507	–0.428388	–0.872557	0.00338355	0.02357914	H
T,4	–0.540231	–0.463919	–0.933596	0.01342848		EH
9,5	–0.539740	–0.464734	–0.934945	0.00338355		EH
8,6	–0.539980	–0.466403	–0.938109	0.00338355	0.02019559	EH
T,5	–0.540147	–0.504049	–1.008098	0.01342848		RH
9,6	–0.540419	–0.505395	–1.010790	0.00338355		RH
8,7	–0.537279	–0.500949	–1.001898	0.00338355	0.02019559	RH
T,6	–0.540827	–0.535975	–1.071950	0.01342848		RH
9,7	–0.537719	–0.536508	–1.073016	0.00338355	0.01681203	RH
T,7	–0.418772	–0.581173	–1.162347	0.01342848		S
9,8	–0.416212	–0.581330	–1.162660	0.00338355	0.01681203	S
T,8	–0.175488	–0.644734	–1.289468	0.01342848	0.01342848	S
T,9	0.067854	–0.727160	–1.454321	0.01342848	0.01342848	S
A,2	–0.540250	–0.102987	–0.509947	0.00339243	0.00339243	H
A,3	–0.540181	–0.137733	–0.510062	0.00339243	0.00339243	H
A,4	–0.540118	–0.173350	–0.511903	0.00339243	0.00339243	H
A,5	–0.540033	–0.209467	–0.516723	0.00339243	0.00339243	H
A,6	–0.419604	–0.196275	–0.455533	0.00339243	0.00339243	H
A,7	–0.179245	–0.143158	–0.343788	0.00339243	0.00339243	H
A,8	0.063271	–0.087760	–0.234060	0.00339243	0.00339243	S
A,9	0.554551	0.023067	–0.013719	0.00339243	0.00339243	S
A,T	1.500000	0.117644	0.174830	0.01346372	0.01346372	S

Hand	Stand	Hit	Double	Spl1	Spl2	Spl3	Freq.	BS
				NDAS				
A,A	–0.539205	–0.067269	–0.508503	0.181408	0.283046	0.300206	0.00164751	P
2,2	–0.541301	–0.287740	–1.082602	–0.481146	–0.508295	–0.512857	0.00163891	H
3,3	–0.541162	–0.338652	–1.063576	–0.527886	–0.554323	–0.558744	0.00163891	H
4,4	–0.541048	–0.248356	–0.745474	–0.577374	–0.623654	–0.631447	0.00163891	H
5,5	–0.540876	0.026671	–0.005214	–0.631734	–0.724670	–0.740373	0.00163891	H
6,6	–0.542207	–0.382074	–0.797936	–0.677781	–0.719336	–0.726327	0.00163891	H
7,7	–0.536807	–0.472124	–0.949744	–0.637698	–0.660505	–0.664265	0.00163891	EH
8,8	–0.537752	–0.536485	–1.072970	–0.494356	–0.488317	–0.487312	0.00163891	EP
9,9	–0.172895	–0.644913	–1.289826	–0.301125	–0.319251	–0.322319	0.00163891	S
T,T	0.557987	–0.847618	–1.695235	0.050680	–0.210638	–0.367458	0.02643732	S
				DAS				
A,A	–0.539205	–0.067269	–0.508503	0.181408	0.283046	0.300206	0.00164751	P
2,2	–0.541301	–0.287740	–1.082602	–0.472045	–0.497894	–0.502234	0.00163891	H
3,3	–0.541162	–0.338652	–1.063576	–0.518528	–0.543629	–0.547822	0.00163891	H
4,4	–0.541048	–0.248356	–0.745474	–0.567838	–0.612755	–0.620316	0.00163891	H
5,5	–0.540876	0.026671	–0.005214	–0.622217	–0.713785	–0.729253	0.00163891	H
6,6	–0.542207	–0.382074	–0.797936	–0.668338	–0.708546	–0.715306	0.00163891	H
7,7	–0.536807	–0.472124	–0.949744	–0.628061	–0.649476	–0.652996	0.00163891	EH
8,8	–0.537752	–0.536485	–1.072970	–0.485033	–0.477667	–0.476435	0.00163891	EP
9,9	–0.172895	–0.644913	–1.289826	–0.292309	–0.309214	–0.312079	0.00163891	S
T,T	0.557987	–0.847618	–1.695235	0.050680	–0.210638	–0.367458	0.02643732	S

TABLE A76
8D H17: Dealer's Upcard A

Hand	Stand	Hit	Double	Freq.	Cum. Freq.	BS
3,2	–0.597078	–0.317444	–1.194156	0.00063275	0.00063275	EH
4,2	–0.596812	–0.343401	–1.181226	0.00063275	0.00063275	EH
5,2	–0.596909	–0.351174	–1.086719	0.00063275		EH
4,3	–0.596558	–0.352320	–1.085971	0.00063275	0.00126550	EH
6,2	–0.596451	–0.265287	–0.832138	0.00063275		H
5,3	–0.596647	–0.265274	–0.831345	0.00063275	0.00126550	H
7,2	–0.595762	–0.126160	–0.455415	0.00063275		H
6,3	–0.596189	–0.124522	–0.454319	0.00063275		H
5,4	–0.596369	–0.124202	–0.453320	0.00063275	0.00189825	H
8,2	–0.596566	0.033623	–0.031676	0.00063275		H
7,3	–0.595501	0.033630	–0.029936	0.00063275		H
6,4	–0.595912	0.034174	–0.028405	0.00063275	0.00189825	H
9,2	–0.597355	0.105943	0.112521	0.00063275		D
8,3	–0.596295	0.106247	0.114904	0.00063275		D
7,4	–0.595214	0.106619	0.117516	0.00063275		D
6,5	–0.595987	0.107162	0.118807	0.00063275	0.00253100	D
T,2	–0.599046	–0.383552	–0.834866	0.00253989		EH
9,3	–0.597096	–0.386577	–0.840433	0.00063275		EH
8,4	–0.596019	–0.386497	–0.838535	0.00063275		EH
7,5	–0.595307	–0.386667	–0.837190	0.00063275	0.00443814	EH
T,3	–0.598788	–0.427288	–0.896293	0.00253989		EH
9,4	–0.596821	–0.425953	–0.893167	0.00063275		EH
8,5	–0.596113	–0.431836	–0.904024	0.00063275		EH
7,6	–0.594859	–0.432808	–0.904992	0.00063275	0.00443814	EH
T,4	–0.598514	–0.468881	–0.959600	0.00253989		EH
9,5	–0.596915	–0.468698	–0.958835	0.00063275		EH
8,6	–0.595665	–0.469880	–0.961330	0.00063275	0.00380539	EH
T,5	–0.598606	–0.508878	–1.025780	0.00253989		RH
9,6	–0.596463	–0.508208	–1.024957	0.00063275		RH
8,7	–0.594991	–0.504033	–1.016322	0.00063275	0.00380539	RH
T,6	–0.598166	–0.540543	–1.081086	0.00253989		RH
9,7	–0.595791	–0.539524	–1.079048	0.00063275	0.00317264	RH
T,7	–0.514228	–0.579048	–1.158097	0.00253989		RS
9,8	–0.512689	–0.577681	–1.155363	0.00063275	0.00317264	RS
T,8	–0.223968	–0.639664	–1.279327	0.00253989	0.00253989	S
T,9	0.190962	–0.732351	–1.464702	0.00253989	0.00253989	S
A,2	–0.597589	–0.100066	–0.585775	0.00061298	0.00061298	H
A,3	–0.597329	–0.134921	–0.585359	0.00061298	0.00061298	H
A,4	–0.597054	–0.171245	–0.587824	0.00061298	0.00061298	H
A,5	–0.597155	–0.207640	–0.592228	0.00061298	0.00061298	H
A,6	–0.514597	–0.220993	–0.549410	0.00061298	0.00061298	H
A,7	–0.225455	–0.159852	–0.417915	0.00061298	0.00061298	H
A,8	0.189122	–0.064133	–0.227994	0.00061298	0.00061298	S
A,9	0.604528	0.031089	–0.039613	0.00061298	0.00061298	S
A,T	1.500000	0.103858	0.107271	0.00246051	0.00246051	S

Hand	Stand	Hit	Double	Spl1	Spl2	Spl3	Freq.	BS
				NDAS				
A,A	–0.597839	–0.064439	–0.586446	0.116549	0.213773	0.229603	0.00028733	P
2,2	–0.597338	–0.292566	–1.194675	–0.491475	–0.519315	–0.523980	0.00030649	EH
3,3	–0.596828	–0.343405	–1.181256	–0.537771	–0.564784	–0.569278	0.00030649	EH
4,4	–0.596279	–0.263777	–0.830517	–0.586614	–0.631960	–0.639586	0.00030649	H
5,5	–0.596461	0.034059	–0.028259	–0.641913	–0.737247	–0.753343	0.00030649	H
6,6	–0.595536	–0.386488	–0.836824	–0.681968	–0.723208	–0.730099	0.00030649	EH
7,7	–0.594187	–0.475418	–0.972494	–0.697639	–0.728102	–0.733099	0.00030649	EH
8,8	–0.595796	–0.539453	–1.078906	–0.521618	–0.518789	–0.518260	0.00030649	RP
9,9	–0.221073	–0.638594	–1.277188	–0.241888	–0.244434	–0.244800	0.00030649	S
T,T	0.600677	–0.857214	–1.714428	0.057411	–0.225393	–0.396752	0.00505771	S
				DAS				
A,A	–0.597839	–0.064439	–0.586446	0.116549	0.213773	0.229603	0.00028733	P
2,2	–0.597338	–0.292566	–1.194675	–0.490138	–0.517747	–0.522366	0.00030649	EH
3,3	–0.596828	–0.343405	–1.181256	–0.536066	–0.562790	–0.567228	0.00030649	EH
4,4	–0.596279	–0.263777	–0.830517	–0.584478	–0.629464	–0.637019	0.00030649	H
5,5	–0.596461	0.034059	–0.028259	–0.639629	–0.734578	–0.750599	0.00030649	H
6,6	–0.595536	–0.386488	–0.836824	–0.679716	–0.720578	–0.727395	0.00030649	EH
7,7	–0.594187	–0.475418	–0.972494	–0.695588	–0.725714	–0.730646	0.00030649	EH
8,8	–0.595796	–0.539453	–1.078906	–0.520178	–0.517134	–0.516566	0.00030649	RP
9,9	–0.221073	–0.638594	–1.277188	–0.241049	–0.243502	–0.243856	0.00030649	S
T,T	0.600677	–0.857214	–1.714428	0.057411	–0.225393	–0.396752	0.00505771	S

TABLE A77
8D H17: Dealer's Upcard 2

Hand	Stand	Hit	Double	Freq.	Cum. Freq.	BS
3,2	–0.286200	–0.125977	–0.572401	0.00088828	0.00088828	H
4,2	–0.286114	–0.138923	–0.552109	0.00088828	0.00088828	H
5,2	–0.284232	–0.108335	–0.427609	0.00088828		H
4,3	–0.286043	–0.110560	–0.430104	0.00091693	0.00180521	H
6,2	–0.284604	–0.023277	–0.201488	0.00088828		H
5,3	–0.284161	–0.023497	–0.201636	0.00091693	0.00180521	H
7,2	–0.284984	0.073498	0.065587	0.00088828		H
6,3	–0.284532	0.074560	0.065231	0.00091693		H
5,4	–0.284073	0.074907	0.065735	0.00091693	0.00272215	H
8,2	–0.285334	0.184433	0.364431	0.00088828		D
7,3	–0.284901	0.184826	0.365041	0.00091693		D
6,4	–0.284433	0.185304	0.365961	0.00091693	0.00272215	D
9,2	–0.285842	0.240483	0.476300	0.00088828		D
8,3	–0.285262	0.241095	0.477494	0.00091693		D
7,4	–0.284813	0.241774	0.478818	0.00091693		D
6,5	–0.282557	0.243091	0.481665	0.00091693	0.00363908	D
T,2	–0.288712	–0.252610	–0.505220	0.00355312		H
9,3	–0.285771	–0.255298	–0.510596	0.00091693		H
8,4	–0.285175	–0.254479	–0.508957	0.00091693		H
7,5	–0.282938	–0.253645	–0.507290	0.00091693	0.00630392	H
T,3	–0.288641	–0.308136	–0.616271	0.00366774		S
9,4	–0.285684	–0.306761	–0.613521	0.00091693		S
8,5	–0.283301	–0.311423	–0.622846	0.00091693		S
7,6	–0.283311	–0.311517	–0.623034	0.00091693	0.00641854	S
T,4	–0.288555	–0.364138	–0.728277	0.00366774		S
9,5	–0.283809	–0.363110	–0.726219	0.00091693		S
8,6	–0.283672	–0.363607	–0.727214	0.00091693	0.00550161	S
T,5	–0.286693	–0.420689	–0.841379	0.00366774		S
9,6	–0.284196	–0.420028	–0.840056	0.00091693		S
8,7	–0.284067	–0.415350	–0.830700	0.00091693	0.00550161	S
T,6	–0.287065	–0.472569	–0.945139	0.00366774		S
9,7	–0.284577	–0.471493	–0.942986	0.00091693	0.00458467	S
T,7	–0.156996	–0.538402	–1.076805	0.00366774		S
9,8	–0.154450	–0.537493	–1.074985	0.00091693	0.00458467	S
T,8	0.109835	–0.624780	–1.249560	0.00366774	0.00366774	S
T,9	0.377972	–0.732024	–1.464047	0.00366774	0.00366774	S
A,2	–0.285456	0.045865	–0.064702	0.00088828	0.00088828	H
A,3	–0.285385	0.022282	–0.065038	0.00091693	0.00091693	H
A,4	–0.285310	–0.000502	–0.067778	0.00091693	0.00091693	H
A,5	–0.283432	–0.021034	–0.069590	0.00091693	0.00091693	H
A,6	–0.153662	–0.000625	–0.005768	0.00091693	0.00091693	H
A,7	0.112397	0.060298	0.115866	0.00091693	0.00091693	D
A,8	0.380082	0.121257	0.237818	0.00091693	0.00091693	S
A,9	0.636989	0.181885	0.359249	0.00091693	0.00091693	S
A,T	1.500000	0.238638	0.471942	0.00366774	0.00366774	S

Hand	Stand	Hit	Double	Spl1	Spl2	Spl3	Freq.	BS
				NDAS				
A,A	–0.284645	0.081466	–0.062173	0.481540	0.589324	0.607476	0.00044414	P
2,2	–0.286259	–0.112626	–0.572519	–0.149313	–0.154268	–0.155068	0.00041638	H
3,3	–0.286129	–0.138949	–0.552136	–0.196884	–0.204877	–0.206198	0.00044414	H
4,4	–0.285956	–0.022792	–0.203115	–0.225436	–0.253931	–0.258729	0.00044414	H
5,5	–0.282186	0.185996	0.367548	–0.248368	–0.309245	–0.319459	0.00044414	D
6,6	–0.282929	–0.253650	–0.507300	–0.268976	–0.270568	–0.270742	0.00044414	H
7,7	–0.283691	–0.368785	–0.737570	–0.212550	–0.202292	–0.200522	0.00044414	P
8,8	–0.284429	–0.471348	–0.942696	–0.042234	–0.007951	–0.002142	0.00044414	P
9,9	0.112141	–0.624180	–1.248359	0.147182	0.152124	0.152959	0.00044414	P
T,T	0.633452	–0.854339	–1.708677	0.355538	0.209971	0.121278	0.00727817	S
				DAS				
A,A	–0.284645	0.081466	–0.062173	0.481540	0.589324	0.607476	0.00044414	P
2,2	–0.286259	–0.112626	–0.572519	–0.084512	–0.080607	–0.079959	0.00041638	P
3,3	–0.286129	–0.138949	–0.552136	–0.131961	–0.130784	–0.130554	0.00044414	P
4,4	–0.285956	–0.022792	–0.203115	–0.160304	–0.179594	–0.182834	0.00044414	H
5,5	–0.282186	0.185996	0.367548	–0.184361	–0.262010	–0.275089	0.00044414	D
6,6	–0.282929	–0.253650	–0.507300	–0.203379	–0.195669	–0.194264	0.00044414	P
7,7	–0.283691	–0.368785	–0.737570	–0.146490	–0.126833	–0.123463	0.00044414	P
8,8	–0.284429	–0.471348	–0.942696	0.021776	0.065100	0.072437	0.00044414	P
9,9	0.112141	–0.624180	–1.248359	0.182620	0.192546	0.194220	0.00044414	P
T,T	0.633452	–0.854339	–1.708677	0.355538	0.209971	0.121278	0.00727817	S

TABLE A78
8D H17: Dealer's Upcard 3

Hand	Stand	Hit	Double	Freq.	Cum. Freq.	BS
3,2	−0.245816	−0.093181	−0.491632	0.00088828	0.00088828	H
4,2	−0.243908	−0.104369	−0.468510	0.00091693	0.00091693	H
5,2	−0.243610	−0.074848	−0.349880	0.00091693		H
4,3	−0.243823	−0.076947	−0.351216	0.00088828	0.00180521	H
6,2	−0.243978	0.007675	−0.130851	0.00091693		H
5,3	−0.243525	0.007457	−0.130965	0.00088828	0.00180521	H
7,2	−0.244337	0.102836	0.127551	0.00091693		D
6,3	−0.243882	0.103892	0.127204	0.00088828		D
5,4	−0.241611	0.104241	0.128239	0.00091693	0.00272215	D
8,2	−0.244844	0.208922	0.416261	0.00091693		D
7,3	−0.244253	0.209576	0.417492	0.00088828		D
6,4	−0.241979	0.210091	0.418702	0.00091693	0.00272215	D
9,2	−0.247778	0.260884	0.520083	0.00091693		D
8,3	−0.244760	0.263825	0.526052	0.00088828		D
7,4	−0.242350	0.264965	0.528507	0.00091693		D
6,5	−0.241681	0.266014	0.530603	0.00091693	0.00363908	D
T,2	−0.248340	−0.232264	−0.464529	0.00366774		H
9,3	−0.247695	−0.236468	−0.472935	0.00088828		H
8,4	−0.242858	−0.233497	−0.466994	0.00091693		H
7,5	−0.242053	−0.232647	−0.465294	0.00091693	0.00638989	H
T,3	−0.248256	−0.290886	−0.581772	0.00355312		S
9,4	−0.245792	−0.290506	−0.581013	0.00091693		S
8,5	−0.242560	−0.293672	−0.587345	0.00091693		S
7,6	−0.242421	−0.293573	−0.587146	0.00091693	0.00630392	S
T,4	−0.246367	−0.350586	−0.701172	0.00366774		S
9,5	−0.245509	−0.350606	−0.701212	0.00091693		S
8,6	−0.242943	−0.349725	−0.699451	0.00091693	0.00550161	S
T,5	−0.246069	−0.410526	−0.821052	0.00366774		S
9,6	−0.245878	−0.410802	−0.821603	0.00091693		S
8,7	−0.243314	−0.404927	−0.809853	0.00091693	0.00550161	S
T,6	−0.246438	−0.465458	−0.930915	0.00366774		S
9,7	−0.246249	−0.465145	−0.930290	0.00091693	0.00458467	S
T,7	−0.120413	−0.534023	−1.068045	0.00366774		S
9,8	−0.120372	−0.533856	−1.067711	0.00091693	0.00458467	S
T,8	0.137888	−0.622641	−1.245282	0.00366774	0.00366774	S
T,9	0.394786	−0.726628	−1.453256	0.00366774	0.00366774	S
A,2	−0.245149	0.073858	0.000384	0.00091693	0.00091693	H
A,3	−0.245077	0.050574	−0.001503	0.00088828	0.00088828	H
A,4	−0.243173	0.029292	−0.002766	0.00091693	0.00091693	H
A,5	−0.242875	0.008799	−0.005661	0.00091693	0.00091693	H
A,6	−0.117176	0.028886	0.056347	0.00091693	0.00091693	D
A,7	0.140675	0.088051	0.174678	0.00091693	0.00091693	D
A,8	0.399077	0.150105	0.298662	0.00091693	0.00091693	S
A,9	0.645407	0.203764	0.405843	0.00091693	0.00091693	S
A,T	1.500000	0.260935	0.519541	0.00366774	0.00366774	S

Hand	Stand	Hit	Double	Spl1	Spl2	Spl3	Freq.	BS
				NDAS				
A,A	−0.244405	0.103750	0.003080	0.528868	0.637630	0.655944	0.00044414	P
2,2	−0.245899	−0.080271	−0.491799	−0.095887	−0.098060	−0.098422	0.00044414	H
3,3	−0.245732	−0.105607	−0.471977	−0.132845	−0.136421	−0.136983	0.00041638	H
4,4	−0.241915	0.008401	−0.130794	−0.156285	−0.179225	−0.183051	0.00044414	H
5,5	−0.241313	0.210620	0.419754	−0.181289	−0.236156	−0.245352	0.00044414	D
6,6	−0.242051	−0.232674	−0.465349	−0.200799	−0.195715	−0.194759	0.00044414	P
7,7	−0.242807	−0.354696	−0.709392	−0.145700	−0.131743	−0.129343	0.00044414	P
8,8	−0.243822	−0.464173	−0.928346	0.018606	0.055731	0.062018	0.00044414	P
9,9	0.135585	−0.623082	−1.246163	0.195505	0.203645	0.204967	0.00044414	P
T,T	0.644190	−0.854062	−1.708124	0.403762	0.277625	0.200658	0.00727817	S
				DAS				
A,A	−0.244405	0.103750	0.003080	0.528868	0.637630	0.655944	0.00044414	P
2,2	−0.245899	−0.080271	−0.491799	−0.019153	−0.010463	−0.008984	0.00044414	P
3,3	−0.245732	−0.105607	−0.471977	−0.055935	−0.048995	−0.047837	0.00041638	P
4,4	−0.241915	0.008401	−0.130794	−0.078560	−0.090467	−0.092418	0.00044414	H
5,5	−0.241313	0.210620	0.419754	−0.105100	−0.178893	−0.191314	0.00044414	D
6,6	−0.242051	−0.232674	−0.465349	−0.118212	−0.101321	−0.098345	0.00044414	P
7,7	−0.242807	−0.354696	−0.709392	−0.063001	−0.037215	−0.032790	0.00044414	P
8,8	−0.243822	−0.464173	−0.928346	0.090592	0.137898	0.145909	0.00044414	P
9,9	0.135585	−0.623082	−1.246163	0.235396	0.249103	0.251356	0.00044414	P
T,T	0.644190	−0.854062	−1.708124	0.403762	0.277625	0.200658	0.00727817	S

TABLE A79
8D H17: Dealer's Upcard 4

Hand	Stand	Hit	Double	Freq.	Cum. Freq.	BS
3,2	–0.201546	–0.056246	–0.403093	0.00091693	0.00091693	H
4,2	–0.201215	–0.067786	–0.383499	0.00088828	0.00088828	H
5,2	–0.200953	–0.038781	–0.268623	0.00091693		H
4,3	–0.199309	–0.039525	–0.266631	0.00088828	0.00180521	H
6,2	–0.201298	0.042057	–0.057600	0.00091693		H
5,3	–0.199041	0.042717	–0.055481	0.00091693	0.00183387	H
7,2	–0.201818	0.132236	0.190685	0.00091693		D
6,3	–0.199397	0.134353	0.192741	0.00091693		D
5,4	–0.198708	0.134693	0.193429	0.00088828	0.00272215	D
8,2	–0.204749	0.232415	0.464830	0.00091693		D
7,3	–0.199918	0.236172	0.472344	0.00091693		D
6,4	–0.199065	0.236916	0.473832	0.00088828	0.00272215	D
9,2	–0.205361	0.284846	0.569693	0.00091693		D
8,3	–0.202850	0.286419	0.572839	0.00091693		D
7,4	–0.199586	0.289790	0.579580	0.00088828		D
6,5	–0.198803	0.290820	0.581640	0.00091693	0.00363908	D
T,2	–0.205889	–0.211461	–0.422923	0.00366774		S
9,3	–0.203461	–0.215228	–0.430457	0.00091693		S
8,4	–0.202518	–0.213931	–0.427861	0.00088828		S
7,5	–0.199324	–0.211388	–0.422775	0.00091693	0.00638989	S
T,3	–0.204003	–0.272906	–0.545812	0.00366774		S
9,4	–0.203143	–0.272555	–0.545111	0.00088828		S
8,5	–0.202270	–0.277039	–0.554078	0.00091693		S
7,6	–0.199695	–0.275388	–0.550776	0.00091693	0.00638989	S
T,4	–0.203670	–0.336170	–0.672340	0.00355312		S
9,5	–0.202881	–0.336164	–0.672329	0.00091693		S
8,6	–0.202627	–0.336396	–0.672792	0.00091693	0.00538699	S
T,5	–0.203408	–0.399814	–0.799627	0.00366774		S
9,6	–0.203238	–0.400046	–0.800092	0.00091693		S
8,7	–0.203148	–0.395297	–0.790594	0.00091693	0.00550161	S
T,6	–0.203765	–0.458008	–0.916016	0.00366774		S
9,7	–0.203760	–0.457833	–0.915665	0.00091693	0.00458467	S
T,7	–0.081488	–0.529657	–1.059314	0.00366774		S
9,8	–0.083846	–0.530273	–1.060546	0.00091693	0.00458467	S
T,8	0.164845	–0.615975	–1.231949	0.00366774	0.00366774	S
T,9	0.413885	–0.725547	–1.451093	0.00366774	0.00366774	S
A,2	–0.202783	0.104016	0.068239	0.00091693	0.00091693	H
A,3	–0.200882	0.082320	0.067545	0.00091693	0.00091693	H
A,4	–0.200550	0.060679	0.064202	0.00088828	0.00088828	D
A,5	–0.200288	0.041090	0.061395	0.00091693	0.00091693	D
A,6	–0.078144	0.060917	0.121835	0.00091693	0.00091693	D
A,7	0.169828	0.120881	0.241762	0.00091693	0.00091693	D
A,8	0.415771	0.175525	0.351051	0.00091693	0.00091693	S
A,9	0.656117	0.229480	0.458960	0.00091693	0.00091693	S
A,T	1.500000	0.284544	0.569088	0.00366774	0.00366774	S

Hand	Stand	Hit	Double	Spl1	Spl2	Spl3	Freq.	BS
				NDAS				
A,A	–0.202098	0.128661	0.070502	0.578111	0.687827	0.706298	0.00044414	P
2,2	–0.203452	–0.045028	–0.406904	–0.035526	–0.034169	–0.033937	0.00044414	P
3,3	–0.199640	–0.066794	–0.380712	–0.058222	–0.056749	–0.056457	0.00044414	P
4,4	–0.198971	0.043972	–0.055286	–0.083653	–0.100797	–0.103551	0.00041638	H
5,5	–0.198447	0.237153	0.474306	–0.107913	–0.156129	–0.164195	0.00044414	D
6,6	–0.199159	–0.211201	–0.422402	–0.126569	–0.115927	–0.114064	0.00044414	P
7,7	–0.200216	–0.340184	–0.680368	–0.073541	–0.055373	–0.052256	0.00044414	P
8,8	–0.206080	–0.458635	–0.917270	0.077718	0.117579	0.124282	0.00044414	P
9,9	0.164860	–0.615969	–1.231937	0.254283	0.266599	0.268630	0.00044414	P
T,T	0.654940	–0.853812	–1.707624	0.454079	0.348250	0.283428	0.00727817	S
				DAS				
A,A	–0.202098	0.128661	0.070502	0.578111	0.687827	0.706298	0.00044414	P
2,2	–0.203452	–0.045028	–0.406904	0.054263	0.068313	0.070693	0.00044414	P
3,3	–0.199640	–0.066794	–0.380712	0.032503	0.046817	0.049284	0.00044414	P
4,4	–0.198971	0.043972	–0.055286	0.008718	0.004281	0.003617	0.00041638	H
5,5	–0.198447	0.237153	0.474306	–0.015205	–0.083984	–0.095553	0.00044414	D
6,6	–0.199159	–0.211201	–0.422402	–0.025706	–0.000656	0.003671	0.00044414	P
7,7	–0.200216	–0.340184	–0.680368	0.028420	0.061138	0.066740	0.00044414	P
8,8	–0.206080	–0.458635	–0.917270	0.157899	0.208996	0.217583	0.00044414	P
9,9	0.164860	–0.615969	–1.231937	0.298133	0.316570	0.319624	0.00044414	P
T,T	0.654940	–0.853812	–1.707624	0.454079	0.348250	0.283428	0.00727817	S

TABLE A80

8D H17: Dealer's Upcard 5

Hand	Stand	Hit	Double	Freq.	Cum. Freq.	BS
3,2	−0.157469	−0.017467	−0.314937	0.00091693	0.00091693	H
4,2	−0.157175	−0.028476	−0.296263	0.00091693	0.00091693	H
5,2	−0.156911	0.000229	−0.186052	0.00088828		H
4,3	−0.156850	−0.000668	−0.185034	0.00091693	0.00180521	H
6,2	−0.157428	0.076942	0.017342	0.00091693		H
5,3	−0.156593	0.078071	0.019971	0.00088828	0.00180521	H
7,2	−0.160373	0.161897	0.252802	0.00091693		D
6,3	−0.157110	0.166315	0.259345	0.00091693		D
5,4	−0.156297	0.167540	0.261824	0.00088828	0.00272215	D
8,2	−0.160988	0.260620	0.521239	0.00091693		D
7,3	−0.160055	0.261967	0.523935	0.00091693		D
6,4	−0.156815	0.265409	0.530817	0.00091693	0.00275080	D
9,2	−0.161530	0.310991	0.621982	0.00091693		D
8,3	−0.160670	0.312262	0.624524	0.00091693		D
7,4	−0.159759	0.313657	0.627314	0.00091693		D
6,5	−0.156557	0.317168	0.634336	0.00088828	0.00363908	D
T,2	−0.162025	−0.189934	−0.379867	0.00366774		S
9,3	−0.161225	−0.193886	−0.387772	0.00091693		S
8,4	−0.160389	−0.192649	−0.385297	0.00091693		S
7,5	−0.159515	−0.191597	−0.383193	0.00088828	0.00638989	S
T,3	−0.161706	−0.255084	−0.510168	0.00366774		S
9,4	−0.160930	−0.254748	−0.509495	0.00091693		S
8,5	−0.160130	−0.259226	−0.518453	0.00088828		S
7,6	−0.160033	−0.259068	−0.518137	0.00091693	0.00638989	S
T,4	−0.161411	−0.321139	−0.642279	0.00366774		S
9,5	−0.160672	−0.321109	−0.642219	0.00088828		S
8,6	−0.160649	−0.321492	−0.642983	0.00091693	0.00547295	S
T,5	−0.161153	−0.387915	−0.775829	0.00355312		S
9,6	−0.161190	−0.388279	−0.776557	0.00091693		S
8,7	−0.163593	−0.384688	−0.769375	0.00091693	0.00538699	S
T,6	−0.161672	−0.449904	−0.899807	0.00366774		S
9,7	−0.164136	−0.450710	−0.901420	0.00091693	0.00458467	S
T,7	−0.046071	−0.520177	−1.040354	0.00366774		S
9,8	−0.046183	−0.520181	−1.040362	0.00091693	0.00458467	S
T,8	0.195388	−0.612793	−1.225586	0.00366774	0.00366774	S
T,9	0.437176	−0.724148	−1.448296	0.00366774	0.00366774	S
A,2	−0.159015	0.136612	0.137558	0.00091693	0.00091693	D
A,3	−0.158697	0.115190	0.136507	0.00091693	0.00091693	D
A,4	−0.158403	0.094321	0.132980	0.00091693	0.00091693	D
A,5	−0.158146	0.074998	0.129483	0.00088828	0.00088828	D
A,6	−0.040456	0.096732	0.193463	0.00091693	0.00091693	D
A,7	0.197978	0.149995	0.299990	0.00091693	0.00091693	D
A,8	0.438984	0.205232	0.410465	0.00091693	0.00091693	S
A,9	0.669897	0.257245	0.514491	0.00091693	0.00091693	S
A,T	1.500000	0.310105	0.620210	0.00366774	0.00366774	S

Hand	Stand	Hit	Double	Spl1	Spl2	Spl3	Freq.	BS
				NDAS				
A,A	−0.160232	0.159763	0.138175	0.628892	0.739843	0.758518	0.00044414	P
2,2	−0.157786	−0.005663	−0.315573	0.041028	0.047745	0.048901	0.00044414	P
3,3	−0.157151	−0.028484	−0.296039	0.015817	0.022349	0.023498	0.00044414	P
4,4	−0.156555	0.079564	0.020767	−0.008624	−0.020809	−0.022825	0.00044414	H
5,5	−0.156040	0.266294	0.532588	−0.032294	−0.072657	−0.079185	0.00041638	D
6,6	−0.157089	−0.189878	−0.379756	−0.051016	−0.035664	−0.033007	0.00044414	P
7,7	−0.162978	−0.327749	−0.655499	−0.007495	0.014394	0.018082	0.00044414	P
8,8	−0.164209	−0.450726	−0.901453	0.145899	0.189436	0.196754	0.00044414	P
9,9	0.195376	−0.612786	−1.225571	0.317223	0.334085	0.336878	0.00044414	P
T,T	0.668790	−0.853473	−1.706946	0.509838	0.425745	0.374050	0.00727817	S
				DAS				
A,A	−0.160232	0.159763	0.138175	0.628892	0.739843	0.758518	0.00044414	P
2,2	−0.157786	−0.005663	−0.315573	0.145037	0.166529	0.170197	0.00044414	P
3,3	−0.157151	−0.028484	−0.296039	0.123736	0.145594	0.149349	0.00044414	P
4,4	−0.156555	0.079564	0.020767	0.102608	0.106234	0.106909	0.00044414	P
5,5	−0.156040	0.266294	0.532588	0.078863	0.017085	0.007022	0.00041638	D
6,6	−0.157089	−0.189878	−0.379756	0.069018	0.101491	0.107070	0.00044414	P
7,7	−0.162978	−0.327749	−0.655499	0.111865	0.150558	0.157078	0.00044414	P
8,8	−0.164209	−0.450726	−0.901453	0.234438	0.290382	0.299781	0.00044414	P
9,9	0.195376	−0.612786	−1.225571	0.365127	0.388680	0.392593	0.00044414	P
T,T	0.668790	−0.853473	−1.706946	0.509838	0.425745	0.374050	0.00727817	S

TABLE A81
8D H17: Dealer's Upcard 6

Hand	Stand	Hit	Double	Freq.	Cum. Freq.	BS
3,2	–0.116691	0.018226	–0.233383	0.00091693	0.00091693	H
4,2	–0.116414	0.008594	–0.215234	0.00091693	0.00091693	H
5,2	–0.116339	0.035424	–0.108590	0.00091693		H
4,3	–0.116121	0.034461	–0.107934	0.00091693	0.00183387	H
6,2	–0.119286	0.106038	0.080627	0.00088828		H
5,3	–0.116046	0.109541	0.087946	0.00091693	0.00180521	H
7,2	–0.119903	0.189030	0.309711	0.00091693		D
6,3	–0.118993	0.191438	0.312261	0.00088828		D
5,4	–0.115774	0.195062	0.319528	0.00091693	0.00272215	D
8,2	–0.120443	0.284853	0.569707	0.00091693		D
7,3	–0.119610	0.286189	0.572377	0.00091693		D
6,4	–0.118720	0.287722	0.575443	0.00088828	0.00272215	D
9,2	–0.120958	0.333887	0.667774	0.00091693		D
8,3	–0.120164	0.335134	0.670268	0.00091693		D
7,4	–0.119352	0.336525	0.673051	0.00091693		D
6,5	–0.118660	0.338026	0.676052	0.00088828	0.00363908	D
T,2	–0.121374	–0.171034	–0.342068	0.00366774		S
9,3	–0.120665	–0.175023	–0.350046	0.00091693		S
8,4	–0.119892	–0.173804	–0.347608	0.00091693		S
7,5	–0.119277	–0.172996	–0.345993	0.00091693	0.00641854	S
T,3	–0.121081	–0.239285	–0.478570	0.00366774		S
9,4	–0.120393	–0.238963	–0.477927	0.00091693		S
8,5	–0.119817	–0.243618	–0.487236	0.00091693		S
7,6	–0.122223	–0.245002	–0.490004	0.00088828	0.00638989	S
T,4	–0.120809	–0.308390	–0.616781	0.00366774		S
9,5	–0.120318	–0.308532	–0.617064	0.00091693		S
8,6	–0.122764	–0.310248	–0.620495	0.00088828	0.00547295	S
T,5	–0.120734	–0.377945	–0.755890	0.00366774		S
9,6	–0.123265	–0.379473	–0.758946	0.00088828		S
8,7	–0.123382	–0.374905	–0.749811	0.00091693	0.00547295	S
T,6	–0.123661	–0.438855	–0.877711	0.00355312		S
9,7	–0.123863	–0.438881	–0.877762	0.00091693	0.00447005	S
T,7	–0.008904	–0.516262	–1.032524	0.00366774		S
9,8	–0.009010	–0.516267	–1.032535	0.00091693	0.00458467	S
T,8	0.221064	–0.610899	–1.221797	0.00366774	0.00366774	S
T,9	0.451179	–0.723323	–1.446646	0.00366774	0.00366774	S
A,2	–0.116815	0.166144	0.202168	0.00091693	0.00091693	D
A,3	–0.116523	0.145750	0.201233	0.00091693	0.00091693	D
A,4	–0.116252	0.126128	0.198770	0.00091693	0.00091693	D
A,5	–0.116167	0.110070	0.200517	0.00091693	0.00091693	D
A,6	–0.005284	0.125698	0.251397	0.00088828	0.00088828	D
A,7	0.222828	0.178233	0.356466	0.00091693	0.00091693	D
A,8	0.452450	0.230918	0.461836	0.00091693	0.00091693	D
A,9	0.677749	0.281742	0.563484	0.00091693	0.00091693	S
A,T	1.500000	0.333231	0.666462	0.00366774	0.00366774	S

Hand	Stand	Hit	Double	Spl1	Spl2	Spl3	Freq.	BS
				NDAS				
A,A	–0.116641	0.188937	0.203740	0.675893	0.787223	0.805945	0.00044414	P
2,2	–0.116983	0.029218	–0.233966	0.108833	0.120218	0.122166	0.00044414	P
3,3	–0.116393	0.008593	–0.215191	0.084605	0.095630	0.097542	0.00044414	P
4,4	–0.115849	0.110966	0.088623	0.061107	0.054339	0.053241	0.00044414	H
5,5	–0.115700	0.290927	0.581855	0.038124	0.002818	–0.003085	0.00044414	D
6,6	–0.121606	–0.174496	–0.348991	0.012205	0.030515	0.033512	0.00041638	P
7,7	–0.122841	–0.315381	–0.630763	0.060889	0.086717	0.091064	0.00044414	P
8,8	–0.123902	–0.438879	–0.877758	0.207225	0.253702	0.261512	0.00044414	P
9,9	0.221002	–0.610903	–1.221807	0.370866	0.391655	0.395107	0.00044414	P
T,T	0.677004	–0.853263	–1.706525	0.558109	0.494813	0.455688	0.00727817	S
				DAS				
A,A	–0.116641	0.188937	0.203740	0.675893	0.787223	0.805945	0.00044414	P
2,2	–0.116983	0.029218	–0.233966	0.230395	0.259047	0.263931	0.00044414	P
3,3	–0.116393	0.008593	–0.215191	0.209085	0.237797	0.242717	0.00044414	P
4,4	–0.115849	0.110966	0.088623	0.188662	0.200043	0.202034	0.00044414	P
5,5	–0.115700	0.290927	0.581855	0.166122	0.107562	0.097687	0.00044414	D
6,6	–0.121606	–0.174496	–0.348991	0.147834	0.184646	0.190663	0.00041638	P
7,7	–0.122841	–0.315381	–0.630763	0.196982	0.241954	0.249525	0.00044414	P
8,8	–0.123902	–0.438879	–0.877758	0.304522	0.364638	0.374737	0.00044414	P
9,9	0.221002	–0.610903	–1.221807	0.422321	0.450299	0.454955	0.00044414	P
T,T	0.677004	–0.853263	–1.706525	0.558109	0.494813	0.455688	0.00727817	S

Appendix B

Basic Strategy Charts for the 1-, 2-, and Multi-Deck Games

Charts depicting allegedly correct basic strategy for various numbers of decks and rules variations are standard fare in virtually all books on blackjack. In addition, there are dozens of Web sites purporting to furnish completely accurate BS tables, and some are actually quite impressive in their thoroughness. Finally, commercially sold, laminated BS cards, that fit conveniently inside a purse or shirt pocket, are commonly found in the gift shops of most major casinos. And yet, careful perusal of these tables almost always leads to finding some sort of error, either of commission or omission, in the supposedly accurate charts.

The first two editions of *BJA* did not feature BS tables. It was felt that there were sufficient other sources for the reader to gather such information, and I deemed it unnecessary to rehash the "obvious." Well, I've changed my mind! After collaborating with Cacarulo, I came to understand and appreciate the rarity of a set of perfectly reliable BS charts. And, in so doing, I decided it was time that such a collection grace the pages of *BJA*. Based on the e.v. tables that you have just read (or at least glanced at), the BS charts that follow are, quite frankly, the most painstakingly accurate depiction of basic strategy (according to the definition that we furnished in Appendix A) that you are likely to find anywhere — be it another blackjack book, a Web site, or a commercially sold product. Rather boldly, I'll stick my neck out and say that, should another source disagree with a BS recommendation provided in our charts, that other source is in error. It's as simple as that. I hope that you will find the following information useful and that even the savviest veteran player may discover a play or two that was hitherto unknown to him or her. I know I did!

How to Read the Charts

The charts you are about to see are unique in the annals of blackjack BS charts. Frankly, I have never seen anything quite like them. We owe Cacarulo a tremendous debt of gratitude not only for conceiving their design, but for producing the prototype, as well. Here's how they work. Down the left-hand side, there are two columns, a darkly shaded "generic," total-dependent one, and a more lightly shaded composition-dependent column to the right. Use the C-D column to locate your original two-card holding versus the dealer's upcard, across the top. Find the correct play in the box under the dealer's upcard. Sometimes, that box has only a single entry. In those instances, the recommended play is identical for both T-D and C-D strategy. Other times, the cell is split: The right-hand play pertains to the C-D, original hand, while the left-hand play is for the generic, T-D total. One important reminder: Use the C-D plays, where, often, the two-card holding is broken down into its component parts, for your *original hand only*. Use the "generic" total and corresponding play (such as "8," "12," or "16") for any multi-card holding, *or for any hand that is the result of a pair-split*. Finally, if you do not see your specific holding listed in the "C-D" section, then simply follow the T-D play for the total of the hand in question.

Here is an example: Suppose we're playing a two-deck, s17, das game (Table B6), and I receive 8,3 for my first two cards. Dealer shows an ace. Across from the "8,3" entry, in the right-hand column of the hard 11 ("H11") section, on the right side of the box under "A," the "H" advises me to hit, while the "Dh" entry in the left-hand part of the box tells me that, in general, for an overall total of hard 11 (obtained, perhaps, via a pair-split, or if I am not distinguishing the cards that comprise the total), the proper play is to double, if permitted. So, later, suppose I split a pair of 8s v. the dealer's A, and I draw a 3 to my first 8. I no longer consult the C-D 8,3 line; instead, I consult the generic, hard 11 entry, and the correct play is to double down, if permitted. Again, the concept is to use the C-D entries *only* for the first two cards you receive.

Let's take another example. For pair-splitting, there are no divided boxes. I follow the C-D advice for the original hand. But, suppose resplitting is not permitted. Here, I simply ascertain the total of the cards in the resulting hand and consult the generic, hard-hand play, above. Thus, with 6,6 v. 2 in the 1D, s17, ndas game (Table B1), suppose I receive another 6, after splitting. If I am not permitted to resplit, I simply add the two sixes, obtaining 12, and go to the hard 12 ("H12") row, above, versus the dealer's 2, where I find that the correct play is to hit.

With these caveats in mind, enjoy the most accurate BS charts you've ever seen!

TABLE B1
1-Deck S17 NDAS

T-D	C-D	2	3	4	5	6	7	8	9	T	A
H4	H4	H	H	H	H	H	H	H	H	H	H
H5	H5	H	H	H	H	H	H	H	H	H	H
H6	H6	H	H	H	H	H	H	H	H	H	H
H7	H7	H	H	H	H	H	H	H	H	H	H
H8	6,2	H	H	H	Dh / H	Dh / H	H	H	H	H	H
H9	H9	Dh	Dh	Dh	Dh	Dh	H	H	H	H	H
H10	H10	Dh	Dh	Dh	Dh	Dh	Dh	Dh	Dh	H	H
H11	H11	Dh	Dh	Dh	Dh	Dh	Dh	Dh	Dh	Dh	Dh
H12	T,2	H	H[2]	S / H	S	S / H	H	H	H	H	H
	8,4	H	H[2] / S	S	S	S	H	H	H	H	H
	7,5	H	H[2] / S	S	S	S	H	H	H	H	H
H13	T,3	S / H	S	S	S	S	H	H	H	H	H
H14	H14	S	S	S	S	S	H	H	H	H	H
H15	T,5	S	S	S	S	S	H	H	H	H / Rh	H
	9,6	S	S	S	S	S	H	H	H	H / Rh	H
H16	9,7	S	S	S	S	S	H	H	H	Rh[1]	Rh / H
H17	H17	S	S	S	S	S	S	S	S	S	S
H18	H18	S	S	S	S	S	S	S	S	S	S
H19	H19	S	S	S	S	S	S	S	S	S	S
H20	H20	S	S	S	S	S	S	S	S	S	S
H21	H21	S	S	S	S	S	S	S	S	S	S
S13	A,2	H	H	Dh	Dh	Dh	H	H	H	H	H
S14	A,3	H	H	Dh	Dh	Dh	H	H	H	H	H
S15	A,4	H	H	Dh	Dh	Dh	H	H	H	H	H
S16	A,5	H	H	Dh	Dh	Dh	H	H	H	H	H
S17	A,6	Dh	Dh	Dh	Dh	Dh	H	H	H	H	H
S18	A,7	S	Ds	Ds	Ds	Ds	S	S	H	H	S
S19	A,8	S	S	S	S	Ds	S	S	S	S	S
S20	A,9	S	S	S	S	S	S	S	S	S	S
S21	A,T	S	S	S	S	S	S	S	S	S	S
PAIRS	A,A	P	P	P	P	P	P	P	P	P	P
	2,2	H	P	P	P	P	P	H	H	H	H
	3,3	H	H	P	P	P	P	H	H	H	H
	4,4	H	H	H	Dh	Dh	H	H	H	H	H
	5,5	Dh	Dh	Dh	Dh	Dh	Dh	Dh	Dh	H	H
	6,6	P	P	P	P	P	H	H	H	H	H
	7,7	P	P	P	P	P	P	H	H	Rs	H
	8,8	P	P	P	P	P	P	P	P	P	P
	9,9	P	P	P	P	P	S	P	P	S	S
	T,T	S	S	S	S	S	S	S	S	S	S

H = Hit S = Stand R = Late Surrender P = Split D = Double | Use lowercase play if uppercase play not permitted.

[1] Stand if 16 is multi-card or the result of a pair-split. [2] Stand if 12 is multi-card, original hand only.

TABLE B2
1-Deck S17 DAS

T-D	C-D	2	3	4	5	6	7	8	9	T	A
H4	H4	H	H	H	H	H	H	H	H	H	H
H5	H5	H	H	H	H	H	H	H	H	H	H
H6	H6	H	H	H	H	H	H	H	H	H	H
H7	H7	H	H	H	H	H	H	H	H	H	H
H8	6,2	H	H	H	Dh | H	Dh | H	H	H	H	H	H
H9	H9	Dh	Dh	Dh	Dh	Dh	H	H	H	H	H
H10	H10	Dh	Dh	Dh	Dh	Dh	Dh	Dh	Dh	H	H
H11	H11	Dh	Dh	Dh	Dh	Dh	Dh	Dh	Dh	Dh	Dh
H12	T,2	H	H[2]	S | H	S	S | H	H	H	H	H	H
	8,4	H	H[2] | S	S	S	S	H	H	H	H	H
	7,5	H	H[2] | S	S	S	S	H	H	H	H	H
H13	T,3	S | H	S	S	S	S	H	H	H	H	H
H14	H14	S	S	S	S	S	H	H	H	H	H
H15	T,5	S	S	S	S	S	H	H	H	H | Rh	H
	9,6	S	S	S	S	S	H	H	H	H | Rh	H
H16	9,7	S	S	S	S	S	H	H	H	Rh[1]	Rh | H
H17	H17	S	S	S	S	S	S	S	S	S	S
H18	H18	S	S	S	S	S	S	S	S	S	S
H19	H19	S	S	S	S	S	S	S	S	S	S
H20	H20	S	S	S	S	S	S	S	S	S	S
H21	H21	S	S	S	S	S	S	S	S	S	S
S13	A,2	H	H	Dh	Dh	Dh	H	H	H	H	H
S14	A,3	H	H	Dh	Dh	Dh	H	H	H	H	H
S15	A,4	H	H	Dh	Dh	Dh	H	H	H	H	H
S16	A,5	H	H	Dh	Dh	Dh	H	H	H	H	H
S17	A,6	Dh	Dh	Dh	Dh	Dh	H	H	H	H	H
S18	A,7	S	Ds	Ds	Ds	Ds	S	S	H	H	S
S19	A,8	S	S	S	S	Ds	S	S	S	S	S
S20	A,9	S	S	S	S	S	S	S	S	S	S
S21	A,T	S	S	S	S	S	S	S	S	S	S
PAIRS	A,A	P	P	P	P	P	P	P	P	P	P
	2,2	P	P	P	P	P	P	H	H	H	H
	3,3	P	P	P	P	P	P	P	H	H	H
	4,4	H	H	P	P	P	H	H	H	H	H
	5,5	Dh	Dh	Dh	Dh	Dh	Dh	Dh	Dh	H	H
	6,6	P	P	P	P	P	P	H	H	H	H
	7,7	P	P	P	P	P	P	P	H	Rs	H
	8,8	P	P	P	P	P	P	P	P	P	P
	9,9	P	P	P	P	P	S	P	P	S	S
	T,T	S	S	S	S	S	S	S	S	S	S

H = Hit S = Stand R = Late Surrender P = Split D = Double | Use lowercase play if uppercase play not permitted.

[1] Stand if 16 is multi-card or the result of a pair-split. [2] Stand if 12 is multi-card, original hand only.

TABLE B3
1-Deck H17 NDAS

T-D	C-D	2	3	4	5	6	7	8	9	T	A
H4	H4	H	H	H	H	H	H	H	H	H	H
H5	H5	H	H	H	H	H	H	H	H	H	H
H6	H6	H	H	H	H	H	H	H	H	H	H
H7	H7	H	H	H	H	H	H	H	H	H	H
H8	6,2	H	H	H	Dh	Dh | H	H	H	H	H	H
H9	H9	Dh	Dh	Dh	Dh	Dh	H	H	H	H	H
H10	H10	Dh	Dh	Dh	Dh	Dh	Dh	Dh	Dh	H	H
H11	H11	Dh	Dh	Dh	Dh	Dh	Dh	Dh	Dh	Dh	Dh
H12	T,2	H	H[2]	S | H	S	S	H	H	H	H	H
	8,4	H	H[2] | S	S	S	S	H	H	H	H	H
	7,5	H	H[2] | S	S	S	S	H	H	H	H	H
H13	H13	S	S	S	S	S	H	H	H	H	H
H14	H14	S	S	S	S	S	H	H	H	H	H
H15	T,5	S	S	S	S	S	H	H	H	H | Rh	Rh
	9,6	S	S	S	S	S	H	H	H	H | Rh	Rh
	8,7	S	S	S	S	S	H	H	H	H	Rh | H
H16	H16	S	S	S	S	S	H	H	H	Rh[1]	Rh
H17	9,8	S	S	S	S	S	S	S	S	S	Rs | S
H18	H18	S	S	S	S	S	S	S	S	S	S
H19	H19	S	S	S	S	S	S	S	S	S	S
H20	H20	S	S	S	S	S	S	S	S	S	S
H21	H21	S	S	S	S	S	S	S	S	S	S
S13	A,2	H	H	Dh	Dh	Dh	H	H	H	H	H
S14	A,3	H	H	Dh	Dh	Dh	H	H	H	H	H
S15	A,4	H	H	Dh	Dh	Dh	H	H	H	H	H
S16	A,5	H	H	Dh	Dh	Dh	H	H	H	H	H
S17	A,6	Dh	Dh	Dh	Dh	Dh	H	H	H	H	H
S18	A,7	S	Ds	Ds	Ds	Ds	S	S	H	H	H
S19	A,8	S	S	S	S	Ds	S	S	S	S	S
S20	A,9	S	S	S	S	S	S	S	S	S	S
S21	A,T	S	S	S	S	S	S	S	S	S	S
PAIRS	A,A	P	P	P	P	P	P	P	P	P	P
	2,2	H	P	P	P	P	P	H	H	H	H
	3,3	H	H	P	P	P	P	H	H	H	H
	4,4	H	H	H	Dh	Dh	H	H	H	H	H
	5,5	Dh	Dh	Dh	Dh	Dh	Dh	Dh	Dh	H	H
	6,6	P	P	P	P	P	H	H	H	H	H
	7,7	P	P	P	P	P	P	H	H	Rs	Rh
	8,8	P	P	P	P	P	P	P	P	P	P
	9,9	P	P	P	P	P	S	P	P	S	S
	T,T	S	S	S	S	S	S	S	S	S	S

H = Hit S = Stand R = Late Surrender P = Split D = Double | Use lowercase play if uppercase play not permitted.

[1] Stand if 16 is multi-card or the result of a pair-split. [2] Stand if 12 is multi-card, original hand only.

TABLE B4
1-Deck H17 DAS

<table>
<tr><th>T-D</th><th>C-D</th><th>2</th><th>3</th><th>4</th><th>5</th><th>6</th><th>7</th><th>8</th><th>9</th><th>T</th><th>A</th></tr>
<tr><td>H4</td><td>H4</td><td>H</td><td>H</td><td>H</td><td>H</td><td>H</td><td>H</td><td>H</td><td>H</td><td>H</td><td>H</td></tr>
<tr><td>H5</td><td>H5</td><td>H</td><td>H</td><td>H</td><td>H</td><td>H</td><td>H</td><td>H</td><td>H</td><td>H</td><td>H</td></tr>
<tr><td>H6</td><td>H6</td><td>H</td><td>H</td><td>H</td><td>H</td><td>H</td><td>H</td><td>H</td><td>H</td><td>H</td><td>H</td></tr>
<tr><td>H7</td><td>H7</td><td>H</td><td>H</td><td>H</td><td>H</td><td>H</td><td>H</td><td>H</td><td>H</td><td>H</td><td>H</td></tr>
<tr><td>H8</td><td>6,2</td><td>H</td><td>H</td><td>H</td><td>Dh</td><td>Dh | H</td><td>H</td><td>H</td><td>H</td><td>H</td><td>H</td></tr>
<tr><td>H9</td><td>H9</td><td>Dh</td><td>Dh</td><td>Dh</td><td>Dh</td><td>Dh</td><td>H</td><td>H</td><td>H</td><td>H</td><td>H</td></tr>
<tr><td>H10</td><td>H10</td><td>Dh</td><td>Dh</td><td>Dh</td><td>Dh</td><td>Dh</td><td>Dh</td><td>Dh</td><td>Dh</td><td>H</td><td>H</td></tr>
<tr><td>H11</td><td>H11</td><td>Dh</td><td>Dh</td><td>Dh</td><td>Dh</td><td>Dh</td><td>Dh</td><td>Dh</td><td>Dh</td><td>Dh</td><td>Dh</td></tr>
<tr><td rowspan="3">H12</td><td>T,2</td><td>H</td><td>H[2]</td><td>S | H</td><td>S</td><td>S</td><td>H</td><td>H</td><td>H</td><td>H</td><td>H</td></tr>
<tr><td>8,4</td><td>H</td><td>H[2] | S</td><td>S</td><td>S</td><td>S</td><td>H</td><td>H</td><td>H</td><td>H</td><td>H</td></tr>
<tr><td>7,5</td><td>H</td><td>H[2] | S</td><td>S</td><td>S</td><td>S</td><td>H</td><td>H</td><td>H</td><td>H</td><td>H</td></tr>
<tr><td>H13</td><td>H13</td><td>S</td><td>S</td><td>S</td><td>S</td><td>S</td><td>H</td><td>H</td><td>H</td><td>H</td><td>H</td></tr>
<tr><td>H14</td><td>H14</td><td>S</td><td>S</td><td>S</td><td>S</td><td>S</td><td>H</td><td>H</td><td>H</td><td>H</td><td>H</td></tr>
<tr><td rowspan="3">H15</td><td>T,5</td><td>S</td><td>S</td><td>S</td><td>S</td><td>S</td><td>H</td><td>H</td><td>H</td><td>H | Rh</td><td>Rh</td></tr>
<tr><td>9,6</td><td>S</td><td>S</td><td>S</td><td>S</td><td>S</td><td>H</td><td>H</td><td>H</td><td>H | Rh</td><td>Rh</td></tr>
<tr><td>8,7</td><td>S</td><td>S</td><td>S</td><td>S</td><td>S</td><td>H</td><td>H</td><td>H</td><td>H</td><td>Rh | H</td></tr>
<tr><td>H16</td><td>H16</td><td>S</td><td>S</td><td>S</td><td>S</td><td>S</td><td>H</td><td>H</td><td>H</td><td>Rh[1]</td><td>Rh</td></tr>
<tr><td>H17</td><td>9,8</td><td>S</td><td>S</td><td>S</td><td>S</td><td>S</td><td>S</td><td>S</td><td>S</td><td>S</td><td>Rs | S</td></tr>
<tr><td>H18</td><td>H18</td><td>S</td><td>S</td><td>S</td><td>S</td><td>S</td><td>S</td><td>S</td><td>S</td><td>S</td><td>S</td></tr>
<tr><td>H19</td><td>H19</td><td>S</td><td>S</td><td>S</td><td>S</td><td>S</td><td>S</td><td>S</td><td>S</td><td>S</td><td>S</td></tr>
<tr><td>H20</td><td>H20</td><td>S</td><td>S</td><td>S</td><td>S</td><td>S</td><td>S</td><td>S</td><td>S</td><td>S</td><td>S</td></tr>
<tr><td>H21</td><td>H21</td><td>S</td><td>S</td><td>S</td><td>S</td><td>S</td><td>S</td><td>S</td><td>S</td><td>S</td><td>S</td></tr>
<tr><td>S13</td><td>A,2</td><td>H</td><td>H</td><td>Dh</td><td>Dh</td><td>Dh</td><td>H</td><td>H</td><td>H</td><td>H</td><td>H</td></tr>
<tr><td>S14</td><td>A,3</td><td>H</td><td>H</td><td>Dh</td><td>Dh</td><td>Dh</td><td>H</td><td>H</td><td>H</td><td>H</td><td>H</td></tr>
<tr><td>S15</td><td>A,4</td><td>H</td><td>H</td><td>Dh</td><td>Dh</td><td>Dh</td><td>H</td><td>H</td><td>H</td><td>H</td><td>H</td></tr>
<tr><td>S16</td><td>A,5</td><td>H</td><td>H</td><td>Dh</td><td>Dh</td><td>Dh</td><td>H</td><td>H</td><td>H</td><td>H</td><td>H</td></tr>
<tr><td>S17</td><td>A,6</td><td>Dh</td><td>Dh</td><td>Dh</td><td>Dh</td><td>Dh</td><td>H</td><td>H</td><td>H</td><td>H</td><td>H</td></tr>
<tr><td>S18</td><td>A,7</td><td>S</td><td>Ds</td><td>Ds</td><td>Ds</td><td>Ds</td><td>S</td><td>S</td><td>H</td><td>H</td><td>H</td></tr>
<tr><td>S19</td><td>A,8</td><td>S</td><td>S</td><td>S</td><td>S</td><td>Ds</td><td>S</td><td>S</td><td>S</td><td>S</td><td>S</td></tr>
<tr><td>S20</td><td>A,9</td><td>S</td><td>S</td><td>S</td><td>S</td><td>S</td><td>S</td><td>S</td><td>S</td><td>S</td><td>S</td></tr>
<tr><td>S21</td><td>A,T</td><td>S</td><td>S</td><td>S</td><td>S</td><td>S</td><td>S</td><td>S</td><td>S</td><td>S</td><td>S</td></tr>
<tr><td rowspan="10">PAIRS</td><td>A,A</td><td>P</td><td>P</td><td>P</td><td>P</td><td>P</td><td>P</td><td>P</td><td>P</td><td>P</td><td>P</td></tr>
<tr><td>2,2</td><td>P</td><td>P</td><td>P</td><td>P</td><td>P</td><td>P</td><td>H</td><td>H</td><td>H</td><td>H</td></tr>
<tr><td>3,3</td><td>P</td><td>P</td><td>P</td><td>P</td><td>P</td><td>P</td><td>P</td><td>H</td><td>H</td><td>H</td></tr>
<tr><td>4,4</td><td>H</td><td>H</td><td>P</td><td>P</td><td>P</td><td>H</td><td>H</td><td>H</td><td>H</td><td>H</td></tr>
<tr><td>5,5</td><td>Dh</td><td>Dh</td><td>Dh</td><td>Dh</td><td>Dh</td><td>Dh</td><td>Dh</td><td>Dh</td><td>H</td><td>H</td></tr>
<tr><td>6,6</td><td>P</td><td>P</td><td>P</td><td>P</td><td>P</td><td>P</td><td>H</td><td>H</td><td>H</td><td>H</td></tr>
<tr><td>7,7</td><td>P</td><td>P</td><td>P</td><td>P</td><td>P</td><td>P</td><td>P</td><td>H</td><td>Rs</td><td>Rh</td></tr>
<tr><td>8,8</td><td>P</td><td>P</td><td>P</td><td>P</td><td>P</td><td>P</td><td>P</td><td>P</td><td>P</td><td>P</td></tr>
<tr><td>9,9</td><td>P</td><td>P</td><td>P</td><td>P</td><td>P</td><td>S</td><td>P</td><td>P</td><td>S</td><td>P</td></tr>
<tr><td>T,T</td><td>S</td><td>S</td><td>S</td><td>S</td><td>S</td><td>S</td><td>S</td><td>S</td><td>S</td><td>S</td></tr>
</table>

H = Hit S = Stand R = Late Surrender P = Split D = Double | Use lowercase play if uppercase play not permitted.

[1] Stand if 16 is multi-card or the result of a pair-split. [2] Stand if 12 is multi-card, original hand only.

TABLE B5
2-Deck S17 NDAS

T-D	C-D	2	3	4	5	6	7	8	9	T	A
H4	H4	H	H	H	H	H	H	H	H	H	H
H5	H5	H	H	H	H	H	H	H	H	H	H
H6	H6	H	H	H	H	H	H	H	H	H	H
H7	H7	H	H	H	H	H	H	H	H	H	H
H8	H8	H	H	H	H	H	H	H	H	H	H
H9	H9	Dh	Dh	Dh	Dh	Dh	H	H	H	H	H
H10	H10	Dh	Dh	Dh	Dh	Dh	Dh	Dh	Dh	H	H
H11	9,2	Dh	Dh	Dh	Dh	Dh	Dh	Dh	Dh	Dh	Dh | H
	8,3	Dh	Dh	Dh	Dh	Dh	Dh	Dh	Dh	Dh	Dh | H
H12	T,2	H	H	S | H	S	S	H	H	H	H	H
H13	H13	S	S	S	S	S	H	H	H	H	H
H14	H14	S	S	S	S	S	H	H	H	H	H
H15	8,7	S	S	S	S	S	H	H	H	Rh | H	H
H16	H16	S	S	S	S	S	H	H	H	Rh[1]	Rh
H17	H17	S	S	S	S	S	S	S	S	S	S
H18	H18	S	S	S	S	S	S	S	S	S	S
H19	H19	S	S	S	S	S	S	S	S	S	S
H20	H20	S	S	S	S	S	S	S	S	S	S
H21	H21	S	S	S	S	S	S	S	S	S	S
S13	A,2	H	H	H	Dh	Dh	H	H	H	H	H
S14	A,3	H	H	H	Dh	Dh	H	H	H	H	H
S15	A,4	H	H	Dh	Dh	Dh	H	H	H	H	H
S16	A,5	H	H	Dh	Dh	Dh	H	H	H	H	H
S17	A,6	H	Dh	Dh	Dh	Dh	H	H	H	H	H
S18	A,7	S	Ds	Ds	Ds	Ds	S	S	H	H	S | H
S19	A,8	S	S	S	S	S	S	S	S	S	S
S20	A,9	S	S	S	S	S	S	S	S	S	S
S21	A,T	S	S	S	S	S	S	S	S	S	S
PAIRS	A,A	P	P	P	P	P	P	P	P	P	P
	2,2	H	H	P	P	P	P	H	H	H	H
	3,3	H	H	P	P	P	P	H	H	H	H
	4,4	H	H	H	H	H	H	H	H	H	H
	5,5	Dh	Dh	Dh	Dh	Dh	Dh	Dh	Dh	H	H
	6,6	P	P	P	P	P	H	H	H	H	H
	7,7	P	P	P	P	P	P	H	H	H	H
	8,8	P	P	P	P	P	P	P	P	P	P
	9,9	P	P	P	P	P	S	P	P	S	S
	T,T	S	S	S	S	S	S	S	S	S	S

H = Hit S = Stand R = Late Surrender P = Split D = Double | Use lowercase play if uppercase play not permitted.

[1] Stand if 16 is multi-card or the result of a pair-split.

TABLE B6
2-Deck S17 DAS

T-D	C-D	2	3	4	5	6	7	8	9	T	A
H4	H4	H	H	H	H	H	H	H	H	H	H
H5	H5	H	H	H	H	H	H	H	H	H	H
H6	H6	H	H	H	H	H	H	H	H	H	H
H7	H7	H	H	H	H	H	H	H	H	H	H
H8	H8	H	H	H	H	H	H	H	H	H	H
H9	H9	Dh	Dh	Dh	Dh	Dh	H	H	H	H	H
H10	H10	Dh	Dh	Dh	Dh	Dh	Dh	Dh	Dh	H	H
H11	9,2	Dh	Dh	Dh	Dh	Dh	Dh	Dh	Dh	Dh	Dh / H
	8,3	Dh	Dh	Dh	Dh	Dh	Dh	Dh	Dh	Dh	Dh / H
H12	T,2	H	H	S / H	S	S	H	H	H	H	H
H13	H13	S	S	S	S	S	H	H	H	H	H
H14	H14	S	S	S	S	S	H	H	H	H	H
H15	8,7	S	S	S	S	S	H	H	H	Rh / H	H
H16	H16	S	S	S	S	S	H	H	H	Rh[1]	Rh
H17	H17	S	S	S	S	S	S	S	S	S	S
H18	H18	S	S	S	S	S	S	S	S	S	S
H19	H19	S	S	S	S	S	S	S	S	S	S
H20	H20	S	S	S	S	S	S	S	S	S	S
H21	H21	S	S	S	S	S	S	S	S	S	S
S13	A,2	H	H	H	Dh	Dh	H	H	H	H	H
S14	A,3	H	H	H	Dh	Dh	H	H	H	H	H
S15	A,4	H	H	Dh	Dh	Dh	H	H	H	H	H
S16	A,5	H	H	Dh	Dh	Dh	H	H	H	H	H
S17	A,6	H	Dh	Dh	Dh	Dh	H	H	H	H	H
S18	A,7	S	Ds	Ds	Ds	Ds	S	S	H	H	S / H
S19	A,8	S	S	S	S	S	S	S	S	S	S
S20	A,9	S	S	S	S	S	S	S	S	S	S
S21	A,T	S	S	S	S	S	S	S	S	S	S
PAIRS	A,A	P	P	P	P	P	P	P	P	P	P
	2,2	P	P	P	P	P	P	H	H	H	H
	3,3	P	P	P	P	P	P	H	H	H	H
	4,4	H	H	H	P	P	H	H	H	H	H
	5,5	Dh	Dh	Dh	Dh	Dh	Dh	Dh	Dh	H	H
	6,6	P	P	P	P	P	P	H	H	H	H
	7,7	P	P	P	P	P	P	P	H	H	H
	8,8	P	P	P	P	P	P	P	P	P	P
	9,9	P	P	P	P	P	S	P	P	S	S
	T,T	S	S	S	S	S	S	S	S	S	S

H = Hit S = Stand R = Late Surrender P = Split D = Double | Use lowercase play if uppercase play not permitted.

[1] Stand if 16 is multi-card or the result of a pair-split.

TABLE B7
2-Deck H17 NDAS

T-D	C-D	2	3	4	5	6	7	8	9	T	A
H4	H4	H	H	H	H	H	H	H	H	H	H
H5	H5	H	H	H	H	H	H	H	H	H	H
H6	H6	H	H	H	H	H	H	H	H	H	H
H7	H7	H	H	H	H	H	H	H	H	H	H
H8	H8	H	H	H	H	H	H	H	H	H	H
H9	H9	Dh	Dh	Dh	Dh	Dh	H	H	H	H	H
H10	H10	Dh	Dh	Dh	Dh	Dh	Dh	Dh	Dh	H	H
H11	H11	Dh	Dh	Dh	Dh	Dh	Dh	Dh	Dh	Dh	Dh
H12	T,2	H	H	S | H	S	S	H	H	H	H	H
	8,4	H	H | S	S	S	S	H	H	H	H	H
	7,5	H	H | S	S	S	S	H	H	H	H	H
H13	H13	S	S	S	S	S	H	H	H	H	H
H14	H14	S	S	S	S	S	H	H	H	H	H
H15	8,7	S	S	S	S	S	H	H	H	Rh | H	Rh | H
H16	H16	S	S	S	S	S	H	H	H	Rh[1]	Rh
H17	H17	S	S	S	S	S	S	S	S	S	Rs
H18	H18	S	S	S	S	S	S	S	S	S	S
H19	H19	S	S	S	S	S	S	S	S	S	S
H20	H20	S	S	S	S	S	S	S	S	S	S
H21	H21	S	S	S	S	S	S	S	S	S	S
S13	A,2	H	H	H	Dh	Dh	H	H	H	H	H
S14	A,3	H	H	Dh	Dh	Dh	H	H	H	H	H
S15	A,4	H	H	Dh	Dh	Dh	H	H	H	H	H
S16	A,5	H	H	Dh	Dh	Dh	H	H	H	H	H
S17	A,6	H	Dh	Dh	Dh	Dh	H	H	H	H	H
S18	A,7	Ds	Ds	Ds	Ds	Ds	S	S	H	H	H
S19	A,8	S	S	S	S	Ds	S	S	S	S	S
S20	A,9	S	S	S	S	S	S	S	S	S	S
S21	A,T	S	S	S	S	S	S	S	S	S	S
PAIRS	A,A	P	P	P	P	P	P	P	P	P	P
	2,2	H	H	P	P	P	P	H	H	H	H
	3,3	H	H	P	P	P	P	H	H	H	H
	4,4	H	H	H	H	H	H	H	H	H	H
	5,5	Dh	Dh	Dh	Dh	Dh	Dh	Dh	Dh	H	H
	6,6	P	P	P	P	P	H	H	H	H	H
	7,7	P	P	P	P	P	P	H	H	H	H
	8,8	P	P	P	P	P	P	P	P	P	Rp
	9,9	P	P	P	P	P	S	P	P	S	S
	T,T	S	S	S	S	S	S	S	S	S	S

H = Hit S = Stand R = Late Surrender P = Split D = Double | Use lowercase play if uppercase play not permitted.

[1] Stand if 16 is multi-card or the result of a pair-split.

TABLE B8
2-Deck H17 DAS

T-D	C-D	2	3	4	5	6	7	8	9	T	A
H4	H4	H	H	H	H	H	H	H	H	H	H
H5	H5	H	H	H	H	H	H	H	H	H	H
H6	H6	H	H	H	H	H	H	H	H	H	H
H7	H7	H	H	H	H	H	H	H	H	H	H
H8	H8	H	H	H	H	H	H	H	H	H	H
H9	H9	Dh	Dh	Dh	Dh	Dh	H	H	H	H	H
H10	H10	Dh	Dh	Dh	Dh	Dh	Dh	Dh	Dh	H	H
H11	H11	Dh	Dh	Dh	Dh	Dh	Dh	Dh	Dh	Dh	Dh
H12	T,2	H	H	S | H	S	S	H	H	H	H	H
	8,4	H	H | S	S	S	S	H	H	H	H	H
	7,5	H	H | S	S	S	S	H	H	H	H	H
H13	H13	S	S	S	S	S	H	H	H	H	H
H14	H14	S	S	S	S	S	H	H	H	H	H
H15	8,7	S	S	S	S	S	H	H	H	Rh | H	Rh | H
H16	H16	S	S	S	S	S	H	H	H	Rh[1]	Rh
H17	H17	S	S	S	S	S	S	S	S	S	Rs
H18	H18	S	S	S	S	S	S	S	S	S	S
H19	H19	S	S	S	S	S	S	S	S	S	S
H20	H20	S	S	S	S	S	S	S	S	S	S
H21	H21	S	S	S	S	S	S	S	S	S	S
S13	A,2	H	H	H	Dh	Dh	H	H	H	H	H
S14	A,3	H	H	Dh	Dh	Dh	H	H	H	H	H
S15	A,4	H	H	Dh	Dh	Dh	H	H	H	H	H
S16	A,5	H	H	Dh	Dh	Dh	H	H	H	H	H
S17	A,6	H	Dh	Dh	Dh	Dh	H	H	H	H	H
S18	A,7	Ds	Ds	Ds	Ds	Ds	S	S	H	H	H
S19	A,8	S	S	S	S	Ds	S	S	S	S	S
S20	A,9	S	S	S	S	S	S	S	S	S	S
S21	A,T	S	S	S	S	S	S	S	S	S	S
PAIRS	A,A	P	P	P	P	P	P	P	P	P	P
	2,2	P	P	P	P	P	P	H	H	H	H
	3,3	P	P	P	P	P	P	H	H	H	H
	4,4	H	H	H	P	P	H	H	H	H	H
	5,5	Dh	Dh	Dh	Dh	Dh	Dh	Dh	Dh	H	H
	6,6	P	P	P	P	P	P	H	H	H	H
	7,7	P	P	P	P	P	P	P	H	H	H
	8,8	P	P	P	P	P	P	P	P	P	P
	9,9	P	P	P	P	P	S	P	P	S	S
	T,T	S	S	S	S	S	S	S	S	S	S

H = Hit S = Stand R = Late Surrender P = Split D = Double | Use lowercase play if uppercase play not permitted.

[1] Stand if 16 is multi-card or the result of a pair-split.

TABLE B9
Multi-Deck S17 NDAS

T-D	C-D	2	3	4	5	6	7	8	9	T	A
H4	H4	H	H	H	H	H	H	H	H	H	H
H5	H5	H	H	H	H	H	H	H	H	H	H
H6	H6	H	H	H	H	H	H	H	H	H	H
H7	H7	H	H	H	H	H	H	H	H	H	H
H8	H8	H	H	H	H	H	H	H	H	H	H
H9	H9	H	Dh	Dh	Dh	Dh	H	H	H	H	H
H10	H10	Dh	Dh	Dh	Dh	Dh	Dh	Dh	Dh	H	H
H11	H11	Dh	Dh	Dh	Dh	Dh	Dh	Dh	Dh	Dh	H
H12	T,2	H	H	S / H[2]	S	S	H	H	H	H	H
H13	H13	S	S	S	S	S	H	H	H	H	H
H14	H14	S	S	S	S	S	H	H	H	H	H
H15	8,7	S	S	S	S	S	H	H	H	Rh / H[3]	H
H16	H16	S	S	S	S	S	H	H	Rh	Rh[1]	Rh
H17	H17	S	S	S	S	S	S	S	S	S	S
H18	H18	S	S	S	S	S	S	S	S	S	S
H19	H19	S	S	S	S	S	S	S	S	S	S
H20	H20	S	S	S	S	S	S	S	S	S	S
H21	H21	S	S	S	S	S	S	S	S	S	S
S13	A,2	H	H	H	Dh	Dh	H	H	H	H	H
S14	A,3	H	H	H	Dh	Dh	H	H	H	H	H
S15	A,4	H	H	Dh	Dh	Dh	H	H	H	H	H
S16	A,5	H	H	Dh	Dh	Dh	H	H	H	H	H
S17	A,6	H	Dh	Dh	Dh	Dh	H	H	H	H	H
S18	A,7	S	Ds	Ds	Ds	Ds	S	S	H	H	H
S19	A,8	S	S	S	S	S	S	S	S	S	S
S20	A,9	S	S	S	S	S	S	S	S	S	S
S21	A,T	S	S	S	S	S	S	S	S	S	S
PAIRS	A,A	P	P	P	P	P	P	P	P	P	P
	2,2	H	H	P	P	P	P	H	H	H	H
	3,3	H	H	P	P	P	P	H	H	H	H
	4,4	H	H	H	H	H	H	H	H	H	H
	5,5	Dh	Dh	Dh	Dh	Dh	Dh	Dh	Dh	H	H
	6,6	H	P	P	P	P	H	H	H	H	H
	7,7	P	P	P	P	P	P	H	H	H	H
	8,8	P	P	P	P	P	P	P	P	P	P
	9,9	P	P	P	P	P	S	P	P	S	S
	T,T	S	S	S	S	S	S	S	S	S	S

H = Hit S = Stand R = Late Surrender P = Split D = Double | Use lowercase play if uppercase play not permitted.

[1] Stand if 16 is multi-card or the result of a pair-split. [2] Stand in 8-deck. [3] Surrender in 7 or more decks.

TABLE B10
Multi-Deck S17 DAS

T-D	C-D	2	3	4	5	6	7	8	9	T	A
H4	H4	H	H	H	H	H	H	H	H	H	H
H5	H5	H	H	H	H	H	H	H	H	H	H
H6	H6	H	H	H	H	H	H	H	H	H	H
H7	H7	H	H	H	H	H	H	H	H	H	H
H8	H8	H	H	H	H	H	H	H	H	H	H
H9	H9	H	Dh	Dh	Dh	Dh	H	H	H	H	H
H10	H10	Dh	Dh	Dh	Dh	Dh	Dh	Dh	Dh	H	H
H11	H11	Dh	Dh	Dh	Dh	Dh	Dh	Dh	Dh	Dh	H
H12	T,2	H	H	S | H[2]	S	S	H	H	H	H	H
H13	H13	S	S	S	S	S	H	H	H	H	H
H14	H14	S	S	S	S	S	H	H	H	H	H
H15	8,7	S	S	S	S	S	H	H	H	Rh | H[3]	H
H16	H16	S	S	S	S	S	H	H	Rh	Rh[1]	Rh
H17	H17	S	S	S	S	S	S	S	S	S	S
H18	H18	S	S	S	S	S	S	S	S	S	S
H19	H19	S	S	S	S	S	S	S	S	S	S
H20	H20	S	S	S	S	S	S	S	S	S	S
H21	H21	S	S	S	S	S	S	S	S	S	S
S13	A,2	H	H	H	Dh	Dh	H	H	H	H	H
S14	A,3	H	H	H	Dh	Dh	H	H	H	H	H
S15	A,4	H	H	Dh	Dh	Dh	H	H	H	H	H
S16	A,5	H	H	Dh	Dh	Dh	H	H	H	H	H
S17	A,6	H	Dh	Dh	Dh	Dh	H	H	H	H	H
S18	A,7	S	Ds	Ds	Ds	Ds	S	S	H	H	H
S19	A,8	S	S	S	S	S	S	S	S	S	S
S20	A,9	S	S	S	S	S	S	S	S	S	S
S21	A,T	S	S	S	S	S	S	S	S	S	S
PAIRS	A,A	P	P	P	P	P	P	P	P	P	P
	2,2	P	P	P	P	P	P	H	H	H	H
	3,3	P	P	P	P	P	P	H	H	H	H
	4,4	H	H	H	P	P	H	H	H	H	H
	5,5	Dh	Dh	Dh	Dh	Dh	Dh	Dh	Dh	H	H
	6,6	P	P	P	P	P	H	H	H	H	H
	7,7	P	P	P	P	P	P	H	H	H	H
	8,8	P	P	P	P	P	P	P	P	P	P
	9,9	P	P	P	P	P	S	P	P	S	S
	T,T	S	S	S	S	S	S	S	S	S	S

H = Hit S = Stand R = Late Surrender P = Split D = Double | Use lowercase play if uppercase play not permitted.

[1] Stand if 16 is multi-card or the result of a pair-split. [2] Stand in 8-deck. [3] Surrender in 7 or more decks.

TABLE B11
Multi-Deck H17 NDAS

T-D	C-D	2	3	4	5	6	7	8	9	T	A
H4	H4	H	H	H	H	H	H	H	H	H	H
H5	H5	H	H	H	H	H	H	H	H	H	H
H6	H6	H	H	H	H	H	H	H	H	H	H
H7	H7	H	H	H	H	H	H	H	H	H	H
H8	H8	H	H	H	H	H	H	H	H	H	H
H9	H9	H	Dh	Dh	Dh	Dh	H	H	H	H	H
H10	H10	Dh	Dh	Dh	Dh	Dh	Dh	Dh	Dh	H	H
H11	H11	Dh	Dh	Dh	Dh	Dh	Dh	Dh	Dh	Dh	Dh
H12	H12	H	H	S	S	S	H	H	H	H	H
H13	H13	S	S	S	S	S	H	H	H	H	H
H14	H14	S	S	S	S	S	H	H	H	H	H
H15	8,7	S	S	S	S	S	H	H	H	Rh | H[2]	Rh
H16	H16	S	S	S	S	S	H	H	Rh	Rh[1]	Rh
H17	H17	S	S	S	S	S	S	S	S	S	Rs
H18	H18	S	S	S	S	S	S	S	S	S	S
H19	H19	S	S	S	S	S	S	S	S	S	S
H20	H20	S	S	S	S	S	S	S	S	S	S
H21	H21	S	S	S	S	S	S	S	S	S	S
S13	A,2	H	H	H	Dh	Dh	H	H	H	H	H
S14	A,3	H	H	H	Dh	Dh	H	H	H	H	H
S15	A,4	H	H	Dh	Dh	Dh	H	H	H	H	H
S16	A,5	H	H	Dh	Dh	Dh	H	H	H	H	H
S17	A,6	H	Dh	Dh	Dh	Dh	H	H	H	H	H
S18	A,7	Ds	Ds	Ds	Ds	Ds	S	S	H	H	H
S19	A,8	S	S	S	S	Ds	S	S	S	S	S
S20	A,9	S	S	S	S	S	S	S	S	S	S
S21	A,T	S	S	S	S	S	S	S	S	S	S
PAIRS	A,A	P	P	P	P	P	P	P	P	P	P
	2,2	H	H	P	P	P	P	H	H	H	H
	3,3	H	H	P	P	P	P	H	H	H	H
	4,4	H	H	H	H	H	H	H	H	H	H
	5,5	Dh	Dh	Dh	Dh	Dh	Dh	Dh	Dh	H	H
	6,6	H	P	P	P	P	H	H	H	H	H
	7,7	P	P	P	P	P	P	H	H	H	H
	8,8	P	P	P	P	P	P	P	P	P	Rp
	9,9	P	P	P	P	P	S	P	P	S	S
	T,T	S	S	S	S	S	S	S	S	S	S

H = Hit S = Stand R = Late Surrender P = Split D = Double | Use lowercase play if uppercase play not permitted.

[1] Stand if 16 is multi-card or the result of a pair-split. [2] Surrender in 7 or more decks.

TABLE B12
Multi-Deck H17 DAS

T-D	C-D	2	3	4	5	6	7	8	9	T	A
H4	H4	H	H	H	H	H	H	H	H	H	H
H5	H5	H	H	H	H	H	H	H	H	H	H
H6	H6	H	H	H	H	H	H	H	H	H	H
H7	H7	H	H	H	H	H	H	H	H	H	H
H8	H8	H	H	H	H	H	H	H	H	H	H
H9	H9	H	Dh	Dh	Dh	Dh	H	H	H	H	H
H10	H10	Dh	Dh	Dh	Dh	Dh	Dh	Dh	Dh	H	H
H11	H11	Dh	Dh	Dh	Dh	Dh	Dh	Dh	Dh	Dh	Dh
H12	H12	H	H	S	S	S	H	H	H	H	H
H13	H13	S	S	S	S	S	H	H	H	H	H
H14	H14	S	S	S	S	S	H	H	H	H	H
H15	8,7	S	S	S	S	S	H	H	H	Rh / H[2]	Rh
H16	H16	S	S	S	S	S	H	H	Rh	Rh[1]	Rh
H17	H17	S	S	S	S	S	S	S	S	S	Rs
H18	H18	S	S	S	S	S	S	S	S	S	S
H19	H19	S	S	S	S	S	S	S	S	S	S
H20	H20	S	S	S	S	S	S	S	S	S	S
H21	H21	S	S	S	S	S	S	S	S	S	S
S13	A,2	H	H	H	Dh	Dh	H	H	H	H	H
S14	A,3	H	H	H	Dh	Dh	H	H	H	H	H
S15	A,4	H	H	Dh	Dh	Dh	H	H	H	H	H
S16	A,5	H	H	Dh	Dh	Dh	H	H	H	H	H
S17	A,6	H	Dh	Dh	Dh	Dh	H	H	H	H	H
S18	A,7	Ds	Ds	Ds	Ds	Ds	S	S	H	H	H
S19	A,8	S	S	S	S	Ds	S	S	S	S	S
S20	A,9	S	S	S	S	S	S	S	S	S	S
S21	A,T	S	S	S	S	S	S	S	S	S	S
PAIRS	A,A	P	P	P	P	P	P	P	P	P	P
	2,2	P	P	P	P	P	P	H	H	H	H
	3,3	P	P	P	P	P	P	H	H	H	H
	4,4	H	H	H	P	P	H	H	H	H	H
	5,5	Dh	Dh	Dh	Dh	Dh	Dh	Dh	Dh	H	H
	6,6	P	P	P	P	P	H	H	H	H	H
	7,7	P	P	P	P	P	P	H	H	H	H
	8,8	P	P	P	P	P	P	P	P	P	Rp
	9,9	P	P	P	P	P	S	P	P	S	S
	T,T	S	S	S	S	S	S	S	S	S	S

H = Hit S = Stand R = Late Surrender P = Split D = Double | Use lowercase play if uppercase play not permitted.

[1] Stand if 16 is multi-card or the result of a pair-split. [2] Surrender in 7 or more decks.

TABLE B13

Basic Strategy for Early Surrender

	1-Deck		2-Deck		Multi-Deck		
	T	A	T	A	9	T	A
5		E		E			E
6		E		E			E
7		E		E			E
8		E[4]		E[4]			
12		E		E			E
13		E		E			E
14	E[1]	E	E[2]	E		E	E
15	E	E	E	E		E	E
16	E	E	E	E	E[3]	E	E
17		E		E			E
2,2		E		E[4]			E[4]
3,3		E		E			E
4,4		E[4]					
6,6		E		E			E
7,7	E	E	E	E		E	E
8,8	E[5]	E	E	E		E	E

[1] Do not surrender T,4 or 9,5. Surrender 8,6 only.

[2] Do not surrender T,4.

[3] Do not surrender T,6 with 3-deck.

[4] Surrender only if dealer hits soft 17.

[5] Surrender only if NDAS.

TABLE B14

Basic Strategy for European No Hole Card

	Any No. of Decks	
	T	A
Hard 11	H	H
A,A	P	H
8,8	H	H

H = Hit P = Split

Appendix C

The Effect of Rules Variations on Basic Strategy Expectations

As we have seen, from both the e.v. tables and the BS charts, variations in the fundamental rules of blackjack have a cumulative effect on the expectations, and hence the basic strategy, of the games we play. Once again, there are many sources, in print and on the Internet, for the values of these different rules, and again, many of those sources are quite accurate. In this final appendix, we present what we feel is a complete treatment of the effect of rules variations on BS expectations. The chart that follows is adapted from a similar one found on Richard Reid's *bjmath.com* Web site, and we thank him for permitting us to use the format in our third edition. However, it was the inimitable Cacarulo, who, upon reviewing the data in the Reid chart, decided to recalculate all of the entries, to make the table you are about to read as precise as possible. *[Text continues on p. 494.]*

TABLE C1

The Effect of Rules Variations on Basic Strategy Expectations

Abbreviations	Description of Rules	1D S17	1D H17	2D S17	2D H17	4D S17	4D H17	6D S17	6D H17	8D S17	8D H17
Base Rules: S17, DOA, NDAS, SPA1, SPL3, NS	Benchmark	0.033		–0.324		–0.491		–0.546		–0.573	
Dealer Choices	**Dealer Choices**										
S17 | H17	Dealer Stands | Hits on Soft 17	0.000	–0.195	0.000	–0.204	0.000	–0.212	0.000	–0.215	0.000	–0.217
Double Downs	**Double Downs**										
NDDA	No Double Down Allowed	–1.592	–1.607	–1.453	–1.466	–1.391	–1.401	–1.372	–1.380	–1.362	–1.370
DOA or DO2	Double on Any First Two Cards	0.000	0.000	0.000	0.000	0.000	0.000	0.000	0.000	0.000	0.000
D11	Double Down only on Hard 11	–0.784	–0.790	–0.689	–0.695	–0.647	–0.652	–0.633	–0.639	–0.626	–0.632
D10	Double Down only on Hard 10 or 11	–0.268	–0.276	–0.206	–0.214	–0.181	–0.189	–0.173	–0.181	–0.170	–0.178
D9	Double Down only on Hard 9, 10, or 11	–0.136	–0.146	–0.104	–0.114	–0.093	–0.102	–0.089	–0.098	–0.087	–0.097
Splits	**Splits**										
SPA0 or NSA	No Splitting of Aces Allowed	–0.164	–0.164	–0.175	–0.175	–0.179	–0.179	–0.180	–0.180	–0.180	–0.180
SPA1 or NRSA (single hit)	Split Aces Once (up to 2 hands)	0.000	0.000	0.000	0.000	0.000	0.000	0.000	0.000	0.000	0.000
SPA2 or RSA1 (single hit)	Split Aces Twice (up to 3 hands)	0.030	0.030	0.047	0.047	0.056	0.056	0.059	0.059	0.061	0.060
SPA3 or RSA2 (single hit)	Split Aces Three Times (up to 4 hands)	0.032	0.031	0.053	0.053	0.065	0.064	0.069	0.068	0.071	0.070
Splits w/DAS*	**Splits w/Double After Split Allowed**										
DAS/SPL0 or NSP	No Splitting of Non-Ace Pairs Allowed	–0.211	–0.206	–0.234	–0.228	–0.244	–0.238	–0.247	–0.241	–0.249	–0.243
DAS/SPL1	Split Non-Ace Pairs Once (up to 2 hands)	0.112	0.115	0.098	0.101	0.090	0.094	0.088	0.091	0.086	0.090
DAS/SPL2	Split Non-Ace Pairs Twice (up to 3 hands)	0.139	0.142	0.137	0.140	0.135	0.138	0.134	0.137	0.134	0.136
DAS/SPL3	Split Non-Ace Pairs Three Times (up to 4 hands)	0.141	0.144	0.142	0.145	0.142	0.145	0.142	0.144	0.142	0.144
DAS/SPA1 (multiple hits)	Split Aces Once (up to 2 hands)	0.282	0.283	0.310	0.311	0.324	0.325	0.329	0.329	0.331	0.331
DAS/SPA2 (multiple hits)	Split Aces Twice (up to 3 hands)	0.307	0.308	0.350	0.351	0.372	0.373	0.380	0.380	0.383	0.383
DAS/SPA3 (multiple hits)	Split Aces Three Times (up to 4 hands)	0.308	0.310	0.355	0.356	0.380	0.380	0.388	0.388	0.392	0.392

* All double downs after splits assume DOA.

Abbreviations	Description of Rules	1D		2D		4D		6D		8D	
		S17	H17	S17	H17	S17	H17	S17	H17	S17	H17
Splits w/NDAS	**Splits w/No Double After Split Allowed**										
NDAS/SPL0 or NSP	No Splitting of Non-Ace Pairs Allowed	–0.211	–0.206	–0.234	–0.228	–0.244	–0.238	–0.247	–0.241	–0.249	–0.243
NDAS/SPL1	Split Non-Ace Pairs Once (up to 2 hands)	–0.017	–0.017	–0.027	–0.027	–0.033	–0.032	–0.034	–0.033	–0.035	–0.034
NDAS/SPL2	Split Non-Ace Pairs Twice (up to 3 hands)	–0.001	–0.001	–0.003	–0.003	–0.004	–0.004	–0.005	–0.005	–0.005	–0.005
NDAS/SPL3	Split Non-Ace Pairs Three Times (up to 4 hands)	0.000	0.000	0.000	0.000	0.000	0.000	0.000	0.000	0.000	0.000
NDAS/SPA1 (multiple hits)	Split Aces Once (up to 2 hands)	0.131	0.130	0.160	0.158	0.175	0.172	0.181	0.177	0.183	0.180
NDAS/SPA2 (multiple hits)	Split Aces Twice (up to 3 hands)	0.156	0.154	0.199	0.197	0.222	0.219	0.230	0.227	0.234	0.230
NDAS/SPA3 (multiple hits)	Split Aces Three Times (up to 4 hands)	0.157	0.156	0.204	0.202	0.230	0.226	0.238	0.235	0.243	0.239
Surrender	**Surrender**										
NS	No Surrender	0.000	0.000	0.000	0.000	0.000	0.000	0.000	0.000	0.000	0.000
ES	Early Surrender	0.617	0.700	0.619	0.703	0.626	0.711	0.629	0.715	0.631	0.717
ESA	Early Surrender v. Dealer Ace	0.430	0.513	0.403	0.487	0.390	0.476	0.386	0.472	0.384	0.471
ES10	Early Surrender v. Dealer Ten	0.187	0.187	0.216	0.216	0.234	0.234	0.241	0.241	0.244	0.244
LS	Late Surrender	0.023	0.039	0.052	0.067	0.067	0.082	0.073	0.088	0.076	0.092
Other Rules	**Other Rules**										
INS	Insurance	–0.226	–0.226	–0.261	–0.261	–0.279	–0.279	–0.284	–0.284	–0.287	–0.287
BJ3:1	Blackjacks Pays 3 to 1	6.974	6.974	6.868	6.868	6.816	6.816	6.798	6.798	6.790	6.790
BJ2:1	Blackjacks Pays 2 to 1	2.325	2.325	2.289	2.289	2.272	2.272	2.266	2.266	2.263	2.263
BJ6:5	Blackjacks Pays 6 to 5	–1.395	–1.395	–1.374	–1.374	–1.363	–1.363	–1.360	–1.360	–1.358	–1.358
BJ1:1	Blackjacks Pays 1 to 1	–2.325	–2.325	–2.289	–2.289	–2.272	–2.272	–2.266	–2.266	–2.263	–2.263
ENHC	European No Hole Card (Dealer takes all bets)	–0.106	–0.112	–0.106	–0.111	–0.108	–0.110	–0.109	–0.110	–0.110	–0.110
NHC (OBO)	No Hole Card (Dealer takes Original Bets Only)	0.000	0.000	0.000	0.000	0.000	0.000	0.000	0.000	0.000	0.000

As virtually all of the entries are self-explanatory, a single example should suffice to explain how to use the tables. Across the top, for each of the 1-, 2-, 4-, 6-, and 8-deck games, we use the dealer action of standing on all 17s ("S17") as the benchmark, off-the-top, BS-player edge (usually a negative value, and, therefore, a player disadvantage). To those values, we add or subtract the effects of all the rules variations that apply to the game we're playing.

Let's suppose we're on the Las Vegas Strip, using basic strategy against the rather popular 6-deck, s17, das, ls, rsa game. What is our BS edge off the top of the pack? Well, we start with the –0.546 (which already incorporates the s17 rule) in the benchmark row, at the top of the chart. To that we add: 1) 0.142, which we find on the "DAS/SPL3" = "Split Non-Ace Pairs Three Times (up to 4 hands)" line, in the "Splits w/Double After Split Allowed" section; 2) 0.073 for "LS" = "Late Surrender"; and 3) 0.069 for "SPA3 or RSA2 (single hit)" = "Split Aces Three Times (up to 4 hands)." Final player BS edge? –0.262.

Maybe you'd like to try one. What is the BS edge for the Northern Nevada basic strategist, who faces a 1D game, where the dealer hits soft 17, doubling is restricted to totals of 10 and 11 only, and there are no other favorable rules? I'll wait a minute, while you figure it out. ... All done? I hope you got: 0.033 + (–0.195) + (–0.276) = –0.438.

Again, enjoy the table!

APPENDIX D

Effects of Removal for the 1-, 2-, 6-, and 8-Deck Games

"Don's new EOR tables extend and complete the pioneering work of Peter Griffin on the subject. One can use them to see how the fluctuating composition of the pack of cards affects each individual strategy decision. The tables are impressive and make a nice addition to *Blackjack Attack*."

— *Ed Thorp, author of* ***Beat the Dealer***

"Effect of removal" (EOR, henceforth) — three little words that form the cornerstone of how we play our hands, or bet, at blackjack and how we quantify the favorability of one playing decision over another. We all know that blackjack is a game of dependent trials and that we are able to obtain an advantage, at certain moments, because cards that have been used in previous rounds are no longer available to be played until after the next shuffle. This simple notion, we agree, is the very foundation of the concept of card counting. But, at the more elementary level, when we consider the basic strategy of play, removal of cards from the deck also plays an important role.

Were we to attempt to catalog all of the basic strategy decisions without removing the specific cards that comprise the situations we are contemplating, we would not always arrive at the same conclusions that we reach by actually removing those cards. That is, so-called "full-deck favorabilities" are not the same as those that we find by judiciously removing our original two cards (or more) and the dealer's upcard from the pack. As a single example, consider our (correct) decision to double a total of 11, against the dealer's ace, in the single-deck game, where the dealer stands on soft 17. Were we to consider that action by simply writing "11 v. A" on a piece of paper and deciding to double or draw (to 17 or better), based on the availability of a full pack of 52 cards, it is interesting to note that drawing would be superior to doubling by a little less than one percent (0.85%). Why, then, do we double? Because, in an actual game of blackjack, we don't conduct such "thought experiments"; rather, we actually extract the dealer's ace and the cards that comprise our total of 11 (say a 6 and a 5) from the pack before we make our strategic decision. And, by so doing, we alter the full-deck favorability for the play in question. Enter EORs!

The removal of any card or cards from the pack has an immediate effect on the probabilities that apply to the decisions we are about to make. We learned, for example, in our previous Appendix B, that composition-dependent basic strategy differed, on occasion, from total-dependent basic strategy. Again, a single example will suffice. In the one-deck game, we stand on our total of 12 versus the dealer's 4, unless that 12 is composed of a T and a 2, in which case we hit. Why? Because the EOR of the T on the play of this hand is so different from the EORs of all the other cards involved in the play (e.g., 9,3; 8,4; 7,5), and has such a positive effect on the favorability of hitting, that removal of this lone card is sufficient to prompt us to hit, rather than stand, with, of course, less chance of breaking our 12, now that a T has been removed from the pack.

It goes without saying that, once we begin to explore all of the strategic variations possible, given the removal of certain ranks of cards from the pack, the possibilities are endless. In his masterwork, *The Theory of Blackjack,* Professor Peter Griffin constructed what he called "virtually complete strategy tables," consisting of the effects of removal of each card rank on almost all of the holdings of interest to the blackjack player. That is, he cataloged the EORs for the player actions of hitting (versus standing), hard doubling (versus hitting), soft doubling (versus hitting or standing), and pair splitting (versus hitting or standing), against all dealer upcards and, in the case of the dealer's 6 or A, for both dealer actions (hit or stand) on a total of soft 17. We urge the reader to consult Griffin's work (pp. 71–86 and 231–33) for many fascinating examples of how to get the most out of such multi-purpose tables.

In addition to the aforementioned playing-EOR charts, we provide several others (Tables D17 and D18, p. 522), furnishing the effects of removing the various ranks on full-deck favorabilities for betting and insurance, and the betting and insurance correlations for several of the most popular count systems.

The Construction of the Tables

It all started quite innocently when one of the most valued contributors to "Don's Domain," who goes by the handle of "Zenfighter," began to post updated, more precise EORs to the *AdvantagePlayer.com* Web site. Long a fan of Griffin's work, Zen (as we'll call him for short) wanted, in his own way, to pay homage to the late professor by continuing and, dare we say it, perfecting the studies that Griffin had done in this area. And, as has often been the case with some of the research performed to create the content of this book, the project sprouted wings and expanded in ways that our new pioneer could not have readily envisioned. Encouraged — indeed urged — by me to "go all the way" with his newfound labor of love, Zen produced the masterful charts you're about to read. Along the way, he enlisted the aid of Cacarulo, to complete the arduous task of calculating the pair-splitting EORs, when resplits are permitted — a computation very difficult to achieve with extreme accuracy — and together, these formidable and innovative blackjack researchers finalized a revision and expansion of Griffin's EORs, to be published (where else?) in this softcover edition of *BJA3.* I am proud to include this impressive collaboration of Zenfighter and Cacarulo in my book, and I thank them both for their dedication to this exciting project.

To respect Griffin's conventions, and for ease of comparison, the present charts resemble quite closely, in their format, the *Theory of Blackjack* tables. Nonetheless, we saw fit to make certain improvements that, we feel, will facilitate usage. In all, there are 16 tables of playing EORs: one for each of the ten possible dealer upcards, and six extra ones to cover the dealer's upcards of A through 6, when he hits soft 17. A new feature lists, across the top, each of the ranks whose EORs are being cataloged, as we often found it annoying, with Griffin's unlabeled tables, to have to count, from left to right (or vice versa), to arrive at the rank whose EOR we were studying. In addition, we saw fit to list the player's holdings down the left-hand column, in front of the EORs, again, to facilitate use of the tables. One final change moved the "Hitting Soft 18" line (now called "Soft Hitting") from underneath the "Doubling" rows (and, therefore, separated from the hard-hitting array) to directly after those hard-hitting EORs (now called "Hard Hitting"), where, we feel, it more logically belonged.

As for the entries themselves, there are several improvements over Griffin's work. Among them are: 1) Greater accuracy. Given today's higher-speed computers, compared to the ones available to Griffin, several of the values differ slightly from (and are more accurate than) those cataloged by the professor. The main discrepancies are in the doubling and splitting entries. Here, our tables are based on exact computations and not on estimation and careful alteration of infinite-deck probabilities, actually the only tool Griffin had at his disposal in the late seventies. In those days, the amount of time required to execute these calculations was staggering, to say the least; 2) Two more decimal places. Griffin's values were truncated to two decimal places, whereas the current EORs are given to four-decimal-place accuracy; 3) Newly created pages (a novelty for which, once again, we may thank Cacarulo) that include EORs for when the dealer hits soft 17, for upcards of 2 through 5. Griffin did not feel that it was necessary to make this distinction with the s17 values, but there are subtle differences, and we opted for the greater accuracy in our revamped tables; 4) Columns newly created to include full-deck favorabilities for the 2-, 6-, and 8-deck games, in addition to the lone 1-deck favorabilities that Griffin furnished (he did, however, include infinite-deck m's and explained how to use them, by interpolation of the reciprocals, to calculate m's for other numbers of decks). The inclusion in our tables of these extra 12th, 13th, and 14th columns, to complement the famous "m," or 11th column full-deck favorabilities of Griffin (more on their meaning in a little while), prompted us to re-baptize that traditional "m" as "m1," to conform with the "m2," "m6," and "m8" that now follow. We trust that the reader will be comfortable with this slightly modified terminology; 5) Pair-splitting EORs for resplitting (SPL3), instead of just the former SPL1 variety, and for DAS (not provided by Griffin); and, finally, 6) Late Surrender EORs, also not furnished by Griffin.

One quick word about the final column, labeled "ss." The "sum of the squares" is obtained by squaring the values in the first nine columns, summing, and then adding four times the square of the tenth column to that sum. Caution: If you try to replicate the ss columns by doing the individual calculations yourself, you may not arrive at the precise values in the charts. This is because all roundings of ss values were done *after* squaring and summing the non-truncated EORs, whereas the EORs in the tables have already been rounded to four decimal places. We shall have a few more words to say about the meaning and use of the ss values a bit later in our presentation.

Checking the Quality of the Tables

An EORs table is supposedly a fair or accurate one if: Sum Ei x Pi = 0; that is, the EOR of a single rank times its probability of being drawn, summed over the ten different values (tens are four times more probable, remember), should be zero. Obviously, by rounding off decimals, what we obtain is a final result as close to zero as is practical. As long as this final sum is no greater than 0.01 in absolute value, we can be satisfied with the accuracy of those EORs. As can be seen, all of the sums for the tables that follow are within a range of –0.0003 to +0.0003 and are, therefore, quite precise.

As an example, consider the Table D10 entry for hitting 16 v. T: (–0.4992)(1/13) + (–0.2903)(1/13) + … + (0.5524)(1/13) + (1.1151)(4/13) = 0.000008, which, of course, is well within our guidelines.

How to Use the Tables

There is little to change from Griffin's directions on his pp. 72–73, and so we summarize the most salient points here. In the "Hitting" section, we are considering the favorability of hitting our holding exactly one time, as opposed to standing. (Shortly, we shall give examples of how to carry out the actual calculations.) The signs of the entries in the 11th–14th columns will indicate, before actual cards are removed, whether hitting (positive) or standing (negative) is the superior full-deck choice. As you may see, it is often the case that what begins as one decision, for full-deck favorability, changes to the alternative, once cards are removed. (Please note that all the values in the tables of this Appendix D are given in percent.)

For "Hard Doubling," we catalog the favorability of doubling over simply hitting. In similar fashion, the "Soft Doubling" EORs examine the preferences for doubling over the other choices of either hitting or standing, depending on the individual holdings. Finally, for "Splitting," EORs are furnished for both the NDAS and DAS rules, while the "SPA1" and "SPL3" designations inform us that aces may be split only once and that pairs may be split a total of three times (to create four hands, maximum).

Griffin reminds us that the entries in the tables, that is, the EORs themselves, are not expectations, but rather *differences* in the expectations for the two separate actions being contemplated (p. 73). And so, we cannot, for example, use the charts to replicate the e.v.s found in our Appendix A, but rather, we may use the tables to ascertain (with great accuracy) the *differences* between the values given in one column of an Appendix A chart (say hitting) and another (say standing).

Now is also a good time to remind you that, although the EORs furnished by Zen and Cacarulo are impressively accurate, the very nature of the endeavor (creating EORs) precludes absolute precision. EORs are, by definition, the best linear *estimates* (albeit darn good ones!), under the criterion of least squares, of the exact effects of removal of a particular rank, or ranks, on the favorabilities of the actions being considered. As such, we shouldn't be surprised, upon carrying out some of the calculations in the examples that follow, if they do not provide identical answers to those obtained by comparing the e.v. of one column to that of another, in our Appendix A.

From Theory to Practice: Some Examples

But enough talk; I know that you want to actually *use* the tables! I don't blame you. Let's get down to cases. In each example, we'll designate a specific holding and decide which of two alternatives is the superior choice. As a start, suppose I'm curious to know why we stand on A,7 v. A in the single-deck, s17 game but hit in 6-deck. Here is how to use the charts to explain the phenomenon:

1. For the single-deck decisions, we go to Table D1, for the dealer's upcard of ace. In the Soft Hitting section, we find our A,7 (Soft 18) holding. Now, m1, the full-deck favorability, tells us that, before any cards are removed from the pack, it's better by 0.0309% to hit rather than to stand. (As we're comparing favorability of hitting to standing, a positive value indicates that hitting is preferable.)

2. Next, we remove the three cards involved in the decision, namely, the dealer's A and our A and 7. The EOR for the A is –0.4197 and that of the 7 is –0.0302. Reckoning the effect of removing all three is simply additive: 2(–0.4197) + (–0.0302) = –0.8696.

3. We now need to make a slight adjustment to the above sum, before proceeding to the final step. As constructed, the tables require no adjustments to ascertain the EOR of a *single* card from a full, single deck. However, when more than one card is removed (as is almost always the case), or when more than one deck is used, we must multiply the sum of the EORs by 51/(52k – n), where k is the number of decks being used, and n is the total number of cards removed from the full pack. It's easy to see, therefore, why no adjustment is necessary when a single card is removed from a single deck. In that case, k = 1, and n = 1, as well. So, our above fraction reduces to 51/(52 – 1), which, of course, is simply 51/51, or 1. But, in our A,7 v. A example, we have removed three cards, and so, multiplication by 51/49 is in order. Thus, (51/49)(–0.8696) = –0.9051.

4. Finally, adding –0.9051 to the initial m1 value of 0.0309 gives –0.8742(%), indicating that hitting is not favorable, and so we stand. (Note that the precise difference in the e.v.s of the two actions can be found in Table A2, on p. 395, where the Stand column's e.v. of –0.101015 is superior to the Hit e.v. of –0.108386 by precisely 0.007371, which can be compared to our –0.008742, found above. The difference is just slightly more than one-tenth of one percent.

But, our job is only half done. We need to examine the 6-deck game, to see why hitting our A,7 now becomes preferable to standing, against the dealer's A. Let's walk through the math together:

1. We now look to the 13th column, m6 value for full-deck (pack) favorability, where we read 0.6152.

2. The same three cards are removed, so there is no change in the sum of our EORs, namely, –0.8696.

3. This time, however, our adjustment factor becomes 51/[(52 x 6) – 3], or 51/309. Now, (51/309)(–0.8696) = –0.1435.

4. We add –0.1435 to the m6 value of 0.6152, obtaining 0.4717(%). The result is positive, so we infer that, this time, hitting our A,7 is 0.4717% better than standing. Again, comparison to the precise expectations, given in Table A50, p. 443, shows

that hitting is actually –0.095345 – (–0.100278) = 0.004933 (or 0.4933%) better than standing, a difference of a minuscule 0.000216, or just over two-hundredths of one percent!

OK, now it's your turn. Can you show me that, in a 2-deck, s17 game, whereas it would be correct to hit our T,6 v. T, we'd be better off standing if our 16 were composed of, say, 8,4,4? Of course you can! I'll wait a minute (or two!) while you do the necessary calculations, and then you can check below to see how you did. All done? Here's what the math should look like:

1. From Table D10, the 12th column m2 value is –0.1915.
2. Removing T,6 and T leads to: 2(1.1151) + 1.6446 = 3.8748.
3. The adjustment factor is now 51/101, and (51/101)(3.8748) = 1.9566.
4. Finally, 1.9566 + (–0.1915) = 1.7651, which is positive, and so we hit.

Now for the second calculation, where we consider what to do with 8,4,4:

1. There is no change for the m2 value of –0.1915.
2. This time, however, we remove 8,4,4 and T: –0.0567 +2(–1.7279) + 1.1151 = –2.3974.
3. The adjustment factor is 51/100, and (51/100)(–2.3974) = –1.2227.
4. Finally, adding –1.2227 to –0.1915 yields –1.4142, and the negative value tells us that we are roughly 1.4% better off to stand than to hit.

Of course, there is really no end to the examples that we may generate with our amazingly complete set of tables. Let's look at a few more. Say we're in A.C., playing an 8-deck game, alone at the table. We're shown the burn card (an A.C. custom), and it's an 8. Our first hand is A,2 v. dealer's 5. Our play? "A marginal double," you say? Well, let's see.

1. From Table D5, the 14th column m8 value is –0.6527.
2. Removing 8, and then A,2 and 5 leads to: –2.8332 + 1.4274 + 1.8206 + 2.0658 = 2.4806.
3. The adjustment factor is 51/412, and (51/412)(2.4806) = 0.3071.
4. Adding 0.3071 to –0.6527 yields –0.3456, and so, by the slimmest of margins (just over three-tenths of one percent), hitting is now superior to doubling.

Note what is happening here. The play (to double) was extremely close to begin with (m8 absolute value of less than 1%). Even in the 8-deck game, removal of a single 8, the most important card for our A,2 double-down, tips the favorability to the hitting side.

Remaining in A.C. for a moment, a similar type of analysis might be made for a sticky surrender question. Many A.C. casinos have both 8-deck and 6-deck games, and it isn't all that rare for someone to play at both, during the course of a day. Suppose we're dealt 8,7 v. dealer's T, early in the day, at an 8-deck game, and then, later on, we have the same hand, this time at 6-deck. What's more, we're at a casino where requesting surrender is honored, and you may obtain this favorable rule, simply by asking (yes, some places offer surrender, but are reluctant to advertise it). The question is: What do I do in each situation?

Here, Table D10 furnishes the two relevant m values in the Late Surrender section: m8 = 0.4551 and m6 = 0.4576. Not surprisingly, they are almost identical. And, of course, the "raw," or unadjusted, EORs will also be the same for the two situations. Removing 8,7 and T yields: –0.9141 + (–0.6826) + (–1.3193) = –2.9160. But, here's where things get interesting. The adjustment factor for the 8-deck game is 51/413, while that of the 6-deck game is 51/309. When we multiply the sum of the EORs by the first factor, for the 8-deck game, we get (51/413)(–2.9160) = –0.3601, while a similar calculation yields (51/309)(–2.9160) = –0.4813, for the 6-deck game. And, do you know what? Adding the first result to 0.4551 (m8) leaves us (barely) on the positive side of zero — and so we surrender, in the 8-deck game — while adding the second result to 0.4576 (m6) leaves us ever so slightly on the negative side (–0.0237), and so surrender is no longer as desirable as hitting!

Let's try a final example, this time for our Northern Nevada brethren. We're playing in a single-deck, h17, NDAS game. Off the top, we get 2,2 v. 3, and we split. On our first deuce, we draw an 8 (can't double), then a 7, for 17. On our second deuce, we draw yet another 2. Split 'em again? Not so fast! Table D13 shows an m1 entry of 0.4509 for our original holding. By the time we contemplate our second pair split, we've seen three deuces, an 8, a 7, and the dealer's 3. EORs are, respectively: 3(0.1562) + (–1.8372) + (0.3552) + (0.3839). They sum to –0.6295. The factor is 51/46, which yields –0.6979, after multiplication. Adding our m1 value of 0.4509 leaves us with –0.2470, and, chastened by the negative value, we eschew the pair split and meekly hit our pathetic total of four!

Further Uses of the Tables

By now, you may have noticed that we haven't made any use of the last column, the ss, or sum of the squares, value. In fact, that calculation plays a role in determining several important blackjack metrics. But, before we see how to use the ss, let's try to understand, intuitively, its meaning. In addition to having a full-deck favorability (the m value), each holding we consider also has a "volatility" associated with the play. Think of this number as an indication of how likely the m value is to change as cards are removed from the pack. A large ss portends a play that is volatile, that is, a play that may easily be affected by removal of a relatively small number of key ranks, whereas a small ss indicates a stable strategic play, one that is unlikely to change or be affected, even with the removal of several cards from the pack.

So, the ss value acts as a kind of barometer, in conjunction with the raw m value, tipping us off as to whether or not a play might be a candidate for a strategy departure, once certain key cards are no longer available to be played. To see how this may work, note, in Table D1, that despite a rather large m1 for splitting A,A v. A (16.9799), there is an astronomical 255.5365 ss value. The culprit is the huge EOR of a single T, in this case –6.4592. In a three-handed single-deck game, it would suffice that the two players to our right stand on their T,T twenties (and that we manage to glimpse their cards) for us to refuse to split our aces and simply hit the hand.

Now, the sum of the squares is used in several other manners, as well. We don't have room to explain them all here, but the reader is urged to consult Griffin, pp. 41–44, 71–72, and 86–94, where ss is used to calculate correlation coefficients, strategic efficiencies, opportunity at various penetration levels, and chances of being behind, among several other fascinating topics. The reader will also note that ss was involved in the creation of the "Illustrious 18," where, in Table 5.1, p. 62, the "Sum of the Squares" constitutes our column 4 in the array.

Finally, for many years, before the advent of our multi-purpose software, which now cranks out index numbers to our heart's content, players who wanted a full matrix of strategy-departure indices often turned to Arnold Snyder's "Algebraic Approximations of Optimum Blackjack Strategy" (RGE Publishing, 1981). We note that Griffin's EORs played a central role in Snyder's algorithm for calculating the indices, and we also note, in passing, that those who would still wish today to use Snyder's methodology could now create even more accurate indices, thanks to the new tables you now have at your disposal.

Betting and Insurance Effects of Removal

We have just spent a good deal of time examining the effects of removal on *playing* decisions, but EORs affect the overall *betting* and *insurance* advantages of the game, as well. Table D17 catalogs the effects of removal on off-the-top BS edges for the four standard varieties of the single-deck game (C-D BS, with the assumption of the rules set given at the top of the table), followed by a full set for when late surrender is offered, plus two extra sets of EORs for Northern Nevada rules (h17, D10), and a final set of insurance EORs.

For betting, the idea is to ascertain how our global advantage is helped or hindered as the various ranks are removed from a full pack, and for insurance, we note how removal of the ranks affects the advisability of taking insurance. Note that, as was the case with the playing EORs, as the number of decks increases, the betting effects become smaller, which is logical. We're much happier to see a single 5 removed from a 2-deck game than we are to see that same 5 removed from eight decks. So, these single-deck EORs need to be adjusted roughly in proportion to the number of decks employed, and we apply the same 51/(52 k – n) factor as was used previously.

A single example should suffice. Suppose, in a double-deck, s17, DAS game, we hit our T,2 v. dealer's 3 with a 5. He flips a 9 in the hole and outdraws us with a 7. What is our advantage before we play the next hand? EORs are (in order): (–0.5121) + 0.3809 + 0.4339 + 0.7274 + (–0.1731) + 0.2823 = 1.1393. Multiplying by 51/98 yields 0.5929. To this value, we add the mean, full-deck player edge (m2) of –0.1820, and we now perceive an advantage of 0.4109% for the upcoming hand. Note that although many counting systems (such as Hi-Lo) do not reckon the 9 or the 7, in fact, removal of one of each rank slightly increases our betting edge, as the last two EORs clearly illustrate.

Insurance EORs are important in determining the insurance correlation and efficiency of a point count (see Griffin's discussions of these concepts on pp. 43 and 71–72), as well as in furnishing insight, through methods such as Snyder's algebraic approximation, into calculation of correct insurance indices. As a final exercise, let's try to see how we could come up with the conditional penalty, in percent, for deviating from basic strategy by insuring our natural A,T, in the 4-deck game that was used to produce the statistics in Table 7.2, on p. 98.

At the bottom of that table, we noted that taking insurance when we held A,T would cost (rounded) 3.90% of our *original* wager. Let's see if we can use our EOR table to reproduce this value. Although we do not list an m4 in our chart, it is trivial to understand that all the full-deck favorabilities for insurance are identical, regardless of the number of decks shuffled, since the ratio of non-tens to tens is not affected by the number of decks used. So, m4, had it been listed, would be the same –7.6923 as all the other entries for m1 through m8. Next, we remove two aces and a ten, yielding: 2(1.8100) + (–4.0724) = –0.4524. Multiplying by 51/205 gives (51/205) (–0.4524) = –0.1125. Adding this value to m gives –0.1125 + (–7.6923) = –7.8048. Now, to express this in terms of our original wager, we need to halve the value (the insurance wager is half of our original bet), producing –3.9024, which, rounded to two decimal places, gives the aforementioned 3.90% listed in Table 7.2.

Betting and Insurance Correlations

Before we conclude, so that you may enjoy the tables, there is one final chart, Table D18, comprising several lines, that needs to be explained. On pp. 43–44, Griffin describes how to use the betting EORs, in conjunction with the tag values of a particular point-count system, to derive the betting correlation (BC) for that count. Later, on p. 71, he uses a similar methodology to obtain the insurance correlation for a given point count. In our new table, we furnish representative betting and insurance correlations for some of the most popular count systems. It is interesting to note minor differences in these present values, when compared to those we are accustomed to reading in the existing literature, as our numbers result directly from the improved betting EORs of Table D17.

It is also fascinating to observe that, without exception, a system's BC is actually *higher* when reckoned against a rules set that features h17, as opposed to s17. Finally, we see that, when it comes to BC, the Rolls Royce of point-count systems is Wong's level-3 Halves, which nears perfection with its set of 99+% BCs against all rules sets.

And now, thanks to Zenfighter and Cacarulo, we invite you to peruse the most accurate EORs ever published anywhere!

TABLE D1
Dealer's Upcard A (S17)

	A	2	3	4	5	6	7	8	9	T	m1	m2	m6	m8	ss
							Hard Hitting								
17	–0.5262	–1.5406	–2.4796	–3.0918	0.4844	0.6114	–1.3937	–0.3744	0.5664	1.9360	–8.8905	–8.4036	–8.0845	–8.0450	36.3618
16	–0.0228	–0.9231	–1.6624	–2.5294	–3.1818	–1.4315	–0.4129	0.5341	1.4016	2.0570	13.7958	14.3919	14.7850	14.8339	41.5325
15	0.3728	–0.1487	–0.8796	–1.7609	–2.6149	–4.9333	–0.2292	0.6666	1.4833	2.0109	16.4852	17.0200	17.3695	17.4128	54.0825
14	0.4133	0.2619	–0.1336	–0.9922	–1.7772	–4.1737	–3.7896	0.7893	1.5561	1.9614	19.0872	19.6030	19.9387	19.9802	54.6148
13	0.4535	0.2775	0.2527	–0.1852	–0.9457	–3.1909	–3.0908	–2.8327	1.6246	1.9093	21.6199	22.1490	22.4950	22.5379	46.2547
12	0.4620	0.2960	0.3163	0.2529	–0.1196	–2.2169	–2.1709	–2.1905	–2.0483	1.8548	24.0898	24.6627	25.0404	25.0874	32.8617
							Soft Hitting								
18	–0.4197	–1.2412	–2.1925	0.0751	–0.1992	1.2171	–0.0302	–0.9112	0.1922	0.8774	0.0309	0.3874	0.6152	0.6432	11.9978
							Hard Doubling								
11	1.7361	1.7877	1.9158	2.1667	2.6460	2.9952	1.7613	0.4467	–0.9006	–3.6388	–0.8517	–2.1412	–2.9804	–3.0842	87.6227
10	–2.0326	1.7373	1.8488	2.1829	2.7190	3.3073	2.2504	0.9406	0.1783	–3.2830	–6.7288	–8.1604	–9.0909	–9.2059	82.7572
							Splitting (NDAS, SPA1, SPL3)								
9,9	–2.8358	–0.9368	0.8877	0.8304	0.7063	1.6724	0.2806	0.5342	2.2501	–0.8472	–3.4196	–3.5255	–3.6128	–3.6246	21.9906
8,8	–2.5509	–0.6832	0.0328	3.0095	3.8097	–2.2701	0.4081	2.1943	3.4028	–1.8383	15.1433	14.8056	14.5768	14.5480	65.7764
7,7	–1.8129	–1.4109	–1.8173	–1.4372	1.7059	1.7631	5.4095	0.9676	1.3039	–1.1679	–21.0342	–21.1755	–21.1581	–21.1596	54.0190
6,6	–1.3292	–1.3369	–2.7691	–4.3162	–4.5026	1.6116	2.6410	4.2047	5.6745	0.0306	–29.1673	–29.7505	–30.1093	–30.1530	109.5817
3,3	–0.9850	0.5786	2.0158	1.8344	0.7891	–3.3220	–3.7071	–2.8820	2.6237	0.7637	–18.0139	–17.9354	–17.8656	–17.8558	51.6567
2,2	–1.0764	0.5644	0.0332	0.7915	1.1205	0.1062	0.1536	–2.5636	–1.8931	0.6909	–18.4224	–18.2430	–18.1128	–18.0969	15.4605
A,A	2.7120	2.8772	3.0661	3.4008	4.0172	4.6017	3.3702	1.8376	–0.0458	–6.4592	16.9799	14.9866	13.6073	13.4435	255.5365
							Splitting (DAS, SPA1, SPL3)								
9,9	–2.5051	–0.7125	1.2592	1.2328	1.1909	2.2252	0.6065	0.6143	2.1161	–1.5068	–3.2745	–3.7654	–3.6128	–3.6246	30.5633
8,8	–2.2157	–0.3378	0.2501	3.4074	4.2918	–1.7140	0.7341	2.2932	3.2675	–2.4941	15.3513	14.5993	14.5768	14.5480	79.4100
7,7	–1.4738	–1.0777	–1.4683	–1.2130	2.1883	2.3252	5.7333	1.0894	1.1743	–1.8194	–20.7428	–21.3378	–21.1581	–21.1596	65.8338
6,6	–1.0010	–1.0059	–2.4204	–3.9245	–4.2445	2.1502	3.0180	4.3352	5.5532	–0.6151	–28.7532	–29.8509	–30.1093	–30.1530	106.1669
3,3	–0.6658	0.9061	2.3663	2.2603	1.3040	–2.7436	–3.3744	–2.8151	2.4526	0.0774	–18.0264	–18.2573	–17.8656	–17.8558	46.5510
2,2	–0.7696	0.8954	0.4073	1.2094	1.6274	0.6668	0.4747	–2.4988	–1.9715	–0.0102	–18.6801	–18.6856	–18.1128	–18.0969	16.4721
A,A	2.7120	2.8772	3.0661	3.4008	4.0172	4.6017	3.3702	1.8376	–0.0458	–6.4592	16.9799	14.9866	13.6073	13.4435	255.5365
							Late Surrender								
17	0.3936	0.1616	–0.0737	–0.2176	–0.5043	0.0436	–2.1615	–1.5291	–0.9218	1.2023	–2.7352	–2.4590	–2.2826	–2.2610	14.1317
16	0.1465	0.9024	1.5801	2.2585	2.6804	–0.6310	–1.0289	–1.3637	–1.6470	–0.7243	2.1175	1.9176	1.7827	1.7658	23.7456
15	–0.0015	0.1794	0.9082	1.6708	2.3355	2.9632	–1.1971	–1.5566	–1.8586	–0.8608	–1.6222	–1.8128	–1.9376	–1.9531	28.1579
14	0.0022	–0.0085	0.2134	1.0180	1.6878	2.5786	2.3601	–1.7545	–2.0746	–1.0056	–5.5022	–5.7548	–5.9187	–5.9389	27.5774
13	–0.0021	0.0510	0.0164	0.2765	0.9968	1.9340	1.9383	1.7710	–2.3100	–1.1680	–9.5960	–9.9537	–10.1897	–10.2190	22.4997
12	0.0253	0.0771	0.0197	0.0111	0.2506	1.2495	1.2559	1.3088	1.1718	–1.3424	–13.8840	–14.4145	–14.7687	–14.8130	13.5030
	A	**2**	**3**	**4**	**5**	**6**	**7**	**8**	**9**	**T**	**m1**	**m2**	**m6**	**m8**	**ss**

TABLE D2
Dealer's Upcard 2 (S17)

	A	2	3	4	5	6	7	8	9	T	m1	m2	m6	m8	ss
								Hard Hitting							
17	−1.9227	−1.8636	−2.3847	−2.9269	−0.0327	−0.7243	−0.2971	0.2089	0.6639	2.3198	−38.2286	−38.2753	−38.3040	−38.3074	44.0485
16	−1.4316	−1.3709	−1.8296	−2.5634	−4.4744	−0.3717	0.0878	0.6075	1.0565	2.5724	−17.6677	−17.7458	−17.7963	−17.8025	61.9684
15	−0.9809	−0.7278	−1.2249	−1.9586	−3.7366	−3.8186	0.3013	0.7912	1.2058	2.5372	−12.1932	−12.2856	−12.3489	−12.3569	63.2928
14	−0.7951	−0.3627	−0.5987	−1.3106	−2.9774	−3.1006	−3.1721	0.9712	1.3522	2.4984	−6.7743	−6.8545	−6.9114	−6.9187	59.1213
13	−0.6442	−0.2611	−0.2084	−0.6480	−2.2023	−2.3651	−2.4848	−2.5371	1.4998	2.4628	−1.3255	−1.4101	−1.4699	−1.4776	50.5128
12	−0.4709	−0.1174	−0.0681	−0.2236	−1.4274	−1.6312	−1.7954	−1.8975	−2.0530	2.4211	4.4902	4.2122	4.0297	4.0071	39.4751
								Hard Doubling							
11	1.6687	0.9996	0.9847	1.0536	1.8552	1.6306	1.0947	0.6496	0.1826	−2.5298	23.7212	23.4705	23.3085	23.2885	39.2174
10	−0.7339	0.9258	0.8070	0.9108	1.7322	1.7609	1.5226	1.2627	0.7088	−2.2242	17.8683	17.7556	17.6811	17.6718	33.1820
9	−0.5841	1.4567	0.4585	0.5337	1.4906	1.5296	1.7734	1.6915	0.8005	−2.2876	−1.4082	−1.3676	−1.3437	−1.3409	35.0992
8	−0.2303	0.9957	1.1536	0.3028	1.3464	1.5585	1.7044	1.5362	0.8485	−2.3039	−18.5822	−18.4227	−18.3197	−18.3070	33.9255
								Soft Doubling							
A,9	−1.8379	1.9264	1.9745	2.1559	3.3492	3.4857	2.9399	1.2064	−0.5299	−3.6676	−27.3388	−27.7238	−27.9782	−28.0099	103.1858
A,8	−1.1884	−2.8164	1.7451	1.8934	3.1840	3.2645	2.6895	1.5897	0.4599	−2.7053	−14.0046	−14.2242	−14.3713	−14.3897	76.0169
A,7	−0.4709	−2.1256	−3.1834	1.6136	2.9333	2.4280	1.9492	2.0901	1.4736	−1.6770	−0.1417	−0.1744	−0.1918	−0.1937	53.5665
A,6	0.4963	−0.7588	−1.2488	−1.6069	1.6176	1.7786	1.7336	1.6966	1.4704	−1.2946	−0.9039	−0.7835	−0.6988	−0.6879	25.4938
A,5	1.3014	−0.2331	−0.7226	−1.0406	−0.5867	1.6964	1.6043	1.5367	1.2839	−1.2100	−5.4728	−5.2637	−5.1242	−5.1068	19.0146
A,4	1.7843	0.5538	−0.1943	−0.4967	0.0505	−0.5968	1.5720	1.4589	1.2036	−1.3338	−7.5261	−7.3299	−7.2054	−7.1902	17.2980
A,3	1.8052	1.0186	0.5499	0.0668	0.6784	0.0278	−0.7579	1.3937	1.1117	−1.4736	−9.7495	−9.5630	−9.4497	−9.4361	17.5022
A,2	1.7915	1.0078	1.0366	0.8344	1.3000	0.6352	−0.1564	−0.9775	1.0170	−1.6221	−12.0483	−11.9244	−11.8531	−11.8448	20.6289
								Splitting (NDAS, SPA1, SPL3)							
9,9	−1.4800	−0.5559	1.6518	2.1317	3.5758	2.4029	1.8778	1.7913	−0.0556	−2.8350	4.0973	3.6432	3.3408	3.3031	67.2181
8,8	−1.4086	−0.9417	−1.0069	1.3225	1.1674	2.0160	2.6332	1.1723	1.3613	−1.5789	30.2962	29.7791	29.4420	29.4004	31.1944
7,7	−0.6880	−0.6604	−0.9574	−1.2130	1.4660	2.2240	1.5856	1.8353	1.1853	−1.1944	9.2113	8.9750	8.8197	8.8004	23.3863
6,6	0.2366	−0.3233	−0.8203	−0.9126	1.4317	3.8285	4.0490	3.8784	3.3735	−3.6854	−3.4422	−3.3768	−3.3295	−3.3234	115.5177
3,3	0.0994	0.4356	1.0592	1.0613	1.5899	−0.0788	−0.7037	−0.8140	1.0117	−0.9151	−7.7350	−7.4063	−7.1900	−7.1631	10.5131
2,2	−0.1159	0.2944	0.4373	0.8571	1.2191	0.8446	0.0795	−1.8665	−2.3853	0.1589	−2.3855	−3.3650	−4.0099	−4.0901	12.5064
A,A	2.5557	2.0211	2.0587	2.2257	3.2217	2.9647	2.2830	1.7012	1.2542	−5.0715	39.7237	39.2951	39.0172	38.9829	151.5370
								Splitting (DAS, SPA1, SPL3)							
9,9	−1.0601	−1.4510	1.9408	2.4303	4.0271	2.8104	2.1853	2.0305	−0.0977	−3.2039	8.5722	7.9772	7.5840	7.5350	86.9870
8,8	−1.0529	−1.3143	−1.6810	1.8669	2.0153	2.8242	3.3088	1.3695	1.7385	−2.2687	38.2173	37.4308	36.9172	36.8536	57.6183
7,7	−0.4374	0.0782	−1.2660	−1.7717	2.5929	3.3370	2.2111	2.7499	1.7451	−2.3098	17.2061	16.5433	16.2987	16.2566	59.6345
6,6	0.5528	0.3485	−0.4552	−1.5117	1.5734	4.8391	5.4253	5.1087	4.2291	−5.0275	4.5285	4.1930	4.1788	4.1547	203.3306
3,3	0.3283	1.2516	1.3018	1.7469	2.7679	0.9854	−0.6361	−1.0881	1.5437	−2.0504	0.2636	0.1563	0.2899	0.2936	35.8406
2,2	0.1341	0.7460	1.0831	1.5554	2.4109	1.9937	1.0025	−1.9297	−2.9864	−1.0024	5.5614	4.1736	3.4504	3.3520	31.6211
A,A	2.5557	2.0211	2.0587	2.2257	3.2217	2.9647	2.2830	1.7012	1.2542	−5.0715	39.7237	39.2951	39.0172	38.9829	151.5370
	A	**2**	**3**	**4**	**5**	**6**	**7**	**8**	**9**	**T**	**m1**	**m2**	**m6**	**m8**	**ss**

TABLE D3
Dealer's Upcard 3 (S17)

	A	2	3	4	5	6	7	8	9	T	m1	m2	m6	m8	ss
								Hard Hitting							
17	−1.9062	−1.8604	−2.3697	−3.6405	−1.0666	−0.6749	−0.2293	0.3110	2.0473	2.3474	−41.4418	−41.4459	−41.4463	−41.4462	53.9376
16	−1.4533	−1.3045	−2.0670	−3.9871	−4.4639	−0.2981	0.2340	0.6833	2.2689	2.5969	−21.3389	−21.2444	−21.1825	−21.1748	76.6443
15	−1.0125	−0.6977	−1.4591	−3.2493	−3.7153	−3.7341	0.4617	0.8772	2.2678	2.5653	−15.5490	−15.4709	−15.4228	−15.4170	74.3944
14	−0.8579	−0.3378	−0.8171	−2.4971	−2.9530	−3.0137	−3.0109	1.0726	2.2675	2.5369	−9.7306	−9.6851	−9.6595	−9.6565	66.6569
13	−0.6807	−0.1950	−0.4074	−1.7471	−2.1898	−2.2955	−2.3388	−2.4457	2.2906	2.5024	−3.5410	−3.7206	−3.8380	−3.8525	55.5300
12	−0.4685	−0.0561	−0.2514	−1.2531	−1.4081	−1.5567	−1.6467	−1.8058	−1.4094	2.4639	2.5671	2.2056	1.9712	1.9423	38.5052
								Hard Doubling							
11	1.5019	0.9383	0.9531	1.4916	2.0135	1.7608	1.3240	0.8464	−1.6174	−2.3031	26.6044	26.0916	25.8611	25.8325	39.7257
10	−0.7617	0.7902	0.7787	1.3224	1.8858	1.9318	1.9590	1.4250	−1.2920	−2.0098	20.9261	20.5310	20.3931	20.3757	34.5417
9	−0.6264	1.3716	0.3639	1.1274	1.7087	1.9316	2.1404	1.5515	−1.3009	−2.0670	2.2975	2.0356	1.9829	1.9760	36.0981
8	−0.2863	0.9088	1.1231	1.0524	1.8105	1.9197	1.7533	1.3116	−1.2740	−2.0798	−14.2767	−14.4086	−14.4170	−14.4182	33.9599
7	0.1440	0.5154	0.8554	2.2751	1.9325	1.6527	1.2815	0.8583	−1.4622	−2.0132	−28.1815	−28.3115	−28.3168	−28.3175	33.3896
								Soft Doubling							
A,9	−1.7992	1.7970	1.9788	3.1923	3.8731	3.9045	2.7714	0.9996	−2.2404	−3.6193	−23.0254	−23.5614	−23.9175	−23.9619	116.9137
A,8	−1.1611	−2.7949	1.8167	2.9934	3.6226	3.0590	2.5436	1.9422	−1.3082	−2.6783	−10.0960	−10.4772	−10.7288	−10.7601	84.5474
A,7	−0.4426	−2.1283	−2.8971	2.7027	2.7493	2.2664	2.3890	2.3889	−0.3188	−1.6774	3.2878	3.1057	2.9902	2.9761	55.8889
A,6	0.4254	−0.7746	−1.2072	−0.9036	1.9745	1.9398	1.9240	1.6735	−0.5717	−1.1200	2.5251	2.5223	2.5818	2.5893	22.5633
A,5	1.2027	−0.2856	−0.6632	−0.1908	−0.2562	1.8173	1.7598	1.5004	−0.7573	−1.0318	−1.8694	−1.7870	−1.6809	−1.6678	15.5522
A,4	1.6565	0.4483	−0.1601	0.4153	0.3481	−0.3905	1.6820	1.4134	−0.8576	−1.1389	−3.8686	−3.7856	−3.6859	−3.6740	14.1672
A,3	1.6390	0.8836	0.5846	1.0079	0.9380	0.1864	−0.5766	1.3182	−0.9679	−1.2533	−5.9489	−5.9018	−5.8340	−5.8262	15.0293
A,2	1.6243	0.9131	1.0612	1.7826	1.5039	0.7403	−0.0444	−0.9931	−1.0616	−1.3816	−7.7371	−7.9837	−8.0840	−8.0967	20.3362
								Splitting (NDAS, SPA1, SPL3)							
9,9	−1.4316	−0.6470	1.8754	3.2596	2.8430	2.3508	2.8797	2.0413	−2.0828	−2.7721	7.3095	6.8401	6.5275	6.4884	77.7552
8,8	−1.3746	−0.8367	−1.2990	0.7878	2.3001	3.0937	3.2178	1.0540	−0.7225	−1.5551	32.4365	31.9802	31.6718	31.6331	41.4197
7,7	−0.7004	−0.5946	−1.0926	−1.1459	2.6497	2.7024	1.8847	1.9363	−0.8873	−1.1881	12.1074	11.8765	11.7199	11.7002	31.4097
6,6	0.2474	−0.3967	−0.5616	0.9512	1.8963	3.9559	4.3267	3.9502	0.5967	−3.7415	2.0618	2.1598	2.2178	2.2247	111.3610
3,3	0.0324	0.3960	0.9041	1.7434	1.8518	0.1776	−0.4239	−0.7687	−1.1944	−0.6796	−3.5124	−3.6008	−3.5859	−3.5841	11.5202
2,2	−0.2060	0.1612	0.3278	0.9116	1.4165	1.0436	0.3562	−1.8358	−4.2726	0.5243	0.7806	−0.6548	−1.5428	−1.6533	26.9538
A,A	2.3406	1.9310	1.9771	2.4273	2.9848	2.5847	2.0144	1.5404	−0.8828	−4.2294	43.7168	41.7933	41.5500	41.5197	113.3567
								Splitting (DAS, SPA1, SPL3)							
9,9	−1.0226	−1.6578	2.1637	3.6659	3.3489	2.8115	3.2757	2.3529	−2.4353	−3.1257	12.2136	11.6195	11.2232	11.1736	102.3111
8,8	−1.0098	−1.3462	−2.0886	1.5497	3.2679	4.0409	4.1137	1.3481	−1.0134	−2.2156	41.3669	40.6089	40.0994	40.0355	76.0051
7,7	−0.6350	−0.3998	−2.0566	−0.9899	4.5325	4.5243	3.1395	3.5019	−1.4002	−2.5542	22.4559	21.6501	21.1169	21.0505	96.9609
6,6	0.4383	0.2769	−0.3059	0.7290	2.2154	5.1726	6.0963	5.4168	0.0577	−5.0243	12.4518	11.9238	11.5726	11.5288	200.0406
3,3	0.3065	1.2394	1.1215	2.8169	3.2521	1.2903	−0.2148	−1.0236	−1.6952	−1.7733	6.3942	5.6821	5.2930	5.2450	39.6090
2,2	0.0736	0.5522	0.9858	1.9701	2.7930	2.4211	1.3739	−1.9066	−5.8928	−0.5926	10.2085	8.3997	7.2624	7.1207	60.4776
A,A	2.3406	1.9310	1.9771	2.4273	2.9848	2.5847	2.0144	1.5404	−0.8828	−4.2294	43.7168	41.7933	41.5500	41.5197	113.3567
	A	**2**	**3**	**4**	**5**	**6**	**7**	**8**	**9**	**T**	**m1**	**m2**	**m6**	**m8**	**ss**

TABLE D4
Dealer's Upcard 4 (S17)

	A	2	3	4	5	6	7	8	9	T	m1	m2	m6	m8	ss
Hard Hitting															
17	−1.8878	−1.7845	−3.0873	−4.6961	−1.0052	−0.6087	−0.0255	1.7070	1.9783	2.3524	−44.6251	−44.6335	−44.6402	−44.6411	68.6778
16	−1.4142	−1.5170	−3.4697	−3.9790	−4.3528	−0.2112	0.3512	1.9327	2.2344	2.6064	−24.4636	−24.4916	−24.5118	−24.5144	87.1893
15	−1.0124	−0.9176	−2.7409	−3.2381	−3.6097	−3.6374	0.5886	1.9759	2.2494	2.5856	−18.3209	−18.3847	−18.4301	−18.4359	82.1757
14	−0.8376	−0.5153	−2.0150	−2.5009	−2.8663	−2.9338	−2.8923	2.0403	2.2873	2.5584	−11.8108	−12.1010	−12.2909	−12.3144	72.0457
13	−0.6274	−0.3609	−1.5273	−1.7447	−2.1049	−2.2090	−2.2297	−1.5992	2.2927	2.5276	−5.3771	−5.8545	−6.1639	−6.2021	53.5515
12	−0.4425	−0.2295	−1.2788	−1.2133	−1.3099	−1.4509	−1.5294	−1.1347	−1.4226	2.5029	0.6524	0.1964	−0.1007	−0.1375	37.8864
Hard Doubling															
11	1.4068	0.8876	1.3780	1.7763	2.2324	2.1044	1.6140	−0.8250	−1.5061	−2.2671	28.7977	28.5477	28.3835	28.3631	43.3454
10	−0.7929	0.7394	1.1704	1.6098	2.1613	2.4893	2.2163	−0.4689	−1.1759	−1.9872	23.3121	23.1788	23.0907	23.0798	38.3144
9	−0.6803	1.3082	0.9311	1.4856	2.2239	2.4504	2.1373	−0.4592	−1.1941	−2.0507	5.2739	5.2860	5.2933	5.2942	39.2246
8	−0.3487	0.8650	1.8739	1.6624	2.2947	2.1373	1.6687	−0.6784	−1.1790	−2.0740	−10.6961	−10.6057	−10.5457	−10.5382	38.8186
7	0.0582	0.5572	2.0324	2.8175	2.0995	1.7662	1.2274	−1.1216	−1.3715	−2.0163	−24.1568	−24.0432	−23.9665	−23.9568	40.8188
Soft Doubling															
A,9	−1.8180	1.7750	3.0051	3.7299	4.2648	3.7386	2.4683	−0.7945	−2.1308	−3.5596	−19.2382	−19.6254	−19.8827	−19.9148	123.5103
A,8	−1.1771	−2.6282	2.9169	3.4448	3.3868	2.9428	2.7933	0.1247	−1.2241	−2.6450	−6.7353	−6.9734	−7.1272	−7.1461	86.0989
A,7	−0.4933	−1.9227	−1.5970	2.5217	2.5476	2.7387	2.6480	0.5247	−0.2757	−1.6730	5.9251	6.0170	6.0816	6.0899	45.3986
A,6	0.3514	−0.7841	−0.5219	−0.3006	2.2451	2.2696	2.0080	−0.2315	−0.4898	−1.1366	5.3588	5.6465	5.8374	5.8612	20.7853
A,5	1.0914	−0.2743	0.1625	0.2499	0.0741	2.1365	1.8486	−0.4033	−0.6790	−1.0516	1.0317	1.4440	1.7127	1.7459	14.3896
A,4	1.5068	0.4203	0.7205	0.8258	0.6411	−0.0107	1.7728	−0.4922	−0.7753	−1.1523	−0.8544	−0.4606	−0.2087	−0.1778	13.3562
A,3	1.4924	0.8815	1.4675	1.3737	1.1819	0.5202	−0.4309	−0.5667	−0.8557	−1.2660	−2.4435	−2.3022	−2.2105	−2.1992	16.3619
A,2	1.5450	0.9289	1.9441	2.0992	1.7344	1.0558	0.0852	−2.8436	−0.9656	−1.3958	−3.9854	−4.2018	−4.3387	−4.3554	32.3781
Splitting (NDAS, SPA1, SPL3)															
T,T	−3.8702	2.9133	5.1918	6.5530	7.5445	6.5342	4.1187	−1.9397	−4.4883	−5.6393	−38.7063	−37.7037	−37.0711	−36.9939	361.0558
9,9	−1.3717	−0.3612	3.3048	2.8281	3.0528	3.7359	3.4439	0.0625	−1.8011	−3.2235	10.1465	9.8909	9.7267	9.7065	100.8800
8,8	−1.2939	−1.1001	−1.5722	2.2323	3.6674	3.8897	3.5033	−0.7590	−0.3253	−2.0606	34.3459	34.1298	33.9836	33.9653	68.8576
7,7	−0.5927	−0.7207	−0.7142	0.3890	3.4439	3.4599	2.1354	0.1124	−0.4351	−1.7695	15.0615	14.8778	14.7517	14.7358	42.6494
6,6	−0.2403	−0.4025	0.0696	0.2949	1.0145	2.7660	2.9215	0.0437	−0.5920	−1.4689	7.8496	7.7598	7.6960	7.6879	26.5092
4,4	0.7793	0.8167	2.5954	1.6446	0.8569	0.2982	−0.0491	−0.6108	−1.0136	−1.3294	−16.4625	−16.3073	−16.2063	−16.1938	20.0108
3,3	0.1155	0.4906	1.7254	2.5456	2.6479	1.1096	0.2938	−2.3776	−0.6488	−1.4755	0.3832	0.2142	0.1021	0.0881	32.8223
2,2	0.0007	0.1292	1.3927	2.3059	2.8617	2.7042	1.8349	−2.5506	−2.9435	−1.4338	0.7230	0.6644	0.6260	0.6212	49.5349
A,A	2.2387	1.8389	2.2193	2.6543	3.0637	2.8321	2.3811	0.0337	−0.7734	−4.1221	45.4299	44.6834	44.1899	44.1285	112.0083
Splitting (DAS, SPA1, SPL3)															
T,T	−3.8702	2.9133	5.1918	6.5530	7.5445	6.5342	4.1187	−1.9397	−4.4883	−5.6393	−38.7063	−37.7037	−37.0711	−36.9939	361.0558
9,9	−0.9877	−1.4783	3.6801	3.2775	3.5861	4.2617	3.8808	0.0312	−2.1592	−3.5231	15.4667	15.1092	14.8766	14.8478	127.8387
8,8	−0.9630	−1.7206	−2.3316	3.0808	4.7104	4.9967	4.4789	−1.1593	−0.5940	−2.6246	44.1086	43.6539	43.3462	43.3075	115.2815
7,7	−0.6484	−0.6420	−1.2806	0.5630	5.4693	5.6157	3.4413	0.2074	−0.8933	−2.9581	27.4479	26.7351	26.2651	26.2066	111.9242
6,6	−0.1942	0.2819	0.5697	0.2125	1.4431	4.1643	4.7795	−0.0056	−1.0794	−2.5429	20.3963	19.6789	19.2075	19.1490	69.7850
4,4	1.2274	1.7366	3.6899	2.4351	2.0770	0.6261	0.2614	−0.7280	−1.5894	−2.4341	−4.6562	−5.3049	−5.7262	−5.7782	55.5971
3,3	0.6247	1.4911	2.3901	4.0058	4.4083	2.3351	0.3546	−3.8028	−1.2797	−2.6318	11.4268	10.9170	10.4946	10.4421	93.1892
2,2	0.5682	0.6129	2.6554	3.8564	4.6703	4.4701	2.7748	−4.2384	−4.8261	−2.6359	10.7981	11.0307	10.9076	10.8921	141.1618
A,A	2.2387	1.8389	2.2193	2.6543	3.0637	2.8321	2.3811	0.0337	−0.7734	−4.1221	45.4299	44.6834	44.1899	44.1285	112.0083

TABLE D5
Dealer's Upcard 5 (S17)

	A	2	3	4	5	6	7	8	9	T	m1	m2	m6	m8	ss
Hard Hitting															
17	-1.8607	-2.5345	-4.1452	-4.6622	-0.9236	-0.4252	1.3916	1.7468	2.0145	2.3497	-47.8542	-47.8295	-47.8128	-47.8107	80.9686
16	-1.3725	-2.9910	-3.5072	-3.9485	-4.2911	0.0070	1.5876	1.9625	2.2404	2.5782	-28.5353	-28.3817	-28.2818	-28.2695	95.1149
15	-0.9701	-2.2895	-2.7854	-3.2191	-3.5571	-3.4418	1.6959	2.0319	2.2862	2.5622	-21.6430	-21.7361	-21.7967	-21.8042	87.2946
14	-0.7623	-1.8232	-2.0573	-2.4701	-2.8025	-2.7490	-1.9105	2.0986	2.3005	2.5439	-14.8075	-15.1205	-15.3218	-15.3465	68.8827
13	-0.5797	-1.5698	-1.5517	-1.6985	-2.0148	-2.0210	-1.3930	-1.5908	2.2925	2.5317	-8.3768	-8.7007	-8.9108	-8.9367	51.6020
12	-0.3999	-1.3173	-1.2778	-1.1679	-1.2221	-1.2746	-0.8795	-1.1312	-1.3942	2.5161	-1.9849	-2.2993	-2.5057	-2.5313	37.3296
Hard Doubling															
11	1.3211	1.3231	1.6396	2.0313	2.5704	2.3652	-0.0585	-0.8626	-1.4837	-2.2115	31.4361	31.0848	30.8514	30.8222	45.0226
10	-0.7906	1.1398	1.4336	1.9235	2.7163	2.7173	0.3260	-0.5196	-1.1701	-1.9441	26.0956	25.8600	25.7038	25.6843	39.3040
9	-0.6822	1.8066	1.2866	2.0480	2.7475	2.4211	0.1069	-0.5048	-1.1895	-2.0100	8.7868	8.6430	8.5490	8.5374	40.8316
8	-0.3709	1.6197	2.4375	2.1879	2.5100	2.0327	-0.3604	-0.7370	-1.1971	-2.0306	-6.5927	-6.6665	-6.7127	-6.7183	42.5210
7	0.0004	1.7663	2.5698	3.0004	2.2124	1.6630	-0.7588	-1.1659	-1.3951	-1.9731	-19.3834	-19.4698	-19.5216	-19.5278	45.8406
Soft Doubling															
A,9	-1.8197	2.8054	3.5354	4.1485	4.0781	3.3585	0.6166	-0.8027	-2.0935	-3.4567	-14.7206	-15.2550	-15.6084	-15.6524	122.0027
A,8	-1.2102	-1.2220	3.3208	3.2268	3.2396	3.1184	0.9371	0.0287	-1.2149	-2.5561	-2.8103	-3.0845	-3.2647	-3.2870	73.1055
A,7	-0.5541	-0.5662	-1.6117	2.3600	3.0036	2.8913	0.8031	0.3916	-0.2942	-1.6058	9.4530	9.5110	9.5482	9.5528	37.3759
A,6	0.3055	-0.0705	0.0663	0.0951	2.5647	2.3365	0.0804	-0.3150	-0.5406	-1.1306	8.5761	8.8515	9.0306	9.0528	17.6594
A,5	0.9729	0.5900	0.5977	0.6199	0.4787	2.1940	-0.0608	-0.4730	-0.7123	-1.0518	4.5999	4.9325	5.1461	5.1723	12.2388
A,4	1.3537	1.3118	1.1141	1.1464	1.0001	0.1106	-0.1148	-0.5341	-0.7791	-1.1522	3.1922	3.3005	3.3674	3.3755	13.3367
A,3	1.4043	1.7367	1.8158	1.6818	1.5323	0.6271	-2.2692	-0.5946	-0.8740	-1.2651	1.8745	1.6363	1.4819	1.4629	26.5232
A,2	1.4274	1.8206	2.2833	2.3898	2.0658	1.1411	-1.7577	-2.8332	-1.0217	-1.3788	-0.0202	-0.3851	-0.6233	-0.6527	41.6120
Splitting (NDAS, SPA1, SPL3)															
T,T	-3.7894	4.8985	6.2769	7.4197	7.2718	5.9003	0.8270	-1.8776	-4.3401	-5.6468	-30.3028	-29.6256	-29.2027	-29.1514	371.0879
9,9	-1.4218	1.1785	2.5959	2.8048	4.1046	3.7899	1.1642	0.0409	-1.9221	-3.0837	14.4963	14.0874	13.8198	13.7866	92.3165
8,8	-1.1860	-1.4641	-0.3492	3.2835	4.0942	3.8943	1.0955	-0.9632	-0.3347	-2.0176	36.4513	36.3732	36.3125	36.3045	64.9047
7,7	-0.5542	-0.4387	0.4479	0.9229	3.7510	3.4759	-0.1447	-0.0193	-0.5221	-1.7297	17.7976	17.8735	17.9155	17.9203	39.9648
6,6	-0.2480	0.5029	0.5698	0.7840	1.5066	2.6706	0.6567	-0.0519	-0.6301	-1.4402	11.7830	11.6162	11.4987	11.4838	19.7835
4,4	0.6473	2.5575	2.6301	1.5101	1.2903	0.4607	-2.1546	-0.6544	-1.0143	-1.3182	-10.6805	-10.9569	-11.1378	-11.1603	31.0845
3,3	0.0502	1.4892	1.9117	2.9569	3.0712	1.2597	-1.8805	-2.3579	-0.6989	-1.4504	4.9649	4.5480	4.2761	4.2425	43.6355
2,2	-0.0349	0.8742	1.8199	2.7071	3.2849	2.7969	-0.3916	-2.5502	-2.8741	-1.4081	5.1088	4.8958	4.7526	4.7346	52.8673
A,A	2.1316	2.0884	2.4194	2.7466	3.2861	3.1610	0.8689	-0.0076	-0.7580	-3.9841	47.4661	46.6417	46.0945	46.0263	107.9128
Splitting (DAS, SPA1, SPL3)															
T,T	-3.7894	4.8985	6.2769	7.4197	7.2718	5.9003	0.8270	-1.8776	-4.3401	-5.6468	-30.3028	-29.6256	-29.2027	-29.1514	371.0879
9,9	-1.0435	-0.0055	3.0348	3.3171	4.7269	4.3804	1.2958	0.0247	-2.2830	-3.3619	20.3733	19.8035	19.4286	19.3820	114.9354
8,8	-0.8437	-2.0654	-1.1291	4.2746	5.3504	5.1189	1.4047	-1.3950	-0.5619	-2.5384	47.3118	46.8977	46.6121	46.5759	109.3618
7,7	-0.7567	0.0073	-0.0377	1.1970	6.1780	5.8163	-0.2128	0.0491	-0.9414	-2.8248	32.3794	31.8901	31.5561	31.5140	106.8560
6,6	-0.4089	2.0724	1.6670	1.3057	2.6521	4.5313	1.0393	-0.2500	-1.3162	-2.8232	25.8857	25.7558	25.1233	25.0449	71.2698
4,4	0.9307	3.9645	4.0262	2.5109	2.6846	0.8223	-3.2133	-0.7685	-1.5541	-2.3508	3.3220	2.2074	1.4745	1.3834	82.4184
3,3	0.2853	3.2461	3.1914	5.1542	5.7292	2.8832	-3.2474	-4.0612	-1.5205	-2.9150	17.4069	17.0924	16.4436	16.3627	151.8459
2,2	0.2662	2.0030	3.8054	4.9775	5.9426	5.1971	-0.9212	-4.4980	-5.0709	-2.9254	16.4133	16.7377	16.4282	16.3889	186.6913
A,A	2.1316	2.0884	2.4194	2.7466	3.2861	3.1610	0.8689	-0.0076	-0.7580	-3.9841	47.4661	46.6417	46.0945	46.0263	107.9128

TABLE D6
Dealer's Upcard 6 (S17)

	A	2	3	4	5	6	7	8	9	T	m1	m2	m6	m8	ss
								Hard Hitting							
17	−2.1224	−3.5600	−4.0790	−4.4970	−0.7479	1.0692	1.3899	1.6095	1.9207	2.2542	−52.0909	−52.0694	−52.0558	−52.0541	84.2791
16	−3.7186	−2.9639	−3.3993	−3.7655	−4.0833	1.3486	1.6915	1.9241	2.2381	2.6821	−27.2913	−27.5102	−27.6526	−27.6702	107.1854
15	−3.1160	−2.1956	−2.7264	−3.0811	−3.3972	−2.2413	1.7971	2.0113	2.2660	2.6708	−20.2660	−20.7472	−21.0593	−21.0978	88.9636
14	−2.7757	−1.6156	−1.9218	−2.3719	−2.6798	−1.7396	−1.8506	2.0441	2.2681	2.6607	−13.7312	−14.2242	−14.5448	−14.5844	70.9056
13	−2.4380	−1.3585	−1.3202	−1.5445	−1.9528	−1.2176	−1.4035	−1.6520	2.2987	2.6471	−7.2365	−7.7203	−8.0365	−8.0757	55.2249
12	−2.1026	−1.1023	−1.0429	−0.9243	−1.1057	−0.7091	−0.9374	−1.1979	−1.3997	2.6304	−0.7782	−1.2336	−1.5337	−1.5710	41.2524
								Hard Doubling							
11	1.6638	1.5889	1.8962	2.2679	2.8809	0.6132	−0.2053	−0.7796	−1.4398	−2.1216	33.8559	33.6128	33.4503	33.4300	43.4347
10	−0.4894	1.4165	1.7302	2.3553	3.0159	0.8720	0.2257	−0.4679	−1.1619	−1.8741	28.9085	28.8456	28.8019	28.7963	36.3121
9	−0.2085	2.1434	1.7675	2.4305	2.7752	0.4725	0.1129	−0.3701	−1.2161	−1.9768	11.9860	12.0467	12.0850	12.0897	38.8546
8	0.2836	2.1548	2.9055	2.2623	2.4518	0.0505	−0.3186	−0.4856	−1.1358	−2.0422	−3.1154	−2.9529	−2.8469	−2.8337	42.6067
7	1.4041	2.2105	2.6729	2.9475	2.1102	−0.3745	−0.7663	−0.9468	−1.2539	−2.0009	−17.2402	−16.9935	−16.8321	−16.8121	46.3523
								Soft Doubling							
A,9	−0.4346	3.2750	3.9320	3.9300	3.8488	1.5833	0.4890	−0.8536	−2.0979	−3.4180	−12.3762	−12.6055	−12.7595	−12.7788	111.2387
A,8	0.3797	−0.7976	3.1678	3.0636	3.5902	1.3637	1.0078	0.0932	−1.3231	−2.6363	−1.7716	−1.7010	−1.6581	−1.6530	65.5256
A,7	1.3130	−0.7855	−1.7135	2.7835	3.3434	1.1909	0.8402	0.6538	−0.3003	−1.8314	8.9958	9.4081	9.6751	9.7080	40.2601
A,6	0.6306	0.3151	0.2923	0.3111	2.7129	0.4685	0.1055	−0.0885	−0.4018	−1.0864	11.8345	12.3276	12.6477	12.6872	13.1602
A,5	1.7255	0.7249	0.7065	0.7188	0.5646	0.3275	−0.0515	−0.2578	−0.5808	−0.9694	7.3031	7.7045	7.9647	7.9969	9.1104
A,4	2.2724	1.4508	1.1758	1.2032	1.0458	−1.7988	−0.0979	−0.3073	−0.6626	−1.0703	6.1660	6.1558	6.1516	6.1512	19.5536
A,3	2.3841	2.0781	1.8992	1.6900	1.5258	−1.3322	−2.2869	−0.4259	−0.8039	−1.1821	4.2971	4.1731	4.0956	4.0862	32.2153
A,2	2.4934	2.1550	2.5284	2.4034	1.9853	−0.8769	−1.8540	−2.6828	−0.9192	−1.3081	2.2525	2.0270	1.8790	1.8606	46.0649
								Splitting (NDAS, SPA1, SPL3)							
T,T	−1.1824	5.8239	7.0580	7.0490	6.8801	2.7110	0.6180	−1.9418	−4.3101	−5.6765	−25.9556	−24.7364	−23.9612	−23.8663	341.1243
9,9	1.3229	0.4410	2.5493	3.7377	4.3307	1.6280	1.1207	0.0588	−2.0338	−3.2888	12.7524	12.7561	12.7577	12.7578	92.4802
8,8	−3.1816	−0.2075	0.8469	3.8560	4.3830	1.8120	1.2605	−0.9233	−0.2579	−1.8970	45.0017	45.0387	45.0599	45.0624	65.1479
7,7	−1.4464	0.6543	0.9497	1.2202	4.0155	1.3165	−0.2904	0.2988	−0.2791	−1.6098	24.2260	24.5836	24.8074	24.8346	33.3858
6,6	0.4886	0.6978	0.7918	0.9819	1.5969	0.5629	0.6486	0.1653	−0.5412	−1.3482	14.6176	14.8177	14.9419	14.9569	13.1945
4,4	3.2915	2.3629	2.3592	1.4279	1.2410	−1.9089	−2.3272	−0.5261	−0.8972	−1.2558	−10.8846	−10.8884	−10.8901	−10.8903	42.0118
3,3	1.1779	1.8172	2.1070	3.1139	3.1033	−1.0905	−1.9818	−2.2370	−0.5831	−1.3568	7.5685	7.4242	7.3269	7.3147	46.2802
2,2	1.0852	1.0203	2.1246	2.9688	3.3578	0.4486	−0.5199	−2.3898	−2.8173	−1.3196	7.7342	7.7903	7.8178	7.8207	47.9066
A,A	2.2854	2.4096	2.6576	2.9920	3.7669	1.6400	0.6789	−0.0116	−0.7928	−3.9065	49.6641	48.8992	48.3939	48.3310	106.0542
								Splitting (DAS, SPA1, SPL3)							
T,T	−1.1824	5.8239	7.0580	7.0490	6.8801	2.7110	0.6180	−1.9418	−4.3101	−5.6765	−25.9556	−24.7364	−23.9612	−23.8663	341.1243
9,9	1.7771	−0.8174	3.0498	4.3154	5.0274	1.8897	1.2367	0.0719	−2.3992	−3.5379	19.1156	18.9528	18.8440	18.8304	117.9519
8,8	−2.5592	−0.9760	0.6600	5.5779	6.4556	2.6082	1.7099	−1.3294	−0.6920	−2.8637	56.6592	56.6056	56.4037	56.3785	125.5019
7,7	−1.1607	1.1318	0.5561	1.6964	6.7029	2.2991	−0.4412	0.4898	−0.6424	−2.6579	40.7357	40.4258	40.2150	40.1884	85.1360
6,6	0.6564	2.5060	2.1315	1.6945	2.8357	0.6932	1.0474	0.1581	−1.1560	−2.6417	31.5492	31.2267	30.8997	30.8589	53.0194
4,4	3.7826	4.0223	4.0252	2.5864	2.6750	−2.9183	−3.5049	−0.5278	−1.3687	−2.1930	5.0339	4.3328	3.8720	3.8144	102.7222
3,3	1.7884	3.9606	3.6268	5.5901	5.7851	−1.2204	−3.4765	−3.8868	−1.3098	−2.7144	22.4607	21.9679	21.6122	21.5673	156.6253
2,2	1.7484	2.3971	4.4942	5.5197	6.1834	1.0747	−1.1885	−4.2711	−5.0328	−2.7313	21.5052	21.7997	21.6535	21.6344	173.6809
A,A	2.2854	2.4096	2.6576	2.9920	3.7669	1.6400	0.6789	−0.0116	−0.7928	−3.9065	49.6641	48.8992	48.3939	48.3310	106.0542

TABLE D7
Dealer's Upcard 7 (S17)

	A	2	3	4	5	6	7	8	9	T	m1	m2	m6	m8	ss
							Hard Hitting								
17	-3.6524	-2.8444	-3.3094	-3.7088	0.7781	1.1655	1.4682	1.6715	2.0083	1.6059	-37.7856	-37.7263	-37.6871	-37.6823	67.4000
16	-1.8829	-1.9267	-2.4386	-2.7835	-2.3262	1.7981	2.1033	2.3217	2.7689	0.5915	6.0652	6.0609	6.0597	6.0597	48.4768
15	-0.9512	-0.8138	-1.8908	-2.3337	-1.9632	-1.9439	2.1151	2.2869	2.6950	0.6999	10.1666	10.1290	10.1061	10.1033	37.1472
14	-0.7621	0.0343	-0.7534	-1.7646	-1.5898	-1.6111	-1.6792	2.2522	2.6480	0.8064	14.2490	14.1888	14.1499	14.1451	26.8917
13	-0.5753	0.1494	0.1181	-0.6083	-1.0966	-1.2918	-1.3989	-1.5339	2.5979	0.9099	18.2951	18.2311	18.1880	18.1826	17.9788
12	-0.3901	0.2643	0.2546	0.2820	-0.0376	-0.8535	-1.0829	-1.2459	-1.2340	1.0108	22.3124	22.2587	22.2212	22.2164	9.4308
							Soft Hitting								
17	-1.9826	-1.6710	-2.1553	-2.6216	-0.2702	0.3088	0.8500	1.2232	1.8090	1.1275	15.6924	15.8818	16.0036	16.0186	28.9858
							Hard Doubling								
11	2.8746	2.2558	2.6199	3.0196	2.2165	0.7950	-0.8996	-1.8196	-2.8857	-2.0441	18.1746	17.6197	17.2550	17.2097	64.0397
10	-0.4120	2.2508	2.7306	3.1455	2.2956	1.6999	0.0449	-1.4623	-2.6529	-1.9100	14.0137	13.7822	13.6277	13.6084	54.5163
9	-0.4582	2.9509	2.5644	2.9570	1.9526	1.1383	0.5041	-0.9861	-2.8540	-1.9423	-6.5065	-6.6339	-6.7191	-6.7298	53.8076
8	0.0988	2.5980	3.2271	2.6770	1.5592	0.6386	-0.0128	-0.6420	-2.5161	-1.9070	-27.3238	-27.1552	-27.0468	-27.0334	48.4682
							Soft Doubling								
A,7	-1.0382	-2.5228	-2.8065	2.9329	2.1369	1.8225	1.4839	1.3442	-0.5596	-0.6983	-19.0765	-18.5125	-18.1433	-18.0975	38.0813
A,6	0.6604	-0.6094	-0.3709	-0.1432	2.2251	1.3530	0.5204	0.0358	-0.8652	-0.7015	-7.0045	-6.8816	-6.7994	-6.7891	10.7361
							Splitting (NDAS, SPA1, SPL3)								
9,9	-3.4598	-1.3464	1.0861	1.2335	0.6996	1.4146	1.6738	1.1634	-0.2672	-0.5494	-7.6185	-7.0912	-6.7362	-6.6917	24.4084
8,8	-0.7239	2.0266	2.4774	5.1666	4.3837	0.1171	0.1565	-2.0246	-0.0222	-2.8893	66.6548	67.3617	67.8208	67.8775	94.2095
7,7	-0.9939	-0.4293	0.1494	1.0557	2.5258	2.9262	2.1585	-0.9674	-0.5743	-1.4627	21.5781	21.5870	21.5828	21.5818	31.7337
6,6	-1.7707	-2.1731	-3.0369	-3.7127	-3.6717	2.5252	3.2310	3.6782	4.3470	0.1459	-8.9474	-9.8339	-10.4098	-10.4810	93.6731
3,3	-0.2378	1.6945	2.1160	1.9796	-0.1904	-3.2325	-3.8949	-3.5676	1.2168	1.0291	4.9532	4.6393	4.4458	4.4225	55.4246
2,2	-0.6903	0.0529	1.1922	2.2533	0.7689	-0.1629	-0.6703	-3.2897	-3.1411	0.9218	4.4679	4.2215	4.0525	4.0312	32.1326
A,A	3.8334	3.2091	3.5593	4.0951	3.7237	2.5329	0.2277	-0.8637	-2.0522	-4.5663	30.1436	29.9397	29.8070	29.7906	163.1277
							Splitting (DAS, SPA1, SPL3)								
9,9	-2.8372	-1.8650	1.6749	1.8933	1.1970	1.6542	1.5918	0.8987	-0.8255	-0.8455	-4.0109	-3.7474	-3.5649	-3.5417	28.9692
8,8	-0.1105	2.1931	2.5849	6.5023	5.3920	0.7463	0.1540	-2.7119	-0.9041	-3.4615	73.2549	73.4230	73.5300	73.5431	139.5385
7,7	-0.3610	0.6054	0.4309	1.2809	3.5462	3.5594	1.8874	-1.4115	-1.4300	-2.0269	28.4100	27.7643	27.3307	27.2764	51.6010
6,6	-1.1326	-1.1430	-1.8492	-3.3335	-3.7144	2.8341	3.2399	3.2407	3.4876	-0.4074	-2.0858	-3.6350	-4.6533	-4.7798	72.7756
3,3	0.3560	2.6725	2.9310	3.2735	0.7945	-2.6504	-4.6099	-4.7345	0.3069	0.4151	10.6084	10.2364	10.0021	9.9735	78.6814
2,2	-0.1196	0.7218	2.3126	3.5415	1.7244	0.4193	-0.7090	-4.3299	-4.6717	0.2776	9.6315	9.5718	9.5262	9.5203	62.9589
A,A	3.8334	3.2091	3.5593	4.0951	3.7237	2.5329	0.2277	-0.8637	-2.0522	-4.5663	30.1436	29.9397	29.8070	29.7906	163.1277
	A	**2**	**3**	**4**	**5**	**6**	**7**	**8**	**9**	**T**	**m1**	**m2**	**m6**	**m8**	**ss**

TABLE D8
Dealer's Upcard 8 (S17)

	A	2	3	4	5	6	7	8	9	T	m1	m2	m6	m8	ss
							Hard Hitting								
17	−2.2913	−2.3387	−2.8600	−2.4754	1.3772	1.6552	1.8399	2.1804	1.7811	0.2829	−12.3434	−12.3755	−12.3944	−12.3966	41.2946
16	−1.0509	−1.3859	−2.4377	−2.1989	−2.3607	1.7246	1.9253	2.3760	0.2019	0.8016	5.2286	5.2155	5.2097	5.2092	34.3136
15	−0.6263	−0.3146	−1.3478	−1.7643	−2.0453	−2.0373	1.9063	2.3179	0.3260	0.8963	9.0277	8.9977	8.9807	8.9787	26.0812
14	−0.4579	0.0425	−0.2557	−0.7419	−1.6091	−1.7712	−1.8992	2.2869	0.4483	0.9893	12.8089	12.7719	12.7491	12.7464	19.5066
13	−0.2912	0.1521	0.1205	0.2822	−0.6081	−1.3863	−1.6222	−1.5329	0.5678	1.0795	16.5563	16.5300	16.5124	16.5101	12.4587
12	−0.1256	0.2619	0.2481	0.5637	0.3941	−0.3871	−1.2289	−1.3007	−3.0940	1.1671	20.2719	20.2706	20.2694	20.2693	18.9919
							Soft Hitting								
18	−2.2063	−1.6901	−2.1999	0.1029	−0.0052	0.4165	0.6498	1.0917	2.0196	0.4553	−6.7751	−6.6991	−6.6508	−6.6449	19.2698
							Hard Doubling								
11	2.7304	2.1334	2.5337	1.6438	1.8828	1.0728	−0.3484	−2.0876	−1.3162	−2.0612	12.6217	12.3461	12.1626	12.1397	49.0291
10	−0.6284	2.2399	2.6234	1.7114	1.9583	1.4555	0.5798	−1.1486	−1.0632	−1.9320	8.8239	8.8498	8.8628	8.8643	38.8938
9	−0.3443	2.8618	2.2786	1.3818	1.5540	0.8047	0.4634	−0.7608	−0.3811	−1.9645	−12.8015	−12.6384	−12.5333	−12.5204	34.8481
							Soft Doubling								
A,7	−1.9843	−2.7976	−3.0834	1.7333	1.8013	1.5080	1.3504	0.9456	2.2150	−0.4221	−13.9703	−13.7758	−13.6493	−13.6336	38.1310
							Splitting (NDAS, SPA1, SPL3)								
9,9	−3.5826	−0.2007	1.8072	1.0815	1.1034	1.6369	1.9174	2.2710	0.6244	−1.6646	8.9625	9.8188	10.3856	10.4562	41.5160
8,8	−0.7146	0.9911	1.8771	3.1496	3.3981	−0.7669	−0.6049	−1.7815	−1.0501	−1.1244	39.3341	39.5717	39.7170	39.7344	36.7715
7,7	−0.4794	−0.9182	−1.0964	−1.5436	1.5229	2.1358	2.6304	−1.8875	−1.5787	0.3037	−4.9545	−5.3742	−5.6524	−5.6871	24.8818
6,6	−0.9015	−1.3855	−2.9361	−4.5470	−4.4732	2.0525	3.0945	3.4863	3.8145	0.4488	−17.7309	−18.4433	−18.9092	−18.9670	93.3376
3,3	−0.1423	1.4763	2.5114	1.7837	0.1434	−3.4816	−4.2192	−3.8473	0.5552	1.3051	−4.8896	−5.0809	−5.1931	−5.2063	63.5558
2,2	−0.3190	0.6278	0.8536	0.9487	1.1170	0.2337	−0.7818	−3.7172	−3.7886	1.2065	−5.1716	−5.3449	−5.4618	−5.4765	38.0307
A,A	3.4873	3.0303	3.5126	2.7708	3.1981	2.7362	1.4079	−0.9598	−0.3189	−4.7161	25.2955	25.4288	25.5152	25.5259	151.0455
							Splitting (DAS, SPA1, SPL3)								
9,9	−3.0097	−0.5015	2.3597	1.4596	1.5229	1.9094	1.9091	1.9318	0.3515	−1.9832	11.5905	12.2196	12.6395	12.6920	46.2050
8,8	−0.2066	1.3367	2.1784	3.9063	4.2390	−0.1710	−0.4521	−2.4213	−1.3918	−1.7544	43.9771	43.7805	43.6419	43.6242	60.1482
7,7	0.0423	0.0384	−0.6547	−1.5259	2.3508	2.7060	2.5950	−2.3796	−1.9084	−0.3159	−0.2989	−1.1545	−1.7228	−1.7937	32.0467
6,6	−0.3895	−0.4359	−1.8448	−4.4001	−4.4331	2.4008	3.2485	2.9981	3.4913	−0.1588	−13.1307	−14.2440	−14.9849	−15.0774	80.3539
3,3	0.3331	2.3755	3.3339	2.4952	0.9527	−2.9103	−4.5503	−4.7648	0.2011	0.6335	−1.1325	−1.3133	−1.4149	−1.4265	77.5262
2,2	0.1321	1.3079	1.9036	1.6564	1.9258	0.8286	−0.6387	−4.5653	−4.6337	0.5208	−1.6126	−1.6750	−1.7158	−1.7208	56.2967
A,A	3.4873	3.0303	3.5126	2.7708	3.1981	2.7362	1.4079	−0.9598	−0.3189	−4.7161	25.2955	25.4288	25.5152	25.5259	151.0455
							Late Surrender								
17	−1.2389	−0.3068	−0.5486	0.2926	0.2814	0.6303	0.8331	1.2070	1.0518	−0.5505	−12.0182	−11.9113	−11.8403	−11.8314	6.9612
16	−0.1246	0.9451	1.7926	2.4317	2.6249	−1.0671	−1.0359	−0.9462	−0.7312	−0.9724	−4.3840	−4.2679	−4.1928	−4.1836	24.3489
15	−0.4069	−0.0175	0.8958	2.1671	2.4924	2.5509	−1.1764	−1.0885	−0.8772	−1.1349	−8.5044	−8.4119	−8.3516	−8.3441	26.8744
14	−0.4328	−0.3054	−0.0823	1.2892	2.2207	2.4292	2.4500	−1.2711	−1.0378	−1.3149	−12.9199	−12.8636	−12.8265	−12.8219	28.3936
13	−0.4682	−0.2981	−0.4213	0.3388	1.3655	2.1568	2.2767	2.3119	−1.2135	−1.5122	−17.6146	−17.6295	−17.6367	−17.6374	28.2651
12	−0.5123	−0.3247	−0.4412	0.0525	0.4193	1.2543	1.9623	2.1258	2.3701	−1.7265	−22.6748	−22.7609	−22.8158	−22.8225	28.2251
	A	**2**	**3**	**4**	**5**	**6**	**7**	**8**	**9**	**T**	**m1**	**m2**	**m6**	**m8**	**ss**

TABLE D9
Dealer's Upcard 9 (S17)

	A	2	3	4	5	6	7	8	9	T	m1	m2	m6	m8	ss
Hard Hitting															
17	−1.4710	−1.8109	−2.1302	−2.5141	1.3072	1.5611	1.7932	1.3971	−0.0977	0.4913	−13.2924	−13.1738	−13.0941	−13.0841	26.5905
16	−0.7875	−0.9039	−1.2507	−2.1007	−2.4822	1.6004	1.9556	−0.2152	0.3878	0.9491	2.9665	3.1746	3.3134	3.3307	23.7604
15	−0.3998	−0.3420	−0.2853	−1.1364	−2.0948	−2.2176	1.9152	−0.0735	0.5000	1.0335	6.5254	6.7119	6.8355	6.8509	19.1516
14	−0.2468	−0.0092	0.2098	−0.1701	−1.1417	−1.8784	−1.9056	0.0665	0.6105	1.1163	10.0666	10.2412	10.3551	10.3692	13.9585
13	−0.0952	0.0967	0.4730	0.3044	−0.1878	−0.9252	−1.5716	−3.5968	0.7181	1.1961	13.5710	13.7518	13.8683	13.8827	22.8717
12	0.0545	0.2029	0.4847	0.5465	0.3204	0.0255	−0.6511	−3.0986	−2.9772	1.2731	17.0357	17.2413	17.3743	17.3908	26.0527
Soft Hitting															
18	−1.1784	−1.4889	−1.8309	0.3518	0.2471	0.6128	0.7507	1.7837	1.0265	−0.0686	8.5777	8.4106	8.2983	8.2842	12.3352
Hard Doubling															
11	2.5860	2.0672	1.1330	1.2477	1.5569	0.8363	−0.1096	0.1675	−1.2392	−2.0614	7.5408	7.2458	7.0501	7.0257	35.4986
10	−0.3784	2.1223	1.0953	1.2720	1.5858	1.1074	0.2955	0.9663	−0.4501	−1.9040	2.7212	2.7536	2.7718	2.7739	26.9313
Splitting (NDAS, SPA1, SPL3)															
9,9	−2.1710	−0.2025	1.1722	0.9128	0.8216	1.3026	1.5320	2.0536	−0.0159	−1.3514	8.7596	9.0170	9.1798	9.1997	23.2027
8,8	−0.3596	−0.2477	−0.7388	2.3954	2.5327	−1.6323	−1.3960	−2.3352	−0.4390	0.5551	11.3310	10.8980	10.6011	10.5636	24.3810
7,7	−0.1382	−0.9243	−2.6828	−2.2617	0.7456	2.0337	3.0479	−1.7717	0.2850	0.4167	−15.8849	−16.1572	−16.3361	−16.3583	31.0823
6,6	−0.6472	−0.7391	−2.9898	−4.6514	−4.7409	1.6363	2.2249	2.9447	4.1780	0.6961	−27.5726	−28.2049	−28.6143	−28.6648	89.7081
3,3	0.1156	1.0733	1.7494	1.5559	0.5395	−3.0887	−4.3907	−4.4920	0.8779	1.5149	−15.9832	−16.0801	−16.1328	−16.1387	65.8848
2,2	−0.0421	1.0659	−0.1499	0.3966	0.5769	0.6230	−0.3823	−4.2944	−3.4945	1.4252	−15.9987	−16.1961	−16.3205	−16.3357	40.9636
A,A	3.3036	3.1591	2.4013	2.3764	2.8980	2.1032	1.5358	1.9083	−0.1529	−4.8832	22.5876	22.6843	22.7436	22.7507	146.5346
Splitting (DAS, SPA1, SPL3)															
9,9	−1.6364	−0.2947	1.4362	1.1991	1.1636	1.4976	1.5485	2.1183	−0.2566	−1.6939	10.4599	10.4921	10.5086	10.5103	28.2901
8,8	0.1305	0.2894	−0.7011	2.9455	3.1866	−1.2128	−1.3024	−2.1828	−0.7013	−0.1129	13.8079	13.0145	12.4790	12.4117	27.8968
7,7	0.3370	−0.0635	−2.4339	−2.2042	1.3638	2.4395	3.0374	−1.5152	0.0338	−0.2487	−13.6016	−14.1306	−14.4867	−14.5313	30.4813
6,6	−0.1650	0.1267	−2.5228	−4.3987	−4.6201	1.9183	2.3194	3.2362	3.9503	0.0389	−25.1502	−26.1034	−26.7388	−26.8182	82.2457
3,3	0.5562	1.9005	2.0867	2.0613	1.1850	−2.6333	−4.4416	−4.5808	0.6152	0.8127	−13.9803	−14.2076	−14.3377	−14.3528	64.5952
2,2	0.4157	1.7517	0.2990	0.9280	1.2433	1.0486	−0.2838	−4.2462	−4.0087	0.7131	−14.0690	−14.3600	−14.5375	−14.5589	43.0522
A,A	3.3036	3.1591	2.4013	2.3764	2.8980	2.1032	1.5358	1.9083	−0.1529	−4.8832	22.5876	22.6843	22.7436	22.7507	146.5346
Late Surrender															
17	−1.0820	−0.2987	0.5435	0.4656	0.4244	0.7474	0.9668	0.8132	−0.7875	−0.4481	−8.0896	−7.8855	−7.7512	−7.7345	5.5302
16	−0.2625	0.4667	1.6938	2.5795	2.8745	−0.8430	−0.8124	−0.5991	−0.8419	−1.0639	0.7992	0.8681	0.9114	0.9166	25.0383
15	−0.5610	−0.0223	0.8288	1.7820	2.6820	2.8437	−0.9360	−0.7266	−0.9860	−1.2261	−2.9273	−2.8842	−2.8561	−2.8526	27.8470
14	−0.5976	−0.3089	0.3535	0.9119	1.9039	2.6581	2.7072	−0.8689	−1.1477	−1.4029	−6.9368	−6.9227	−6.9124	−6.9111	29.3734
13	−0.6454	−0.3255	0.0777	0.4577	1.0533	1.8352	2.4795	2.7766	−1.3243	−1.5962	−11.2087	−11.2496	−11.2735	−11.2763	31.0189
12	−0.7001	−0.3520	0.1222	0.1591	0.5681	0.9526	1.6438	2.5466	2.2866	−1.8067	−15.7560	−15.8788	−15.9590	−15.9690	29.3577
	A	**2**	**3**	**4**	**5**	**6**	**7**	**8**	**9**	**T**	**m1**	**m2**	**m6**	**m8**	**ss**

TABLE D10
Dealer's Upcard T (S17)

	A	2	3	4	5	6	7	8	9	T	m1	m2	m6	m8	ss
Hard Hitting															
17	–0.8804	–0.6577	–1.7455	–2.5230	1.2408	1.4685	1.0388	–0.5832	0.0559	0.6465	–16.9323	–16.7023	–16.5501	–16.5311	17.4100
16	–0.4992	–0.2903	–0.8042	–1.7279	–2.5683	1.6446	–0.7109	–0.0567	0.5524	1.1151	–0.4459	–0.1915	–0.0233	–0.0023	19.0546
15	–0.1699	0.1873	–0.3168	–0.7279	–1.7532	–2.2287	–0.5370	0.0853	0.6642	1.1992	3.1103	3.3550	3.5158	3.5358	15.2239
14	–0.0802	0.4374	0.1709	–0.2614	–0.7670	–1.4149	–4.2072	0.2249	0.7738	1.2809	6.6386	6.8887	7.0509	7.0709	27.7989
13	0.0085	0.4564	0.4040	0.2055	–0.2645	–0.4334	–3.2152	–3.4800	0.8802	1.3596	10.1258	10.4000	10.5781	10.6002	31.2874
12	0.0978	0.4526	0.4041	0.4655	0.2360	0.0352	–2.0550	–2.5215	–2.8569	1.4355	13.5765	13.8916	14.0985	14.1242	27.6372
Soft Hitting															
18	–0.2835	–0.9367	–1.9253	0.5569	0.4682	0.3019	1.4495	0.6916	–0.3391	0.0041	3.5343	3.4909	3.4595	3.4554	7.9792
Hard Doubling															
11	1.6097	0.7649	0.7748	0.7855	1.1062	0.8410	1.5305	0.7412	–0.6131	–1.8852	5.6953	5.8641	5.9733	5.9867	23.8074
10	–1.9307	0.5811	0.5844	0.6370	0.8799	0.8810	1.6360	0.8464	–0.0494	–1.0164	–2.7487	–3.0727	–3.2852	–3.3116	13.8909
Splitting (NDAS, SPA1, SPL3)															
9,9	–3.0619	–1.4883	0.6728	0.7757	0.6932	0.2196	1.3540	–0.1661	–0.5454	0.3866	–13.9296	–14.4623	–14.8245	–14.8701	15.9295
8,8	–1.4109	–1.4436	–1.3290	1.9926	2.7534	–1.5951	–2.4457	–0.4875	1.2084	0.6893	5.7702	5.3063	4.9842	4.9432	29.5168
7,7	–1.3494	–1.7914	–2.3888	–1.9201	0.8294	1.3089	3.4790	0.0182	0.5131	0.3253	–19.5320	–19.8690	–20.0891	–20.1164	29.6147
6,6	–1.2319	–1.2950	–1.9873	–2.9617	–2.8521	0.6899	1.7487	2.9147	4.1436	0.2078	–34.1248	–34.3894	–34.5584	–34.5791	53.4209
3,3	–0.9099	0.0881	1.4008	0.7955	–0.0438	–2.3475	–4.5631	–4.1345	1.4010	2.0784	–22.2494	–22.4693	–22.5966	–22.6115	66.1000
2,2	–0.9428	0.5045	–0.1847	0.4488	0.4804	0.1069	–0.2365	–3.4591	–3.2636	1.6365	–22.5471	–22.7234	–22.8386	–22.8529	35.0064
A,A	2.5672	1.8655	2.1075	2.0011	2.4676	2.1204	2.7994	2.6887	1.0571	–4.9186	23.9539	24.4736	24.8097	24.8511	142.0570
Splitting (DAS, SPA1, SPL3)															
9,9	–2.7267	–1.6437	0.8532	0.9561	0.9199	0.3961	1.6548	–0.0114	–0.6740	0.0689	–12.7828	–13.3420	–13.7243	–13.7726	15.9934
8,8	–1.0730	–1.2669	–1.4920	2.1617	2.9814	–1.4196	–2.1425	–0.3796	1.1371	0.3733	6.9595	6.4474	6.0912	6.0458	27.1445
7,7	–1.0105	–1.6154	–2.2253	–2.1008	1.0561	1.4864	3.7200	0.2040	0.4412	0.0111	–18.3177	–18.7117	–18.9760	–19.0091	30.3957
6,6	–0.8897	–1.1307	–1.8237	–2.7992	–2.9907	0.8075	2.0855	3.1009	4.0710	–0.1077	–32.8738	–33.2157	–33.4403	–33.4681	53.4119
3,3	–0.5841	0.2496	1.5126	0.9833	0.2046	–2.1531	–4.2323	–4.2892	1.3223	1.7465	–21.0653	–21.3370	–21.4938	–21.5122	58.5959
2,2	–0.6271	0.6148	–0.0051	0.6285	0.7158	0.2887	0.0713	–3.2974	–3.5663	1.2942	–21.6084	–21.7104	–21.7749	–21.7828	32.0574
A,A	2.5672	1.8655	2.1075	2.0011	2.4676	2.1204	2.7994	2.6887	1.0571	–4.9186	23.9539	24.4736	24.8097	24.8511	142.0570
Late Surrender															
17	–0.0222	0.7628	0.6475	0.6427	0.5068	0.7309	0.5643	–1.1726	–0.8044	–0.4639	–8.4515	–8.2384	–8.0978	–8.0803	5.4071
16	0.2830	0.9290	1.3769	2.3179	3.0681	–0.7401	–0.5087	–0.7721	–1.0130	–1.2353	4.0662	4.0248	3.9968	3.9932	26.1571
15	–0.0462	0.4514	0.8895	1.3179	2.2529	3.1332	–0.6826	–0.9141	–1.1248	–1.3193	0.5101	0.4783	0.4576	0.4551	27.1559
14	–0.0699	0.1817	0.4234	0.9063	1.4063	2.4936	3.0279	–1.0644	–1.2924	–1.5031	–3.2492	–3.3097	–3.3467	–3.3512	30.2429
13	–0.0811	0.1819	0.1645	0.4498	0.9525	1.6172	2.3683	2.6122	–1.4744	–1.6977	–7.2862	–7.3806	–7.4403	–7.4476	29.9269
12	–0.0931	0.2159	0.1636	0.1534	0.4646	1.1835	1.4716	1.9184	2.1708	–1.9122	–11.5784	–11.7358	–11.8389	–11.8517	26.9064
	A	**2**	**3**	**4**	**5**	**6**	**7**	**8**	**9**	**T**	**m1**	**m2**	**m6**	**m8**	**ss**

TABLE D11
Dealer's Upcard A (H17)

	A	2	3	4	5	6	7	8	9	T	m1	m2	m6	m8	ss
Hard Hitting															
17	−0.6065	−1.6715	−2.6588	−3.3239	0.4286	0.3851	−1.2694	−0.1991	0.7928	2.0307	−7.4421	−6.9046	−6.5502	−6.5062	40.3852
16	−0.1131	−0.8735	−1.7441	−2.6258	−3.1932	−0.0528	−0.5774	0.4192	1.3316	1.8573	4.6978	5.1807	5.4985	5.5380	36.9917
15	0.1060	−0.2386	−0.8596	−1.7530	−2.5430	−3.6047	−0.3815	0.5559	1.4129	1.8264	7.9043	8.3298	8.6068	8.6411	39.1350
14	0.1674	0.0007	−0.2442	−0.8780	−1.6268	−2.8709	−3.8961	0.6841	1.4901	1.7935	11.0499	11.4461	11.7036	11.7354	42.4814
13	0.2280	0.0434	−0.0236	−0.2001	−0.7164	−1.9162	−3.1152	−2.8928	1.5627	1.7575	14.1218	14.5239	14.7873	14.8199	37.1502
12	0.2584	0.0887	0.0736	0.0739	−0.0497	−0.9709	−2.1098	−2.1694	−2.0705	1.7189	17.1245	17.5677	17.8597	17.8960	26.2940
Soft Hitting															
18	−0.4493	−1.4653	−2.3834	−0.0613	−0.2653	0.3075	0.0804	−0.7953	0.3532	1.1697	5.7613	6.2712	6.6030	6.6441	14.4348
Hard Doubling															
11	1.5731	1.6718	1.8905	2.2710	2.3956	2.5078	2.0313	0.6094	−0.8573	−3.5233	2.6135	1.3228	0.4820	0.3780	80.9165
10	−1.8717	1.5697	1.8365	2.3159	2.4334	2.6763	2.2873	1.1186	0.2338	−3.1500	−4.3921	−5.6740	−6.5060	−6.6088	74.0140
Splitting (NDAS, SPA1, SPL3)															
9,9	−3.0254	−1.1896	0.9682	1.0662	0.6397	1.4665	0.1650	0.4542	2.4747	−0.7549	−2.5379	−2.5918	−2.6447	−2.6522	23.8391
8,8	−2.3846	−0.5983	0.2113	3.3730	3.6935	−0.4907	−0.2152	2.0474	2.6479	−2.0710	2.4940	2.1451	1.9038	1.8732	59.7548
7,7	−1.5471	−1.2710	−1.8713	−1.1679	1.5368	2.7563	5.3435	0.4392	0.8795	−1.2745	−26.7102	−26.8348	−26.9146	−26.9243	54.8499
6,6	−1.2869	−0.9889	−2.4219	−3.9325	−4.4594	2.2424	2.2851	3.8846	5.3950	−0.1793	−33.4553	−34.2049	−34.6822	−34.7406	98.4248
3,3	−0.9852	0.5094	2.2221	2.1508	1.0836	−2.3793	−3.8029	−3.1328	2.3098	0.5061	−22.8808	−22.9684	−23.0196	−23.0256	48.2649
2,2	−1.0659	0.8401	0.1371	0.9756	1.0758	1.0352	0.1400	−2.7266	−2.1497	0.4346	−23.2382	−23.2434	−23.2502	−23.2512	17.8722
A,A	2.5775	2.8666	3.2216	3.4686	3.7432	3.9019	3.6997	2.1993	0.0943	−6.4431	20.6057	18.6283	17.3307	17.1696	251.0962
Splitting (DAS, SPA1, SPL3)															
9,9	−2.7129	−1.1226	1.3474	1.5050	1.0908	1.9439	0.5539	0.5770	2.3360	−1.3796	−1.7925	−2.2186	−2.5134	−2.5508	31.3790
8,8	−2.0668	−0.2614	0.2833	3.8056	4.1428	−0.0115	0.1749	2.1609	2.5340	−2.6905	3.3218	2.5604	2.0493	1.9853	76.1414
7,7	−1.2240	−0.9404	−1.5108	−1.0781	1.9882	3.2379	5.7016	0.6069	0.7730	−1.8886	−25.7723	−26.3616	−26.7492	−26.7973	68.0060
6,6	−0.9745	−0.6673	−2.0660	−3.5096	−4.3688	2.6898	2.7193	4.0525	5.2895	−0.7912	−32.4799	−33.7111	−34.5096	−34.6081	98.6031
3,3	−0.6806	0.8294	2.5552	2.6114	1.5635	−1.8813	−3.4019	−3.1928	2.1622	−0.1413	−22.2452	−22.6531	−22.9082	−22.9392	47.0048
2,2	−0.7758	1.1390	0.5198	1.4272	1.5460	1.5219	0.5272	−2.6171	−2.3683	−0.2299	−22.8718	−23.0609	−23.1827	−23.1977	21.8602
A,A	2.5775	2.8666	3.2216	3.4686	3.7432	3.9019	3.6997	2.1993	0.0943	−6.4431	20.6057	18.6283	17.3307	17.1696	251.0962
Late Surrender															
17	0.1296	−0.1207	−0.3720	−0.5624	−0.5281	−0.4753	−1.9248	−1.2362	−0.5680	1.4145	0.7780	1.1752	1.4347	1.4668	14.5497
16	0.1186	0.6760	1.3811	2.0636	2.6737	−0.9509	−0.8984	−1.2235	−1.4892	−0.5878	4.3820	4.2969	4.2398	4.2327	20.5936
15	−0.0119	0.1027	0.6280	1.3929	2.2646	2.5944	−1.0521	−1.3979	−1.6819	−0.7097	0.8258	0.7543	0.7101	0.7048	22.1091
14	−0.0112	−0.0665	0.0786	0.6527	1.5469	2.1346	2.4869	−1.5780	−1.8824	−0.8404	−2.8925	−3.0024	−3.0710	−3.0794	22.4298
13	−0.0156	−0.0196	−0.1014	0.0502	0.7812	1.4102	2.0096	1.9284	−2.0998	−0.9858	−6.7879	−6.9913	−7.1237	−7.1401	18.6653
12	0.0083	0.0034	−0.1057	−0.2110	0.1938	0.6415	1.2633	1.4089	1.3656	−1.1420	−10.8711	−11.2309	−11.4694	−11.4991	11.1677
	A	**2**	**3**	**4**	**5**	**6**	**7**	**8**	**9**	**T**	**m1**	**m2**	**m6**	**m8**	**ss**

TABLE D12
Dealer's Upcard 2 (H17)

	A	2	3	4	5	6	7	8	9	T	m1	m2	m6	m8	ss
Hard Hitting															
17	-1.9551	-1.8770	-2.4017	-2.9664	-0.0361	-0.7210	-0.2862	0.2247	0.6858	2.3332	-38.1185	-38.1487	-38.1671	-38.1693	44.8124
16	-1.2417	-1.3716	-1.8523	-2.4219	-4.5084	-0.4005	0.0656	0.5901	1.0451	2.5239	-18.5092	-18.5921	-18.6463	-18.6531	60.1317
15	-0.8115	-0.7409	-1.2379	-1.8161	-3.7620	-3.8428	0.2801	0.7740	1.1941	2.4908	-12.9998	-13.0909	-13.1534	-13.1613	61.8762
14	-0.6364	-0.3906	-0.6233	-1.1706	-2.9927	-3.1159	-3.1888	0.9552	1.3414	2.4555	-7.5286	-7.6099	-7.6674	-7.6747	57.9793
13	-0.4930	-0.2870	-0.2481	-0.5246	-2.2089	-2.3722	-2.4942	-2.5495	1.4890	2.4221	-2.0427	-2.1230	-2.1796	-2.1869	49.5737
12	-0.3269	-0.1413	-0.1044	-0.1186	-1.4486	-1.6307	-1.7968	-1.9027	-2.0608	2.3827	3.8092	3.5415	3.3661	3.3443	38.7143
Hard Doubling															
11	1.6461	0.9999	0.9922	1.0640	1.8458	1.6347	1.1104	0.6567	0.1779	-2.5319	23.7507	23.4976	23.3341	23.3139	39.2441
10	-0.7224	0.9106	0.8179	0.9340	1.7204	1.7423	1.5180	1.2723	0.7062	-2.2248	17.8483	17.7195	17.6350	17.6245	33.1092
9	-0.5771	1.4507	0.4659	0.5500	1.4831	1.5121	1.7484	1.6824	0.8011	-2.2791	-1.3830	-1.3616	-1.3498	-1.3485	34.7496
8	-0.2281	0.9980	1.1637	0.2951	1.3450	1.5456	1.6812	1.5068	0.8317	-2.2847	-18.4856	-18.3456	-18.2554	-18.2444	33.3556
Soft Doubling															
A,9	-1.8941	1.9269	1.9802	2.1009	3.3332	3.4609	2.9464	1.2396	-0.5175	-3.6441	-27.0397	-27.4303	-27.6881	-27.7202	102.3280
A,8	-1.2754	-2.7966	1.7382	1.8063	3.1705	3.2431	2.6604	1.5950	0.4917	-2.6583	-13.5334	-13.7559	-13.9052	-13.9239	74.4304
A,7	-0.5961	-2.1089	-3.1552	1.4979	2.9216	2.4095	1.9242	2.0623	1.4782	-1.6084	0.4854	0.4601	0.4474	0.4460	51.8315
A,6	0.5107	-0.7222	-1.2089	-1.5831	1.5964	1.7631	1.7115	1.6707	1.4381	-1.2941	-0.8480	-0.7640	-0.7034	-0.6956	24.8950
A,5	1.2580	-0.1888	-0.6697	-1.0476	-0.5645	1.6803	1.5949	1.5234	1.2638	-1.2125	-5.2075	-5.0363	-4.9222	-4.9080	18.6487
A,4	1.7151	0.5865	-0.1360	-0.5011	0.0791	-0.5828	1.5503	1.4464	1.1840	-1.3354	-7.2539	-7.0803	-6.9704	-6.9570	16.9320
A,3	1.7269	1.0216	0.5968	0.0584	0.7151	0.0498	-0.7483	1.3699	1.0947	-1.4713	-9.4388	-9.2737	-9.1738	-9.1619	17.1926
A,2	1.7062	1.0101	1.0522	0.8098	1.3432	0.6643	-0.1397	-0.9703	0.9886	-1.6161	-11.7056	-11.5967	-11.5346	-11.5274	20.3250
Splitting (NDAS, SPA1, SPL3)															
9,9	-1.6620	-0.5681	1.6835	1.9847	3.5649	2.3927	1.8634	1.8186	-0.0364	-2.7603	4.8871	4.4524	4.1625	4.1262	65.5504
8,8	-1.0764	-0.9283	-1.0148	1.5801	1.0689	1.9207	2.5394	1.1366	1.3004	-1.6316	28.9795	28.4348	28.0807	28.0370	30.4590
7,7	-0.4677	-0.6347	-0.9427	-1.0554	1.4350	2.1616	1.5562	1.7716	1.1140	-1.2344	8.4856	8.1990	8.0105	7.9871	22.2527
6,6	0.0952	-0.2547	-0.7311	-0.9896	1.4705	3.8040	4.0265	3.8598	3.3539	-3.6586	-2.6640	-2.6507	-2.6384	-2.6366	114.1220
3,3	0.0758	0.4371	1.0634	1.0634	1.6123	-0.0615	-0.6832	-0.8031	1.0142	-0.9296	-7.5970	-7.2923	-7.0918	-7.0669	10.6585
2,2	-0.0550	0.2966	0.4210	0.8998	1.2001	0.8499	0.0798	-1.8799	-2.3960	0.1459	-2.6596	-3.6402	-4.2865	-4.3669	12.6068
A,A	2.5125	2.0128	2.0565	2.2093	3.1967	2.9499	2.3069	1.7214	1.2660	-5.0580	39.8656	39.4410	39.1664	39.1325	150.6188
Splitting (DAS, SPA1, SPL3)															
9,9	-1.2463	-1.4644	1.9739	2.2854	4.0145	2.8012	2.1740	2.0594	-0.0794	-3.1296	9.3672	8.7912	8.4102	8.3627	84.9307
8,8	-0.7227	-1.3019	-1.6877	2.1312	1.9127	2.7262	3.2171	1.3367	1.6761	-2.3219	36.9014	36.0844	35.5519	35.4861	57.2088
7,7	-0.2176	0.0994	-1.2465	-1.6079	2.5572	3.2688	2.1788	2.6879	1.6730	-2.3483	16.4890	15.8255	15.5588	15.5140	58.2500
6,6	0.4066	0.4195	-0.3563	-1.5778	1.6045	4.8077	5.3980	5.0882	4.2042	-4.9986	5.3284	4.9203	4.8667	4.8379	201.2944
3,3	0.3046	1.2493	1.3106	1.7595	2.7844	0.9963	-0.6175	-1.0779	1.5451	-2.0636	0.4060	0.2686	0.3845	0.3859	36.1774
2,2	0.1953	0.7444	1.0718	1.6088	2.3854	1.9921	1.0012	-1.9414	-2.9994	-1.0145	5.2893	3.8956	3.1701	3.0712	31.8729
A,A	2.5125	2.0128	2.0565	2.2093	3.1967	2.9499	2.3069	1.7214	1.2660	-5.0580	39.8656	39.4410	39.1664	39.1325	150.6188
	A	**2**	**3**	**4**	**5**	**6**	**7**	**8**	**9**	**T**	**m1**	**m2**	**m6**	**m8**	**ss**

TABLE D13
Dealer's Upcard 3 (H17)

	A	2	3	4	5	6	7	8	9	T	m1	m2	m6	m8	ss
							Hard Hitting								
17	−1.9370	−1.8708	−2.4045	−3.6589	−1.0696	−0.6720	−0.2197	0.3256	2.0676	2.3598	−41.3359	−41.3286	−41.3220	−41.3210	54.7211
16	−1.2785	−1.3185	−1.9181	−4.0263	−4.4933	−0.3239	0.2150	0.6686	2.2597	2.5538	−22.1102	−22.0128	−21.9496	−21.9418	75.2454
15	−0.8565	−0.7234	−1.3107	−3.2781	−3.7368	−3.7551	0.4435	0.8626	2.2583	2.5240	−16.2880	−16.2017	−16.1486	−16.1422	73.3095
14	−0.7085	−0.3779	−0.6847	−2.5157	−2.9666	−3.0274	−3.0253	1.0582	2.2579	2.4975	−10.4372	−10.3781	−10.3439	−10.3399	65.7282
13	−0.5377	−0.2325	−0.2925	−1.7765	−2.1955	−2.3022	−2.3467	−2.4563	2.2809	2.4648	−4.2162	−4.3762	−4.4812	−4.4942	54.7489
12	−0.3351	−0.0901	−0.1436	−1.2953	−1.4266	−1.5557	−1.6470	−1.8092	−1.4153	2.4295	1.9392	1.5951	1.3718	1.3443	37.8728
							Hard Doubling								
11	1.5009	0.9352	0.9596	1.5078	2.0020	1.7631	1.3369	0.8511	−1.6232	−2.3084	26.5831	26.1113	25.8826	25.8543	39.8986
10	−0.7387	0.7832	0.7998	1.3405	1.8715	1.9131	1.9529	1.4318	−1.2964	−2.0144	20.8475	20.4938	20.3501	20.3320	34.5348
9	−0.6073	1.3615	0.3838	1.1481	1.6987	1.9143	2.1160	1.5416	−1.3021	−2.0636	2.2639	2.0347	1.9753	1.9677	35.8210
8	−0.2719	0.9086	1.1218	1.0750	1.8055	1.9064	1.7307	1.2830	−1.2914	−2.0669	−14.2469	−14.3393	−14.3592	−14.3619	33.6053
7	0.0888	0.5273	0.7891	2.3029	1.9484	1.6588	1.2768	0.8461	−1.4824	−1.9890	−27.8043	−27.9046	−27.9297	−27.9328	33.1273
							Soft Doubling								
A,9	−1.8513	1.8017	1.9237	3.2001	3.8577	3.8814	2.7768	1.0296	−2.2297	−3.5975	−22.7475	−23.2931	−23.6551	−23.7003	116.0726
A,8	−1.2399	−2.7681	1.7216	3.0043	3.6087	3.0384	2.5156	1.9456	−1.2804	−2.6364	−9.6717	−10.0559	−10.3095	−10.3410	83.0011
A,7	−0.5613	−2.0997	−3.0011	2.7193	2.7383	2.2494	2.3652	2.3623	−0.3155	−1.6143	3.8806	3.6899	3.5696	3.5549	55.3811
A,6	0.4490	−0.7429	−1.1803	−0.8689	1.9651	1.9246	1.9026	1.6483	−0.6030	−1.1236	2.5237	2.5384	2.5770	2.5820	22.2176
A,5	1.1741	−0.2436	−0.6687	−0.1407	−0.2380	1.8135	1.7496	1.4864	−0.7779	−1.0387	−1.6802	−1.5821	−1.4991	−1.4889	15.4411
A,4	1.6049	0.4802	−0.1634	0.4707	0.3722	−0.3778	1.6727	1.4000	−0.8778	−1.1454	−3.6773	−3.5602	−3.4743	−3.4640	14.1116
A,3	1.5828	0.8865	0.5685	1.0687	0.9680	0.2057	−0.5694	1.3058	−0.9868	−1.2574	−5.7373	−5.6457	−5.5872	−5.5805	15.0634
A,2	1.5650	0.9150	1.0201	1.8326	1.5393	0.7651	−0.0311	−0.9892	−1.0793	−1.3843	−7.5156	−7.6999	−7.8010	−7.8138	20.4499
							Splitting (NDAS, SPA1, SPL3)								
9,9	−1.6014	−0.6441	1.7395	3.2818	2.8324	2.3408	2.8651	2.0655	−2.0662	−2.7033	8.0543	7.5842	7.2719	7.2329	76.2540
8,8	−1.0667	−0.8461	−1.0424	0.7658	2.2114	3.0075	3.1332	1.0231	−0.7771	−1.6022	31.2178	30.7542	30.4415	30.4023	39.1969
7,7	−0.4961	−0.5842	−0.9359	−1.1413	2.6217	2.6462	1.8595	1.8784	−0.9520	−1.2241	11.4359	11.1687	10.9885	10.9659	30.5279
6,6	0.1173	−0.3191	−0.6496	1.0478	1.9305	3.9328	4.3052	3.9319	0.5768	−3.7184	2.7733	2.8178	2.8411	2.8437	110.4625
3,3	0.0303	0.3945	0.8958	1.7734	1.8677	0.2015	−0.4088	−0.7624	−1.1959	−0.6991	−3.4739	−3.5081	−3.5008	−3.4999	11.7663
2,2	−0.1324	0.1562	0.3839	0.8888	1.3971	1.0486	0.3552	−1.8372	−4.2845	0.5061	0.4509	−0.9064	−1.7929	−1.9033	26.9138
A,A	2.3478	1.9075	1.9841	2.4239	2.9508	2.5668	2.0313	1.5520	−0.8792	−4.2213	43.6116	41.9202	41.6829	41.6533	112.8427
							Splitting (DAS, SPA1, SPL3)								
9,9	−1.1961	−1.6557	2.0280	3.6920	3.3364	2.8021	3.2637	2.3782	−2.4196	−3.0573	12.9612	12.3671	11.9715	11.9221	100.4506
8,8	−0.7020	−1.3564	−1.8304	1.5348	3.1754	3.9516	4.0300	1.3194	−1.0694	−2.2632	40.1436	39.3786	38.8649	38.8005	73.3510
7,7	−0.4696	−0.3859	−1.9114	−0.9710	4.5003	4.4627	3.1073	3.4432	−1.4637	−2.5780	21.8735	21.0362	20.4823	20.4134	95.3716
6,6	0.3053	0.3557	−0.3898	0.8437	2.2424	5.1429	6.0705	5.3969	0.0324	−5.0000	13.1761	12.5846	12.1910	12.1419	198.5416
3,3	0.3058	1.2331	1.1165	2.8595	3.2622	1.3088	−0.2014	−1.0179	−1.6980	−1.7922	6.4302	5.7704	5.3726	5.3235	40.1992
2,2	0.1489	0.5428	1.0458	1.9601	2.7670	2.4192	1.3716	−1.9064	−5.9067	−0.6105	9.8741	8.1426	7.0062	6.8647	60.6574
A,A	2.3478	1.9075	1.9841	2.4239	2.9508	2.5668	2.0313	1.5520	−0.8792	−4.2213	43.6116	41.9202	41.6829	41.6533	112.8427
	A	**2**	**3**	**4**	**5**	**6**	**7**	**8**	**9**	**T**	**m1**	**m2**	**m6**	**m8**	**ss**

TABLE D14
Dealer's Upcard 4 (H17)

	A	2	3	4	5	6	7	8	9	T	m1	m2	m6	m8	ss
								Hard Hitting							
17	–1.9162	–1.8156	–3.0996	–4.7130	–1.0078	–0.6058	–0.0166	1.7202	1.9973	2.3642	–44.5219	–44.5224	–44.5241	–44.5244	69.4776
16	–1.2621	–1.3624	–3.5009	–4.0136	–4.3790	–0.2339	0.3340	1.9203	2.2271	2.5676	–25.1777	–25.2035	–25.2235	–25.2261	86.1752
15	–0.8747	–0.7752	–2.7643	–3.2639	–3.6289	–3.6567	0.5714	1.9628	2.2409	2.5471	–19.0213	–19.0697	–19.1061	–19.1109	81.3527
14	–0.7049	–0.3897	–2.0489	–2.5178	–2.8782	–2.9459	–2.9061	2.0271	2.2785	2.5215	–12.4833	–12.7520	–12.9288	–12.9508	71.3263
13	–0.5031	–0.2429	–1.5734	–1.7711	–2.1093	–2.2139	–2.2363	–1.6083	2.2844	2.4935	–6.0062	–6.4638	–6.7612	–6.7980	52.9509
12	–0.3265	–0.1190	–1.3205	–1.2516	–1.3263	–1.4492	–1.5291	–1.1367	–1.4274	2.4716	0.0669	–0.3711	–0.6574	–0.6929	37.3933
								Hard Doubling							
11	1.4063	0.8906	1.3800	1.7910	2.2213	2.1062	1.6257	–0.8204	–1.5113	–2.2724	28.7732	28.5248	28.3614	28.3411	43.5079
10	–0.7706	0.7533	1.1737	1.6258	2.1472	2.4712	2.2101	–0.4632	–1.1802	–1.9918	23.2283	23.0945	23.0061	22.9951	38.2605
9	–0.6589	1.3121	0.9442	1.5031	2.2128	2.4329	2.1138	–0.4693	–1.1961	–2.0487	5.2206	5.2317	5.2381	5.2389	39.0304
8	–0.3328	0.8600	1.8796	1.6829	2.2895	2.1241	1.6472	–0.7052	–1.1954	–2.0624	–10.6742	–10.5951	–10.5425	–10.5359	38.6223
7	0.0133	0.4869	2.0545	2.8423	2.1133	1.7710	1.2224	–1.1339	–1.3910	–1.9947	–23.8108	–23.7217	–23.6607	–23.6530	40.7688
								Soft Doubling							
A,9	–1.8632	1.7188	3.0139	3.7363	4.2497	3.7164	2.4731	–0.7672	–2.1206	–3.5393	–18.9795	–19.3758	–19.6389	–19.6717	122.6488
A,8	–1.2489	–2.7045	2.9192	3.4557	3.3744	2.9241	2.7680	0.1280	–1.1986	–2.6043	–6.3130	–6.5692	–6.7337	–6.7539	85.5189
A,7	–0.5953	–2.0286	–1.5616	2.5354	2.5364	2.7220	2.6255	0.4991	–0.2734	–1.6149	6.4780	6.5601	6.6197	6.6274	44.8271
A,6	0.3747	–0.7601	–0.4909	–0.2687	2.2360	2.2552	1.9880	–0.2550	–0.5190	–1.1400	5.3550	5.6160	5.7896	5.8113	20.6021
A,5	1.0680	–0.2845	0.2073	0.2962	0.0903	2.1324	1.8389	–0.4167	–0.6984	–1.0584	1.2052	1.5865	1.8350	1.8658	14.4313
A,4	1.4637	0.4043	0.7689	0.8765	0.6624	–0.0004	1.7632	–0.5058	–0.7951	–1.1594	–0.6899	–0.3085	–0.0640	–0.0340	13.4784
A,3	1.4477	0.8435	1.5059	1.4284	1.2085	0.5367	–0.4256	–0.5799	–0.8749	–1.2726	–2.2756	–2.1326	–2.0394	–2.0278	16.6241
A,2	1.4930	0.8827	1.9555	2.1450	1.7666	1.0788	0.0976	–2.8406	–0.9826	–1.3990	–3.7839	–3.9986	–4.1341	–4.1506	32.5902
								Splitting (NDAS, SPA1, SPL3)							
T,T	–3.9521	2.8104	5.2144	6.5701	7.5214	6.4974	4.1339	–1.8811	–4.4630	–5.6128	–38.2116	–37.2364	–36.6204	–36.5452	359.2233
9,9	–1.5392	–0.5369	3.3246	2.8569	3.0475	3.7272	3.4309	0.0852	–1.7888	–3.1518	10.9375	10.6848	10.5238	10.5040	99.7662
8,8	–1.0439	–0.8664	–1.5882	2.1935	3.5929	3.8142	3.4291	–0.7877	–0.3704	–2.0933	33.3291	33.1070	32.9566	32.9377	66.6755
7,7	–0.4335	–0.5921	–0.7032	0.4043	3.3969	3.4108	2.1107	0.0640	–0.4904	–1.7919	14.5606	14.3443	14.1962	14.1776	41.9118
6,6	–0.2450	–0.3983	0.1161	0.3446	1.0278	2.7441	2.9026	0.0257	–0.6135	–1.4760	7.9606	7.8424	7.7600	7.7495	26.4537
4,4	0.6046	0.6478	2.6613	1.7005	0.9211	0.3513	–0.0077	–0.5962	–1.0215	–1.3153	–15.6388	–15.5093	–15.4247	–15.4142	20.0508
3,3	0.0815	0.4489	1.7322	2.5851	2.6676	1.1343	0.3092	–2.3696	–0.6718	–1.4793	0.5871	0.4053	0.2851	0.2702	33.2093
2,2	–0.0316	0.0895	1.4142	2.3269	2.8686	2.7158	1.8434	–2.5377	–2.9393	–1.4375	0.9297	0.8542	0.8048	0.7986	49.7712
A,A	2.2440	1.8445	2.1989	2.6513	3.0322	2.8155	2.3967	0.0451	–0.7691	–4.1148	45.3296	44.6026	44.1214	44.0615	111.4862
								Splitting (DAS, SPA1, SPL3)							
T,T	–3.9521	2.8104	5.2144	6.5701	7.5214	6.4974	4.1339	–1.8811	–4.4630	–5.6128	–38.2116	–37.2364	–36.6204	–36.5452	359.2233
9,9	–1.1553	–1.6524	3.7000	3.3084	3.5789	4.2532	3.8695	0.0545	–2.1475	–3.4523	16.2537	15.8991	15.6697	15.6414	126.8600
8,8	–0.7103	–1.4814	–2.3463	3.0470	4.6312	4.9185	4.4057	–1.1861	–0.6411	–2.6593	43.0754	42.6133	42.3004	42.2610	112.6442
7,7	–0.5185	–0.5219	–1.2563	0.5902	5.4157	5.5600	3.4105	0.1574	–0.9491	–2.9720	27.0024	26.2630	25.7750	25.7143	110.5990
6,6	–0.1912	0.2915	0.6258	0.2792	1.4502	4.1352	4.7539	–0.0274	–1.1079	–2.5523	20.4836	19.7306	19.2349	19.1734	69.6793
4,4	1.0490	1.5693	3.7667	2.5071	2.1405	0.6793	0.2971	–0.7159	–1.6027	–2.4226	–3.8401	–4.5124	–4.9489	–5.0028	55.7261
3,3	0.5870	1.4481	2.4056	4.0629	4.4244	2.3589	0.3712	–3.7963	–1.3079	–2.6385	11.6218	10.8469	10.6479	10.5945	93.9822
2,2	0.5311	0.5719	2.6816	3.8927	4.6742	4.4779	2.7860	–4.2220	–4.8251	–2.6421	10.9982	11.1897	11.0565	11.0398	141.6444
A,A	2.2440	1.8445	2.1989	2.6513	3.0322	2.8155	2.3967	0.0451	–0.7691	–4.1148	45.3296	44.6026	44.1214	44.0615	111.4862

TABLE D15
Dealer's Upcard 5 (H17)

	A	2	3	4	5	6	7	8	9	T	m1	m2	m6	m8	ss
							Hard Hitting								
17	−1.8810	−2.5372	−4.1497	−4.6683	−0.9245	−0.4240	1.3950	1.7518	2.0214	2.3542	−47.8143	−47.7811	−47.7586	−47.7558	81.2924
16	−1.2461	−3.0013	−3.5188	−3.9614	−4.3010	−0.0016	1.5811	1.9575	2.2375	2.5635	−28.8045	−28.6883	−28.6127	−28.6034	94.7591
15	−0.8485	−2.3043	−2.7941	−3.2287	−3.5645	−3.4491	1.6894	2.0266	2.2829	2.5476	−21.9079	−22.0317	−22.1112	−22.1209	86.9967
14	−0.6479	−1.8427	−2.0697	−2.4761	−2.8067	−2.7535	−1.9155	2.0936	2.2973	2.5303	−15.0565	−15.3984	−15.6174	−15.6444	68.6290
13	−0.4726	−1.5878	−1.5687	−1.7082	−2.0162	−2.0226	−1.3954	−1.5944	2.2895	2.5191	−8.6098	−8.9609	−9.1876	−9.2156	51.3939
12	−0.3000	−1.3337	−1.2932	−1.1820	−1.2281	−1.2737	−0.8792	−1.1322	−1.3959	2.5045	−2.2019	−2.5418	−2.7637	−2.7912	37.1625
							Hard Doubling								
11	1.3247	1.3215	1.6402	2.0366	2.5662	2.3657	−0.0541	−0.8608	−1.4858	−2.2136	31.4253	31.0740	30.8408	30.8117	45.0719
10	−0.7764	1.1368	1.4352	1.9296	2.7113	2.7107	0.3238	−0.5172	−1.1715	−1.9456	26.0680	25.8259	25.6651	25.6451	39.2627
9	−0.6738	1.8027	1.2920	2.0548	2.7438	2.4149	0.0983	−0.5084	−1.1899	−2.0086	8.7747	8.6239	8.5248	8.5125	40.7774
8	−0.3738	1.6207	2.4401	2.1958	2.5086	2.0283	−0.3681	−0.7467	−1.2029	−2.0255	−6.5749	−6.6570	−6.7095	−6.7159	42.4987
7	−0.0588	1.7742	2.5787	3.0103	2.2183	1.6654	−0.7600	−1.1698	−1.4018	−1.9641	−19.2416	−19.3251	−19.3773	−19.3836	45.8992
							Soft Doubling								
A,9	−1.8650	2.8096	3.5394	4.1512	4.0727	3.3505	0.6184	−0.7920	−2.0895	−3.4488	−14.6187	−15.1452	−15.4943	−15.5378	121.8978
A,8	−1.2811	−1.2084	3.3214	3.2305	3.2351	3.1115	0.9278	0.0302	−1.2049	−2.5405	−2.6497	−2.9095	−3.0814	−3.1028	72.8470
A,7	−0.6486	−0.5516	−1.5988	2.3652	2.9995	2.8852	0.7949	0.3825	−0.2932	−1.5838	9.6627	9.7455	9.7986	9.8051	37.0945
A,6	0.3119	−0.0593	0.0777	0.1068	2.5612	2.3311	0.0729	−0.3236	−0.5513	−1.1319	8.5745	8.8383	9.0084	9.0293	17.6505
A,5	0.9486	0.6059	0.6142	0.6366	0.4842	2.1920	−0.0648	−0.4782	−0.7198	−1.0547	4.6587	4.9902	5.2017	5.2276	12.2892
A,4	1.3285	1.3233	1.1315	1.1647	1.0074	0.1137	−0.1190	−0.5396	−0.7870	−1.1559	3.2412	3.3588	3.4322	3.4411	13.4496
A,3	1.3737	1.7379	1.8299	1.7020	1.5421	0.6331	−2.2675	−0.5995	−0.8812	−1.2677	1.9347	1.7078	1.5609	1.5428	26.6370
A,2	1.3908	1.8220	2.2871	2.4066	2.0778	1.1496	−1.7533	−2.8322	−1.0281	−1.3801	0.0528	−0.2992	−0.5287	−0.5571	41.6864
							Splitting (NDAS, SPA1, SPL3)								
T,T	−3.8747	4.9090	6.2866	7.4268	7.2639	5.8871	0.8328	−1.8548	−4.3301	−5.6366	−30.1081	−29.4201	−28.9917	−28.9398	371.1814
9,9	−1.5628	1.1882	2.6032	2.8152	4.1026	3.7870	1.1599	0.0500	−1.9175	−3.0565	14.7967	14.4306	14.1908	14.1610	92.1214
8,8	−1.0019	−1.4727	−0.3554	3.2690	4.0665	3.8662	1.0682	−0.9737	−0.3509	−2.0288	36.0784	35.9383	35.8370	35.8239	64.1451
7,7	−0.4593	−0.4358	0.4521	0.9290	3.7339	3.4582	−0.1539	−0.0365	−0.5414	−1.7366	17.6245	17.6523	17.6603	17.6608	39.7505
6,6	−0.2611	0.5179	0.5865	0.8015	1.5111	2.6623	0.6495	−0.0586	−0.6379	−1.4428	11.8209	11.6499	11.5279	11.5122	19.8534
4,4	0.5038	2.5745	2.6542	1.5304	1.3140	0.4799	−2.1393	−0.6487	−1.0171	−1.3129	−10.3746	−10.6158	−10.7754	−10.7953	31.1520
3,3	0.0153	1.4938	1.9143	2.9712	3.0783	1.2688	−1.8750	−2.3544	−0.7071	−1.4513	5.0447	4.6317	4.3617	4.3282	43.7932
2,2	−0.0698	0.8797	1.8286	2.7154	3.2870	2.8013	−0.3884	−2.5453	−2.8722	−1.4091	5.1902	4.9794	4.8362	4.8182	52.9689
A,A	2.1453	2.0790	2.4117	2.7451	3.2745	3.1547	0.8749	−0.0030	−0.7565	−3.9814	47.4283	46.6069	46.0627	45.9948	107.6949
							Splitting (DAS, SPA1, SPL3)								
T,T	−3.8747	4.9090	6.2866	7.4268	7.2639	5.8871	0.8328	−1.8548	−4.3301	−5.6366	−30.1081	−29.4201	−28.9917	−28.9398	371.1814
9,9	−1.1838	0.0045	3.0422	3.3285	4.7240	4.3775	1.2923	0.0340	−2.2786	−3.3351	20.6714	20.1445	19.7976	19.7545	114.5693
8,8	−0.6564	−2.0735	−1.1347	4.2622	5.3207	5.0893	1.3776	−1.4046	−0.5790	−2.5504	46.9316	46.4545	46.1276	46.0863	108.6183
7,7	−0.6722	0.0113	−0.0292	1.2075	6.1585	5.7959	−0.2241	0.0316	−0.9610	−2.8296	32.2183	31.6908	31.3297	31.2841	106.4303
6,6	−0.4185	2.0861	1.6871	1.3304	2.6546	4.5197	1.0286	−0.2594	−1.3271	−2.8254	25.9194	25.7773	25.1359	25.0562	71.4357
4,4	0.7886	3.9809	4.0535	2.5363	2.7072	0.8411	−3.1999	−0.7634	−1.5586	−2.3464	3.6230	2.5446	1.8344	1.7460	82.6433
3,3	0.2510	3.2493	3.1972	5.1757	5.7336	2.8904	−3.2434	−4.0599	−1.5315	−2.9156	17.4852	17.1738	16.5289	16.4484	152.2098
2,2	0.2318	2.0071	3.8163	4.9924	5.9427	5.1984	−0.9191	−4.4939	−5.0709	−2.9262	16.4928	16.8203	16.5139	16.4751	186.9148
A,A	2.1453	2.0790	2.4117	2.7451	3.2745	3.1547	0.8749	−0.0030	−0.7565	−3.9814	47.4283	46.6069	46.0627	45.9948	107.6949

TABLE D16

Dealer's Upcard 6 (H17)

	A	2	3	4	5	6	7	8	9	T	m1	m2	m6	m8	ss
Hard Hitting															
17	-2.3174	-3.6129	-4.1605	-4.6046	-0.7715	1.0783	1.4345	1.6789	2.0213	2.3135	-51.3769	-51.3409	-51.3182	-51.3155	89.0646
16	-2.7236	-3.0404	-3.4975	-3.8859	-4.1544	1.2981	1.6744	1.9309	2.2752	2.5308	-31.7856	-31.9843	-32.1146	-32.1308	100.2665
15	-2.2263	-2.3425	-2.7747	-3.1467	-3.4266	-2.2711	1.7799	2.0141	2.2963	2.5244	-24.5998	-25.0157	-25.2854	-25.3187	82.9321
14	-1.9377	-1.8447	-2.0358	-2.3827	-2.6643	-1.7280	-1.8490	2.0459	2.2958	2.5251	-17.8062	-18.2379	-18.5187	-18.5534	65.4444
13	-1.6518	-1.5689	-1.5147	-1.6193	-1.8953	-1.1664	-1.3653	-1.6317	2.3239	2.5224	-11.0524	-11.4789	-11.7580	-11.7926	50.4351
12	-1.3684	-1.2938	-1.2170	-1.0781	-1.1270	-0.6211	-0.8605	-1.1412	-1.3593	2.5166	-4.3337	-4.7366	-5.0027	-5.0358	37.0693
Hard Doubling															
11	1.6602	1.5705	1.9133	2.3604	2.8179	0.6281	-0.1295	-0.7493	-1.4703	-2.1503	33.7190	33.4765	33.3144	33.2941	44.0250
10	-0.3518	1.3752	1.7619	2.4629	2.9424	0.7735	0.2000	-0.4204	-1.1764	-1.8918	28.4117	28.3307	28.2758	28.2689	36.3576
9	-0.0821	2.0823	1.8557	2.5441	2.7167	0.3729	-0.0248	-0.4250	-1.2192	-1.9551	11.7029	11.7295	11.7466	11.7487	38.7362
8	0.3812	2.1568	2.9386	2.3809	2.4153	-0.0340	-0.4552	-0.6541	-1.2377	-1.9729	-2.9824	-2.8904	-2.8288	-2.8211	42.6728
7	1.1121	2.2833	2.7646	3.0546	2.1482	-0.3909	-0.8419	-1.0647	-1.4144	-1.9127	-15.1054	-14.9957	-14.9232	-14.9141	46.6688
Soft Doubling															
A,9	-0.7313	3.3065	3.9561	3.9363	3.7262	1.4178	0.4873	-0.7152	-2.0667	-3.3292	-10.7543	-11.0420	-11.2330	-11.2569	107.8632
A,8	-0.0757	-0.6414	3.1200	3.0719	3.4601	1.1958	0.8012	0.0655	-1.2125	-2.4462	0.7719	0.7851	0.7913	0.7920	59.0413
A,7	0.6582	-0.6295	-1.5779	2.7982	3.2044	1.0194	0.6353	0.4345	-0.3490	-1.5484	12.3760	12.7742	13.0332	13.0652	32.7601
A,6	0.8122	0.4903	0.4708	0.4950	2.6508	0.3747	-0.0211	-0.2346	-0.5805	-1.1144	11.6320	12.0493	12.3196	12.3530	13.8941
A,5	1.6148	0.9600	0.9487	0.9669	0.6303	0.2739	-0.1385	-0.3649	-0.7231	-1.0420	8.1781	8.4931	8.6967	8.7218	10.8546
A,4	2.0247	1.6239	1.4446	1.4828	1.1469	-1.7633	-0.1803	-0.4121	-0.8039	-1.1408	7.0501	7.0392	7.0351	7.0348	21.5010
A,3	2.0926	2.0816	2.1094	2.0000	1.6640	-1.2561	-2.2785	-0.5241	-0.9371	-1.2379	5.3700	5.2480	5.1721	5.1629	33.9831
A,2	2.1544	2.1593	2.5690	2.6534	2.1553	-0.7647	-1.8068	-2.6894	-1.0428	-1.3469	3.5398	3.3134	3.1649	3.1465	47.0154
Splitting (NDAS, SPA1, SPL3)															
T,T	-1.7159	5.9195	7.1365	7.0900	6.6830	2.4303	0.6505	-1.6426	-4.2177	-5.5834	-22.8902	-21.8240	-21.1440	-21.0606	335.3573
9,9	0.2147	0.4893	2.5708	3.8144	4.2031	1.4852	0.9565	0.1137	-2.0458	-2.9504	17.7654	17.7629	17.7606	17.7602	81.2491
8,8	-1.5458	-0.2387	0.8565	3.7373	4.0478	1.4707	0.9315	-0.9721	-0.4062	-1.9702	38.5184	38.5741	38.6097	38.6141	53.1997
7,7	-0.3904	0.7437	1.0590	1.3561	3.7816	1.0709	-0.3815	0.0615	-0.5510	-1.6875	20.9200	21.1558	21.3002	21.3175	30.9571
6,6	0.5023	0.9281	1.0478	1.2490	1.6450	0.4156	0.5112	0.0350	-0.6882	-1.4114	15.0967	15.2292	15.3074	15.3166	15.3551
4,4	2.1743	2.5430	2.6632	1.6669	1.5248	-1.6991	-2.1874	-0.5385	-1.0453	-1.2755	-5.7917	-5.9220	-6.0065	-6.0170	38.9519
3,3	0.9561	1.8609	2.1278	3.3327	3.2003	-0.9693	-1.9221	-2.2172	-0.7454	-1.4060	8.8274	8.6127	8.4700	8.4522	48.2657
2,2	0.8845	1.0833	2.2314	3.0751	3.3743	0.4968	-0.4937	-2.3428	-2.8223	-1.3716	8.9920	8.9602	8.9304	8.9262	49.2472
A,A	2.3021	2.2785	2.5529	2.9861	3.5914	1.5561	0.7931	0.0750	-0.7546	-3.8452	49.1555	48.4501	47.9835	47.9254	101.5897
Splitting (DAS, SPA1, SPL3)															
T,T	-1.7159	5.9195	7.1365	7.0900	6.6830	2.4303	0.6505	-1.6426	-4.2177	-5.5834	-22.8902	-21.8240	-21.1440	-21.0606	335.3573
9,9	0.6679	-0.7653	3.0735	4.4091	4.8865	1.7490	1.0862	0.1321	-2.4158	-3.2058	24.0997	23.9330	23.8217	23.8077	104.9973
8,8	-1.0669	-0.9622	0.6830	5.5055	6.0816	2.2324	1.3671	-1.3736	-0.8269	-2.9100	50.4360	50.1552	49.9729	49.9504	113.1223
7,7	-0.2805	1.2495	0.7403	1.9011	6.4202	2.0020	-0.5808	0.2338	-0.9236	-2.6905	37.7607	37.3758	37.1157	37.0830	81.2305
6,6	0.7472	2.7313	2.4513	2.0863	2.8422	0.4908	0.8465	-0.0241	-1.3608	-2.7027	31.8972	31.4398	31.0345	30.9838	58.4865
4,4	2.6493	4.2037	4.3954	2.9204	2.9530	-2.7089	-3.3995	-0.5544	-1.5482	-2.2277	10.0790	9.2626	8.7275	8.6607	102.7075
3,3	1.5611	3.9817	3.7065	5.9387	5.8535	-1.1211	-3.4371	-3.9085	-1.5212	-2.7634	23.7056	23.1457	22.7591	22.7105	162.7667
2,2	1.5344	2.4389	4.6386	5.7445	6.1751	1.0858	-1.1741	-4.2355	-5.0757	-2.7830	22.7519	22.9841	22.8035	22.7802	178.1915
A,A	2.3021	2.2785	2.5529	2.9861	3.5914	1.5561	0.7931	0.0750	-0.7546	-3.8452	49.1555	48.4501	47.9835	47.9254	101.5897

TABLE D17

Betting and Insurance Effects of Removal

	A	2	3	4	5	6	7	8	9	T	m1	m2	m6	m8	ss
Betting EORs															
Base rules: DOA, SPA1, SPL3															
S17, NDAS	–0.5944	0.3828	0.4377	0.5582	0.7095	0.4131	0.2898	0.0053	–0.1698	–0.5081	0.0325	–0.3243	–0.5458	–0.5731	2.8224
S17, DAS	–0.5794	0.3809	0.4339	0.5680	0.7274	0.4118	0.2823	–0.0033	–0.1731	–0.5121	0.1735	–0.1820	–0.4041	–0.4316	2.8491
H17, NDAS	–0.5311	0.3950	0.4614	0.5960	0.7135	0.4450	0.2856	0.0145	–0.1920	–0.5470	–0.1623	–0.5287	–0.7610	–0.7898	3.0286
H17, DAS	–0.5173	0.3931	0.4584	0.6053	0.7319	0.4439	0.2785	0.0061	–0.1949	–0.5512	–0.0187	–0.3837	–0.6166	–0.6455	3.0623
Base rules: DOA, SPA1, SPL3, LS															
S17, NDAS	–0.5901	0.4130	0.4876	0.6365	0.8105	0.4534	0.2618	–0.0325	–0.2169	–0.5558	0.0554	–0.2728	–0.4730	–0.4972	3.3765
S17, DAS	–0.5751	0.4110	0.4839	0.6462	0.8284	0.4521	0.2543	–0.0412	–0.2202	–0.5599	0.1963	–0.1305	–0.3313	–0.3557	3.4109
H17, NDAS	–0.5299	0.4257	0.5138	0.6793	0.8208	0.4914	0.2498	–0.0321	–0.2471	–0.5930	–0.1235	–0.4622	–0.6726	–0.6981	3.6337
H17, DAS	–0.5161	0.4238	0.5109	0.6886	0.8392	0.4903	0.2426	–0.0404	–0.2500	–0.5972	0.0201	–0.3172	–0.5283	–0.5539	3.6756
Base rules: D10, SPA1, SPL3															
H17, NDAS	–0.5170	0.3616	0.4316	0.5502	0.6598	0.4209	0.2663	–0.0012	–0.1802	–0.4980	–0.4387	–0.7430	–0.9424	–0.9673	2.5950
H17, DAS	–0.5028	0.3605	0.4304	0.5563	0.6743	0.4209	0.2596	–0.0124	–0.1840	–0.5007	–0.3208	–0.6201	–0.8171	–0.8418	2.6136
Insurance EORs															
	1.8100	1.8100	1.8100	1.8100	1.8100	1.8100	1.8100	1.8100	1.8100	–4.0724	–7.6923	–7.6923	–7.6923	–7.6923	95.8211

TABLE D18

Betting and Insurance Correlations

	Hi-Lo	Hi-Opt I	Hi-Opt I/A[1]	RPC	AOII	AOII/A[1]	Halves
Betting Correlations							
Base rules: DOA, SPA1, SPL3							
S17, NDAS	0.9652	0.8735	0.9480	0.9813	0.9161	0.9873	0.9909
S17, DAS	0.9648	0.8775	0.9475	0.9804	0.9197	0.9867	0.9915
H17, NDAS	0.9685	0.8947	0.9494	0.9834	0.9364	0.9888	0.9924
H17, DAS	0.9676	0.8979	0.9492	0.9822	0.9391	0.9883	0.9926
Base rules: D10, SPA1, SPL3							
H17, NDAS	0.9684	0.8899	0.9505	0.9840	0.9312	0.9893	0.9930
H17, DAS	0.9679	0.8933	0.9502	0.9829	0.9342	0.9887	0.9935
Insurance Correlations[2]							
	0.7885	0.8536	0.8701	0.8096	0.8536	0.8701	0.7509

[1] A side count of aces is employed [2] After removing dealer's ace

Epilogue

As I write these closing lines, there is much unrest in the blackjack-playing community. It seems that, in their continuing quest to offer blackjack games to only those players who cannot win in the long run, casino owners are redoubling their efforts to find efficient and technologically advanced methods of systematically "eliminating" card counters from their premises. Whether or not these tactics will prove to be pervasive or, more likely, just another short-term fit of hysteria on the part of the unenlightened managers, remains to be seen. What is clear, however, is that, for the time being, increased vigilance on the part of the advantage player is the order of the day. In short, playing winning blackjack is not getting any easier!

Along these lines, one of the most frequently asked questions that I receive is: "Can I make a living at this game?" My answer is almost always in the following vein: "You *can*, but you probably shouldn't try." As outlined in several of my previous chapters, playing blackjack for continual profit is a very difficult endeavor. Many liken it to the "fast lane" professions of securities, derivatives, or commodities trading. Although there are both similarities and differences, it is clear that each occupation takes a very special mindset. For all but a select few, high-stakes casino 21 is best left, in my opinion, as an avocation, rather than a full-time job. And, having to camouflage one's blackjack play serves only to exacerbate the problem.

Indeed, when I was first approached by the individual who, ultimately, was successful in luring me away from teaching and onto Wall Street, his sales pitch went something like this: "Don, it's the biggest casino in the world, and … they can't throw you out!"

It is my hope that the material you have just read will have provided you with many keen insights into how best to formulate your own, personal "Blackjack Attack" on the casinos. I wish that it could be a full, frontal attack, rather than one where the art of subterfuge is of such paramount importance; but I'm afraid that wishing won't make it so. On the other hand, we live in an age where rapid dissemination of information is reaching mind-boggling proportions.

Once upon a time, I would write a "Gospel" for *Blackjack Forum* a month or two before the column appeared in print, and then would have to wait yet another three months before being able to furnish a comment or correction on the piece. Today, via e-mail and the Internet, discussions and exchanges of information proceed at breakneck speed, on a daily, or even hourly, basis. In this respect, never before has the opportunity to become an informed and knowledgeable player been greater than it is right now.

I hope that I have been successful in conveying to you some of the finer points of this passion that you and I share for the fascinating and challenging game of casino blackjack. I wish you all the best in your future endeavors, as you continue, or begin, "playing the pros' way."

Good luck, and … good cards!

Don Schlesinger
New York, March 1997

Postscript: It's been seven years since I wrote the above lines, and, unfortunately, the anti-counter casino tactics to which I alluded have proven to be far from "short-term." If anything, the climate for playing professionally or otherwise has gotten considerably worse in the interim. Surveillance measures have become, in some venues, quite oppressive, with new facial recognition software and fancy, hi-tech gizmos in place to more readily detect the presence of counters. Once apprehended, players are facing rude barrings and sometimes even ejections from hotels at which they are staying. As they move on to a new casino, counters often find that through SIN (Surveillance Information Network) — which truly is a sin! — their photo, which has been faxed around town, is already awaiting their arrival in the next blackjack pit.

As if all of this weren't bad enough, game conditions have deteriorated, as well. What was once strictly an Atlantic City curse — the 8-deck shoe game, has now reared its ugly head along the Las Vegas Strip. And, the policy of "no mid-shoe entry," aimed at thwarting back-counters, has proliferated in A.C. and is creeping, ever so slowly, into the Nevada games as well. In addition, good penetration is extremely hard to come by, as unenlightened casino managers "throw the baby out with the bath water," shuffling so frequently that they lose volume to the masses, simply to stymie the occasional card counter in their midsts. Finally, insane rules variations, such as short-changing players by offering 6 to 5 (120%) bonuses for naturals, instead of the customary 3 to 2 (150%), rob players, in this case, of that extra 30% return on the blackjacks that they receive in the single-deck games at which this horrible rule is offered.

No, sad to say, twenty-first century blackjack isn't exactly a picnic! My hat is off to those who continue to ply their trade successfully in this never-ending cat-and-mouse game. There was a time when I felt an ebb and flow to these kinds of shenanigans. We could always count on the pendulum's swinging back the other way, towards more favorable conditions. Well, it's been quite some time now, and, truth be told, the pendulum is stuck! All we can do is wait and hope for better times ahead.

New York, March 2004

Selected References

There have been several hundred books written on the game of blackjack, yet only a handful are worthy of the serious player's attention. Below, I have listed those works that I feel deserve a place in your library. Finally, if I were forced to limit my holdings to only the most essential references — those books or resources to which I refer most often, or those that are of major historical import — I would recommend the titles preceded by an asterisk.

I. Books on Blackjack

*Andersen, Ian. *Burning the Tables in Las Vegas.* Las Vegas: Huntington Press, 1999; second edition, 2003.

*———. *Turning the Tables on Las Vegas.* New York: The Vanguard Press, 1976.

Auston, John. *The World's Greatest Blackjack Simulation* (series of four booklets). Oakland: RGE Publishing, 1997.

Blaine, Rick. *Blackjack Blueprint: How to Operate a Blackjack Team.* Oakland: RGE Publishing, 2000.

———. *Blackjack in the Zone.* Oakland: RGE Publishing, 2000; revised, 2002.

*Braun, Julian H. *How to Play Winning Blackjack.* Chicago: Data House Publishing Co., Inc., 1980.

Canfield, Richard Albert. *Blackjack Your Way to Riches.* Scottsdale: Expertise Publishing Co., 1977.

*Carlson, Bryce. *Blackjack for Blood.* Santa Monica: CompuStar Press, 1992; revised, La Jolla: Pi Yee Press, 2001.

*Chambliss, Carlson R., and Thomas C. Roginski. *Fundamentals of Blackjack.* Las Vegas: GBC, 1990.

Chambliss, C. R., and T. C. Roginski. *Playing Blackjack in Atlantic City.* Las Vegas: GBC, 1981.

*Dalton, Michael. *Blackjack: A Professional Reference.* Merritt Island, FL: Spur of the Moment Publishing, 1991; third edition, 1993.

Forte, Steve. *Read the Dealer.* Berkeley: RGE Publishing, 1986.

*Griffin, Peter A. *The Theory of Blackjack.* Las Vegas: GBC, 1979; sixth edition, Las Vegas: Huntington Press, 1999.

Humble, Lance, and Carl Cooper. *The World's Greatest Blackjack Book.* New York: Doubleday, 1980.

Meadow, Barry. *Blackjack Autumn: A True Tale of Life, Death, and Splitting Tens in Winnemucca.* Anaheim, CA: TR Publishing, 1999.

Mezrich, Ben. *Bringing Down the House: The Inside Story of Six M.I.T. Students Who Took Vegas for Millions.* New York: Free Press, 2002.

Perry, Stuart. *Las Vegas Blackjack Diary.* New York: self published, 1995; third edition, Pittsburgh: ConJelCo, 1997.

Renzey, Fred. *Blackjack Bluebook II.* Elk Grove Village, IL: Blackjack Mentor Press, 2003.

*Revere, Lawrence. *Playing Blackjack as a Business.* Secaucus, NJ: Lyle Stuart, 1969; last revised, 1980.

Rose, I. Nelson, and Robert A. Loeb. *Blackjack and the Law.* Oakland: RGE Publishing, 1998.

*Snyder, Arnold. *Beat the X-Deck Game* (series of five books). Oakland: RGE Publishing, 1987.

*———. *Blackbelt in Blackjack: Playing 21 as a Martial Art.* Berkeley: RGE Publishing, 1983; revised 1998.

———. *Blackjack Wisdom.* Oakland: RGE Publishing, 1998.

Stricker, Ralph. *Silver Fox Blackjack System.* North Brunswick, NJ: Silver Fox Enterprises, 1981; revised, 1996.

*Thorp, Edward O. *Beat the Dealer: A Winning Strategy for the Game of Twenty-One.* New York: Random House, 1962; revised edition, New York: Vintage Books, 1966.

Tilton, Nathaniel. *The Blackjack Life.* Las Vegas: Huntington Press, 2012.

*Uston, Ken. *The Big Player: How a Team of Blackjack Players Made a Million Dollars.* New York: Holt, Rinehart and Winston, 1977.

———. *Ken Uston on Blackjack.* Secaucus, NJ: Lyle Stuart, 1986.

———. *Million Dollar Blackjack.* Hollywood: SRS Enterprises, 1981.

———. *Two Books on Blackjack.* Wheaton, MD: The Uston Institute of Blackjack, 1979.

*Vancura, Olaf, and Ken Fuchs. *Knock-Out Blackjack: The Easiest Card-Counting System Ever Devised.* Las Vegas: Huntington Press, 1996; revised 1998.

*Walker, Katarina. *The Pro's Guide to Spanish 21 and Australian Pontoon.* New York: Maven Press, 2008.

*Wattenberger, Norman. *Modern Blackjack: An Illustrated Guide to Blackjack Advantage Play.* Volumes I and II. New York: QFIT, 2009; second edition, 2010.

*Wong, Stanford. *Basic Blackjack.* La Jolla: Pi Yee Press, 1992; revised 1993.

———. *Blackjack Secrets.* La Jolla: Pi Yee Press, 1993.

*———. *Professional Blackjack.* La Jolla: Pi Yee Press, 1975; last revised 1994.

II. Periodicals or Newsletters on Blackjack

Curtis, Anthony. *Las Vegas Advisor.* Las Vegas: Huntington Press, monthly since 1983.

Dalton, Michael. *Blackjack Review.* Merritt Island, FL: Spur of the Moment Publishing, quarterly, from 1992–1998.

*Snyder, Arnold. *Blackjack Forum.* Berkeley: RGE Publishing, quarterly, from 1981–2002; Las Vegas: Huntington Press, quarterly, from 2002–04.

*Wong, Stanford. *Current Blackjack News.* La Jolla: Pi Yee Press, monthly from 1979–2002; Las Vegas: Pi Yee Press, since 2002 .

*———. *Stanford Wong's Blackjack Newsletter. Vols. 1–5.* La Jolla: Pi Yee Press, 1979–83.

III. Blackjack Software

*Auston, John. *Blackjack Risk Manager 2002 (BJRM).*

Carlson, Bryce. *Omega II Blackjack Casino.*

*Janecek, Karel. *Statistical Blackjack Analyzer 5.5 (SBA 5.5).*

Marcus, Hal. *Blackjack 6•7•8.* Chicago: Stickysoft Corp.

Pronovost, Dan. *Blackjack Audit.*

*Wattenberger, Norman. *Casino Vérité (CV), CVSIM, CVData, and CVCX.* New York: QFit.

*Wong, Stanford. *Blackjack Count Analyzer (BCA) and Professional Blackjack Analyzer (PBA).* La Jolla: Pi Yee Press.

IV. Books on Casino Gambling and Mathematics

Curtis, Anthony. *Bargain City.* Las Vegas: Huntington Press, 1993.

*Epstein, Richard A. *The Theory of Gambling and Statistical Logic.* New York: Academic Press, 1977; revised 1995.

*Ethier, Stewart N. *The Doctrine of Chances: Probabilistic Aspects of Gambling.* Berlin: Springer, 2010.

Ethier, Stewart N., and William R. Eadington, eds. *Optimal Play: Mathematical Studies of Games and Gambling.* Reno: Institute for the Study of Gambling and Commercial Gaming, 2007.

Griffin, Peter. *Extra Stuff: Gambling Ramblings.* Las Vegas: Huntington Press, 1991.

*Grosjean, James. *Beyond Counting: Exploiting Casino Games from Blackjack to Video Poker.* Oakland: RGE Publishing, 2000.

Orkin, Mike. *Can You Win?* New York: W. H. Freeman and Company, 1991.

Ortiz, Darwin. *Gambling Scams: How They Work, How to Detect Them, How to Protect Yourself.* New York: Dodd, Mead & Company, 1984; reissue edition, Secaucus, NJ: Lyle Stuart, 1990.

Poundstone, William. *Fortune's Formula.* New York: Hill and Wang, 2005.

Rubin, Max. *Comp City: A Guide to Free Las Vegas Vacations.* Las Vegas: Huntington Press, 1994; second edition, 2001.

*Scarne, John. *Scarne's Complete Guide to Gambling.* New York: Simon and Schuster, 1961.

*Thorp, Edward O. *The Mathematics of Gambling.* Secaucus, NJ: Lyle Stuart, 1984.

*Vancura, Olaf. *Smart Casino Gambling: How to Win More and Lose Less.* San Diego: Index Publishing Group, Inc., 1996; revised 1999.

*Wilson, Allan N. *The Casino Gambler's Guide: Enlarged Edition.* New York: Harper & Row, 1970.

*Wong, Stanford. *Casino Tournament Strategy.* La Jolla: Pi Yee Press, 1992.

Index

About the Author

Don Schlesinger has been involved with blackjack, in several capacities, for over 30 years. As a writer, lecturer, teacher, and player, Schlesinger has "done it all," in casinos worldwide.

Schlesinger was born in 1946 and graduated from the City College of New York (CCNY) with a B.S. degree in mathematics. In addition, he holds M.A. and M.Phil. degrees in French from the City University of New York.

A former teacher of mathematics and French in the New York City school system, Don changed professions in 1984 and, until 1998, enjoyed a successful career at one of Wall Street's most prestigious investment banks, where he was a principal (executive director) of the firm. Now in semi-retirement, he continues to pursue business ventures in the financial community.

While Schlesinger still contributes on a daily basis to several online blackjack forums, he also enjoys coaching horizontal jumping and is a major track and field enthusiast.

He and his wife live in a suburb of New York City. They have two children.

Who Is Don Schlesinger?

"Don Schlesinger is recognized as one of the foremost experts in blackjack."
— Ralph Stricker, author of the Silver Fox Blackjack System

"Schlesinger is, in my opinion, one of the most knowledgeable blackjack players in the world. He has perhaps the finest perspective on the game today."
— George C., author of Advanced Card Counting

"I've come to the conclusion that 'Don Schlesinger' cannot possibly be one person. It must be a trade name, like 'Mr. Goodwrench.' The so-called Don Schlesinger has been making too many insightful contributions for too long for it to be just one person."
— "Dunbar," well-known player and member of the Internet blackjack community

"Don Schlesinger is not only as knowledgeable as any human could expect to be, he's a terrific writer as well."
— "Grimy Fellow," player

"Don S. [is] the best overall blackjack authority."
— "DD'," noted blackjack and team-play authority

"Don Schlesinger is the best blackjack expert right now. He knows more about the game intuitively than most know by experimentation."
— Tom Turcich, author of Vegas on a Chip

"His writing is clear, his mathematics are sound, and his insights are eminently worthwhile. Schlesinger cements his reputation as a master teacher of the game."
— Nick Christenson, on jetcafe.org

Praise for *Blackjack Attack*

First and Second Editions

"Every few years a great, must-have, blackjack book is published. This is one of those books. Don's intelligent and eloquent writing style alone is worth the price."
— *George C., author of Advanced Card Counting*

"If you want to know what is really important to become a winning player, you need to read and study this material. Schlesinger has successfully answered many of the toughest questions about this game and has made some major discoveries along the way. *Blackjack Attack* is one of the best blackjack books that have ever been published! Outstanding. Highly recommended."
— *Michael Dalton, Blackjack Review*

"Everybody assumes that *Blackjack Attack* is going to be recognized as a classic, and I share that opinion of the book."
— *Stanford Wong, author of Professional Blackjack*

"The best teacher of statistics and probability for me over the years I've been involved in blackjack and casino gambling is Donald Schlesinger. [He] should feel good about the excitement [his] book has generated amongst players."
— *Henry Tamburin, author of Blackjack: Take the Money and Run*

"In *Blackjack Attack,* Don Schlesinger writes about virtually everything that is important in the game of blackjack for a card counter. No matter how much experience he has, a counter can learn from this book."
— *Stuart Perry, author of Blackjack Diary*

"Get this book; you won't be sorry."
— *Bryce Carlson, author of Blackjack for Blood*

"Schlesinger joins the ranks of Peter Griffin, Stanford Wong, and a handful of other authors who have cracked the so-called 'inner circle' of blackjack superstars."
— *Howard Schwartz, Gambler's Book Club*

"*Blackjack Attack* is one of the best blackjack books to hit the bookshelves in the last 35 years."
— *Eddie Olsen, editor of Blackjack Confidential magazine*

"*Blackjack Attack* is not a book, it's the *Bible.* I say that because it is the only book besides the *Bible* that I've read over and over and over. It's one of the greatest blackjack books ever written. Trust me, you'll be taken aback. It's like that commercial: 'Don't leave home without it.'"
— *"Hollywood," high-stakes player*

"It gives me great pleasure to announce the publication of the best modern book on blackjack. Period! This is offered with no hesitation on my part. *Blackjack Attack* is the pinnacle, the cream of knowledge spawned in this incredibly imaginative time. More than just a summary or potpourri of the experiences of others, Schlesinger has linked arms with the ranks of his mentors, joining them in the pantheon of the great blackjack mathematicians and probability experts. He has crafted the final word. I say this because after reading the new edition of his book, I can't imagine what there is left to say concerning the game of blackjack. I will tell you that if you can't find the answer to your questions and prayers concerning blackjack amidst the myriad charts, chapters, and tables, I can't help you. Don Schlesinger has truly written the final word on the game of blackjack as we know it. *Blackjack Attack* is a book for the ages, one that serious students of the game will be referring to in years hence. I can't give it a higher recommendation."
— *Peter Ruchman, former manager of the Gambler's Book Club*

"A masterful presentation of the latest research for every player. If Stanford Wong's *Professional Blackjack* is considered the blackjack bible, then Don Schlesinger's *Blackjack Attack* is the New Testament."
— *Jake Smallwood, noted blackjack authority*

"*Blackjack Attack* is one of the few must-have books on the game. This book is one of the elite. The prose is perfect, the content incontestable, the mathematics magnificent. There is no other work that addresses these topics with the same insight, clarity, and ease of comprehension. It is certainly required reading for the serious player of all levels."
— "anon," well-known player and member of the Internet blackjack community

"No other blackjack book contains more valuable information and tips on how to beat the game. If you could have only one book about blackjack, *Blackjack Attack* would be it. The information in it has made me more money than any of the other 20-odd blackjack books I own. Thank you, Don, for your wonderful book."
— "SOTSOG," well-known player and member of the Internet blackjack community

"The book is too damn good!"
— "Scarce," player

Third Edition

"The more I use *BJA3* the more I'm impressed by it."
— Ed Thorp, author of Beat the Dealer

"With more than 100 new pages (the last edition was published in 2000), Schlesinger's work moves to the top of the 'best-bet/must-read' list for serious players."
— Howard Schwartz, Gambler's Book Club

"It really is a wonderful book, Don. It's great to have a readable, in-depth mathematical resource such as *Blackjack Attack* right there any time I need it. There is just nothing else out there like it."
— Bryce Carlson, author of Blackjack for Blood

"I received my brand-new *Blackjack Attack* yesterday, and all I can say is WOW WOW WOW! This is mandatory reading for any real-deal BJ players. Don, I can't tell you how impressed I am. For all the math people, they should make this a mandatory text for students. For all the non-math people (like myself), who are not greedy and want only to pick up an extra couple of 100 thousand a year (!), all I can say is: 'Go buy the damn book.'"
— "Hollywood," high-stakes player

"It looks like a classic!"
— Hal Jordan, developer of Blackjack 6•7•8

"The new charts are fantastic, the basic strategy charts by Cacarulo are amazing, ... the presentation is excellent. All in all, I love the book."
— "AdvantageRay," on advantageplayer.com

"First I would like to say thanks for the great book. I'm new to [advantage play] and this book really is one of a kind! It has lots of insights that other books do not provide. Thanks again!
— Michael Y., e-mail correspondence

Find a
Wide Selection
of the Very Best
Gambling Products at:
The World's Greatest
GAMBLiNG CATALOG
GamblingCatalog.com
877-798-7743